Current Advances in Fern Research

Helena Fernández
Editor

Current Advances in Fern Research

Springer

Editor
Helena Fernández
Department of Organisms and Systems Biology
Oviedo University
Oviedo, Spain

ISBN 978-3-030-09151-4 ISBN 978-3-319-75103-0 (eBook)
https://doi.org/10.1007/978-3-319-75103-0

Printed on acid-free paper

This Springer imprint is published by the registered company Springer International Publishing AG part of Springer Nature.
The registered company address is: Gewerbestrasse 11, 6330 Cham, Switzerland

To everybody feeling lost,
exhausted and desperate,
living a life,
none of us can even imagine....

Preface

This is the second book focussing on ferns, of our particular collection, given that the majority of researchers who have participated in the first book, titled *Working with Ferns: Issues and Applications*, decided to engage this new project. The first book had so many entries (around 20,000 since its publication in December 2010) that Springer considered a good idea to face a new venture, which was welcome by all the authors.

There is other taxa, far away *Arabidopsis* and all those important crops we cultivate, hiding interesting clues that are waiting to be deciphered, so that we could finally understand how plants are, as we know them today. Ferns are the pioneering vascular plants that started living on land and still continue colonizing and spreading to all the corners of the Earth. Around 1200 species of ferns exist at the present time.

Research on ferns is scarce. That is true. We are a small scientist community in reality but not for that less important. Extant ferns are representative of a legacy of great value, which can be admired and debated from different angles, such as evolutionary, taxonomical, genomic, genetic, physiological, biochemistry, environmental or merely by the beauty that their fronds have and the important market created around them.

Finally, and beyond a purely scientific reasoning, ferns have also joined all of us, and have contributed to bridge the gap existing among our routines, our countries, our big efforts to go ahead with our goals.

Nothing of this sort would have been possible, without the cooperation and generosity of the almost 70 authors, who have prepared their chapters, and put their experiments to the dream that this second book to see light.

Thanks to the editorial Springer for having been committed to us this challenge of giving birth to a new book on ferns.

The editor

P.D. The funds I will receive as editor will be entirely donated to the NGO Médecins sans Frontiers.

Oviedo, Spain　　　　Helena Fernández

Contents

Contributors

Maria do Rosario Anjos University of Trás-os-Montes and Alto Douro, Department of Biology and Environment, Centre for the Research and Technology of Agro-Environmental and Biological Sciences (CITAB), Vila Real, Portugal

Daniel Ballesteros Center for Conservation and Research of Endangered Wildlife, Cincinnati Zoo & Botanical Garden, Cincinnati, OH, USA

Comparative Seed Biology Group, Comparative Plant and Fungal Biology Department, Royal Botanic Gardens Kew, Richmond, UK

Linh Thuy Bui Department of Biology, The University of Iowa, Iowa City, IA, USA

Department of Biology, Indiana University, Bloomington, IN, USA

Alberto Cabezuelo Department of Plant Sciences I, Universidad Complutense, Madrid, Spain

Ashley E. Cannon BioDiscovery Institute, Department of Biology, The University of North Texas, Denton, TX, USA

Araceli Cantero Department of Molecular Biosciences, The University of Texas, Austin, TX, USA

María Jesús Cañal Area of Plant Physiology, Department of Organisms and Systems Biology (BOS), Oviedo University, Oviedo, Spain

Francisco Carrapiço Centro de Ecologia, Evolução e Alterações Ambientais e Centro de Filosofia das Ciências, Departamento de Biologia Vegetal, Faculdade de Ciências, Universidade de Lisboa, Lisboa, Portugal

Myriam Catalá Department of Biology and Geology, Physics and Inorganic Chemistry, ESCET, Rey Juan Carlos University, Madrid, Spain

Yue Cao State Key Laboratory of Pollution Control and Resource Reuse, School of the Environment, Nanjing University, Jiangsu, China

Yanshan Chen State Key Laboratory of Pollution Control and Resource Reuse, School of the Environment, Nanjing University, Jiangsu, China

Chi-Lien Cheng Department of Biology, The University of Iowa, Iowa City, IA, USA

Wen-Liang Chiou Taiwan Forestry Research Institute, Taipei, Taiwan

Paula Conde Area of Plant Physiology, Department of Organisms and Systems Biology (BOS), Oviedo, Spain

Angela R. Cordle Department of Biology, The University of Iowa, Iowa City, IA, USA

F. B. Cuevas-Fernández Department of Plant Sciences I, Universidad Complutense, Biodiversity and Taxonomy of Cryptogamic Plants Research Group, UCM, Madrid, Spain

David Rodríguez de la Cruz Department of Botany and Plant Physiology, University of Salamanca, Salamanca, Spain

Institute Hispano-Luso de Investigaciones Agrarias (CIALE), University of Salamanca, Salamanca, Spain

Jan de Vries Department of Biochemistry and Molecular Biology, Dalhousie University, Halifax, NS, Canada

Sophie de Vries Department of Biochemistry and Molecular Biology, Dalhousie University, Halifax, NS, Canada

Alicja Dolzblasz Department of Plant Developmental Biology, Institute of Experimental Biology, University of Wrocław, Wrocław, Poland

Bhupinder Dhir Department of Genetics, University of Delhi South Campus, New Delhi, India

Marta Esteban Area of Environmental Toxicity, National Center of Environmental Health, Instituto de Salud Carlos III, Majadahonda, Spain

Helena Fernández Area of Plant Physiology, Department of Organisms and Systems Biology (BOS), Oviedo University, Oviedo, Spain

Pablo Fernández Department of Plant Sciences I, Universidad Complutense, Biodiversity and Taxonomy of Cryptogamic Plants Research Group, UCM, Madrid, Spain

Beatriz Fernández-Marín Department of Plant Biology and Ecology, University of the Basque Country (UPV/EHU), Vizcaya, Spain

Jose María Gabriel y Galán Department of Plant Sciences I, Universidad Complutense, Biodiversity and Taxonomy of Cryptogamic Plants Research Group, UCM, Madrid, Spain

Helena García-Cortés National Centre for Environmental Health, National Health Institute Carlos III, Madrid, Spain

Jose Ignacio García-Plazaola Department of Plant Biology and Ecology, University of the Basque Country (UPV/EHU), Vizcaya, Spain

Edyta M. Gola Department of Plant Developmental Biology, Institute of Experimental Biology, University of Wrocław, Wrocław, Poland

Aránzazu Gómez-Garay Department of Plant Sciences I, Universidad Complutense, Madrid, Spain

Ueli Grossniklaus Department of Plant and Microbial Biology, University of Zürich, Zürich, Switzerland

Małgorzata Grzyb Department of Experimental Botany, Polish Academy of Sciences, Botanical Garden – Center for Biological Diversity Conservation in Warsaw, Warsaw, Poland

Yao-Moan Huang Taiwan Forestry Research Institute, Taipei, Taiwan

Erin E. Irish Department of Biology, The University of Iowa, Iowa City, IA, USA

Deepali Johari Pteridology Laboratory, Plant Diversity, Systematics & Herbarium Division, CSIR-National Botanical Research Institute, Lucknow, UP, India

Pei-Hsuan Lee Taiwan Forestry Research Institute, Taipei, Taiwan

Holly Long Department of Biology, The University of Iowa, Iowa City, IA, USA

Eugenia López-López Laboratorio de Evaluación de la Salud de los Ecosistemas Acuáticos, Escuela Nacional de Ciencias Biológicas, Instituto Politécnico Nacional, Ciudad de México, Mexico

Marina López-Pozo Department of Plant Biology and Ecology, University of the Basque Country (UPV/EHU), Vizcaya, Spain

Lena Ma State Key Laboratory of Pollution Control and Resource Reuse, School of the Environment, Nanjing University, Jiangsu, China

Soil and Water Science Department, University of Florida, Gainesville, FL, USA

Luisa Martín Department of Plant Sciences I, Universidad Complutense, Madrid, Spain

Kelly K. S. Matsunaga Department of Earth and Environmental Sciences, University of Michigan, Ann Arbor, MI, USA

Anna Mikuła Department of Experimental Botany, Polish Academy of Sciences, Botanical Garden – Center for Biological Diversity Conservation in Warsaw, Warsaw, Poland

Antonio Murciano Department of Applied Mathematics (Biomathematics), Faculty of Biology, Universidad Complutense, Madrid, Spain

Neural Plasticity Research Group, IdISSC" and "Neuro-computing and Neuro-robotics Research Groups, UCM, Madrid, Spain

Elżbieta Myśkow Department of Plant Developmental Biology, Institute of Experimental Biology, University of Wrocław, Wrocław, Poland

Valerie C. Pence Center for Conservation and Research of Endangered Wildlife, Cincinnati Zoo & Botanical Garden, Cincinnati, OH, USA

Teresa Pinto University of Trás-os-Montes and Alto Douro, Department of Biology and Environment, Centre for the Research and Technology of Agro-Environmental and Biological Sciences (CITAB), Vila Real, Portugal

Beatriz Pintos Department of Plant Biology I, Universidad Complutense, Madrid, Spain

Carmen Prada Department of Plant Sciences I, Universidad Complutense, Madrid, Spain

Luis G. Quintanilla Department of Biology and Geology, Physics and Inorganic Chemistry, University Rey Juan Carlos, Móstoles, Spain

Bala Rathinasabapathi Horticultural Sciences Department, University of Florida, Gainesville, FL, USA

Alejandro Rivera Area of Plant Physiology, Department of Organisms and Systems Biology (BOS), Oviedo University, Oviedo, Spain

José Luis Rodríguez-Gil Department of Biology, University of Ottawa, Ontario, Canada

Alexis Joseph Rodríguez-Romero Laboratorio de Evaluación de la Salud de los Ecosistemas Acuáticos, Escuela Nacional de Ciencias Biológicas, Instituto Politécnico Nacional, Ciudad de México, Mexico

Stanley J. Roux Department of Molecular Biosciences, The University of Texas, Austin, TX, USA

Jan Jarosław Rybczyński Department of Experimental Botany, Polish Academy of Sciences, Botanical Garden – Center for Biological Diversity Conservation in Warsaw, Warsaw, Poland

Mari L. Salmi Department of Molecular Biosciences, The University of Texas, Austin, TX, USA

José Ángel Sánchez-Agudo Department of Botany and Plant Physiology, University of Salamanca, Salamanca, Spain

Instituto Hispano-Luso de Investigaciones Agrarias (CIALE), University of Salamanca, Salamanca, Spain

Estefanía Sánchez-Reyes Department of Botany and Plant Physiology, University of Salamanca, Salamanca, Spain

Instituto Hispano-Luso de Investigaciones Agrarias (CIALE), University of Salamanca, Salamanca, Spain

José Sánchez-Sánchez Department of Botany and Plant Physiology, University of Salamanca, Salamanca, Spain

Instituto Hispano-Luso de Investigaciones Agrarias (CIALE), University of Salamanca, Salamanca, Spain

Jacinto Elías Sedeño-Díaz Coordinación Politécnica para la Sustentabilidad. Instituto Politécnico Nacional. Av. Instituto Politécnico Nacional s/n, Ciudad de México, Mexico

Andrea Seral Department of Plant Sciences I, Universidad Complutense, Biodiversity and Taxonomy of Cryptogamic Plants Research Group, UCM, Madrid, Spain

Emily B. Sessa Department of Biology, University of Florida, Gainesville, FL, USA

Ajit Pratap Singh Pteridology Laboratory, Plant Diversity, Systematics & Herbarium Division, CSIR-National Botanical Research Institute, Lucknow, UP, India

Liliana Cristina Soare Department of Natural Science, University of Piteşti, Piteşti, Romania

Nicoleta Anca Şuţan Department of Natural Science, University of Piteşti, Piteşti, Romania

Alexandru M. F. Tomescu Department of Biological Sciences, Humboldt State University, Arcata, CA, USA

Karolina Tomiczak Department of Experimental Botany, Polish Academy of Sciences, Botanical Garden – Center for Biological Diversity Conservation in Warsaw, Warsaw, Poland

Janos Vetter Department of Botany, University of Veterinary Sciences, Budapest, Hungary

Part I
Biology and Biotechnology in Ferns

Chapter 1 The Gametophyte of Fern: Born to Reproduce

Alejandro Rivera, María Jesús Cañal, Ueli Grossniklaus, and Helena Fernández

1.1 Introduction

I would like to confess with some degree of disappointment, that a not small part of my teaching activity on Plant Physiology at the Faculty, must be dedicated to seduce my students, underlining the enormous and not well recognized yet, importance the plants have for all the organisms living in the Earth. Plants transform the electromagnetic energy into the biosphere, in energy of chemical bounding, providing food to the rest of organisms. There are no doubts about the need to apply the most advanced technologies available so far to their study, in order to face the three major challenges coming up for human being such as food, energy, and environmental changes (Ehrhardt and Frommer 2012). In the order of preferences, our students place plants' interest behind animals and microorganisms and consider the plant research as a poor relative that does not deserve major attention. I have often to remind that plants and animals share many homologous genes, metabolic pathways, and cellular processes. Moreover, plants can go even far away in life spans, environmental resources management, and cellular plasticity. All these aspects make plants good experimental systems to understand the fundamental principles of life and also how we are able to cope with the continuous changes in the planet, inviting us to explore their great adaptive capacity under very stressful situations, some of them derived from the anthropogenic activity such as ultraviolet light, high temperature, etc.

A. Rivera · M. J. Cañal · H. Fernández (✉)
Area of Plant Physiology, Department of Organisms and Systems Biology (BOS), Oviedo University, Oviedo, Spain
e-mail: fernandezelena@uniovi.es

U. Grossniklaus
Department of Plant and Microbial Biology, University of Zürich, Zürich, Switzerland

H. Fernández (ed.), *Current Advances in Fern Research*,
https://doi.org/10.1007/978-3-319-75103-0_1

1.2 Why Is It Important to Lead Studies on Mosses, Lycophytes, and Ferns?

Certainly, a great success in the history of life, especially for green plants, was the acquisition from a single eukaryotic ancestor of a photosynthetic cyanobacterium (the ancestral plastid), according to the endosymbiont theory of Lynn Margulis (Schwartz 2007). Later on, no minus repercussions had the conquest of land from the fresh water, streptophyte green algae, during the Ordovician period, over 470 million years ago, colonizing and dominating terrestrial sceneries (Kernick and Crane 1997; Becker and Marin 2009). Flora expansion between the Silurian and Permian periods leads to the origin of plant groups represented today by bryophytes, lycophytes, and euphyllophytes, which include monilophytes (ferns) and spermatophytes (seed plants).

Bryophytes (which include hornworts, mosses, and liverworts), lycophytes, and monilophytes are remains of the invasion of land by plants, hiding clues capable to shade light on how plant development evolved. The terrestrial environment imposed variations in water availability and temperature, as well as increased exposure to radiation, thus demanding changes in body plan and modifications to cellular, physiological, and regulatory processes (Rensing et al. 2008; Pires and Dolan 2012). Certainly, ferns are among the pioneer vascular plants coping with the new environmental conditions and represent a critical clade for comparative evolutionary studies in land plants. They keep traits of an ancestral life history such as the lack of secondary growth, homospory, motile sperm, and independent free-living gametophyte and sporophyte generations. Moreover, ferns represent an unexplored genetic diversity that could be taken advantage for improving plants by means of genetic transfer technologies (Rathinasabapathi 2006).

Ferns have received minor attention, and to date, only few fern species such as *Adiantum capillus-veneris*, *Anemia phyllitidis*, *Blechnum spicant*, *Dryopteris affinis* ssp. *affinis*, *Ceratopteris richardii*, *Marsilea vestita*, *Matteuccia struthiopteris*, *Onoclea sensibilis*, and *Pteridium aquilinum* have been used to study basic developmental processes such as photomorphogenesis (Wada 2007), germination (Salmi et al. 2005, 2007; Suo et al. 2015), cell polarity (Salmi and Bushart 2010), cell wall composition (Eeckhout et al. 2014), or asexual and sexual reproduction as the gametophyte is an autonomous living organism, for in vitro culture and sample collection (Whittier and Steeves 1960, 1962; von Aderkas 1984; Wen et al. 1999; Fernández and Revilla 2003; Cordle et al. 2007, 2010, 2012; Kezmierczak 2010; López and Renzaglia 2014; Valledor et al. 2014; DeVries et al. 2016; Grossmann et al. 2017).

Apart from the abovementioned processes, we assist recently to the use of ferns to resolve interesting problems in the plant world caused by abiotic and biotic stress. Drought is one of the most severe abiotic stress factors affecting plant growth and productivity and has caused considerable reduction in crop yield worldwide. Several fern and fern-ally species of *Actiniopteridaceae*, *Sinopteridiaceae*, *Pteridaceae*, and *Selaginellaceae* have been associated to desiccation tolerance. Concretely, the fern-ally *Selaginella* is one of the most primitive vascular resurrection plants, which can survive a desiccated state and recover when water becomes available, by morphological adaptations, hormonal regulation, antioxidant protection, and accumulation of osmolytes, which could serve to cope with drought in crops (Wang et al. 2010). Other important adaptations of ferns to extreme environments such as salinity, heavy metal, epiphytism, or invasiveness tolerance are summarized by Rathinasabapathi (2006). More recently, it was published in a fascinating paper based upon the feature that ferns and mosses are rarely infested by phytophagous insect in the field (Hendrix 1980; Markham et al. 2006) and in which an insecticidal protein from the fern species *Tectaria macrodonta* (Fee) C. Chr. was identified and expressed in transgenic cotton lines, conferring protection against whitefly, a sap-sucking pest (Shukla et al. 2016).

However, ferns are reported to have higher chromosome numbers and larger genomes than mosses and seed plants (Barker and Wolf 2010), making it difficult to establish genetic resources for them such as genomic and transcript sequence data. Recently, the advent of the next-generation sequencing (NGS) technologies, such as Roche'454 GS-FLX Titanium and Illumina HiSeq sequencers, by means of which is possible to characterize the transcriptome in plants, representing a small but information reach-target compared to complete genome (Ward et al. 2012). The variation in gene expression induced by whatever environmental or inner condition can be examined in non-model organisms because these techniques have become more feasible as automation and efficiency has reduced the cost. Until present and recently, some NGS transcriptome data sets have been published for ferns, which include the species *Pteridium aquilinum* (Der et al. 2011), *Ceratopteris richardii* (Bushart et al. 2013), *Lygodium japonicum* (Aya et al. 2015), *Dryopteris affinis* ssp. *affinis* (Grossmann et al. 2017), and some others resulting from the oneKP project (Matasci et al. 2014). Moreover, one of the major genomics centers in the world, BGI (in Beijing) and the China National GeneBank (CNGB), have announced 10KP, their plan to sequence 10,000 genomes or more, crossing every major plant clade and eukaryotic microbes, which will build on oneKP project.

Omics technologies based on comprehensive biochemical and molecular characterizations of an organism, tissue, or cell type and next-generation omics approaches facilitate the analyses of non-model organisms owing to the rapid generation of large amounts of de novo systems biology data, making them attractive options for studying plant development and evolution. In our lab, during the last years, they have published a significant number of reports, dealing with different topics carried out either in model species such as *Arabidopsis* and *Chlamydomonas* or non-model species, the case of *Pinus radiate*, focusing on different aspects of plant development (Valledor et al. 2010; Romero-Rodríguez et al. 2014; Jorrín-Novo et al. 2015; Sánchez-Lucas et al. 2016).

1.3 The Gametophyte: Born to Reproduce Either Sexually or Asexually

During its life cycle, a fern exists in two distinct forms: the small, sample, haploid gametophyte and the large, morphologically complex, diploid sporophyte. In ferns, the gametophyte lives separately from sporophyte (Klekowski and Baker 1966; Haufler et al. 2016) excluding some time after fertilization takes place, supporting the beginning of sporophyte development. In fact, as sporophyte develops, the gametophyte disappears in most cases, reflecting to have a role purely involved on reproduction. In *Thelypteris palustris* (Salisb.) Schott, sporophytes are supported by gametophytes until the first leaves develop; then sporophytes grow on the organic matter they produce themselves (Sakamaki and Ino 2007).

The evolution of sexual reproduction represents a major transition in the evolution of life, and occurred well before plants first ventured onto land some 470 million years ago, when the ancestors already had differentiated male and female gametes (spermatozoids and eggs), inside the sex organs archegonia and antheridia (Pannell 2017). The first land plants had also already evolved an alternation of haploid and diploid generations, gametophyte and sporophyte (Kenrick and Crane 1997), being the gametophyte, the generation when the differentiation of sex organs into egg- or sperm-producing tissues occurs. Actually, in seedless plants, gametophytes may be male or female, with separate individuals producing either antheridia or archegonia, or they may be hermaphroditic, with both antheridia and archegonia present in the same individuals (Okada et al. 2001).

Sex determination is a matter of the differentiation of cells and tissues in different parts of the same individuals. In many bryophytes, sex is determined in gametophytes by (U and V) sex chromosomes (Okada et al. 2001). To my knowledge, no sex chromosomes have been described for any ferns or lycophytes so far. In these plant groups, sex determination differs between "homosporous" species, in which sporophytes produce spores of the same size (as in bryophytes), and "heterosporous" species, in which sporophytes produce both small "microspores" and larger "megaspores". In some homosporous ferns, gametophytes are all functionally hermaphroditic, with both antheridia and archegonia. In these cases, sex determination is a question of cellular and tissue differentiation into different male and female "gametangia" (Klekowski 1969). In other homosporous species, gametophytes may develop a unisexual function (producing either only antheridia or only archegonia) (Klekowski 1969). In "heterosporous" ferns and lycophytes, sex determination acts through the size of the spore from which the gametophytes germinate. The gametophytes do not differ genetically, and the sporophyte controls the sex of its gametophytes by regulating the spore-producing "sporangia" to produce either small "microspores", which develop into male (micro-) gametophytes, or larger "megaspores", which develop into female (mega-) gametophytes. Heterospory has evolved several times in vascular land plants, in ferns and lycophytes, as well as independently in the lineage that gave rise to seed plants (Bateman and Dimichele 1994).

In angiosperms (flowering plants) and gymnosperms, the gametophytic phase has become even more highly modified and reduced, and differentiation between the male and female structures takes place in sporophytic structures in which the gametophytes form. In contrast to non-seed plants, seed-plant gametophytes are always either male or female, and it is the sporophyte that determines their sex (Pannell 2017).

1.4 Blechnum spicant: A Sexual Species

Blechnum spicant L. belongs to one of the most ancient groups of ferns. Sexual development in the gametophyte of this species is carried out by the action of the unknown antheridium-inducing substances (Fig. 1.1). When cultured in vitro, initially the gametophyte develops female sexual organs or archegonia and produces and excretes antheridiogens into the medium that induces the formation of male sexual organs or antheridia in the youngest filamentous or spatulate-shaped gametophytes (Klekowski 1969; Cousens 1979; Fernández et al. 1997, 1999; Menéndez et al. 2006a).

In Pteridophyta, sporophyte formation occurs by the means of the fusion of sexual cells, either of the same or different prothalli, and we can distinguish between intro- and intergametophytic crosses, respectively (Klekowski 1969). Natural systems have been reported to favor intergametophytic crosses, promoting genetic interchange at different levels, such as morphological, population, or genetic. In many ferns, a chemical messenger, antheridiogen, controls the onset of antheridium formation in young gametophytes, and it has been also found in *B. spicant* L. (Cousens 1979; Fernández et al. 1997, 1999).

It is clear that this induction should be reflected in a complete reprogramming of the cell fate and function, since there are not only functional but structural changes during this process. Proteomic analyses have been conducted in flowering plants in relation to reproduction, and, concretely, several reports have been published on male gametophyte development in higher plants (Kerim et al. 2003; Dai et al. 2007; Sheoran et al. 2007; Mayank et al. 2012; Ischebeck et al. 2014; Chaturvedi et al. 2016a, b). Valledor et al. (2014) compared the proteomic profiles between female and male gametophytes of *B. spicant.* Female gametophytes were obtained in MS medium (Murashige and Skoog 1962) directly from spores, and the males ones were obtained growing the gametophytes in the same medium supplemented with an extract of mature gametophytes showing antheridiogenic effects. The induction of male gametophytes by antheridiogens correlated to a global increase in stress- and defense-related pathways, which affects also flavonoid signaling and cell division while reducing protein biosynthesis pathways (Fig. 1.2). On the other hand, the photosynthesis and other energy-related pathways are affected also during this induction, being an indicator of how stressed are the gametophytes during induction process.

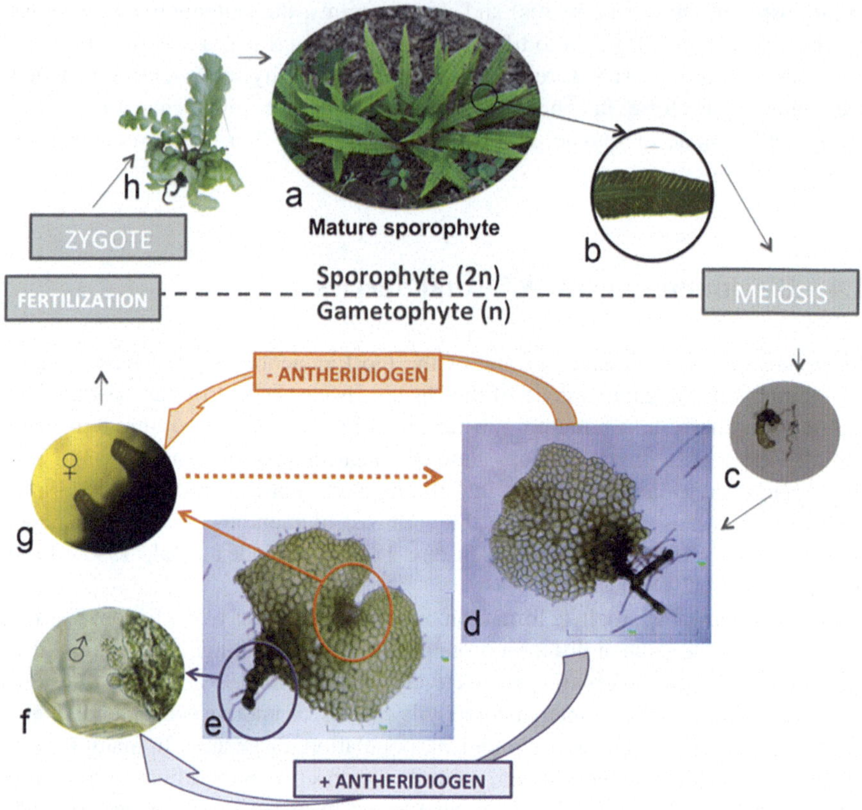

Fig. 1.1 *Blechnum spicant* sexual life cycle. Mature sporophytes (**a**) have reproductive leaves with sori (**b**). Each sorus is a cluster of sporangia in which spores are produced via meiosis. The fern spores develop into a small gametophyte (**c**) that sustains itself by photosynthesis. Young gametophytes are spatulate-shaped (**d**) and quickly maturate into a heart shape (**d**). The gametophytes with fastest growth differentiate as a female (archegonia, **e**), and, along with this maturation, they produce and excrete the antheridiogen to the media. The presence of this pheromone will induce the male differentiation of the surrounding immature gametophytes (antheridia, **e**). Anteridium produces fern sperm (**f**) that uses flagella to swim through moisture from antheridia to fertilize the eggs in the archegonia (**g**). After fertilization a zygote develops into a new young sporophyte (**h**)

1.5 *Dryopteris affinis* ssp. *affinis*: An Apomictic Species

Sexual reproduction generates new genetic individuals by combining the genetic material of two parental individuals, while asexual reproduction is limited to one genetic entity. Most angiosperms reproduce sexually through seeds, where a single generative cell (archesporial cell or megaspore mother cell) undergoes meiosis to produce four chromosomally reduced cells (megaspores). After significant cellular enlargement, the nucleus of the functional megaspore usually undergoes three

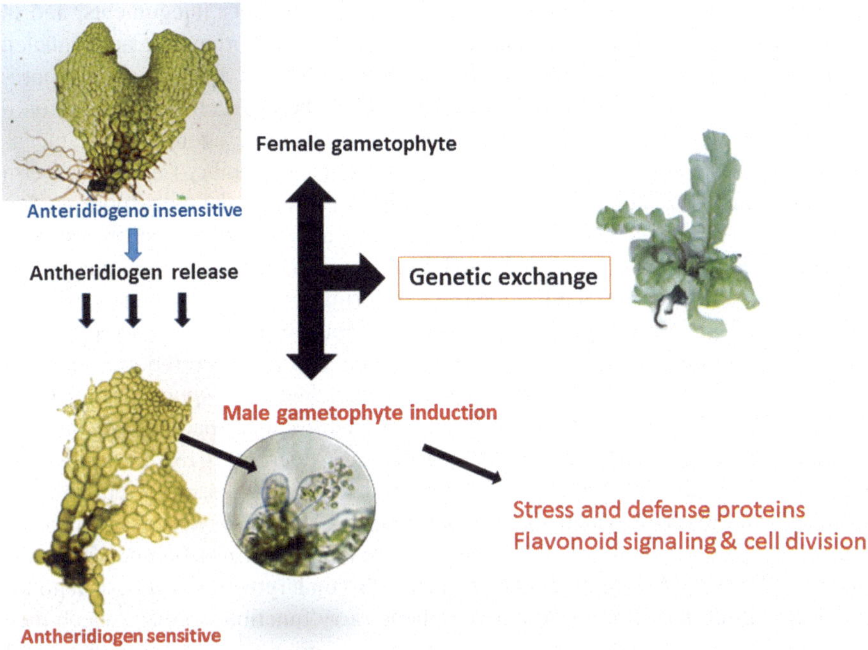

Fig. 1.2 Scheme illustrating the antheridiogen-mediated maleness operating in *Blechnum spicant* L. and the proteins induced

rounds of mitosis before giving rise to a gametophyte composed of seven cells: two companion synergids, the egg cell, a binucleated central cell, and three antipodals. Double fertilization of both the egg and central cell is necessary to trigger embryogenesis and endosperm development, respectively (Taiz and Zeiger 2015; Rodriguez-Leal and Vielle-Calzada 2012).

In plants, as in animals, transition to asexuality may occur; however, despite of its importance, little is known about the mechanisms that cause these transitions and why some taxa experience much higher transition rates than others, especially in plants, where more intensively it has been studied, due to the potential use of asexuality in crop plants to improving agriculture (Neiman et al. 2014). Asexual reproduction in plants can occur either through budding and vegetative growth (e.g., shoots and runners) or by seeds, referred as apomixis, which was a great obstacle for Mendel, when he decided to corroborate his results of genetic inheritance in peas (*Pisum*) in hawkweed (*Hieracium*) and was astonished that *Hieracium* offspring were morphologically identical to their mother plants (Neiman et al. 2014).

There are examples of asexual seed formation (apomixis), where seeds form without meiosis and fertilization (Nogler 1984; Koltunow et al. 1995; Koltunow and Grossniklaus 2003; Bicknell and Koltunow 2004; Ozias-Akins 2006; Rodriguez-Leal and Vielle-Calzada 2012; Barcaccia and Albertini 2013). In its simplest form (adventitious embryony, also called sporophytic apomixis), apomictic plants form embryos directly from a somatic cell from nucellus or inner integuments, and in species with endosperm development, the endosperm is formed by polar nuclei. Apomictic plants can also form embryos directly from a chromosomally unreduced female gametophyte (apomeiosis) in which the egg cell develops autonomously into an embryo by parthenogenesis (gametophytic apomixis), by means of two types, apospory and diplospory. In apospory, the embryo and endosperm develop in unreduced embryo sac in the ovule. In this case, the megaspore mother cell in the sexual ovule starts to develop but stops at some stage, and one or more somatic cells in the ovule and their nuclei start to develop, resembling megaspore mother cells. Before the mature embryo sac formation, the megaspore or young embryo is aborted and replaced by developing aposporous sacs. Apospory is by far the most common mechanism in higher plants and has been reported in *Beta*, *Brachiaria*, *Cenchrus*, *Chloris*, Compositae, *Eriochloa*, *Heteropogon*, *Hieracium*, *Hyparrhenia*, *Hypericum*, *Panicum*, *Paspalum*, *Pennisetum*, Poaceae, *Ranunculus*, *Sorghum*, *Themeda*, and *Urochloa* (Barcaccia and Albertini 2013). In diplospory, the embryo and endosperm develop in an unreduced embryo sac derived from the megaspore mother cell, which differentiates as in sexual ovules but does not undergo meiosis. Diplospory is found in *Tripsacum*, *Eragrostis*, and *Taraxacum* (Kandemír and Saygili 2015). Apomictic female gametes (2n) undergo embryogenesis autonomously, without fertilization. Apomictic plants carry functional pollen, which they sometimes need for endosperm formation as the formation of endosperm is still dependent on fertilization of the central cell (pseudogamy) (Rodriguez-Leal and Vielle-Calzada 2012; Kandemír and Saygili 2015). Observation of apomixis is difficult since it is generally accompanied by sexual reproduction or facultative apomixis (Kandemír and Saygili 2015).

In ferns, apomixis is an important mode of asexual reproduction (Döpp 1939; Manton 1950), which has evolved several times independently and being its frequency at least 3%, a value much higher than in other major plant groups (Ekrt and Koutecký 2016). However, most apomictic fern species are concentrated in just four families (Liu et al. 2012). Apomixis in ferns (Fig. 1.1) includes apogamy, the formation of sporophytes from somatic cells of the prothallium, and agamospermy (or diplospory), the production of unreduced (diplo) spores (Manton 1950; Ekrt and Koutecký 2016).The archesporial cell of sexual fern species usually undergoes 4 mitoses to produce 16 spore mother cells that undergo regular meiosis, resulting in 64 reduced spores in 16 tetrads. Under the prevailing type of agamospermy (Döpp-Manton scheme), the last (premeiotic) mitosis fails, resulting in 8 spore mother cells that undergo regular meiosis, producing 32 diplospores in 8 tetrads (Döpp 1939; Manton 1950). Rarely, the first meiotic division fails, which results in 32 diplospores in 16 diads (Braithwaite 1964). Genetic variation among apomictic offspring has been documented (Peredo et al. 2013).

Apogamy in ferns may be obligate, when gametophytes produce nonfunctional gametes, facultative, or induced by exogenous factors Menéndez et al. 2006b; Cordle et al. 2007). Contrary to ferns, apogamy does not occur naturally in angiosperms (Yang and Zhou 1992), but apogamous sporophytes can be induced by the culture of pollen or embryo sacs, indicating that the developmental plasticity necessary to overcome meiosis and fertilization barriers is not restricted to ferns (Seguí-Simarro 2010; Germanà 2011). In apogamy, somatic cells of the gametophyte are reprogrammed to start the sporophytic developmental program (Okano et al. 2009).

Because apomixis allows the fixation of complex genotypes, including that of highly productive F1 hybrids, many researchers have extolled the tremendous potential that apomixis holds for plant improvement, whose benefits could surpass those of the green revolution (Grossniklaus et al. 1998; Spillane et al. 2004; Marimuthu et al. 2011). Apomixis combines the advantages of propagation by seed (higher multiplication rate, easier storage and planting, suitability for machine planting, less seed material use, and less bearing of diseases) with those of propagation by clone (maintaining genetic structure and hence fixing superior genotypes after crossing) (Kandemír and Saygili 2015).

Over the last decades, several studies focusing on apomixis in model species of angiosperms concluded that sexual and apomictic pathways share gene expression profiles and, thus, common molecular regulatory features, indicating that they are not distinct pathways (Grossniklaus et al. 2001; Tucker et al. 2003). The initiation of apomixis invariably occurs during early ovule ontogeny; sexual and apomictic development can coexist within the same ovule, or within different ovules of a same individual, suggesting that apomixis could have originated as a modified form of sexual reproduction that has undergone deregulation of key developmental steps during gametogenesis (Koltunow and Grossniklaus 2003). So far it is a truth that very few crop species are apomictic, and attempts to introduce this trait by crossing have failed. The alternative would be to de novo engineer apomixis, but for this strategy to be applied, the genes that confer elements of apomixis must be identified.

Apomixis research can be faced by different approaches in each of its three major stages: apomeiosis, parthenogenesis, and seed formation. In most species under study, the basic components of apomixis can be explained by a few genes that control unreduced gamete formation and parthenogenesis, respectively; however, polyploidy, segregation distortion, suppressed recombination, epistatic interactions, naturally active modifiers, and environmental effects complicate their genetic analyses (Rodriguez-Leal and Vielle-Calzada 2012).

There are various ways of converting crop plants into apomictic ones: (a) wide crosses with apomictic wild relatives, (b) mutation, and (c) genetic transformation. Transfer of apomixis from wild relatives via sexual hybridization depends on the presence of relatives with which interspecific hybridizations can be made but this is not possible for most cultivated species. Although significant developments have been done recently, it is generally accepted that apomixis transfer via wide crosses has been unsuccessful so far (Spillane et al. 2004). Cloning of LOA and LOP genes of *Hieracium* is underway, and, after their cloning and transfer, apomixis could be introgressed into other crop species (Kotani et al. 2014). On the other hand, studies in the model plant Arabidopsis revealed that apomixis can also be achieved through artificial mutations. One of them involves the gen Osd1, which controls the entering into the second meiotic division (d'Erfurth et al. 2009). By combining a mutation in this gene with two other mutations, one that eliminates recombination and pairing (Atspo11-1) and another that modifies chromatid segregation (Atrec8), a new genotype was created in which meiosis is totally replaced by mitosis without affecting subsequent sexual processes and called MiMe for "mitosis instead of meiosis" (d'Erfurth et al. 2009; Marimuthu et al. 2011). The induction of apomeiosis by the creation of the MiMe genotype is an important step toward understanding and engineering apomixis, giving place to a genotype called MiMe in which meiosis was replaced with mitosis. In another mechanism, a mutation in the Arabidopsis SWI1 gene leads to apomeiosis and diploid egg formation (Ravi et al. 2008). Chaudhury et al. (1997) reported seed development in mutations of Arabidopsis FIS1, FIS2, and FIS3 genes in the absence of fertilization.

Dryopteris affinis (Lowe) Fraser-Jenkins ssp. *affinis* is a diploid fern with an apomictic life cycle, and it probably originates from the crossing of *D. wallichiana* and *D. oreades*, being broadly distributed in the Mediterranean, Macaronesia, and western Eurosiberian regions (Salvo 1990). When cultured in vitro, apogamy in this species is evident as gametophyte develops a brown meristematic area near the apical indentation that evolves into a new sporophyte. Sexual reproduction is not possible due to the lack of archegonia (Fernández et al. 1996) (Fig. 1.3). Apogamy in ferns can be seen as an opportunity to investigate on embryogenesis, one of the more powerful tools in plant biotechnology. On the other hand, in both apomixis and apogamy, unreduced cells form an embryo without fertilization, and it might be expected that they share some common features. Moreover, the mechanism of asexual reproduction in lower and higher plants appears to be controlled by overlapping sets of genes (Cordle et al. 2012). How somatic cells, either of sporophytic or gametophytic origin, become and develop as embryogenic is still poorly understood (Radoeva and Weijers 2014). Somatic embryogenesis research in ferns is scarce so far and limited to a few reports (Mikuła et al. 2015; Domżalska et al. 2017) and even minor is the number of reports involving embryogenesis linked to apogamy (Bui et al. 2017; Grossmann et al. 2017). In general terms, we can assume that the molecular basis of embryogenesis by sexual or asexual means, in plant kingdom, is far to be understood (De Smet et al. 2010). In seed plants, several groups of genes, mostly encoding transcription factors, such as BABY BOOM, AINTEGUMENTA-like 5, FUSCA3, LEAFY COTYLEDON, etc., have been associated to embryogenesis even though they are required either for embryogenesis itself or plan cell viability and would need still further explanation (Radoeva and Weijers 2014).

Performing proteomic studies on non-model organisms with little or no genomic information is still difficult. In our lab, recently, both transcriptomic and proteomic analyses were performed to increase our knowledge on the molecular basis of apogamy in *D. affinis* ssp. *affinis*, by using next-generation sequencing (NGS) and shotgun proteomics by tandem mass spectrometry (Grossmann et al. 2017). This way 1397 protein clusters with 5865 unique peptide sequences were identified, identifying homologs of proteins involved in several activities (Fig. 1.4a) from several taxa (Fig. 1.4b) and in particular on reproduction of higher plants, including proteins with a potential role in apogamy. Furthermore, in this study, we detected some fern

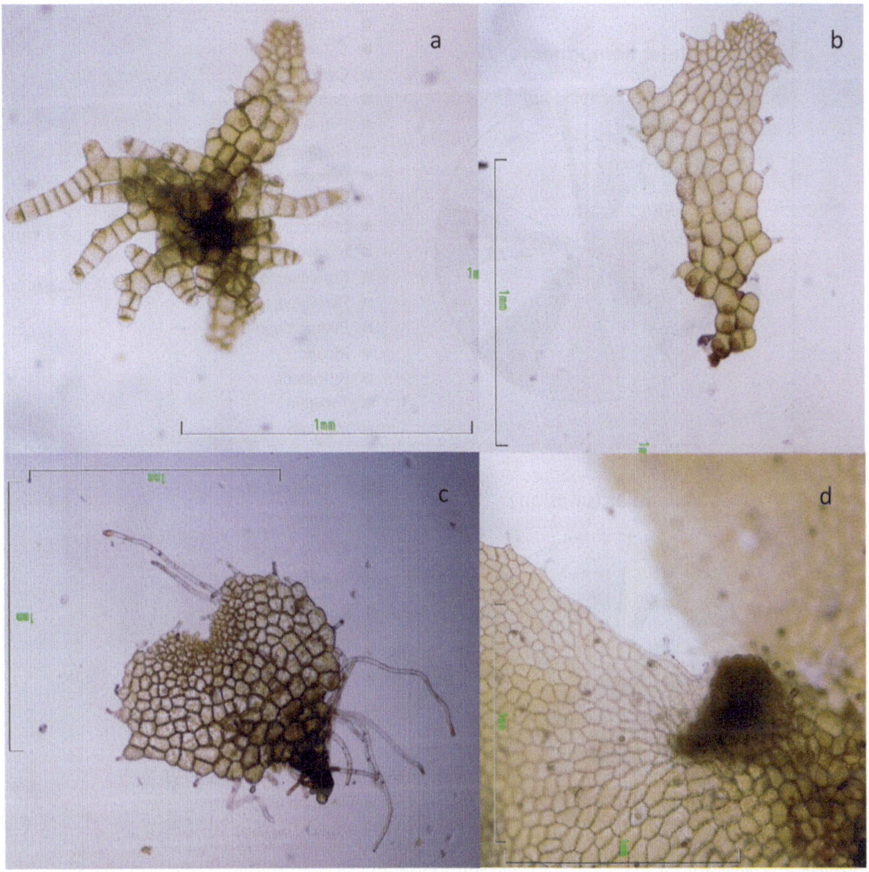

Fig. 1.3 Vegetative developmental and apogamy in the gametophyte of the fern *Dryopteris affinis* ssp. *affinis*. Successive morphological appearance of gametophyte: (**a**) filamentous, (**b**) spatulate- and (**c**) heart-shaped, and (**d**) apogamous embryo

protein homologous to ARGONAUTE10/PINHEAD/ZWILLE and also to the *A. thaliana* SERRATE (SE) RNA effector protein which could participate in the meristematic activity of the incipient apogamic embryo or unknown roles in the switch between sexual and asexual reproduction and perhaps in the regulation of apogamy in ferns (Grossmann et al. 2017). Additionally, proteins involved in gene silencing or enzymes involved in cell wall modifications such as pectinecterases were also identified, which could hide some role on apogamy (Li et al. 2011). Recently, we have done a RNA-Seq approach to compare gene expression profiles of one- and two-dimensional gametophytes of this species, finding several thousands of genes differentially expressed and related to different aspects of either vegetative or reproductive behavior of the gametophyte (Fernández, personal communication).

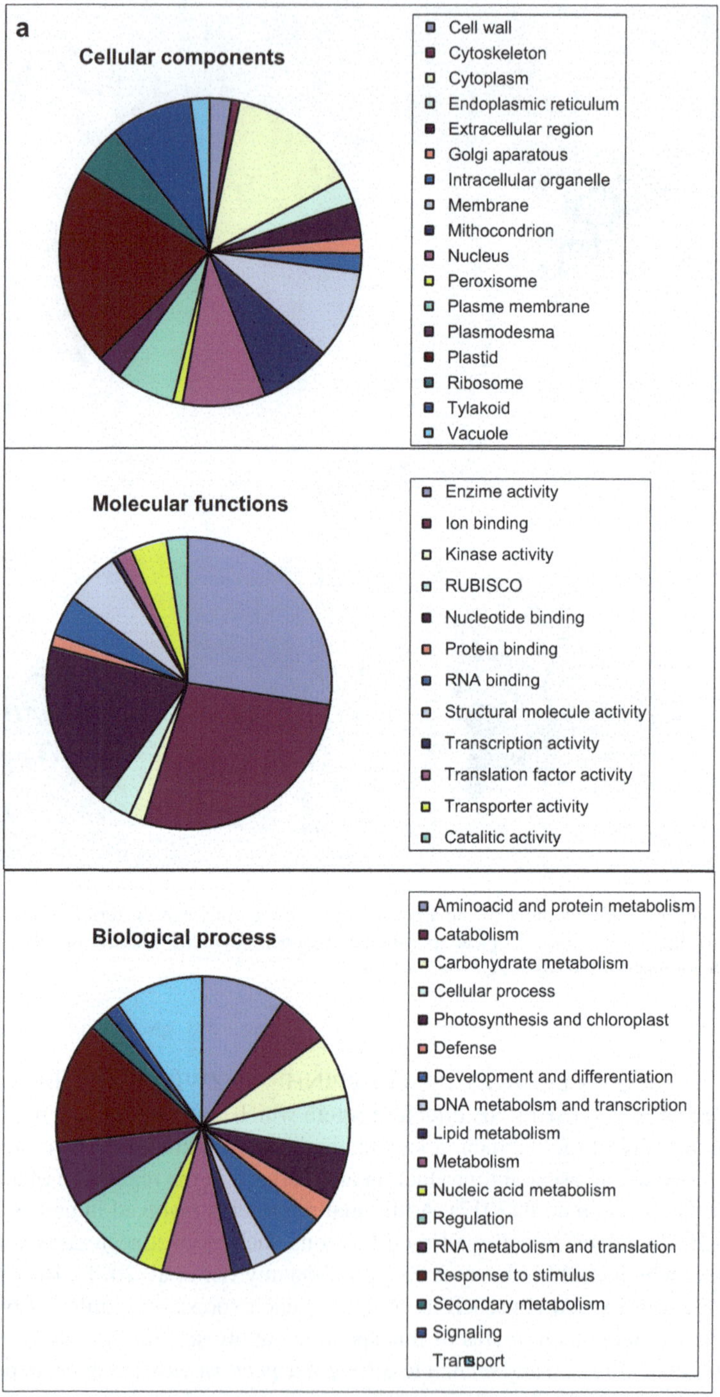

Fig. 1.4 Proteomic profile from the gametophyte of *Dryopteris affinis* ssp. *affinis*. (**a**) Distribution of plant GO-slim functional categories; (**b**) matching species

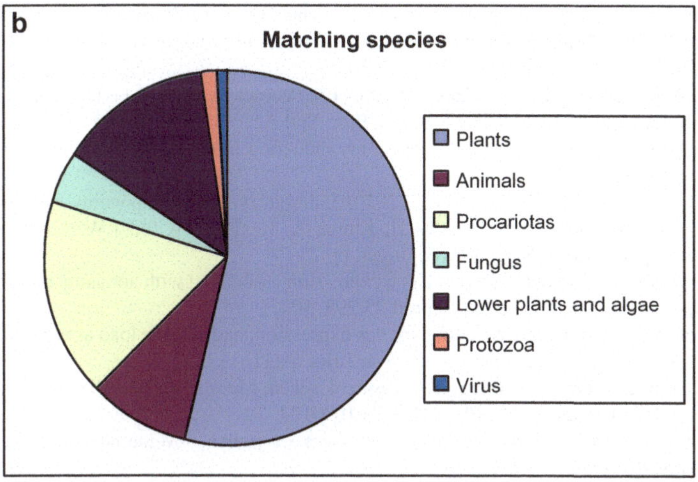

Fig. 1.4 (continued)

In summary, with the increasing availability of genomic data from non-model species, similar proteogenomics approaches will improve the sensitivity in protein identification for species only distantly related to models.

References

Aya K, Kobayashi M, Tanaka J, Ohyanagi H, Suzuki T, Yano K, Takano T, Yano K, Matsuoka M (2015) *De novo* transcriptome assembly of a fern, *Lygodium japonicum*, and a web resource database, Ljtrans DB. Plant Cell Physiol 56:e5

Barcaccia G, Albertini E (2013) Apomixis in plant reproduction: a novel perspective on an old dilemma. Plant Reprod 26:159–179

Barker MS, Wolf PG (2010) Unfurling fern biology in the genomics age. Bioscience 60:177–185

Bateman RM, Dimichele WA (1994) Heterospory: the most iterative key innovation in the evolutionary history of the plant kingdom. Biol Rev 69:345–417

Becker B, Marin B (2009) Streptophyte algae and the origin of embryophytes. Ann Bot 103:999–1004

Bicknell RA, Koltunow AM (2004) Understanding apomixis: recent advances and remaining conundrums. Plant Cell 16(Suppl):S228–S245

Braithwaite AF (1964) A new type of apogamy in ferns. New Phytol 63:293–305

Bui LT, Pandzic D, Youngstrom CE, Wallace S, Irish EE, Szövényi P, Cheng C-L (2017) A fern *AINTEGUMENTA* gene mirrors *BABY BOOM* in promoting apogamy in *Ceratopteris richardii*. Plant J 90:122–132

Bushart TJ, Cannon AE, Ul Haque A, San Miguel P, Mostajeran K, Clark GB, Porterfield DM, Roux SJ (2013) RNA-seq analysis identifies potential modulators of gravity response in spores of *Ceratopteris* (Parkeriaceae): evidence for modulation by calcium pumps and apyrase activity. Am J Bot 100:161–174

Chaturvedi P, Ghatak A, Weckwerth W (2016a) Pollen proteomics: from stress physiology to developmental priming. Plant Reprod 29:119–132

Chaturvedi P, Selymesi M, Ghatak A, Mesihovic A, Scharf K-D, Weckwerth W, Simm S, Schleiff E (2016b) The membrane proteome of male gametophyte in *Solanum lycopersicum*. J Proteome 131:48–60

Chaudhury AM, Ming L, Miller C, Craig S, Dennis ES, Peacock WJ (1997) Fertilization-independent seed development in *Arabidopsis thaliana*. Proc Natl Acad Sci U S A 94(8):4223–4228

Cordle AR, Irish EE, Cheng CL (2007) Apogamy induction in *Ceratopteris richardii*. Int J Plant Sci 168:361–369

Cordle AR, Bui LT, Irish EE, Cheng CL (2010) Laboratory-induced apogamy and apospory in *Ceratopteris richardii*. In: Fernández H, Kumar A, Revilla MA (eds) Working with Ferns. Issues and applications, New York, Springer, pp 25–36

Cordle AR, Irish EE, Cheng CL (2012) Gene expression associated with apogamy commitment in *Ceratopteris richardii*. Sex Plant Reprod 25:293–304

Cousens MI (1979) Gametophytic ontogeny, sex expression, and genetic load as measures of population divergence in *Blechnum spicant*. Am J Bot 66:116–132

d'Erfurth I, Jolivet S, Froger N, Catrice O, Novatchkova M, Mercier R (2009) Turning meiosis into mitosis (GP Copenhaver, Ed.) PLoS Biol 7:e1000124

Dai S, Wang T, Yan X, Chen S (2007) Proteomics of pollen development and germination. J Proteome Res 6:4556–4563

De Vries J, Fischer AM, Roettger M, Rommel S, Schluepmann H (2016) Cytokinin-induced promotion of root meristem size in the fern *Azolla* supports a shoot-like origin of euphyllophyte roots. New Phytologist, 209:705–720

Der JP, Barker MS, Wickett NJ, dePamphilis CW, Wolf PG (2011) De novo characterization of the gametophyte transcriptome in bracken fern, *Pteridium aquilinum*. BMC Genomics 12:99

DeSmet I, Lau S, Mayer U, Jürgens G (2010) Embryogenesis – the humble beginnings of plant life. Plant J 61:959–970

Domżalska L, Kędracka-Krok S, Jankowska U, Grzyb M, Sobczak M, Rybczyński JJ, Mikuła A (2017) Proteomic analysis of stipe explants reveals differentially expressed proteins involved in early direct somatic embryogenesis of the tree fern *Cyathea delgadii* Sternb. Plant Sci 258:61–76

Döpp W (1939) Cytologische und genetische Untersuchungen innerhalb der Gattung Dryopteris. Planta 29:481

Eeckhout S, Leroux O, Willats WGT, Popper ZA, Viane RLL (2014) Comparative glycan profiling of *Ceratopteris richardii* C-Fern` gametophytes and sporophytes links cell-wall composition to functional specialization. Ann Bot 114:1295–1307

Ehrhardt DW, Frommer WB (2012) New technologies for 21st century plant science. Plant Cell 24:374–394

Ekrt L, Koutecký P (2016) Between sexual and apomictic: unexpectedly variable sporogenesis and production of viable polyhaploids in the pentaploid fern of the Dryopteris affinis agg. (Dryopteridaceae). Ann Bot 117:97–106

Fernández H, Revilla MA (2003) *In vitro* culture of ornamental ferns. Plant Cell Tissue Organ Cult 73:1–13

Fernández H, Bertrand AM, Sánchez-Tamés R (1996) Influence of tissue culture conditions on apogamy in *Dryopteris affinis sp. affinis*. Plant Cell Tissue Organ Cult 45:93–97

Fernández H, Bertrand AM, Feito I, Sánchez-Tamés R (1997) Gametophyte culture in vitro and antheridiogen activity in *Blechnum spicant*. Plant Cell Tissue Organ Cult 50:71–77

Fernández H, Bertrand AM, Sierra MI, Sánchez-Tamés R (1999) An apolar GA-like compound responsible for the antheridiogen activity in *Blechnum spicant*. Plant Growth Regul 28:143–144

Germanà MA (2011) Gametic embryogenesis and haploid technology as valuable support to plant breeding. Plant Cell Rep 30:839–857

Grossmann J, Fernández H, Chaubey PM, Valdés AE, Gagliardini V, Cañal MJ, Russo G, Grossniklaus U (2017) Proteogenomic analysis greatly expands the identification of proteins related to reproduction in the apogamous fern *Dryopteris affinis ssp. affinis*. Front Plant Sci 8

Grossniklaus U, Koltunow A, van Lookeren Campagne M (1998) A bright future for apomixis. Trends Plant Sci 3:415–416

Grossniklaus U, Nogler GA, van Dijk PJ (2001) How to avoid sex: the genetic control of developmental aspects. Plant Cell 13:1491–1497
Haufler CH, Pryer KM, Schuettpelz E, Sessa EB, Farrar DR, Moran R, Schneller JJ, Watkins JE, Windham MD (2016) Sex and the single gametophyte: revising the homosporous vascular plant life cycle in light of contemporary research. Bioscience 66:928–937
Hendrix SD (1980) An evolutionary and ecological perspective of the insect fauna of ferns The American Society of Naturalists An Evolutionary and Ecological Perspective of the Insect Fauna of Ferns. Am Nat Source Am Nat 115:171–196
Ischebeck T, Valledor L, Lyon D, Gingl S, Nagler M, Meijón M, Egelhofer V, Weckwerth W (2014) Comprehensive cell-specific protein analysis in early and late pollen development from diploid microsporocytes to pollen tube growth. Mol Cell Proteomics 14(:295–310
Jorrín-Novo JV, Pascual J, Sánchez-Lucas R, Romero-Rodríguez MC, Rodríguez-Ortega MJ, Lenz C, Valledor L (2015) Fourteen years of plant proteomics reflected in *Proteomics*: moving from model species and 2DE-based approaches to orphan species and gel-free platforms. Proteomics 15:1089–1112
Kandemír N, Saygili İ (2015) Turkish Journal of Agriculture and Forestry Apomixis: new horizons in plant breeding
Kazmierczak A (2010) Gibberellic acid and ethylene control male sex determination and development of *Anemia phyllitidis* gametophytes. In: Fernández H, Kumar A, Revilla MA (eds) Working with ferns. Issues and applications. Springer, New York, pp 49–65
Kenrick P, Crane PR (1997) The origin and early evolution of plants on land. Nature 389:33–39
Kerim T, Imin N, Weinman JJ, Rolfe BG (2003) Proteome analysis of male gametophyte development in rice anthers. Proteomics 3:738–751
Klekowski EJ (1969) Reproductive biology of the Pteridophyta. III. A study of the Blechnaceae. Bot J Linn Soc 62:361–377
Klekowski EJ, Baker HG (1966) Evolutionary significance of polyploidy in the pteridophyta. Science 153:305–307
Koltunow AM, Grossniklaus U (2003) Apomixis: a developmental perspective. Annu Rev Plant Biol 54:547–574
Koltunow AM, Bicknell RA, Chaudhury AM (1995) Apomixis: molecular strategies for the generation of genetically identical seeds without fertilization. Plant Physiol 108:1345–1352
Kotani Y, Henderson ST, Suzuki G, et al (2014) The LOSS OF APOMEIOSIS (LOA) locus in Hieracium praealtum can function independently of the associated large-scale repetitive chromosomal structure. New Phytologist 201:973–981
Li JJ, Liu L, Ouyang YD, Yao JL (2011) Sexual reproduction development in apomictic *Eulaliopsis binata* (Poaceae). Genet Mol Res 10:2326–2339. https://doi.org/10.4238/2011.October.5.3
Liu H-M, Dyer RJ, Guo Z-Y, Meng Z, Li J-H, Schneider H (2012) The evolutionary dynamics of apomixis in ferns: a case study from polystichoid ferns. J Bot 2012:1–11
Lopez RA, Renzaglia KS (2014) Multiflagellated sperm cells of *Ceratopteris richardii* are bathed in arabinogalactan proteins throughout development. Am J Bot 101:2052–2061
Manton I (1950) Problems o cytology and evolution in the pteridophyta. Cambridge University Press
Marimuthu MPA, Jolivet S, Ravi M et al (2011) Synthetic clonal reproduction through seeds. Science 331:876–876
Markham K, Chalk T, Stewart CN Jr (2006) Evaluation of fern and moss protein-based defenses against phytophagous insects. Int J Plant Sci 167:111–117
Matasci N, Hung L-H, Yan Z et al (2014) Data access for the 1,000 Plants (1KP) project. 3:17
Mayank P, Grossman J, Wuest S, Boisson-Dernier A, Roschitzki B, Nanni P, Nühse T, Grossniklaus U (2012) Characterization of the phosphoproteome of mature *Arabidopsis* pollen. Plant J 72:89–101
Menéndez V, Revilla MA, Bernard P, Gotor V, Fernández H (2006a) Gibberellins and antheridiogen on sex in *Blechnum spicant* L. Plant Cell Rep 25:1104–1110
Menéndez V, Villacorta NF, Revilla MA, Gotor V, Bernard P, Fernández H (2006b) Exogenous and endogenous growth regulators on apogamy in *Dryopteris affinis (Lowe) Fraser-Jenkins sp. affinis*. Plant Cell Rep 25:85–91

Mikuła A, Pożoga M, Tomiczak K, Rybczyński JJ (2015) Somatic embryogenesis in ferns: a new experimental system. Plant Cell Rep 34:783–794
Murashige T, Skoog F (1962) A revised medium for rapid growth and bioassays with tobacco tissue cultures. Physiol Plant 15:73–497
Neiman M, Sharbel TF, Schwander T (2014) Genetic causes of transitions from sexual reproduction to asexuality in plants and animals. J Evol Biol 27:1346–1359
Nogler GA (1984) Embryology of angioesperms. In: Johri B (ed) Embryology of angiosperms. Springer, Berlín, pp 475–518
Okada S, Sone T, Fujisawa M et al (2001) The Y chromosome in the liverwort *Marchantia polymorpha* has accumulated unique repeat sequences harboring a male-specific gene. Proc Natl Acad Sci U S A 98:9454
Okano Y, Aono N, Hiwatashi Y, Murata T, Nishiyama T, Ishikawa T (2009) A polycomb repressive complex 2 gene regulates apogamy and gives evolutionary insights into early land plant evolution. Proc Natl Acad Sci U S A 106:16321–16326
Ozias-Akins P (2006) Apomixis: Developmental characteristics and genetics. Crit Rev Plant Sci 25:199–214
Pannell JR (2017) Plant sex determination. Curr Biol 27:R191–R197
Peredo EL, Méndez-Couz M, Revilla MA, Fernández H (2013) Mating system in *Blechnum spicant* and *Dryopteris affinis ssp. affinis* correlates with genetic variability. Am Fern J 103:27
Pires ND, Dolan L (2012) Morphological evolution in land plants: new designs with old genes. Philos Trans R Soc Biol Sci 367:508
Radoeva T, Weijers D (2014) A roadmap to embryo identity in plants. Trends Plant Sci 19:709–716
Rathinasabapathi B (2006) Ferns represent an untapped biodiversity for improving crops for environmental stress tolerance. New Phytol 172:385–390
Ravi M, Marimuthu MPA, Siddiqi I (2008) Gamete formation without meiosis in *Arabidopsis*. Nature 451:1121–1124
Rensing SA, Lang D, Zimmer AD et al (2008) The Physcomitrella genome reveals evolutionary insights into the conquest of land by plants. Science (New York, NY) 319:64–69
Rodriguez-Leal D, Vielle-Calzada J-P (2012) Regulation of apomixis: learning from sexual experience. Curr Opin Plant Biol 15:549–555
Romero-Rodríguez MC, Pascual J, Valledor L, Jorrín-Novo J (2014) Improving the quality of protein identification in non-model species. Characterization of Quercus ilex seed and Pinus radiata needle proteomes by using SEQUEST and custom databases. J Proteome 105:85–91
Sakamaki Y, Ino Y (2007) Gametophyte contribution to sporophyte growth on the basis of carbon gain in the fern *Thelypteris palustris*: effect of gametophyte organic-matter production on sporophytes. J Plant Res 120:301–308
Salmi ML, Bushart TJRS (2010) Cellular, molecular, and genetic changes during the development of *Ceratopteris richardii* gametophytes. In: Fernández AK H, Revilla MA (eds) Working with ferns. Issues and applications. Springer, New York, pp 11–24
Salmi ML, Bushart TJ, Stout SC, Roux SJ (2005) Profile and analysis of gene expression changes during early development in germinating spores of *Ceratopteris richardii*. Plant Physiol 138:1734–1745
Salmi ML, Morris KE, Roux SJ, Porterfield DM (2007) Nitric oxide and cGMP signaling in calcium-dependent development of cell polarity in *Ceratopteris richardii*. Plant Physiol 144:94–104
Salvo E (1990) *Guía de helechos de la Península Ibérica y Baleares*. Ediciones Pirámide S.A.
Sanchez-Lucas R, Mehta A, Valledor L et al (2016) A year (2014-2015) of plants in *Proteomics* journal. Progress in wet and dry methodologies, moving from protein catalogs, and the view of classic plant biochemists. Proteomics 16:866–876
Schwartz W (2007) Lynn Margulis, Origin of eukaryotic cells. Evidence and research implications for a theory of the origin and evolution of microbial, plant, and animal cells on the precambrian earth. XXII u. 349 S., 89 Abb., 49 Tab. New Haven-London 1970: Yale University. Z Allg Mikrobiol 13:186–186
Seguí-Simarro JM (2010) Androgenesis revisited. Bot Rev 76:377–404

Sheoran IS, Ross ARS, Olson DJH, Sawhney VK (2007) Proteomic analysis of tomato (Lycopersicon esculentum) pollen. J Exp Bot 58:3525–3535
Shukla AK, Upadhyay SK, Mishra M et al (2016) Expression of an insecticidal fern protein in cotton protects against whitefly. Nat Biotechnol 34:1046–1051
Spillane C, Curtis MD, Grossniklaus U (2004) Apomixis technology development—virgin births in farmers' fields? Nat Biotechnol 22:687–691
Suo J, Zhao Q, Zhang Z, Chen S, Cao J, Liu G, Wei X, Wang T, Yang C, Dai S (2015) Cytological and proteomic analyses of *Osmunda cinnamomea* germinating spores reveal characteristics of fern spore germination and rhizoid tip growth. Mol Cell Proteomics 14:2510–2534
Taiz L, Zeiger E (2015) Plant physiology and development. Sinauers Associates, Inc., Sunderland
Tucker MR, Araujo A-CG, Paech NA, Hecht V, Schmidt EDL, Rossell J-B, De Vries SC, Koltunow AMG (2003) Sexual and apomictic reproduction in *Hieracium subgenus pilosella* are closely interrelated developmental pathways. Plant Cell 15:1524–1537
Valledor L, Jorrín JV, Rodríguez JL, Lenz C, Meijón M, Rodríguez R, Cañal MJ (2010) Combined proteomic and transcriptomic analysis identifies differentially expressed pathways associated to *Pinus radiata* needle maturation. J Proteome Res 9:3954–3979
Valledor L, Menéndez V, Canal MJ, Revilla A, Fernández H (2014) Proteomic approaches to sexual development mediated by antheridiogen in the fern *Blechnum spicant* L. Proteomics 14:1–11
von Aderkas P (1984) Promotion of apogamy in *Matteuccia struthiopteris*, the Ostrich Fern. Am Fern J 74(1):1–6
Wada M (2007) The fern as a model system to study photomorphogenesis. J Plant Res 120:3–16
Wang X, Chen S, Zhang H, Shi L, Cao F, Guo L, Xie Y, Wang T, Yan X, Dai S (2010) Desiccation tolerance mechanism in resurrection Fern-Ally *Selaginella tamariscina* revealed by physiological and proteomic analysis. J Proteome Res. 9:6561–6577
Ward JA, Ponnala L, Weber CA (2012) Strategies for transcriptome analysis in nonmodel plants. Am J Bot 99:267–276
Wen CK, Smith R, Banks JA (1999) ANI1. A sex pheromone-induced gene in *Ceratopteris* gametophytes and its possible role in sex determination. Plant Cell 11:1307–1318
Whittier DP, Steeves T (1960) The induction of apogamy in the bracken fern. Can J Bot 38:925–930
Whittier D, Steeves T (1962) Further studies on induced apogamy in ferns. Can J Bot 40:1525–1531
Yang HY, Zhou C (1992) Experimental plant reproductive biology and reproductive cell manipulation in higher plants: now and the future. Am J Bot 79:354–363

Chapter 2
Azolla: A Model System for Symbiotic Nitrogen Fixation and Evolutionary Developmental Biology

Sophie de Vries and Jan de Vries

Abbreviations

ABA	Abscisic acid
BR	Brassinosteroids
CK	Cytokinin
DGGE	Denaturing gradient gel electrophoresis
ET	Ethylene
GA	Gibberellin
HIF	Hormogonia-inducing factor
IAA	Indole-3-acetic acid
IRRI	International Rice Research Institute
JA	Jasmonic acid
LCOs	Lipochitooligosaccharides
MAMPs	Microbial-associated molecular patterns
RLK	Receptor-like kinase
SA	Salicylic acid
SL	Strigolactones

2.1 Introduction

Only prokaryotic nitrogen fixation makes inorganic nitrogen available for biological usage, such as amino acid or nucleotide biosynthesis. Hence, nitrogen is frequently among the limiting factors in plant growth. Many plants engage in a symbiotic interaction with nitrogen-fixing bacteria. Naturally, nitrogen-fixing symbioses gain much attention, as they are both ecologically and agriculturally relevant.

S. de Vries (✉) · J. de Vries (✉)
Department of Biochemistry and Molecular Biology, Dalhousie University, Halifax, NS, Canada
e-mail: sophie.devries@dal.ca; jan.devries@dal.ca

H. Fernández (ed.), *Current Advances in Fern Research*,
https://doi.org/10.1007/978-3-319-75103-0_2

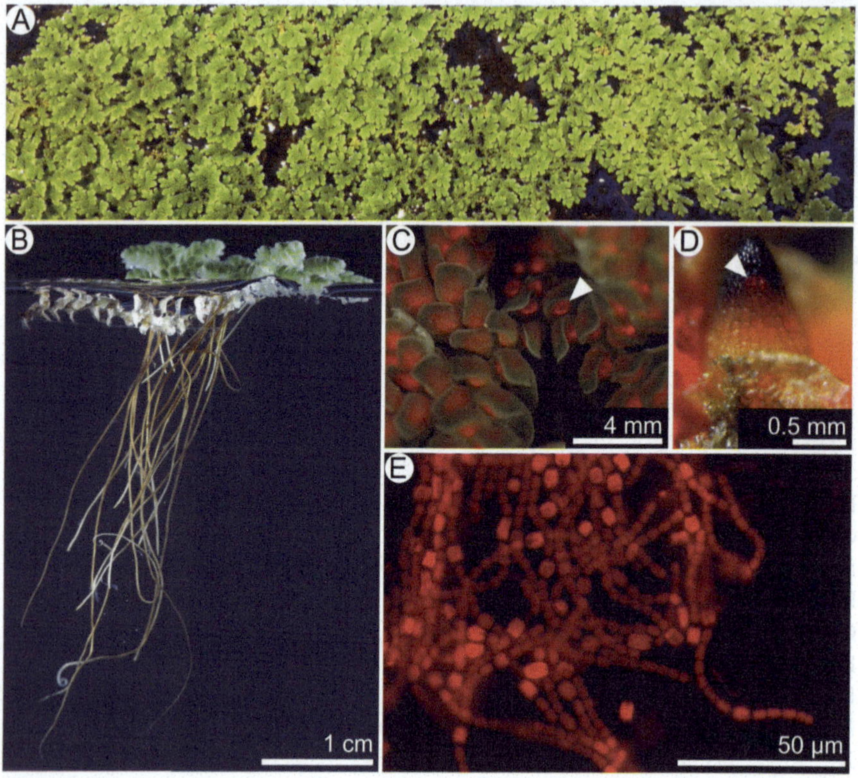

Fig. 2.1 *Azolla filiculoides.* (**a**) Dense culture of *A. filiculoides* growing in the greenhouse. (**b**) Lateral view of *A. filiculoides.* (**c**) Autofluorescencing cyanobionts (arrowhead) fill the leaf cavities of each dorsal leaf lobe. (**d**) Autofluorescencing cyanobionts (arrowhead) within the indusium chamber of the megasporocarps. (**e**) Tightly packed autofluorescencing cyanobionts within the leaf cavity

The symbiotic interaction between the water fern *Azolla* (Fig. 2.1a, b) and its nitrogen-fixing cyanobacterial symbiont (cyanobionts; Fig. 2.1c, d, and e) is unique: *Azolla*'s cyanobiont is vertically inherited from generation to generation (e.g., Perkins and Peters 1993; Fig. 2.1d). This suggests a very intricate communication forged over 66–100 million years (my) of coevolution between the two organisms (Collinson 2002; Carrapiço 2006). Classical studies have characterized this interaction in great detail (Peters and Meeks 1989). With the lowered costs and increased throughput of DNA sequencing techniques, a new, genomic era of *Azolla* research is on the horizon.

In this chapter, we begin by providing an overview of the symbiosis between *Azolla* sp. and its cyanobionts and synthesize its importance in the context of general insights into the molecular communication between plants and microbial symbionts. We then highlight why *Azolla* is a promising system for gaining insights into fern biology and evolutionary (developmental) plant biology.

2.2 *Azolla* and Symbiotic Nitrogen Fixation

Across the biological world, the nitrogen-fixing reaction from N_2 to NH_3 (or NH_4^+) is catalyzed by the multi-subunit bacterial enzyme nitrogenase. The most important components of the enzyme complex are the nitrogenase reductase and the nitrogenase. The nitrogenase reductase consists of two identical subunits encoded by *nifH* (Eady 1996), while the nitrogenase consists of four subunits, of which the two identical α-subunits are encoded by *nifD* and the two identical β-subunits are encoded by *nifK* (Eady 1996). Yet, nitrogen fixation requires more than only a nitrogenase: it requires high amounts of energy in the form of ATP. Low amounts of oxygen can inhibit nitrogenase activity. Consequently, in many organisms (including the nitrogen-fixing symbionts), tight regulation is necessary; oxygen is needed to produce ATP, but nitrogen fixation requires an anaerobic environment. Moreover, these organisms are able to react to cues, such as altered levels of fixed nitrogen, and adapt the expression of several *nif* genes and the abundance of their corresponding proteins accordingly (Poza-Carrión et al. 2014).

Nitrogen fixation is regulated both on a transcriptional and protein level, i.e., by the availability of cofactors, protein-protein interactions, or protein modification (cf. Eady 1996; Martinez-Argudo et al. 2004; Ekman et al. 2008; Poza-Carrión et al. 2014). NifA transcriptionally regulates the expression of several *nif* genes in some nitrogen-fixing bacteria (Martinez-Argudo et al. 2004). NifL is a negative regulator interacting with nifA and responsive to environmental cues such as oxygen or fixed nitrogen (Hill et al. 1981; Merrick et al. 1982; Naberhaus et al. 1995). Other nif proteins, such as nifU and nifS, regulate the rate of nitrogen fixation via the provision of cofactors or iron-sulfur clusters important for nitrogenase activity (Zhao et al. 2007). These regulatory steps all take place within the bacterium. In symbioses, regulation of nitrogen fixation in the symbiont by the host is also important: this starts with recognition and attraction of an effective nitrogen-fixing symbiont (if applicable) and requires a fine-tuned communication of host and symbiont to regulate the exchange of fixed nitrogen from the symbiont with the nutrients from the host during all stages of the symbiosis (Rai et al. 2000). Both of these aspects depend on the nature of the symbiotic interaction in question.

2.2.1 *Diversity of Symbiotic Nitrogen Fixation*

Symbiotic nitrogen fixation evolved multiple times independently in various plant lineages. Notwithstanding, it sometimes recruits similar molecular pathways of the hosts (Markmann and Parniske 2008; Op den Camp et al. 2011). Several types of bacteria, namely, rhizobia, actinobacteria, and cyanobacteria, are known to have become nitrogen-fixing symbionts (Santi et al. 2013; Gourion et al. 2015). Nitrogen fixation by rhizobia in legumes (a group of plants from the family Fabaceae) is one of the best-studied systems, not least because several of their host plants have been sequenced (Schmutz et al. 2010; Young et al. 2011; Schmutz et al. 2014) and are genetically accessible (Chee et al. 1989; Trieu et al. 2000). Interestingly, the plant genus *Parasponia*, which does not belong to the Fabaceae, but to the Cannabaceae, can also undergo a symbiosis with rhizobia (Trinick 1973). This symbiosis is suggested to be evolutionary much younger than its counterpart in the legumes (Op den Camp et al. 2011). The nitrogen-fixing symbiosis of *Parasponia* therefore makes it possible to study how the dialogue between symbiont and host in such symbioses first evolved (Behm et al. 2014).

Actinobacteria infect several plant species of the order Fagales (Pawlowski 2009). Like rhizobia, they colonize the roots of their hosts. Their infection mechanism resembles that of rhizobia in many ways (Santi et al. 2013), yet the formation of nodule primordia is different (Pawlowski and Bisseling 1996), and the initial communication between host and symbiont appears distinct at the molecular level (Pawlowski and Bisseling 1996; Svistoonoff et al. 2010). The third major group of nitrogen-fixing plant symbionts is cyanobacteria (cyanobionts). Cyanobionts are able to initiate symbioses with a wide range of plants. Their host species include representatives from liverworts, hornworts, mosses, ferns, gymnosperms, and angiosperms (Bergman et al. 2007). In contrast to the diversity of the host species, the cyanobacteria involved belong primarily to the Nostocaceae (Bergman et al. 2007).

Nostocaceae interact with various types of plants. They associate with mosses only epiphytically, but still fix nitrogen for their hosts (Bay et al. 2013). In other associations, the cyanobacteria are kept in specific plant structures: in the symbiosis with the liverwort *Blasia pusilla*, cyanobionts colonize specific cavities on the upper side of the liverwort's thallus, while the hornwort *Anthoceros punctatus* forms the cavities on the underside of its thallus (Rodgers and Stewart 1977). This is similar to the water fern *Azolla*, which hosts its cyanobiont in mucilage-filled leaf cavities (Peters and Mayne 1974, Fig. 2.1c, e). In contrast to these interactions, cycads (which are gymnosperm) and the angiosperm *Gunnera* house their cyanobionts within organ tissues. Cycads keep the cyanobacteria between cells, in mucilage-filled zones in coralloid roots (Milindasuta 1975; Obukowicz et al. 1981). In *Gunnera*, *Nostoc* is attracted to a mucilage-filled tract and comes to reside within specially developing cells, surrounded by the host's plasma membrane (Schaede 1951; Silvester and McNamara 1976; Towata 1985; Nilsson et al. 2006). *Nostoc* spp. isolated from cycads and *Anthoceros* sp. were able to colonize *Gunnera* (Johansson and Bergman 1994). Similarly, *A. punctatus* was able to form associations with cyanobionts isolated from cycads and *Gunnera* (Enderlin and Meeks 1983), suggesting that host range in these cyanobacteria is very broad. In contrast, the cyanobacterial symbiont of *Azolla* spp., not able to live isolated from its host, seems restricted to *Azolla* spp. (cf. Adams et al. 2013).

2.2.2 *Diversity of* Azolla

The *Azolla* genus can be divided into two sections: the *Rhizosperma* section, consisting of two species, *Azolla nilotica* and *Azolla pinnata*, and the *Azolla* section, consisting of the five species, *Azolla caroliniana*, *Azolla filiculoides* (for which we have the most molecular data (Ran et al. 2010; Li and Pryer 2014; de Vries et al. 2016a; Brouwer et al. 2017)), *Azolla mexicana*, *Azolla microphylla*, and *Azolla rubra* (Reid et al. 2006; Metzgar et al. 2007; Pereira et al. 2011). It should be noted that this classification is still being debated. For example, Evrard and Van Hove (2004) proposed that the *Azolla* section contains only *Azolla cristata* and *Azolla filiculoides*, while all other species are mere synonyms.

Azolla's cyanobiont coevolved with its host. This is supported by short tandemly repeated repetitive sequence fingerprinting and 16S rRNA phylogenies (Zheng et al. 1999; Papaefthimiou et al. 2008; reviewed by Pereira and Vasconcelos 2014). Despite this observation, more than one band corresponding to a cyanobacterium was detected for some *Azolla* species using denaturing gradient gel electrophoresis (DGGE) profiling of the 16S rRNA gene (Papaefthimiou et al. 2008). Molecular studies of cyanobacteria in the leaf cavities of *Azolla* showed that the cyanobionts of all *Azolla* species have one common ancestor, i.e., this particular symbiosis happened once (Zheng et al. 1999; Papaefthimiou et al. 2008). Further the cyanobiont and the plant host show patterns of coevolution: cyanobionts of hosts that are more closely related are also more similar to each other than they are to a cyanobiont from a more distantly related *Azolla* species (Zheng et al. 1999; Papaefthimiou et al. 2008). Despite this coevolution it is possible to colonize one *Azolla* species with the cyanobiont of another (van Cat et al. 1989). This was achieved by transferring parts of *Azolla*'s reproductive organ from one *Azolla* species to another later in the reproductive cycle (i.e., when akinetes germinate or when the cyanobiont differentiates into vegetative filaments; Adams et al. 2013).

The *Azolla-N. azollae* coevolution scenario is also supported by phylogeny-independent data such as the fatty acid profiles of the cyanobionts (Caudales et al. 1995). Also, the *Azolla* cyanobionts show a strong conservation of their *nif* gene cluster (Franche and Cohen-Bazire 1987). Additionally, free-living, heterocyst-forming cyanobacteria differ from *N. azollae* by having an 11 kb segment located between the *nifK* gene and the genes *nifD* and *nifH* (Franche and Cohen-Bazire 1987; Rice et al. 1982). This 11 kb fragment is excised from the chromosome in mature heterocyst (Golden et al. 1985). Golden et al. (1985) hypothesized that the *nif* genes can only be transcribed when the fragment is excised, suggesting that vegetative cells, which contain the 11 kb segment, cannot express a functional nitrogenase. This 11 kb segment was not detected in the cyanobionts of *Azolla* (Franche and Cohen-Bazire 1987), suggesting that i) functional nitrogenase enzymes could theoretically be produced independent of the cell type and ii) that the synthesis of the enzyme may be faster in the cyanobiont compared to its free-living relatives. Regardless of some taxonomic ambiguities, these data support the hypothesis that the *N. azollae* cyanobiont has become an integral part of the *Azolla* plant – one might even say that it is on its way to becoming the plant's third symbiotically derived organelle.

2.3 *Azolla* Has a Heritable Cyanobiont

For the host plant, the first step in ensuring the availability of fixed nitrogen is to attract their nitrogen-fixing symbionts. In nonpermanent nitrogen-fixing symbioses of other plants, this step is the first point to enrich for symbionts compared to non-nitrogen-fixing bacteria. For example, rhizobia are recruited to their host plants via plant-derived signaling molecules such as flavonoids (Muñoz Aguilar et al. 1988). Exuding flavonoids may also be important in attracting *Frankia* sp., the actinobacteria involved in some nitrogen-fixing symbioses, although whether or not they promote infection is less clear and may be strain dependent (Hughes et al. 1999; Popvici et al. 2010). Mosses, *Gunnera*, and cycads first secrete a non-specific factor that induces hormogonia, motile cyanobacterial filaments, in such species that are able to form them, independent of whether or not they participate in a nitrogen-fixing symbiosis with the host (Johansson and Bergman 1994; Rasmussen et al. 1994; Santi et al. 2013). These factors are called hormogonia-inducing factors (HIFs), but, in most interactions, it is not known what this secreted substance actually is.

The differentiation into hormogonia is part of the developmental cycle of many filamentous cyanobacteria belonging to sections III to IV (Rippka et al. 1979). Cyanobacterial differentiation into hormogonia is influenced by many external cues such as nitrogen availability (Herdman and Rippka 1988), solidity of the growth substrate (Hernandez-Muñiz and Stevens 1987), and red light (Robinson and Miller 1970; Damerval et al. 1991). It is hence a relic of the cyanobacterium's free-living state. Yet, in many of the plant-cyanobiont symbioses, the ability to differentiate into a motile hormogonium is co-opted to enable the cyanobacterium's movement toward its host. As described above, the plant exudes HIFs that promote the differentiation of hormogonia in cyanobacteria. This process is best understood in the interaction between *Gunnera* and its cyanobiont, where a heat-labile protein smaller than 12 kDa has been found to be the HIF (Rasmussen et al. 1994). HIFs are also produced by cycads and the hornwort *Anthoceros* (Campbell and Meeks 1989; Bergman et al. 1996).

HIFs induce motility of cyanobacteria in a non-specific manner; hence to attract the symbionts, additional molecules are secreted by their hosts (Johansson and Bergman 1994; Rasmussen et al. 1994). These attractants seem to be more specific than the HIFs (Rasmussen et al. 1994; Santi et al. 2013). However, the nature of the attractant is presently unknown. In the *Gunnera-Nostoc* symbioses, it was noticed that the mucilage is rich in polysaccharides (Khamar et al. 2010). In addition, Nilsson et al. (2006) showed that *Nostoc* sp. was attracted to seven different sugars. This makes polysaccharides good candidates for attracting cyanobacterial symbionts. Yet, the *Nostoc* spp. isolates tested in the above studies have all been isolated from facultative interactions, i.e., symbionts must be taken up from the environment anew with every generation. As explained above, in *Azolla*, the situation is quite different; next we explore how symbiont inheritance is ensured in this instance.

2.3.1 Continuity of the **Azolla-Nostoc** *Symbiosis*

Azolla's leaves are never without their cyanobionts. This is achieved by mechanisms that transmit a cyanobiont inoculum during sexual and asexual (vegetative) reproduction (Peters and Meeks 1989; Rai et al. 2000). This is unique among land plants. Yet, vertical inheritance of a nitrogen-fixing cyanobiont in a photosynthetic eukaryote does occur in some rhopalodiacean diatoms, such as *Rhopalodia* and *Epithemia* (Kneip et al. 2008; Nakayama et al. 2014). While the transmission of the cyanobiont is already quite complex in the context of a unicellular diatom – it "simply" requires a certain level of synchronization of the division of host and endosymbiont as it occurs in plastids (Sumiya et al. 2016; de Vries and Gould 2017) – it gets even more complex in the context of a multicellular organism such as *Azolla*. There are two major mechanisms that facilitate the transmission of a cyanobiont inoculum during *Azolla*'s life cycle.

The first mechanism mediates transmission during vegetative growth of the sporophyte and hence assures that there is no leaf cavity that lacks cyanobionts. This is achieved by trichomes that inoculate newly formed leaf cavities from old ones (Calvert et al. 1985; Hill 1989). Calvert and Peters (1981) distinguished between branched and simple trichomes involved in the interaction with the cyanobiont. The branched trichomes occur only twice per cell (Calvert and Peters 1981; Calvert et al. 1985); it is branched trichomes that facilitate the transfer from one cavity to the other (Peters and Meeks 1989). Both the branched and the simple trichome are considered to be relevant for the nutrient exchange between cyanobionts and the *Azolla* host (Calvert et al. 1985).

The second mechanism occurs during sexual reproduction of *Azolla*. Here, cyanobionts are co-transmitted to the next generation of ferns by resting in a special chamber, the indusium chamber, of the female megasporocarps (Perkins and Peters 1993). According to Becking (1987), the cyanobacteria enter this indusium chamber by slipping in through a pore. Hence, it seems that in contrast to the vegetative propagation (where the process is facilitated by the trichomes), the cyanobacteria are not directly injected into the indusium chamber; instead (as described above), this mechanism is based on a means of attracting the cyanobionts. No HIFs have so far been identified in *Azolla*, but during sexual reproduction the cyanobiont forms hormogonia (Zheng et al. 2009). There are also no data on what the attractant of *Azolla* is, but since hormogonia are attracted toward the sporocarp (Zheng et al. 2009), one has to exist. The holdout of the cyanobiont is the indusium that seems to be reached not by chemotaxis, but by sheer force of the growing and maturing sporangium, which serves to push the cells to their final destination (Zheng et al. 2009).

Zheng et al. (2009) showed the presence of hormogonia during megasporocarp development using microscopy. Moreover, they demonstrated that the cyanobacterial differentiation (including a hormogonia stage) is synchronized with the entire process of the megasporocarp development of *Azolla*. Indeed, work by Zheng et al. (2009) reveals that after the cyanobacterial inoculum has been successfully attracted to the chamber, the cyanobiont differentiates into akinetes, the resting spores of Nostocaceae (cf. e.g., Rippka et al. 1979). Interestingly, Zheng et al. (2009) showed formation of a biofilm with other bacteria that seemed to be induced during sexual reproduction of the fern. This led the authors to suggest that these additional bacteria play a third part in *Azolla*'s unique symbiosis. This is in agreement with Papaefthimiou et al. (2008), who found that for leaf cavity extracts of some *Azolla* species, more than one 16S amplicon was visible in the DGGE gel, suggesting the presence of other bacteria; the presence of a plethora of bacteria was recently confirmed by Dijkhuizen et al. (2018). Additionally, a diversity of non-cyanobacterial bacteria has been found in sporocarps and leaf cavities of *A. microphylla*, with *Bacillus cereus* being a major isolate (Zheng et al. 2008).

2.3.2 Cross Talk in Symbiotic Nitrogen Fixation

Nitrogenase is an oxygen-sensitive enzyme (Haysted et al. 1970). To allow for nitrogen fixation in bacteria, an oxygen-poor environment has to be achieved. This is achieved by different means in different symbiotic interactions, partly provided by the host and partly by the bacterium. All plant cyanobionts (and many of their free-living relatives) are capable of forming heterocysts (Tyagi 1975; Papaefthimiou et al. 2008). Heterocysts are morphologically and biochemically distinct from the remaining vegetative cells. Their thick cell walls prevent oxygen influx (Tyagi 1975). Further, heterocysts are suggested to not carry out the photosynthetic reaction of photosystem II (Donze et al. 1972), which would otherwise lead to high oxygen production. Cyanobionts tend to have high frequencies of heterocystous cells (Rai et al. 2000); in *Azolla*'s cyanobiont the heterocyst frequency is between 18% and 30% (Hill 1975; Grilli Caiola et al. 1989). This is three to six times more than observed in their free-living relatives (Fay 1969), suggesting that heterocyst formation may be specifically induced during these symbioses.

The availability of sugar from the host is also known to trigger the cyanobiont to increase its rate of nitrogen fixation. Indeed, certain sugars (that are produced in the gland tissue of *Gunnera*) inhibit secondary hormogonia formation in *Gunnera* symbionts (Khamar et al. 2010). In *N. azollae*, the cyanobiont of *Azolla*, the amount of heterocysts is promoted by the presence of fructose, as is nitrogen fixation (Rozen et al. 1986). Cyanobiont surface sugars, interacting with sugar-binding lectins of *Azolla*, have also been proposed to mediate adherence of cyanobionts in the leaf cavity (Kobiler et al. 1981). It is noteworthy though that the studies by Rozen et al. (1986) and Kobiler et al. (1981) were done in a culture of the potential cyanobiont identified as *Anabaena azollae*. Later it became apparent that these cultures of *Azolla* cyanobionts are rather less strongly associated cyanobacteria and not the actual cyanobiont (cf. Adams et al. 2013). Despite this, analysis of a proteome from the cyanobiont (isolated directly from the *Azolla* cavity) showed that a potential hexose transporter is one of the most strongly expressed proteins in the *Azolla* cyanobacterial symbiont (Ekman et al. 2008). What is more, this transporter even shows enhanced levels compared to the cultured strain *Nostoc* PCC 73102 (Ekman et al. 2008). This points to sugars as a common signal for cyanobionts to induce heterocyst formation (independent of their type of association). Expression data on nitrogen symbiosis-associated factors generally tell the same tale: key genes in nitrogen fixation, such as the nitrogen regulator *NtcA*, a gene involved in heterocyst differentiation (*hetR*), or the nitrogenase reductase protein nifH, are upregulated in *Nostoc* sp. during infection of young *Gunnera* stem tissue (Wang et al. 2004). Likewise, the cyanobiont of *Azolla* showed enhanced nifH and nifK protein abundance compared to the cultured *Nostoc* PCC 73102 (Ekman et al. 2008). In addition to enhanced differentiation into heterocysts, the leaf cavities in which *Azolla*'s cyanobiont resides show a decreased concentration of oxygen (Grilli Caiola et al. 1989), suggesting that some regulatory mechanism is present to deprive this extracellular but protected environment of the cyanobiont from oxygen.

Sugar is more than a signaling component of the host to communicate that the cyanobiont is supposed to differentiate heterocysts. It also seems to be important to provide a reductant for the nitrogenase. Studies have shown a strong upregulation of the oxidative pentose phosphate pathway in *N. azollae* (*Azolla's* cyanobiont) (Rozen et al. 1986; Ekman et al. 2008). It is noteworthy that this pattern is not only true for the very intricate association of *Azolla* and *N. azollae* but also in other symbioses with nitrogen-fixing cyanobacteria: a *Nostoc* strain infecting *Gunnera manicata* was shown to upregulate 6-phoshogluconate dehydrogenase, a key enzyme of the oxidative pentose phosphate pathway, in heterocysts (Ekman et al. 2006).

Besides the host's influence on committing the cyanobiont to nitrogen fixation, additional cross talk between the two partners must happen. Environmental factors such as temperature, light, or plant growth have an influence on the rate of nitrogen fixation (Hechler and Dawson 1995). In addition, availability of fixed nitrogen in the environment may impact the nitrogen fixation rate of *Azolla*'s cyanobiont. Other hosts, which do not inherit the nitrogen-fixing symbiont every generation, could modulate this by restricting access to nutrients to the symbiont (cf. Usher et al. 2007).

Azolla does not kill off its symbiont in the presence of available fixed nitrogen – if it did, their coevolution would have found a swift end. Yet somehow the host must react to such a strong environmental change. Indeed, treatment with NH_4 and NO_3 resulted in reduced nitrogen fixation rates (Hechler and Dawson 1995). Furthermore, NH_4Cl reduced the expression of *nifH* at 10 days after continuous treatment (Brouwer et al. 2017). Additionally, pathogens allocate fixed nitrogen from the host to themselves (Horst et al. 2010). Hence, *Azolla* has to protect its symbiotically acquired nitrogen from pathogens. Simultaneously, *Azolla* needs to fine-tune its defense responses against its pathogens such that the symbiont is not harmed. This and more details on the intimate relationship of *Azolla* and its symbiont will be discussed in the following subsection.

2.3.3 Communication and Dependency

Symbiont recruitment alone will not lead to a successful symbiosis. The host needs to further exert some control over the symbiont. In many cases, such coevolution resulted in infection of the host by the symbiont (Santi et al. 2013). This is a difficult balancing act. All bacterial symbionts, like any other bacteria, including pathogens, present microbial-associated molecular patterns (MAMPs) to their hosts. These include conserved molecules, or parts of these, such as flagellin or the peptidoglycan layer in the bacterial cell wall (Felix et al. 1999; Gust et al. 2007). These MAMPs induce a first layer of immunity to hinder infection of the plant by the microbes (Goméz-Goméz and Boller 2000; Gust et al. 2007). Hence there must be a mechanism by which the symbionts can evade the immune response.

In the interactions of rhizobia and legumes, one of the key factors is the secretion of lipochitooligosaccharides (LCOs), molecules with a backbone similar to chitin (Lerouge et al. 1990; Schultze et al. 1992). These LCOs are recognized via membrane-spanning receptors that often, but not always, possess a kinase domain located in the cytoplasm, named receptor-like kinases (RLKs, Radutoiu et al. 2003; Fliegmann et al. 2013; Zipfel and Oldroyd 2017). RLKs operate as dimers and can also associate with those receptors that do not possess the kinase domain, extending their possible combinations even further (Radutoiu et al. 2003; Fliegmann et al. 2016). The receptor domains of these RLKs constitute either of leucine-rich repeats or a sequence of three lysin motif domains, called LysM receptors (Madsen et al. 2003; Petutschnig et al. 2010; Fliegmann et al. 2013). LCOs but also similar chemical compounds such as peptidoglycan and chitin are bound and recognized by LysM receptors (Petutschnig et al. 2010; Fliegmann et al. 2013; Zipfel and Oldroyd 2017). The different possible combinations of LysM interactions seem to be a critical point in distinguishing between these molecules and hence between symbiont and pathogen (Zipfel and Oldroyd 2017).

Symbiont or pathogen perception by receptors induces a signaling cascade. This signaling readily results in changes in phytohormone-associated pathways (Oldroyed and Downie 2008), of which one of the most prominent is auxin. Additionally, for example, *Frankia* strains are able to produce indole-3-acetic acid (IAA), an auxin (Berry et al. 1989; Perrine-Walker et al. 2010). This leads to local changes in auxin maxima at the site of infection (Perrine-Walker et al. 2010). Likewise, in the rhizobium symbiosis with the legume *Lotus japonicus*, auxin transport is enhanced, and auxin was shown to accumulate in small nodule primordia after LCO treatment (Pacios-Bras et al. 2003). It was suggested that such enhanced auxin levels are associated with proliferating cortical cells (from which nodule primordia arise) in the rhizobium-legume symbioses (Suzaki et al. 2012). Whether auxin is also involved in gland development in *Gunnera* and coralloid root development in cycads is not known. Yet, 83% of symbiotic cyanobacteria, among them a *Nostoc* isolate from *Gunnera*, are able to produce IAA (Sergeeva et al. 2002). Both endogenous and released IAA levels of this particular *Nostoc* isolate increased upon addition of tryptophan (Sergeeva et al. 2002). Enhanced local auxin or IAA production seems to be a recurrent and important theme in most nitrogen-fixing symbiosis; whether enhanced cell division and cell enlargement in association with enhanced IAA levels play a role during any time in the life cycle of the *Azolla* symbiosis is however not known.

While under natural conditions *Azolla* is never without its cyanobionts, many studies have successfully removed the cyanobiont by applying harsh antibiotic treatment, mainly erythromycin (Forni et al. 1991; Schor-Fumbarov et al. 2005; Brouwer et al. 2017). Yet, once *Azolla* is completely devoid of cyanobionts, it cannot be reinfected with cyanobacteria from the environment (highlighted, e.g., in Quispel 1991). *Azolla* plants with empty leaf cavities visibly experience stress (Schor-Fumbarov et al. 2005) and are dependent on externally supplied nitrogen for growth (Forni et al. 1991; Brouwer et al. 2017). These observations underpin the interwoven relationship of *Azolla* and its cyanobionts: that *Azolla* cannot be reinfected is explained by the fact that the cyanobiont has become an integral part of the plant. *Azolla* is beyond the stage where it attracts and communicates with its cyanobiont *Nostoc azollae* in the environment. The fact that *Azolla*'s cyanobionts are hardly cultivatable outside of the plant likely rests with the cyanobiont genome itself. Hence, in the next section, we will explore how coevolution and dependency has shaped the cyanobiont's genome.

2.3.4 Reductive Evolution in the Cyanobiont of Azolla

Some of the best insights into how its obligate symbiotic lifestyle has shaped the cyanobiont were gained from its genome sequence obtained by Ran et al. in 2010. The authors found a high degree of genome erosion as evident by i) reduced number of protein-encoding genes, while the genome size was similar compared to other closely related cyanobacteria and ii) pseudogenization. Pseudogenization in two DNA repair enzyme-encoding genes (*recD* and *alkA*, Ran et al. 2010) has probably served to increase the rate of erosion of this genome by reducing the symbiont's ability to repair mutations. Additionally, the DNA replication initiator *dnaA*-encoding gene and other genes involved in DNA replication are also good examples, as their pseudogenization suggests that the *Azolla* cyanobiont may have problems with replication and growth (Ran et al. 2010). The authors hypothesized that due to these occurrences, *Azolla* is able to control its symbionts' growth. Among other molecular mechanisms, pseudogenization also impacts glycolysis, as well as phosphate and nitrate uptake (Ran et al. 2010). The latter is interesting, because (as mentioned above) treatment with fixed nitrogen reduces nitrogen fixation rates (Hechler and Dawson 1995). This further supports the hypothesis that the host is communicating with its cyanobiont and can influence its nitrogen fixation. Moreover, the genome analyses of Ran et al. (2010) point to the use of the oxidative pentose phosphate pathway as a source of energy metabolism. The entire *nif* gene operon remains intact (Ran et al. 2010), yet some portion of the nifH protein was suggested to be nonfunctional due to a posttranslational protein modification (Eckman et al. 2008). What this protein modification means for the symbiosis with *Azolla* is not clear. All in all, the entire genome points to a streamlining towards nitrogen fixation and photosynthate import by the cyanobiont.

2.4 *Azolla*'s Potential for Agriculture and Science

2.4.1 Azolla *Grows Like a Weed!*

Azolla has a remarkable growth rate – it can double its biomass within just 2 days (Peters et al. 1980; Wagner 1997). Indeed, *Azolla*'s growth rate is even documented by earth's geochemistry. Arctic Ocean sediments from the Eocene (approximately 50 my old) contain up to 300,000 *Azolla* remains (mainly from the spores) per gram of dry sediment (Brinkhuis et al. 2006). These (*Azolla*-based) inferences not only highlight that there seem to have been (episodic) freshwater conditions in the Arctic basin but also *Azolla*'s phenomenal growth rate. Even today, "green tides" of *Azolla* can be observed, e.g., outlined by Carrapiço (2010), who vividly described an *Azolla* bloom in the Guadiana River in Portugal in 1993. *Azolla*'s growth rate leads to a high rate of biomass production, which has found its useful applications in, for example, agriculture.

Azolla is a classical fertilizer that has been used in rice fields for thousands of years (Shi and Hall 1988; Wagner 1997). Its application can lead to more than a 200% increase in yield (reviewed in Wagner 1997). Furthermore, *Azolla* has been used as fodder, for example, for pigs (Leterme et al. 2009) and fish (Abou et al. 2007). Given the agricultural potential *Azolla* bears, Brouwer et al. (2014) embarked upon the task to laying the groundwork for its domestication. This was further substantiated in Brouwer et al. (2017), where experiments to evaluate the biomass yield that can be obtained in a continuously harvested system were performed. The authors found that *Azolla* can be a reliably harvested as a high-protein biomass crop. In addition to making it suitable for agricultural applications, its fast growth rate makes *Azolla* an attractive laboratory system.

2.4.2 The Azolla *Nuclear Genome*

Extant vascular plants are deeply split in two lineages: the lycophytes and the euphyllophytes. The euphyllophytes consist of the ferns and seed plants (to which gymnosperms, such as conifers, and angiosperms, the flowering plants, belong).

These phylogenetic patterns are long-standing (e.g., Pryer et al. 2001) and have been further supported by recent large-scale phylogenomic analyses (Wickett et al. 2014). Ferns are hence essential for our understanding of euphyllophyte evolution. Despite their important phylogenetic position, fern molecular biology has been vastly understudied. Indeed, as of yet, there is only one lycophyte genome sequenced – that of *Selaginella moellendorffii* (Banks et al. 2011) – and not a single fern genome. But things are about to change. There are several sequencing projects aimed at a broader coverage of land plants and their algal relatives, the streptophyte algae (Delwiche 2016; Sessa and Der 2016; Rensing 2017). These projects include fern genomes.

Ferns have among the largest known genomes of all land plants (Bennett and Leitch 2001; Leitch et al. 2005; Sessa et al. 2014). For example, some homosporous ferns have genomes averaging around 10 gigabases in size (Leitch et al. 2005; Sessa et al. 2014); hence they are often larger than the human genome (cf. International Human Genome Sequence Consortium 2001; cf. Venter et al. 2001). Indeed, very recent data by Hidalgo et al. (2017) suggest that the fern *Tmesipteris obliqua* has a genome of similar size as the current record holder in plant genome sizes, *Paris japonica*, which has an estimated genome of 150 Gbp (Pellicer et al. 2010). While the cost of sequencing drops from year to year and new and improved sequencing techniques appear at mind-boggling speed, sequencing of such enormous genomes is (at the moment) still a herculean task. Fortunately, *A. filiculoides* with its roughly 750 Mbp (cf. Li and Pryer 2014) ranks among the ferns with the smallest known genomes. The only fern that has a significantly smaller genome is another Salviniales, the (namegiving) *Salvinia* with a genome of 250 Mbp (Li et al. 2016; see also Clark et al. 2016). But this does not mean that only these small heterosporous ferns are being sequenced. The homosporous fern *Ceratopteris* is currently being sequenced, too, and its genome is ~11 Gbp in size (Wolf et al. 2015; Sessa and Der 2016). Genomic insights into fern biology will enrich our understanding of multiple key vascular plant characters. In the next section, we explore how *Azolla* has been used as a tool for understanding plant evolution.

2.4.3 **Azolla** *Is a Promising System for Plant Evo-devo*

Ceratopteris richardii is one of the best-studied ferns in terms of molecular biology. After *Pteridium aquilinum* (Der et al. 2011), it was among the first ferns for which RNAseq data have been generated (Bushart et al. 2013). Moreover, it is the only fern for which there are stable transformation protocols (Plackett et al. 2014). There have been some key insights into fern biology using *C. richardii* as a model system. These include, for example, the finding that abscisic acid (ABA) signaling plays a key role in fern sex determination (McAdam et al. 2016), early development, and fern spore germination (Salmi et al. 2005). Research on *Azolla* is catching up.

In 2014, Brouwer et al. generated the first large-scale sequencing data on *A. filiculoides*. The conditions for establishing *Azolla* as a model system for the lab could not be better. Foremost among them is the aforementioned growth rate;

experiments using hundreds of *Azolla* plants can be carried using a negligible amount of resources (including space; Fig. 2.1a). *Azolla* can be cultivated under controlled conditions using the well-established IRRI medium (Watanabe et al. 1992), and its entire life cycle can be completed under laboratory and greenhouse conditions. Furthermore, Brouwer et al. (2014) showed that in vitro fertilization of *Azolla* megaspores occurs at a decent success rate of up to 27%. This means that controlled crossing experiments with *Azolla* are feasible. Brouwer et al. (2014) went further and assessed possible strategies for preserving spores. They found that spores were viable after months of storage at −80 °C. In summary, *Azolla* is an easy-to-rear lab system that can be subjected to crossing experiments and has storable germplasm.

Despite being a floating fern, the sporophyte of *Azolla* has roots. *Azolla*'s roots arise at the nodes from which the ventral and dorsal leaves emerge. This highlights "primary homorhizy," the growth pattern that shapes the roots of not only *Azolla* but ferns in general (Goebel 1930). What this means is that ferns form a root system that is mainly post-embryogenic and characterized by secondarily emerging roots (Schneider 2013). This is a stark contrast to root development of angiosperms. Regardless of whether the root system is characterized by a dominating taproot (allorhizy; e.g., in *Arabidopsis* or carrot) or fibrous roots ([secondary] homorhizy; e.g., in rice), they have a permanent primary root that is formed during embryogenesis. Ferns lack such a permanent embryonic root (Schneider 2013). The reason for this likely is found with their embryo development.

The embryos of angiosperms are bipolar. This bipolar arrangement is achieved by a concentration gradient of key developmental regulators such as auxin (facilitated by positioning of the PIN auxin efflux carrier proteins; Friml et al. 2003; Petrasek and Friml 2009). This bipolar arrangement early establishes the root apical meristem that will form the primary embryonic root – a process again mediated by auxin (Friml et al. 2002; Blilou et al. 2005; Schlereth et al. 2010). Ferns however do not form such a bipolar embryo, although the molecular mechanisms underpinning fern embryo development have not been studied. It is, however, very likely that they will also involve auxin and PIN proteins, which have radiated in ferns independent of angiosperms (Bennett et al. 2014a; Bennett 2015). The involvement of auxin in fern root development however warrants attention.

In *Arabidopsis* roots, auxin accumulates in the root tip and is necessary for activity of the root apical meristem (Petrasek and Friml 2009; Petricka et al. 2012). Another phytohormone, cytokinin, restricts this meristematic zone along the longitudinal axis of the root by promoting differentiation (Dello-Ioio et al. 2007). The two phythormones are hence acting antagonistically in establishing the size of the meristematic zone: auxin promotes the size of the meristematic zone while cytokinin restricts it (Dello Ioio et al. 2007). This pattern largely holds true for exogenously applied auxin and cytokinin (cf. Dello Ioio et al. 2007) – very high exogenous auxin can have a slight inhibitory effect on root meristem size (Růžička et al. 2009). The roots of *Azolla*, however, respond very differently to exogenous application of these two classical phytohormones: auxin reduced the size of the *Azolla* root meristematic zone while cytokinin promoted it (de Vries et al. 2016a). Sequencing of mRNAs collected from the tips of *Azolla* roots that have been exposed to exogenous auxin and cytokinin shows that the differential responses are likely governed by a chassis of protein cell wall modifiers (de Vries et al. 2016a).

The different responses of the *Azolla* root toward auxin raise questions about the role of auxin in fern root development as in contrast to *Arabidopsis*. Are fern roots very different from those of other euphyllophytes? What does this mean for the evolutionary origin of the euphyllophyte root? The aforementioned insights gained from studying the response of the *Azolla* root toward these key phytohormones support the idea that the root apical meristem evolved from a shoot apical meristem (cf. Jiang and Feldman 2005; de Vries et al. 2016a; Harrison 2017), as development of the latter is known to be promoted by cytokinin (Kurakawa et al. 2007). Comparative analyses between molecular mechanisms that underpin primary and secondary homorhizy as well as detailed analyses of the fern embryo are bound to provide more comprehensive answers to these outstanding questions. Fortunately, the RNAseq data included large parts of the canonical auxin and cytokinin signaling pathways (de Vries et al. 2016a), and future studies – hopefully soon aided by a genome sequence – can now thoroughly dissect these pathways to gain further insights into molecular developmental biology of ferns. In summary, *Azolla* has been proven to be a suitable system for evo-devo studies.

2.4.4 Evolution of Phytohormone Signaling as an Example for Azolla*'s Potential for Plant Evolutionary Studies*

Although auxin is best known for its role in plant development (Box 2.1), it plays a key role during symbioses with cyanobacteria and other nitrogen-fixing symbionts. This leads to the question how plants like *Azolla*, which participate in nitrogen-fixing symbioses, balance the level of this hormone without harming either its symbioses or its development. For auxin, we know that many aspects of plant development are regulated based upon the creation of local auxin maxima or gradients (Bennett et al. 2014b). But this question is also valid for other plant hormones, for example, those required for defense responses against microbial pathogens. We have already touched upon the similarities between pathogens and symbionts and how plants may distinguish between them. But what if a pathogen occurs in parallel with a symbiont? Defense responses toxic to microbial pathogen may also be toxic to a symbiont. Given that *Azolla* is unable to grow without its cyanobiont in the absence of fixed nitrogen (Brouwer et al. 2017), the loss of its cyanobiont would be detrimental. How does *Azolla* protect its resources that should be allocated to its symbiont or the fixed nitrogen from the symbiont from pathogens? Fine-tuning mechanisms likely involve spatiotemporal regulation of phytohormone levels. Here we briefly discuss which phytohormones may be present in *Azolla* and how these could conceivably influence the symbiosis with its cyanobiont.

Some phytohormones are most closely associated with plant growth and development, while others are more associated with defense responses and biotic stress. All of their signaling pathways are, however, integrated with each other (Robert-Seilaniantz et al. 2011; Berens et al. 2017, Box 2.1). What this means is that these phytohormones do not act in a binary fashion. Phytohormone action is a matter of fine-tuning the levels of a phytohormone present as well as their interaction in the phytohormone signaling networks. For example, JA signaling alone is required for defense against herbivores, while the integration of JA and ET is required to trigger defense responses against necrotrophic pathogens (i.e., pathogens that actively induce plant cell death; Lorenzo et al. 2003; Glazebrook 2005). Another example is ABA, which funnels into the JA and SA signaling pathway. It was found to negatively regulate systemic acquired resistance via the SA signaling pathway in *Arabidopsis thaliana* (Yasuda et al. 2008) and to negatively regulate both SA and JA during defense responses against a nematode parasite of rice (Nahar et al. 2012).

Phytohormone regulatory networks are at the heart of various plant functions. The nine best-studied phytohormones (Box 2.1) are mainly explored in angiosperms. This warrants attention, as not all plant lineages possess all of these nine hormones. Some biosynthesis and signaling pathways have evolved earlier in plant evolution than others (Han 2017). Therefore, different networks are presumably present in the different plant lineages. Auxin and CK are ancient. Their biosynthesis and downstream signaling machineries were found in streptophyte algae (Hori et al. 2014; Ju et al. 2015). IAA, an auxin derivative, is produced by streptophyte algae (Hori et al. 2014; Beilby et al. 2015), which are the closest algal relatives to land plants (Delwiche and Cooper, 2015; de Vries et al. 2016b; de Vries and Archibald, 2018). To find the antagonism somewhat inverted in terms of root development in *Azolla* (de Vries et al. 2016a) is a good example of why we need to broaden the range of physiological investigations.

ABA, JA, SA, and ET have been measured in various streptophyte algae (Beilby et al. 2015; Ju et al. 2015). Further, streptophyte algae have the genetic tools for using the phenylpropanoid pathway (de Vries et al. 2017), which some plants use as alternative route for SA production (Chen et al. 2009). With the exception of the JA pathway, homologs of most of the signaling components downstream of these hormones can be found as early on as streptophyte algae (Hori et al. 2014). Theoretically, all four of these plant hormones should also exist in *Azolla.* Despite being present early in plant evolution, current data suggests that the JA and SA antagonism evolved after the split of gymnosperms and angiosperms (Kozlowski et al. 1999; Arnerup et al. 2013). Additionally, JA seems to have been lost several times during evolution. The model moss *Physcomitrella patens* does not produce JA, but only a closely related derivative (Stumpe et al. 2010). In agreement with this observation, Záveská Drábková et al. (2015) showed that within the bryophytes JA concentrations fluctuate very strongly. The lycophyte *Selaginella moellendorffii* was recently shown to possess the JA biosynthesis pathway and synthesize measurable levels of the hormone (Pratiwi et al. 2017). This makes ferns a possible “missing link” in terms of JA/SA evolution. Analyses of published *Azolla* RNAseq data show that most of the biosynthesis, signal transduction, and downstream pathways of both JA and SA are present in *A. filiculoides* (de Vries et al. 2018). However, the data also suggest that the fern responds only to SA (de Vries et al. 2018).

SL induces branching of arbuscular mycorrhizal fungi, which is critical for the colonization of host plants (Akiyama et al. 2005). These symbioses are as old as land plants themselves, suggesting that they may be a crucial driver of terrestrialization (Redecker et al. 2000). There is no direct evidence of how ancient SL is, but its importance in mycorrhizal symbioses suggests that it may be produced by streptophyte algae. In agreement with this, signaling components of the SL pathway can also be found in some streptophyte algae (cf. Hori et al. 2014). Hence SL may also be present in *Azolla.*

GAs seem to not be as old as auxin, CK, ABA, SA, JA, ET, and SL. While genes relevant for GA perception and signaling were found in *P. patens* and *S. moellendorffii*, only the homologs of *S. moellendorffii* GA receptors were able to bind to GA (Hirano et al. 2007). Further, Hirano et al. (2007) showed that *S. moellendorffii* can produce GAs. Lycophytes evolved before ferns, and hence GAs should be produced by *Azolla.* Homologs of many components of the BR perception and signaling are present in streptophyte algae, *P. patens* and *S. moellendorffii*, although only angiosperms seem to have the entire perception module (Wang et al. 2015; cf. Ju et al. 2015). These results hence suggest that at least some components for BR perception and signaling are present, but whether *Azolla* produces or is able to sense BRs is not known. It is, however, noteworthy that the pteridophyte *Equisetum arvense* L. was shown to produce BRs (Takatsuto et al. 1990), suggesting an alternative BR perception mechanism outside of the angiosperms.

Pathogens, like plants, require fixed nitrogen. They, therefore, try to exploit the plant not only for the photosynthates that were originally allocated for the symbiont but also for the fixed nitrogen (Horst et al. 2010) the symbiont produces for them. While sometimes higher nitrogen levels can also improve resistance against pathogens, infection success of other pathogens is reduced with lower amounts of fixed nitrogen (Hoffland et al. 2000). Hence this association highly depends on the plant-pathogen interactions. Interestingly, in *Azolla*, application of the defense-associated hormone SA reduces *nifE* expression (de Vries et al. 2018), thus directly impacting nitrogen fixation. Interestingly, SA signaling is highly regulated by nitric oxide (Tada et al. 2008; Lindermayr et al. 2010). It seems that SA not only regulates fixed nitrogen availability via a direct or indirect route (de Vries et al. 2018), but that NO is also a regulator of SA signaling (Tada et al. 2008; Lindermayr et al. 2010).

In summary, *Azolla* is an interesting case for studying phytohormone action. Every alteration in *Azolla*'s physiology is bound to have a great impact on the cyanobiont. The cyanobionts' nitrogen fixation rate, for example, is highly sensible to drought, because reduced humidity leads to a drop in the fixation rate (Hechler and Dawson 1995). The stress hormone ABA could therefore be of vital importance to

Box 2.1: A Snapshot of Phytohormones

The major plant hormones are auxin, cytokinin (CK), abscisic acid (ABA), gibberellin (GA), brassinosteroids (BR), ethylene (ET), strigolactones (SL), salicylic acid (SA), and jasmonic acid (JA). Auxin – probably the most studied plant growth factor – functions in plant development, from plant embryo development to the development of roots and shoots (Petrášek and Friml 2009). CK can act as its antagonist (cf. Bishopp et al. 2011), and that seems to apply to some aspects of *Azolla*'s development, too (de Vries et al. 2016a).

(continued)

Box 2.1 (continued)
However, in case of the effects of auxin and CK observed on *Azolla* root development, the phenotypic outcome differed from that known from *Arabidopsis* (de Vries et al. 2016a). ABA has an impact on many physiological roles in the plant, including its regulatory role in abiotic and biotic stress responses (Cutler et al. 2010; Lee and Luan 2012). GA regulates plant growth from the seed to the development of flowers (Swain and Singh 2005), and BRs are, for example, involved in cell elongation, plant growth, and photomorphogenesis (Haubrick and Assmann 2006). ET on the other hand regulates ripening and senescence responses in plants (Koyama 2014; Liu et al. 2015), but is also involved in disease responses (Broekaert et al. 2006). SL, SA, and JA are the major hormones involved in plant-microbe interactions. SL is a plant signaling phytohormone that is crucial for infection of the plant by mycorrhizal fungi (Xie et al. 2010). SA and JA on the other hand are involved in the defense against microbial pathogens (bacteria, fungi, and oomycetes) and parasites, such as insects and nematodes (Glazebrook 2005). In the canonical model, JA and SA are antagonists (Glazebrook 2005).

regulate drought and other abiotic stresses to act against a loss in fixed nitrogen. *Azolla* ought to be a very instructive case for understanding the integration of a permanent symbiont into plant molecular physiology.

Acknowledgment JdV thanks the German Research Foundation (DFG) for a Research Fellowship (VR132/1-1). SdV gratefully acknowledges support through a Killam Postdoctoral Fellowship. We thank John Archibald for many useful and helpful comments on the manuscript.

References

Abou Y, Fiogb ED, Micha J-C (2007) A preliminary assessment of growth and production of Nile Tilapia, *Oreochromis niloticus* L., Fed *Azolla*-Based-Diets in Earthen Ponds. J Appl Aquac 19:55–69

Adams DG, Bergman B, Nierzwicki-Bauer SA et al (2013) Cyanobacterial-plant symbioses. In: Rosenberg E, DeLong EF, Lory S et al (eds) The prokaryotes. Springer, Berlin/Heidelberg, pp 359–400

Akiyama K, K-i M, Hayashi H (2005) Plant sesquiterpenes induce hyphal branching in arbuscular mycorrhizal fungi. Nature 435:824–827

Arnerup J, Nemesio- Gorriz M, Lundén K, Asiegbu FO, Stenlid J, Elfstrand M (2013) The primary module in Norway spruce defence signalling against *H. annosum s.l.* seems to be jasmonate-mediated signalling without antagonism of salicylate-mediated signalling. Planta 237:1037–1045

Banks JA, Nishiyama T, Hasebe M et al (2011) The *Selaginella* genome identifies genetic changes associated with the evolution of vascular plants. Science 332:960–963

Bay G, Nahar N, Oubre M et al (2013) Boreal feather mosses secrete chemical signals to gain nitrogen. New Phytol 200:54–60

Becking JH (1987) Endophyte transmission and activity in the *Anabaena-Azolla* association. Plant Soil 100:183–212

Behm JE, Geurts R, Kiers ET (2014) *Parasponia*: a novel system for studying mutualism stability. Trends Plant Sci 19:757–763

Beilby MJ, Turi CE, Baker TC et al (2015) Circadian changes in endogenous concentrations of indole-3-acetic acid, melatonin, serotonin, abscisic acid and jasmonic acid in Characeae (*Chara australis* Brown). Plant Signal Behav 10:e1082697

Bennett T (2015) PIN proteins and the evolution of plant development. Trends Plant Sci 20:498–507

Bennett MD, Leitch IJ (2001) Nuclear DNA amounts in pteridophytes. Ann Bot 87:335–345

Bennett T, Brockington SF, Rothfels C, Graham SW, Stevenson D, Kutchan T, Rolf M, Thomas P, Wong GK, Leyser O, Glover BJ, Harrison CJ (2014a) Paralogous radiations of PIN proteins with multiple origins of noncanonical PIN structure. Mol Biol Evol 31:2042–2060

Bennett T, Hines G, Leyser O (2014b) Canalization: what the flux? Trends Genet 30:41–48

Berens ML, Berry HM, Mine A et al (2017) Evolution of hormone signaling networks in plant defense. Annu Rev Phytopathol 55:8.1–18.25

Bergman B, Matveyev A, Rasmussen U (1996) Chemical signalling in cyanobacterial-plant symbioses. Trends Plant Sci 1:191–197

Bergman B, Rai AN, Rasmussen U (2007) Cyanobacterial associations. In: Elmerich C, Newton WE (eds) Associative and endophytic nitrogen-fixing bacteria and cyanobacterial associations. Springer, Berlin, pp 257–301

Berry AM, Kahn RKS, Booth MC (1989) Identification of indole compounds secreted by *Frankia* HFPArI3 in defined culture medium. Plant Soil 118:205–209

Bishopp A, Help H, El-Showk S, Weijers D, Scheres B, Friml J, Benková E, Mähönen AP, Helariutta Y (2011) A mutually inhibitory interaction between auxin and cytokinin specifies vascular pattern in roots. Curr Biol 21:917–926

Blilou I, Xu J, Wildwater M et al (2005) The PIN auxin efflux facilitator network controls growth and patterning in Arabidopsis roots. Nature 433:39–44

Brinkhuis H, Schouten S, Collinson ME et al (2006) Episodic fresh surface waters in the Eocene Arctic Ocean. Nature 441:606–609

Broekaert WF, Delauré SL, De Bolle MFC et al (2006) The role of ethylene in host-pathogen interactions. Annu Rev Phytopathol 44:393–416

Brouwer P, Bräutigam A, Külahoglu C et al (2014) *Azolla* domestication towards a biobased economy? New Phytol 202:1069–1082

Brouwer P, Bräutigam A, Buijs VA et al (2017) Metabolic adaptation, a specialized leaf organ structure and vascular responses to diurnal N2 fixation by *Nostoc azollae* sustain the astonishing productivity of *Azolla* ferns without nitrogen fertilizer. Front Plant Sci 8:15–16

Bushart TJ, Cannon AE, ul Haque A et al (2013) RNA-seq analysis identifies potential modulators of gravity response in spores of Ceratopteris (Parkeriaceae): Evidence for modulation by calcium pumps and apyrase activity. Am J Bot 100:161–174

Calvert HE, Peters GA (1981) The *Azolla-Anabaena azollae* relationship. New Phytol 89:327–335

Calvert HE, Pence MK, Peters GA (1985) Ultrastructural ontogeny of leaf cavity trichomes in *Azolla* implies a functional role in metabolite exchange. Protoplasma 129:10–27

Campbell EL, Meeks JC (1989) Characteristics of hormogonia formation by symbiotic *Nostoc* spp. in response to the presence of *Anthoceros punctatus* or its extracellular products. Appl Environ Microbiol 55:125–131

Carrapiço F (2006) Is the *Azolla-Anabaena* symbiosis a co-evolution case? In: Sitnykov A (ed) General botany: traditions and perspectives. Kazan University, Kazan, pp 193–195

Carrapiço F (2010) *Azolla* as a superorganism. Its implication in symbiotic studies. In: Seckbach J, Grube M (eds) Symbioses and stress: joint ventures in biology, Cellular origin, life in extreme habitats and astrobiology 17. Springer, Dordrecht, pp 225–241

Caudales R, Wells JM, Antoine AD et al (1995) Fatty acid composition of symbiotic cyanobacteria from different host plant (*Azolla*) species: evidence for coevolution of host and symbiont. Int J Syst Bacteriol 45:364–370
Chee PP, Fober KA, Slightom JL (1989) Transformation of soybean (*Glycine max*) by infecting germinating seeds with *Agrobacterium tumefaciens*. Plant Physiol 91:1212–1218
Chen Z, Zheng Z, Huang J, Lai Z, Fan B (2009) Biosynthesis of salicylic acid in plants. Plant Signal Behav 4:493–496
Clark J, Hidalgo O, Pellicer J, Liu H, Marquardt J, Robert Y, Christenhusz M, Zhang S, Gibby M, Leitch IJ, Schneider H (2016) Genome evolution of ferns: evidence for relative stasis of genome size across the fern phylogeny. New Phytol 210:1072–1082
Collison ME (2002) The ecology of Cainozoic ferns. Rev Palaeobot Palynol 119:51–68
Cutler SR, Rodriguez PL Finkelstein RR, Abrams SR (2010) Abscisic acid: emergence of a core signaling network. Annu Rev Plant Biol 61:651–679
Damerval T, Guglielmi G, Houmard J et al (1991) Hormogonium differentiation in the cyanobacterium *Calothrix*: a photoregulated developmental process. Plant Cell 3:191–201
de Vries J, Fischer AM, Roettger M et al (2016a) Cytokinin-induced promotion of root meristem size in the fern *Azolla* supports a shoot-like origin of euphyllophyte roots. New Phytol 209:705–720
de Vries J, Stanton A, Archibald JM et al (2016b) Streptophyte terrestrialization in light of plastid evolution. Trends Plant Sci 21:467–476
de Vries J, Gould SB (2017) The monoplastidic bottleneck in algae and plant evolution. J Cell Sci. (in press)
de Vries J, de Vries S, Slamovits CH, Rose LE, Archibald JM (2017) How embryophytic is the biosynthesis of phenylpropanoids and their derivatives in streptophyte algae? Plant Cell Physiol 58:934–945
de Vries J, Archibald JM (2018) Plant evolution: landmarks on the path to terrestrial life. New Phytol 217:1428–1434
de Vries S, de Vries J, Teschke H, von Dahlen JK, Rose LE, Gould SB (2018) Jasmonic and salicylic acid response in the fern *Azolla filiculoides* and its cyanobiont. Plant Cell Environ in press https://doi.org/10.1111/pce.13131
Dello Ioio R, Linhares FS, Scacchi E et al (2007) Cytokinins determine *Arabidopsis* root-meristem size by controlling cell differentiation. Curr Biol 17:678–682
Delwiche CF (2016) The genomes of charophyte green algae. Adv Bot Res 78:255–270
Delwiche CF, Cooper ED (2015) The evolutionary origin of a terrestrial flora. Curr Biol 25:R899–R910
Der JP, Barker MS, Wickett NJ et al (2011) De novo characterization of the gametophyte transcriptome in bracken fern, *Pteridium aquilinum*. BMC Genomics 12:99
Dijkhuizen LW, Brouwer P, Bolhuis H, Reichart G-J, Koppers N, Huettel B Bolger AM, Li F-W, Cheng S, Liu X, Wong GK-S, Pryer K, Weber A, Bräutigam A, Schluepmann H (2018) Is there foul play in the leaf pocket? The metagenome of floating fern Azolla reveals endophytes that do not fix N2 but may denitrify. New Phytol 217:453–466
Donze M, Haveman J, Schiereck P (1972) Absence of photosystem 2 in heterocysts of blue-green alga *Anabaena*. Biochim Biophys Acta 256:157–161
Eady RR (1996) Structure-function relationships of alternative nitrogenases. Chem Rev 96:3013–3030
Ekman M, Tollbäck P, Klint J et al (2006) Protein expression profiles in an endosymbiotic cyanobacterium revealed by a proteomic approach. Mol Plant Microbe Interact 19:1251–1261
Ekman M, Tollbäck P, Bergman B (2008) Proteomic analysis of the cyanobacterium of the *Azolla* symbiosis: identity, adaptation, and NifH modification. J Exp Bot 59:1023–1034
Enderlin CS, Meeks JC (1983) Pure culture and reconstitution of the *Anthoceros-Nostoc* symbiotic association. Planta 158:157–165
Evrard C, Van Hove C (2004) Taxonomy of the American *Azolla* species (Azollaceae): a critical review. Syst Geogr Plants 74:301–318

Fay P (1969) Cell differentiation and pigment composition in *Anabaena cylindrica*. Arch Mikrobiol 67:62–70

Felix G, Duran JD, Volko S et al (1999) Plants have a sensitive perception system for the most conserved domain of bacterial flagellin. Plant J 18:265–276

Fliegmann J, Canova S, Lachaud C et al (2013) Lipo-chitooligosaccharidic symbiotic signals are recognized by LysM receptor-like kinase LYR3 in the legume *Medicago truncatula*. ACS Chem Biol 8:1900–1906

Fliegmann J, Jauneau A, Pichereaux C et al (2016) LYR3, a high-affinity LCO-binding protein of *Medicago truncatula,* interacts with LYK3, a key symbiotic receptor. FEBS Lett 590:1477–1487

Forni C, Tel-Or E, Bar E et al (1991) Effects of antibiotic treatments on *Azolla-Anabaena* and *Arthrobacter*. Plant Soil 137:151–155

Franche C, Cohen-Bazire G (1987) Evolutionary divergence in the nifHDK gene region among nine symbiotic *Anabaena azollae* and between *Anabaena azollae* and some free-living heterocystous cyanobacteria. Symbiosis 3:159–178

Friml J, Benková E, Blilou I et al (2002) AtPIN4 mediates sink-driven auxin gradients and root patterning in *Arabidopsis*. Cell 108:661–673

Friml J, Vieten A, Sauer M et al (2003) Efflux-dependent auxin gradients establish the apical–basal axis of *Arabidopsis*. Nature 426:147–153

Glazebrook J (2005) Contrasting mechanisms of defense against biotrophic and necrotrophic pathogens. Annu Rev Phytopathol 43:205–227

Goebel K (1930) Organographie der Pflanzen, insbesondere der Archegoniaten und Samenpflanzen. Gustav Fischer Verlag, Berlin

Golden JW, Robinson SJ, Haselkorn R (1985) Rearrangement of nitrogen fixation genes during heterocyst differentiation in the cyanobacterium *Anabaena*. Nature 314:419–423

Goméz-Goméz L, Boller T (2000) FLS2: an LRR receptor-like kinase involved in the perception of the bacterial elicitor flagellin in *Arabidopsis*. Mol Cell 5:1003–1011

Gourion B, Berrabah F, Ratet P, Stacey G (2015) Rhizobium legume symbioses: the crucial role of plant immunity. Trends Plant Sci 20:186–194

Grilli Caiola M, Canini A, Moscone D (1989) Oxygen concentration, nitrogenase activity and heterocyst frequency in the leaf cavities of *Azolla filiculoides* Lam. FEMS Microbiol Lett 59:283–288

Gust AA, Biswas R, Lenz HD et al (2007) Bacteria-derived peptidoglycans constitute pathogen-associated molecular patterns triggering innate immunity in *Arabidopsis*. J Biol Chem 282:32338–32348

Han G-Z (2017) Evolution of jasmonate biosynthesis and signaling mechanisms. J Exp Bot 68:1323–1331

Harrison CJ (2017) Development and genetics in the evolution of land plant body plans. Philos T Roy Soc B 372:20150490

Haubrick LL, Assmann SM (2006) Brassinosteroids and plant function: some clues, more puzzles. Plant Cell Environ 29:446–457

Haystead A, Robinson R, Stewart WDP (1970) Nitrogenase activity in extracts of heterocystous and non-heterocystous blue-green algae. Arch Mikrobiol 74:235–243

Hechler WD, Dawson JO (1995) Factors affecting nitrogen fixation in *Azolla caroliniana*. Trans Illinois State Acad Sci 88:97–107

Herdman M, Rippka R (1988) Cellular differentiation: Hormogonia and baeocytes. Methods Enzymol 167:232–242

Hernandez-Muñiz W, Stevens SE (1987) Characterization of the motile hormogonia of *Mastigocladus laminosus*. J Bacteriol 169:218–223

Hidalgo O, Pellicer J, Christenhusz M et al (2017) Genomic gigantism in the whisk-fern family (Psilotaceae): *Tmesipteris obliqua* challenges record holder *Paris japonica*. Bot J Linn Soc 183:509–514

Hill DJ (1975) The pattern of development of *Anabaena* in the *Azolla-Anabaena* symbiosis. Planta 122:179–184

Hill DJ (1989) The control of the cell cycle in microbial symbionts. New Phytol 112:175–184
Hill S, Kennedy C, Kavanagh E et al (1981) Nitrogen fixation gene (*nifL*) involved in oxygen regulation of nitrogenase synthesis. Nature 290:424–426
Hirano K, Nakajima M, Asano K et al (2007) The GID1-mediated gibberellin perception mechanism is conserved in the lycophyte *Selaginella moellendorffii* but not in the bryophyte *Physcomitrella patens*. Plant Cell 19:3058–3079
Hoffland E, Jeger MJ, van Beusichem ML (2000) Effect of nitrogen supply rate on disease resistance in tomato depends on the pathogen. Plant Soil 218:239–247
Hori K, Maruyama F, Fujisawa T et al (2014) *Klebsormidium flaccidum* genome reveals primary factors for plant terrestrial adaptation. Nat Commun 5:3978
Horst RJ, Doehlemann G, Wahl R et al (2010) *Ustilago maydis* infection strongly alters organic nitrogen allocation in maize and stimulates productivity of systemic source leaves. Plant Physiol 152:293–308
Hughes M, Donnelly C, Crozier A et al (1999) Effects of the exposure of roots of *Alnus glutinosa* to light on flavonoids and nodulation. Can J Bot 77:1311–1315
International Human Genome Sequencing Consortium (2001) Initial sequencing and analysis of the human genome. Nature 409:860–921
Jiang K, Feldman LJ (2005) Regulation of root apical meristem development. Annu Rev Cell Dev Biol 21:485–509
Johansson C, Bergman B (1994) Reconstitution of the symbiosis of *Gunnera manicata* Linden: cyanobacterial specificity. New Phytol 126:643–652
Ju C, Van de Poel B, Cooper ED, Thierer JH, Gibbons TR, Delwiche CF, Chang C (2015) Conservation of ethylene as a plant hormone over 450 million years of evolution. Nat Plants 1:14004
Khamar HJ, Breathwaite EK, Prasse CE et al (2010) Multiple roles of soluble sugars in the establishment of *Gunnera-Nostoc* endosymbiosis. Plant Physiol 154:1381–1389
Kneip C, Voß C, Lockhart PJ et al (2008) The cyanobacterial endosymbiont of the unicellular algae *Rhopalodia gibba* shows reductive genome evolution. BMC Evol Biol 8:30
Kobiler D, Cohen-Sharon A, Tel-Or E (1981) Recognition between the N_2-fixing *Anabaena* and the water fern *Azolla*. FEBS Lett 133:157–160
Koyama T (2014) The roles of ethylene and transcription factors in the regulation of onset of leaf senescence. Front Plant Sci 5:650
Kozlowski G, Buchala A, Métraux J-P (1999) Methyl jasmonate protects Norway spruce [*Picea abies* (L.) Karst.] seedlings against *Pythium ultimum* Trow. Physiol Mol Plant Pathol 55:53–58
Kurakawa T, Ueda N, Maekawa M et al (2007) Direct control of shoot meristem activity by a cytokinin-activating enzyme. Nature 445:652–655
Lee SC, Luan S (2012) ABA signal transduction at the crossroad of biotic and abiotic stress responses. Plant Cell Environ 35:53–60
Leitch IJ, Soltis DE, Soltis PS, Bennett MD (2005) Evolution of DNA amounts across land plants (Embryophyta). Ann Bot 95:207–217
Lerouge P, Roche P, Faucher C et al (1990) Symbiotic host-specificity of *Rhizobium meliloti* is determined by a sulphated and acylated glucosamine oligosaccharide signal. Nature 334:781–784
Leterme P, Londoño AM, Muñoz JE, Súarez J, Bedoya CA, Souffrant WB, Buldgen A (2009) Nutritional value of aquatic ferns (*Azolla filiculoides* Lam. and *Salvinia molesta* Mitchell) in pigs. Anim Feed Sci Tech 149:135–148
Li F-W, Pryer KM (2014) Crowdfunding the *Azolla* fern genome project: a grassroots approach. GigaScience 3:16–14
Li F-W, Kuo LY, Simenc M (2016) *Salvinia cucullata*: The smallest fern genome known. Plant and Animal Genome XXV, San Diego
Lindermayr C, Sell S, Müller B et al (2010) Redox regulation of the NPR1-TGA1 system of *Arabidopsis thaliana* by nitric oxide. Plant Cell 22:2894–2907
Liu M, Pirrello J, Chervin C et al (2015) Ethylene control of fruit ripening: revisiting the complex network of transcriptional regulation. Plant Physiol 169:2380–2390

Lorenzo O, Piqueras R, Sánchez-Serrano JJ et al (2003) ETHYLENE RESPONSE FACTOR1 integrates signals from ethylene and jasmonate pathways in plant defense. Plant Cell 15:165–178

Madsen EB, Madsen LH, Radutoiu S et al (2003) A receptor kinase gene of the LysM type is involved in legume perception of rhizobial signals. Nature 425:637–640

Markmann K, Parniske M (2008) Evolution of root endosymbiosis with bacteria: how novel are nodules. Trends Plant Sci 14:77–86

Martinez-Argudo I, Little R, Shearer N et al (2004) The NifL-NifA System: a multidomain transcriptional regulatory complex that integrates environmental signals. J Bacteriol 186:601–610

McAdam SA, Brodribb TJ, Banks JA, Hedrich R, Atallah NM, Cai C, Geringer MA, Lind C, Nichols DS, Stachowski K, Geiger D, Sussmilch FC (2016) Abscisic acid controlled sex before transpiration in vascular plants. Proc Natl Acad Sci U S A 113:12862–12867

Merrick M, Hill S, Hennecke H et al (1982) Repressor properties of the *nifL* gene product in *Klebsiella pneumoniae*. Mol Gen Genet 185:75–81

Metzgar JS, Schneider H, Pryer KM (2007) Phylogeny and divergence time estimates for the fern genus *Azolla* (Salviniaceae). Int J Plant Sci 168:1045–1053

Milindasuta B-E (1975) Developmental anatomy of coralloid roots in cycads. Amer J Bot 62:468–472

Muñoz Aguilar M, Ashby AM, Richards AJM et al (1988) Chemotaxis of *Rhizobium leguminosarum* biovar *phaseoli* towards flavonoid inducers of the symbiotic nodulation genes. J Gen Microbiol 134:2741–2746

Naberhaus F, Lee H-S, Schmitz RA et al (1995) The C-terminal domain of NifL is sufficient to inhibt NifA activity. J Bacteriol 177:5078–5087

Nahar K, Kyndt T, Nzogela YB et al (2012) Abscisic acid interacts antagonistically with classical defense pathways in rice – migratory nematode interaction. New Phytol 196:901–913

Nakayama T, Kamikawa R, Tanifuji G et al (2014) Complete genome of a nonphotosynthetic cyanobacterium in a diatom reveals recent adaptations to an intracellular lifestyle. P Natl Acad Sci USA 111:11407–11412

Nilsson M, Rasmussen U, Bergman B (2006) Cyanobacterial chemotaxis to extracts of host and nonhost plants. FEMS Microbiol Ecol 55:382–390

Obukowicz M, Schaller M, Kennedy GS (1981) Ultrastructure and phenolic histochemistry of the *Cycas revoluta-Anabaena* symbiosis. New Phytol 87:751–759

Oldroyd GED, Downie JA (2008) Coordinating nodule morphogenesis with rhizobial infection in legumes. Annu Rev Plant Biol 59:519–546

Op den Camp R, Streng A, De Mita S et al (2011) LysM-type mycorrhizal receptor recruited for rhizobium symbiosis in nonlegume *Parasponia*. Science 331:909–911

Pacios-Bras C, Schlaman HRM, Boot K et al (2003) Auxin distribution in *Lotus japonicus* during root nodule development. Plant Mol Biol 52:1169–1180

Papaefthimiou D, Van Hove C, Lejeune A et al (2008) Diversity and host specificity of *Azolla* cyanobionts. J Phycol 44:60–70

Pawlowski K (2009) Induction of actinorhizal nodules by *Frankia*. In: Pawlowski K (ed) Prokaryotic symbionts in plants, Microbiol Monogr, vol 8. Springer, Berlin, pp 127–154

Pawlowski K, Bisseling (1996) Rhizobial and actinorhizal symbioses: what are the shared features? Plant Cell 8:1899–1913

Pellicer J, Fay MF, Leitch IJ (2010) The largest eukaryotic genome of them all? Bot J Linn Soc 164:10–15

Pereira AL, Vasconcelos V (2014) Classification and phylogeny of the cyanobiont *Anabaena azollae* Strasburger: an answered question? Int J Syst Evol Microbiol 64:1830–1840

Pereira AL, Martins M, Oliveira MM et al (2011) Morphological and genetic diversity of the family Azollaceae inferred from vegetative characters and RAPD markers. Plant Syst Evol 297:213–226

Perkins SK, Peters GA (1993) The *Azolla–Anabaena* symbiosis: endophyte continuity in the *Azolla* life-cycle is facilitated by epidermal trichomes. New Phytol 123:53–64

Perrine-Walker F, Doumas P, Lucas M et al (2010) Auxin carriers localization drives auxin accumulation in plant cells infected by *Frankia* in *Casuarina glauca* actinorhizal nodules. Plant Phys 154:1372–1380

Peters GA, Mayne BC (1974) The *Azolla Anabaena azollae* relationship. Plant Physiol 53:813–819

Peters GA, Meeks JC (1989) The *Azolla-Anabaena* symbiosis: basic biology. Annu Rev Plant Biol 40:193–210

Peters GA, Toia RE, Evans WR et al (1980) Characterization and comparisons of five N2-fixing *Azolla-Anabaena* associations, I. Optimization of growth conditions for biomass increase and N content in a controlled environment. Plant Cell Environ 3:261–269

Petrášek J, Friml J (2009) Auxin transport routes in plant development. Development 136:2675–2688

Petricka JJ, Winter CM, Benfey PN (2012) Control of *Arabidopsis* root development. Annu Rev Plant Biol 63:563–590

Petutschnig EK, Jones AME, Serazetdinova L et al (2010) The lysin motif receptor-like kinase (LysM-RLK) CERK1 is a major chitin-binding protein in *Arabidopsis thaliana* and subject to chitin-induced phosphorylation. J Biol Chem 285:28902–28911

Plackett ARG, Huang L, Sanders HL, Langdale JA (2014) High-efficiency stable transformation of the model fern species *Ceratopteris richardii* via microparticle bombardment. Plant Phys 165:3–14

Popovici J, Comte G, Bagnarol E et al (2010) Differential effects of rare specific flavonoids on compatible and incompatible strains in the *Myrica gale – Frankia* actinorhizal symbiosis. Appl Environ Microbiol 76:2451–2460

Poza-Carrión C, Jiménez-Vincente E, Navarro-Rodríguez M, Echavarri-Erasun C, Rubio LM (2014) Kinetics of *nif* gene expression in a nitrogen-fixing bacterium. J Bacteriol 196:595–603

Pratiwi P, Tanaka G, Takahashi T et al (2017) Identification of jasmonic acid and jasmonoyl-isoleucine, and characterization of AOS, AOC, OPR and JAR1 in the model lycophyte *Selaginella moellendorffii*. Plant Cell Physiol 58:789–801

Pryer KM, Schneider H, Smith AR et al (2001) Horsetails and ferns are a monophyletic group and the closest living relatives to seed plants. Nature 409:618–621

Quispel A (1991) A critical evaluation of the prospects for nitrogen fixation with non-legumes. Plant Soil 137:1–11

Radutoiu S, Madsen LH, Madsen EB et al (2003) Plant recognition of symbiotic bacteria requires two LysM receptor-like kinases. Nature 425:585–592

Rai AN, Söderbäck E, Bergman B (2000) Cyanobacterium-plant symbioses. New Phytol 147:449–481

Ran L, Larsson J, Vigil-Stenman T et al (2010) Genome erosion in a nitrogen-fixing vertically transmitted endosymbiotic multicellular cyanobacterium. PLoS One 5:e11486

Rasmussen U, Johansson C, Bergman B (1994) Early communication in the *Gunnera-Nostoc* symbiosis: plant-induced cell differentiation and protein synthesis in the cyanobacterium. Mol Plant Microbe Interact 7:696–702

Redecker D, Kodner R, Graham LE (2000) Glomalean fungi from the Ordovician. Science 289:1920–1921

Reid JD, Plunkett GM, Peters GA (2006) Phylogenetic Relationships in the heterosporous fern genus *Azolla* (Azollaceae) based on DNA sequence data from three noncoding regions. Int J Plant Sci 167:529–538. https://doi.org/10.1086/501071

Rensing SA (2017) Why we need more non-seed plant models. New Phytol. https://doi.org/10.1111/nph.14464

Rice D, Mazur BJ, Haselkorn R (1982) Isolation and physical mapping of nitrogen fixation genes from the cyanobacterium *Anabaena* 7120. J Biol Chem 257:13157–13163

Rippka R, Deruelles J, Waterbury JB et al (1979) Generic assignments, strain histories and properties of pure cultures of cyanobacteria. J Gen Microbiol 111:1–61

Robert-Seilaniantz A, Grant M, Jones JDG (2011) Hormone crosstalk in plant disease and defense: more than just jasmonate-salicylate antagonism. Annu Rev Phytopathol 49:317–343

Robinson BL, Miller JH (1970) Photomorphogenesis in the blue-green alga *Nostoc* commune. Physiol Plant 23:461–472

Rodgers A, Stewart WDP (1977) The cyanophyte-hepatic symbiosis. New Phytol 78:441–458

Rozen A, Arad H, Schönfeld M et al (1986) Fructose supports glycogen accumulation, heterocysts differentiation, N2 fixation and growth of the isolated cyanobiont *Anabaena azollae*. Arch Microbiol 145:187–190

Růžička K, Simásková M, Duclercq J et al (2009) Cytokinin regulates root meristem activity via modulation of the polar auxin transport. P Natl Acad Sci USA 106:4284–4289

Salmi ML, Bushart TJ, Stout SC et al (2005) Profile and analysis of gene expression changes during early development in germinating spores of *Ceratopteris richardii*. Plant Physiol 138:1734–1745

Santi C, Bogusz D, Franche C (2013) Biological nitrogen fixation in non-legume plants. Ann Bot 111:743–767

Schaede R (1951) Über die Blaualgensymbiose von *Gunnera*. Planta 39:154–170

Schlereth A, Möller B, Liu W, Kientz M, Flipse J, Rademacher EH, Schmid M, Jürgens G, Weijers D (2010) MONOPTEROS controls embryonic root initiation by regulating a mobile transcription factor. Nature 464:913–916

Schmutz J, Cannon SB, Schlueter J et al (2010) Genome sequence of the palaeopolyploid soybean. Nature 463:178–183

Schmutz J, McClean PE, Mamidi S et al (2014) A reference genome for common bean and genome-wide analysis of dual domestications. Nat Gen 46:707–713

Schneider H (2013) Evolutionary morphology of ferns (monilophytes). Annu Plant Rev 45:115–140

Schor-Fumbarov T, Goldsbrough PB, Adam Z et al (2005) Characterization and expression of a metallothionein gene in the aquatic fern *Azolla filiculoides* under heavy metal stress. Planta 223:69–76

Schultze M, Quiclet-Sire B, Kondorosi É et al (1992) *Rhizobium meliloti* produces a family of sulfated lipooligosaccharides exhibiting different degrees of plant host specificity. P Natl Acad Sci USA 89:192–196

Sergeeva E, Liaimer A, Bergman B (2002) Evidence for production of the phytohormone indole-3-acetic acid by cyanobacteria. Planta 215:229–238

Sessa EB, Der JP (2016) Evolutionary genomics of ferns and lycophytes. Adv Bot Res 78:215–254

Sessa EB, Banks JA, Barker MS et al (2014) Between two fern genomes. GigaScience 3:15–17

Shi D-J, Hall DO (1988) The *Azolla-Anabaena* association: historical perspective, symbiosis and energy metabolism. Bot Rev 54:353–386

Silvester WB, McNamara PJ (1976) The infection process and ultrastructure of the *Gunnera-Nostoc* symbiosis. New Phytol 77:135–141

Stumpe M, Göbel C, Faltin B et al (2010) The moss *Physcomitrella patens* contains cyclopentenones but no jasmonates: mutations in allene oxide cyclase lead to reduced fertility and altered sporophyte morphology. New Phytol 188:740–749

Sumiya N, Fujiwara T, Era A et al (2016) Chloroplast division checkpoint in eukaryotic algae. P Natl Acad Sci USA 113:E7629–E7638

Suzaki T, Yano K, Ito M et al (2012) Positive and negative regulation of cortical cell division during root nodule development in *Lotus japonicus* is accompanied by auxin response. Development 139:3997–4006

Svistoonoff S, Sy M-O, Diagne N et al (2010) Infection-specific activation of the *Medicago truncatula Enod11* early nodulin gene promoter during actinorhizal root nodulation. Mol Plant Microbe Interact 23:740–747

Swain SM, Singh DP (2005) Tall tales from sly dwarves: novel functions of gibberellins in plant development. Trends Plant Sci 10:1360–1385

Tada Y, Spoel SH, Pajerowska-Mukhtar K et al (2008) Plant immunity requires conformational charges of NPR1 via S-nitrosylation and thioredoxins. Science 321:952–956

Takatsuto S, Abe H, Gamoh K (1990) Evidence for brassinosteroids in strobilus of *Equisetum arvense* L. Agric Biol Chem 54:1057–1059

Towata EM (1985) Mucilage glands and cyanobacterial colonization in *Gunnera kaalensis* (Haloragaceae). Bot Gaz 146:56–62

Trieu AT, Burleigh SH, Kardailsky IV et al (2000) Transformation of *Medicago truncatula* via infiltration of seedlings or flowering plants with *Agrobacterium*. Plant J 22:531–541

Trinick MJ (1973) Symbiosis between *Rhizobium* and the non-legume, *Trema aspera*. Nature 244:459–460

Tyagi VVS (1975) The heterocysts of blue-green algae (myxophyceae). Biol Rev 50:247–248

Usher KM, Bergman B, Raven JA (2007) Exploring cyanobacterial mutualisms. Annu Rev Ecol Evol Syst 38:255–273

van Cat D, Watanabe I, Zimmerman WJ et al (1989) Sexual hybridization among *Azolla* species. Can J Bot 67:3482–3485

Venter JC, Adams MD, Myers EW et al (2001) The sequence of the human genome. Science 291:1304–1351

Wagner GM (1997) Azolla: a review of its biology and utilization. Bot Rev 63:1–26

Wang C-M, Ekman M, Bergman B (2004) Expression of cyanobacterial genes involved in heterocyst differentiation and dinitrogen fixation along a plant symbiosis development profile. Mol Plant Microbe Interact 17:436–443

Wang C, Liu Y, Li S-S et al (2015) Insights into the origin and evolution of the plant hormone signaling machinery. Plant Phys 167:872–886

Watanabe I, Roger PA, Ladha JK, Van Hove C (1992) Azolla. In: Watanabe I, Roger PA, Ladha JK, Van Hove C (eds) Biofertilizer germplasm collections at IRRI. International Rice Research Institute, Manila, pp 5–15

Wickett NJ, Mirarab S, Nguyen N et al (2014) Phylotranscriptomic analysis of the origin and early diversification of land plants. P Natl Acad Sci USA 111:E4859–E4868

Wolf PG, Sessa EB, Marchant DB et al (2015) An exploration into fern genome space. Genome Biol Evol 7:2533–2544

Xie X, Yoneyama K, Yoneyama K (2010) The strigolactone story. Annu Rev Phytopathol 48:93–117

Yasuda M, Ishikawa A, Jikumaru Y et al (2008) Antagonistic interaction between systemic acquired resistance and the abscisic acid-mediated abiotic stress response in *Arabidopsis*. Plant Cell 20:1678–1692

Young ND, Debellé F, Oldroyd GE et al (2011) The *Medicago* genome provides insight into the evolution of rhizobial symbioses. Nature 480:520–524

Záveská Drábková L, Dobrev PI, Motyka V (2015) Phytohormone profiling across the bryophytes. PLoS One 10:e0125411

Zhao D, Curatti L, Rubio LM (2007) Evidence for *nifU* and *nifS* participation in the biosynthesis of the Iron-Molybdenum cofactor of nitrogenase. J Biol Chem 282:37016–37025

Zheng WW, Nilsson M, Bergman B et al (1999) Genetic diversity and classification of cyanobacteria in different *Azolla* species by the use of PCR fingerprinting. Theor Appl Genet 99:1187–1193

Zheng S-P, Bin C, Xiong G et al (2008) Diversity analysis of endophytic bacteria within *Azolla microphylla* using PCR-DGGE and electron microscopy. Chinese J Agri Biotechnol 5:269–276

Zheng W, Bergman B, Chen B et al (2009) Cellular responses in the cyanobacterial symbiont during its vertical transfer between plant generations in the *Azolla microphylla* symbiosis. New Phytol 181:53–61

Zipfel C, Oldroyd GED (2017) Plant signaling in symbiosis and immunity. Nature 543:328–336

Chapter 3
Meristems of Seedless Vascular Plants: The State of the Art

Alicja Dolzblasz, Elżbieta Myśkow, and Edyta M. Gola

3.1 Introduction

The indeterminate growth and proper development of a plant are ensured by the continuous activity of specific groups of undifferentiated, dividing, and potentially immortal stem cells. They perpetuate organ growth and simultaneously maintain their meristematic condition. Depending on their location in the plant and morphological characteristics, meristems are generally categorized as (i) apical meristems, when located at the tips of axial organs; (ii) lateral meristems, when forming the inner cylinder parallel to the stem surface; and (iii) intercalary meristems, when meristematic cells are separated from the apical meristem by mature tissues (Evert 2006). Because of the significance of these structures for proper plant function, it is increasingly necessary to study representatives of all major clades of vascular plants, specifically in the context of the evolutionary origins and developmental regulation. The shoot and root apical meristems (SAM and RAM) of the model angiosperm plant *Arabidopsis thaliana* are the best known and most intensively studied. Data on the function of meristems in seedless vascular plants are however scarce, scattered, and refer mostly to structural aspects and, on the molecular level, to the sequence homologies, where these are available.

Seedless vascular plants (SVPs) is a non-taxonomic name for a heterogeneous group of early-diverged vascular plants containing lycophytes (Lycopodiopsida), a sister lineage to all other vascular plants, and ferns (Polypodiopsida). Extant lycophytes are represented by three clades: clubmosses (Lycopodiales), quillworts (Isoëtales), and spikemosses (Selaginellales), whereas ferns include horsetails (subclass Equisetidae), ophioglossid ferns (Ophioglossidae) with whisk ferns

A. Dolzblasz · E. Myśkow · E. M. Gola (✉)
Department of Plant Developmental Biology, Institute of Experimental Biology, University of Wrocław, Wrocław, Poland
e-mail: alicja.dolzblasz@uwr.edu.pl; elzbieta.myskow@uwr.edu.pl; edyta.gola@uwr.edu.pl

H. Fernández (ed.), *Current Advances in Fern Research*,
https://doi.org/10.1007/978-3-319-75103-0_3

(Psilotales), marattioid ferns (Marattiidae), and leptosporangiate ferns (Polypodiidae) (PPG I 2016).

Since the nineteenth century, the meristem structure in ferns and fern allies has been the subject of intensive study; the first descriptions of shoot and root apical meristems come from these early days (e.g., Nägeli 1846; Bower 1889, 1923; Farmer 1890). Traditionally, fern apical meristems have been recognized as having a distinct pyramidal apical cell (AC), which is the main source of all cells in the meristem and, thus, the entire plant body. An alternative organization of the apical meristems, with a group of morphologically similar initial cells instead of the single AC, has been described in the Isoëtales and Lycopodiales (Farmer 1890; Härtel 1937) as well as in some ophioglossid and marattioid ferns (Bower 1889, 1923; references in Popham 1951; Ogura 1972; Philipson 1990). In addition, because of specific growth of the *Isoëtes* corm, the vascular cambium in this genus has been described as an exceptional example of a secondary meristem in extant SVPs (Stokey 1909; West and Takeda 1915; Stewart 1947). Currently, the traditional description of the meristems in SVPs is generally accepted, and the data obtained with use of modern techniques of visualization and improved tools to analyze the meristem structure at the cellular and subcellular levels have not substantially changed it. In-depth analysis of the data concerning meristem development and activity in a particular genus is still not fully available, and sometimes confusing or even contradictory details occur. Recently, new data has been published for *Ceratopteris* and *Selaginella* as representatives of these genera have been selected as model organisms in evolutionary developmental research, but other groups are barely studied. We summarize in this review the data on meristem organization and its genetic regulation in all groups of the SVPs.

3.2 Shoot Apical Meristem (SAM) Structure

3.2.1 Ferns (Polypodiopsida)

In the majority of ferns, regardless of their taxonomic position (basal vs. advanced taxa), the structure of the SAM is still recognized as relating to the presence and activity of a single apical cell (AC). The AC is usually large and morphologically distinct from other meristematic cells. Mostly, it is of an inverted-pyramidal shape with the triangular or rectangular base directed toward the outside. The best-known example of such a tetrahedral AC is the SAM of horsetails (Golub and Wetmore 1948a, b; Gifford and Kurth 1983; Gifford and Foster 1989; Evert 2006); and it has also been exhaustively documented for numerous genera and species of the leptosporangiate ferns (Hirsch and Kaplan 1974; Lucansky and White 1976; Bierhorst 1977; Hébant-Mauri 1975, 1977, 1984, 1990, 1993; Hébant-Mauri and Veillon 1989; Imaichi 1980, 1982, 1988, 2008; White and Turner 1995).

Interestingly, in several fern species, for example, in *Platycerium alcicorne*, *Pteridium aquilinum*, *Polypodium vulgare*, and water ferns (Salviniaceae and Azollaceae), instead of the typical tetrahedral AC, the AC is wedge-like or lenticular-shaped (Bower 1889; Campbell 1893; Ogura 1972; Foster and Gifford 1974; White and Turner 1995). In *Salvinia*, a lenticular AC with two alternating cutting faces has been visualized in confocal microscope (CLSM). The precise and regular sequence of its divisions and fixed fates of derivative segments are undoubtedly related to the dorsiventral symmetry of shoots and exemplify the specialization and adaptation to the environment (Croxdale 1978, 1979; Lemon and Posluszny 1997). The interrelationship of the lenticular shape and the dorsiventral symmetry of organs have also been suggested for some rhizomatous fern species; however, detailed structural analyses confirmed the presence of the typical tetrahedral AC with the regular segmentation pattern as in *Microgramma* (Hirsch and Kaplan 1974), *Trichomanes* (Hébant-Mauri 1984), *Lomogramma* (Hébant-Mauri and Gay 1993), *Hypolepis*, and *Histiopteris* (Imaichi 1980, 1982). In addition, the question of the AC shape is still more complicated as it can ontogenetically change from wedge-shaped in young plants to the typical pyramidal AC in adults as in *Polypodium vulgare* (Ogura 1972) and *Pteridium aquilinum* (Dasanayake 1960).

Regardless of the shape, the AC undergoes regular divisions, cutting off derivative cells (segments, merophytes) precisely along the sidewalls. There are no divisions parallel to the external wall; thus, the typical tetrahedral AC has three cutting faces giving rise to spirally arranged segments, either clockwise or counterclockwise when viewed from the top; and the lenticular AC has only two alternating planes of division. Subsequent divisions within each merophyte are according to the stable species-dependent sequence and result in a segmentation pattern of the whole apex. When the apex is convex and relatively tall, e.g., as in the textbook SAM of horsetails (Bierhorst 1959; Gifford and Kurth 1983) and *Ceratopteris thelypteris* (Hébant-Mauri 1977, 1993), the AC is pronounced and only few segments are traceable; when the meristem is relatively flat, more segments can be traced, although their inner borders demarking stelar tissue and the AC itself are less evident. Instead, the superficial layer of large anticlinally elongated so-called prismatic cells becomes more visible at the meristem as in *Trichomanes*, *Cardiomanes* (Hébant-Mauri 1984, 1993), *Microgramma* (Hirsch and Kaplan 1974), and Filicales (Bierhorst 1977). As prismatic cells are formed in each newly formed merophyte due to uneven periclinal divisions, they surround the AC, which becomes almost indiscernible. Formation of the prismatic cell layer is probably why the single AC becomes obscure at the meristem of ophioglossid ferns (Bower 1923; Bhambie and Madan 1980). Detailed studies on the meristem structure in the genus *Botrychium* demonstrated, however, that the SAM changes upon development and a single AC is present at the apex (Imaichi and Nishida 1986; Imaichi 1989; Stevenson 1976a, c), at least in young plants.

This view on the SAM structure and especially on the role of the AC was challenged in the 1970s and 1980s, when intensified research on the meristem structure in plants, especially in angiosperms, prompted researchers to compare fern meristems with those in seed plants. Subsequent extensive studies on AC activity, potential polyploidy, segmentation pattern, and histogenesis within the apex covered a wide range of fern species. The proposed histological zonation of fern SAMs was similar to that in seed plants with the following zones: the surface (initial) layer, containing the AC and prismatic cells; small subsurface cells, which give rise to procambial (stelar) tissue with more-or-less identifiable pith; and the peripheral zone where the initiation of leaf primordia takes place (e.g., in *Botrychium*, *Dennstaedtia* (Stevenson 1976b, c), and *Matteuccia* (Ma and Steeves 1994, 1995). However, developmental analyses proved that both the cell-lineage-dependent segmentation pattern and the cytological properties of derivative cells, their further divisions, and differentiation are integral parts of the proper functioning of the meristem (White and Turner 1995).

3.2.2 Lycophytes

In lycophytes, the SAM structure differs between orders. In the Selaginellales, the SAM organization is most widely recognized as related to the activity of a single AC (e.g., *S. willdenowii*, *S. martensii*, *S. rupestris*, and *S. apoda*; Barclay 1931; Dengler 1983; Jacobs 1988; Schulz et al. 2010), although two functioning apical cells (*S. grandis*, *S. wallichii*, *S. kraussiana*; Wand 1914) or a layer of initial cells (*S. oregana*, *S. uncinata*, *S. lyallii*, *S. willdenowii*; Wand 1914; Popham 1951 and references therein; Cusick 1953; Ogura 1972) have also been reported. Similar to that in ferns, the AC shape shows interspecific diversity. Typically, they are either lenticular or wedge-shaped with two planes of division (e.g., in *S. kraussiana*, *S. umbrosa*) or pyramidal with three to (occasionally) five cutting faces (e.g., in *S. martensii* and *S. apoda*; Wand 1914; Popham 1951; Ogura 1972; Dengler 1983; Jacobs 1988; Philipson 1990; Schulz et al. 2010). In addition, their shape can differ between shoots of the same plant, dependent on the age and size of the dichotomous branch, e.g., in *Selaginella martensii* (Wand 1914; Popham 1951; Dengler 1983; Lu and Jernstedt 2005).

This variability in the number and shape of the functioning AC is inconsistent even in the most current literature data. For instance, the SAM organization in the intensively studied species *S. kraussiana* remains disputable. Using CLSM, Jones and Drinnan (2009) demonstrated the presence of one AC with two cutting faces. This finding is contradictory to that of Harrison et al. (2007) who, based on clonal analysis of the mutagenized shoots (with albino sectors), inferred the presence of two ACs at the meristem. Furthermore, the latter authors argued that based on the apex geometry this arrangement is more probable (Harrison and Langdale 2010). However, the SAM structure in *Selaginella* appears quite changeable during the ontogeny due to frequent dichotomous divisions of the apex. Geometric changes of the apex both prior to and during the division, as well as the selection of the new initials for each meristem of twin dichotomous branches, significantly affect the meristem structure. Even given a single AC, which in other *Selaginella* species has been proven to be short-lived (Lu and Jernstedt 2005), it undergoes the segmentation prior to dichotomous division, giving rise to a pool of meristematic cells that are larger, anticlinally elongated, and of transient rectangular shape in frontal longitudinal view.

In the order Isoëtales, in both extant genera *Isoëtes* and *Stylites*, one, two, or several functioning ACs have been mentioned (e.g., reviewed in Popham 1951, Paolillo 1963 and Philipson 1990). The single AC of lenticular or tetrahedral shape is found, if at all, in young plants (e.g., in *I. lacustris*; Scott and Hill 1900), and further ontogenetic changes of the SAM result in the presence of a group of undistinguishable meristematic cells (Farmer 1890; Scott and Hill 1900; Paolillo 1963). The group of initial cells can also be found throughout the plant life, even in sporelings, of *I. japonica, I. hystrix, I. velata*, and *I. taiwanensis* (West and Takeda 1915; Chiang 1976).

In Lycopodiales, the meristem contains a group of superficial multiple initial cells throughout the entire life of the plant (Härtel 1937; Wetmore and Smith 1942; Freeberg and Wetmore 1967; Stevenson 1976a; Gola and Jernstedt 2011). Generally, current authors do not distinguish a single AC in the meristems of lycopods, although such records were mentioned in the early literature of the subject (Treub 1886 after Popham 1951). The initial cells are indistinguishable based on morphology from the remaining meristematic cells (Härtel 1937; Wetmore and Smith 1942; Freeberg and Wetmore 1967; Stevenson 1976a), but it is possible to identify them and infer their function based on the position at the meristem and divisional behavior. Based on clonal analysis, four superficial impermanent initial cells were described to function at the SAM. Before the dichotomous division, their number significantly increases forming a pool of meristematic cells, from which new initial cells for both meristems of twin branches are selected (Gola and Jernstedt 2011). The histological zonation of lycopod meristems has also been related to that of the seed plants. These studies showed the presence of distinct zones of superficial and subsuperficial initials, with the rib meristem, the source of procambial tissue, formed below these, and a peripheral organogenic zone (Stevenson 1976a).

Interestingly, in an evolutionarily distinct and independently diverged clade of lycophytes, structural solutions for the SAM organization show much higher diversity and variability than in ferns or seed plants, posing questions about the developmental and/or evolutionary origin of this structural diversification. Perhaps one of the driving factors is the mode of shoot branching, which in all lycophytes is dichotomous. A meristem with multiple initial cells, as in lycopods, appears beneficial for dichotomous branching. In apices with a single AC, this process requires the initiation of new ACs for both twin branches by the segmentation of the original AC and/or additional formative divisions in the derivative merophyte (Gola 2014). When two initials function at the apex, twin branches are formed in each of two sagittal sectors generated by them as in *S. kraussiana* (Harrison et al. 2007). In either case, the new extra initial cells must be initiated upon the meristem split. Conversely, in ferns terminal dichotomous branching is rarely described (Hagemann and Schultz 1978; *Lygodium*, Mueller 1982). Mostly, branching is acrogenous or related to the formation of non-axillary phyllogenous or cauline buds, for which the ACs are formed on a regular basis and traceable early within the segmented apex (e.g., Bierhorst 1977; Hébant-Mauri 1993; White and Turner 1995). Therefore, it seems likely that meristem structures preserved and fixed in ferns and diversified in lycophytes are two examples of different adaptation and specialization to the dominant form of growth.

The question of which meristem structure, with a single AC or multiple initial cells, is ancestral is discussed in the literature (Imaichi 2008; Tomescu 2011). The view that the ancestral structure is a single AC is based primarily on the similarities to the bryophyte meristem structure and is supported by the lack of the single AC-based meristem in advanced lineages of seed plants. Conversely, the structure of SAM with numerous initial cells is considered as an advanced feature based on its convergence to that in gymnosperms. Lycophytes are the first group of extant vascular plants, which experimented with the meristem organization, and each structural solution was then adapted and tuned either in ferns or seed plants. Unfortunately, based on available fossil data, it is difficult to conclude which structure was first in the plant phylogeny (compare Tomescu 2011).

3.3 Root Apical Meristem (RAM) Structure

Root emergence was probably an independent innovation in the phylogeny of lycophytes and ferns (Pires and Dolan 2012). In addition, in both groups, the primary embryonic root is formed laterally in respect to the primary SAM and is only short-lived. In adult plants, roots are adventitious and originated endogenously close to the apex or in relation to leaves. Regardless of RAM origin, its organization, with a single AC or multiple initial cells, is generally similar to that of their SAM.

3.3.1 Ferns

In ferns, RAM is primarily based on the activity of the single AC, usually tetrahedral in shape, but in contrast to that in the SAM it cuts off segments parallel to all the walls. The base of the pyramid is externally oriented, and the derivative cell cut off parallel to it gives rise to the root cap. The AC and its derivatives undergo divisions according to a fixed sequence, giving rise to particular tissues of the root, which have been shown for different species of ferns, e.g., *Azolla* (Gunning et al. 1978; Nitayangkura et al. 1980), *Equisetum* (Gifford and Kurth 1982; Gifford 1993), *Asplenium* (Gifford 1991), and *Ceratopteris* (Hou and Hill 2002, 2004). The segmentation of the RAM is even more pronounced than the SAM as here the lateral organs (roots) are formed endogenously below the root apex and do not disturb the regularity of segments, while in the SAM there is intensive organogenesis within the apex itself. The same pattern is also observed during the formation of the lateral roots. The ACs of lateral roots are endogenously initiated due to the fixed sequence of divisions, from a single endodermal cell (most leptosporangiate ferns, e.g., *Marsilea*, Lin and Raghavan 1991; *Ceratopteris*, Hou and Hill 2002, 2004;

Adiantum, and *Pteris*, Chiang and Chou 1974) or below it, in the pericycle, if it consists of more than one cell layer (e.g., *Osmunda* and *Angiopteris*, Chiang and Chou 1974).

In contrast to the typical one-AC-based structure of the RAM, multiple initial cells have been reported in the roots of *Osmunda regalis*. Although initially one AC is present, the rate of its divisions reduces in older plants and the single AC becomes obscure. Similarly, multiple initial cells have been reported in RAMs of the families Osmundaceae and Marattiaceae (Bower 1923; Bhambie and Rao 1972; Freeberg and Gifford 1984; Ogura 1972).

3.3.2 Lycophytes

The organization of the RAM in lycophytes is highly diversified and typically related to the presence of a group of initial cells. In lycopods, roots are initiated in the cortex and/or in the stele below the apex and run in the cortex through the stem before emerging to the surface. Meristematic cells are similar to each other and show a tendency to form layers in a genus-dependent manner (Fujinami et al. 2017). In *Lycopodium* and *Diphasiastrum*, one group of unstratified initial cells is present, whereas in the genera *Lycopodiella* and *Huperzia*, three discrete meristematic tiers are specified for the root cap, for the protoderm, and for the ground tissue and stele (Fujinami et al. 2017). In *Isoëtes*, two separate layers were selected – the outermost one for the root cap and the epidermis and the internal one for the root cortex and stelar tissue (Paolillo 1963; Yi and Kato 2001); this seems to be slightly exaggerated as the most recent data suggest the presence of an unstratified group of meristematic cells forming the root meristem in this genus (Fujinami et al. 2017).

The RAM structure in the third order of the extant lycophytes, the Selaginellales, is more complicated as roots are initiated at the tips of specific root-bearing organs (rhizophores); thus, discussion of rhizophores and roots is given separately below.

3.3.3 **Selaginella***: Rhizophore and Root Apical Meristems and the Angle Meristem*

Rhizophores are leafless cylindrical axes with the emerging roots at their tips, unique to the genus *Selaginella*. Due to combined shoot and root characteristics, the identity of rhizophores was widely disputed and historically interpreted as (i) transformed shoots (Schoute 1938), (ii) roots in the aerial phase of their development (Webster and Steeves 1964, 1967; Webster 1992), or (iii) distinctive organs, characteristic of *Selaginella*, with no homologies to any other structures of extant vascular plants (Imaichi and Kato 1989, 1991; Imaichi 2008; Otręba and Gola 2011). The first, marginal, interpretation was supported by the developmental plasticity of rhizophores, which, as well as typically being the root-bearing organs, have the potential to develop as leafy shoots (Schoute 1938) when the hormonal balance is changed (compare e.g., Imaichi and Kato 1989, 1991; Webster 1992; Jernstedt et al. 1992, 1994). The other interpretations were much more intensively discussed, with the structure and ontogenetic changes of the rhizophore/root apex decisive in both viewpoints. Rhizophores develop in the branching point of the dichotomizing shoot, where another meristem characteristic exclusive to *Selaginella*, i.e., the angle meristem, is located (Webster 1969, 1992; Jernstedt et al. 1992, 1994). The angle meristem is defined as a meristematic residue (detached meristem), which remains temporarily dormant between twin branches after dichotomous division of the shoot apex. One meristematic cell within the angle meristem undergoes a formative division, giving rise to a single tetrahedral AC with three cutting faces, typical of the SAM, whose further divisions cause the outgrowth of the rhizophore. Rhizophores, with preserved AC, can remain unbranched or divide dichotomously several times depending on the species, always continuing the same organization of resultant rhizophore apices and the manner of the AC divisions (Imaichi and Kato 1989, 1991; Jernstedt et al. 1992, 1994; Lu and Jernstedt 1996; Otręba and Gola 2011). Ultimately, at the rhizophore tips, the root primordia are formed; however, the manner of the initiation of root ACs caused discrepancies in the interpretation of the rhizophore identity. Researchers questioned whether the functioning tetrahedral AC changes the division pattern to cut off an additional segment for a root cap parallel to the pyramid base. In such case, the same AC functions throughout the organ development, and the entire rhizophore-root system is a developmental continuum; thus, the rhizophore should be interpreted only as an aerial phase of root development (Webster and Steeves 1964, 1967; Webster 1992). The alternative interpretation states that the initiation of ACs for root primordia is independent and starts after the segmentation and disappearance of the original rhizophore AC. The sequence of developmental events leading to the root primordia formation was documented in detail in several *Selaginella* species, proving the structural and functional distinctiveness of rhizophores and roots; thus, the currently accepted interpretation is that the rhizophores are the unique axial organs incomparable to roots or shoots of vascular plants (Imaichi and Kato 1989, 1991; Lu and Jernstedt 1996; Imaichi 2008; Otręba and Gola 2011). Furthermore, the results of more recent molecular analyses, which will be presented later, support this (Kawai et al. 2010).

3.4 Intercalary Meristems

Apart from the apical meristems, plant growth, especially the elongation of axial organs, can be related to the activity of the intercalary meristems (IM). The best-known intercalary meristems occur in monocots, and only rarely are present in dicotyledonous plants (Evert 2006). In SVPs, the intercalary meristems are rare and usually found only in relation to specific organs or growth forms. The most studied was the IM in the genus *Equisetum,* characterized by noticeable internodes in the aerial shoots (Golub and Wetmore 1948a, b; French 1984; Vasco et al. 2013). Within the distinctly segmented SAM of horsetails, the precursors of the IM are determined already during the initiation of the third node and the internode below it. The cell, which will give rise to the internode, divides transversely, with the basal derivative cell being the incipient IM. Repeated transverse divisions of this founder cell produce a pile of short derivatives, which elongation starts from the first formed (basal) cells, leading to the rapid growth of the entire internode and thus plant. Simultaneously with differentiation of the IM cells in the fifth or sixth internode, successive intercalary meristems develop in the internodes above (Golub and Wetmore 1948a, b). On the other hand, in rhizomes two different patterns of internode development can occur: with a normal IM (subgenus *Hippochaetae*) or with an uninterrupted meristem, non-disjunctive to the apical meristem (subgenus *Equisetum*; French 1984).

The presence of IM was also shown in the distal part of the rachis, just below the sporangial zone of the fertile spike, and in the sterile part of the frond of *Ophioglossum petiolatum* (Peterson and Cutter 1969; Vasco et al. 2013).

The IM is also the driving force for the growth of rhizophores of *Selaginella kraussiana*, and an adaptation for rapid delivery of roots to the soil, as the ACs of root primordia have initially only limited divisional activity. This meristem develops below both root primordia partly from their cells and partly from the cells of the rhizophore. The mixed origin and obscure disjunction from the RAM (by meristematic cells of different fixed fate) result in the specificity of the intercalary meristem in *Selaginella* rhizophore (Otręba and Gola 2011).

3.5 Lateral Meristems: Vascular Cambium and Phellogen

In the extant plants, there are two types of lateral meristems, the vascular cambium and the phellogen (cork cambium), the activity of which results in the secondary thickening of the stem. These meristems are typically related to the woody phenotype, increasing mechanical properties and plants' diameter. Both meristems are formed inside the plant, at the surface parallel to the stem circumference. Typically, their initial cells divide parallel to the stem surface (periclinally) giving daughter cells both inward and outward; thus, these meristems are called bidirectional or bifacial. Periclinal divisions of the cambial initials produce secondary phloem outside and secondary xylem inside, whereas phellogen gives rise to the phellem (cork) and phelloderm, respectively (Evert 2006).

Although lateral meristems were present in many representatives of extinct lycophytes and ferns, among extant SVPs their presence is reported only exceptionally, in *Isoëtes* and *Botrychium/Botrypus*. As data is scattered and often contradictory, detailed descriptions of both cases will be given below.

3.5.1 Isoëtes

Vascular Cambium

Isoëtes is a perennial plant with a relatively short axis called "corm" divided into two parts: the upper with the SAM located in the small depression and giving rise to the leaves and the lower with numerous spirally arranged roots (Stewart 1947; Bhambie 1963; Paolillo 1963; Karrfalt and Eggert 1977b, 1978; Yi and Kato 2001). The central part of the corm is filled with the primary xylem composed of tracheids and parenchyma, with scarce irregularly arranged elements of primary phloem located outside of them (Paolillo 1963; Yang et al. 1975; Chiang 1976; Kruatrachue and Evert 1977; Chiang and Chen 1986); however, the presence of the latter tissue was also negated (Stokey 1909).

The thickening of the corm is derived from the activity of the cambium, consisting of initial cells organized in two- or three-celled radial files (Yang et al. 1975; Chiang 1976; Kruatrachue and Evert 1977; Yi and Kato 2001). All cambial cells are similar in shape (Yang et al. 1975), thus in *Isoëtes*, two distinct types of cambial initials (fusiform and ray initials), which typically occur in seed plants (Evert 2006), are not distinguished. The mechanism of cambium function and characteristics of the secondary tissues produced are differently interpreted. According to some authors (Stokey 1909; Yang et al. 1975; Chiang 1976; Chiang and Chen 1986), in *Isoëtes* cambium is one bifacial vascular meristem dividing periclinally and giving rise to the inner prismatic layer (secondary vascular tissue) and the outer secondary parenchymatic cortex. Other authors suggest that the cambium is composed of two parts, the lateral meristem (corresponding structurally and functionally to the earlier-mentioned cambium) and the basal meristem, producing secondary xylem inside of the stem and root-bearing tissue outside (Paolillo 1963; Kruatrachue and Evert 1977; Yi and Kato 2001). The unity and continuity of both parts are evidenced in the central part of the corm and are additionally proved by the ability of cells to transform from one part of the meristem to another (Yi and Kato 2001).

Due to divisions of the lateral meristem, the inner prismatic layer and the outer secondary cortex are formed. The secondary cortex is composed of thin-walled parenchyma cells with abundant starch grains, and importantly, the phloem cells expected on this side of (typical) cambium have never been detected here. The thickness of the secondary cortex increases seasonally, but its external parts are sloughed off (Stokey 1909; Kruatrachue and Evert 1977). The identity of the cells in the prismatic layer is more controversial. Its location at the inner side of the cambium suggests that these cells should have characteristics of secondary xylem; accordingly, xylem parenchyma cells and tracheids were recognized here (Stokey 1909). Currently, the inner secondary tissues are accepted to contain the secondary phloem and parenchyma cells. Interestingly, the formation of the secondary xylem on this side can be delayed or even lacking depending on the species and the corm age (Paolillo 1963; Kruatrachue and Evert 1977; Chiang and Chen 1986). In addition, since *Isoëtes* is an aquatic plant, the presence of the secondary xylem may be redundant for the proper plant functioning (Chiang and Chen 1986). The reason why the secondary vascular tissues are deposited only inward despite the presence of the bifacial cambium remains unsolved.

The basal meristem, interpreted as a functionally separate part of the cambium, forms a ribbonlike structure in the lower part of the *Isoëtes* corm and is a root-producing or root-bearing meristem (Paolillo 1963; Kruatrachue and Evert 1977; Yi and Kato 2001). The initial cells of this meristem form radial files of two or three cells, similar to those in the lateral meristem. They also divide periclinally producing derivative cells to both sides (bifacial meristem): parenchyma cells and tracheids inside and cortical cells outside, increasing the thickness of the corm (Kruatrachue and Evert 1977; Yi and Kato 2001). These similarities to and continuity with the lateral meristem led to the basal meristem being considered as a secondary meristem (Paolillo 1963; Kruatrachue and Evert 1977). Importantly, due to divisions of the cortical cells generated by the basal meristem, the primordia of adventitious roots are formed (Yi and Kato 2001). The continuous organogenic activity of the basal meristem, leading to the repetitive formation of roots, was comparable to that of the SAM indicating that the basal meristem is a primary meristem (Stewart 1947; Karrfalt and Eggert 1977a, 1978; Karrfalt 1982). Consequently, the xylem and parenchyma cells produced inward by this meristem should be interpreted as primary tissues (Chiang 1976). Currently, the basal meristem is assumed to be a distinct, unique, unparalleled meristem, characteristic only of *Isoëtes* due to its mixed primary (branching, organ formation) and secondary (corm thickening, bifacial activity) characteristics (Yi and Kato 2001).

The origin of vascular meristem is also disputable; with two hypotheses initially postulated to explain its development (reviewed in Paolillo 1963). The first suggests that the procambium is located outside of the primary xylem as the founder tissue (Scott and Hill 1900), as in conifers and dicotyledonous trees (Evert 2006), whereas in the second the vascular cambium originates from the parenchyma cells located outside of the primary phloem (Stokey 1909; Lang 1915; West and Takeda 1915). Currently, it is assumed that the vascular cambium originates from the procambial cells, which are atypically located outside of the primary phloem (Paolillo 1963; Yang et al. 1975; Chiang 1976). In addition, although the lateral and basal meristems are functionally bifacial, derivative cells of both vascular types are formed only inward. Furthermore, the lateral meristem is continuous with the basal meristem (Yi and Kato 2001). Therefore, regarding all these issues and what we know now about the vascular cambium development and activity in different plant groups, in *Isoëtes* neither the basal nor the lateral meristem are the true vascular cambium. So-called vascular cambium in *Isoëtes* is a specific meristem incomparable to any other meristem in lycophytes (Stewart 1947) and plants in general.

Phellogen and the Secondary Protective Tissue (Periderm)

Data on the secondary covering tissue formation and morphology in *Isöetes* are extremely scarce and affected by interpretation of the nature of lateral and basal meristems. As mentioned above, the outermost parts of the secondary cortex are sloughed off every year (Stokey 1909; Kruatrachue and Evert 1977). Beneath these layers, the corm is covered by dead cells with suberized cell walls, forming the corky layers (West and Takeda 1915; Paolillo 1963; Kruatrachue and Evert 1977). Due to the controversial secondary nature of the cortex, it is unclear whether this corky layer is a primary or secondary protective tissue, especially as proliferative divisions of the cells in the primary cortex can also form the secondary cortex (Karrfalt 1982).

3.5.2 Botrychium/Botrypus

Botrychium and *Botrypus* are the other genera in which the presence of cambium was reported, although its activity ceases after several years, when all the cambial initials transform to the parenchymatous cells (Lee and Soh 1995). Also in the case of *Botrychium/Botrypus*, cambium features are differently regarded. This cambium was considered as unifacial, producing exclusively the secondary xylem to the inside of the stem (Stevenson 1980), or as bifacial with the secondary xylem deposited toward the center and only parenchymatous cells to the outside. Furthermore, its structure is characterized by uniseriate, or sometimes biseriate, rays (Takahashi and Kato 1988). However, several developmental peculiarities have caused the identity of this vascular meristem to be challenged (Rothwell and Karrfalt 2008). The continuous cylinder of cambium originates before the first vascular elements of protoxylem and protophloem undergo differentiation, and metaxylem elements and secondary phloem are not formed in these genera at all (Stevenson 1980; Takahashi and Kato 1988). This indicates that there is no developmental continuity of successive phases (these should be proto- and metaxylem and phloem, followed by the secondary tissues). In addition, the formation of the continuous cylinder of vascular meristem and the presence of rays are not obligatory markers of the cambium, and they can emerge even in the functioning procambium (Evert 2006; Myśkow 2010). Based on these features, the vascular meristem in *Botrychium/Botrypus* should be interpreted throughout as the primary (procambium) and not the secondary (cambium) meristem (Rothwell and Karrfalt 2008). The only feature related to the activity of the secondary meristem is the formation of radial rows of the cells in the cortex. This characteristic phenotype is related to the phellogen activity, which gives rise to the cork cells and phelloderm (Rothwell and Karrfalt 2008).

3.6 Maintenance of the Meristem Identity

Molecular genetic studies are important to properly understand organism structure and development. In plant research, molecular mechanisms accompanying various processes were mostly investigated in a model organism, a representative of angiosperms, *Arabidopsis thaliana*. Conversely, SVP lineages have been little studied at the molecular level, and most recent research is focused on and discussed only from the evolutionary perspective. Limiting factors of research advancement were the lack of full sequences of nuclear genomes, none for fern species (Barker and Wolf 2010), and only one available for a lycophyte (*Selaginella moellendorffii*; Banks et al. 2011), and the lack of the efficient reliable protocols for stable transformation. Importantly, in *Ceratopteris richardii*, there are promising preliminary studies presenting transformation protocols (Plackett et al. 2014, 2015b), and the sequencing of whole nuclear genome is anticipated. Unfortunately, development of a transformation protocol is still incomplete in *Selaginella*, and, in addition, although *S. moellendorffii* has been sequenced, it neither has yet been extensively used in molecular biology nor sufficiently developmentally characterized. Instead, *S. kraussiana* has been the object of several studies despite not yet being sequenced (Korall and Kenrick 2002; Harrison et al. 2005, 2007; Durantini 2009). In addition, SVPs encompass not only spikemosses (*Selaginella*) and leptosporangiate ferns (e.g., *C. richardii*), which are studied at the gene level, but thousands of genera and species, which are neglected in terms of molecular studies.

The central focus of developmentally oriented studies are the stem-cell-bearing shoot and root apical meristems, which maintain the complex signaling network and ensure indeterminate plant growth and development. Most research concerning genetic regulation of meristem activity in SVPs is derivative of knowledge already available for more studied plants, mostly *Arabidopsis*, and thus was initially concentrated on the search for key regulators from the best-known gene families. Due to insufficient methodological options, their probable functional relevance was assessed based on limited resources, typically by comparison of gene sequences and expression patterns, often accompanied by far-reaching interpretations and speculations. This is currently often supplemented by ectopic expression or complementation studies performed in the model angiosperm plant *A. thaliana*, as transgenic SVPs are not yet created routinely and, thus, it lacks a native context. Below we present two examples where the orthologous genes from two different families of HD (homeodomain) transcription factors (TFs), *knotted-related homeobox* (*KNOX*) genes and *WUS/WOX5* from *Wuschel-related homeobox* (*WOX*), which are key regulators of apical meristem maintenance in *Arabidopsis*, were identified and preliminarily functionally characterized in *Ceratopteris* and *Selaginella* (for broader discussion, see Plackett et al. 2015a).

3.6.1 KNOX *Family*

The *KNOX* genes are a class of HD TFs found in all green plant lineages. In monocots and eudicots, the *KNOX* family is divided into two subclasses (Kerstetter et al. 1994; Mukherjee et al. 2009): class I expressed in overlapping domains within the SAMs (Hake et al. 2004; Jackson et al. 1994) and class II showing diverse expression patterns (Zhong et al. 2008). Members of class I are important regulators of SAM development in angiosperms. For example, KNOTTED1 (KN1) activity in maize and related protein SHOOT MERISTEMLESS (STM) in *Arabidopsis* are required to prevent pluripotent cells from differentiation; thus, *stm* and *kn1* mutants fail to establish and maintain a SAM (Smith et al. 1992; Long et al. 1996; Vollbrecht et al. 2000).

The *Arabidopsis* genome contains four class I *KNOX* gene expressions of which in specific domains maintains the activity of the SAM, its lateral organs, and stem boundaries. The first *KNOX* gene expressed during early embryogenesis is *STM*, which expression marks the whole indeterminate SAM (Long et al. 1996). The established boundary between KNOX-expressing and non-expressing cells enables proper lateral organ development (Hake et al. 2004). Collectively called ARP proteins (the name comes from ASYMMETRIC LEAVES1 (AS1) in Arabidopsis, ROUGH SHEATH2 (RS2) in maize, and PHANTASTICA (PHA) in snapdragon) in various plant species are expressed in a mutually distinct domain to that of KNOX proteins, distinguishing leaf founder cells from the meristematic cells, and repress KNOX expression during leaf development (e.g., see a review by Hay and Tsiantis 2010).

In lycophytes, two class I (*SkKNOX1* and *SkKNOX2*) and one class II (*SkKNOX3*) *KNOX* genes were found in *S. kraussiana*. Interestingly, in situ hybridization experiments have shown that both *SkKNOX1* and *SkKNOX2* genes were expressed respectively in the shoot apices, in the area directly beneath the large apparent AC, and in internodal regions and were not detected in the leaf primordia, resembling that of the class I *KNOX* genes in seed plant apices (Harrison et al. 2005). Interestingly, a recent study by Frank et al. (2015) has shown the presence of a class I *KNOX* gene also in the AC of *S. moellendorffii*. In *S. kraussiana* (and *S. viticulosa*), a single *ARP* gene (*SkARP1*) was found to be expressed in leaf primordia, stem vasculature, leaf traces, but also in the meristem. Using a heterologous system, *SkARP1* was able to repress *KNOX* gene expression in *Arabidopsis* and thus is a functional equivalent to *AS1*. In *Selaginella*, *ARP* seems to disallow *KNOX* expression in microphylls and some parts of the stem, but *KNOX* genes seem not to repress *ARP* in the meristem. It is thus speculated that leaf development is regulated similarly in *S. kraussiana* and euphyllophytes and that the presence of ARP in the *Selaginella* meristem facilitates dichotomous division (bifurcation) by repressing *SkKNOX* at the midline of the apex (Harrison et al. 2005). Nonetheless, the exact mechanisms leading to the leaf initiation and the SAM dynamic regulation are yet to be determined, as there seem to be significant differences to evolutionarily younger plants.

In addition, aside from the shoot apical region, the expression of the class I *SuKNOX* gene, homolog found and analyzed in *S. uncinata*, was detected in a rhizophore of *Selaginella*, where it was found in the outer layer around the apex and weakly in vascular bundles, but not in root tips (Kawai et al. 2010). As already described, the identity of the rhizophore is controversial, assumed as either the aerial or stemlike root (Webster and Steeves 1964, 1967; Webster 1992). The expression of *SuKNOX* provided the first molecular evidence that the rhizophore has developmental mechanisms different from the roots, supporting the rhizophore concept as a unique organ type (Imaichi and Kato 1989, 1991; Lu and Jernstedt 1996; Kawai et al. 2010).

In the fern *Ceratopteris richardii*, three *KNOX* genes (two class I and one class II) were characterized. Both class I *KNOX* genes were expressed, as assessed with in situ hybridization, not only in the shoot AC but also in leaf primordia, marginal parts of the leaves, and vascular bundles. Such an expression pattern resembles that of class I *KNOX* genes in angiosperms with compound leaves (Sano et al. 2005). Interestingly, the expression of KNOX in leaf primordia, in addition to SAM, was also observed in other fern species: *Osmunda regalis* (Harrison et al. 2005) and *Anogramma chaeophylla* (Bharathan et al. 2002). Ectopic overexpression of fern *KNOX* in *Arabidopsis* resulted in plant phenotypes similar to that of overexpressing seed plant class I *KNOX* genes. Thus, it is speculated that class I *KNOX* genes may function similarly during meristem development in seed plants and ferns, but until the relevant mutants of ferns are obtained, the exact gene function will not be resolved. Interestingly, the data suggest that the sporophyte and gametophyte meristems of seedless plants are not regulated by the same molecular mechanisms. This conclusion was drawn from studies with a fern *C. richardii* and a model moss species *Physcomitrella patens.* Class I *KNOX* genes from both species seem not to be required for the development of the gametophyte (Sano et al. 2005; Sakakibara et al. 2008). Nevertheless, studies on *P. patens* suggest that *PpKNOX* genes do not regulate the same as in *Arabidopsis* hormone pathways; the target genes diverge although the biochemical activity is retained (Sakakibara et al. 2008).

This example also highlights the potential pitfalls of interpreting gene transfer experiments across very distant lineages. Genetic tools for the studied organisms are required to properly assess endogenous gene function, to avoid misinterpretations drawn from different developmental context. Studies to identify *KNOX* target genes in evolutionarily older plants will enable the investigation of the native context and degree of divergence and conservation of the KNOX pathway and to decipher how this relates to morphological transitions during land plant evolution (for more see Hay and Tsiantis 2010).

3.6.2 WOX *Family*

WOX genes can be divided into three clades by the time of their appearance in plant phylogeny: an ancient clade (*WOX13*, *WOX14*) and two derived clades, *WUS*/modern (*WUS* and *WOX1–7*) and *WOX9*/intermediate (*WOX8*, *WOX9*, *WOX11*, *WOX12*) (Haecker et al. 2004; van der Graaff et al. 2009); the names of the clades vary depending on the publication. With the exception of the HD common to all *WOX* family members, other unique motifs are shared only within a clade (e.g., Deveaux et al. 2008; van der Graaff et al. 2009; Lian et al. 2014). In *Arabidopsis*, *WUS*, a founding member of that family, and its close homolog *WOX5* are expressed in the organizing center (OC) of the SAM and the quiescent center (QC) of the RAM, respectively, where they are essential for stem-cell maintenance. The WOX family proteins, almost exclusively of the WUS clade, were shown to prevent cell differentiation and therefore determinate cell fate in various meristems of *Arabidopsis*: WUS in the shoot and floral meristems, WOX1 and WOX3 in the leaf marginal and plate meristems, WOX4 in the vascular cambium, and WOX5 in the RAM. Due to the WUS clade relevance for meristem development, that family was selected for experimental examination in SVPs. Based on the HD structure, unicellular green algae and the earliest land plants such as mosses cluster into a single ancestral orthologous group: WOX13. Lycophyte representatives, including *S. moellendorffii*, aside from the *WOX13*-like genes, also possess potential genes from *WOX9*-like clade; and gymnosperms and angiosperms additionally possess the genes from the *WUS*-like clade. The number of *WOX* family genes increased substantially with the emergence of the vascular plant lineages. Five *WOX* genes have so far been identified in the incomplete genome sequence of the fern *C. richardii*; it is likely that more are present in the full genome. It has been suggested that the meristem's complexity and the organization of plant stem cells in various niches have significantly changed during the evolution of the vascular plants. This might have been enabled by *WOX* gene family evolution from ancient to WOX9-like, and to WUS clade, with strong expansion and diversification of WUS clade members in gymnosperms and angiosperms (Haecker et al. 2004; Mukherjee et al. 2009; van der Graaff et al. 2009; Lian et al. 2014; Zeng et al. 2015; Ge et al. 2016).

WOX genes have been identified and phylogenetically analyzed in both *S. moellendorffii* (Mukherjee et al. 2009; Lian et al. 2014) and *S. kraussiana* (Harrison et al. 2005, 2007; Ge et al. 2016). *S. kraussiana* was shown to possess at least five ancient clade *WOX* genes, *SkWOX13A–E*, and three *WOX* genes that do not group with the others. One non-*WOX13*-like gene was typical of the WOX9 clade (*SkWOX11C*), and two transitional between the ancient and WOX9 clades (*SkWOX11A*, *B*) seemed specific to lycophytes and vascular plants in general (Ge et al. 2016). The *S. kraussiana* genome does not encode WUS clade genes (Nardmann and Werr 2012; Ge et al. 2016). Ge et al. (2016) also analyzed the expression pattern of the eight genes (*SkWOX13A–E* and *SkWOX11A–C*) in microphylls, shoot tips, rhizophore tips, and stems of *S. kraussiana*. The only preferentially expressed gene was *SkWOX13B*, which was linked to the rhizophore tip. The remaining genes could be detected in all four tissues analyzed, though the expression levels varied: *SkWOX13A* and *SkWOX11A* were barely expressed, *SkWOX13D* was expressed at low level, and *SkWOX13C*, *SkWOX13E*, *SkWOX11B*, and *SkWOX11C* showed a relatively high expression level (Ge et al. 2016). These data suggested that each of the *SkWOX* genes may play specific roles in different tissues. To distinguish their specific expression patterns, it seems important however to analyze other structures, diverse ontogenetic stages, and/or growth conditions. In addition, experiments showing spatial regulation of their expression using in situ hybridization instead of RNAseq/qRT-PCR are needed to assess these genes' relevance in particular tissues for more adequate functional interpretations. Interestingly, the expression pattern for three *WOX* genes in *S. moellendorffii* and *S. kraussiana* was assessed by in situ hybridization in a PhD dissertation (Durantini 2009). In addition, Frank et al. (2015) have lately shown, also by in situ hybridization, that one of the *WOX* genes is expressed in the core domain and primordia of the shoot apex. Nevertheless, a comprehensive analysis of the whole family members is lacking.

The fern *C. richardii* was found to have *WOX* genes not only from ancient and WOX9 clades but interestingly also one *WUS*-like (*CrWUL*) gene, which grouped with the *WUS* lineage, and possessed a canonical WUS box, the characteristic motif of the WUS clade and an EAR motif also found in *WUS/WOX5*. The *CrWUS* gene was shown to be related to the meristematic tissues, as it was transcribed in the whole root tip (Nardmann and Werr 2012). In *Arabidopsis*, the stem-cell factors WUS and WOX5 establish respectively the shoot and root apical stem-cell niche (Mayer et al. 1998; Schoof et al. 2000; Sarkar et al. 2007), for which cell-to-cell movement is essential (Yadav et al. 2011; Pi et al. 2015). In addition, WOX5 and WUS are functionally interchangeable (Sarkar et al. 2007), and most of the WUS clade genes can functionally substitute for WUS, when expressed in the OC (Dolzblasz et al. 2016). Zhang et al. (2017) studied the functional potential of *CrWUL* gene in the evolutionary context. The fern CrWUL has a much longer protein sequence than the WUS/WOX5 proteins of seed plants, and thus was immobile, which prevented it from complementing *Arabidopsis wus-1* and *wox5* mutants' defects, when expressed from *WUS* and *WOX5* promoters (Zhang et al. 2017). Importantly, *CrWUL* can perform stem-cell-related functions in *Arabidopsis* (as *WUS*/*WOX5* in SAM/RAM, respectively), when expressed in a broader domain encompassing not only OC/QC but also the adjacent stem cells or when a truncated version expressed in OC/QC was able to migrate into the L1 and L2 layers and to the columella, respectively. In both cases, *wus-1* and *wox5* defects in *Arabidopsis* were complemented (Zhang et al. 2017). The results suggest that the shoot/root apical stem-cell maintenance activity of the *WUS*-like gene was present in the common ancestor of ferns and seed plants and was independent of intercellular protein mobility. This new feature was gained after the split of the fern lineage, when the WUS/WOX5 predecessor became shorter, which seems crucial for stem-cell maintenance in flowering plants. Mitotic activity of the single AC in the root meristem and broader expression of CrWUL in the whole root tip (in contrast to QC of seed plants) suggest a cell-autonomous action of this gene in the regulation of meristematic potential (Nardmann and Werr 2012; Zhang et al. 2017). Beyond an inability for intercellular movement, CrWUL may still behave differently from seed plant WUS, as its broader expression in the SAM in *Arabidopsis* (from the *CLV3* promoter) did not cause the enlargement of the meristem with increased stem-cell population, as *Arabidopsis WUS* gene does (Brand et al. 2002; Yadav et al. 2010; Zhang et al. 2017).

Moss *WOX* genes (*PpWOX13Ls*) have been shown to play a role in apical stem-cell formation and regeneration (Mukherjee et al. 2009; Sakakibara et al. 2014). Thus, although the WOX13 lineage seems to have no function in apical stem-cell maintenance in flowering plants, it is highly probable that some genes of that clade in SVPs, despite not being sequentially similar to the WUS clade genes of evolutionarily younger plants, might also perform stem-cell-related function. Progress in developing molecular techniques in SVPs will hopefully enable future investigations.

Interpretations of gene function based on such limited data must be taken cautiously. Gene homologs, selected based on their function in stem-cell regulation in *Arabidopsis*, are usually members of a larger family, which underwent sub- and neofunctionalization; thus, the direct comparison of possible functions between *Arabidopsis* and SVPs can be misleading. In addition, the apices of lycophytes and ferns depend on apical initial function, whereas that of more studied angiosperms are layered, with multiple initials, and maintain a stem-cell population (Evert 2006). It is tempting to speculate that, with growing interest of the scientific community in SVPs like *Selaginella* and *Ceratopteris*, advances in experimental techniques will occur and enable more in-detail functional characterization of genes.

3.7 Mechanisms Protecting the Meristem Identity

The significance of the meristems for proper plant growth and development is indisputable, posing the question of how their genetic information is protected against potential mutations during DNA synthesis in successive cell cycles. It is especially important in meristems with a single initial cell (so-called structural meristems), such as in the majority of SVPs, as any change in the genetic information of the AC is passed to all derivative cells and may possibly affect the development of the entire organism (Klekowski and Kazarinova-Fukshansky 1984). A particularly interesting strategy for preferential inheritance and transfer of the template DNA strand has been described for *Marsilea* (Kuligowski-Andrés 1975a, b; Kuligowski-Andrés and Tourte 1978). During successive divisions, the original template DNA always remains in the AC, whereas the DNA molecule synthesized based on the complementary strand is passed to the derivative cell. This preferential maintenance of one DNA strand is related to the establishment of the AC in the embryo and is retained throughout the meristem life (Kuligowski-Andrés 1975a, b).

An alternative mechanism for protecting the genetic information is the impermanency of the initial cells, postulated for the stochastic meristems (with multiple initials). In such meristems, unaffected cells may be selected from the pool of existing meristematic cells, preventing the uncontrolled passing of the mutational changes (Klekowski and Kazarinova-Fukshansky 1984; Ruth et al. 1985; Zagórska-Marek and Turzańska 2000). Importantly, initial cell replacement has been documented in the SAM of the lycopod *Huperzia*, which structurally resemble those of gymnosperms. Here, four initial cells typically function at the meristem surface, though they are transient and can be shifted from the apex center, losing their function. Concurrently, new meristematic cells at the apex take over their initial role. Recurrent dichotomous divisions additionally affect the geometry of the apex and the number of functioning initial cells facilitating their regular replacements (Gola and Jernstedt 2011). Interestingly, even meristems classified as structural demonstrate the potential to behave as stochastic meristems upon dichotomous division. Formation of new meristems for both twin apices during dichotomy requires de novo initiation and/or selection of additional (new) ACs; thus, the meristem bifurcation indirectly forces the replacement of functioning initial cells. This process of recurrent initiation and selection of new ACs has been shown, e.g., during the branching of the rhizophore/root complex in *Selaginella* (Imaichi and Kato 1989, 1991; Otręba and Gola 2011).

Formation of new potential ACs within the functioning meristem can occur on a regular basis during ontogeny as in rhizomes of the whisk fern *Psilotum nudum* (Takiguchi et al. 1997). Here, apart from the single functioning typical AC, numerous new ACs are regularly initiated in successive merophytes. Consequently, a pool of potential initial cells, temporarily dormant, is constantly available at the meristem surface, although the authors stated that their selection and activation are not comparable to those in stochastic meristems nor have to be related to mechanical damage of the AC (Takiguchi et al. 1997). However, regardless of whether the selection of additional ACs in *Psilotum* is the strategy to maintain meristem integrity, it still leads to the exchange of functioning initial cells and, when viewed from the organismal perspective, to protection of the genet (regardless of the ramets' fates).

In all fern taxa, the regular formation of the leaves at the apex is related to the repeated initiation of the apical cells, which will initiate the growth of leaf primordia (compare, e.g., White and Turner 1995; Imaichi 2008; Vasco et al. 2013). Interestingly, experimental studies on *Osmunda* and *Dryopteris* showed that, at the early stages of development, the leaf primordia are not fully determined and maintain totipotency. Their programmed developmental fates can be changed, and primordia can revert into shoots, when isolated from the influences of the SAM and/or older leaves (for the review see Vasco et al. 2013). Whether this ability of apical cells of the leaf can be utilized to replace the original functioning AC of the shoot, for example, in case of its mechanical or mutational damage, requires further analyses. The temporary maintenance of the undetermined characteristics of the leaf AC is further evidenced in the ferns *Lygodium* and *Salpichlaena*, where transient identity and prolonged apical growth of the leaf are adaptive mechanisms to climbing behavior (Mueller 1983; Vasco et al. 2013). Importantly, the leaf AC's totipotency may be related to the broader expression pattern of genes relevant for the maintenance of the meristem function and undetermined stem cells. For example, the abovementioned *Ceratopteris CrWUL* gene has been detected in both the single AC and the adjacent meristematic stem cells (Nardmann and Werr 2012; Zhang et al. 2017), possibly determining the totipotency of the apex, including the leaf ACs, in contrast to precisely delimited domains of WUS/WOX5 expression in seed plants and determined fates of leaf primordia.

3.8 Conclusions

The early-diverged lineages of land plants present an interesting mixture of ancestral and progressive features, illustrating a million years of independent evolution. Unfortunately, even their structure, seemingly so well known at the tissue, cellular, and sometimes ultra-cellular levels, is not really fully acknowledged nor developmentally understood. Moreover, the genetic and molecular mechanisms behind the extreme diversity of forms and underlying developmental processes remain largely unrevealed. Thus, although ferns and fern allies have been studied for more than a hundred years, they still pose new challenges.

Acknowledgments This work was supported by the University of Wrocław grant No. 1068/S/IBE/2017.

References

Banks JA, Nishiyama T, Hasebe M, Bowman JL, Gribskov M, dePamphilis C et al (2011) The Selaginella genome identifies genetic changes associated with the evolution of vascular plants. Science 332(6032):960–963. https://doi.org/10.1126/science.1203810

Barclay BD (1931) Origin and development of tissues in stem of *Selaginella wildenovii*. Bot Gaz 91(4):452–461. https://doi.org/10.1086/334168

Barker MS, Wolf PG (2010) Unfurling fern biology in the genomics age. Bioscience 60(3):177–185. https://doi.org/10.1525/bio.2010.60.3.4

Bhambie S (1963) Studies in Pteridophytes. IV. The development structure and organization of root in *Isoëtes coromandelina* L. Proc Indian Acad Sci Sec B 58(3):153–164

Bhambie S, Madan P (1980) Studies in Pteridophytes. XVII. Ontogenetic study on the shoot apex of *Botrychium lanuginosum* (Wall.) Proc Indian Acad Sci (Pl Sci) 89(1):29–35

Bhambie S, Rao CGP (1972) Studies in Pteridophytes. IX. The root apex organization in some pteridophytes. Proc Indian Natl Sci Acad 39:150–156

Bharathan G, Goliber TE, Moore C, Kessler S, Pham T, Sinha NR (2002) Homologies in leaf form inferred from *KNOXI* gene expression during development. Science 296:1858–1860. https://doi.org/10.1126/science.1070343

Bierhorst DW (1959) Symmetry in *Equisetum*. Am J Bot 46:170–179. https://doi.org/10.2307/2439274

Bierhorst DW (1977) On the stem apex, leaf initiation and early leaf ontogeny in filicalean ferns. Am J Bot 64:125–152. https://doi.org/10.2307/2442101

Bower FO (1889) The comparative examination of the meristems of ferns, as a phylogenetic study. Ann Bot 3(11):305–392

Bower FO (1923) The ferns (Filicales). University Press, Cambridge, UK

Brand U, Grünewald M, Hobe M, Simon R (2002) Regulation of *CLV3* expression by two homeobox genes in *Arabidopsis*. Plant Physiol 129(2):565–575. https://doi.org/10.1104/pp.001867
Campbell DH (1893) On the development of *Azolla filiculoides* Lam. Ann Bot 7:155–184. https://doi.org/10.1093/aob/os-7.2.155
Chiang SHT (1976) The growth cycle of cambium and the structure of the vascular tissue in the corm of *Isoëtes taiwanensis*. Taiwania 21(1):14–26
Chiang SHT, Chen SY (1986) Components of vascular tissue in the corm of *Isoëtes taiwanensis*. Taiwania 31(1):1–14
Chiang SHT, Chou T (1974) Tissue differentiation in the roots of some ferns. Taiwania 19(1):7–18. https://doi.org/10.6165/tai.1974.19.7
Croxdale JG (1978) *Salvinia* leaves. I. Origin and early differentiation of floating and submerged leaves. Can J Bot 56:1982–1991. https://doi.org/10.1139/b78-237
Croxdale JG (1979) *Salvinia* leaves. II. Morphogenesis of the floating leaf. Can J Bot 57:1951–1959. https://doi.org/10.1139/b79-245
Cusick F (1953) Experimental and analytical studies on Pteridophytes. XXII. Morphogenesis in *Selaginella willdenovii* Baker. I. Preliminary morphological analysis. Ann Bot 17:369–383. https://doi.org/10.1093/oxfordjournals.aob.a083356
Dasanayake MD (1960) Aspects of morphogenesis in a dorsiventral fern, *Pteridium aquilinum* (L.) Kuhn. Ann Bot 24(3):317–329. https://doi.org/10.1093/oxfordjournals.aob.a083706
Dengler NG (1983) The developmental basis of anisophylly in *Selaginella martensii*. I. Initiation and morphology of growth. Am J Bot 70:181–192. https://doi.org/10.2307/2443262
Deveaux Y, Toffano-Nioche C, Claisse G, Thareau V, Morin H, Laufs P, Moreau H, Kreis M, Lecharny A (2008) Genes of the most conserved WOX clade in plants affect root and flower development in *Arabidopsis*. BMC Evol Biol 8:291. https://doi.org/10.1186/1471-2148-8-291
Dolzblasz A, Nardmann J, Clerici E, Causier B, van der Graaff E, Chen J, Davies B, Werr W, Laux T (2016) Stem cell regulation by genes. Mol Plant 9(7):1028–1039. https://doi.org/10.1016/j.molp.2016.04.007
Durantini D (2009) Plant stem cell homeostasis: Phylogeny and expression pattern of different components. PhD Thesis, Universität zu Köln
Evert RF (2006) Esau's plant anatomy. Meristems, cells, and tissues of the plant body – their structure, function, and development, 3rd edn. John Wiley & Sons. Inc, Hoboken, NJ
Farmer JB (1890) On *Isoëtes lacustris* L. Ann Bot 5(17):37–62. https://doi.org/10.1093/oxfordjournals.aob.a090626
Foster AS, Gifford EM (1974) Comparative morphology of vascular plants, 2nd edn. W. H. Freeman, San Francisco
Frank MH, Edwards MB, Schultz ER, McKain MR, Fei Z, Sørensen I, Rose JKC, Scanlon MJ (2015) Dissecting the molecular signatures of apical cell-type shoot meristems from two ancient land plant lineages. New Phytol 207:893–904. https://doi.org/10.1111/nph.13407
Freeberg JA, Gifford EM Jr (1984) The root apical meristem of *Osmunda regalis*. Am J Bot 71:558–563. https://doi.org/10.2307/2443331
Freeberg JA, Wetmore RH (1967) The Lycopsida – a study in development. Phytomorphology 17:78–91
French J (1984) Occurrence of intercalary and uninterrupted meristems in *Equisetum*. Am J Bot 71(8):1099–1103. https://doi.org/10.2307/2443385
Fujinami R, Yamada T, Nakajima A, Takagi S, Idogawa A, Kawakami E, Tsutsumi M, Imaichi R (2017) Root apical meristem diversity in extant lycophytes and implications for root origin. New Phytol 215(3):1210–1220. https://doi.org/10.1111/nph.14630
Ge Y, Liu J, Zeng M, He J, Qin P, Huang H, Xu L (2016) Identification of *WOX* family genes in *Selaginella kraussiana* for studies on stem cells and regeneration in lycophytes. Front Plant Sci 7:93. https://doi.org/10.3389/fpls.2016.00093
Gifford EM, Foster AS (1989) Morphology and evolution of vascular plants, 3rd edn. W. H. Freeman and Co., New York

Gifford EM Jr (1991) The root apical meristem of *Asplenium bulbiferum*: structure and development. Am J Bot 78:370–376. https://doi.org/10.2307/2444959
Gifford EM Jr (1993) The root apical meristem of *Equisetum diffusum*: structure and development. Am J Bot 80:468–473. https://doi.org/10.2307/2445393
Gifford EM Jr, Kurth E (1982) Quantitative studies of the root apical meristem of *Equisetum scirpoides*. Am J Bot 69:464–473. https://doi.org/10.2307/2443152
Gifford EM Jr, Kurth E (1983) Quantitative studies of the vegetative shoot apex of *Equisetum scirpoides*. Am J Bot 70(1):74–79. https://doi.org/10.2307/2443206
Gola EM (2014) Dichotomous branching: the plant form and integrity upon the apical meristem bifurcation. Front Plant Sci 5:263. https://doi.org/10.3389/fpls.2014.00263
Gola EM, Jernstedt JA (2011) Impermanency of initial cells in *Huperzia lucidula* (Huperziaceae) shoot apices. Int J Plant Sci 172:847–855. https://doi.org/10.1086/660878
Golub SJ, Wetmore RH (1948a) Studies on development in the vegetative shoot of *Equisetum arvense* L. II. The mature shoot. Am J Bot 35:767–781. https://doi.org/10.2307/2438158
Golub SJ, Wetmore RH (1948b) Studies on development in the vegetative shoot of *Equisetum arvense* L. I. The shoot apex. Am J Bot 35:755–767. https://doi.org/10.2307/2438157
Gunning BES, Hughes JE, Hardham AR (1978) Formative and proliferative cell divisions, cell differentiation, and developmental changes in the meristem of *Azolla* roots. Planta 143:121–144. https://doi.org/10.1007/BF00387786
Haecker A, Gross-Hardt R, Geiges B, Sarkar A, Breuninger H, Herrmann M, Laux T (2004) Expression dynamics of *WOX* genes mark cell fate decisions during early embryonic patterning in *Arabidopsis thaliana*. Development 131:657–668. https://doi.org/10.1242/dev.00963dev.00963
Hagemann W, Schulz U (1978) Wedelanlegung und Rhizomverzweigung bei einigen Gleicheniaceen. Bot Jahrb Syst 99(2/3):380–399
Hake S, Smith HM, Holtan H, Magnani E, Mele G, Ramirez J (2004) The role of *KNOX* genes in plant development. Annu Rev Cell Dev Biol 20:125–151. https://doi.org/10.1146/annurev.cellbio.20.031803.093824
Harrison CJ, Langdale JA (2010) Comment: the developmental pattern of shoot apices in *Selaginella kraussiana* (Kunze) A. Brown. Int J Pl Sci 171(6):690–692. https://doi.org/10.1086/653134
Harrison CJ, Corley SB, Moylan EC, Alexander DL, Scotland RW, Langdale JA (2005) Independent recruitment of a conserved developmental mechanism during leaf evolution. Nature 434:509–514. https://doi.org/10.1038/nature03410
Harrison CJ, Rezvani M, Langdale JA (2007) Growth from two transient apical initials in the meristem of *Selaginella kraussiana*. Development 134:881–889. https://doi.org/10.1242/dev.001008
Härtel K (1937) Studien an Vegetationspunkten einheimischer Lycopodien. Beitr Biol Pflanzen 25(2):124–169
Hay A, Tsiantis M (2010) *KNOX* genes: versatile regulators of plant development and diversity. Development 137:3153–3165. https://doi.org/10.1242/dev.030049
Hébant-Mauri R (1975) Apical segmentation and leaf initiation in the tree fern *Dicksonia squarrosa*. Can J Bot 53:764–772. https://doi.org/10.1139/b75-092
Hébant-Mauri R (1977) Segmentation apicale et initiation foliaire chez *Ceratopteris thalictroides* (Fougère leptosporangiée). Can J Bot 55:1820–1828. https://doi.org/10.1139/b77-208
Hébant-Mauri R (1984) Branching patterns in *Trichomanes* and *Cardiomanes* (hymenophyllaceous ferns). Can J Bot 62:1336–1343. https://doi.org/10.1139/b84-181
Hébant-Mauri R (1990) The branching of *Trichomanes proliferum* (Hymenophyllaceae). Can J Bot 68:1091–1097. https://doi.org/10.1139/b90-137
Hébant-Mauri R (1993) Cauline meristems in leptosporangiate ferns: structure, lateral appendages, and branching. Can J Bot 71:1612–1624. https://doi.org/10.1139/b93-196
Hébant-Mauri R, Gay H (1993) Morphogenesis and its relation to architecture in the dimorphic clonal fern *Lomagramma guianensis* (Aublet) Ching (Dryopteridaceae). Bot J Linn Soc 112(3):257–276. https://doi.org/10.1111/j.1095-8339.1993.tb00319.x

Hébant-Mauri R, Veillon JM (1989) Branching and leaf initiation in the erect aerial system of *Stromatopteris moniliformis* (Gleicheniaceae). Can J Bot 67:407–414. https://doi.org/10.1139/b89-056

Hirsch AM, Kaplan DR (1974) Organography, branching, and the problem of leaf versus bud differentiation in the vining epiphytic fern genus *Microgramma*. Am J Bot 61:217–229. https://doi.org/10.2307/2441600

Hou GC, Hill JP (2002) Heteroblastic root development in *Ceratopteris richardii* (Parkeriaceae). Int J Plant Sci 163:341–351. https://doi.org/10.1086/339156

Hou GC, Hill JP (2004) Developmental anatomy of the fifth shoot-borne root in young sporophytes of *Ceratopteris richardii*. Planta 219(2):212–220. https://doi.org/10.1007/s00425-004-1225-6

Imaichi R (1980) Developmental studies on the leaf and the extra-axillary bud of *Histiopteris incisa*. Bot Mag 93:25–38. https://doi.org/10.1007/BF02489484

Imaichi R (1982) Developmental study on *Hypolepis punctata* (Thunb.) Mett. Initiation of the first and the second petiolar buds in relation to early leaf ontogeny. Bot Mag 95:435–453. https://doi.org/10.1007/BF02489479

Imaichi R (1988) Developmental anatomy of the shoot apex of leptosporangiate ferns. II. Leaf ontogeny of *Adiantum capillus-veneris* (Adiantaceae). Can J Bot 66:1729–1733. https://doi.org/10.1139/b88-237

Imaichi R (1989) Early leaf development and leaf sheath formation of *Botrychium strictum* and *B. virginiana* (Ophioglossaceae). Ann Bot 63:249–256. https://doi.org/10.1093/oxfordjournals.aob.a087739

Imaichi R (2008) Meristem organization and organ diversity. In: Ranker TA, Haufler CH (eds) Biology and evolution of ferns and lycophytes. Cambridge University Press, Cambridge, pp 75–103

Imaichi R, Kato M (1989) Developmental anatomy of the shoot apical cell, rhizophore and root of *Selaginella uncinata*. Bot Mag 102:369–380. https://doi.org/10.1007/BF02488120

Imaichi R, Kato M (1991) Developmental study of branched rhizophores in three *Selaginella* species. Am J Bot 78:1694–1703. https://doi.org/10.2307/2444848

Imaichi R, Nishida M (1986) Developmental anatomy of the three-dimensional leaf of *Botrychium ternatum* (Thunb.) Sw. Bot Mag 99:85–106. https://doi.org/10.1007/BF02488625

Jackson D, Veit B, Hake S (1994) Expression of maize *KNOTTED1* related homeobox genes in the shoot apical meristem predicts patterns of morphogenesis in the vegetative shoot. Development 120:405–413

Jacobs WP (1988) Development of procambium, xylem, and phloem in the shoot apex of *Selaginella*. Bot Gaz 149:64–70. https://doi.org/10.1086/337692

Jernstedt JA, Cutter EG, Gifford EM, Lu P (1992) Angle meristem origin and development in *Selaginella martensii*. Ann Bot 69:351–363. https://doi.org/10.1093/oxfordjournals.aob.a088352

Jernstedt JA, Cutter EG, Lu P (1994) Independence of organogenesis and cell pattern in developing angle shoots of *Selaginella martensii*. Ann Bot 74:343–355. https://doi.org/10.1006/anbo.1994.1127

Jones CS, Drinnan AN (2009) The developmental pattern of shoot apices in *Selaginella kraussiana* (Kunze) A. Brown. Int J Plant Sci 170(8):1009–1018. https://doi.org/10.1086/605118

Karrfalt EE (1982) Secondary development in the cortex of *Isoëtes* L. Bot Gaz 143(4):439–445. https://doi.org/10.1086/337319

Karrfalt EE, Eggert DA (1977a) The comparative morphology and development of *Isoëtes* L. II. Branching of the base of the corm in *I. tuckermanii* A.Br. Bot Gaz 138(2):357–368. https://doi.org/10.1086/336935

Karrfalt EE, Eggert DA (1977b) The comparative morphology and development of *Isoëtes* L. I. Lobe and furrow development in *I. tuckermanii* A.Br. Bot Gaz 138(2):236–247. https://doi.org/10.1086/336919

Karrfalt EE, Eggert DA (1978) The comparative morphology and development of *Isoëtes* L. III. The sequence of root initiation in three- and four-lobed plants of *I. tuckermanii* A.Br. and *I. nuttallii* A.Br. Bot Gaz 139(2):271–283. https://doi.org/10.1086/337001

Kawai J, Tanabe Y, Soma S, Ito M (2010) Class 1 KNOX gene expression supports the *Selaginella* rhizophore concept. J Plant Biol 53:268–274. https://doi.org/10.1007/s12374-010-9113-z

Kerstetter R, Vollbrecht E, Lowe B, Veit B, Yamaguchi J, Hake S (1994) Sequence analysis and expression patterns divide the maize *knotted1*-like homeobox genes into two classes. Plant Cell 6:1877–1887. https://doi.org/10.2307/3869914

Klekowski EJ Jr, Kazarinova-Fukshansky N (1984) Shoot apical meristems and mutation: fixation of selectively neutral cell genotypes. Am J Bot 71:22–27. https://doi.org/10.2307/2443619

Korall P, Kenrick P (2002) Phylogenetic relationships in Selaginellaceae based on *rbcL* sequences. Am J Bot 89:506–517. https://doi.org/10.3732/ajb.89.3.506

Kruatrachue M, Evert RF (1977) The lateral meristem and its derivatives in the corm of *Isoëtes muricata*. Am J Bot 64:310–325. https://doi.org/10.2307/2441975

Kuligowski-Andrés J (1975a) Contribution à l'étude d'une fougère, le *Marsilea vestita* (Marsiléacées), du stade embryon au stade sporophyte adulte. 1. Étude morphologique, histochimique et histoautoradiographique de l'embryon. Premiere partie: mise en place des territoires organogenes. Ann Sci Nat Bot 16:151–216

Kuligowski-Andrés J (1975b) Contribution à l'étude d'une fougère, le *Marsilea vestita* (Marsiléacées), du stade embryon au stade sporophyte adulte. 1. Étude morphologique, histochimique et histoautoradiographique de l'embryon. Deuxieme partie: l'embryon organise. Ann Sci Nat 16:249–308

Kuligowski-Andrés J, Tourte Y (1978) First study of a Pteridophyte fertilization, using radioautography. Biol Cell 31(1):101–108

Lang WH (1915) Studies in the morphology of *Isoëtes*. II. The analysis of the stele of the shoot of *Isoëtes lacustris* in the light of mature structure and apical development. Memoirs and Proceedings of the Manchester Literary and Philosophical Society 59(8):29–57

Lee KB, Soh WY (1995) Comparative anatomy and ultrastructure of active and dormant cambium in rhizome of *Botrychium ternatum*. J Plant Res 108:149–159. https://doi.org/10.1007/BF02344339

Lemon GD, Posluszny U (1997) Shoot morphology and organogenesis of the aquatic floating fern *Salvinia molesta* D.S. Mitchell, examined with the aid of laser scanning confocal microscopy. Int J Plant Sci 158:693–703. https://doi.org/10.1086/297481

Lian G, Ding Z, Wang Q, Zhang D, Xu J (2014) Origins and evolution of WUSCHEL-related homeobox protein family in plant kingdom. Sci World J 2014:534140. https://doi.org/10.1155/2014/534140

Lin BL, Raghavan V (1991) Lateral root initiation in *Marsilea quadrifolia*. I. Origin and histogenesis of lateral roots. Can J Bot 69:123–135. https://doi.org/10.1139/b91-018

Long JA, Moan EI, Medford JI, Barton MK (1996) A member of the KNOTTED class of homeodomain proteins encoded by the *SHOOTMERISTEMLESS* gene of Arabidopsis. Nature 379:66–69. https://doi.org/10.1038/379066a0

Lu P, Jernstedt JA (1996) Rhizophore and root development in *Selaginella martensii*: meristem transitions and identity. Int J Plant Sci 157(2):180–194. https://doi.org/10.1086/297337

Lu P, Jernstedt JA (2005) Meristem development and dichotomous branching in *Selaginella martensii*. Proceedings of the XVII International Botanical Congress, 17–23 July 2005, Vienna, Austria, p 336

Lucansky TW, White RA (1976) Comparative ontogenetic studies in young sporophytes of tree ferns. I. A primitive and an advanced taxon. Am J Bot 63:463–472. https://doi.org/10.2307/2441914

Ma Y, Steeves TA (1994) Vascular differentiation in the shoot apex of *Matteuccia struthiopteris*. Ann Bot 74:573–585. https://doi.org/10.1006/anbo.1994.1157

Ma Y, Steeves TA (1995) Characterization of stelar initiation in shoot apices of ferns. Ann Bot 75:105–117. https://doi.org/10.1006/anbo.1995.1001

Mayer KF, Schoof H, Haecker A, Lenhard M, Jürgens G, Laux T (1998) Role of WUSCHEL in regulating stem cell fate in the *Arabidopsis* shoot meristem. Cell 95:805–815. https://doi.org/10.1016/S0092-8674(00)81703-1

Mueller RJ (1982) Shoot morphology of the climbing fern *Lygodium* (Schizaeaceae): general organography, leaf initiation, and branching. Bot Gaz 143:319–330. https://doi.org/10.1086/337306

Mueller RJ (1983) Indeterminate growth and ramification of the climbing leaves of *Lygodium japonicum* (Schizaeaceae). Am J Bot 70:682–690. https://doi.org/10.2307/2443122

Mukherjee K, Brocchieri L, Burglin TR (2009) A comprehensive classification and evolutionary analysis of plant homeobox genes. Mol Biol Evol 26:2775–2794. https://doi.org/10.1093/molbev/msp201

Myśkow E (2010) Procambium-cambium transition during vascular meristem development in *Diospyros lotus*. Botany 88:985–993. https://doi.org/10.1139/B10-070

Nägeli C (1846) Über das Wachstum des Gefäßstammes. Zschr f wiss Bot 3:129–152

Nardmann J, Werr W (2012) The invention of WUS-like stem cell-promoting functions in plants predates leptosporangiate ferns. Plant Mol Biol 78:123–134. https://doi.org/10.1007/s11103-011-9851-4

Nitayangkura S, Gifford EM Jr, Rost TL (1980) Mitotic activity in the root apical meristem of *Azolla filiculoides* Lam., with special reference to the apical cell. Am J Bot 67:1484–1492. https://doi.org/10.2307/2442877

Ogura Y (1972) Comparative anatomy of vegetative organs of the pteridophytes. In: Encycl plant anatomy, VII. 3. Gebrüder Borntraeger, Berlin, Stuttgart

Otręba P, Gola EM (2011) Specific intercalary growth of rhizophores and roots in *Selaginella kraussiana* (Selaginellaceae) is related to unique dichotomous branching. Flora 206:227–232. https://doi.org/10.1016/j.flora.2010.07.001

Paolillo DJ (1963) The developmental anatomy of *Isoëtes*. Illinois Biological Monographs 31. University of Illinois Press, Urbana, pp 1–131. https://doi.org/10.5962/bhl.title.50237

Peterson RL, Cutter EG (1969) The fertile spike of *Ophioglossum petiolatum*. I. Mechanism of elongation. Am J Bot 56(5):473–483. https://doi.org/10.2307/2440640

Philipson WR (1990) The significance of apical meristems in the phylogeny of land plants. Plant Syst Evol 173(1):17–38. https://doi.org/10.1007/BF00937760

Pi L, Aichinger E, van der Graaff E, Llavata-Peris CI, Weijers D, Hennig L, Groot E, Laux T (2015) Organizer-derived WOX5 signal maintains root columella stem cells through chromatin-mediated repression of CDF4 expression. Dev Cell 33(5):576–588. https://doi.org/10.1016/j.devcel.2015.04.024

Pires ND, Dolan L (2012) Morphological evolution in land plants: new designs with old genes. Philos Trans R Soc Lond Ser B Biol Sci 367:508–518. https://doi.org/10.1098/rstb.2011.0252

Plackett ARG, Huang L, Sanders HL, Langdale JA (2014) High-efficiency stable transformation of the model fern species *Ceratopteris richardii* via microparticle bombardment. Plant Physiol 165(1):3–14. https://doi.org/10.1104/pp.113.231357

Plackett ARG, Di Stilio VS, Langdale JA (2015a) Ferns: the missing link in shoot evolution and development. Front Plant Sci 6:972. https://doi.org/10.3389/fpls.2015.00972

Plackett ARG, Rabbinowitsch EH, Langdale JA (2015b) Genetic transformation of the fern *Ceratopteris richardii* through microparticle bombardment. Plant Methods 11:37. https://doi.org/10.1186/s13007-015-0080-8

Popham RA (1951) Principal types of vegetative shoot apex organization in vascular plants. Ohio J Sci 51:249–270

PPG I, Schuettpelz E, Schneider H, Smith AR, Hovenkamp P, Prado J, Rouhan G et al (2016) A community-derived classification for extant lycophytes and ferns. J Syst Evol 54(6):563–603. https://doi.org/10.1111/jse.12229

Rothwell GW, Karrfalt EE (2008) Growth, development, and systematics of ferns: does *Botrychium* s.l. (Ophioglossales) really produce secondary xylem? Am J Bot 95:414–423. https://doi.org/10.3732/ajb.95.4.414

Ruth J, Klekowski EJ Jr, Stein OL (1985) Impermanent initials of the shoot apex and diplontic selection in a juniper chimera. Am J Bot 72:1127–1135. https://doi.org/10.2307/2443459

Sakakibara K, Nishiyama T, Dequchi H, Hasebe M (2008) Class 1 KNOX genes are not involved in shoot development in the moss *Physcomitrella patens* but do function in sporophyte development. Evol Dev 10:555–566. https://doi.org/10.1111/j.1525-142X.2008.00271.x

Sakakibara K, Reisewitz P, Aoyama T, Friedrich T, Ando S, Sato Y et al (2014) *WOX13-like* genes are required for reprogramming of leaf and protoplast cells into stem cells in the moss *Physcomitrella patens*. Development 141:1660–1670. https://doi.org/10.1242/dev.097444

Sano R, Juárez CM, Hass B, Sakakibara K, Ito M, Banks JA, Hasebe M (2005) KNOX homeobox genes potentially have similar function in both diploid unicellular and multicellular meristems, but not in haploid meristems. Evol Dev 7:69–78. https://doi.org/10.1111/j.1525-142X.2005.05008.x

Sarkar AK, Luijten M, Miyashima S, Lenhard M, Hashimoto T, Nakajima K, Scheres B, Heidstra R, Laux T (2007) Conserved factors regulate signalling in *Arabidopsis thaliana* shoot and root stem cell organizers. Nature 446:811–814. https://doi.org/10.1038/nature05703

Schoof H, Lenhard M, Haecker A, Mayer KF, Jürgens G, Laux T (2000) The stem cell population of *Arabidopsis* shoot meristems is maintained by a regulatory loop between the *CLAVATA* and *WUSCHEL* genes. Cell 100:635–644. https://doi.org/10.1016/S0092-8674(00)80700-X

Schoute JC (1938) Morphology. In: Verdoorn F (ed) Manual of pteridology. Martinus Nijhoff, Hague, pp 1–104

Schulz C, Little DP, Stevenson DW, Bauer D, Moloney C, Stützel T (2010) An overview of the morphology, anatomy, and life cycle of a new model species: the lycophyte *Selaginella apoda* (L.) Spring. Int J Plant Sci 171(7):693–712. https://doi.org/10.1086/654902

Scott DH, Hill TG (1900) The structure of *Isoëtes hystrix*. Ann Bot 14:413–454. https://doi.org/10.1093/oxfordjournals.aob.a088787

Smith L, Greene B, Veit B, Hake S (1992) A dominant mutation in the maize homeobox gene, *Knotted-1*, causes its ectopic expression in leaf cells with altered fates. Development 116:21–30

Stevenson DW (1976a) Observations on phyllotaxis, stelar morphology, the shoot apex and gemmae of *Lycopodium lucidulum* Michaux (Lycopodiaceae). Bot J Linn Soc 72:81–100. https://doi.org/10.1111/j.1095-8339.1976.tb01353.x

Stevenson DW (1976b) Shoot apex organisation and origin of the rhizome-born roots and their associated gaps in *Dennstaedtia cicutaria*. Am J Bot 63:673–678. https://doi.org/10.2307/2441830

Stevenson DW (1976c) The cytohistological and cytohistochemical zonation of the shoot apex of *Botrychium multifidum*. Am J Bot 63:852–856. https://doi.org/10.2307/2442045

Stevenson DW (1980) Ontogeny of the vascular system of *Botrychium multifidum* (S. G. Gmelin) Rupr. (Ophioglossaceae) and its bearing on stelar theories. Bot J Linn Soc 80:41–52. https://doi.org/10.1111/j.1095-8339.1980.tb01656.x

Stewart WN (1947) A comparative study of *Stigmarian* appendages and *Isoëtes* roots. Am J Bot 34:315–324. https://doi.org/10.2307/2437143

Stokey AG (1909) The anatomy of *Isoëtes*. Bot Gaz 47(4):311–335. https://doi.org/10.1086/329874

Takahashi A, Kato M (1988) Developmental anatomy of vascular cambium and periderm of *Botrypus virginianus* and its bearing on the systematic position of Ophioglossaceae. Bot Mag 101:373–385. https://doi.org/10.1007/BF02488083

Takiguchi Y, Imaichi R, Kato M (1997) Cell division patterns in the apices of subterranean axis and aerial shoot of *Psilotum nudum* (Psilotaceae): morphological and phylogenetic implications for the subterranean axis. Am J Bot 84(5):588–596. https://doi.org/10.2307/2445894

Tomescu AMF (2011) The sporophytes of seed-free vascular plants – major vegetative developmental features and molecular genetic pathways. In: Fernández H, Kumar A, Revilla MA (eds) Working with ferns: issues and applications. Springer Science+Business Media LLC, New York, pp 67–94. https://doi.org/10.1007/978-1-4419-7162-3_6

Treub M (1886) Études sur les Lycopodiacées. Annales du Jardin Botanique de Buitenzorg 5:87–139. (cited after Popham 1951)

van der Graaff E, Laux T, Rensing SA (2009) The WUS homeobox-containing (WOX) protein family. Genome Biol 10:248. https://doi.org/10.1186/gb-2009-10-12-248

Vasco A, Moran RC, Ambrose BA (2013) The evolution, morphology, and development of fern leaves. Front Plant Sci 4:345. https://doi.org/10.3389/fpls.2013.00345

Vollbrecht E, Reiser L, Hake S (2000) Shoot meristem size is dependent on inbred background and presence of the maize homeobox gene, *knotted1*. Development 127:3161–3172

Wand A (1914) Beiträge zur Kenntnis des Scheitelwachstums und der Verzweigung bei *Selaginella*. Flora 106:237–263

Webster TR (1969) An investigation of angle-meristem development in excised stem segments of *Selaginella martensii*. Can J Bot 47:717–722. https://doi.org/10.1139/b69-102

Webster TR (1992) Developmental problems in *Selaginella* (Selaginellaceae) in an evolutionary context. Ann Mo Bot Gard 79:632–647. https://doi.org/10.2307/2399757

Webster TR, Steeves TA (1964) Developmental morphology of the root of *Selaginella kraussiana* A. Br. and *S. wallacei* Hieron. Can J Bot 42:1665–1676. https://doi.org/10.1139/b64-165

Webster TR, Steeves TA (1967) Developmental morphology of the root of *Selaginella martensii* Spring. Can J Bot 45:395–404. https://doi.org/10.1139/b67-039

West C, Takeda H (1915) On *Isoëtes japonica* A. Br. Trans Linn Soc London. Ser Bot 8:333–376

Wetmore RH, Smith AC (1942) Development of the shoot system in *Lycopodium*. Am J Bot 29(suppl):19

White R, Turner M (1995) Anatomy and development of the fern sporophyte. Bot Rev 61:281–305. https://doi.org/10.1007/BF02912620

Yadav RK, Tavakkoli M, Reddy GV (2010) WUSCHEL mediates stem cell homeostasis by regulating stem cell number and patterns of cell division and differentiation of stem cell progenitors. Development 137:3581–3589. https://doi.org/10.1242/dev.054973

Yadav RK, Perales M, Gruel J, Girke T, Jönsson H, Reddy GV (2011) WUSCHEL protein movement mediates stem cell homeostasis in the *Arabidopsis* shoot apex. Genes Dev 25:2025–2030. https://doi.org/10.1101/gad.17258511

Yang TY, Chiang SHT, DeVol CE (1975) Vascular elements in the corm of *Isoëtes taiwanensis*. Taiwania 20:99–106

Yi SY, Kato M (2001) Basal meristem and root development in *Isoëtes asiatica* and *Isoëtes japonica*. Int J Plant Sci 162:1225–1235. https://doi.org/10.1086/322942

Zagórska-Marek B, Turzańska M (2000) Clonal analysis provides evidence for transient initial cells in shoot apical meristems of seed plants. J Plant Growth Regul 19:55–64. https://doi.org/10.1007/s003440000007

Zeng M, Hu B, Li J, Zhang G, Ruan Y, Huang H, Wang H, Xu L (2015) Stem cell lineage in body layer specialization and vascular patterning of rice root and leaf. Sci Bull 61(11):847–858. https://doi.org/10.1007/s11434-015-0849-1

Zhang Y, Jiao Y, Jiao H, Zhao H, Zhu YX (2017) Two-step functional innovation of the stem-cell factors WUS/WOX5 during plant evolution. Mol Biol Evol 34(3):640–653. https://doi.org/10.1093/molbev/msw263

Zhong R, Lee C, Zhou J, McCarthy RL, Ye ZH (2008) A battery of transcription factors involved in the regulation of secondary cell wall biosynthesis in Arabidopsis. Plant Cell 20:2763–2782. https://doi.org/10.1105/tpc.108.061325

Chapter 4
Biotechnology in Clone Gametophytes: Future Perspectives in Homosporous Ferns

Deepali Johari and Ajit Pratap Singh

4.1 Introduction

Pteridophytes (ferns and fern allies) are ancient vascular plants and exist from the Silurian period. They are spore-bearing land plants and comprise 10,000 species under 250 genera (Smith 1972). They grow in moist, cool climatic condition of tropical and temperate forest of the world, from lowland to alpine zones (Tryon and Tryon 1982), but some species are confined to certain locality in small population as endemic and endangered (Chiou et al. 2006). On the basis of spore morphology, they are divided into homosporous and heterosporous ferns. Homosporous ferns produce only one kind of spores, but the life cycle of heterosporous fern is more analogue to the seed plants, where it produces megaspore and microspore (Raghvan 1989). The megaspore and microspore give rise to independent gametophytes that bear female and male sex gametes, respectively. All spores of a homosporous fern give rise to bisexual gametophyte by the mitotic division of single-celled spore. Ferns represent a crucial, unexplored genetic diversity that may be exploited for understanding evolutionary trends in plants. Gametophytes are usually haploid and produce gametes for sexual fertilization. Gametophyte of fern is comparatively long-lived than to the gametophytes of other plant groups; therefore, they could be best organism for the study of development biology, reproductive mechanisms, genetic barriers, homozygous and heterozygous traits (Singh et al. 2017) and polyploidization (Fauzi et al. 2016), as their genome usually remains pure and haploid. The haploid, pure genome of gametophytes could be used in improving plants via gene-transfer technologies. Genome sizes of ferns get increased by the substitution of an additional set of chromosomes (Otto 2007) that leads to polyploidization.

D. Johari · A. P. Singh (✉)
Pteridology Laboratory, Plant Diversity, Systematics & Herbarium Division,
CSIR-National Botanical Research Institute, Lucknow, UP, India

H. Fernández (ed.), *Current Advances in Fern Research*,
https://doi.org/10.1007/978-3-319-75103-0_4

Polyploidy in ferns can occur spontaneously or by error in meiotic or mitotic division, including the fusion of unreduced gametes (Comai 2005). Sometimes, meiotic failure in sporogenesis results diploid spores that produces apogamous sporophyte. Thus, in both the stages, i.e. gametophyte and sporophyte, the fern species remain pure diploid cytotype, which in other word would be the most important organism for biotechnological investigations. Application of biotechnology in fern gametophytes began prevalent, while people attempted to produce ferns by means of asexual or sexual methods in culture conditions. The in vitro culture of spore, gametophytes and sporophytes were carried out (Dyer 1979; Sheffield et al. 2001; Mikula and Rybczynski 2006; Rybczynski and Mikula 2011; Behera et al. 2011; Singh et al. 2013, 2017) with different aims, using a variety of culture combination. Except the contributions (Chiou and Farrar 1997; Fernandez et al. 1999; Singh et al. 2012, 2013), nothing is known about production of clone gametophytes in ferns. Fern gametophytes were used as germplasm for gene banks, genome-wide analysis of reproductive barriers (Nakazato et al. 2007), genetic diversity (Sigel 2016), cloning and expression of genes (Huang et al. 2014; Ruiz-Estevez et al. 2017), unique model for somatic embryogenesis (Mikuła et al. 2015), model for exploration of fern genome space (Wolf et al. 2015; Fauzi et al. 2016), genetic transformation (Plackett et al. 2015), perspective of apomixes (Grusz 2016), mating systems (Haufler et al. 2016) and barcoding for species identification (Pinson et al. 2017). Clone gametophytes could be used for elucidating the progresses in the genetics and evolutionary biology, as well as resolving the issues of fern biology in this genomic age (Baker and Wolf 2010). Gametophyte produces the sporophytes, which could be harvested for prospection of potential molecules. *Tectaria macrodonta*, reported for insecticidal protein (Tma12) to whitefly (Shukla et al. 2016), was required in bulk during prospection; the mass multiplication helped to achieve the goal. Possible function of such proteins in the ferns is still to be discovered, which can be achieved by exploring the expression of proteins and genes in different developmental stages of this species. Production of clone gametophyte can provide different developmental stages. This would help to elucidate presence of specific proteins and genes, as these molecules express at a certain age of the gametophytes or sporophytes. The biotechnology in higher plants has advanced, but such researches in ferns are lagged behind. There is only one sequenced genome of fern allies, i.e. *Selaginella moellendorffii*, and none from any homosporous ferns (Sessa and Der 2016). In spite of above uniqueness of ferns, there is inappropriate efforts to take the fern gametophyte as a model organism to understand evolution pattern, genetic barriers, cloning potential, sexual expression, mating system, genomic spectrum and response of silenced or expressed half set genome in haploid and diploid stage and prospection of active biomolecules. In view of above, present chapter provides information about the problem in homosporous ferns, genetic barriers, gametophyte as a model organism, gametophyte clone formation, in vitro methods for gametophyte development and explant culture, regeneration of clones, sexual expression and matings in clone, sporophyte production, transplantation, hardening and acclimatization and prospection of active biomolecules for transgenic crops.

4.2 Problems in the Gametophytes of Homosporous Ferns

Majority of the ferns (Fig. 4.1a–c) are under threat of survival. Population of such ferns are reducing either by edaphic or genetic factors. Amongst these factors, the genetic factor plays a significant role, because the establishment potential is determined by the sexual efficiency of gametophytes. Genetic traits, viz. spore germination, gametophyte development, sexual expression, functional gametes (eggs, sperms) and gamete fusion, are inherited traits of the gametophyte. To investigate the sexual potential of ferns, it is important to apply the biotechnological approaches, studying the sexual compatibility of different genotypes, functional gametes and lethal alleles inhibiting fertilization, though homosporous ferns are usually bisexual and mates either through intragametophytic selfing, intergametophytic selfing or intergametophytic crossing (Nayar and Kaur 1971). Nevertheless, it happens to happen that the inbreeding depression, gap in sexual expression, incompatibility of different genotype and their alleles are the subsidiary factors (Banks 1997) in the success of fern establishment. Changing climate, habitat fragmentation and anthropogenic intervention are also exerting selection pressure in the species to colonize new areas. Disturbance reduces their reproduction ability or forces the species to colonize in new areas over long distances (De Groot et al. 2012). Amongst the unknown reasons, it is being hypothesized and speculated that the set of lethal genes altogether causes failure of fertilization and population establishment (Goodenough 1978). Most important and challenging issues in the biotechnology are to understand the genetic cause of reproductive bottlenecks using the fern gametophytes to ensure endurance for their sustainable utilization.

4.3 Genetic Barriers in Gametophytes

Homozygous gametophytes in ferns exhibit two copies of similar genetic alleles. Both of these alleles are responsible for similar traits. These alleles are inherited from parent gametophytes during fertilization and are randomly united during pairing of the homologous chromosomes. In fertilization, one chromosome in each pair comes from male and another from female parents. Alleles on these chromosomes usually determine the traits in sporophytes. Homozygous alleles could be dominant or recessive for a particular trait. Fusion between two gametes exhibiting similar alleles for a trait could be lethal for some other specific desired traits. In heterozygous conditions, the gametophytes exhibit variable set of alleles and express different phenotypes. Since the homozygous genotype express one phenotype, therefore, usually, the heterozygous dominant allele may mask the recessive allele phenotype, which usually happens to be lethal for the individual. Ganders (1972) subjected to breeding tests in homosporous ferns to reveal the presence of recessive lethal alleles in their genotypes and suggested that natural selection operates to strongly favour heterozygous sporophytes. In heterozygous genotype, the traits of two different

Fig. 4.1 Threatened fern species. (**a**) *Anemia rotundifolia* Schrad. (**b**) *Cyathea spinulosa* Wall. *ex* Hook. (**c**) *Pronephrium nudatum* (Roxb.) Holtt

alleles co-express, which minimize the chance of lethality for required traits. In addition, the genetic traits altogether with their homozygous or heterozygous genotypic alleles determine the fate of species population (Wubs et al. 2010). The allele in homozygous genotype become lethal in inbreeding (Wubs et al. 2010), which also has a huge impact on population structure. In addition to the biotic as well as abiotic factors, the allele in both the homozygous and heterozygous genotype has significant ability in establishing the ferns population. Studies have evidence that if some lethal genes are favoured in the heterozygous state, the level of genetic load maintained in the population will be even higher (Goodenough 1978). The sources of genetic load include somatic mutation. These mutations may be transmitted through next generation of gametophytes; particularly, outcrossing populations may accumulate some of this genetic load. Presence of any genetic load would prevent the establishment of viable sporophyte from most isolated gametophytes. For the reason of existence of homozygous, heterozygous and homoeozygous genotypes, the gametophyte of ferns could serve as a best model to discover the mating mechanism and elucidate the genetic barriers.

4.4 Gametophyte as a Model Organism for Biotechnological Studies

In ferns, the gametophyte is multicellular, green, autotrophic, haploid phase (with half set of chromosome), which alternates with a diploid sporophyte phase. Gametophyte development in fern is more distinct, as it is only terrestrial organism, which has comparatively long-lived haploid stage for study of the phenotypic, genotypic and physio-biochemical processes. Fern gametophyte provides a novel source of germplasm that can circumvent many of the problems associated with development, evolution, sexual expression, genetics of matings, inheritance of genetic traits, genetic barriers, fertilization, apogamy, apospory and polyploidization (Nakazato et al. 2007; Haufler et al. 2016). They can be best source of genetic manipulation, as the trait(s) expressing alleles and genes can be easily handled for the reason of their pure line haploid homozygous, heterozygous or homoeozygous genotypes. From a genetic perspective, they are interesting because their polyploids represent the union of divergent parent genome (Sigel 2016). *Ceratopteris richardii* has proven to be useful in identification and characterization of mutations affecting the key development process in gametophyte (Banks 1999). Different stages of gametophytes can be easily and quickly multiplied (Fernández and Revilla 2003), as they also possess the ability to regenerate from gametophyte tissues (Fernández et al. 1999). Gametophytic explant's raised regenerates usually comprise of homozygous, heterozygous or homoeozygous genotypic alleles. For the reason of exhibiting different genotypic alleles, the mating in ferns often varies from species to species. It is the genetic make-up of these peculiar genotypes, which can help in identifying the range of genetic variation between the extant ferns and their progenitors. In addition to above, the ferns are useful in studying the natural polyploidization (Fauzi et al. 2016); therefore, they could be as a best organism for biotechnological approaches.

4.5 Clone Gametophyte Formation

Apogamous sporophyte is known to have low (Peredo et al. 2013) as well as high level of genotypic diversity (Ootsuki et al. 2011) and resulted from frequent clone formation of sexual progenitors. For the reason of maintenance of similar genotype in both the gametophytic and sporophytic phases, the apomicts may be assumed to produce genotypically clone offspring (Grusz 2016). Clonal potentiality is not confined to apomicts, but each cells of the gametophyte being totipotent has ability to regenerate a new gametophyte. Albaum (1938) described the clone formation from the pieces of prothallus of *Pteris aquilina* and *Pteris longifolia*. There were many methods, viz. explant culture, cell culture, homogenization of gametophytes, protoplast culture of gametophytes and callus and cell suspension culture of gametophytes (Fernandez et al. 1999; Maeda and Ito 1981) for clone formation. To demonstrate the total totipotency of any cell, the cell is to be isolated and cultured in suitable culture media to produce numerous regenerates. Ito (1962) isolated the single cells from the gametophyte and cultured in a mineral salt medium for regeneration of clone gametophytes. Isolated single cells from the gametophyte enzymatically dissociate from cells of their walls and follow the isolated protoplasts through steps involving reformation of the cell wall, growth and regeneration of the clone gametophyte. Fernandez et al. (1999) mechanically triturated the gametophyte, and resulting homogenates were culture on the culture media. These homogenized pieces of gametophytes gave new cloned gametophytes. Robertson (2002) used the fern gametophytes to produce arrays of genetically identical clones. These clones helped in determining the ability of clones for sporophyte production. Such identical clones of individual gametophyte or different gametophyte origin were combined in pairs in all possible ways to produce the possible mating scheme. Singh et al. (2012) described that culturing the apical, median and basal explants of gametophytes of *Anemia rotundifolia* on the P&T culture media produces numbers of clone gametophytes. Apical and median explants show higher clone formation potentiality than to basal explants; basal explants lost their viability on culturing in extended period. In extended culture, the gametophytic explants produce numbers of primary and secondary regenerate gametophytes, which bears male and female gametes, respectively. The clone gametophytes are offspring with pure line of homozygous genotype than to the parents. Genotypic expression of the clone gametophytes achieved in successive generation (T0 to T∞) may provide information about the range of genetic variation to elucidate the process of genome evolution. Since the clones are the pure genotypic lines of the progenitors, therefore, they show the sequence of genetic substitution or deletion over the period of extant ferns evolution. Clone can provide an understanding on the heritable gene expression amongst the progenitors, early stages of evolutionary divergence in sexual as well as apomictic ferns (Ferreira et al. 2016). Clone gametophytes could serve as best resource with pure line genetic traits to handle the biotechnological implications, besides their significant characteristics feature of multiplication.

4.6 In Vitro Gametophyte Development

Gametophytes can be developed either in nature or in vitro culture conditions (Nayar and Kaur 1971) with the spore germination. They can also be multiplied by culturing the gametophyte explants (Singh et al. 2012; Fernandez et al. 1999) on a variety of culture media, viz. Murashige and Skoog (MS) medium (Somer et al. 2010) and Parker and Thomson's (P&T) media (Klekowski 1969a, b; Klekowski and Lloyd 1968). Authors have successfully developed the gametophytes of many homosporous ferns, viz. *Adiantum peruvianum*, *Anemia rotundifolia*, *Cyathea spinulosa*, *Lepisorus nudus*, *Pronephrium nudatum* and *Pteris biaurita*, growing the spores on P&T culture media (Singh et al. 2012, 2013, 2017). For in vitro development of gametophyte, the single fertile frond containing spores was collected in paper packet and kept in desiccators for a week to release the spores. Spores were sterilized with 2% sodium hypochlorite solution for 2 min and rinsed with sterilized water, thrice. Culture media comprising macro (Parker's) and micro (Thompson's) elements were prepared in 1 L distilled water, solidified with agar and autoclaved (Klekowski 1969a, b; Klekowski and Lloyd 1968), maintaining the pH at 5.6. Sterilized spores were sown on the petri dishes containing culture media and kept in laboratory at 50 μmol m^{-2} s^{-1} light intensity (photoperiod of 10:14 h light and dark), temperature 25 ± 1° C, 50–75% relative humidity. After germination, spore produces green, autotrophic, different, viz. two celled, filamentous, spatulate, semicordate, cordate, sexual gametophytes (Fig. 4.2a–i) for fertilization as well as sporophyte production (Fig. 4.2j–l). Nayar and Kaur (1971) defined pattern of gametophyte development, viz. *Adiantum* type, *Aspidium* type, *Ceratopteris* type, *Drynaria* type, *Kaulinia* type, *Marattia* type and *Osmunda* type. These types of gametophyte development differ in the sequence of cell divisions and the region at which meristem developed. In vitro studies on *Dipteris wallichii* and many other ferns (Behera et al. 2011; Singh et al. 2012, 2017; Khare et al. 2005; Nair et al. 2013; Srivastava and Uniyal 2014; De Groot et al. 2012) were carried out on different culture media to investigate the developmental pattern, sexual expression, mating systems, genetic barriers in gametophytes and ability as well as efficiency of these culture media to produce high numbers of gametophytes and sporophytes. Nutritional requirements, including other environmental sex determination (ESD) factors for the development of the gametophyte in ferns, differ from species to species (De Soto et al. 2008). Gametophyte growth is also affected by the water, sugar, light, temperature and micro and macro elements in in vitro conditions (Brown 1923). Different steps of in vitro gametophyte development could be understood by the provided flow chart (Fig. 4.3).

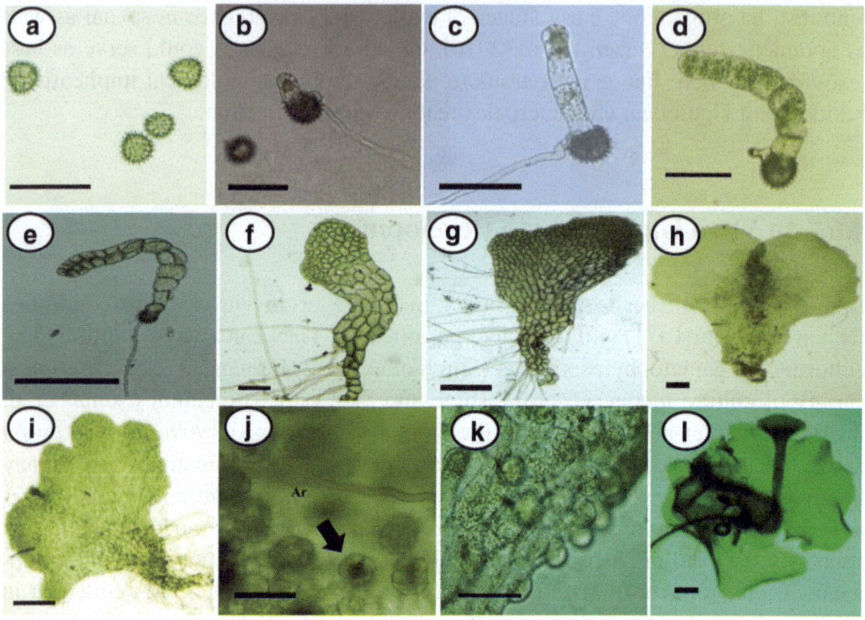

Fig. 4.2 Morphological structures observed when cultured ferns in vitro. (**a**) Spores. (**b**) First protonemal cell. (**c, d**) Filamentous gametophytes. (**e**) Two-dimensional gametophyte. (**f**) Semi-spatulate gametophyte. (**g**) Spatulate gametophyte. (**h, i**) Cordate and sexual gametophytes. (**j**) Archegonia. (**k**) Antheridia. (**l**) Sporophyte. Scale bar for (**a, e**) = 0.25 mm, (**b–d, f, j, k**) = 0.1 mm, (**g–i, l**) = 0.5 mm

4.7 In Vitro Gametophyte Explant Culture

Typically, the single spore gives rise to one gametophyte, but an individual gametophyte can result numerous regenerates subject to availability of favourable axenic conditions, rationale nutritional supply and feasible ESD factors. Majority of the ferns exhibit notched apical meristem region, but the ferns having pluri-meristematic cells are not uncommon (Nayar and Kaur 1971). Therefore, the tissues excised from different regions of a cordate gametophyte, culturing on suitable media, may response variably. The gametophytes of ferns are known to have a great multiplication potentiality, possibly for the reason of their totipotent cells. Unfortunately, the studies on explant culture in ferns sprint around the sporophytes, using frond (Makowski et al. 2016), rhizome meristem (Thakur et al. 1998), scales (Ambrozic-Dolinsek and Camloh 1997), apices and stolon (Lazar et al. 2010) and green sporangia (Soare 2008; Somer et al. 2010), including somatic embryogenesis (Mikuła et al. 2015). Investigations on culture of gametophyte tissues, callus, cells, protoplasm, cell suspension and homogenized tissues show that the culture medium can contribute to increase sporophyte production in ferns (Fernández et al. 1999; Soare 2008; Somer et al. 2010), but the efforts to propagate ferns from gametophyte were

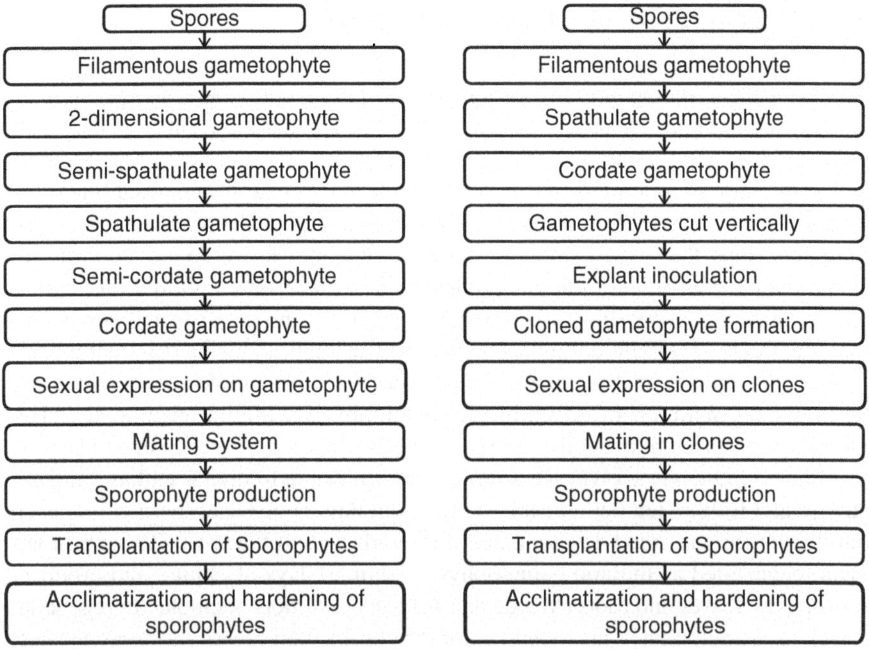

Fig. 4.3 Flow chart showing in vitro propagation of gametophytes through spores and gametophyte explant

least attempted. Attempts to raise multiple gametophyte and sporophyte through homogenized gametophytes of *Woodwardia virginica*, *Dryopteris affinis*, *Osmunda regalis*, *Pteris ensiformis* and other ferns were randomly made (Fernández et al. 1993; Fernández et al. 1999; Soare 2008; Somer et al. 2010). In the process, homogenized tissues were cultured on the culture medium, which regenerated new gametophytes and produced almost thousand sporophytes. Culturing the gametophyte of *Pteris vittata*, *Ampelopteris prolifera* and *Woodwardia orientalis* (Khare 2001) determined two different patterns of regeneration. In *A. prolifera*, regeneration begins only from the apical meristem region, while in other two species it started from the margin of the wings, showing variable pattern of sexual expression in the parent as well as regenerate gametophytes. Gametophyte explant cultures in *Anemia rotundifolia*, *Adiantum peruvianum* (Singh et al. 2012, 2013) and *Cyathea spinulosa*, and *Pronephrium nudatum* led to investigate the viability of explants, regeneration of clone gametophyte, sexual expression in clone, mating, sporophyte production and possible mode of conservation, besides providing numbers of pure line homozygous genotypes. This can be used to promote mating between pure line contrasting genotypes and to elucidate pattern of genetic variation between progenitors and recent species, predicting the genome evolution pattern. In vitro gametophyte explant culture could be easily explained by the provided flow chart (Fig. 4.3).

4.8 Regeneration of Clone Gametophytes

Clones of the gametophytes are genotypic and phenotypic identical replica of the parent gametophyte. Clone production is reliant on regeneration ability of the parent gametophyte. Regeneration is ability of the explants, tissues, cells, homogenized cell suspension or protoplast of the gametophyte to reconstitute a normal functional identical gametophyte from the part of the original ones. Cells of gametophytes are totipotent, thus they have ability to produce clones in favourable conditions. Ito (1962) isolated cells from the gametophyte of *Pteris vittata,* cultured on medium and summed that every single isolated cell regenerates and grows to a mature gametophyte. Reynolds (1979) cultured substance diffusing from the meristematic region of the gametophyte of *Anemia phyllitidis* and found that indole acetic acid (IAA) promotes regeneration of adventitious gametophytes while abscisic acid (ABA) inhibits regeneration from the meristem cells (Reynolds 1981). Protoplasts of *Lygodium japonicum* grown in a culture medium began to divide within 8 days of culture and formed ten-cell clusters on the 30th day. Transfer of these cell clusters into fresh media resulted development of rhizoids and protonemata. These protonemata regenerated a common gametophyte within 50 days of culture and produced sporophytic leaves and roots (Maeda and Ito 1981). Authors attempted to regenerate the clone gametophytes in *Anemia rotundifolia*. In the process of clone development, the mature gametophytes were cut into small pieces (1–2 mm) from three regions (apical, median, basal). Culturing of these explants on P&T culture media revealed (Fig. 4.4a) that the cells in apical explants near the meristem tissues divided transversely and vertically to form numerous semi-cordate secondary gametophytes on the 10th day after inoculation (Fig. 4.4b). Semi-cordate secondary gametophyte develops laterally and grew into an elongated unusual gametophyte with undulate margins (Fig. 4.4c). After 45 days, numerous regions in the unusual gametophytes performed meristematic activity. Meristematic regions cut the cells tangentially and vertically and led to shuttle the vegetative tissues obliquely to emerge regenerates. This led to form tertiary regenerates on secondary regenerate gametophyte (Fig. 4.4d). Each secondary gametophyte bears an average numbers of eight to nine clone gametophytes. Clone formation process assorts the heterozygosity and forwards homozygous alleles in offspring regenerates. Comparative analysis on regeneration ability of the gametophyte explants (apical, median, basal) revealed that after 1 year of culture, the apical explants produced an average of 75.5 regenerates per 15 explants; however, the median and basal explants produced 70 and 64.3 regenerates, respectively (Singh et al. 2012). Authors also observed the clone formation in *Cyathea spinulosa*, using gametophyte explants and growing on P&T media. On culturing, the excised region of the explants became brown; however, the intrinsic heterogeneous chlorophyllous cells formed a thick area by their active division. The viable cells divided subsequently, resulting development of secondary gametophyte. Some of the meristem cells in the secondary gametophyte produced secondary regenerates on or after 20 day of inoculation. Comparative analysis revealed that in extended period of inoculation, highest regeneration potentiality of

Fig. 4.4 Regeneration in clone gametophytes. (**a**) Gametophyte explant grown on P&T media. (**b**) Secondary regenerates with thickened medial region. (**c**) Multilobed secondary regenerates. (**d**) Tertiary regenerates develop on secondary regenerates. (**e**) Archegonia on tertiary regenerate. (**f**) Antheridia present on tertiary regenerate. (**g**, **h**) Developmental stages of sporophytes. (**i**, **j**) Mass propagation of sporophytes in in vitro conditions. (**k**) Acclimatization of sporophytes in earthen pots. (**l**, **m**) Hardening and transplantation of sporophytes. Scale bar for (**b**) = 0.25 mm, (**e**, **f**) = 0.1 mm, (**c**, **d**, **g**, **h**) = 0.5 mm

gametophytes was achieved in apical region (40.5 regenerates) followed by the median (25.3 regenerates). All the basal gametophyte explants lost their viability after 2 months of inoculation; as a result the basal explants could not produce any regenerates. Authors also grew the gametophytic explants of *Pronephrium nudatum* in P&T culture media and observed that the outer cells of the excised region became brown, but the inner one to two layers of cell remained brownish green and viable. These viable cells acted as meristem and divided to give rise secondary gametophytes. Few of the vegetative cells in the secondary gametophyte became thickened with dense protoplasm. These distinct cells divided to form a regenerate, developing from the secondary gametophyte after 20 days of inoculation. Numbers of secondary regenerate gametophyte emerged on parent gametophyte. Cells of these secondary regenerates were green, homogeneous throughout, except in the apical region, where cells were smaller. Frequent divisions occur in small cells to produce juvenile secondary regenerates after 42 days. These regenerates developed into cordate gametophyte and called as secondary regenerates. Analysis of regenerate production revealed that apical explants have highest regeneration potentiality as it shown an average numbers of 68.2 regenerates per fifteen explants. Median explants produced an average number of 52.5 regenerates per fifteen explants, whereas the basal explants lost their viability as well as regeneration potentiality after 6 months. Authors observed that usually the new proliferation arises from the margins of the prothallus, although it was not uncommon to find regenerates on the midrib or the region of the apical meristem. It is assumed that as long as the cells are capable of mitotic activity, the gametophyte exhibit high regeneration potentiality. Authors could not observe the differences between the pattern and growth of the gametophytes arises either from the spore or gametophyte, itself. Regeneration of clone gametophytes is significant for investigation of the genetic variations, amongst the different populations. Since the clones are offspring resulted from subsequent mitotic division in the parent gametophytes, thus, they inherit and move the trait towards homozygous genotype due to continuous genetic assortment in successive generation. Clone may provide genotype that can serve as germplasm with pure genotypic traits for biotechnological implications.

4.9 Sexual Expression and Mating in Clones

Gametophytes in fern have potential to generate sexual gametangia. Most of the homosporous ferns produce bisexual gametophytes that promote inbreeding and attainment of homozygosity in the progeny. The clones, which are replicas of parent gametophyte, inherit the parent genotypic traits. They bring about the sexual expression, but the stage and pattern of sexual expression may vary because of their move towards homozygosity. Sex of gametophyte can vary depending on its social environment and by the pheromone antheridiogen (Atallah and Banks 2015); therefore, besides homozygous alleles, these factors may also influence the sexual expression in clone gametophytes. There were many observations on the regeneration of the gametophytes (Khare 2001; Somer et al. 2010), but observations on the sexual expression in clone gametophytes were never made. Authors regenerated the clones by culturing the explant gametophytes on P&T culture media and observed the pattern of sexual expression and mating system in secondary or tertiary regenerates, whether the regenerates are homothallic (self-compatible) or heterothallic (self-incompatible). Mating in fern depends on the timing and sequence of the sexual expression and can be determined on the basis of phenotypic traits or using the microsatellite markers and evaluating genetic diversity and structure of different population of clone gametophytes (Imai et al. 2016). Observation on *A. rotundifolia* revealed that numerous cells on the secondary clones became distinct and matured into archegonia. These secondary clones did not produce any antheridia during and after the emergence of archegonia (Fig. 4.4e). Some cells in the middle region of tertiary clones were distinct and turned into archegonia. Few cells near the archegonia on tertiary regenerate became globose, elevate out of the cells and matured into antheridia (Fig. 4.4f). The secondary gametophyte bears only archegonia, but not antheridia. This showed lethality effect in the first clone gametophyte that could be due to the impact of moderate heterozygosity in explant gametophyte. Tertiary regenerates bear both the gametes with overlap expression. This led option for both intra- and inter-gametophytic selfing, i.e. (i) intra-gametophytic selfing in a tertiary regenerates and (ii) intergametophytic selfing between the secondary and tertiary regenerate gametophytes. This could happen for the reason because all the tertiary regenerates were comprised of pure line homozygous allelic genotype. Homozygosity has accelerated the expression of both the gametes to perform inter- and intra-gametophytic selfing for sporophyte production on the 85th day (Fig. 4.4g, h). Homozygosity of isolate gametophyte was carried forward to the regenerates that compassionated for the intra- or inter-gametophytic selfing. Analysis revealed that the sporophyte production through gametophyte explants reached about 135.5%. Authors also observed sexual expression in the clone gametophytes of *Cyathea spinulosa* by culturing the gametophytic explants on P&T culture media. Both the male and female gametangia developed on the same secondary regenerate gametophytes on different days showing monoecious tendency. Nevertheless, the determinants of sexual phenotype in ferns are diverse (Brown 1923; De Soto et al. 2008) ranging from sex chromosomes to cross talk between individuals of *Ceratopteris richardii* (Tanurdzic and Banks 2004) and many other ferns. Secondary regenerates exhibit both the gametes showing maximum chances of intra-gametophytic selfing. This resulted maximum possibility of fusion of the gametes (produced on secondary regenerates that had derived from common parent gametophyte explants) to demeanour fertilization and sporophyte production. Analysis revealed that 19 sporophyte were produced by an average numbers of 40.5 homozygous regenerates that had derived itself from apical explants only. Average sporophyte production was achieved 190%. Sexual expression in *Pronephrium nudatum* by culturing the gametophytic explants on P&T culture media revealed that both the antheridia and archegonia were present on different regenerate gametophytes showing dioecious tendency. Archegonia developed on the secondary regenerate on the 55th day; however, antheridia developed on small lanceolate tertiary regenerate. Both antheridia and archegonia derived from secondary and tertiary regenerate gametophyte were available for inter-gametophytic selfing, as both secondary and tertiary regenerate gametophyte were derived from the same explant gametophyte. The secondary and tertiary regenerate gametophyte served as clone of the parent gametophyte. Through the three generations, they have got opportunity for assortment of homozygous or heterozygous genotype, if any. After maturation of the sex gametes, numerous sporophytes were produced by fertilization on 113th day. Average sporophyte production through apical, median and basal explant culture reached about 88.8%. Result asserted that single gametophyte produced from a single spore can produce multiple regenerate gametophytes, which after fertilization, may produce numbers of sporophytes. Study on sexual expression in clone gametophytes can help in manipulation of different genotypes for pure line culture of required traits, multiplication of the targeted genotypes, inheritance and expression of required traits, analysis of genetic variations across the filial generations and multiplication for conservation.

4.10 Propagation of Sporophyte

Conventional method of ferns propagation is spores, but the vegetative propagation can also be an effective technique for propagation of ferns using a variety of culture media and culturing the vegetative portion of sporophyte under feasible in vitro culture conditions. Propagation of sporophyte may be achieved by the methods as discussed in heading 4.6 of the chapter. Similar to that the sporophyte can be propagated for various purposes. Sporophyte propagation under in vitro culture condition was carried out by many worker (Fernandez and Revilla 2003; Somer et al. 2010); reporting the regeneration through the frond (Makowski et al. 2016), rhizome (Thakur et al. 1998), scales (Ambrozic-Dolinsek and Camloh 1997), stolon (Lazar et al. 2010), green sporangia (Somer et al. 2010), somatic embryos (Mikuła et al. 2015) and homogenization process. The micropropagation using shoot tip as initial explants was first reported by Hennen and Sheenan (1978) for *Platycerium stemaria*, who reported that the bud development occurred at the base of the explants and on the leaves connected with the culture media. Wee et al. (1992) used rhizome pieces as explants and cultured on MS media supplemented with some growth regulators to initiate the in vitro culture of *Platycerium coronarium* and *P. ridleyi*, where sporophytes were produced after 2 months. Authors realize that the use of scales as explants to propagate sporophyte is difficult task because their tissues become dead as soon as scale reaches to maturity. Nevertheless, Ambrozic-Dolinsek and Camloh (1997) used scale for culturing on nutrient medium and observed that regeneration begins on or after 30 days of culture and new proliferated group of cells formed which later turned into buds and aposporous gametophytes. Sporophytes were also propagated using green globular bodies (GGB) on culture media supplemented with cytokinin. These GGBs later developed into sporophytes (Amaki and Higuchi 1991). Besides, the somatic embryogenesis in ferns and fern allies has rarely been described; it was reported for *Huperzia selago* (Szypula et al. 2005), including

Mikuła et al. (2015). Another micropropagation process (Cooke 1979) was studied by homogenization of cell and culture, who was the first to use the term homogenization and used homogenized sporophytes of *Platycerium* species for multiplication. Hegde et al. (2006) obtained callus from the rhizome tips of *Drynaria quercifolia* and cultured on Knop's salt solution in different concentrations of auxins, to harvest the sporophyte. Besides, there were many studies (Singh et al. 2013, 2017; Ravi et al. 2015) on mass propagation of sporophytes using in vitro gametophyte cultures. Survey of literatures as well as studies by the authors (Behera et al. 2011; Singh et al. 2012, 2013, 2017) suggests that the gametophyte is the best material for propagation of fern sporophyte. There, the spores can be sown on a variety of culture media (P&T, MS) under in vitro culture conditions (50 μmol m^{-2} s^{-1} light intensity with photoperiod of 10:14 h light and dark, 25 ± 1° C temperature, 50–75% relative humidity) to achieve different developmental stages, viz. filamentous, spathulate, semi-cordate, cordate and sexual of gametophytes for large-scale propagation of sporophytes (Figs. 4.2a–l and 4.4a–h) through mating of gametophytes either produced by the spore germination or explant culturing. Newly produced sporophytes were left to grow in the petri dishes (containing culture media) either for 4–5 weeks or till they achieve about 12–15 cm height so that they can be maintained in the glass chimney (Fig. 4.4i, j).

4.11 Transplantation of Sporophytes

In in vitro practices, the transplantation may be referred as shifting of the in vitro growing ferns from one medium to another medium, where the medium may be either artificial culture media (MS, P&T) or the soil, in addition to the variable climatic conditions (light, temperature, relative humidity, moisture, air, etc.). Although sporophytes in many form were produced (Singh et al. 2012, 2013, 2017; Ravi et al. 2015) and were transplanted from one petri dish to another petri dish, none of them described the transplantation method in the contributions. Oh et al. (2014) stated that the in vitro-raised sporophyte of *Isoetes coreana* may be subjected to acclimatization and transplantation to a test tube containing vermiculite and sand mixture (1:1) after 4 weeks of incubating light. After another 2 weeks of incubation, the sporophytes with roots and five leaves were suggested to be transferred to potting soil. Plackett et al. (2015) merely stated that the T0 sporophytes of the fern *Ceratopteris richardii* may be transferred for a period of 8 weeks on medium containing the antibiotics, but this suggestion was for transgenic model. Authors suggest that the in vitro-raised sporophytes may be allowed to grow on the culture media containing petri dish, until they attain a height of approximately 12–15 cm (Fig. 4.4i, j). It is to remember that the sporophytes are repeatedly transferred from one culture containing petri dish to another culture containing petri dish, once the media is utilized. Shifting of sporophytes from petri dish to petri dish is done after 15 days, but it can also vary, as many species require high amount of culture media;

however, a few use it in less quantity. Authors transplanted many ferns for their conservation, after they were grown in in vitro culture conditions (Fig. 4.4j). Transfers of sporophyte were done under axenic condition to avoid any contamination, as algae often have been noticed to make a green film all around the culture media, which inhibits the growth of the sporophytes. Conventionally, the ferns were collected from the forests and directly introduced to captivity centre for long-term conservation. Authors collected many ferns, including *Helminthostachys zeylanica* from nature and transplanted in the fern house of the institute. Direct transplantation revealed that this method of multiplication and conservation of ferns in the captivity was not satisfactory, as in the captivity they are difficult to grow due to their specific requirements. Transplantation through in vitro-raised sporophyte may become more worthy, as it maintains axenic climatic condition, temperature, light, humidity and nutritional requirements for proper growth and development of ferns.

4.12 Hardening and Acclimatization of Sporophytes

Hardening is a process of giving exposure of a variety of climatic factors, viz. relative humidity, light irradiance, temperature, air and pressure to the in vitro-raised plants in increasing order to strengthen their physio-biochemical, morphological, anatomical as well as cellular response, if any. Shukla and Khare (2014) and Shekhawat and Manokari (2015) provided information about the hardening and acclimatization procedure for the sporophytes of *Cyathea spinulosa*, *Nephrolepis biserrata* and *Marsilea quadrifolia*, respectively. Authors hardened many fern sporophytes in the laboratory and suggest that it can be achieved by given exposure of increasing relative humidity, temperature as well as irradiance of light to the in vitro-raised sporophytes. The well-developed and rooted sporophyte (with 3–5 small fronds up to 12–15 cm) can be removed out from the petri dishes along with culture media and transferred to the earthen pots previously filled with the soil, peat, moss and compost (1:1:1:1). The earthen pots (containing newly emerged sporophytes) are brought to the plastic tray (partially filled with water), where the pots are to be placed on a platform, which is partially dipped into the water of plastic tray (Fig. 4.4k). The earthen pots containing the sporophytes were kept in the laboratory conditions with controlled light, temperature and humidity for 1–2 weeks. Sporophyte containing pots were covered with a transparent, closed plastic hood so that the sporophyte can receive the light and increasing humidity to get hardened and ready for acclimatization (Fig. 4.4l, m). These plants were regularly sprayed with water, as soil should be kept moist-wet at all times. During process, the sporophytes gradually response to the factors and harden the cells, tissues and cuticles to acclimatize. Proper hardening and acclimatization reduces the mortality rate of in vitro-raised sporophytes. It enhances the multiplication rate of those ferns, which were produced by sophisticated in vitro culture techniques and whose survivals were always on risk. Authors observed that the sporophytes were hardening and acclimatization shown 80% of survival rate. After hardening, the ferns were successfully transferred to the fern house of institute, where they were growing in captivity under controlled light, temperature, humidity and water, portrayed by net-lon shade and functional sprinkler as well as water mists.

4.13 Biotechnology in Ferns: Biomolecules, Transgenic and Future Perspectives

Lower plants, particularly the ferns and fern allies, got little attention of biotechnological implications for prospection of protein, gene, bioactive compounds and secondary metabolites (Schaufelberger and Hostettmann 1983; Jones and Firn 1978; Nakazato et al. 2006; Leroux et al. 2013), including genetic inheritance and genomic evolution. There is urgent requirement for sequenced genomes (Cronk 2009) in ferns. Studies on *Azolla* and *Ceratopteris* (Sessa et al. 2014) have revealed nuclear genome structure only in two ferns; however, they are exemplars of polyploidy and a variety of mating system, particularly inbreeding. Ferns can be helpful in understanding the genetics as well as inheritance of genetic traits (Sigel 2016). It can generate DNA sequence data to understand the phylogeny, evolution of plastid and nuclear genome, duplication of genome, speciation and gene function (Barker and Wolf 2010). There are no full sequenced fern genomes (Wolf et al. 2015); the NCBI short read archive (SRA) database has incomplete genome data for *Ceratopteris richardii* and *Pteridium aquilinum* (Der et al. 2011; Leroux et al. 2013). Okamoto et al. (1997) transformed phytochrome gene (PHY1) cDNA from the *Adiantum capillus-veneris* into *Arabidopsis* ecotype *Landsberg erecta* and found that the fern phytochrome gene can incorporate in the chromophore of the host plants. Similarly, MADS-box genes were cloned and analysed in *Ceratopteris richardii* and *Ceratopteris pteridoides*, but a new MADS-box gene, DfMADS1(GU385475), was cloned from *Dryopteris fragrans* to understand the role of MADS genes in the evolution (Huang et al. 2014). Identification and characterization of TALE homeobox genes in *Vandenboschia speciosa* (Ruiz-Estevez 2017) was also made recently. It's not time to only clone and express the gene, but there is urgent need to develop methods for genetic transformation of the ferns through a variety of vectors. There is a known fern, *Tectaria macrodonta*, which has potential Tma 12 proteins and genes (Shukla et al. 2016) that require genetic transformation to develop transgenic lines for extended investigations. Though, Plackett et al. (2015) developed a protocol for genetic transformation of the *Ceratopteris richardii* through microparticle bombardment to achieve transgenic progeny for antibiotic resistance, but possibly it does not meets robust transformation. Ferns exhibit insecticidal proteins causing 70–100% morality of *Spodoptera frugipedra* and *Helicoverpa zea* (Markham et al. 2006). Recently, a protein (Tma 12) from *Tectaria macrodonta* that is insecticidal to whitefly and interferes with its life cycle at sublethal doses was discovered (Shukla et al. 2016). A transgenic cotton lines using the Tma 12 gene was developed (Fig. 4.5). Transgenic cotton lines were resistant to whitefly infestation in contained field trials, with no detectable yield penalty. Transgenic cotton lines were also

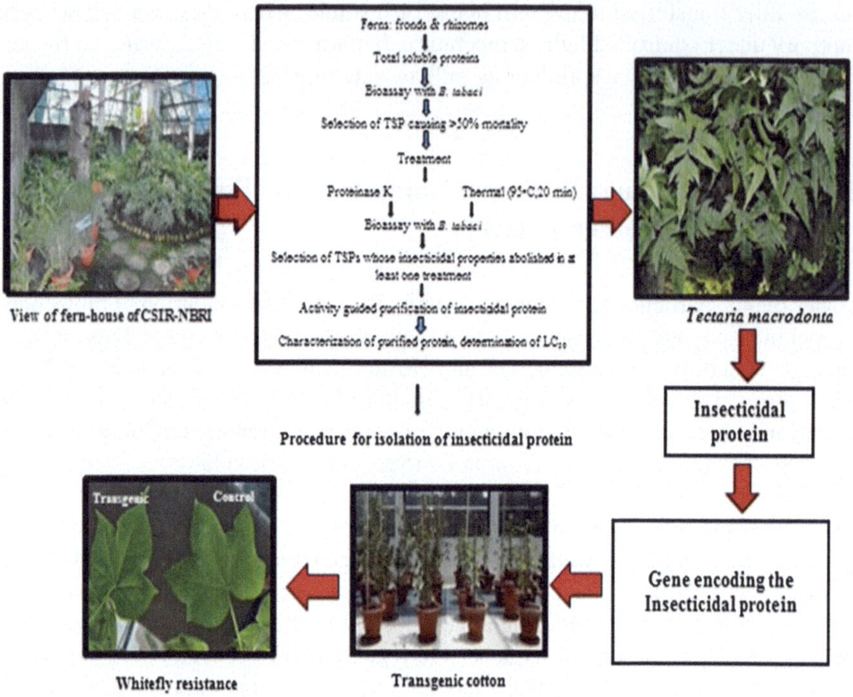

Fig. 4.5 Schematic representation for prospection of *T. macrodonta* to develop transgenic cotton against whitefly (Part of photo from Shukla et al. (2016))

protected from a whitefly-borne cotton leaf curl viral disease, showing that Tma 12 might be well suited for deployment in GM crops to control whitefly and the viruses it carries (Shukla et al. 2016). Such potentiality of ferns suggests that the knowledge about the complete genome sequence may provide insights into genome evolution, genetic and genomic features that characterize ferns and also opens the door for transgenic development. Authors believe that the biotechnological approaches may produce clones of a variety of genotypes, viz. homozygous, heterozygous and homoeozygous. It could differentiate the extent of homoeozygous genotype constituent in the ferns by the cytogenetic testing. It can also assist in elucidating the genetic variation between the progenitors and divergent extant ferns, in addition to the substitution and subtraction of genetic constitutions during polyploidization and genome evolution (Sigel 2016) in sexual and apomict (Grusz 2016) ferns. Ferns have potential molecules, proteins and genes that can help to develop the transgenic plants. Only thing is that we have to investigate those ferns of tremendous potentiality and infuse their ability to combat against various challenges in plant diseases, proteomics, genomics and genome evolution.

4.14 Conclusion

Gametophytes of homosporous ferns are usually bisexual with undecided sexual expression or occasionally apomicts behaviour. Gametophyte exhibits genetic barriers at different developmental stages and may be of homozygous, heterozygous or homoeozygous genotype. They are short-lived, autotrophous independent, haploid organism and can serve as source of model organism for molecular study. They have tremendous ability of clone formation and, thus, can be grown and multiplied either through spores or explant culture in in vitro conditions. Clone is a replica of parent gametophytes with homozygous alleles; thus, they could be helpful in investigating mating system and genetic variation between progenitors and divergent extant ferns, including genome evolution. Gametophytes can produce sporophyte by sexual fusion between the gametes produced on parent or clone gametophytes. Proper transplantation of sporophytes for hardening and acclimatization may enhance the possibility of sporophyte production in higher rate. Mass propagation and multiplication of ferns make availability of their biomass for future prospection, particularly for isolation of biomolecules, proteins and gene to be used to develop transgenic crop plants in future.

Acknowledgements Authors are indebted to Director, CSIR-National Botanical Research Institute, Lucknow, for extending the facilities of research and development activities. Thanks also to Prof. Eelena M. F. Gonzalez, the editor of the book, for inviting us to contribute in this volume and also for her generous support during preparation of the chapter. Council of Scientific and Industrial Research (CSIR), New Delhi, is greatly acknowledged for financial support.

References

Albaum HG (1938) Normal growth, regeneration and adventitious outgrowth formation in fern. Am J Bot 25:37–44

Amaki W, Higuchi H (1991) A possible propagation system of *Nephrolepis*, *Asplenium*, *Pteris*, *Adiantum* and *Rumohra* (Arachniodes) through tissue culture. Acta Hortic 300:237–243

Ambrozic-Dolinsek J, Camloh M (1997) Gametophytic and sporophytic regeneration from bud scales of the fern *Platycerium bifurcatum* (Cav.) C. Chr. *in vitro*. Ann Bot 80:23–28

Atallah NM, Banks JA (2015) Reproduction and the pheromonal regulation of sex type in fern gametophytes. Front Plant Sci 6:1–6

Banks JA (1997) Sex determination in the fern *Ceratopteris*. Trends Plant Sci 2:175–180

Banks JA (1999) Gametophyte development in ferns. Ann. Rev. Plant Physiol Plant Mol Biol 50:163–186

Barker MS, Wolf PG (2010) Unfurling fern biology in the genomics age. Bioscience 60(3):177–185

Behera SK, Rawat VK, Singh AP, Khare PB (2011) Studies on the spore germination, developmental pattern and sexuality of gametophytes in *Dipteris wallichii* (R. Br. *ex* Hook. *et* Grev.) T. Moore. Indian Fern J 28:172–178

Brown EDW (1923) Apogamy in *Phegopteris polypodioides*. Bull Torrey Bot Club 50:17–34

Chiou WL, Farrar DR (1997) Comparative gametophyte morphology of selected species of the family Polypodiaceae. Am Fern J. 87:77–86

Chiou WL, Huang YM, Chen CM (2006) Conservation of two endangered ferns, *Angiopteris somai* and *A. itoi* (Marattiaceae: Pteridophyta), by propagation from stipules. Fern. Gazette 17:271–278

Comai L (2005) The advantages and disadvantages of being polyploid. Nat Rev Genet 6(11):836–846

Cooke RC (1979) Homogenization as an aid in tissue culture propagation of *Platycerium* and *Davallia*. Hort Sci 14:21–22

Cronk QCB (2009) The molecular organography of plants. Oxford University Press, Oxford

De Groot GA, Verduyn B, Wubs ERJ, Erkens RHJ, During HJ (2012) Inter and intra-specific variation in fern matins systems after long-distance colonization: the importance of selfing. BMC Plant Biol 12:1–12

De Soto L, Quintanilla LG, Mendez M (2008) Environmental sex determination in ferns: effects of nutrient availability and individual density in *Woodwardia radicans*. J Ecol 96:1319–1327

Der JP, Barker MS, Wickett NJ, dePamphilis CW, Wolf PJ (2011) *De novo* characterization of the gametophyte transcriptome in bracken fern, *Pteridium aquilinum*. BMC Genomics 12:99

Dyer AF (1979) The culture of fern gametophytes for experimental investigation. In: Dyer AF (ed) The experimental biology of ferns. Academic, London, pp 253–305

Fauzi A, Aloysius DC and Zubaidah S (2016) The utilization of ferns as a model organism for studying natural polyploidization concept in genetic course. In: International conference on education. Universitas Negeri Malang, Indonesia

Fernandez H, Revilla MA (2003) *In vitro* culture of ornamental ferns. Plant Cell Tissue Organ Cult 73:1–13

Fernandez H, Bertrand AM, Sanchez-Tames R (1993) *In-vitro* regeneration of *Asplenium nidus* L. from gametophytic and sporophytic tissue. Sci Hortic 56:71–77

Fernandez H, Bertrand AM, Sanchez-Tames R (1999) Biological and nutritional aspects in fern multiplication. Plant Cell Tissue Organ Cult 56:211–214

Ferreira de Carvahlo J, Oplaat C, Pappas N, Derks M, de Ridder D, Verhoeven KJF (2016) Heritable gene expression differences between apomictic clone members in *Taraxacum officinale*: insights into early stages of evolutionary divergence in asexual plants. BMC Genomics 17:203

Ganders FR (1972) Heterozygosity for recessive lethal in homosporous fern populations: *Thelypteris palustris* and *Onoclea sensibilis*. Bot J Linn Soc 65(2):211–221

Goodenough U (1978) Genetics, 2nd edn. Holt, Rinehart & Winston, New York

Grusz AL (2016) A current perspective on apomixes in ferns. J Syst Evol 54(6):656–665

Haufler CH, Pryer KM, Schuettpelz E, Sessa EB, Farrar DR, Moran R, Schneller J Jr, Watkins JE, Windham MD (2016) Sex and the single gametophyte: revising the homosporous vascular plant life cycle in light of contemporary research. Bioscience 66:928–937

Hegde S, Kumar Menon V, Noronha R, D'Souza L (2006) Callus culture and an unconventional pattern of sporophyte regeneration in *Drynaria quercifolia* – a medical fern. In Vitro Cell Dev Biol Plant 42:508–513

Hennen GR, Sheenan TJ (1978) *In vitro* propagation of *Platycerium stemaria* (Beauvois) Desv. Hort Sci 13:245

Huang Q, Li W, Fan R, Chang Y (2014) New MADS-box gene in Fern: cloning and expression analysis of DfMADS1 from *Dryopteris fragrans*. PLoS One 9(1):1–6

Imai R, Tsuda Y, Matsumoto S, Ebihara A, Watano Y (2016) The relationship between mating system and genetic diversity in diploid sexual populations of *Cyrtomium falcatum* in Japan. PLoS One 11(10):1–17

Ito M (1962) Studies on the differentiation of fern gametophytes I. Regeneration of single cells isolated from cordate gametophytes of *Pteris vittata*. Bot. Mag. Tokyo 75:19–27

Jones CG, Firn RD (1978) The role of phytoecdysteroids in bracken fern *Pteridium aquilinum* Kuhn. As a defense against phytophagous insect attack. J Chem Ecol 4:117–138

Khare PB (2001) Regeneration of gametophytes in homosporous ferns. Indian Fern J. 18:11–14

Khare PB, Behera SK, Srivastava R, Shukla SP (2005) Studies on reproductive biology of a threatened tree fern, *Cyathea spinulosa* Wall. ex. Hook. Curr Sci 89(1):173–177
Klekowski EJ Jr (1969a) Reproductive biology of the Pteridophyta II. Theoretical considerations. Bot J Linn Soc 62:347–359
Klekowski EJ Jr (1969b) Reproductive biology of the Pteridophyta III. A study of Blechenaceae. Bot J Linn Soc 62:361–377
Klekowski EJ Jr, Lloyd RM (1968) Reproductive biology of the Pteridophyta I, general considerations and a study of *Onoclea sensibilis* L. J Linn Soc Bot 60:315–324
Lazar A, Cerasela P, Sorina P (2010) Studies concerning the "*in-vitro*" multiplication of *Nephrolepis exaltata* "Fluffy Ruffles.". J Hortic For Biotechnol 14(3):191–193
Leroux O, Eeckhout S, Viane RLL, Popper ZA (2013) *Ceratopteris richardii* (C-Fern): a model for investigating adaptive modification of vascular plant cell walls. Front Plant Sci 4:1–8
Maeda M, Ito M (1981) Isolation of protoplasts from fern prothallia and their regeneration to gametophytes. Bot Mag Tokyo 94(1):35–40
Makowski D, Tomiczak K, Rybczynski JJ, Mikula A (2016) Integration of tissue culture and cryopreservation methods for propagation and conservation of the fern *Osmunda regalis* L. Acta Physiol Plant 38:19
Markham K, Chalk T, Stewart N Jr (2006) Evaluation of fern and moss protein based defences against phytophagous insects. Int J Plant Sci 167:111–117
Mikula A, Rybczynski JJ (2006) Preliminary studies on cryopreservation and strategies of *Cyathea australis* gametophyte development *in vitro*. Bot Guidebook 29:133–142
Mikuła A, Pożoga M, Tomiczak K, Rybczyński JJ (2015) Somatic embryogenesis in ferns: a new experimental system. Plant Cell Rep 34(5):783–794
Nair AG, Pradessh S, Nikhila GS, Sangeetha G, Mini I, Swapna TS (2013) *In vitro* propagation of a rare medicinal fern of western Ghats – *Diplazium esculentum* (Retz.) Sw. Ind J Exp Biol 51:919–923
Nakazato T, Jung M-K, Housworth EA, Rieseberg LH, Gastony GJ (2006) Genetic map-based analysis of genome structure in the homosporous fern *Ceratopteris richardii*. Genetics 173:1585–1597
Nakazato T, Jung MK, Housworth EA, Rieseberg LH, Gastony GJ (2007) A genomewide study of reproductive barriers between allopatric populations of a homosporous ferns, *Ceratopteris richardii*. Genetics 177:1141–1150
Nayar BK, Kaur S (1971) Gametophytes of homosporous ferns. Bot Rev 37:295–396
Oh MJ, Kim C, Na HR, Shin H, Liu JR, Choi HK, Kim SW (2014) High frequency sporophytes regeneration from the spore culture of the endangered aquatic Fern *Isoetes coreana*. Am J Plant Sci 4:14–20
Okamoto H, Sakamoto K, Tomizawa K-I, Nagatani A, Wada M (1997) Photoresponses of transgenic *Arabidopsis* overexpressing the fern *Adiantum capillus-veneris* PHY1. Plant Physiol 115:79–85
Ootsuki R, Shinohara W, Suzuki T, Murakami N (2011) Genetic variation in the apogamous fern *Cyrtomium fortune* (Dryopteridaceae). Acta Phytotaxonomica Geobotanica 62:1–14
Otto SP (2007) The evolutionary consequences of polyploidy. Cell 131(3):452–462
Peredo EL, Mendez-Couz M, Revilla M, Fernandez H (2013) Mating system in *Blechenum spicant* and *Dryopteris affinis* ssp. *affinis* correlates with genetic variability. Am. Fern J. 103:27–39
Pinson JB, Chambers SM, Nitta JH, Kuo LY, Sessa EB (2017) The separation of generations: biology and biogeography of long-lived sporophytes less fern gametophytes. Int J Plant Sci 178(1):1–18
Plackett ARG, Rabbinowitsch EH, Langdale JA (2015) Protocol: genetic transformation of the fern *Ceratopteris richardii* through microparticle bombardment. Plant Methods 11(37):1–10
Raghvan V (1989) Developmental biology of fern gametophytes. Cambridge University Press, Cambridge, Great Britain, pp 1–352

Ravi BX, Varuvel GVA, Kilimas R, Robert J (2015) Apogamous sporophyte development through spore reproduction of a South Asia's critically endangered fern: *Pteris tripartita* Sw. Asian Pac J Reprod 4(2):135–139
Reynolds TL (1979) Apical dominance in *Anemia phyllitidis* gametophytes. Am Fern J 69:92–94
Reynolds TL (1981) Effects of auxin and abscisic acid on adventitious gametophyte formation by *Anemia phyllitidis* (L.) Sw. Z Pflanzenphysiol 103:9–13
Robertson FW, (2002) Reproductive Behavior of Cloned Gametophytes of Pteridium aquilinum (L.) Kuhn. American Fern Journal 92 (4):270–287
Ruiz-Estevez M, Bakkali M, Martin-Blazquez R, Garriso-Ramos MA (2017) Identification and characterization of *TALE* homeobox genes in the endangered fern *Vandenboschia speciosa*. Genes 8(10):1–14
Rybczynski JJ, Mikula A (2011) Tree fern biotechnology: from spores to sporophytes. In: Fernandez H et al (eds) Working with ferns: issues and applications. Springer, New York, pp 135–147
Schaufelberger D, Hostettmann K (1983) On the molluscicidal activity of tannin containing plants. Planta Med 48:105–107
Sessa EB, Der JP (2016) Evolutionary genomics of ferns and lycophytes. Adv Bot Res 78:215–254
Sessa EB, Banks JA, Barker MS, Der JP, Duffy AM, Graham SW, Hasebe M, Langdale J, Li FW, Marchant DB, Pryer KM, Rothfels CJ, Roux SJ, Salmi ML, Sigel EM, Soltis DE, Soltis PS, Stevenson DW, Wolf PG (2014) Between Two Fern Genomes. Giga Science 3:15
Sheffield E, Douglas GE, Hearne SJ, Huxham S, Wynn JM (2001) Enhancement of fern spore germination and gametophyte growth in artificial media. Am Fern J 91:179–186
Shekhawat MS, Manokari M (2015) Direct organogenesis from rhizome explants in *Marsilea quadrifolia* L.: a threatened fern species. Adv Biol 2015:1–6
Shukla SP, Khare PB (2014) *In vitro* conservation of some threatened and economically important ferns belonging to the Indian subcontinent. J Bot 2014:1–8
Shukla AK, Upadhyay SK, Mishra M, Saurabh S, Singh R, Singh H, Thakur N, Rai P, Pandey P, Hans AL, Srivastava S, Rajapure V, Yadav SK, Singh MK, Kumar J, Chandrashekhar K, Verma PC, Singh AP, Nair KN, Bhadauria S, Wahajuddin M, Singh S, Sharma S, Omkar Upadhyay RS, Ranade SA, Tuli R, Singh PK (2016) Expression of an insecticidal fern protein in cotton protects against whitefly. Nat Biotechnol 34:1046–1051
Sigel EM (2016) Genetic and genomic aspects of hybridization in ferns. J Syst Evol 54(6):638–655
Singh AP, Johari D, Singh A, Behera SK, Khare PB (2012) Studies on regeneration of gametophytes and mass multiplication of *Anemia rotundifolia* Schrad. (Pteridophyte). Int J Curr Res 4(12):235–240
Singh AP, Johari D, Singh A, Khare PB (2013) Reproductive biology of a rare fern species *Adiantum peruvianum* (Adiantaceae: Pteridophyta). Phytomorphology 63(1&2):105–114
Singh AP, Johari D, Khare PB (2017) Studies on ontogeny and reproductive behaviour of *Lepisorus nudus* (Hook.) Ching (Polypodiaceae). J Plant Res 130(2):281–290
Smith AR (1972) Comparison of fern and flowering plant distributions with some evolutionary interpretations for ferns. Biotropica 4:4–9
Soare LC (2008) *In vitro* development of gametophyte and sporophyte in several fern species. Not Bot Hort Agrobot Cluj 36(1):13–19
Somer M, Arbesu R, Menendez V, Revilla MA, Fernandez H (2010) Sporophytic induction studies in ferns *in-vitro*. Euphytica 171:203–210
Srivastava R, Uniyal PL (2014) Developmental studies in an ornamental fern *Polystichum lentum* (D. Don) Moore. Natl Acad Sci Lett 37(3):281–284
Szypula W, Pietrosiuk A, Suchocki P, Olszowska O, Furnamowa M, Kazimierska O (2005) Somatic embryogenesis and *in vitro* culture of *Huperzia selago* shoots as a potential source of huperzine A. Plant Sci 168:1443–1452
Tanurdzic M, Banks JA (2004) Sex-determining mechanisms in land plants. Plant Cell 16:S61–S71
Thakur RC, Hosoi Y, Ishii K (1998) Rapid *in-vitro* propagation of *Matteuccia struthiopteris* (L.) Todaro – an edible fern. Plant Cell Rep 18:203–208

Tryon RM, Tryon AF (1982) Ferns and allied plants with special reference to tropical America. Springer, New York/Heidelberg/Berlin, pp 1–357

Wee YC, Kwa SH, Loh CS (1992) Production of sporophytes from *Platycerium coronarium* and *P. ridleyi* frond strips and rhizome pieces cultured *in vitro*. Am Fern J. 82:75–79

Wolf PG, Sessa EB, Marchant DB, Li FW, Rothfels CJ, Sigel EM, Gitzendanner MA, Visger CJ, Banks JA, Soltis DE, Soltis PS, Pryer KM, Der JP (2015) An exploration into fern genome space. Genome Biol Evol 7(9):2533–2544

Wubs ERJ, De Groot GA, During HJ, Vogel JC, Grundmann M, Bremer P, Schneider H (2010) Mixed mating system in the fern *Asplenium scolopendrium*: implications for colonization potential. Ann Bot 106:175–183

Chapter 5
Morphogenic Events in Ferns: Single and Multicellular Explants In Vitro

Jan Jarosław Rybczyński, Karolina Tomiczak, Małgorzata Grzyb, and Anna Mikuła

Abbreviations

0.5MS medium	½ Strength MS medium
2,4,5-T	2,4,5-Trichlorophenoxyacetic acid
2,4-D	2,4-Dichlorophenoxyacetic acid
2-iP	N^6-(2-Isopentenyl adenine)
AC	Active charcoal
BA	6-Benzylaminopurine
CH	Casein hydrolysate
CM	Coconut milk
CPW solution	Mineral salt complex (Frearson et al. 1973)
GA_3	Gibberellic acid
JA	Jasmonic acid
MFMM	Murashige Fern Multiplication Medium
MS medium	Murashige and Skoog medium (1962)
NAA	α-Naphthaleneacetic acid
SE	Somatic embryo
SERK	*Somatic embryogenesis receptor kinase* gene
TDZ	Thidiazuron-1-phenyl-3-(1,2,3-thiadiazol-5-yl) urea

5.1 Introduction

Ferns undergo both sexual and asexual (apogamy or apospory) reproduction, and in vitro culture conditions extend these possibilities. Under these conditions, the fern body provides a source of various explants, both single-celled and

J. J. Rybczyński (✉) · K. Tomiczak · M. Grzyb · A. Mikuła
Department of Experimental Botany, Polish Academy of Sciences, Botanical Garden – Center for Biological Diversity Conservation in Warsaw, Warsaw, Poland
e-mail: jjryb@obpan.pl

H. Fernández (ed.), *Current Advances in Fern Research*,
https://doi.org/10.1007/978-3-319-75103-0_5

Table 5.1 Comparison of the involvement of plant parts in both in vivo and in vitro vegetative propagation

No.	In vivo/in the habitat[a]	In vitro	References for in vitro studies
1.	Stolons subterranean and aerial	Somatic embryos	Mikuła et al. (2015), Li et al. (2017)
2.	Bulbils on leaf rachis	Green globular bodies	Amaki and Higuchi (1991), Bertrand et al. (1999), Liao and Wu (2011), Yu et al. (2017)
3.	Creeping rhizomes	Gametophytes	Goller and Rybczyński (1995, 2007)
4.	Proliferous leaves	Leaf primordium	Shukla and Khare (2012a, b)
5.	Division of the leafy crown of the plant	Rhizome	Hegde et al. (2006)
6.	Tassels	Shoots and their organs[b]	Mikuła et al. (2015)
7.	Auricles	Callus	Kwa et al. (1995, 1997), Hegde et al. (2006), Packett et al. (2014)
8.	Tubers	Callus tissue on homogenized gametophytes	Menéndez et al. (2006)
9.	Propagation from spores	Meristemoids	Shukla and Khare (2012a)
10.	Aerial layering	Cell suspensions	Bryne and Caponett (1992), Teng (1997), Teng and Teng (1997)
11.	Root buds	Single cells	Kato (1964)
12.		Protoplasts	Redford et al. (1987), Maeda et al. (1990)
13.		Leaves	Bristow (1962)
14.		Stolon pieces	Ambrósio Melo (2004)
15.		Scales	Ambrozič-Dolenisek et al. (1999)
16.		Knots	Davidson et al.(2008)
17.		Epidermal cells of excised leaves	Hirsch (1975, 1976)
18.		Root culture	Munroe and Sussex (1969)
19.		Epidermal hairs developed gametophyte	Martin et al. (2006)
20.		Shoot apices	Hicks and Aderkas (1986)
21.		Disinfected rhizomes	Bertrand et al. (1999)
22		Polyembryony	Ravi (2016)
23.		Circinate part of young leaves	Salome and Pais (1987)
24.		Gemmae	Emigh and Farrar (1977)
25		Green sporangia (sori)	Soare et al. (2010)
26		Frond, frond micro-cuttings	Avila-Peréz (2011)
		Shoot apex	Plackett et al. (2014, 2015)

[a]The content of this part of the table was prepared based on Chap 16 of *Encyclopedia of ferns* by D.L. Jones

[b]Stipe, internode, shoot apical meristem

Fig. 5.1 Bulbils regenerated on the midrib of *Tectaria cicutaria* leaf (Image was taken by the authors in a greenhouse at the Polish Academy of Sciences, Botanical Garden-CBDC in Warsaw, Poland)

multicellular (Table 5.1), and these have been explored in regeneration experiments and in more advanced experiments involving plant genome manipulation (Plackett et al. 2015). Many similarities can be recognized between spermatophytes and spore-producing plants; however one significant difference is the absence of secondary growth in the latter.

The majority of published papers concerning fern tissue culture and biotechnology stress the importance of an alternation of generations between a haploid gametophyte and a diploid sporophyte in the natural life cycle of a fern, and the egg cell is fertilized within the archegonium. Each generation commences with the development of a single cell; in the case of the gametophyte, this is the haploid spore, and in the case of the sporophyte, it is the diploid zygote. Both generations of the fern are able to survive independently of each other (Bui et al. 2012). They are very adaptable to a wide range of climatic conditions and thus vary greatly in their morphology, from annual and herbaceous to perennial tree ferns. A number of structures may be produced by the sporophyte that enable vegetative propagation. These include modification of the shoot rhizomes, or the leaves, which may bear bulbils. These bulbils are formed on the main midrib of the leaf (Fig. 5.1). Once disseminated, on reaching the ground, and under favorable conditions, they continue to grow with further development of the initial leaf and the formation of roots, resulting in the establishment of a new individual and independent plant.

In vitro conditions reveal the additional morphogenetic potential of both generations. Thera are numerous papers describing the fern multiplication system in vitro basing on culture of spores, multiplication of gametophyte (Camloh and Gogala 1992), and spontaneous syngamy finally giving sporophytes. Various organs of the sporophyte produce green globular bodies (see later). Very rarely, scales from the juvenile sporophyte (*Platycerium bifurcatum*) have been used for morphogenesis experiments in ferns.

When whole sporophytes of *Asplenium nidus* and *Pteris ensiformis* were cultured on MS medium supplemented with 4.4 μM BA, the rhizomes became swollen due to bud proliferation. Isolated and homogenized rhizomes were plated on MS hormone-free medium. In *Pteris ensiformis*, the rhizome explants developed into sporophytes, but for *Asplenium nidus* both gametophyte and sporophyte regeneration occurred. The system was quite effective giving approximately 500 sporophytes from homogenates of 0.5 g of BA-treated rhizomes (Fernández et al. 1997a). Isolated shoot apices of *Matteuccia struthiopteris* cultured in the presence of kinetin developed in multibud structures giving the first step of the species multiplication on Knudson's medium (Hicks and von Aderkas 1986). Ambrozič-Dolenisek et al. (1999) have reported the development of sporophytes following adventitious shoot formation when scales were cultured on MS medium in the absence of plant growth regulators. Also, epidermal cells of the excised leaf of the fern *Microgramma vacciniifolia* produced both aposporous gametophytes and sporophytes (Hirsch 1975, 1976). Sporophytes were produced by initiation of cell division in the adaxial epidermis cells and the internal proliferation of epidermal derivatives, resulting in the formation of an embryo-like structure. The first leaf of this structure emerged precociously through the abaxial epidermal surface (Hirsch 1976). Later histological analysis of indirect somatic embryogenesis in pteridophytes was given by Atmane and co-workers (2000) for *Lycopodiella inundata*.

On the other hand, large numbers of adventitious gametophytes and single-celled gemmae are produced on a very simple medium, being these gemmae initiated by meristematic activity of marginal cells of the gametophyte (Emigh and Farrar 1977; Fernández et al. 1997b).

The callus tissue of ferns can contain cells representing both generations and, if derived from cell suspensions and protoplasts that have the morphogenic capacity to regenerate both gametophyte and sporophyte, may be used as the initial stage for plant cell manipulation (Kwa et al. 1997; Byrne and Caponetti 1992; Teng and Teng 1997; Kwa et al. 1997; Joyce et al. 2014). More complex regeneration systems appeared in in vitro cultures of ferns during sporophytic bud regeneration in aposporous gametophytes. Regeneration of transformants was observed for shoot apex-derived callus (Plackett et al. 2014, 2015) modified by microparticle bombardments using 35S:*GUS* plasmid pCAMBIA1305 (Cambia) (Plackett et al. 2014).

The in vitro phenomenon of somatic embryogenesis has intrigued the experimental biologist since the 1950s, firstly in terms of the structural changes and morphogenic processes involved, but, more recently, in terms of genetic control (Ledwoń and Gaj 2011; Gaj 2011) and proteomics (Domżalska et al. 2016). The beginning of the present century saw the publication of the description of somatic embryogenesis in the cryptogamic plants *Lycopodiella inundata* (Atmane et al. 2000) and *Huperzia selago* (Szypuła et al. 2005). Later, mitotic activity of one cell was indicated in the culture of *Argyrochosma nivea* gametophyte, giving proliferative area from which

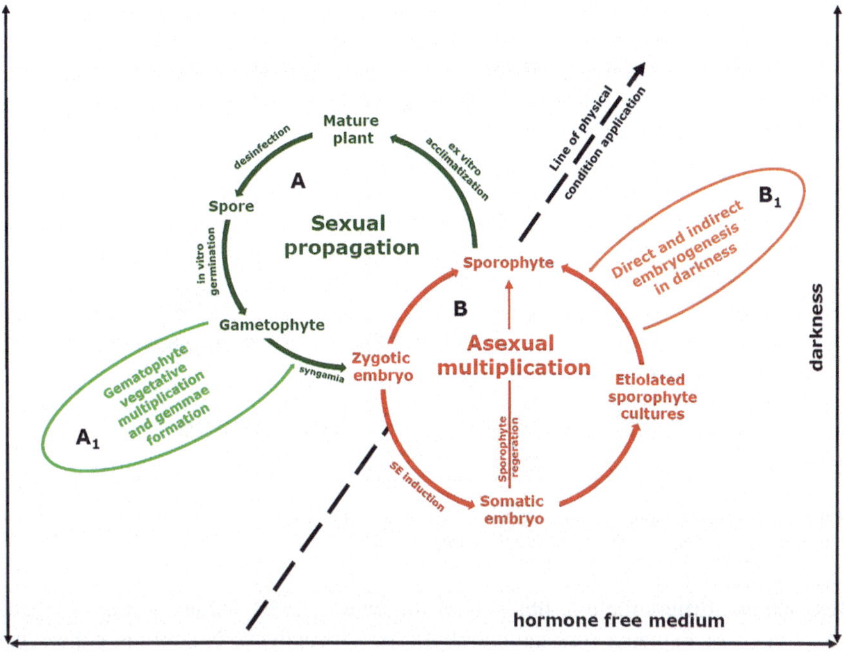

Fig. 5.2 Diagram showing somatic embryogenesis position in sexual and asexual reproduction in fern

finally apogamous sporophytes were formed (Galan 2011). In our lab, more recently somatic embryogenesis of tree ferns (*Cyathea delgadii*) has been described (Fig. 5.2) involving plant regeneration from a single epidermal cell on hormone-free medium in darkness. In the near future, this may provide a new avenue for fern somatic cell genetic manipulation (Mikuła et al. 2015).

5.2 Callus Induction, Maintenance, and Its Morphogenic Potential

Information about spontaneous callus formation in gametophytes was first provided by Morel and Wetmore (1951) for *Osmunda cinnamomea* grown on agar medium containing inorganic salts, sugar, and vitamins. Gametophytes of *Pteridium aquilinum* were the second to show experimental induction of callus formation in the presence of sugar and yeast extract (Sussex and Steeves 1953).

The gametophytes and sporophytes that develop from these are very different in their morphology. Recognizing these main differences is fundamental to the creation of culture conditions for inducing and maintaining callus tissue and for the expression of its morphogenic potential (Fig. 5.3). Both generations appear not to undergo secondary growth. In order to better convey the details of callus induction, its main-

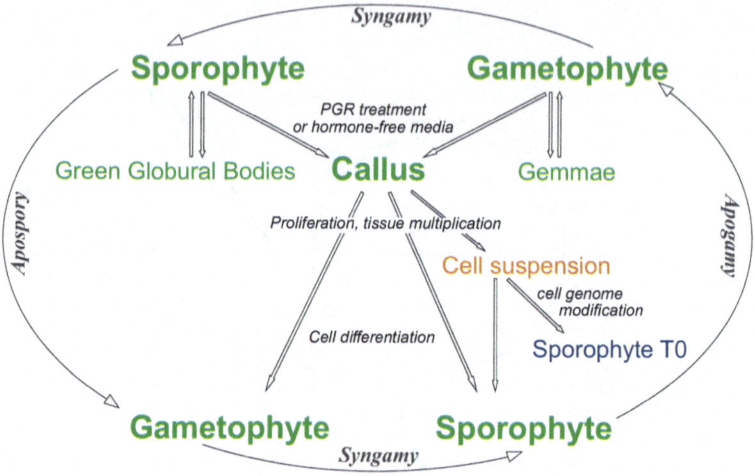

Fig. 5.3 Diagram indicates the various relationships that exist between generations, the callus being the central subject for all fern cell manipulations

tenance and differentiation, this part of the review is divided according to whether the source of explants was gametophytic or sporophytic. Numerous papers have appeared in the literature regarding the development of particular species, with special emphasis on morphogenic events, the initiation of spore germination, prothallial development and gametophyte proliferation, and the development of a relatively undifferentiated multicellular body. The photosynthetic surface comprises active cells that are able to relinquish their prime function and start to self-replicate, resulting in the formation of callus tissue and the potential to form a new generation.

5.2.1 *Gametophytic Source of Explants*

Pteris vittata (Joyce et al. 2014) and *Adiantum capillus-veneris* (Maridass et al. 2010) were the fern species selected for callus induction and plant regeneration from callus tissue initiated in fully developed prothalli. Prothalli of *P. vittata* were transferred onto callus induction medium consisting of 0.5MS medium supplemented with 20 g l^{-1} of sucrose, 1.4 μM gibberellic acid (GA_3), and 2.2 μM BA, known as Yang medium. The callus tissue contained two types of cells: the first were actively dividing whereas the other contained a single large vacuole. Sporophytes regenerated from callus asynchronously form clusters of cells in a manner apparently analogous to direct organogenesis. Extracellular matrices were observed in actively growing callus and at the base of regenerating sporophytes (Joyce et al. 2014). Callus induction of *A. capillus-veneris* prothalli occurred when grown on MS medium supplemented with 1.5 μM 2,4-D + 2.0 μM BA, i.e., a commonly used combination of plant growth hormones. A combination of two auxins, namely,

2,4-D and IBA, at concentrations of 0.75 and 1.50 μM, respectively, produced unexpected and rather peculiar results (Maridass et al. 2010).

5.2.2 *Sporophytic Source of Explants*

For *Cyathea spinulosa*, Parker and Thompson medium supplemented with 9.04 μM 2,4-D and 2.21 μM BA was found to be the most suitable for the induction of deep-yellow to greenish callus tissue from the cut surface of leaf primordium explants excised from in vitro-raised sporophytes. After 2 weeks, excellent growth and development of callus and differentiation occurred on complex media such as P&T supplemented with 2,4-D, BA, and NAA. The callus induction medium (with 9.04 μM 2,4-D and 2.21 μM BA), during subculture, promoted the formation of meristemoid zones instead of shoot differentiation (Shukla and Khare 2012a). The best responses (80% and 70%) for green and compact callus induction in *Pteris vittata* were obtained for explants grown for 2 weeks on full strength MS and P&T media, respectively, supplemented with 2.26 μM 2,4-D and 2.22 μM BA (Shukla and Khare 2012b). The use of lower concentrations of mineral salts in both of the media mentioned above resulted in the formation of a completely different type of callus of pale green coloration, characterized by GGB regeneration (see below). Immature, excised leaf blades of greenhouse-grown *Adiantum capillus-veneris* regenerated callus tissue on MS medium supplemented with 2.2 μM BA+2,4-D that was capable of shoot formation. Shoots appeared from meristemoid formations on the callus in the presence of 2.2 μM BA in 0.5 MS medium (Li et al. 2015).

Following the subculture of pieces of calli on Parker and Thompson medium containing different concentrations of 2,4-D, BA, and NAA, both alone and in combinations, 12.5 ± 0.45 shoots measuring 8.3 ± 0.6 cm in length were produced with 4.43 μM BA and 5.36 μM NAA after 4 weeks, as had previously been reported for the same combinations of growth regulators (Shukla and Khare 2012a). These results were confirmed by culturing *Pteris vittata*, with BA, which was found to be the most useful cytokinin for the experimental regeneration of tree ferns.

Palta and Mehra (1983) described remarkable callus function in experiments, thereby demonstrating the possibilities of ploidy manipulation with the help of tissue induction and plant regeneration for consecutive ploidy levels (2*x*, 4*x*, 8*x*) for *Pteris vittata* grown in vitro. The morphogenic potential of callus was explored at various ploidy levels for the production of gametophytes in the order from sporophytes via callus to gametophyte with increasing chromosome number for each cycle of callus induction and gametophyte regeneration. Figure 5.3 summarizes the various relationships that exist between generations, the callus being the central subject for all fern cell manipulations.

5.3 Sporophytic Source of Explants

Published results provide a full description of the morphogenic events that occur in culture, but that never happen in nature, for example, the formation of globular masses of cells or direct embryo formation by proliferation of photosynthetic gametophyte cells.

An exception is the formation of sporophytes from spores of *Equisetum arvense*, as described by Kuriyama and Maeda (1999). The spores, cultured as a haploid explant in the presence of a low concentration of BA, developed into globular cell masses after 2–3 weeks of culture. Each globular mass was composed of various types of cells, but predominantly consisted of small cells with dense cytoplasm. After 2 months of culture in the presence of 1 μM BA, the masses underwent morphogenic changes via sporophyte formation, and regenerants were obtained. In the case of *Dryopteris affinis*, the culture initiated by the transfer of sori onto hormone-free 0.5 MS medium resulted in spore germination, prothallus formation, and direct apogamous embryo regeneration by the gametophyte. The embryo was very easily recognized due to the lack of chloroplasts in its cells and, hence, its pale coloration, a characteristic and specific trait of embryonic tissue (Soare et al. 2010).

5.4 Proliferation of Secondary and Tertiary Gametophytes

Single-celled spores are excellent experimental material, but have limited shelf-life, and therefore, their percentage germination and percentage gametophyte survival under natural, unfavorable conditions are very limited. In vitro culture conditions, however, have opened up for numerous fern species the possibility of producing an unlimited supply of gametophytes under aseptic conditions. Since the spores germinate easily in vitro, this technique has been used to study different aspects of fern spore germination and growth and development of both gametophytes and sporophytes. Our experiences of in vitro culture of ferns indicate that gametophyte culture and multiplication of both tree ferns and herbaceous ferns do not pose any particular experimental problems (Goller and Rybczyński 1995; Makowski et al. 2016). However, the gametophyte is an excellent subject for this work, since it is haploid, arises from a single cell, and has a marginal meristem comprising a chain of single cells that under specific conditions can undergo cell division, finally producing a new gametophyte, for example, numerous species of *Cyathea* taxa (Goller and Rybczyński 2007). This phenomenon has been described among others for *Phlebodium aureum* and *Pronephrium articulatum*, with excellent photographic documentation. Under control conditions, secondary and tertiary gametophyte formation followed the same path as the initial gametophyte (parent) obtained directly by spore germination, in terms of initiation and development. However, in the case of *Phlebodium aureum*, these generations can be distinguished by their cordate and irregularly lobed prothalli (Johnson and Manickam 2006, 2011).

5.5 Tissue Culture in Liquid Media

Over many years of scientific inquiry, tissue culture techniques have developed so as to enable researchers to carry out experiments on small pieces (fragments) of plant body and even single cells. The introduction and use of liquid media in culturing helped to establish cell suspension cultures, which for many species of seed plants permitted an alternative approach, namely, the use of protoplasts. The culture of protoplasts having morphogenic potential appeared to be the main objective for various plant cell manipulation techniques, including the transformation and modification of the genome, especially with the exploration of electroporation (Wójcik and Rybczyński 2015), electrofusion (Tomiczak et al. 2017), or genome editing (Weinthal and Gürel 2016). In ferns, alternation of generations of distinct morphology resulted in the early abandonment of such studies, and nowadays, progress in fern biotechnology is limited to only a very few species. Not all published data fulfill the conditions required to characterize cell suspension cultures of ferns. Numerous papers describing the development and establishment of a tissue culture system for ferns emphasize the pharmacological value of studying cryptogams (Bienaime et al. 2015). A long history of the use of liquid media in the culture of cell suspensions has resulted in the establishment of two main types of cell suspensions defined as differentiated (completed by cells possessing a different shape, large vacuoles, and high rates of secondary metabolite production) and embryogenic (composed of small cells having dense cytoplasm with prominent starch grains and that form aggregates and carrying embryogenesity). Developing differentiated cell suspensions of particular species helps us to protect those species from the brutal exploitation of their habitats in response to increased demands by the market for plant material such as ingredients. Embryogenic cell suspension, on the other hand, as an alternative system of cell culture, could be used for somatic cell genetic manipulation of ferns. The number of species used and the level of plant cell manipulations employed for ferns are very limited. Nevertheless, disseminating information about the present status of our knowledge concerning cell suspensions of ferns is worthwhile since, in contrast to considerable progress made in this field with regard to seed plants, only scattered reports are currently available for ferns on callus induction, the establishment of cell suspensions, protoplast culture, and fern regeneration from suspensions.

5.5.1 Callus and Cell Suspensions

The earliest paper published that concentrated on regeneration from isolated cells of fern gametophytes was that of Meyer (1953), followed, almost 10 years later, by that of Ito (1962).

Based on a study of the growth and differentiation of single cells isolated from callus tissue of *Pteris vittata* L., Kato (1964) developed a complex medium that was supplemented with various ingredients, depending on the developmental stage of the culture. Gametophytes cultured in the darkness proliferated and produced various types of callus, only one of which, a very friable one, enabled the establishment of a cell suspension. Numerous small pieces of tissue and a suspension of cell aggregates consisting of a few to hundreds of cells, as well as individual cells, constituted the suspension culture. The supernatant of the culture contained many isodiametric and small, free cells. These small cells contained abundant cytoplasm, starch grains, plastids, and other organelles. In order to investigate regeneration capacity, the suspension was placed on agar medium and maintained under white light. Green callus tissue differentiated and slowly grew to form gametophytes. A fraction of free cells formed normal gametophytes. The process of regeneration from isolated cells essentially resembled that from spores, with rhizoids developing first. Differentiation into gametophytes occurred on defined medium not supplemented with organic substances.

A paper describing the process of fern differentiation from cell suspension in *Nephrolepis exaltata* was published (Bryne and Caponetti 1992). For culture initiation, apical explants of stolon tips were selected to develop callus and to establish cell suspensions. Callus culture was initiated and established on agar medium containing 2 μM 2,4-D and 3% sucrose and subcultured many times. Due to the soft texture of callus tissue, its agitation in liquid medium produced a suspension of individual cells, and small multicellular aggregates were produced in the presence of Murashige Fern Multiplication Medium (MFMM) supplemented with 2 μM 2,4-D and 3% sucrose. The morphology of single cells varied both in shape and size, but small multicellular aggregates were mostly gourd-shaped. Single cell and cell aggregates formed numerous cell masses and callus clones when transferred to agar medium. Medium supplemented with kinetin and NAA supported the regeneration processes, eventually resulting in the formation of plantlets. Callus cultures, as the main source of cell suspensions, were derived from 2-month-old gametophytes of *Platycerium coronarium* cultured on MS medium supplemented with 20 μM 2,4-D and 2% sucrose as described by Kwa et al. (1995). Cell suspensions were prepared by transferring 0.2 g of callus tissue to 20 ml of liquid medium comprising 5 μM 2,4-D. Manipulation of medium ingredients resulted in the establishment of two kinds of callus, characterized by differences in their morphogenic potential and their capacity to regenerate. Sporophytes were regenerated by culturing pale-green-colored callus tissue in the presence of kinetin, and, following proliferation, maintaining the tissue on MS medium containing a reduced concentration of synthetic auxin (2 μM 2,4-D). Subculturing onto basal MS medium resulted in the development of sporophytes which contrasted with the dark-green, gametophyte-producing callus tissue. Gametophytes have also been used for callus induction in *Anogramma leptophylla* (Chemma 1983), *Nephrolepis cordifolia* (Sulklyan and Mehra 1977), and *Ceratopteris thalictroides* (Cheema and Sharma 1991). Furthermore, roots (Mehra and Palta 1971) and rhizome apices (Cheema 1983) of *Cyclosorus dentatus* and *Regnellidium diphyllum*, respectively, have been used to develop callus cultures with auxins and kinetin in the presence of sucrose.

Hegde and co-workers (2006) demonstrated an excellent system for the induction of callus and the regeneration of sporophytes from rhizomes of *Drynaria quercifolia* by the use of cytokinins and auxins. The callus was subcultured using solid and liquid media, and significant sporophyte regeneration was observed only on solid medium supplemented with 48.2 μM 2-iP. An atypical pattern of development followed, resulting in the formation of one-cell-thick, leafy structures, together with microrhizomes. A prolonged period of culture caused the above to produce in vitro a juvenile sporophyte. Cell suspension cultures were the only source of plant biomass used for pharmaceutical purposes, but these were not used as plant material for morphogenesis induction experiments.

A simple culture suspension of *Platycerium bifurcatum* was developed (Teng and Teng 1997; Teng 1997), where sporophytes could be regenerated directly from cells of macerated leaf blade or indirectly via an aposporous gametophyte stage, under the same culture conditions in the presence of MS medium supplemented with 0.54 μM NAA. A single cell is able to form regenerated gametophytes by means of two separate routes. Some single cells produce a bipolar protonema early on during culture, whereas most form a gametophyte cluster. During protonema development, distally located cells branch out in various directions and then grow into armlike structures. These structures also grow to form the next branches, which finally result in the formation of ribbonlike gametophytes that form apogamous sporophytes. Single cells or small aggregates (about 100 cells) are able to develop aposporous gametophytes which later gave rise to sporophytes. This phenomenon, like the natural life cycle of a fern, is limited to the regeneration of one sporophyte per gametophyte. Larger aggregates composed of 500–1000 cells form apogamous sporophytes directly. This system demonstrates that the regeneration patterns employed are closely related to the size of cell aggregates.

Bearing in mind the various cell suspensions referred to in the literature relating to fern biotechnology, we would like to contribute to all these previous publications by presenting here figures of our own work on the liquid culture of cell aggregates, the subject of our current research interest. Our purpose in doing this is to help others understand how new plants can arise from both cell suspension aggregates and single cells (Fig. 5.4).

The use of plant protoplast culture manipulation is well established for seed plants. In the case of ferns, for the last two decades of the twentieth century, the 1980s and 1990s, there was a revival of interest in spore-bearing plants, since their possession of two different generations made it possible to investigate two separate sources of various characters. Plant material comprising two generations with various ploidy levels for the same genome seemed an excellent subject for experimenting with plant genome expression along two morphogenic pathways. However, eventually, there was a decline in such work because the biochemical properties of fern cells complicated the release of the protoplast. Sometimes, mechanical treatment of plant tissue helped release protoplasts from the green, single-layered gametophyte and from the mesophyll of sporophyte leaves. It is also worth considering the increased efforts made to develop embryogenic cell suspensions. Recently published papers relating to the somatic embryogenesis of tree ferns represent a first

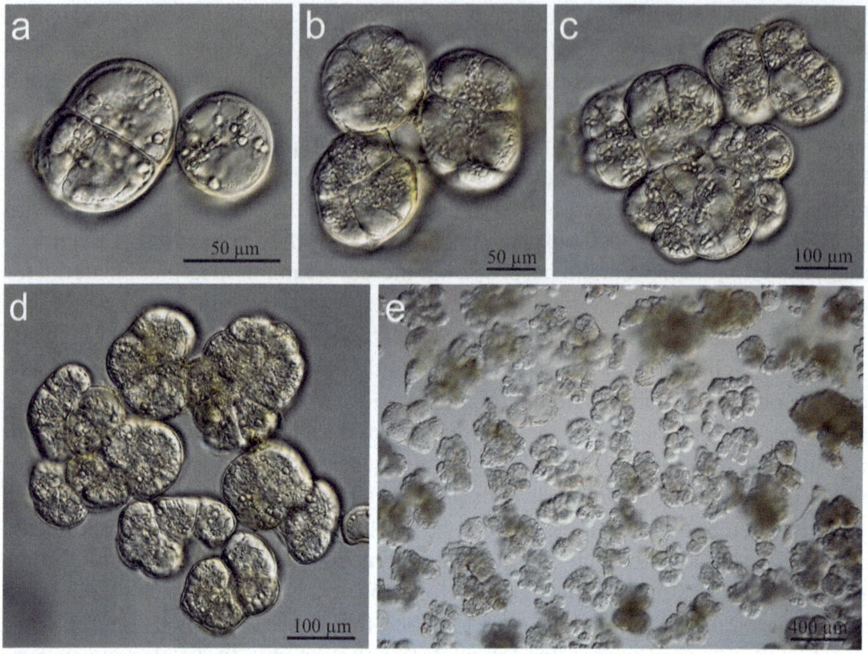

Fig. 5.4 Image of initial cell divisions of developing embryogenic cell suspension derived from root callus tissue of *Cyathea australis* in liquid medium 0.5 MS supplemented with 0.2 mg/l BA and 2.0 mg/l 2,4-D. Pictures of living specimens were taken with the help of Nomarski contrast microscopy: (**a**) Image of undivided cell and two sister cells (**b**) Image of three four to six cell aggregates showing cell division leading to form embryo-like structure of 2-month-old cell suspension culture (**c**) Storage materials cells of various shape cell suspension aggregates (**d**) Numerous cell aggregates of (embryogenic) cell suspension (**e**) General image of cell suspension aggregates of 2-week-old subculture

step in this direction. Our hope is that researchers working on ferns might find here important information and help for studying the development of in vitro cultures and the biotechnology of cereals, where the embryogenic character of cultures significantly advanced the biotechnology.

5.5.2 *Protoplast*

Protoplasts are the best model for studying morphogenic events leading to the acquisition of fully formed plants from single cells. In the case of ferns, the number of papers involving protoplast isolation and their subsequent culture is scant, and therefore, in this part of the chapter, protoplast experiments involving fern allies will also be considered. It is very interesting, and very characteristic, that fern protoplasts derived from young green prothalli of *Ceratopteris richardii* behave like single spores in terms of their first cellular division, which give rise to primary rhizoids after a period of at least 10–24 days of culture on agarose. Finally, after 29 days, planar growth of the prothallus is observed, and within 60 days of culture, the gametophytes reach sexual maturity (Edwards and Roux 1998).

Protoplast isolation was carried out for the first time by Klercker (1892) using a mechanical method on the plasmolyzed cells of tobacco leaf. Cocking (1960), however, used biochemical cell wall degradation with the help of cellulase to obtain plant protoplasts. Later, mostly enzymatic cell wall digestion was used for protoplast isolation. In the case of ferns, the first isolation of fern protoplasts was obtained for a callus-like mutant gametophyte of *Pteridium aquilinum* (Partanen et al. 1980). Microscope studies (light, scanning, and transmission electron microscopy) were used to describe protoplasts derived from the gametophyte and from gametophyte rhizoids of *Onoclea sensibilis* and *Pteridium aquilinum*, respectively. The age of the gametophyte appears to be a critical factor for protoplast isolation, the optimal age for protoplast isolation being between 10 and 12 weeks of in vitro culture (Breznovits 1994). The ultrastructure of studied protoplasts showed numerous structural differences related to their origin (Huckaby et al. 1982; Attree and Shefield 1985). Protoplasts were successfully isolated for the first time from the filamentous protonemal cells of ferns *Adiantum capillus-veneris* and *Pteris vittata*. Protoplasts were easily isolated from these protonemal cells by digestion of the cell wall with cellulase and pectinase. A total of $0.8–1.9 \times 10^4$ and $0.6–2.0 \times 10^4$ protoplasts were obtained of *Adiantum* and of *Pteris*, respectively. Viability, as judged by staining with fluorescein diacetate, was more than 90% for both species. Staining with 4′6-diamidino-2-phenylindole (DAPI) revealed that about half of the protoplasts of both species contained a nucleus (Kadota and Wada 1989). The gametophyte of *Lygodium japonicum* was the subject of a study that investigated the viable culture of protoplasts derived from green prothallial cells (Maeda et al. 1990) and 15–20-day-old gametophytes (Nakamura and Maeda 1995). For protoplast isolation, Cellulase Onozuka and Macerozyme R-10 in 0.6 M mannitol and CPW solution was used in both cases (Frearson et al. 1973). The protoplasts, grown in a culture medium containing 0.6 M mannitol and 0.05 M sucrose supplemented with NAA and BA, began to divide within a few days of culture, and after 30 days, cell clusters were observed. Sequential reduction in mannitol concentration resulted in the development of rhizoids and protonemata. Later, regeneration of a common gametophyte occurred, and subsequently the regenerated gametophytes produced sporophytic leaves and roots (Maeda and Ito 1981). The manipulation of sugars, a reduction in mannitol concentration, and protoplast-derived cell clusters cultured on hormone-free medium led to the formation of rhizoids and protonema-like regenerants from protoplasts derived from the juvenile leaflets of sporophytes of *Lygodium japonicum* (Maeda et al. 1990). Manipulation of BA and NAA concentrations aimed at stimulating the growth and development of cultured protoplasts is supported by supplementing the culture with inorganic and organic compounds. Jasmonic acid was the PGR used for sporophytic protoplast culture of the fern *Platycerium bifurcatum.* The lowest concentration of JA used, 0.01 μM, stimulated protoplast division giving significantly better results than controls after 8 days of culture. The higher concentrations used in the experiment did not contradict results obtained for the lowest JA concentration, suggesting that low concentrations of JA stimulate cell division. Nonorganic active charcoal (AC) is used to stimulate the growth of cultures of various plant tissues by absorbing harmful metabolites produced by tissues, cells, and protoplasts growing in vitro. By placing liquid medium over the surface of agar medium containing charcoal, the number of dividing green cells was increased. Active charcoal at 1.0% was sufficient for increasing both cell division and maintaining cell viability. Protoplasts growing in the presence of AC in the medium retained their chloroplasts and remained green in color, regardless of whether or not they divided. Stimulatory effects were exerted by AC only during the initial stages of protoplast culture, but AC is not necessary for the subsequent development of regenerated cells during the later stages of culture. Subculturing resulted in the formation of rhizoids, which developed into young gametophyte. On reaching maturity, the gametophytes formed sporophytic shoots in the presence of low concentrations of BA (Kuriyama et al. 1999).

Protoplasts (plant cells deprived of their cell walls) are considered ideal experimental models for increasing the genetic variation of somatic cells for gene transfer and somatic hybridization. Literature sources for fern somatic cell hybridization are extremely poor and, to the best of our knowledge, represented by a single paper presented by Breznovits (1994). This paper describes somatic hybridization using electrofusion experiments in *Pteridium aquilinum* streptomycin-resistant x-sensitive lines and other fern species (*Pteris vittata*, *Pteris cretica* cv. *albolineata*, *Polystichum aculeatum*, *Polystichum setiferum*, *Phyllitis scolopendrium*, *Asplenium ruta-muraria*). Interspecific hybridization of the species mentioned above resulted in hybrids that failed to survive due to chromosome incompatibility. The Fraction I protein analysis revealed that the fusion products selected for streptomycin resistance were viable hybrids containing chloroplasts obtained from the donor *Pteridium aquilinum*. These electrofusion experiments showed that cytoplasmic hybrids in ferns can be readily generated by protoplast fusion, in contrast to nuclear hybrids that are nonviable for most interspecific combinations (Breznovits 1994).

5.6 Green Globular Bodies (GGBs)

Table 5.1 compares two different systems of vegetative fern propagation. The life cycle in ferns is highly variable and depends on the habitat occupied. Generally, fern morphology varies little under in vitro conditions, regardless of species, as the plants are exposed to the same degree of stress; however, when deflasked, they assume the form characteristic of the species under consideration. The left side of Table 5.1 shows the natural potential for vegetative propagation of each fern when grown naturally, i.e., in vivo (Jones 1987), whereas the right side shows how each of these ferns propagates vegetatively in vitro. It seems that bulbils formed by leaves under natural conditions can be compared with green globular bodies (GGBs), the topic discussed here. Bulbils are formed naturally on sporophyte leaves and occur only in some species, but GGBs are formed in response to cytokinin in vitro and produced by unlimited cell division of sporophytes. The aim of the GGB experiments is commercial, namely, to develop an efficient system for vegetative propagation.

The term GGB was first introduced into experimental work on the life cycle of ferns by Higuchi and co-workers (1987) who worked on *Nephrolepis cordifolia*. Subsequently, authors extended the list of species used for the experimental initiation of GGB formation as a simple means for propagating ferns vegetatively in vitro. The system was developed for *Asplenium nidus*, *Pteris ensiformis*, *Adiantum raddianum*, and *Rumohra adiantiformis*, with special emphasis on the cultivation of two cultivars of *Nephrolepis cordifolia* and nine cultivars of *Nephrolepis exaltata* for commercial purposes (Amaki and Higuchi 1991). The main factor involved in GGB initiation was the cytokinin BA at a concentration of 4.4 μM MS medium which was also supplemented with 2% sucrose. There was a great difference in GGB growth rate between species. The various requirements for in vitro cultivation were reflected by the degree of proliferation and plantlet regeneration in particular species. Fresh weight as an indicator of variation revealed that GGBs of *Pteris ensiformis* (Fernandez et al. 1996) grew rapidly, whereas those of *Asplenium nidus* and *Adiantum raddianum* grew very slowly and did not exceed a twofold increase in weight. Variation in morphology of the GGB surface was surprising, as it was characteristic for each individual species. GGB development required hormone-free medium that promoted sporophyte regeneration (Liao and Wu 2011).

As mentioned above, GGBs have been developed for *Polypodium cambricum* in studies of various explants that also originated from young sporophyte. Therefore, experiments were extended to the use of two cytokinins (BA and kinetin) in combination, with or without NAA and sucrose. Homogenization of the sporophyte body for a range of experiments is quite common in the vegetative propagation of ferns, not only in vitro. The method, however, has a number of drawbacks, for example, the size of the explants necessary, as it requires relatively young donor plants. Nevertheless, it appears to be effective in promoting vegetative propagation. This procedure was used for the species discussed. Segments of rhizome, petiole, or the basal part of frond and root tips of 6-month-old sporophytes were of particular interest. The experimental conditions created were useful in determining the various morphogenetic potentials of the investigated explants, each of which (rhizome, petiole, and frond explants – 50%) showed a similar response in terms of GGB formation, but no GGB regeneration occurred on root explants (Bertrand et al. 1999).

Alternatively, epidermal or green mesophyll cells of young *Adiantum capillus-veneris* leaves were used as a source of GGB regeneration in the presence of BA (Li et al. 2015). For *Platycerium bifurcatum* GGB regeneration using both BA and the next purine-derived cytokinin, 2-iP [N^6-(2-isopentenyl adenine)] was employed. In these experiments, explants were derived from juvenile leaves of in vitro-grown sporophytes. As in the other studies mentioned, sporophyte regeneration required hormone-free medium in order to implement a program of plantlet differentiation. During plant regeneration, rooting usually required additional special treatments, which in this case were omitted, and shoot regenerants were directly transferred onto sphagnum substrate in order to promote rooting (Liao and Wu 2011).

In order to determine the optimal conditions for *Cibotium barometz* multiplication, mineral medium supplemented with the artificial urine-derived cytokinin, thidiazuron, was used in GGB induction experiments. Three-leaf plantlets (sporophytes) responded by shoot apical dome multiplication and differentiation. Multiplication of GGB was very rapid because both mother and daughter GGBs were involved in the simultaneous formation of the next vegetative generation. A single GGB, during the multiplication phase, had a flat or dome-shaped upper surface (Yu et al. 2017). Stolon tips are the next possible explants for use in the development of a rapid propagation system using BA (2.2 μM) via GGB formation. GGBs easily regenerate plants of *Nephrolepis cordifolia* (Higuhi et al. 1987). The green globular body system fully met our expectation for multiplication, not only for commercial reasons but also for the propagation and conservation of endangered species. However, this system of multiplication was developed simultaneously as an alternative to the shoot regeneration of *Pteris vittata* in the presence of 2,4-D and BA (Shukla and Khare 2012a, b).

Papers published to date do not precisely address the cyto- and histological origin of GGBs. One published analysis indicates that GGB formation depends on the organ from which explants are derived. GGB structure was compared to that of a protocorm body (Yu et al. 2017). However, protocorms are freely growing multicellular complexes, whereas GGBs form no part of the explant tissue until initiation of liquid culture occurs, which, in the case of some species, does not stimulate GGB overgrowth. Only the independence of multi-shoot meristem formation as the initial stage in shoot formation is similar. In the case of runner explants of *Nephrolepis cordifolia*, the initial GGBs arise from the peripheral zone of the vascular bundle of tip explants. The swollen GGB usually has many meristems, and each of these is directly connected to the main vascular bundle system of explants. The efficiency of propagation depends on the size of the GGB, a very small piece of plant body giving better results than a large one in the presence of only 2.2 μM BA. Completion of plant regeneration from GGB via the development of a particular meristem occurs on hormone-free medium (Higuchi et al. 1987). The processes of GGB induction and multiplication were described for juvenile leaf explants of *Platycerium bifurcatum* (Liao and Wu 2011) and also for juvenile sporophytes of *Cibotium barometz* (Yu et al. 2017). It is necessary to stress that, for rhizomes, the GGB structures originate by explant tissue proliferation, namely, from the epidermis or primary cork, but not by single cell proliferation. Its single cell origin has been proven for only three species of fern that form somatic embryos, with cytohistological evidence provided for *Microgramma vacciniifolia* (Hirsch 1976), *Argyrochosma nivea* (Galan 2011), and the tree fern *Cyathea delgadii* grown on hormone-free medium (Mikuła et al. 2015). Histological analysis of *Cibotium barometz* GGB multiplication revealed the multicellular organization of meristematic activity involved in the formation of primary meristems and a second meristem (based on the description of daughter bodies) which represents the initial stage of multiplication. As the process proceeds, the formation of protrusions becomes visible and these later give rise to new GGBs

(Higuchi and Amaki 1989). This phenomenon is not limited to experimental systems, and each GGB develops into a single shoot via frond regeneration (Yu et al. 2016). For *Neottopteris nidus*, both systems of GGB growth, namely, agar and suspension medium, were developed so as to provide experimental material for conserving plant genetic resources (Mikuła et al. 2011). For the cryopreservation of GGBs, the droplet-vitrification method has been adopted (Li et al. 2013). The medium used for suspension culture has a lower concentration of BA, but the same concentration of NAA and sucrose. The method used involves a loading solution and PVS2 solution. Over a period of 30–90 days following the freezing of GGB cultures, the structures passed through various stages of development. Initially they turned black due to phenolic oxidation, and later, new body tissue was regenerated, and new meristems became organized that finally resulted in the formation of new leaves. The extension of the culture period resulted in the production of many more new leaves. GGBs in suspension culture showed a similar trend of regrowth (Li et al. 2013).

5.7 Media and Plant Growth Hormones Used

The following media were employed in the tissue culture of ferns: Knopp, Knudson, Murashige and Skoog (with various concentrations of inorganic salts), Gamborg B5, modified Prague, Hoagland and Arnon, Dryer, Moor, and Parker-Thompson, Miller, Yang. Their use closely follows the history of plant tissue culture, commencing with very simple media and becoming increasingly complex, such as Gamborg B5 and Murashige and Skoog media. At present, the medium mainly used is that of Murashige and Skoog. Its mineral salt content is mostly halved in the majority of experiments for initiation, establishing and maintaining cultures. Some media were modified by changing nitrogen concentrations. Fern cultures are maintained on solidified medium, very rarely liquid medium. Liquid culture of ferns is not popular and is mainly used for plant cell suspension cultures. Generally, lack of this type of culture reflects a lack of progress in plant cell genetic manipulation in fern biotechnology.

The set of plant growth hormones used does not differ much from that used for cell and tissue culture of seed plants, both monocots and dicots. The list includes cytokinins, Kin, 2-iP, zeatin, TDZ, BA and auxins, 2,4-D, 2,4,5-T, IAA, NAA, and IBA. Gibberellic acid (GA_3), JA, and ABA are plant growth hormones that are very rarely used in the in vitro culture of ferns (Raynolds 1981). Organic supplements, coconut milk (CM) and casein hydrolysate (CH), were used as a source of organic compounds for stimulating tissue growth and cell proliferation (Bharati et al. 2013). Active charcoal is used for greatly improving the growth efficiency of the sporophyte regeneration (Teng 1997; Avila-Perez 2011).

5.8 Conclusions

The number of fern species used for developing plant cell, tissue, and organ culture manipulation is limited. Some of these, however, are recognized and used as model systems (e.g., *Ceratopteris thalictroides*). Morphological variation is reflected by the choice of species used for propagation by vegetative multiplication (Bharati et al. 2013). Pharmaceutical value, horticultural value, environmental degradation, and species protection were the criteria considered in selecting these species and introducing them to in vitro propagation.

The review presented indicates that some of the techniques result only in the formation of initials, and these usually commence with the sowing of spores onto agar medium so as to have sufficient experimental material for PGR manipulation and to induce morphogenic events. For ferns, the cell, not the protoplast, is usually the level at which manipulation and the production of modified regenerants occurs. In the majority of fern species, the effect of PGRs on the regeneration of new plants is studied on well-organized explant gametophytes and sporophytes or their various elements. Green globular bodies, formed as a result of cellular proliferation of explants, originated from various elements of the juvenile plant body and are of interest, not only as a vegetative propagule but also as the structure in which cell multiplication and meristem organization occur.

It is a great pleasure to work with ferns, plants without secondary growth, and to develop a new means of regenerating plants via the single cell origin of somatic embryos (see Chap 6) and by cell suspensions (see figures for this chapter) that carry characters typical of the embryogenic cell suspensions described for seed plants. These two elements are very promising for future experimental work on ferns. Our work in this field commenced many years ago with the initiation of spore germination and gametophyte multiplication on a simple medium, with the aim of obtaining sporophytes. We performed our work like every other fern tissue culturist. Now, however, we have much evidence to indicate that this work on such variable experimental material could prove to be very rewarding.

References

Amaki W, Higuchi H (1991) Possible propagation system of *Nephrolepis, Asplenium, Pteris, Adiantum,* and *Rumohra* (Arachiniodes) through tissue culture. Acta Horticulture 300:237–243

Ambrósio ST, de Melo NF (2004) Interaction between sucrose and pH during in vitro culture of Nephrolepis biserrata (Sw.) Schott (Pteridophyta). Acta Bot Bras 18:13–18

Ambrozic-Dolinse J, Camloh M, Žel J (1999) Plant regeneration from scales of the fern *Platycerium bifurcatum* (Cav.) C. Chr. Phyton (Austria) 39:297–300

Atmane N, Blervacq AS, Michaux-Ferriere N, Vasseur J (2000) Histological analysis of indirect somatic embryogenesis in the marsh clubmoss *Lycopodiella inundata* (L.) Holub (Pteridophytes). Plant Sci 156:159–167

Attree SM, Scheffield E (1985) Plasmolysis of *Pteridium* protoplasts: a study using light and scanning-electron microscopy. Planta 165:151–157

Avila- Peréz MCR (2011) In vitro regeneration of leathererleaf fern [*Rumohra adiantiformis* (G.Forest.) Ching]. Amer Fern J 101:25–35

Bertrand MA, Albuerne H, Fernández A, González R, Sánchez-Tamés R (1999) In vitro organogenesis of *Polypodium cambricum*. Plant Cell Tissue Organ Cult 57:65–69
Bharati SK, Manabendra DC, Behari MP (2013) In vitro propagation in Pteridophytes: a review. Int J Res Ayurveda Pharm 4:297–203
Bienaime C, Melin A, Bensaddek L, Attoumbre J, Nava-Saucedo E, Baltora-Rosse S (2015) Effects of plant growth regulators on cell growth and alkaloids production by cell cultures of *Lycopodiella inundata*. Plant Cell Tissue Organ Cult 123:523–533
Breznovits A. (1994) Somatic Hybridization in Ferns. In: Bajaj YPS (eds) Somatic Hybridization in Crop Improvement I. Biotechnology in Agriculture and Forestry. Springer, Berlin, Heidelberg. 27:518–528
Bristow JM (1962) The controlled *in vitro* differentiation of callus derived from fern, *Pteris cretica* L., into gametophytic or sporophytic tissues. Dev Biology 4:361–375
Bui LT, Hurst A, Irish EE, Cheng C-L (2012) The effect of sugars and ethylene on apospory and regeneration in *Ceratopteris richardii*. Amer J Plant Sci 3:953–961
Byrne TE, Caponett JD (1992) Morphogenesis in three cultivars of Boston fern. II. Callus production from stolon tips and plantlet differentiation from callus. Amer Fern J 82(1):11
Camloh M, Gogala N (1992) *In vitro* culture of *Platycerium bifurcatum* gametophytes. Sci Hortic 51:343–346
Cheema HK (1983) In vitro studies on callus induction and differentiation of gametophytes and sporophytes in ferns. In: National Seminar on Progress in Botanical Research. Penjab University. pp 44–46
Cheema HK, Sharma MB (1991) In vitro induction and differentiation of gametophyte callus in *Ceratopteris thalictroides*. Indian Fern J 8:74–77
Cocking EC (1960) The method for the isolation of plant protoplasts and vacuoles. Nature 187:962–963
Davidson C, Prusinkiewicz P, von Aderkas P (2008) Description of a novel organ in the gametophyte of the fern *Schizaea pusilla* and its contribution to overall plant architecture. Botany 86:1217–1223
Domżalska L, Mikuła A, Rybczyński JJ (2016) Protein extraction from Ca-alginate encapsulated plant material for comparative proteomic analysis. Protein Expr Purif 126:55–61
Edwards ES, Roux SJ (1998) Gravity and light control of the developmental polarity of regenerating protoplasts isolated from prothallial cells of the fern *Ceratopteris richardii*. Plant Cell Rep 17:711–716
Emigh VD, Farrar DR (1977) Gemmae: a role in sexual reproduction in the fern genus *Vittaria*. Science 198(4314):297–298
Fernández H, Bertrand A, Sanchez-Tamés R (1996) Micropropagation and phase change in *Blechnum spicata* and *Pteris ensiformis*. Plant Cell Tissue Organ Cult 44:261–265
Fernández H, Bertrand A, Sanchez-Tamés R (1997a) Gemmation in cultured gametophytes of *Osmunda regalis*. Plant Cell Rep 16:358–362
Fernández H, Bertrand A, Sánchez-Tamés R (1997b) Plantlet regeneration in *Asplenium nidus* L. and *Pteris ensiformis* by homogenization of BA treated rhizomes. Sci Hort 68:243–247
Frearson EM, Power JB, Cocking EC (1973) The isolation, culture and regeneration of petunia leaf protoplasts. Dev Biol 33:130–137
Gaj MD (2011) Somatic embryogenesis and plant regeneration in the culture of *Arabidopsis thaliana* (L.) Heynh. Immature zygotic embryos. Meth Mol Biol 710:257–265
Galán y JMG (2011) Gametophyte development and reproduction of *Argyrochosma nivea* (Pteridaceae). Biologia 66:50–54
Goller K, Rybczyński JJ (1995) *In vitro* culture used for woody fern *Cyathea australis* (R.Br) Domin vegetative propagation. Acta Soc Bot Pol 64:13–17
Goller K, Rybczyński JJ (2007) Gametophyte and sporophyte of tree ferns in vitro culture. Acta Soc. Bot Pol 76:193–199
Hedge S, Menon VK, Noronha R, D'Souza L (2006) Callus culture and an unconventional pattern of sporophyte regeneration in *Drynaria quercifolia* – medicinal fern. In Vitro Cell Dev Biol-Planta 42:508–513

Hicks G, von Aderkas P (1986) A tissue culture of the ostrich fern *Matteucia struthiopteris* (L.) Todaro. Plant Cell Tissue Organ Cult 5:199–204

Higuchi H, Amaki W (1989) Effects of 6-benzylaminopurine on the organogenesis of *Asplenium nidus* L. through *in vitro* propagation. Sci. Hort. 37:351–359

Higuhi H, Amak W, Suzuki S (1987) In vitro propagation of Nephrolepis Cordifolia Prel. Sci Hort 32:105–113

Hirsch AM (1975) The effect of sucrose on the differentiation of excised fern leaf tissue into either gametophytes and sporophytes. Plant Physiol 56:390–393

Hirsch AM (1976) The development of aposporous gametophytes and regenerated sporophytes from epidermal cells of excised fern leaves: an anatomical studies. Amer J Botany 63:263–271

Huckaby CS, Bassel AR, Miller IH (1982) Isolation of rhizoid and prothallial protoplasts from gametophytes of the fern, *Onoclea sensibilis*. Plant Sci Lett 25:203–208

Ito M (1962) Studies on the differentiation of fern gametophytes. Bot Mag Tokyo 75:19–27

Johnson M, Manickam VS (2006) Adventitious proliferation of secondary and tertiary prothalli from the primary prothalli of *Pronephrium articulatum* (Houlst. & Moore) Holit. Eth J Sci Technol 3(2):93–96

Johnson M, Manickam VS (2011) *In vitro* proliferation of secondary and tertiary gametophytes from the spore derived gametophytes of the fern *Phlebodium aureum*. J Basic Appl Biol 5:352–356

Jones DL (1987) Chapter 16. Encyclopedia of ferns. In: Vegetative propagation. Luthian Publish Company PTY. LTD, Melbourne, pp 125–139

Joyce BL, Eda S, Dunlap J, Stewart C Jr (2014) Morphology and ploidy level determination of *Pteris vittata* Callus during induction and regeneration. BMC Biotechnol 14:96

Kadota A, Wada M (1989) Enzymatic isolation of protoplasts from fern protonemal cells stainable with a fluorescent brightener. Plant Cell Physiol 30:1107–1113

Kato Z (1964) Physiological and morphological studies of fern gametophyte by aseptic culture. III. Growth and differentiation of single cells isolated from callus tissue of *Pteris vittata*. Cytologia (Tokyo) 29:79–85

Klercker JA (1892) Eine methone sur isoleirung labender protoplasten. Swenka, Vet Akad Forh Stockholm 9:463–471

Kuriyama A, Maeda M (1999) Direct production of sporophytic plants from spores of *Equisetum arvense*. Plant Cell Tissue Organ Cult 58:77–79

Kwa S-H, Wee Y-C, Lim T-M, Kumar PP (1995) Establishment and physiological analyses of photoautotrophic callus cultures of the fern *Platycerium coronarium* (Koenbing) Dev. under CO_2 enrichment. J Exp Bot 46:1535–1542

Kwa S-H, Wee Y-C, Lim T-M, Kumar PP (1997) Morphogenetic plasticity of callus reinitiated from cell suspension cultures of the fern *Platycerium coronarium*. Plant Cell Tissue Organ Cult 48:37–44

Ledwoń A, Gaj MD (2011) *LEAFY COTYLEDON1, FUSCA3* expression and auxin treatment in relation to somatic embryogenesis induction in *Arabidopsis*. Plant Growth Reg 65:157–167

Li T, Xu L, Li Z, Panis B (2013) Cryopreservation of *Neottopteris nidus* prothallus and green globular bodies by droplet-vitrification. CryoLetters 34:481–489

Li X, Fang YH, Han JD, Bai S-N, Rao G-Y (2015) Isolation and characterization of a novel somatic embryogenesis receptor kinase gene expressed in the fern *Adiantum capillus-veneris* during shoot regeneration *in vitro*. Plant Mol Biol Rep 33:638–647

Li X, Han J-D, Fang Y-H, Bai S-N, Rao G-Y (2017) Expression analyses of embryogenesis-associatedgenes during gomatic gmbryogenesis of *Adiantum capillus-veneris* L. In vitro: New insights into the evolution ofreproductive organs in land plants. Front Plant Sci 8:658. doi:10.3389/fpls.2017.00658

Liao YK, Wu Y, H. (2011) *In vitro* propagation of *Platycerium bifurcatum* (Cav.) C. *via* green globular body initiation. Botanica Studies 52:456–463

Maeda M, Ito M (1981) Isolation of protoplasts from fern prothallia and their regeneration to gametophytes. Botanical Magazine (Tokyo) 94:35–40

Maeda M, Sugimoto Y, Nakamura M, Masuda K, Kaneko H, Sugai M (1990) Division and gametophytic tissue formation from protoplasts of young sporophytes in fern *Lygodium japonicum*. Plant Cell Rep 9:113–118

Makowski D, Tomiczak K, Rybczyński JJ, Mikuła A (2016) Integration of tissue culture and cryopreservation methods for propagation and conservation of the fern *Osmunda regalis* L. Acta Physiol Plant 38:19

Maridass M, Mahesh R, Raju G, Muthuchelian K (2010) Clonal propagation of *Adiantum capillus - veneris*. Inter J Biol Tech 1:33–37

Martin KP, Sini S, Zhang C-L, Slater A, Madhusoodana PV (2006) Efficient induction of apospory and apogamy in vitro in silver fern (Pityrogramma calomelanos L.). Plant Cell Rep 25:1300–1307.

Mehra PN, Palta HK (1971) In vitro control differentiation of the root callus of *Cyclosorus dentatus*. Phytomorphology 21:367–375

Menéndez V, Revilla MA, Bernard P, Gotor V, Fernández H (2006) Gibberellins and antheridiogen on sex in Blechnum spicant L. Plant Cell Rep 25(10):1104–10

Meyer DT (1953) Uber das verhalteneinzelner isolierter prothalliumzellen und dessen bedeuntung correlation und regeneration. Planta 41:642–645

Mikuła A, Tomiczak K, Wójcik A, Rybczyński JJ (2011) Encapsulation-dehydration method of cryopreservation elevates embryogenesity of *Gentiana kurroo* cell suspension and carrying on genetic stability of its regenerants after cryopreservation. Acta Hortic 908:143–154

Mikuła A, Pożoga M, Tomiczak K, Rybczyński JJ (2015) Somatic embryogenesis in ferns: a new experimental system. Plant Cell Rep 34:783–794

Morel G, Wetmore RN (1951) Fern callus tissue culture. Am J Bot 38:141–143

Munroe MH, Sassex IM (1969) Gametophyte formation in bracken fern root cultures. Can J Bot 47:617–621

Nakamura M, Maeda M (1995) Gametophytes derived from sporophytic tissue in a fern, *Lygodium japonicum* L. 1. Induction of the gamatophytes and their protoplast isolation. J Plant Physiol 145:185–188

Palta HK, Mehra PN (1983) *In vitro* induction of polyhaploid and octoploid *Pteris vittata* L. and their meiosis. Caryologia 36:325–332

Partanen CR, Power JB, Cocking EC (1980) Isolation and division of protoplasts of *Pteridium aquilinum*. Plant Sci Lett 17:333–338

Plackett ARG, Huang L, Sanders HL, Langdale JA (2014) High-efficiency stable transformationof the model fern species *Ceratopteris richardii* via microparticle bombardment. Plant Physiol 165:3–14

Plackett ARG, Rabbinowitsch EH, Langdale JA (2015) Protocol: genetic transformation of the fern *Ceratopteris richardii* through microparticle bombardment. Plant Methods 11:37

Ravi BX (2016) In vitro polyembryony induction in a critically endangered fern, *Pteris tripartita* Sw. Asian Pacific J Reprod 5:345–350

Redford K, Berlinew MD, Gates JE, Fisher RW, Matthews BF (1987) Protoplast induction from sporophyte tissues of the heterosporous fern A. zolla Plant Cell, Tiss Org Cult 10:187–196

Reynolds TL (1981) Effects of auxins and abscisic acid on adventitious gametophyte formation by *Anemia phyllitidis*. (L.) Sw. Z Pflanzenphysiol 103:9–13

Salome M, Pais S (1987) Propagation of the fern *Adiantum capillus-veneris* through tissue culture of the circinate part of young leaves. Acta Horticulture 212:651–654

Shukla SP, Khare PB (2005) In-vitro mass multiplication of a threatened tree fern, Cyathea Spinulosa wall. Ex hook. Indian Fern J 22:157–161

Shukla SP, Khare PB (2012a) In vitro mass multiplication of a threatened tree fern, Cyathea Spinulosa wall. Ex hook. Int J Genet Engin Biotech 3:15–23

Shukla SP, Khare PB (2012b) In vitro shoot regeneration via caulogenesis in fern, *Pteris vittata* L. J Environ Biol 33:683–687

Soare LC, Visoiu E, Dobrescu CM (2010) In vitro culture and regeneration of the fern Dryopteris Affinis, species growing in protected area. Romanian Biotechnology Letters 15:45–54

Sulkyan DS, Mehra PN (1977) In vitro morphogenetic studies *Nephrolepis cordifolia*. Phytomortph 27:396–407

Sussex IM, Steeves TA (1953) Experimental and analytical studies of pteridophyta XXI. Sterile culture of abnormal growth frm the gametophyte of bracken fern. ANN Botany 17:395–401

Teng W-L (1997) Activated charcoal affects morphogenesis and enhances sporophyte regeneration during leaf cell suspension culture of *Platycerium bifurcatum*. Plant Cell Rep 17:77–83

Teng W-L, Teng M-C (1997) In vitro regeneration patterns of *Platycerium bifurcatum* leaf cell suspension culture. Plant Cell Rep 16:820–824

Tomiczak K, Sliwińska E, Rybczyński JJ (2017) Protoplast fusion in the genus Gentiana: genomic compositionand genetic stability of somatic hybrids between *Gentiana kurroo* Royle and *G. cruciata*. Plant Cell Tiss Organ Cult 131:1–14

Weinthal DM, Gürel F (2016) Chapter 4. Plant genome editing and its applications in cereals. In: Jamal F (ed) Agricultural and biological sciences "Genetic engineering – an insight into the strategies and applications" InTechPublisher, open access.

Wójcik AI, Rybczyński JJ (2015) Electroporation and morphogenic potential of *Gentiana kurroo* (Royle) embryogenic cell suspension protoplasts. BioTechnology 96:19–29

Yu R, Zhangfei Z, Li H, Cao H, Mo X, Gui M, Zhou X, Jiang Y, Li S, Wang J (2017) *In vitro* propagation of the endangered tree fern *Cibotium barometz* through formation of green globular bodies. Plant Cell Tissue Organ Cult 128:369–379

Chapter 6
Experimental and Practical Application of Fern Somatic Embryogenesis

Anna Mikuła, Małgorzata Grzyb, Karolina Tomiczak, and Jan Jarosław Rybczyński

Abbreviations

ABA	Abscisic acid
ACC	Subunit acetyl-CoA carboxylase
ACS	Acetyl-coenzyme A synthetase
APX	L-ascorbate peroxidase
CKs	Cytokinins
DAHPS1	Phospho-2-dehydro-3-deoxyheptonate aldolase 1
DHQ synthase	3-dehydroquinate synthase
GDH	Glutamate dehydrogenase
GST	Glutathione S-transferase
IAA	Indole-3-acetic acid
MED37	The mediator of RNA polymerase II transcription subunit 37
RGP1	UDP-arabinose mutase
RHM1	UDP-4-keto-L-rhamnose-reductase RHM1
SE	Somatic embryogenesis

6.1 Introduction

Somatic embryogenesis (SE) is a process by which bipolar structures called somatic embryos develop from non-zygotic cells. It was first described for carrot callus cells 60 years ago (Reinert 1958; Steward et al. 1958). By 2015, this unique developmental process had already been recognized for many spermatophytes, but for only two cryptogamic species, i.e. *Lycopodiella inundata* (L.) Holub and *Huperzia selago*

A. Mikuła (✉) · M. Grzyb · K. Tomiczak · J. J. Rybczyński
Department of Experimental Botany, Polish Academy of Sciences, Botanical Garden – Center for Biological Diversity Conservation in Warsaw, Warsaw, Poland
e-mail: amikula@obpan.pl

H. Fernández (ed.), *Current Advances in Fern Research*,
https://doi.org/10.1007/978-3-319-75103-0_6

(L.) Bernh. ex Schrank. (Atmane et al. 2000; Szypuła et al. 2005). In 2015, the first report of SE in a fern was published for *Cyathea delgadii* Sternb., a plant species belonging to Monilophyta, a clade that contains the closest living relatives of spermatophytes (Mikuła et al. 2015b). Although the life cycle of ferns is completely different from that of spermatophytes, the ability to switch the developmental program of mature cells to that of somatic embryo formation seems to be universal and needs to be explained (Kennedy and Norman 2005). The switch in cell fate, both in seed plants and cryptogamic plants, is characterized by changes in phytohormone content and abundance of protein, by patterns of gene expression, as well as structural rearrangement of explant cells. Determining the changes by which somatic cells are able to reset a new developmental pathway is important if we are to understand the capacity of plant cells for asexual embryogenesis. Our recent work shows that the use of *C. delgadii* is useful in understanding somatic-to-embryogenic transition.

6.2 Story of Somatic Embryogenesis in the Tree Fern *Cyathea delgadii*

The sequence of events leading to termination of zygotic embryogenesis under in vitro culture conditions (Fig. 6.1) provides not only information about sexual reproduction in *C. delgadii* but is also a source of experimental material for SE. It is worth emphasizing that maintaining cultures in darkness inhibits the development of gametophytes, and the latter are prevented from reaching sexual maturity. Under 16/8 h photoperiod conditions, gametophytes achieve maturity within 6 months following spore sowing on 1/2 strength Murashige and Skoog (MS) medium supplemented with 1% sucrose. Details of sex organ formation in this species were described by Rybczyński and Mikuła (2011). Following fertilization, early development of the zygotic embryo takes place inside an archegonium. Following rupture of the archegonial envelope by the first leaf (Fig. 6.1a), subsequent embryonic growth is rapid. Within 4–5 weeks, the main organographic regions are determined, and the leaf-root axis is visible (Fig. 6.1b). Subsequently, the first leaf elongates, and the development of the second follows shortly thereafter (Fig. 6.1c, d). The lamina of the first leaf, grown under photoperiod conditions, is entirely different from that of mature adult leaves (Fig. 6.1e, f). The first leaf shows dichotomous venation, whereas the second is more complex and pinnate (Fig. 6.1f). Zygotic embryo-derived sporophytes were readily induced on gametophytes growing on culture medium lacking ammonium nitrate (NH_4NO_3) and with the mineral salt concentration of the medium reduced to 1/8 MS (Fig. 6.1e).

These findings confirm that requirements for nutrients for zygotic embryo initiation in the tree fern *C. delgadii* are low. Indeed, starvation has previously been used for the initiation of fern sporophytes in *Pteris ensiformis* Burm. f., *Adiantum reniforme* var. *sinense* Y. X. Lin., and *Osmunda regalis* L. (Fernández et al. 1999; Wu et al. 2010; Makowski et al. 2016). Parts of very young zygotic embryos and zygotic

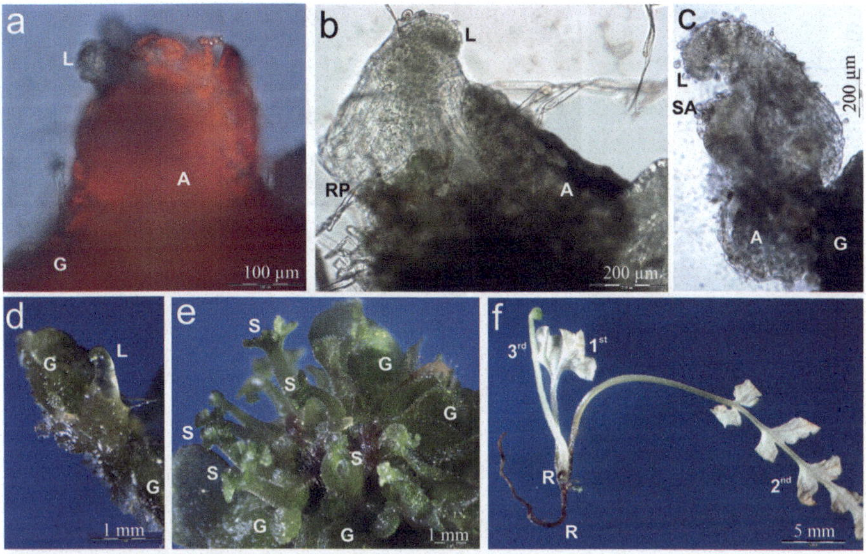

Fig. 6.1 Details of zygotic embryo development in *Cyathea delgadii* maintained in vitro at a 16 h/8 h photoperiod on 1/2 MS medium. (**a**) Archegonial envelope ruptured by the first emerging leaf of the zygotic embryo; red autofluorescence of chlorophyll was induced by exposure to blue-violet light (BV filter: 400–440 nm). (**b**) Well-developed zygotic embryo showing first leaf and a root primordium; thus, the leaf-root axis is visible. (**c**) The first crosier of a zygotic embryo. (**d**) An elongating leaf of a zygotic embryo. (**e**) Numerous zygotic embryo-derived sporophytes formed on 1/8 MS medium lacking ammonium nitrate (NH_4NO_3). (**f**) Zygotic embryo-derived sporophyte with first juvenile leaf, second pinnate leaf, and third young crozier. *A* archegonium, *G* gametophyte, *L* leaf, *RP* root primordium, *R* root, *S* sporophyte, *SA* shoot apex, *1st* first leaf, *2nd* second leaf, *3rd* third leaf

embryo-derived sporophytes that were allowed to develop under reduced light conditions (3.5 µE m^{-2} s^{-1}) were capable of producing somatic embryos (19% efficient) (Mikuła et al. 2015b). Ultimately, the capacity of *C. delgadii* explants to undergo SE depends on the light conditions that prevail during the development of source plantlets. No embryogenic induction was observed in stipes excised from sporophytes cultured at a light intensity of 50 µM m^{-2} s^{-1} (Mikuła et al. 2015a). Long-term etiolation of somatic embryo-derived sporophytes has proved to be the main factor controlling induction of SE on hormone-free medium. Later studies revealed that the course of SE is dependent on the type of explant used for culture initiation. Stipe explant is important for inducing a single-celled source of somatic embryos (Fig. 6.2a–d) (Mikuła et al. 2015b), whereas an internodal explant is required to induce a multicellular SE pathway (Fig. 6.2e–g). When stipe explants were used for culture initiation, the first cell divisions typically commenced at 10 days of culture (Fig. 6.2a). During the next 4–6 days, the epidermal cells divided several times (Fig. 6.2b). Early (Fig. 6.2c) and late embryonic leaf stages (Fig. 6.2d), with numerous trichomes on the embryo surface, were clearly visible at about 21–28 days following culture initiation. Numerous divided cells were visible on the surface of internodal explants after 7 days of culture. These divisions led to the formation of

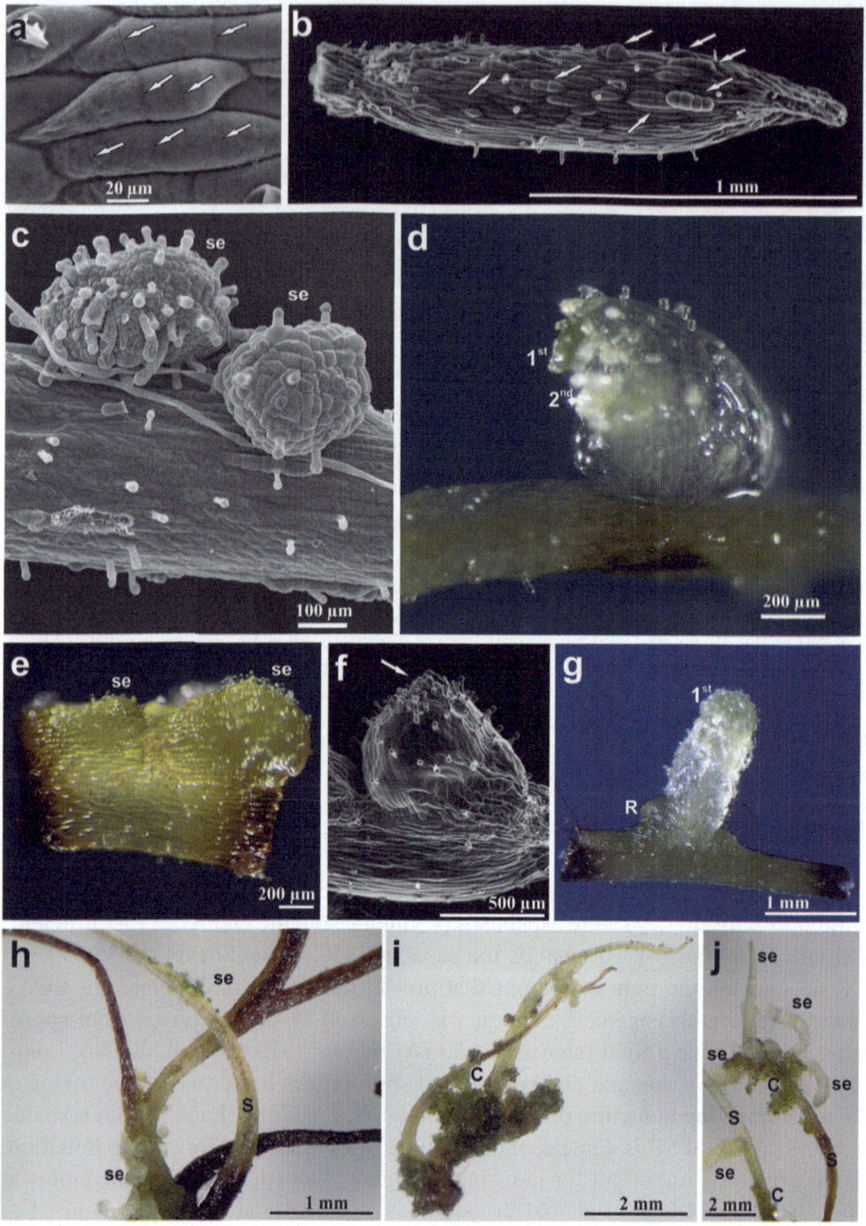

Fig. 6.2 Different methods of inducing somatic embryogenesis in *Cyathea delgadii*. (**a**–**d**) Direct unicellular pathway method induced on stipe explants. (**a**) First divisions (arrows) of epidermal cells of stipe explant; day 12 of culture. (**b**) Numerous proembryos at linear stage of development; day 14 of culture. (**c**) Somatic embryos at early (day 16 of culture) and (**d**) late embryonic leaf stages (day 21 of culture). (**e**–**g**) Direct multicellular pathway method induced on internodal explants. (**e**) Emerging nodular proembryos on internodal explant; day 13 of culture.

compact, nodular structures of multicellular origin (Fig. 6.2e), and these subsequently developed very quickly. Differentiation of the leaf primordium was observed at about day 16 of culture (Fig. 6.2f). Elongation of both embryo body and root then occurred (Fig. 6.2g). This method of somatic embryo formation in fern species is shown here for the first time. Sporophyte cultures maintained in darkness without any subculture for more than 5 months became an additional source of somatic embryos (Mikuła et al. 2015b). These embryos were spontaneously formed on various parts of donor plantlets by direct (Fig. 6.2h) or indirect SE (Fig. 6.2i, j). Both processes have been observed to occur simultaneously under the same tissue culture conditions. The course of SE in *C. delgadii*, commencing with spores followed by the culturing of gametophytes and formation of zygotic embryo-derived sporophytes, to the formation of somatic embryos and, ultimately, the derivation of mature plantlets from these, is summarized in Fig. 6.3.

Somatic embryo-derived sporophytes that have developed three leaves are the best source of initial explants. The process of SE from stipe explants of sporophytes (Fig. 6.3; in gray) continue in a cyclic manner, thus providing a continuous supply of embryogenic material. As an experimental system, it has certain advantages compared to those systems of SE described for the majority of spermatophytes. The system is induced (1) on hormone-free medium, (2) from a single cell of the stipe epidermis or from several cells of internodal segments, (3) during a short-time period, and (4) with very high multiplication and replication rate. Additionally, embryogenicity can be maintained for prolonged periods independent of a primary explant source. There are a few examples of species for which SE is driven by growth regulator-free medium (Raemakers et al. 1995). Although progress has undeniably been made in characterizing the underlying mechanisms of SE, thanks to the use of model plants such as *Arabidopsis* and *Daucus carota* (Raghavan 2006), the unique system that occurs in *C. delgadii* is extremely helpful in advancing this research.

6.3 Insight into Early Somatic Embryogenesis Using the *C. delgadii* Model System

The acquisition of embryogenic capacity occurs with greater or lesser ease depending on the plant species and the in vitro regeneration system employed. For most, the response of explant cells and the method of somatic embryo formation are

Fig. 6.2 (continued) (**f**) Further embryo development showing location of differentiation of leaf primordium (arrow) at the apex of the embryonic structure; day 16 of culture. (**g**) Elongation of embryo body and differentiation of both the first leaf and root of somatic embryo. (**h**–**j**) Direct or indirect pathway method in aging cultures of sporophytes. (**h**) Numerous somatic embryos induced directly from stipe cells of aging sporophyte. (**i**) Proliferation of embryogenic callus and (**j**) differentiation of somatic embryos. Images are based on scanning electron (**a**–**c**, **f**) and light (**d**–**e**, **g**–**j**) microscopy studies. *C* callus, *R* root, *S* sporophyte, *SE* somatic embryo, *1st* first leaf, *2nd* second leaf

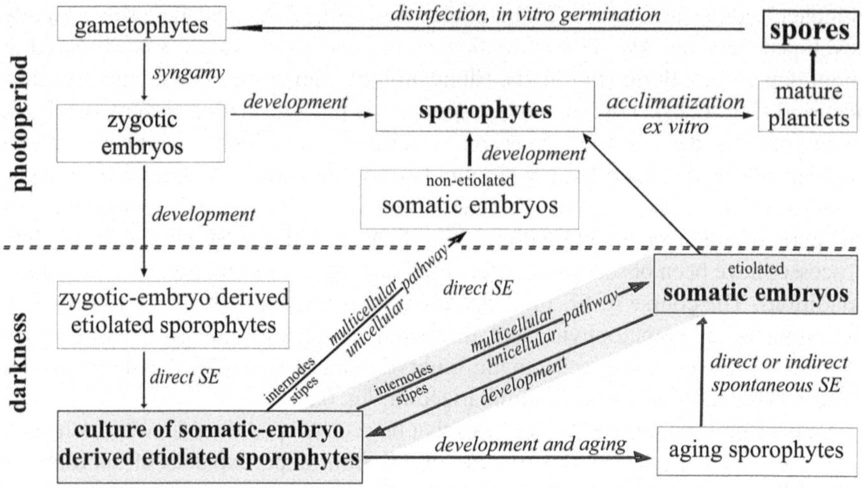

Fig. 6.3 Schematic representation of somatic embryogenesis pathways during micropropagation of *Cyathea delgadii*. Recurrent embryogenesis and continuous production of somatic embryos are shown in gray. This is achieved by maintaining the culture of somatic embryo-derived sporophytes in the dark. The method of somatic embryo formation is dependent on the explant type used for culture initiation. *SE* somatic embryogenesis

modified by the interaction of natural hormones and synthetic plant growth regulators (PGR) (Gaj 2004). Exogenously applied PGRs alter the synthesis, activation, transport, and destruction of endogenous hormones and regulate the morphogenetic response of explant cells (Gaspar et al. 1996). Conversely, SE can be induced in the absence of PGRs by using different stress treatments, such as osmotic stress, high or low temperature, or salinity (Nic-Can et al. 2016). Research on *C. delgadii* showed that the ability of explants to undergo somatic embryo production can also be controlled by the light conditions under which the donor plantlets are grown (Mikuła et al. 2015a). This occurs by the regulation of both endogenous hormone contents and hormonal balance (Grzyb et al. 2017). A summary of hormonal changes resulting from the 5-month-long culture of donor plantlets in constant darkness and during early SE is shown in Fig. 6.4.

Etiolation reduces by almost 12-fold the concentration of the phytohormone abscisic acid (ABA), and it is this PGR that is considered to be the main factor involved in inhibiting the explant to undergo SE (Ivanova et al. 1994; Jiménez and Bangerth 2001). It also reduces the ABA/cytokinin (CK) and ABA/ indole-3-acetic acid (IAA) ratios by as much as 107-fold and 29-fold, respectively (Fig. 6.4). In the absence of photosynthesis, there is also a clear reduction in sugar content. However, light per se does not appear to be associated with the acquisition of embryogenic totipotency by cells of etiolated explants since both dark- and light-cultured explants are highly efficient in producing somatic embryos (Mikuła et al. 2015a). However, light is nevertheless one of the most important external factors in that it affects both the type of morphogenetic response and also fern sporophyte development. In *C.*

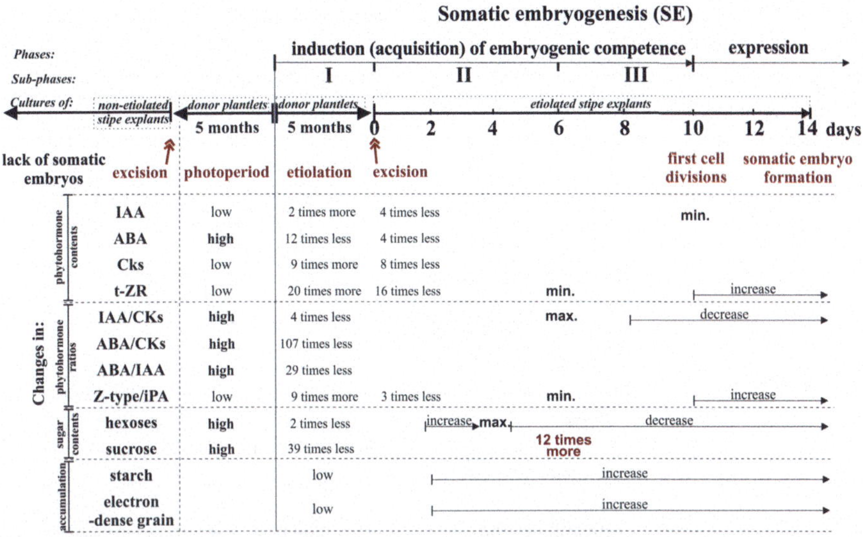

Fig. 6.4 Physiological events leading to the formation of single-cell origin of somatic embryos in the tree fern *Cyathea delgadii*

delgadii, stipe explants cultured under photoperiod conditions are able to form both somatic embryos and gametophytes next to each other (Mikuła et al. 2015a).

Although there has been considerable progress in the identification of genes implicated in SE, our understanding of the mechanisms by which stresses influence the acquisition of embryogenic capacity still remains unclear (Elhiti et al. 2013). In *C. delgadii*, the events that lead to single-cell somatic embryogenesis have been analyzed at the level of hormonal and cytomorphological responses. These results enable one to speculate on how SE started (Table 6.1). It is proposed that the excision of a stipe explant from an etiolated donor plantlet of *C. delgadii* leads to the loss of the resources required for hormone biosynthesis, resulting in a dramatic reduction in the concentration of all phytohormones, yet the hormonal balance remains almost unchanged (Fig. 6.4; Grzyb et al. 2017). The leakage of cytoplasm from mechanically disrupted cells probably results in a fall in turgor pressure (Nonami and Schulze 1989). One of the earliest responses to excision is the production of reactive oxygen species (ROS) that promote redifferentiation (Rose et al. 2010). Studies in cotton have indicated the involvement of ROS homeostasis in the initiation of SE by modifying auxin signaling (Zhou et al. 2016). Alternatively, a shortage of water activates the osmoregulation process (Pelleschi et al. 1997) that functions in accordance with acid growth theory (Wang and Ruan 2013). As a result, fructose and glucose accumulate over the 4 days of *C. delgadii* stipe explant culture (Grzyb et al. 2017). The influx of glucose and fructose into the cells, followed by an almost 12-fold increase in sucrose concentration, may represent a short-term osmotic signal required for the somatic-to-embryogenic transition. Associations between acid invertase activity, sucrose metabolism, and the cell water status, as

Table 6.1 Proposed and described physiological changes preceding the formation of somatic embryos in *Cyathea delgadii*

Exogenous treatments	Proposed changes	Described changes
1. Long-term darkness treatment of donor plantlets	Etiolation	Changes in the content of phytohormones Establishment of a new hormonal balance that is suitable for the induction of SE
2. Explant excision	Loss of endogenous hormone sources Leakage of cytoplasm Reduction in turgor pressure[a] Reactive oxygen species (ROS) generation[b]	Reduction in the content of IAA, ABA, and cytokinins without disturbing the hormonal balance
3. Explant culture	Increase in acid invertase activity[c,d]	Accumulation of fructose and glucose
	Increase in sucrose phosphate synthase activity[c,d]	Accumulation of sucrose
	Activation of the osmoregulation process[c,d]	Accumulation of starch in amyloplasts Reduction in sucrose content Accumulation of intravacuolar electron-dense grains

[a]Nonami and Schulze (1989)
[b]Rose et al. (2010)
[c]Wang and Ruan (2013)
[d]Pelleschi et al. (1997)

shown for carrot suspension cultures (Ikeda et al. 1999), support our hypothesis. Sucrose also appears to be involved in the morphogenetic response of juvenile leaves of the fern *Microgramma vacciniifolia* (Langsd. & Fisch.) Copel. (Hirsch 1976). In the presence of low level of sucrose (1%) in the induction medium, both embryo-like structures and gametophytes were produced. On the higher sucrose levels (3–4%), the development of aposporous gametophytes was inhibited and the sporophytes developed on a large number of excised leaves.

The significance of auxin as the main controlling factor in SE induction was recognized by Fehér (2015). The role of endogenous cytokinins, however, is still poorly understood. Using the *C. delgadii* model system, it has been shown that establishing an equilibrium between endogenous IAA and cytokinin content is essential both for the acquisition of embryogenic competence and the expression phase of SE (Grzyb et al. 2017). The embryogenic process may be completely arrested by disrupting this balance (Grzyb et al. 2018). A thorough examination of the effect of transport and hormone biosynthesis inhibitors on SE in *C. delgadii* showed that the role of phytohormones in somatic embryo induction is a highly complex process. Under the influence of different inhibitor substances (such as the auxin polar transport inhibitor 2, 3, 5-triiodobenzoic acid, the ABA biosynthesis inhibitor fluridone, or ethylene biosynthesis inhibitor salicylic acid), the concentrations of both indole-3-acetic acid (IAA) and cytokinins became strongly modified. Moreover, imbalances in phytohormone levels are responsible for the modification of soluble sugar concentrations, including that of sucrose – the main factor for triggering embryogenesis in *C. delgadii*. A survey of the literature has shown that a synthetic auxin is initially required to induce SE, but that its withdrawal drives embryo formation (Halperin 1966). Similar relationships were also found for changes in phytohormone concentrations in the *C. delgadii* model system. More endogenous auxin than cytokinin was specifically related to the acquisition of embryogenic competence, whereas cytokinins favored SE expression during which frequent cell divisions occur resulting in the formation and appearance of somatic embryos (Grzyb et al. 2017). Of the different cytokinins, the trans-zeatin riboside was clearly associated with this phase of SE (Fig. 6.4). This agrees with results obtained by Centeno et al. (1997), who observed an increase in the level of Z-type cytokinin during the maturation of *Corylus avellana* L. zygotic embryos.

With reference to cytomorphological studies, only two differences were found to occur between initial explants and those cultured on hormone-free 1/2 MS medium for about 16 days (Domżalska et al. 2017). These were a massive accumulation of starch grains in amyloplasts and an accumulation of electron-dense intravacuolar granules, both of which appear to be necessary for the early SE of *C. delgadii*. Increased levels of starch have been reported at the onset of several in vitro development processes, including the induction of SE (Ho and Vasil 1983; Stamp 1987; Barciela and Vieitez 1993; Canhoto et al. 1996; Pinto et al. 2010). By contrast, a dramatic reduction in the amount of starch has been documented at the commencement of SE and organogenesis (Martin et al. 2000; Fortes and Pais 2000; Puigderrajols 2001; Mikuła et al. 2004). Starch is considered a primary source of energy and carbon for cell proliferation during plant morphogenetic events (Martin et al. 2000). Research conducted on *C. delgadii* suggests that the starch accumulation in early SE may play a role in the osmoregulatory response to a sudden increase in endogenous sucrose level. The importance of starch was emphasized by Fortes and Pais (2000) for *Humulus lupulus* var. Nugget. By means of the cyclic process of starch accumulation/mobilization, they demonstrated the potential role of this carbohydrate in the initiation of organs and later in their development. The accumulation of starch in *C. delgadii* stipe explants was reflected in the proteomic analysis. Comparative analysis performed by Domżalska et al. (2017) showed that many proteins were involved in carbohydrate metabolism (33%). The results of this study are in accordance with the data obtained for *Picea glauca*, *Zea mays*, and *Medicago truncatula* (Lippert et al. 2005; Almeida et al. 2012; Varhaníková et al. 2014). Changes in this protein category, represented by enolase and 6-phosphogluconate dehydrogenase, and decarboxylation of 6PGDH, can be directly related to starch accumulation.

6.4 Molecular Aspects of Fern Somatic Embryogenesis

Acquisition of embryogenic capacity and the process of somatic embryo differentiation are a consequence of physiological changes induced by stress treatment (Fehér 2015; Grzyb et al. 2017). It promotes changes in molecular regulation and the proteomic response, especially in proteins that play integral roles in hormone perception and signaling (Elhiti et al. 2013). Evaluation of proteomic changes allows the characterization of different events that occur during SE, and these can be used as markers for monitoring the process (Rosas et al. 2016). In *C. delgadii*, the protein expression patterns associated with early SE were investigated using two-dimensional gel electrophoresis (2-DE) and mass spectrometry (Domżalska et al. 2017). The study revealed the presence of several proteins that are triggered by induction of SE and the embryo-forming process. Of the 114 differentially regulated proteins identified in *C. delgadii* stipe explants at day 16 of culture, it is thought that 16 and 20 are markers of the induction and expression phase of SE, respectively.

Proteins associated with the induction phase play an essential role in the protection of plant cells against oxidative stress (chalcone synthase, chaperone protein ClpB3, HSP70, chaperonin GroEL, MED37), in the interruption of cell communication (RHM1, RGP1, ACC, ACS), and in cellular reprogramming (GST, APX, DAHPS1, DHQ synthase, GDH). These proteins have been found to be associated with the early SE of many seed plants (Moriguchi et al. 1999; Tchorbadjieva 2005; Almeida et al. 2012; Gallego et al. 2014; Heringer et al. 2015; Zhou et al. 2016). Moreover, the activities of two enzymes involved in malate metabolism, i.e., malate dehydrogenase (MDH) and NADP-dependent malic enzyme (NADP-ME), increased more than fourfold. According to Crecelius et al. (2003), malate serves not only as an additional carbon sink but may also function as a vacuolar osmolyte balancing increased concentrations of soluble sugars.

Nine of the proteins characterized for the expression phase of SE in *C. delgadii* are involved in the conversion of sugars and the production of metabolically usable energy. Next, 11 proteins are associated with the high growth rate of embryogenic cells, protein synthesis, and degradation. It is worth noting that enolase is thought to be a molecular marker of late embryogenesis in spermatophytes (Andriotis et al. 2010; Tonietto et al. 2012; Vale et al. 2014), whereas in the tree fern *C. delgadii*, it accumulates during early SE (Domżalska et al. 2017). The study provided a list of early SE-related proteins that are potential biomarker candidates for future investigations of somatic embryogenic development in this species of tree fern.

6.5 Application of Somatic Embryogenesis for the Propagation of Ferns

Ferns (around 12,000 species) are a very ancient group of plants that are distributed worldwide, from Greenland to the Antarctic and from sea level to high altitudes. Of these, tree ferns are the second most diverse group with 600–660 species, distributed mainly throughout the tropics (Large and Braggins 2004). Most fern species are habitat- and niche- specific because of their sensitivity to wet or cold conditions. In recent years, ferns have become increasingly popular in horticulture. Many fern species are grown for their ornamental value, both for indoor and garden culture (Fernández and Revilla 2003). Ferns are used for food (Liu et al. 2012), medicine (Ho et al. 2011), and phytoremediation (Rathinasabapathi 2011). Tree ferns are also used for the production of handicrafts (Eleutério and Pérez-Salicrup 2006). Edible ferns are some of the most common wild food plants collected by people around the world (Liu et al. 2012). Habitat destruction and fragmentation, deforestation, and overexploitation of fern fronds, stipes, rhizomes/trunks, and scales are the main factors related to the reduction in fern populations. Nearly 70% of fern species are threatened with extinction (Barnicoat et al. 2011).

The conventional methods of fern propagation by spores or vegetative propagules (such as bulbils, aerial growths, stolons, tubers, offsets, stipules, and root buds) are time-consuming procedures that are comparatively slow and inefficient. Furthermore, most tree ferns cannot be multiplied vegetatively, as they do not produce offsets from their trunk or roots (Large and Braggins 2004). Therefore, in vitro propagation and collecting systems have been developed for many ferns (Fernández and Revilla 2003; Rybczyński and Mikuła 2011; Bharati et al. 2013). These systems are primarily based on the vegetative reproduction of gametophytes from which the formation of sporophytes occurs through sexual (Rybczyński and Mikuła 2011; Makowski et al. 2016) or asexual mechanisms (Kuriyama et al. 2004; Camloh and Ambrožič-Dolinšek 2011; Cordle et al. 2011). During in vitro culture, the time required for the development of the sporophyte from spore varies from 1 (*Woodwardia virginica*) to 8 months (*Asplenium nidus*) depending on the species of fern and the length of its life cycle (Fernández et al. 1999). Favorable conditions for in vitro culture can both increase the number of plants produced and shorten the time required for sexual sporophyte production to occur, as documented for *O. regalis* (Makowski et al. 2016), *Cyathea cooperi*, *C. cunninghamii*, and *Dicksonia antarctica* (Moura et al. 2012). However, this method of sporophyte production can also result in failure, as was shown for *Platycerium andinum* and *P. wandae* (Pérez-García et al. 2010). Other effective in vitro propagation methods are based on the culture of homogenized gametophytic (Fernández et al. 1999) and sporophytic tissue (Teng and Teng 1997; Somer et al. 2010; Camloh and Ambrožič-Dolinšek 2011). Homogenization of gametophytes can be considered an excellent method for the propagation of fern species, one that is currently not available for the in vitro culture of seed plants. This system is most effective for species with a short life cycle, such as *Woodwardia virginica* and *Dryopteris affinis* sp. *affinis*, but not for *O. regalis* (Fernández et al. 1999). Numerous reports also document the propagation of sporophytes by apogamy, the use of globular green bodies, and direct or indirect shoot organogenesis (Fernández and Revilla 2003; Camloh and Ambrožič-Dolinšek 2011). However, although these systems work successfully for herbaceous ferns, our knowledge of the in vitro propagation of tree ferns is still in its infancy. Researchers working on this group of ferns have mainly focused on spore germination and gametophyte development (Huang et al. 2001; Hiendlmayer and Randi 2007; Du et al. 2009; Rechenmacher et al. 2010; Moura et al. 2012; Das et al. 2013;

Vargas and Droste 2014). Progress in the production of tree fern sporophytes was made by Goller and Rybczyński (2007) and Rybczyński and Mikuła (2011). In their studies, 12 of 16 tree fern species reached sporophyte stage and were successfully acclimatized to ex vitro conditions (Goller and Rybczyński 2007). Recently, Yu et al. (2017) reported the successful induction of globular green bodies in *Cyathea barometz* (L.) J.Sm. by culturing juvenile sporophytes in the presence of thidiazuron (TDZ) and α-naphthaleneacetic acid (NAA). This is the first species of tree fern for which this rapid regeneration pathway has been described. Another quite efficient system of sporophyte production has also been shown for the tree fern *Cyathea spinulosa* Wall. ex Hook (Shukla and Khare 2013). Here, media supplemented with plant growth regulators promoted indirect shoot production, each having as many as 12.5 shoots per leaf primordium callus.

The excellent multiplication system by direct somatic embryogenesis recently developed for *C. delgadii* (Mikuła et al. 2015a, b) contrasts markedly with the low numbers of sporophytes obtained by the methods described above. To date, it is the fastest and most economically efficient system for the in vitro propagation of ferns. Following 2 months of culture, each responding explant measuring 2.5 mm in length produced as many as 42 somatic embryos (Mikuła et al. 2015a). The efficiency of this process can be improved upon even further by the spontaneous formation of somatic embryos in aging cultures (Mikuła et al. 2015b). This system allows more rapid propagation and may be extended to other fern species, in particular tree ferns. Large-scale propagation of *C. delgadii* plants is one practical application of the process described here, and this may prove beneficial for the conservation of threatened ferns, as well as for the commercial production of ornamental fern species.

6.6 Future Prospects

The tree fern *C. delgadii* is a good experimental model for studying SE at both basic and applied levels. It is still not clear, however, how embryogenic cells differentiate inside explants and what mechanisms control this process. Somatic embryogenesis of *C. delgadii* offers a model system for understanding the physiological and biochemical events that occur during the somatic-to-embryogenic transition. Moreover, the superficial and unicellular origin of somatic embryos makes this process easy to observe in terms of changes to intercellular communication. Symplasmic isolation between neighboring cells is considered an important factor in determining cell fate.

In recent years, substantial progress has been made in the field of SE in seed plants and in various aspects of this process. However, *C. delgadii* is the first fern species for which this process has been described. Consequently, our studies open the door for comparing developmental biology, especially with respect to embryogenesis, in spermatophytes and cryptogamic plants.

References

Almeida AM, Parreira JR, Santos R, Duque AS, Francisco R, Tomé DFA, Ricardo CP, Coelho AV, Fevereiro P (2012) A proteomics study of the induction of somatic embryogenesis in *Medicago truncatula* using 2DE and MALDI-TOF/TOF. Physiol Plant 146:236–249. https://doi.org/10.1111/j.1399-3054.2012.01633.x

Andriotis VM, Kruger NJ, Pike MJ, Smith AM (2010) Plastidial glycolysis in developing *Arabidopsis* embryos. New Phytol 185:649–662. https://doi.org/10.1111/j.1469-8137.2009.03113.x

Atmane N, Blervacq AS, Michaux-Ferriere N, Vasseur J (2000) Histological analysis of indirect somatic embryogenesis in the Marsh clubmoss *Lycopodiella inundata* (L.) Holub (Pteridophytes). Plant Sci 156:159–167. https://doi.org/10.1016/S0168-9452(00)00244-2

Barciela J, Vieitez AM (1993) Anatomical sequence and morphometric analysis during somatic embryogenesis on cultured cotyledon explants of *Camellia japonica* L. Ann Bot 71:395–404. https://doi.org/10.1006/anbo.1993.1050

Barnicoat H, Cripps R, Kendon J, Sarasan V (2011) Conservation *in vitro* of rare and threatened ferns-case studies of biodiversity hotspot and island species. In Vitro Cell Dev Biol Plant 47:37–45. https://doi.org/10.1007/s11627-010-9303-x

Bharati SK, Dutta Choudhury M, Mazumder BP (2013) *In vitro* propagation in Pteridophytes: a review. Int J Res Ayurveda Pharm 4:297–303. https://doi.org/10.7897/2277-4343.04245

Camloh M, Ambrožič-Dolinšek J (2011) *In vitro* regeneration systems of *Platycerium*. In: Fernández H, Kumar A, Revilla MA (eds) Working with ferns: issues and applications. Springer Science+Business Media, LLC, New York, pp 111–126. https://doi.org/10.1007/978-1-4419-7162-3_8

Canhoto J, Mesquita JF, Cruz GS (1996) Ultrastructural changes in cotyledons of pineapple guava (Myrtaceae) during somatic embryogenesis. Ann Bot 78:513–521. https://doi.org/10.1006/anbo.1996.0149

Centeno ML, Rodríguez R, Berros B, Rodríguez A (1997) Endogenous hormonal content and somatic embryogenic capacity of *Corylus aveliana* L. cotyledons. Plant Cell Rep 17:139–144. https://doi.org/10.1007/s002990050367

Cordle AR, Bui LT, Irish EE, Cheng CL (2011) Laboratory-induced apogamy and apospory in *Ceratopteris richardii*. In: Fernández H, Kumar A, Revilla MA (eds) Working with ferns: issues and applications. Springer Science+Business Media, LLC, New York, pp 25–36. https://doi.org/10.1007/978-1-4419-7162-3_3

Crecelius F, Streb P, Feierabend J (2003) Malate metabolism and reactions of oxidoreduction in cold-hardened winter rye (*Secale cereale* L.) leaves. J Exp Bot 54:1075–1083. https://doi.org/10.1093/jxb/erg101

Das S, Dutta Choudhury M, Mazumder BP (2013) *In vitro* propagation of *Cyathea gigantea* (Wall ex. Hook)—a tree fern. Int J Rec Sci Res 4:211–224

Domżalska L, Kędracka-Krok S, Jankowska U, Grzyb M, Sobczak M, Rybczyński JJ, Mikuła A (2017) Proteomic analysis of stipe explants reveals differentially expressed proteins involved in early direct somatic embryogenesis of the tree fern *Cyathea delgadii* Sternb. Plant Sci. https://doi.org/10.1016/j.plantsci.2017.01.017

Du H, Li Y, Li D, Dai S, Jiang C, Shi L (2009) Effect of light, temperature and pH on spore germination and early gametophytic development of *Alsophila metteniana*. Bio Sci 17(2):182–187. https://doi.org/10.3724/SP.J.1003.2009.08262

Eleutério AA, Pérez-Salicrup D (2006) Management of tree ferns (*Cyathea* spp.) for handicraft production in Cuetzalan, Mexico. Econ Bot 60:182–186. http://www.jstor.org/stable/4257089

Elhiti M, Stasolla C, Wang A (2013) Molecular regulation of plant somatic embryogenesis. In Vitro Cell Dev Biol-Plant 49:631–642. https://doi.org/10.1007/s11627-013-9547-3

Fehér A (2015) Somatic embryogenesis—stress-induced remodeling of plant cell fate. Biochim Biophys Acta 1849:385–402. https://doi.org/10.1016/j.bbagrm.2014.07.005

Fernández H, Revilla MA (2003) *In vitro* culture of ornamental ferns. Plant Cell Tissue Organ Cult 73:1–13. https://doi.org/10.1023/A:1022650701341

Fernández H, Bertrand AM, Sánchez-Tamés R (1999) Biological and nutritional aspects involved in fern multiplication. Plant Cell Tissue Organ Cult 56:211–214. https://doi.org/10.1023/A:1006277229136

Fortes AM, Pais MS (2000) Organogenesis from internode-derived nodules of *Humulus lupulus* var. Nugget (Cannabinaceae): histological studies and changes in the starch content. Am J Bot 87:971–979

Gaj MD (2004) Factors influencing somatic embryogenesis induction and plant regeneration with particular reference to *Arabidopsis thaliana* (L.) Heynh. Plant Growth Regul 43:27–47. https://doi.org/10.1023/B:GROW.0000038275.29262.fb

Gallego P, Martin L, Blazquez A, Guerra H, Villalobos N (2014) Involvement of peroxidase activity in developing somatic embryos of *Medicago arborea* L. Identification of an isozyme peroxidase as biochemical marker of somatic embryogenesis. J Plant Physiol 171:78–84. https://doi.org/10.1016/j.jplph.2013.09.017

Gaspar T, Kevers C, Penel C, Greppin H, Reid DM, Thorpe T (1996) Plant hormones and plant growth regulators in plant tissue culture. In Vitro Cell Dev Biol-Plant 32:272–289. https://doi.org/10.1007/BF02822700

Goller K, Rybczyński JJ (2007) Gametophyte and sporophyte of tree ferns in vitro culture. Acta Soc Bot Pol 76:193–199. https://doi.org/10.5586/asbp.2007.022

Grzyb M, Kalandyk A, Waligórski P, Mikuła A (2017) The content of endogenous hormones and sugars in the process of early somatic embryogenesis in the tree fern *Cyathea delgadii* Sternb. Plant Cell Tissue Organ Cult 123:467–478. https://doi.org/10.1007/s11240-017-1185-8

Grzyb M, Kalandyk A, Mikuła A (2018) Effect of TIBA, fluridone and salicylic acid on somatic embryogenesis and endogenous hormone and sugar contents in the tree fern *Cyathea delgadii* Sternb. Acta Physiol Plant 40:1. https://doi.org/10.1007/s11738-017-2577-4

Halperin W (1966) Alternative morphogenetic events in cell suspensions. Am J Bot 53:443–453

Heringer AS, Barroso T, Macedo AF, Santa-Catarina C, Souza GHMF, Floh EIS, Souza-Filho GA, Silveira V (2015) Label-Free quantitative proteomics of embryogenic and non-embryogenic callus during sugarcane somatic embryogenesis. PLoS One 10:23. https://doi.org/10.1371/journal.pone.0127803

Hiendlmayer R, Randi AM (2007) Response of spores and young gametophytes of *Cyathea delgadii* Sternb. (Cyatheaceae) and *Blechnum brasiliense* Desv. (Blechnaceae) to different light levels. Acta Bot Bras 21:909–915. https://doi.org/10.1590/S0102-33062007000400015

Hirsch AM (1976) The development of aposporous gametophytes and regenerated sporophytes from epidermal cells of excised fern leaves: an anatomical study. Am J Bot 63:263–271

Ho W, Vasil IK (1983) Somatic embryogenesis in sugarcane (*Saccharum officinarum* L.) I. The morphology and physiology of callus formation and the ontogeny of somatic embryos. Protoplasma 118:169–180. https://doi.org/10.1007/BF01281800

Ho R, Teai T, Bianchini JP, Lafont R, Raharivelomanana P (2011) Ferns: from traditional uses to pharmaceutical development, chemical identification of active principles. In: Fernández H et al (eds) Working with ferns: issues and applications. Springer Science+Business Media, LLC, New York, pp 321–346. https://doi.org/10.1007/978-1-4419-7162-3_23

Huang YM, Chiou WL, Lee PH (2001) Morphology of the gametophytes and young sporophytes of Cyatheaceae native to Taiwan. Taiwania 46(3):274–283. https://doi.org/10.6165/tai.2001.46(3).274

Ikeda T, Taji A, Hirai M, Yamashita I, Enomoto S, Noguchi M (1999) Association between water status and sucrose metabolism in cell suspension culture of carrot. Plant Biotechnol 16:375–379. https://doi.org/10.5511/plantbiotechnology.16.375

Ivanova A, Velcheva M, Denchev P, Atanassov A, Van Onckelen HA (1994) Endogenous hormone levels during direct somatic embryogenesis in *Medicago falcata*. Physiol Plant 92:85–89. https://doi.org/10.1111/j.1399-3054.1994.tb06658.x

Jiménez VM, Bangerth F (2001) Hormonal status of maize initial explants and of the embryogenic and non-embryogenic callus cultures derived from them as related to morphogenesis in vitro. Plant Sci 160:247–257. https://doi.org/10.1016/S0168-9452(00)00382-4

Kennedy D, Norman C (2005) What don't we know? Science 309(5731):75–102

Kuriyama A, Kobayashi T, Maeda M (2004) Production of sporophytic plants of *Cyathea lepifera*, a tree fern, from in vitro cultured gametophyte. J Jpn Soc Hortic Sci 73:140–142. https://doi.org/10.2503/jjshs.73.140

Large MF, Braggins JE (2004) Tree ferns. Timber Press, Portland, pp 1–359

Lippert D, Zhuang J, Ralph S, Ellis DE, Gilbert M, Olafson R, Ritland K, Ellis B, Douglas CJ, Bohlmann J (2005) Proteome analysis of early somatic embryogenesis in *Picea glauca*. Proteomics 5:461–473. https://doi.org/10.1002/pmic.20040098

Liu Y, Wujisguleng W, Long C (2012) Food uses of ferns in China: a review. Acta Soc Bot Pol 81:263–270. https://doi.org/10.5586/asbp.2012.046

Makowski D, Tomiczak K, Rybczyński JJ, Mikuła A (2016) Integration of tissue culture and cryopreservation methods for propagation and conservation of the fern *Osmunda regalis* L. Acta Physiol Plant 38:19. https://doi.org/10.1007/s11738-015-2037-y

Martin AB, Cuadrado Y, Guerra H, Gallego P, Hita O, Martin L, Dorado A, Villalobos N (2000) Differences in the contents of total sugars, reducing sugars, starch and sucrose in embryogenic and nonembryogenic calli from *Medicago arborea* L. Plant Sci 29:143–151. https://doi.org/10.1016/S0168-9452 (99) 00251-4

Mikuła A, Tykarska T, Zielinska M, Kuraś M, Rybczyński JJ (2004) Ultrastructural changes in zygotic embryos and somatic embryogenesis of *Gentiana punctata* L. during callus formation and somatic embryogenesis. Acta Biol Cracov Ser Bot 46:109–120. https://doi.org/10.1079/IVP2005678

Mikuła A, Pożoga M, Grzyb M, Rybczyński JJ (2015a) An unique system of somatic embryogenesis in the tree fern *Cyathea delgadii* Sternb. – the importance of explant type, and physical and chemical factors. Plant Cell Tiss Org Cult 123:467–478. https://doi.org/10.1007/s11240-015-0850-z

Mikuła A, Pożoga M, Tomiczak K, Rybczyński JJ (2015b) Somatic embryogenesis in ferns: a new experimental system. Plant Cell Rep 34(5):783–794. https://doi.org/10.1007/s00299-015-1741-9

Moriguchi T, Kita M, Tomono Y, EndoInagaki T, Omura M (1999) One type of chalcone synthase gene expressed during embryogenesis regulates the flavonoid accumulation in citrus cell cultures. Plant Cell Physiol 40:651–655. https://doi.org/10.1093/oxfordjournals.pcp.a029589

Moura IR, Simões-Costa MC, Garcia J, Silva MJ, Duarte MC (2012) *In vitro* culture of tree fern spores from Cyatheaceae and Dicksoniaceae families. Acta Hortic 937:455–461. https://doi.org/10.17660/ActaHortic.2012.937.55

Nic-Can GI, Avilez-Montalvo JR, Aviles-Montalvo RN, Márquez-López RE, Mellado-Mojica E, Galaz-Ávalos RM, Loyola-Vargas VM (2016) The relationship between stress and somatic embryogenesis. In: Loyola-Vargas VM, Ochoa-Alejo N (eds) Somatic embryogenesis: fundamental aspects and applications. Springer International Publishing, Switzerland, pp 151–170. https://doi.org/10.1007/978-3-319-33705-0_9

Nonami H, Schulze ED (1989) Cell water potential, osmotic potential, and turgor in the epidermis and mesophyll of transpiring leaves. Combined measurements with the cell pressure probe and nanoliter osmometer. Planta 177:35–46. https://doi.org/10.1007/BF00392152

Pelleschi S, Rocher JP, Prioul JL (1997) Effect of water restriction on carbohydrate metabolism and photosynthesis in mature maize leaves. Plant Cell Environ 20:493–503. https://doi.org/10.1046/j.1365-3040.1997.d01-89.x

Pérez-García B, Mendoza-Ruiz AC, Espinosa-Matías S, Gómez-Pignataro LD (2010) Gametophyte morphology of *Platycerium andinum* Baker and *Platycerium wandae* Racif. Micron 41:806–813. https://doi.org/10.1016/j.micron.2010.05.005

Pinto G, Silva S, Neves L, Araujo C, Santos C (2010) Histocytological changes and reserve accumulation during somatic embryogenesis in *Eucalyptus globulus*. Trees-Struct Funct 24:763–769. https://doi.org/10.1007/s00468-010-0446-5

Puigderrajols P (2001) Ultrastructure of early secondary embryogenesis by multicellular and unicellular pathways in cork oak (*Quercus suber* L.) Ann Bot 87:179–189. https://doi.org/10.1006/anbo.2000.1317

Raemakers CJJM, Jacobsen E, Visser RGF (1995) Secondary somatic embryogenesis and applications in plant breeding. Euphytica 81:93–107. https://doi.org/10.1007/BF00022463

Raghavan V (2006) Can carrot and *Arabidopsis* serve as model systems to study the molecular biology of somatic embryogenesis? Curr Sci 90:1336–1343

Rathinasabapathi B (2011) Arsenic hyperaccumulator fern *Pteris vittata*: utilities for arsenic phytoremediation and plant biotechnology. In: Fernández H et al (eds) Working with ferns: issues and applications. Springer Science+Business Media, LLC, New York, pp 261–269. https://doi.org/10.1007/978-1-4419-7162-3_19

Rechenmacher C, Schmitt JL, Droste A (2010) Spore germination and gametophyte development of *Cyathea atrovirens* (Langsd. & Fisch.) Domin (Cyatheaceae) under different pH conditions. Braz J Biol 70:1155–1160

Reinert J (1958) Untersuchungen über die Morphogenese an Gewebekulturen. Ber Dtsch Bot Ges 71:15

Rosas MM, Quiroz-Figueroa F, Shannon LM, Ruiz-May E (2016) The current status of proteomic studies in somatic embryogenesis. In: Loyola-Vargas VM, Ochoa-Alejo N (eds) Somatic embryogenesis: fundamental aspects and applications. Springer International Publishing, Cham, pp 103–119. https://doi.org/10.1007/978-3-319-33705-0_7

Rose RJ, Mantiri FR, Kurdyukov S, Chen S-K, Wang X-D, Nolan KE, Sheahan MB (2010) Developmental biology of somatic embryogenesis. In: Pua E-C, Davey MR (eds) Plant developmental biology – biotechnological perspectives, vol 2. Springer, Berlin/Heidelberg. https://doi.org/10.1007/978-3-642-04670-4_1

Rybczyński JJ, Mikuła A (2011) Tree ferns biotechnology: from spores to sporophytes. In: Fernández H, Kumar A, Revilla MA (eds) Working with ferns: issues and applications. Springer Science+Business Media, LLC, New York, pp 135–148. https://doi.org/10.1007/978-1-4419-7162-3_10

Shukla SP, Khare PB (2013) *In vitro* mass multiplication of a threatened tree fern, *Cyathea spinulosa* Wall. ex Hook. Int J Genet Eng Biotechnol 3:15–23. https://doi.org/10.13140/2.1.2282.8168

Somer M, Arbesú R, Menéndez V, Revilla MA, Fernández H (2010) Sporophyte induction studies in ferns in vitro. Euphytica 171:203–210. https://doi.org/10.1007/s10681-009-0018-1

Stamp JA (1987) Somatic embryogenesis in cassava: the anatomy and morphology of the regeneration process. Ann Bot 57:451–459

Steward FC, Mapes MO, Mears K (1958) Growth and organized development of cultured cells. II. Organization in cultures grown from freely suspended cells. Am J Bot 45:705–708

Szypuła W, Pietrosiuk A, Suchocki P et al (2005) Somatic embryogenesis and *in vitro* culture of *Huperzia selago* shoots as a potential source of huperzine A. Plant Sci 168:1443–1452. https://doi.org/10.1016/j.plantsci.2004.12.021

Tchorbadjieva MI (2005) Protein markers for somatic embryogenesis. In: Mujib A, Šamaj J (eds) Somatic embryogenesis. Plant cell monograph, 2nd edn. Springer, Berlin/Heidelberg, pp 215–233. https://doi.org/10.1007/7089

Teng WL, Teng MC (1997) In vitro regeneration patterns of *Platycerium bifurcatum* leaf cell suspension culture. Plant Cell Rep 16:820–824. https://doi.org/10.1007/s002990050327

Tonietto Â, Sato JH, Teixeira JB, Souza EM, Pedrosa FO, Franco OL, Mehta A (2012) Proteomic analysis of developing somatic embryos of *Coffea arabica*. Plant Mol Biol Rep 30:1393–1399. https://doi.org/10.1007/s11105-012-0425-7

Vale EM, Heringer AS, Barroso T, Ferreira ATS, Costa MN, Perales JEA, Santa-Catarina C, Silveira V (2014) Comparative proteomic analysis of somatic embryo maturation in *Carica papaya* L. Proteome Sci 12:37. https://doi.org/10.1186/1477-5956-12-37

Vargas IB, Droste A (2014) *In vitro* propagation of *Cyathea atrovirens* (Cyatheaceae): spore storage and sterilization conditions. Rev Biol Trop 62:299–308

Varhaníková M, Uvackova L, Skultety L, Pretova A, Obert B, Hajduch M (2014) Comparative quantitative proteomic analysis of embryogenic and non-embryogenic calli in maize suggests the role of oxylipins in plant totipotency. J Proteome 104:57–65. https://doi.org/10.1016/j.jprot.2014.02.003

Wang L, Ruan YL (2013) Regulation of cell division and expansion by sugar and auxin signaling. Frontiers in Plant Sci 4:163. https://doi.org/10.3389/fpls.2013.00163
Wu H, Xiu-Qun L, Hua J, Long-Qing C (2010) Effect of light, macronutrients, and sucrose on germination and development of the endangered fern *Adiantum reniforme* var. *sinense* (Adiantaceae). Sci Hort 125:417–421. https://doi.org/10.1016/j.scienta.2010.03.004
Yu R, Zhang G, Li H, Cao H, Mo X, Gui M, Zhou X, Jiang Y, Li S, Wang J (2017) *In vitro* propagation of the endangered tree fern *Cibotium barometz* through formation of green globular bodies. Plant Cell Tiss Org Cult 128:369–379. https://doi.org/10.1007/s11240-016-1116-0
Zhou T, Yang X, Guo K, Deng J, Xu J, Gao W, Lindsey K, Zhang X (2016) ROS homeostasis regulates somatic embryogenesis via the regulation of auxin signaling in cotton. Mol Cell Proteomics 15:2108–2124. https://doi.org/10.1074/mcp.M115.049338

Chapter 7
Biotechnology and Apogamy in *Dryopteris affinis* spp. *affinis*: The Influence of Tissue Homogenization, Auxins, Cytokinins, Gibberellic Acid, and Polyamines

Alejandro Rivera, Paula Conde, Ma Jesús Cañal, and Helena Fernández

Abbreviations

BA	6-benzyladenine
CHA	Bromohydrate-cyclohexylamine
F	Flurprimidol
GA_3	Gibberellic acid
IBA	Indole-3-butyric acid
MS	Murashige and Skoog mineral medium (1962)
TIBA	2,3,5-triiodobenzoic acid

7.1 Introduction

Much of the research on ferns has focused on angiosperms, particularly the requirements for some species to become models. Traditionally, ferns have been cultured for their ornamental qualities, with certain species being used for micropropagation, such as *Nephrolepis* (Fernández and Revilla 2003). *Ceratopteris reinhardii* and *Pteridium aquilinum* (see Chap. 1, Sect. 1.2) also have been used to study basic developmental processes such as photomorphogenesis (Wada 2007), germination (Salmi et al. 2005, 2007; Suo et al. 2015), cell polarity (Salmi and Bushart 2010), cell wall composition (Eeckhout et al. 2014), and reproduction. The gametophyte is an autonomous organism that is well suited for *in vitro* cultures, sample collection, and observation under a microscope (Whittier and Steeves 1960, 1962; von Aderkas 1984; Wen et al. 1999; Fernández and Revilla 2003; Cordle et al. 2007, 2010, 2012;

A. Rivera · P. Conde · M. J. Cañal · H. Fernández (✉)
Area of Plant Physiology, Department of Organisms and Systems Biology (BOS), Oviedo, Spain
e-mail: fernandezelena@uniovi.es

H. Fernández (ed.), *Current Advances in Fern Research*,
https://doi.org/10.1007/978-3-319-75103-0_7

Kazmierczak 2010; López and Renzaglia 2014; Valledor et al. 2014; de Vries et al. 2015; Grossmann et al. 2017).

Most angiosperms reproduce sexually through seeds. However, reproduction by seeds is also possible by asexual means in a process known as apomixis (Grossniklaus et al. 2001; Koltunow and Grossniklaus 2003; Barcaccia and Albertini 2013; Kandemir and Saygili 2015). Even though ferns are seedless plants, apomixis is an important mode of asexual reproduction for them. Apomixis in ferns combines apogamy (the formation of sporophytes from somatic cells of the prothallium) and agamospermy (the formation of unreduced diplo spores) to properly adjust the life cycle (Manton 1950, Lovis 1977; Gastony and Windham 1989). Apomixis in ferns has evolved independently several times and occurs at a much higher frequency than in other major plant groups (Ekrt and Koutecký 2016), although it is concentrated in only four families (Liu et al. 2012). Moreover, apomictic ferns are usually obligate (Lovis 1977). Unlike in ferns, apogamy does not occur naturally in angiosperms (Yang and Zhou 1982). However, apogamous sporophytes might be induced by culturing pollen or embryo sacs (Seguí-Simarro 2010; Germaná 2011).

Dryopteris affinis spp. *affinis* is a diploid fern with an obligated apomictic life cycle. It originated from the crossing of *Dryopteris oreades* and a sexual ancestor of the species *Dryopteris walichiana* or *Dryopteris caucasica* (Salvo 1990). This fern is widely distributed in the Mediterranean region, Macaronesia, and west of Eurosiberia. Because the gametophyte of this subspecies is male and no female reproductive organs are formed, apogamy becomes obligate. Usually, when the gametophyte takes the typical heart shape, it is observed near the apical notch, with a group of small and dark brown cells evolving to form an embryo (see Chap. 1, Fig. 1.3) (Menéndez et al. 2006a). Fern embryos do not develop in a bipolar manner (de Vries et al. 2015); the roots arise adventitiously from a shoot.

Apogamy also can be regarded as a somatic embryogenesis process from gametophytic cells. This process can be induced *in vitro* by manipulating factors such as carbohydrates (Whittier and Steeves 1960, 1962; Whittier 1964, 1965; Kawakami et al. 2003), osmotic conditions (Whittier 1975; von Aderkas 1984), and growth regulators if supplied with sucrose (Whittier 1966; Mehra and Sulklyan 1969; Kato 1970; Elmore and Whittier 1975a, b). However, it is unclear how these factors affect the gametophyte development of an apogamous fern species.

Auxins and cytokinins are the most common phytohormones for *in vitro* culture experiments on embryogenesis and organogenesis in seeds and seedless plants. These plant growth regulators have been reported to affect direct or indirect organogenesis by forming shoots, roots, green globular bodies, and gemmae, among others, from sporophytic or gametophytic tissue (Harper 1976; Emigh and Farrar 1977; Loescher and Alberch 1979; Beck and Caponetti 1983; Hicks and von Aderkas 1986; Higuchi et al. 1987; Higuchi and Amaki 1989; Amaki and Higuchi 1991; Fernández et al. 1996, 1997; Somer et al. 2010; see also the extended and detailed revision by Rybcynski et al. in Chap. 5). In addition, auxins and cytokinins have been connected to embryogenesis in ferns (Menéndez et al. 2006b; Mikula et al. 2015; Grossmann et al. 2017). Other plant growth regulators (e.g., gibberellins, abscisic acid, jasmonic acid, brassinoesteroids, polyamines) have received minor attention to date. Some compounds of the gibberellin family have been regarded as possible candidates for chemical messengers to control the onset of male sexual organs in some species of ferns, known as antheridiogens (Fernández et al. 1999; Menéndez et al. 2006b; Kazmierczak 2010; Valledor et al. 2014). The role of brassinosteroids in fern development is also of interest, as discussed in the chapter by Gómez-Garay et al. A few other reports have focused on the effect of phytohormone inhibitors on this plant group, such as 2,3,5-triiodobenzoic acid (TIBA), which is an inhibitor of the polar transport of auxins (Lee and Seo 2014); flurprimidol, which inhibits gibberellin biosynthesis by blocking the conversion of ent-kaurene to ent-kaurenoic acid (Rademacher 2000); and cyclohexylamine, an inhibitor of spermidine synthase that acts on polyamine spermidine biosynthesis, which is considered to be essential for sexual and somatic embryogenesis (Bouchereau et al. 1999; De la Peña et al. 2008; Deeb et al. 2010; Dutra et al. 2013; Schmidt et al. 2014).

In this study, we cultured spores and homogenized gametophytes in several media to analyze the effects of phytohormones and their inhibitors on the vegetative and asexual development of gametophytes, as well as the regeneration capacity of gametophytes and sporophytes, in the fern *D. affinis* spp. *affinis*.

7.2 Material and Methods

7.2.1 Collection, Cleaning, and Culture of Spores and Gametophytes

Spores of *D. affinis* spp. *affinis* were collected from mature fronds obtained in the Valley of Turón (Asturias, Spain; coordinates 43°12′N and 5°43′W). The removed fronds were placed between sheets of paper and stored in a dry environment to release the spores. Spores and sporangia were sieved to separate spores from the rest of plant material, then kept in vials and stored at 4°C until use. Spores (5 mg) were soaked in water for 2 h and then washed for 10 min with a solution of NaClO (0.5% w/v) containing Tween 20 (0.1% w/v). They were then rinsed three times with sterile distilled water. The spores were centrifuged at 700 g for 3 min between rinses. Prior to the *in vitro* culture of spores, the density was adjusted using an optical microscope (Nikon Eclipse E-600) and a camera (Füsch-Rosenthal).

Spores were cultured separately in 100-mL flasks containing 20 mL Murashige and Skoog (1962) medium (MS) supplemented with 2% (w/v) sucrose and 0.7% (w/v) agar. Unless otherwise noted, the pH was adjusted to 5.7 with 1 or 0.1 N NaOH. The cultures were maintained at 25 °C under cool-white fluorescent light (70 µmol m^{-2} s^{-1}) with a 16:8 h light:dark photoperiod. The culture media of spores were supplemented with the following phytohormones and inhibitors of phytohormones, at the noted concentrations: indole-3-butyric acid (IBA; 0.5, 5, and 25 µM), gibberellic acid (GA_3; 0.3, 3, and 15 µM), 6-benzyladenine (BA; 0.4, 4.4, and 22 µM), TIBA (0.2, 2, and 10 µM), flurprimidol (F; 0.3, 3, and 15 µM), and cyclohexylamine (CHA; 0.06–0.6, 5.6, and 55.5 μM). The range of concentrations was broader for CHA because this was our initial investigation with this compound.

7.2.2 Homogenized Gametophyte and Sporophyte Cultures

Two-month-old gametophytes (0.1–0.25 g fresh weight) or 9-month-old sporophytes (0.2–0.4 g.) were mechanically fragmented using a Waring blender for 15 s under aseptic conditions. The homogenized tissue samples were cultured separately in 500-mL Erlenmeyer flasks that contained 100 mL of liquid MS medium with 2% (w/v) sucrose and supplementation with the following phytohormones and inhibitors of phytohormones, either alone or in combination: IBA 0.5 μM + BA 4.4 μM; IBA 2.5 μM + BA 2.2 μM; IBA 5 μM + BA 0.4 μM; naphthalenacetic acid (NAA) 0.5 μM + BA 4.4 μM; NAA 2.7 μM + BA 2.2 μM; NAA 5.3 + BA μM 0.4; GA_3 1.5 and 3 μM; TIBA 1 and 2 μM; and CHA 0.6 and 2.8 μM. The cultures were placed in an orbital shaker and kept at the same conditions described previously for spore cultures.

7.2.3 Microscopic Analyses

Gametophytes from the mentioned origins were observed in fresh under an optical microscope (Nikon Eclipse E-600) or a stereomicroscope (Nikon, SMZ-U) after 60 days. Data regarding cellular regeneration, vegetative development, gametophyte shape (filamentous, spatula, or heart), length/width ratio, and apogamy (percentage of apogamous gametophytes and the apogamous embryo size) were collected.

The frequency of apogamy was categorized in two ways: 1) gametophytes showing a spot, which is indicative of a two-dimensional organizer; and 2) gametophytes showing a "button," which demonstrates a more developed degree in which the embryo has achieved volume and three-dimensionality. One hundred individuals from each treatment were selected randomly to determine the frequency of morphotypes and the length/width ratios. Apogamy data were based on 50 heart-shaped gametophytes, in which the embryo is usually well defined; however, some spatula-shaped cases were also considered for treatments that accelerated the apogamic process.

7.2.4 Statistical Analysis

The chi-square (χ^2) test was used to analyze non-parametric data, such as the frequency of regeneration, morphotype, and the number of apogamous embryos. For parametric data including the rate, length/width, and size of the embryo, we used an analysis of variance (ANOVA) and ad-hoc testing, such as the Levene and Bartlett test for the homogeneity of variances and the Shaphiro-Wilk test for normality. The data were transformed into logarithms so that they would not follow a normal distribution. In cases where there was no equality between treatments in an ANOVA test, we used post-hoc Bonferroni-Tukey and Duncan tests to obtain the data and observe any similarities that existed in the treatments. For every test, we set a significance level of $\alpha = 0.05$. We used the software programs SPSS (IBM) and Microsoft Excel to carry out the analyses.

7.3 Results

7.3.1 Effects of Phytohormones and Inhibitors on Spore Cultures

Table 7.1 shows data on the effects of the phytohormones (IBA, GA_3, BA) and their inhibitors (TIBA, F and CHA) on the parameters of vegetative development (the percentage of gametophytes showing two-dimensional development and the length/width ratio) and other factors associated with apogamy (the percentage of apogamous gametophytes and the size of embryo). Spore germination was inhibited at the highest tested concentrations of IBA, BA, F, and CHA (Rivera, personal communication). Regarding the percentage of two-dimensional gametophytes, the addition of the phytohormones IBA, BA, or GA_3 to the culture medium did not cause any significant changes compared with the control results. This parameter was increased by F at the concentration of 0.3 μM ($\chi^2 = 19.42$, $p \leq 0.001$) and inhibited by TIBA and CHA ($\chi^2 = 19.42$, $p = <0.001$; $\chi^2 = 95.67$, $p \leq 0.0001$, respectively). Concerning the length/width ratio, no differences were observed with respect to the control; however, TIBA at 10 μM was associated with an increase in width ($\chi^2 = 33.51$, $p \leq 0.01$). The

Table 7.1 Concentrations of phytohormones and their inhibitors used to culture spores of *Dryopteris affinis* spp. *affinis* on MS medium

	Spore cultures			
Treatments	Two-dimensional	Length/width ratio	Apogamy	Embryo size
Control	63%	0.62	67%	0.49 mm²
IBA (0.5, 5, and 25 μM)	No	No	No	No
GA_3 (0.3, 3, and 15 μM)	No	No	*0.3 μM >	No
BA (0.4, 4.4, and 22 μM)	No	No	No	No
TIBA (0.2, 2, and 10 μM)	*2 μM <	*10 μM <	No	No
F (0.3, 3, and 15 μM)	*0.3 μM >	No	No	No
CHA (0.06, 0.6, 5.6, and 55.5 μM)	*5.6 μM <	No	*0.06 μM >	No

Data indicate if there were any changes with respect to the control for vegetative and reproductive development after 60 days.

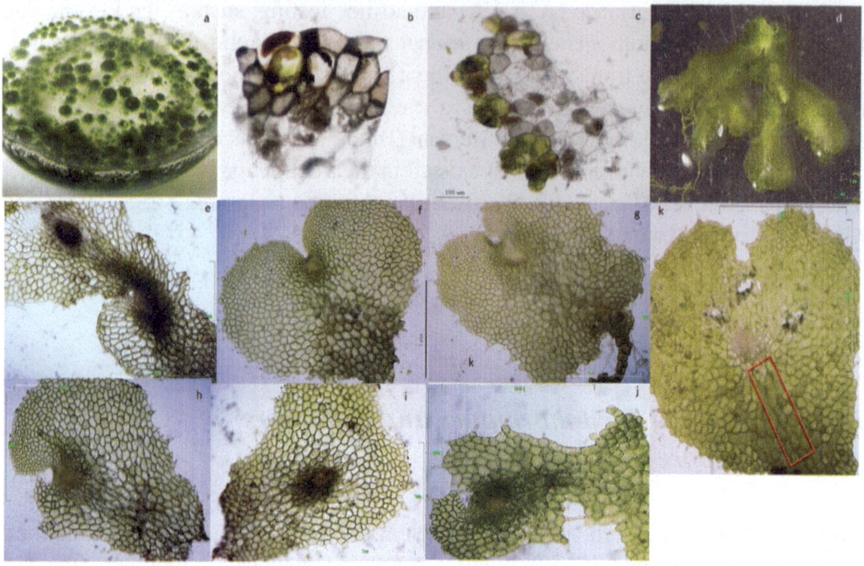

Fig. 7.1 Gametophyte regeneration from homogenized gametophytes of *Dryopteris affinis* spp. *affinis*. (**a**) Aspect of liquid cultures. (**b**, **c**) Recovery of cellular division capacity by one cell or several cells in the explant. (**d**) Gametophyte branching. (**e**) Apogamous embryos in different lobes of the same gametophyte. (**f**) Heart-shaped gametophyte. (**g**) Square-shaped gametophyte cultured in NAA 2.6 μM + BA 2.2 μM. (**h**) Asymmetric gametophyte forming an apogamous embryo, grown with GA_3 3 μM. (**i**, **j**) Apogamous embryo in a spatula-shaped gametophyte, grown in GA_3 3 μM or cyclohexylamine 0,3 μM, respectively. (**k**) A possible route of transport to feed the embryo, highlighted in red

percentage of gametophytes with apogamous embryos increased in the presence of GA_3 at 0.29 μM ($\chi^2 = 6.35$; $p = 0.01$), was inhibited by F at 3 μM ($\chi^2 = 26.83$, $p \leq 0.0001$), and was promoted by CHA at a lower concentration of 0.06 μM ($\chi^2 = 132.2$, $p \leq 0.001$) but inhibited at doses higher than 0.06 μM. No differences were detected in the size of embryos.

Homogenate Cultures from Gametophyte Tissue

The regeneration of gametophytes from the explants was evident in all assayed media (Fig. 7.1a–c). Furthermore, a high formation of secondary gametophytes was noted through the branching from prothallic cells (Fig. 7.1d). The results obtained were not significantly better than those of the controls, with the exception of CHA at 2.8 μM ($\chi^2 = 3.04$; $p = 0.08$; Table 7.2). The two tested formulations of auxin/cytokinin had different effects. The addition of NAA at the highest dose was lethal for the gametophyte survival, but it was not so drastic when using IBA.

Concerning the two-dimensional development of gametophytes, the three tested combinations of BA and IBA gave contrasting results. BA at the highest concentration

Table 7.2 Phytohormones and their inhibitors used to culture homogenized gametophytes of *Dryopteris affinis* spp. *affinis* on MS liquid medium

	Homogenized gametophyte cultures				
Treatments	Regeneration	Two-dimensional	Length/width ratio	Apogamy	Embryo size
Control	60%	60%	1.05	60–70%	0.13 mm^2
IBA 5 μM + BA 0.4 μM	<	+	nd	nd	nd
IBA 2.5 μM + BA 2.2 μM	<	No, callus	nd	nd	nd
IBA 0.5 μM + BA 4.4 μM	<	Callus	nd	nd	nd
NAA 5.3 μM + BA 0.4 μM	–	–	–	–	–
NAA 2.7 μM + BA 2.2 μM	nd	>	No	>	No
NAA 0.5 μM + BA 4.4 μM	nd	>	No	>	No
GA_3 1.5 μM	<	No	nd	>	nd
GA_3 3 μM	NO	<	No	>	No
TIBA 1 μM	NO	<	nd	nd	nd
TIBA 2 μM	<	<	nd	nd	nd
CHA 0.6 μM	>	No	>	>	No
CHA 2.8 μM	<	<	nd	>	nd

Data indicate if there were any changes with respect to the control regarding regeneration, vegetative development, and reproductive development after 70 days of culture. *nd,* not determined.

stopped gametophyte development (χ^2 = 10.17; p < 0.001), whereas a balanced combination of BA/IBA favored a filamentous appearance (χ^2 = 0.42; p < 0.51). In these cases, the formation of cellular aggregates, like a callus, with a solid texture was observed in the cultures. The two-dimensional growth of gametophytes was significantly enhanced by adding a balanced combination of NAA/BA, which was favorable to BA (χ^2 = 77.29; p = 0.001). The presence of the inhibitor CHA at 0.6 μM in the culture medium increased the percentage of gametophytes exhibiting two-dimensional growth (χ^2 = 0.78, p = 0.037). The length/width ratio was only evaluated for the treatments of GA_3 at 3 μM and CHA at 0.6 μM because, at the end of the culture period, there were not enough suitable gametophytes to assess both this parameter and embryo size for the other treatments. The data revealed significant differences in these cases, with the gametophyte shape being longer than wider (F = 135.6, p≤0.001).

After 70 days of culture, apogamy was favored for the two non-lethal combinations of NAA/BA (NAA 2.7 μM + BA 2.2 μM; χ^2 = 19.04; p < 0.001), GA_3 (1.4 and 2.8 μM; χ^2 = 0.008, p = 0.92 and χ^2 = 19.04, p < 0.001, respectively), and CHA (0.6 and 2.8 μM; χ^2 = 13.27, p < 0.001 and χ^2 = 19.04, p < 0.001, respectively). The number of analyzed gametophytes was lowest for the latter case.

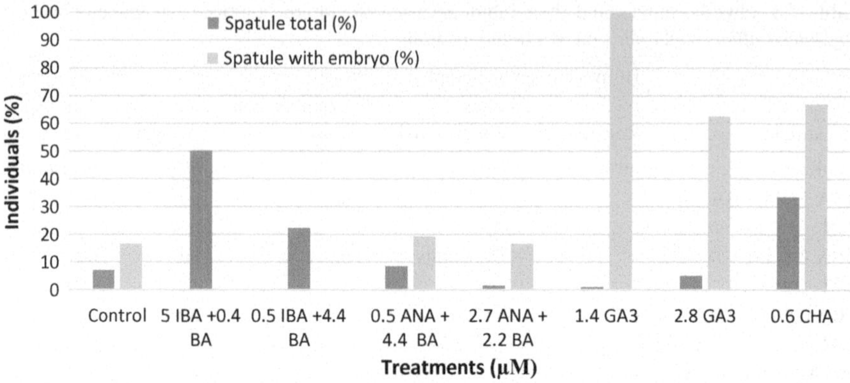

Fig. 7.2 Effects of homogenization, NAA, IBA, BA, GA_3, and CHA inducing apogamy in spatulate-shaped gametophytes of *Dryopteris affinis* spp. *affinis*. Data were collected after 70 days of culture

With the great branching activity in homogenates cultures, apogamous embryos were found to develop in different lobes belonging to the same individual (Fig. 7.1e). Usually, the apogamous embryos become evident when the gametophyte was taking a heart shape (Fig.7.1f); however, they were also found in quadrangles (Fig. 7.1g), asymmetric gametophytes (Fig. 7.1h), and spatulate-shaped gametophytes (Fig. 7.1i, j; Fig. 7.2), without signs of an apical notch or meristematic area. In Fig. 7.1k, the orientation of cell divisions shows the possible transport paths supporting embryo development.

As mentioned previously, the area of the embryos was measured only when adding GA_3 or CHA. The embryo size did not differ significantly from the controls. Finally, in homogenate cultures from sporophyte tissue, no cellular regeneration was observed for any of the tested treatments after 70 days of culture.

7.4 Discussion

Apogamy can be regarded as a case study of somatic embryogenesis, even though the embryo originates from gametophyte cells rather than sporophytes as in seed plants (Schmidt et al. 2015); thus, it occurs without gamete intervention. Ferns are considered to be seedless plants or seed-free plants. Somatic embryogenesis usually denotes the induction of embryos from sporophyte tissue, because sporophytic explants are often studied. However, even though male and female gametophytes are very reduced in seed plants, it is possible to induce somatic embryogenesis *in vitro* through the processes of androgenesis and gynogenesis (Seguí-Simarro 2010; Germaná 2011), which have not been reported to occur naturally in seed plants.

The gametophyte of ferns is a very simple structure known as prothalli, which has its own functional identity that lives spatially and temporally split from the sporophyte. The gametophyte passes from one-dimensional development to grow in length and width. The male and female sexual organs are formed (in most cases but not all) in the same individual, which becomes hermaphrodite. The lack of gametes in the gametophyte impedes a sexual reproduction scheme. This occurs in a significant number of fern species and is known as apogamy.

Apogamy is a highly fixed characteristic. Consequently, this type of embryogenesis does not adhere to an induction system like those in seed plants. Moreover, few studies have reported on either sexual or asexual embryogenesis in seedless plants (Valledor et al. 2014; Mikula et al. 2015; Domzalska et al., 2017; Grossmann et al. 2017). However, because the gametophyte of a fern is a free-living organism, separate from the sporophyte, it might be a good option to enhance plant development, especially for those aspects dealing with reproduction.

In this work, the principal goal was to modify the persistent response of apogamy in *D. affinis* spp. *affinis* through plant growth regulators and their inhibitors. To achieve our objective, we started by culturing single-celled spores because they are simple experimental materials for studying gametophyte development. We expected that germination would not be strongly inhibited at the concentrations used in this study, which ranged from 0.01/0.1 to 5 mg/L. (Here, we use the concentration unit mg/L rather than micromolar numbers to illustrate the similarities among compounds regarding the quantities added to the culture media). In general, it was true that either germination or gametophyte expansion were possible except at the higher doses of IBA, BA, F, and CHA. This was expected because the regulation of growth in plants depends somewhat on the amount of phytohormones present in the cells, tissues, and organs. From these preliminary results, we noted slight variations with respect to the controls, such as when adding GA_3 or the inhibitors TIBA and CHA; thus, it may not be easy to modify growth rates after spore germination has occurred.

Homogenate cultures of both gametophyte and sporophyte tissue were then conducted *in vitro*, with the same goal of analyzing the effects of these compounds on vegetative and reproductive development in the gametophyte. However, we started from explants composed of a few gametophyte cells, which to some extent lost their connections from mechanical disruptions caused by the homogenization process. Because the availability of instruments and space were required to perform *in vitro* homogenate cultures, we reduced the number of conditions tested but still aimed to gather as much data as possible. The homogenate cultures displayed a higher degree of plasticity than did the spores, which allowed us to investigate the extent to which the assayed compounds exogenously disturbed the normal vegetative and reproductive progress as gametophyte regeneration occurred.

Initially, this type of culture was used as an alternative to propagation, particularly fpr ornamental species with great commercial interest (Knauss 1976; Cooke 1979; Finnie and van Staden 1987). Later, in our laboratory, we used this culture many times to explore the effects of phytohormones on phase change—either from gametophyte to sporophyte or vice versa, in what are known as apogamy and apospory events (Fernández et al. 1993, 1997; Fernández and Revilla 2003, Somer et al. 2010). It is well accepted that factors such as the age of plant material, nutrients, growth regulators, and the physiological isolation of cells can affect cell division and differentiation and consequently morphogenesis (Camloh 2006).

In homogenate cultures of *D. affinis* spp. *affinis*, regeneration from gametophyte tissue was theoretically possible for all treatments, but regeneration was never observed when we cultured sporophytic explants. The sporophyte of *D. affinis* spp. *affinis* behaves as a recalcitrant to retain the capacity of cell division and differentiation, unlike what has been reported previously for other fern species such as *Adiantum capillus-veneris, Asplenium adiantum-nigrum, Asplenium nidus-avis, Dryopteris dilatata,* and *Polypodium cambricum* (Somer et al. 2010). In these species, regeneration from sporophytes to sporophytic tissue or aposporous gametophytes was observed with the addition (or not) of the cytokinin BA to the culture media (Somer et al. 2010). It is unclear why the sporophytic cells of *D. affinis* spp. *affinis* are not able to recover the divisional activity; this should be investigated in future studies.

In the homogenate cultures of *D. affinis* spp. *affinis*, the cytokinin BA induced a callus when applied at the highest and intermediate concentrations used in this study. In a previous study with this species (Fernández et al. 1996), gametophytes derived from spores were cultured on a solid MS medium with NAA 0.5 μM + BA 4.4 μM. Calluses occurred with great morphogenic capacity, producing a massive number of sporophytes, when transferred to a hormone-free medium. In our study, neither the auxin IBA nor BA when added separately to the culture medium induced calluses in the spore cultures. Moreover, in the homogenate cultures, callus proliferation was evident with the addition of two combinations of IBA + BA but not NAA + BA. It is not easy to explain the differences found within the same species, or even the same tissue. For more details, the reader is referred to the extensive list of references on the effect of auxins and cytokinins on callus induction in Chap. 5 (Sects. 5.2 and 5.6).

Usually, the formation of an apogamous embryo occurs as the gametophyte reaches the typical heart shape, with a well-defined apical notch or meristematic area from which (in most cases) two symmetric wings or lobes spread. Apogamy has never been reported in a spatula-shaped gametophyte, before the meristematic zone has been outlined. To the best of our knowledge, this work is the first report of the presence of apogamous centers in gametophytes that lacked the typical heart-shaped appearance by the addition of the spermidine biosynthesis inhibitor CHA and GA_3. This compound promoted the regeneration of explants, the transition from one- to two-dimensional growth, and the formation of an apogamous embryo. Our results should be confirmed in future studies.

The essential role of polyamines in the growth and development of plants has been reported previously (Çetin et al. 2000). Furthermore, the synthesis of spermidine is considered to be essential for embryonic development in *Arabidopsis* and in somatic embryogenesis (De la Peña et al. 2008; Dutra et al. 2013). However, transcriptional levels of compounds related to the biosynthesis of the polyamines spermine and spermidine, as well as the catabolism of arginine, have been observed to be upregulated in the pre-apogamic stages of gametophytes of this species (Fernández, personal communication). The effects of gibberellic acid on accelerating apogamy were consistent with previous analyses (Menéndez et al. 2006a), which showed increased levels of GAs as apogamy progresses and the inhibition of gametophytes in developing embryos with the GA inhibitor flurprimidol in the culture medium.

Finally, TIBA induced high regeneration when added to the medium at a concentration of 1 μM, although the gametophytes were mainly filamentous. TIBA is an inhibitor of auxin transport; its function is to prevent the distribution of auxins from the apical meristem to the rest of the plant (Lee and Seo 2014). Proteins involved in the transport of auxins have been identified. The existence of a transport system in the gametophytes of *D. affinis* spp. *affinis* has been revealed (Grossmann et al. 2017), which otherwise might be presumed by the orientation of cell divisions, outlining possible routes to feed the embryos. The possible block of protein transport by TIBA may somehow affect the transition toward a two-dimensional stage from the filamentous stage; however, further investigations are needed.

7.5 Conclusions

In this study, homogenate cultures of gametophytes were used to investigate the regeneration capacity of ferns. In principle, the gametophytes of ferns are an interesting experimental system to study the development of these plants, serving as an alternative to spores as the starting plant material of the apogamous species *Dryopteris affinis*. Using homogenate cultures from homogenized gametophytes, our results showed that factors such as cellular disruption or the presence of GA_3 and CHA can speed up apogamy in the regenerated gametophytes of *D. affinis* spp. *affinis*. Developing embryos were visible in a spatula shape, before the meristematic area and lobulated wings were defined, which is typical of a heart-shaped gametophyte. Homogenate cultures from gametophytic tissue may be a good option for investigating apogamy by manipulating the endogenous and exogenous factors that have a role in this process.

References

von Aderkas P (1984) Promotion of Apogamy in *Matteuccia struthiopteris*, the Ostrich Fern. Am Fern J 74(1):1–6

Amaki W, Higuchi H (1991) A possible propagation system of *Nephrolepis, Asplenium, Pteris, Adiantum* and *Rumora* through tissue culture. Acta Hortic 300:237–243

Barcaccia G, Albertini E (2013) Apomixis in plant reproduction: a novel perspective on an old dilemma. Plant Reprod 26:159–179. https://doi.org/10.1007/s00497-013-0222-y

Beck MJ, Caponetti JD (1983) The effects of kinetin and naphthalenacetic acid *in vitro* shoot multiplication and rooting in the fishtail fern. Am J Bot 70:1–7

Bouchereau A, Aziz A, Larher F, Martin-Tanguy J (1999) Polyamines and environmental challenges: recent development. Plant Sci 140:103–125

Camloh M (2006) *In vitro* culture of the fern *Platycerium bifurcatum* as a tool for various physiological and developmental studies. In: Floriculture, ornamental and plant biotechnology advances and topical issues. Springer, New York, pp 163–170

Cetin E, Yildirim C, Palavan Unsal N, Unal M (2000) Effect of spermine and cyclohexylamine on *in vitro* pollen germination and tube growth in *Helianthus annuus*. Can J Plant Sci 80:241–245

Cooke RC (1979) Homogenization as an aid in tissue cultura propagation of *Platycerium* and *Davallia*. HortSci 14:21–22

Cordle AR, Irish EE, Cheng CL (2007) Apogamy induction in *Ceratopteris richardii*. Int J Plant Sci 168:361–369. https://doi.org/10.1086/511049

Cordle AR, Bui LT, Irish EE, Cheng CL (2010) Laboratory-induced apogamy and apospory in Ceratopteris Richardii. In: Fernández H, Kumar A, Revilla MA (eds) Working with ferns: issues and applications. Springer, New York/Dordretch-Heidelberg/London, pp 25–36

Cordle AR, Irish EE, Cheng CL (2012) Gene expression associated with apogamy commitment in *Ceratopteris richardii*. Sex Plant Reprod 25:293–304

Deeb F, van der Weele CM, Wolniak SM (2010) Spermidine is a morphogenetic determinant for cell fate specification in the male gametophyte of the Water Fern *Marsilea vestita*. The Plant Cell Online 22:3678–3691. https://doi.org/10.1105/tpc.109.073254

De-la-Peña C, Galaz-Avalos RM, Loyola-Vargas VM (2008) Possible role of light and polyamines in the onset of somatic embryogenesis of *Coffea canephora*. Mol Biotechnol 39:215–224

Domżalska L, Kędracka-Krok S, Jankowska U, Grzyb M, Sobczak M, Rybczyński JJ, Mikuła A (2017) Proteomic analysis of stipe explants reveals differentially expressed proteins involved in early direct somatic embryogenesis of the tree fern *Cyathea delgadii* Sternb. Plant Sci 258:61–76. https://doi.org/10.1016/j.plantsci.2017.01.017

Dutra N, Silveira V, Gonçalves I, Ribeiro Gomes-Neto J, Façanha A, Steiner N, Santa-Catarina C (2013) Polyamines affect the cellular growth and structure of pro-embryogenic masses in *Araucaria angustifolia* embryogenic cultures through the modulation of proton pump activities and endogenous levels of polyamines. Physiol Plant 148(1):121–132

Eeckhout S, Leroux O, Willats WGT, Popper ZA, Viane RLL (2014) Comparative glycan profiling of *Ceratopteris richardii* C-Fern` gametophytes and sporophytes links cell-wall composition to functional specialization. Ann Bot 114:1295–1307

Ekrt L, Koutecký P (2016) Between sexual and apomictic: unexpectedly variable sporogenesis and production of viable polyhaploids in the pentaploid fern of the *Dryopteris affinis* agg. (Dryopteridaceae). Ann Bot 117:97–106. https://doi.org/10.1093/aob/mcv152

Elmore HW, Whittier DP (1975a) The involvement of ethylene and sucrose in the inductive and developmental phases of apogamous bud formation in *Pteridium* gametophytes. Can J Bot 53:375–381

Elmore HW, Whittier DP (1975b) Ethylene and carbohydrate requirements for apogamous bud induction in *Pteridium* gametophytes. Can J Bot 52:2089–2096

Emigh VD, Farrar DR (1977) Gemmae: a role in sexual reproduction in the fern gnus Vittaria. Science 198:297–298

Fernández H, Revilla MA (2003) *In vitro* culture of ornamental ferns. Plant Cell Tissue Org Cult 73:1–13. https://doi.org/10.1023/A:1022650701341

Fernández H, Bertrand AM, Sánchez-Tamés R (1993) *In vitro* regeneration of *Asplenium nidus* L. from gametophytic and sporophytic tissue. Sci Hortic 56:71–77

Fernández H, Bertrand AM, Sánchez-Tamés R (1996) Influence of tissue culture conditions on apogamy in *Dryopteris affinis ssp. Affinis*. Plant Cell Tissue Organ Cult 45(1):93–97

Fernández H, Bertrand A, Sánchez-Tamés R (1997) Plantlet regeneration in *Asplenium nidus* l. and *Pteris ensiformis* l. by homogenization of Ba treated rhizomes. Sci Hortic 68:243–247. https://doi.org/10.1016/S0304-4238(96)00986-7

Fernández H, Bertrand AM, Sierra MI, Sánchez-Tamés R (1999) An apolar GA-like compound responsible for the antheridiogen activity in *Blechnum spicant*. Plant Growth Regul 28:143–144. https://doi.org/10.1023/A:1006263326546

Finnie JF, Van Staden J (1987) Multiplication of the tree fern *Cyathea degrei*. HortSci 22:665

Gastony GJ, Windham MD (1989) Species concepts in pteridophytes: the treatment and definition of agamosporous species. Amer Fern J 79:65–77

Germanà MA (2011) Gametic embryogenesis and haploid technology as valuable support to plant breeding. Plant Cell Rep 30:839–857

Grossmann J, Fernández H, Chaubey PM, Valdés AE, Gagliardini V, Cañal MJ, Russo G, Grossniklaus U (2017) Proteogenomic analysis greatly expands the identification of proteins related to reproduction in the Apogamous Fern *Dryopteris affinis* ssp. *affinis*. Front Plant Sci 8. https://doi.org/10.3389/fpls.2017.00336

Grossniklaus U, Nogler GA, van Dijk PJ (2001) How to avoid sex: the genetic control of developmental aspects. Plant Cell 13(7):1491–1497

Harper K (1976) Asexual multiplication of Leptosporangiatae ferns through tissue culture. Dissertation, University of California.

Hicks G, von Aderkas P (1986) A tissue culture of the ostrich fern *Matteuccia struthiopteris* L. Todaro. Plant Cell Tissue Organ Cult 5:199–204

Higuchi H, Amaki W (1989) Effects of 6-benzyladenine on the organogenesis of *Asplenium nidus* L. through *in vitro* propagation. Sci Hortic 37:351–359

Higuchi H, Amaki W, Suzuki S (1987) *In vitro* propagation of *Nephrolepis cordifolia* Prsel. Sci Hortic 32:105–113

Kandemír N, Saygili I (2015) Apomixis: new horizons in plant breeding. Turk J Agric For 39(4):549–556. http://journals.tubitak.gov.tr/havuz/tar-1409-74.pdf

Kato Y (1970) Physiological and morphogenetic studies of fern gametophytes and sporophytes in aseptic culture. XII. Sporophyte formation in the dark cultured gametophyte of *Pteris vittata* L. Bot Gaz 121:205–210

Kawakami, S.M., Kawakami, S., Kondo, K., Kato, J, and Ito, M. 2003. Sporogenesis in haploid sporophytes ofOsmunda japonica (Osmundaceae). Int. J. Plant Sci. 164:527–534

Kazmierczak A (2010) Gibberellic acid and ethylene control male sex determination and development of *Anemia phyllititdis* gametophytes. In: Fernández H, Kumar A, Revilla MA (eds) Working with ferns. Issues and applications. Springer, New York, pp 49–65

Knauss JF (1976) A partial tissue culture method for pathogen-free propagation of selected ferns from spores. Proc Fla State Hort 89:363–365

Koltunow AM, Grossniklaus U (2003) Apomixis: a developmental perspective. Annu Rev Plant Biol 54:547–574

Lee K, Seo PJ (2014) The Arabidopsis E3 ubiquitin ligase HOS1 contributes to auxin biosynthesis in the control of hypocotyl elongation. Plant Growth Regul 76:157–165. https://doi.org/10.1007/s10725-014-9985-x

Liu HM, Dyer RJ, Guo ZY, Meng Z, Li JH, Schneider H (2012) The evolutionary dynamics of apomixis in ferns: a case study from Polystichoid ferns. J Bot 11:510478. https://doi.org/10.1155/2012/510478

Loescher WH, Alberch CN (1979) Development in vitro of *Nephrolepsis exaltata* cv. *Bostoniennsis* runner tissues. Physiol Plant 47:250–254

Lopez RA, Renzaglia KS (2014) Multiflagellated sperm cells of *Ceratopteris richardii* are bathed in arabinogalactan proteins throughout development. Am J Bot 101:2052–2061

Lovis JD (1977) Evolutionary patterns and processes in ferns. Adv Bot Res 4:229–415

Manton I (1950) Problems of cytology and evolution in the Pteridophyta. Cambridge University Press, Cambridge

Mehra PN, Sulklyan DS (1969) *In vitro* studies on apogamy, apospory, and controlled differentiation of rhizome segments of the fern, Ampelopteris prolifera (Retz.) Copel. Bot J Linn Soc 62:431–443

Menéndez V, Revilla MA, Bernard P, Gotor V, Fernández H (2006b) Gibberellins and antheridiogen on sex in *Blechnum spicant* L. Plant Cell Rep 25(10):1104. https://doi.org/10.1007/s00299-006-0149-y

Menéndez V, Villacorta NF, Revilla MA, Gotor V, Bernard P, Fernández H (2006a) Exogenous and endogenous growth regulators on apogamy in *Dryopteris affinis* (Lowe) Fraser-Jenkins sp. *affinis*. Plant Cell Rep 25:85–91. https://doi.org/10.1007/s00299-005-0041-1

Menéndez V, Revilla MA, Fal MA, Fernández H (2009) The effect of cytokinins on growth and sexual organ development in the gametophyte of *Blechnum spicant* L. Plant Cell Tissue Org Cult 96:245–250. https://doi.org/10.1007/s11240-008-9481-y

Mikuła A, Pożoga M, Tomiczak K, Rybczyński JJ (2015) Somatic embryogenesis in ferns: a new experimental system. Plant Cell Rep 34:783–794

Murashige T, Skoog F (1962) A revised medium for rapid growth and bioassays with tobacco tissue culture. Physiol Plant 15:473–497

Rademacher 2000. Growth retardants: Effects on gibberellin biosynthesis and other metabolic pathways. Ann.Rev. Plant Physiol. and Mol. Biol. 51: 501–531

Salmi ML, Bushart TJRS (2010) Cellular, molecular, and genetic changes during the development of *Ceratopteris richardii* gametophytes. In: Fernández AK H, Revilla MA (eds) Working with Ferns. Issues and applications. Springer, New York, pp 11–24

Salmi ML, Bushart TJ, Stout SC, Roux SJ (2005) Profile and analysis of gene expression changes during early development in germinating spores of *Ceratopteris richardii.* Plant Physiol 138:1734–1745

Salmi ML, Morris KE, Roux SJ, Porterfield DM (2007) Nitric oxide and cGMP signaling in calcium-dependent development of cell polarity in *Ceratopteris richardii.* Plant Physiol 144:94–104

Salvo E (1990) Guía de helechos de la Península Ibérica y Baleares. Ediciones Pirámide S.A.

Schmidt A, Schmid MW, Klostermeier UC, Qi E, Guthörl D, Saliler C, Waller M, Rosenstiel P, Grossniklaus U (2014) Apomictic and sexual germline development differ with respect to cell cycle, transcriptional, hormonal and epigenetic regulation. PLoS Genet 10:e1004476

Schmidt A, Schmid MW, Grossniklaus U (2015) Plant germline formation: common concepts and developmental flexibility in sexual and asexual reproduction. Development 142:229–241

Seguí-Simarro JM (2010) Androgenesis Revisited. Bot Rev 76:377–404

Somer M, Arbesú R, Menéndez V, Revilla MA, Fernández H (2010) Sporophyte induction studies in ferns *in vitro*. Euphytica 171:203–210. https://doi.org/10.1007/s10681-009-0018-1

Suo J, Zhao Q, Zhang Z, Chen S, Cao J, Liu G, Wei X, Wang T, Yang C, Dai S (2015) Cytological and proteomic analyses of *Osmunda cinnamomea* germinating spores reveal characteristics of Fern spore germination and rhizoid tip growth. Mol Cell Proteomics 14:2510–2534

Valledor L, Menéndez V, Canal MJ, Revilla A, Fernández H (2014) Proteomic approaches to sexual development mediated by antheridiogen in the fern *Blechnum spicant* L. Proteomics 14(17–18):1–11

de Vries J, Fischer AM, Roettger M, Rommel S, Schluepmann H (2015) Cytokinin-induced promotion of root meristem size in the fern *Azolla* supports a shoot-like origin of euphyllophyte roots. New Phytol 209(2):705–720. https://doi.org/10.1111/nph.13630

Wada M (2007) The fern as a model system to study photomorphogenesis. J Plant Res 120:3–16

Wen CK, Smith R, Banks JA (1999) ANI1. A sex pheromone-induced gene in *Ceratopteris* gametophytes and its possible role in sex determination. Plant Cell 11:1307–1318

Whittier DP (1964) The influence of cultural conditions on the induction of apogamy in *Pteridium* gametophytes. Am J Bot 51:730–736

Whittier DP (1965) Obligate apogamy in *Cheilantes tomentosa* and *C. alabamiensis*. Bot Gaz 126:275–281

Whittier DP (1966) The influence of growth substances on the induction of apogamy in *Pteridium* gametophytes. Am J Bot 53:882–886

Whittier DP (1975) The influence of osmotic conditions on induced apogamy in *Pteridium* gametophytes. Phytomorphology 25:246–249

Whittier DP, Steeves TA (1960) The induction of apogamy in the bracken fern. Can J Bot 38:925–930

Whittier DP, Steeves TA (1962) Further studies on induced apogamy in ferns. Can J Bot 40:1525–1531

Yang HY, Zhou C (1982) *In vitro* induction of haploid plants from unpollinated ovaries and ovules. Theor Appl Genet 63:97–104

Chapter 8
Scope of Ferns in Horticulture and Economic Development

Ajit Pratap Singh and Deepali Johari

8.1 Introduction

The term “horticulture” consists of two words, viz., *hortus* (meaning garden) and *colere* (meaning to cultivate), and was first conceived by Peter Laurenberg. Fern horticulture could be defined as the involvement of science and art of growing the ferns for produce and economic development. Ferns are nonflowering, vascular, seedless plants that are produced by spores and less known for horticulture and economic perspectives. They have a long geological history of evolution (Bennici 2008) and are represented by 10,000 species over the world. Ferns are the important foliage that could be used in establishing, designing, and beautifying the gardens. They grow in very specific climatic conditions (Singh et al. 2017), but the practice of direct introduction, in vitro propagation, and mass multiplication may also help in conserving the ferns (Fig. 8.1a). Ferns can be grown as one of the horticultural crop of ornamental value on unutilized geographical regions to bring the unused soil under cultivation. It may provide additional foliose materials for development of a variety of products as well as societal commodity. Now questions arise: What could be done by the promotion of fern horticulture? How is it beneficial for the people residing in those areas, where the opportunity of employment, food security, medication, and economic crunch is more prevalent? How can fern horticulture contribute in economic development of an individual or the whole country? Is there any scope and application of fern horticulture that can bring the economic development and better livelihood for farmers, women, rural, and unprivileged people? Ferns do not give grains, fruits, vegetables, and flowers, but they can

A. P. Singh (✉) · D. Johari
Pteridology Laboratory, Plant Diversity, Systematics & Herbarium Division, CSIR-National Botanical Research Institute, Lucknow, UP, India

H. Fernández (ed.), *Current Advances in Fern Research*,
https://doi.org/10.1007/978-3-319-75103-0_8

Fig. 8.1 The cultivation and uses of ferns. (**a**) A landscape of fern house. (**b**) *Drynaria quercifolia* growing on tree trunk. (**c**) *Microsorum scolopendrium* growing on coal balls. (**d**) *Doryopteris ludens* growing in earthen pot. (**e**) Use of 'Ghora patti fern" and other raw materials for floral articles at a florist's shop in Lucknow. (**f**) Poor selling *Selaginella bryopteris* for livelihood and showing two bunches of resurrected "sanjeevani booti." (**g**) Florist attracting the customers by "Ghora patti ferns." (**h**) A big package of bunches of "Ghora patti fern" at the florist's shop in Lucknow. (**i**) *Microsorum alternifolium* grows upward and forms a rosette in the basket. (**j**) *Nephrolepis exaltata* in hanging basket. (**k**) *Microlepia strigosa* in hanging basket

provide other necessity of our life, on the cost of their characteristic features of biomass, ornamental value, physio-biochemical properties, metabolic activity, potential molecules, and the ecosystem services they render. Fern horticulture can provide enormous opportunity in the areas of nurseries, indoor and outdoor decoration, foliose and floral products, ornamentals, food, medicines, fertilizer, in vitro micro-propagation, hybrid development, job opportunities, academic and research organizations, and government and nongovernmental developmental projects to sustain the livelihood and economy of rural people.

A variety of ferns grow on tree trunk, rough coal balls, calcareous rock pieces, and soil and in earthen pots (Fig. 8.1b–d), whereas a few grow in water and hanging baskets and as climber and provide aesthetic beauty. *Cyrtomium falcatum* is liked by people for its thick texture and color (Mickel 2003). Use of ferns in floral works shows their implication and worth. A variety of ferns can be used as cut foliose if they can be properly produced, processed, and maintained for post-harvest characteristics (Singh et al. 2003). Survey in Lucknow, India, revealed that the florists use *Nephrolepis tuberosa*, locally known as "Ghora patti fern," for floral works (Fig. 8.1e, g). Ghora patti fern is highly priced, and one bunch (comprised of 50 fronds) cost ranges from 80 to 200 Indian rupees (Fig. 8.1h). Besides, few of the ferns are plants of "curiosity" and become source of employment. *Selaginella bryopteris* "sanjeevani booti" is such a curious plant, whose literal meaning is the "herb which revives the life in dead." It has high demand for the belief that it can "revive the dead." Dried plant (a bunch of ten fronds) of *S. bryopteris* is sold in the name of "sanjeevani booti" for livelihood (Fig. 8.1f). People in different socio-practices use a variety of edible ferns. *Ophioglossum reticulatum*, *Diplazium esculentum*, and *Ampelopteris prolifera*, locally known as "ekpatiya," "dhekiya," and "macchinure," respectively, are delicious foods of the tribes (Singh and Johari 2015). The demand of *Matteuccia struthiopteris* has expanded beyond the boundary of the USA, and the harvest rate reached up to 200 tons per year (Aderkas 1984). *Helminthostachys zeylanica* and *Huperzia serrata* are known for medicinal significance (Singh and Khare 2011; Singh and Johari 2015; Ma et al. 2005). *Azolla* is efficient for mineralized organic nitrogen (Bocchi and Malgioglio 2010) and therefore could be mulched to increase soil fertility (Brouwer et al. 2017). Variable forms and variety of ferns are required by the people. Establishment of a hi-tech fern horticultural infrastructure is the only option to deliver to the demand. Aesthetic beauty of ferns in hanging baskets, potteries, plastic planters, terrariums, containers, and vertical-wall wells (Fig. 8.1i–k) varies from species to species, and such ferns can be sold in high price. They could be multiplied using the spores or tissues of gametophyte and sporophyte (Singh et al. 2012, 2013, 2017) using different culture media to provide thousands of individuals for sustainable uses (Fig. 8.2a–i). The pinnae or pinnules of some ferns could be harvested, properly dried, and used individually or with the flowers to prepare dried foliose scenes (Fig. 8.3a). The entire process of fern horticulture is labor incentive; thus it may guarantee job opportunity and human resource development. Recent studies (Soni et al. 2012; Singh et al. 2017; Shukla et al. 2016) have revealed that the ferns are useful in understanding the basic and applied researches. The market demand of ferns is increasing day by day, but the supply is completely reliant on natural forests, as there is no any horticulture practice. Nowadays, various developmental projects of government and nongovernmental organizations need a variety of indoor and outdoor ferns for the development of green belts. Keeping in mind that the ferns have become an inevitable component in our life, they should be multiplied through horticulture practices and supplied for various purposes. The present paper discourses about the scope of ferns in horticulture, their usefulness in various walks of life, and contribution in economic develop-

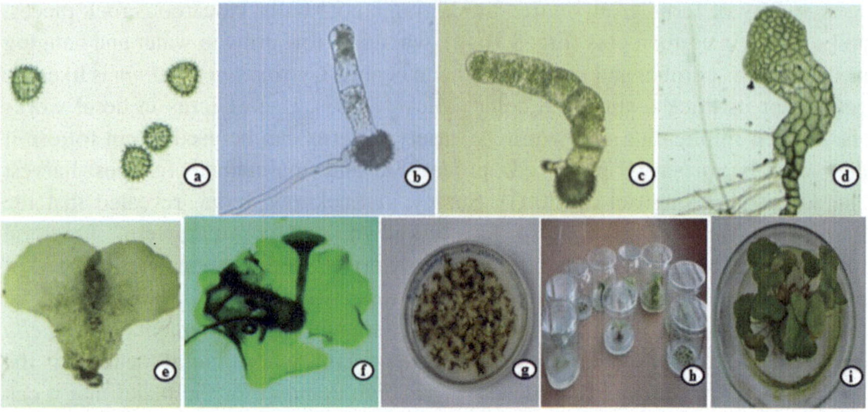

Fig. 8.2 In vitro culture of spores and gametophytes of *Anemia rotundifolia* for multiplication and conservation. (**a**) Spores. (**b**, **c**) Filamentous gametophyte. (**d**) Semi-spathulate gametophyte. (**e**) Cordate gametophyte. (**f**) Sporophyte. (**g**) Sporophytes produced in petri dishes. (**h**) Sporophytes growing under glass chimneys for growth and development. (**i**) Magnified view of sporophyte

ment and future perspectives, which can lure the common people to turn up toward fern horticulture for economic development and better livelihood.

8.2 Horticulture and Pteridophytes

Feathery fronds of fern form a variety of structure that could incline the people to use the ferns for indoor decoration and landscaping. Fern has a huge scope in landscape horticulture, as it involves the production, marketing, and maintenance of landscape ferns. Previous information suggests that the ferns were conserved in the gardens for display and to enhance the scenic beauty of the gardens (Kawano 2015; Huang et al. 2008). Ferns were used as indoor plants (Jones 1987) and are grown in gardens and backyards to add beauty (Kawano 2015). A variety of ferns, including *Microsorum scolopendrium*, *Nephrolepis tuberosa*, and *Pronephrium nudatum*, are conserved in the garden of CSIR-NBRI, India. The Third World countries in Southeast Asia are exploring various ways to enhance their economic growth (Yhome and Chaturvedy 2017), where ferns can be sustainably utilized. A variety of ferns could be used for decorating the rooms, pavilion, flower events, shops, etc. (Fig. 8.3b), in addition to festooning the dais and banquets (Jones 1987). We think of ferns as being suitable for horticulture practices, but the major problem is in identifying the suitable species for production, durability of fronds, and indoor uses. Unless there is no information about horticultural usefulness of ferns, it is hardly easy to select a fern that could be of diverse potential, including florist industry and indoor or outdoor decoration. There is need to demonstrate wide application and usefulness of suitable fern species that could be utilized for horticultural practices.

Fig. 8.3 Ornamental compositions of ferns. (**a**) Scene on greeting cards by the use of pinnae. (**b**) Display of fern canopy and *Selaginella bryopteris* "sanjeevani booti" in annual flower event. (**c**) Fern and white flower wedding bouquet. (**d**) Arrangement of three "button flowers" using *Nephrolepis exaltata* var. *fluffy-ruffles* and pink rose. (**e**) Arrangement of fern, yellow-white gladioli, and red rose for a theme of goddesses "Saraswati"

8.3 Significance of Fern Horticulture

Horticulture substantially contributes in agriculture production and economic development. Survey on state-wise value of output of floriculture in India revealed that the total output of the floriculture during the year 2010–2011 reached around 139.7 billion rupees. Unfortunately the floriculture data did not provide any information about ferns. The ferns may be valued because of their eye-catching foliage, decorative usefulness, and demand as food and medicine and as an essential component for product development. Ferns could be a popular horticultural crop as many species of ferns are often grown as indoor ornamentals for decorative purposes (Adam et al. 2012). Food of people from far-flung regions are also obtained from a variety of ferns. The croziers are used as delicious vegetable and highly prized in Japan, the Philippines, and the USA (Jones 1987; Singh and Johari 2015). Some of the ferns, viz., *Nephrolepis tuberosa*, are used as intrinsic articles in preparation of variety of floral crafts such as bouquet (Fig. 8.3c). Some ferns, viz., *Huperzia selago*, are used for treatment of neurodegenerative diseases; however, many drugs are developed from their chemicals (Talukdar et al. 2011). Horticultural practice is the ultimate option that may propagate the ferns up to the extent of their demand. Fern horticulture may provide high yield per unit area, employment throughout the year, raw materials for fern-based industries, and best utilization of wasteland, where agronomical crops cannot be cultivated and enough potential ferns are needed for humankind.

8.4 Demand of Fern Horticulture

Horticultural practice has increased about 18% than other crop groups in India, and the total output of the floriculture (flower and foliose) reached highest compared to the previous records (Horticulture Statistics at a Glance 2015). Unfortunately, the share of ferns in foliose industry for their demand, supply, as well as economic share could not be quantified due to lack of appropriate knowledge; however, the numbers of ferns are chief constituent in foliose industry. *Adiantum capillus-veneris* and *Huperzia serrata* are medicinally important (Singh and Khare 2011), whereas *Diplazium esculentum* and *Ophioglossum reticulatum* are delicious food in Pacific Asia and serve as alternative food materials. A study entitled "Indian Floriculture Industry: The Way Ahead" of ASSOCHAM claimed that "to provide demands of raw materials related to floriculture (flowers and foliose), the shift toward horticulture has enhanced the productivity of horticulture crops (287.3 million tons in 2016–2017)." Due to changes in the people's liking, the use pattern of foliose plants in India has been shifting toward ferns. Such shifting toward fern is also occurring globally, which opens a wide scope of fern horticulture. An estimate made by the authors in Lucknow, India, has shown that the florists use a variety of ferns for different purposes. Florists informed that the species of *Nephrolepis* "Ghora patti fern" has multifarious application in foliose industry (e.g., Fig. 8.1). It is supplied in the entire country from Darjeeling, West Bengal. *Selaginella bryopteris* "sanjeevani booti" is sold in the market by the poor for their livelihood (Fig. 8.1f); the cost of one bunch of "sanjeevani booti" ranged from 30 to 50 rupees. There is imposed high risk on natural endurance of such significant ferns as there are no horticultural practices for production of ferns. Survey revealed that the cost of one bunch (comprised of 50 fronds) of "Ghora patti fern" ranges from 80 to 200 Indian rupees, and the price of indoor decorative ferns in earthen pots also reaches up to thousand rupees. Ferns are inevitable foliose in preparation of floral works including bouquets, garland of flowers, and artistic arrangement of cut flowers including preparation of arrangement of different flowers (Fig. 8.3c–e). Such huge demand of ferns signifies its economic worth. The increasing demand for ferns means that the more knowledge about horticultural practices of propagation techniques is to be developed to guarantee the continuous supply of ferns (Fernandez and Revilla 2003).

8.5 Different Fern Horticulture Practices

Fern horticultural practice involves the cultivation of ferns by application of the best known techniques for plant selection, plantation, breeding, artificial culture, genetic engineering, rearing, propagation, management, and harvesting. Before initiation of horticulture, one should keep in mind that the ferns grow in different habitats, i.e., epiphyte, aquatic, lithophytes, and saxicolous; thus they require a variety of substratum for growth. It is also necessary to know about the characteristic features of ferns, as different species of ferns are required for different purposes. Fern horticulture requires selection of a particular or group of ferns to grow individually or together for large-scale production. One should also analyze the market demand of respective fern species. If the market demands single species, then mono-foliar horticultural practice is suggestive (Fig. 8.4a). If the market demands a variety of ferns, then multi-species horticultural practices are preferred (Fig. 8.4b). Horticulture should be opted on the basis of the growth habits of ferns, whether they grow on soil, rocks, and tree trunk or in water. As *Nephrolepis tuberosa* grows on composted soil under shade and moist conditions, therefore, such area may be considered for mono-foliar horticulture practices (Fig. 8.4a). *Nephrolepis tuberosa* can also be grown in the mango orchard, as the orchard provides composted soil, shade, and moist conditions. Such practice may extend opportunity for enhanced fern production in suitable plantations. In multi-foliar horticultural practices, a variety of ferns may be grown together (Fig. 8.4b). Besides, different habitats may be established in proposed horticultural land to grow a variety of ferns. Spores, tubers, suckers, and rhizomes, including buds and tissues of gametophyte and sporophyte, may be used for propagation of selected fern species. It is wise to select those ferns, which are most commonly used for ornamental and economic benefits.

8.6 Climate Factors that Aid in Fern Horticulture

Fern horticulture requires a variety of favorable climatic factors. Fern occurs worldwide (Moran 2008), in different habitats (Page 1979), but more factual they grow in moist, shade, and cool conditions with moderate humidity, low temperature, as well as air (Nondorf et al. 2003). Singh et al. (2017) recorded growth of *Lepisorus nudus* at 50 μmol m^{-2} sec^{-1} light intensity with a photoperiod of 10:14 h (light and dark) and temperature of 25 ± 1 °C. Many of the contributors also observed the role of climatic factors, such as the influence of soil (Jones et al. 2011), climatic

Fig. 8.4 Fern horticulture practices. (**a**) Mono-foliar horticulture of *Microsorum alternifolium*. (**b**) Multi-foliar horticulture of ferns. (**c**) Development of new fronds of *Microsorum scolopendrium*, after plantation

variables (de Gasper et al. 2015), relative humidity (Dzwonko and Kornas 1994), precipitation (Cardelús et al. 2006), and the period of dry and wet season (Jones et al. 2013) for fern establishment. For instances, in the fern house of CSIR-NBRI, the agro-shade materials avoid up to 65% irradiance light. Besides, the water is sprinkled for 5 min after each 2 h in summer to provide moisture, water, and relative humidity (Fig. 8.4b). The congenial climatic factors that include optimal water, light, temperature, relative humidity, and air are required that may aid in fern horticultural practices.

8.7 Pre-horticultural Conditions

The pre-horticultural conditions are a set of combined physical, chemical, and environmental conditions that provides basic needs for plantation, growth, and production of ferns. The climatic conditions, spatial land, and kind of soils are a major determinant of kind and quantity of fern cultivation. For example, the *Nephrolepis tuberosa* "Boston fern" thrives on soil in warm, humid conditions, whereas the *Platycerium bifurcatum* "Staghorn fern" thrives as epiphyte and derives moisture and nutrients from the air and rain. Therefore, if a fern grower selects a site for establishing a new horticultural area, he must assess the physical, environmental, climatic, and chemical conditions before planting the ferns. The soils in selected sites are to be loosened, aerated, and mixed with litter or dung compost to become fertile and mulched with peat mosses to conserve soil moisture and to control the weed growth. The mulching keeps the soil cool in day and warm at night and winter, reduces surface run-off, adds humus, prevents soil erosion, allows the absorption water, and reduces irrigation frequency (Kumar 1997). Ferns require 22–25 °C temperature and 70–80% relative humidity, but they don't like the wind; so they should be sheltered at open spaces. The site is to be fenced with a brick wall up to the height of 6 feet, from where the steel structure welded mesh roof extends centrally up to the height of 20 feet or so. The mesh roof may be covered with agro-shade materials to provide shelters and extra warmth for cold periods and also to avoid high intensity (up to 65%) of irradiance light. The mesh roof is to be made in different shapes, viz., gothic, tunnel, gable, sawtooth, gambrel, or skillion, as it accrues benefits, e.g., protection from temperature, light, and winds, maintenance of humidity, and an enhancement of growth (Jones 1987). An incessant waterway, sprinkler, or mist is to be established to provide continuous water supply, moisture, and relative humidity, for better produce of fern.

8.8 Fern Horticultural Conditions

Once, the pre-horticultural condition resumes, the targeted ferns could be selected for plantation. The area of fern plantation is to be in accordance of preferred habitat of the ferns. The common characteristics of the ferns are that they require specific climatic conditions and spatial land as well as soils and can be grown individually or together (Fig. 8.4a, b). In multi-foliar culture, numbers of ferns with diverse habitat may be planted at the same site(s), subject to preparation of extra small beds for each species (Fig. 8.4b). Plantation may be made on a number of small (20 × 6 feet) soil beds, which is a slightly elevated area of soil above the ground level. Beds are to be established between the wall and interconnected waterway so that the soil could retain moisture for underground rhizomes. Plantation can be easily done by making a small pit (5 × 5 × 5 cm) and putting the vegetative portions of the ferns into it. For epiphytic ferns, numbers of pieces of tree trunk are established, and a mat of peat mosses (mixed with litter, compost and soil) are swathe around trunks. Pieces of clumping pads and creeping rhizomes of ferns are placed on peat moss. The mat of peat moss and vegetative portion of ferns are firmed with steel wire mesh (Fig. 8.1b). The beds as well as trunks are continuously irrigated so that the moisture reached up to the vegetative portion of ferns. New fronds gradually appear and turn into mature fronds within months or so (Fig. 8.4a, c).

8.9 Mono-foliar Horticultural Practices

Mono-foliar horticultural practices refer to growing only one fern species on a plantation site(s) year after year. The mono-foliar horticultural practice is recommended due to the suitability of climatic condition, socio-economic condition, or specialty of the plant growers in growing particular fern species. Looking at the climatic factors as one of the major limitations, some most demanding ferns (Fig. 8.1j) can be largely produced on minimum risks of climatic conditions, socioeconomic defeat, and easy mono-foliar culture practices with low investment. Besides, the mango orchard in tropical regions may be also utilized for production of Boston fern, as the mango orchard provides shade, moisture, and low intense light; however the litter and leaves add humus for the fern. Orchard tree has a distance of about 10 m between the two standing tree plants; therefore, authors suggest that *Nephrolepis* may be planted between two trees. Proper irrigation of fern plantation line in the orchard may produce high amount of fronds. In that way the fern grower can produce high amount of ferns through mono-foliar cultivation, in addition to the orchard-produced fruits.

8.10 Multi-foliar Horticultural Practices

In multi-foliar fern horticulture, two or more ferns are grown on different soil beds under one roof in a piece of land in a span of time, as done in other horticultural crops (Bunnett 2002). Many terrestrial ferns grow in moist, shade, and cool climate at optimal temperature and may be produced at large scale through multi-foliar horticulture practices. For such culture, a number of soil beds are built up in the identified site(s) under one roof. The plantation of ferns is to be made in the pattern of mono-foliar culture, except that each species is to be planted on individual soil bed. For instances, in an identified area (60 × 30 feet) of land, about 12 different small beds (15 × 5 feet) may be established, providing extra 5 feet area for each water drain as well as path. Each bed is to be planted by the use of vegetative portions of 12 different species at the same time, followed by continuous irrigation. The climatic conditions are to be maintained by altering the agro-shed and frequency of watering. After 1 month or so, numbers of frond of 12 different fern species from different plantation beds will be produced.

8.11 Fern Horticulture that Aid in Orchard Productivity

The mango orchard is a major cover in low-altitude areas of subtropical and tropical regions of India (Horticultural Statistics at a Glance 2015). Fern horticulture necessitates selection of land, composting of soil and mulching that maintains the soil fertility, moisture, temperature, and water holding capacity and protects the orchards from weeds. Ferns can use the humus produced by trees but also could supplement the soil by death and decay of dried fronds and rhizomes. The tubers, suckers, stolons, and stoloniferous rhizoids run under the soils and loosen it for gaseous exchanges of tree roots. Ferns also retain water and make it available for underground roots of the orchard tree to promote water-mediated metabolism. The normal spacing (10 × 10 m) of trees in mango orchards (100 standing tree/hectare) provides low productivity (1.12 t/ha) and low net returns (Rs. -8000/hectare) to the farmers (Dalvi et al. 2010). To overcome this constraint, authors are of opinion that a trial to accommodate plantation of some ornamental and medicinally valuable ferns is to be made to get higher yield from the mango plantation. Estimate of the authors revealed that the total linear distance between the 100 mango trees/hectare will be equal to 980 m. Estimates revealed that about 313 (±10) fronds of *Nephrolepis tuberosa* arise from 1 m linear area, which can be harvested twice in years. About 626 (±10) fronds can be harvested annually from 1 m linear area. A total of 613,480 fronds may arise out of the 980 m linear area. Since one bunch of the "Ghora patti fern" comprises 50 fronds, therefore, the entire harvested fern from 1 ha orchard area will give rise to 12,270 bunches. The cost of one bunch of "Ghora patti fern" ranges from 80 to 200 rupees. Thus, the additional annual income of the farmers by "Ghora patti fern" may reach around 981,600–2,454,000 rupees per hectare, which is many times higher than the income provided by major tree crops.

8.12 Harvesting of Foliose Ferns

Once the ferns of medicinal, food, and ornamental value are matured, they need to be harvested for a variety of purposes. Medicinal ferns are harvested in accordance to the use of their parts. The edible ferns, viz., *Diplazium esculentum*, are best picked when they produce new curled fronds called "croziers" or "fiddleheads." The ornamental ferns are invariably used for preparation of bouquet, garland, floral works and variety of themes, artistic arrangement of cut flowers, decoration, event organization, and floral and foliose industry (Fig. 8.3c–e). It is important to keep in mind which species is highly demanded and how it can reach to the buyers within firmed time so that harvesting may not be interrupted. Harvesting is to be carried out considering the age and numbers of produced fronds. It is important to measure the impact of harvesting, before ferns are utilized for commercial purposes. Milton (1987) studied the effects of harvesting fronds of ferns and found that all four species are slow growing and have 3–13 fronds per plant. Milton (*l.c.*) suggested for annual harvesting, as these species produce only 3–4 fronds per year, and also emphasized the effects of the harvesting season on frond production. The spatial differences among harvesting sites of ferns are also important, as the study on three edible ferns using multi-criteria evaluation (MCE) and global information system (GIS) has determined the spatial characteristics as one parameter for harvesting and ensuring management of ferns (Matsuura et al. 2014). Ferns are preferred to harvest when fronds become lush green and mature with highest numbers of small pinnae. Authors' estimate suggests that the crown of *Nephrolepis tuberosa* comprises about 133 fronds. The average value of counting of single crowns revealed that out of 133 fronds, about 113 fronds were mature and 20 fronds were immature. It is wiser to harvest only mature frond of the ferns carefully in moderate amounts to maintain the health and vigorness of the ferns (Forest Practices Authority, Tasmania 2012). Before, the ferns are harvested, it is important to ensure their sustainable harvesting, because sustainable harvesting ensures species are not lost and are available for future generations to harvest. The fronds are cut off with a curved sickle that fits in the hands. The harvested fronds are collected in a large-sized (1 × 1 m) polythene bag. Once the harvested fronds are collected, the mouth of the polythene bags is to be tightly held. The airtight polythene bags are to be brought in a cool place to sort out the fronds.

8.13 Foliar Preservation

Foliar preservation refers to "ensuring long-lasting living foliose" by any means. Thus, the best method to ensure long-lasting green foliose is keeping the fronds in refrigerators (Nowak and Rudnicki 1990; Bhattacharjee 1999). Harvested fronds rely on their own sources of energy and water to survive until its final consumption. Furthermore, the compositions of these elements decrease and the green cells and the entire frond start degradation, resulting in complete drying of fronds. Fronds are perishable living products that require coordinated actions by growers, storage operators, processors, and retailers to maintain their quality and reduce loss and waste (Mahajan et al. 2017). Stamps and Chase (1984) reported that the storage of leather leaf fern fronds at 4.5 °C for 1 month did not affect their subsequent vase life. Hvoslef-Eide (1992) performed in vitro culture of *Nephrolepis exaltata* storing it at 5°, 9°, or 13°C temperature and low irradiance light (3–5 μmol m^{-2} s^{-1}). Before storage, the cultures of *Nephrolepis* were subjected to 18°, 21°, 24°, or 27°C and 15, 30, or 45 μmol m^{-2} s^{-1} light. Hvoslef-Eide (*l.c.*) concluded that the preconditioning of ferns before the storage was most favorable. Singh et al. (2003) studied the wet and dry storage using the fronds of two cultivars of *Nephrolepis exaltata* "Bostoniensis" and "Golden Boston." Singh et. al. (*l.c.*) noticed that fronds of *Bostoniensis* exhibited a longer vase life than those of *Golden Boston* on zero day of wet storage. Fronds of *Bostoniensis* dry stored in polypropylene (PP) sleeves continue to show a longer vase life than those kept in polyethylene (PE). Hence, Singh et al. (2003) suggested the storage of fronds in PP sleeves to prolong the vase life of fronds. The above observation suggested that the ferns can be better preserved, provided a few of the variables, which affect the storage life of ferns, are given consideration. Therefore, better understanding on the physical characteristics of fronds and their compatibility with the climatic conditions can assist in taking decisions for storage and minimizing the damage of fronds.

8.14 Packaging of Foliar

The harvested fronds of the ferns are to be packaged in a systematic manner so that it can be avoided from spoilage and damage. Another concern is that the respiration and too much moisture barrier will cause an excessive high temperature as well as humidity in the package and result in accelerated spoilage due to increased metabolism and contamination. To overcome the temperature and chances of contamination, the harvested ferns are subjected for pre-cooling, whereby the temperature is quickly lowered by drenching the fronds with cold water, by immersion, flooding, spraying, or vacuum cooling. Prior to packaging, the ferns are subjected to a cleaning process dipping in soaking tank, where dirt or dust is washed by warm or cold water. The fern produce are segregated for grading into quality groups, i.e., grades A and B. The uniform size of graded (grade A, grade B) fronds are ready for packaging. There are many shapes of packing containers, but a packaging device and method for fern was discovered by Dellecker (1997). Ferns should be wrapped in durable, moisture-proof polyethylene bags. Depending upon the feasibility of handling, transportation, and end users, the ferns may be packaged in consumer unit packaging and transport packaging. Before packaging, the ferns are to be tied together in flat bunches with a number of 50 fronds. A large polyethylene is unrolled on packing board, and bunches are to be laid flat and stacked one by one avoiding damage to any bunch. The bunch is to be wrapped in polyethylene to maintain freshness and avoid dehydration of the fronds (Reid 2009). Lay the packaging of bunches flat on a pack board and carry out carefully (Plastic in Food Packaging 2003) to keep in large boxes for transportation and marketing.

8.15 Market-Linked Demand of Ferns

A survey in Lucknow City of India has informed that the market demand of the *Nephrolepis tuberosa* is increasing day by day. Peoples have fondness in ferns for its aesthetic beauty and decorative significance. Ferns have invaded herbal pharmaceutics, bio-fertilizer, and food domain, too. A variety of ferns, viz., *Helminthostachys zeylanica*, *Dryopteris sparsa*, and *Selaginella bryopteris*, are sold on herbal shops in the name of "Kamraj," "Jatashankari," "sanjeevani booti," etc. with a huge economic turnover (Singh and Khare 2011). Bio-fertilizer industries consume a huge amount of *Azolla* to produce bio-fertilizers for farmers to enhance and sustain the nitrogen content in the agricultural land (Bocchi and Malgioglio 2010). Croziers of ferns are supplied as delicious food by many industries (Walliser 2016). Fronds give feelings of a variety of bird's feather; for that reason, ferns bestow natural scenes for sculpturing on utensils, pots, cloths, and other items. They are utilized in decoration and are an inevitable component of flower and foliar industry. *Matteuccia struthiopteris* has a global demand and market price tag of $10–$15 per pound (Walliser 2016). A survey in Lucknow City also informed that the market cost of one bunch (comprised of 50 fronds) of "Ghora patti fern" ranges from 80 to 200 Indian rupees. The decorative ferns planted in earthen pots, terrarium, and hanging basket also cost up to 1000 rupees. In that way, the return from only *Nephrolepis tuberosa* may reach around 981,600–2,454,000 rupees per hectare, which is higher than the income provided by mango trees in an orchard. Thus the market of herbal, floral, and fertilizer industry has high demands of ferns to develop the product and deliver to the society as well as end users. Now by the imposition of legislative rules of central as well as state governments (The Biological Diversity Act 2002), ferns cannot be harvested from protected or natural forests. Henceforth, the fern supply is to be reliant on the horticultural produce. Such high economic return may remunerate in compounded rates and benefit to the farmers and rural, unprivileged people connecting them in horticulture and product development and meeting the market demand.

8.16 Modes of Supply of Fern Foliose

Mode of supply of ferns or fern produce depends on the quantum of demand in the market and their rational production, harvesting, processing, packaging, transportation, and consumption by the end users. Local inhabitants facilitate supply of ferns from natural forests. The local inhabitants in Darjeeling, West Bengal of India, are under pressure of poverty and dependent on forest resources for their livelihood (Ching 2016). They collect the fronds of *Nephrolepis tuberosa* and *Diplazium esculentum* by cutting with iron sickle and depositing in large-sized polyethylene bags to avoid excessive water loss. They stack the fronds in a bundle of 50 fronds and laid on each other until it resumes 25 kg weight or above and wrapped in polyethylene bags. Local people bring the small-weighted package to the local markets or to the New Jalpaiguri from where it is transported to other major cities like Lucknow, Patna, Bhubaneshwar, Kolkata, Delhi, Bhopal, etc. of the country by trucks or trains and by air as well as ships. The florists and beauticians visit the local foliose markets in the city and bring the fronds to their shops by rickshaw or auto vehicles. The time duration between purchasing from the local foliose markets and bringing to their own shops is more crucial, because long duration may cause excessive water loss in the fronds. The florists and beauticians keep the bunches of fronds in a small thermocol container with ice cubes to sustain their lush greenness (Fig. 8.1h). The mode of supply of ferns may be gauzed by the supply chain management (SCM), which is the process of planning, execution, controlling, and managing the process of the supply chain of ferns with the purpose to satisfy the customers (Chandra et al. 2013). It covers all movement, storage of ferns, and processing of fern produce from point of origin to point of consumption. Thus for maintenance of a better mode of supply of ferns, it is very important to establish a superb quality SCM, because it involves coordinating and integrating these flows both within and among flower, foliose, herbal, and bio-fertilizer industries.

8.17 Market-Linked Demand and Application in Product Preparation

People across the world exhibit diverse culture, belief, and social practices; therefore, the food and cuisine liking and disliking vary from person to person. Tharu tribes in India consume *Diplazium esculentum* and *Ophioglossum reticulatum* as delicious food (Singh and Johari 2015), whereas the Canadian and American people eat *Matteuccia struthiopteris* as vegetable; therefore, demand of its fiddleheads has expanded many times (Aderkas 1984). The fern and fern produce have significant link with market from point of origin to point of consumption, as it involves various steps of selling and buying by the producer, wholesaler, retailer, or consumer. The benefit in terms of returns to producers, wholesalers, and retailers and even to consumers is linked with the demand and supply chain of the ferns. A slight change in the market-linked demand of ferns may tremble the entire system of the fern-based industries. Demand of the ferns will not only strengthen the market operation, but it will also ascend the economic status of the people, who were deployed to facilitate fern-based services, either in horticulture or in the process of product development as well as consumption. The fern product either is purchased by an individual or a group of people, but in both cases, the demand is affirmed paradigm of elevated economy. Thus, the market-linked demand signifies the collective demand of each individual at a certain price and at a certain time. The market-linked demand of ferns helps in product development, because the market is an appropriate platform, where the service providers come to know about the variety of taste and demands of the customers. Ferns are used for preparation of a variety of floret article, including dried flower sceneries (Fig. 8.3a, c). Thus, it is the customers, who can suggest for adding or excluding ferns or flowers to beautify the fern-based product. The suggestion of the customers to improve the quality and buy the products of own choice would enhance the value and demand of ferns. Florists prepare these products using fronds of *Nephrolepis tuberosa*, few more ferns, and leaves or flowers of other higher plants (e.g., Fig. 8.1). Such products may be of different shapes and styles, viz., nosegay, cascading, crescent bouquets, hand-tied, pomander, flower spray, and Biedermeier (Rye 2000; Lee 2014). Few of the wedding bouquets, viz., bouquet of ferns and white flower, summery fern bouquet, green and white bouquet, all-greenery bridal bouquet, woodsy bridal bouquet, and botanical bridal bouquets (Fig. 8.3c), are on high demand, especially for ferns (Lee 2014). Each piece of the bouquet costs from 200 to 15,000 Indian rupees, showing potentiality of ferns in economic development and sustainable livelihood.

8.18 Key Foliose in Florist and Bouquet Shops

Florist and bouquet shops usually intend to provide a variety of flower and foliose-based products to celebrate our special moments, viz., birthday, ceremony, and other rituals. Usually, florists keep leaves and flowers of a variety of plants in the shops, viz., *Asparagus* sp., *Callistemon* sp. "bottle brush," *Caryota* sp. "fishtail palm," *Cycas* sp., *Dianthus caryophyllus* "carnation," *Dypsis lutescens* "areca palm," *Gerbera* sp., *Gladiolus* sp., *Lilium* sp. "lily," *Murraya paniculata* "Manokamna" orchids, *Polianthes tuberosa* "Rajnigandha" rose, *Solidago* sp. "goldenrod," and *Trachycarpus fortunei* "China palm" (Fig. 8.1e, g). Nevertheless, the ferns are key-foliose that beautifies and enhances the splendor of flower articles. Ferns are available all year round; therefore, they have become choice foliose plants for florists. India's biggest florist's chain Ferns N Petals (FNS) provides a variety of foliose-based products that include the lush-green fronds of ferns, viz., *Adiantum capillus-veneris*, *Gleichenia dicarpa*, *Nephrolepis tuberosa*, and *Polystichum lentum*, in preparation of bouquets. The fronds of *Adiantum capillus-veneris* add prettiness to the white- and red-flowered bouquets; therefore, florists use it in preparation of the wedding bouquets such as "bouquet of ferns and white flowers" (Fig. 8.3c). Fronds of *Gleichenia dicarpa* are used to give lush-green color to the "white- or yellow-flowered bouquets," but sometimes the fronds of *Gleichenia dicarpa* and *Polypodium lachnopus* are used together. The "summer fern bouquet" is an on-demand bridal bouquet and is a combination of textured ferns, viz., *Nephrolepis exaltata*, with cheerful daisies, hydrangeas, and feverfew. The "all-greenery bridal bouquet" is usually comprised of fronds of *Nephrolepis* and flowers of dianthus, celosia, or green hydrangeas. The fronds of *Sphenomeris immersa* are mixed with flowers of *Gerbera*, white roses, and a variety of lilies to prepare bridal bouquets, whereas the fronds of *Thelypteris dentata* are used in preparation of woodsy bridal and botanical bridal bouquet (Lee 2014). The fronds as well as pinnae of *Thelypteris dentata* including *Adiantum capillus-veneris* are used to prepare sceneries and pictorial cards (Fig. 8.3a). Thus, the florists and bouquet shops usually store those ferns, whose frond has lush-green color with long durable life, so that it can be used for long term.

8.19 Market Linkage of Ferns for Sustaining Livelihood

Market linkage often serves to link the rural people with urban or cities for direct trade of their produces as well as transaction of economic gains. For that reason, there is necessity to develop market linkages for rural products as well as rural entrepreneur (Islam 2011). Ferns can contribute to enhance the livelihood of rural people, provided they know about their potential value and market-linked economic network. Large-scale production of such ferns and their market linkage may open opportunity to enhance self-reliance. Fern production may be augmented by horticultural practices, and the produce may directly be supplied by the producer to the consumers. Direct selling and buying between the producers and consumers and transaction of economic benefit will help in sustaining the livelihood of farmers (Wiggins and Keats 2013). Execution of such market linkaging may ensure transparent transaction, idle transaction mode, and proper linkage of transportation as well as communication and geographical location of the transaction. Nevertheless, the main purpose of market linkages of ferns in present contribution is to facilitate the flow of produce between the different levels of the marketing system. To achieve the tasks of market linkage of ferns for sustaining livelihood, the smallholder farmers owing small-based plots of land shall be promoted to take initiative for fern horticulture with aim to sustain livelihood.

8.20 Role of Fern Horticulture in Economic Development

Fern horticulture may contribute as an integral and important component in the economy of a country. The significance of fern horticulture can be evaluated by its benefits, viz., best utilization of lands, high yield per unit area, increased supply against demand, development of a variety of products, high export value, high returns, availability of fern raw materials for industries, and increased employment to fern growers and labors. The importance of fern horticulture can be also gauzed by the provision of employment opportunities to the rural people, farmers, and women through fern cultivation, harvesting, processing, transportation, product preparation, marketing, and economic returns. In addition, the ferns are potential sources of medicines, foods, and bio-fertilizers. They constitute important medicines and food-giving materials (Srivastava 2007; Singh and Khare 2011) which are of great economic significance (Berruezo et al. 2017). Some of the ferns are significantly important and required for various purposes, e.g., the croziers of *Matteuccia struthiopteris* are used for food, *Adiantum philippense* is also used as medicine, and *Lycopodium cernuum*, *Nephrolepis exaltata*, and *Platycerium bifurcatum* are frequently sold in the market by some local people in Africa, but in spite of the above usefulness, people do not make it as their primary source of living (Mustacisa 2016). Because of their multifarious uses, ferns can be used to improve the economic condition of middle class farmers and unemployed and economically deprived people. Eventually, the fern cultivation, production, and fern-based product development and its marketing could be highly remunerative for women in rural areas (Ravichandra 2014). *Nephrolepis* is usually used in preparation of all kinds of flower-based bouquets; thus, it would not be hyperbole to rank it as the most widely used fern of ornamental value. It can be considered as "wealth-making fern" and bring revolution to the foliose industry. The potentiality of *Nephrolepis tuberosa* can be presumed as one can earn an additional income of 0.982–2.454 million Indian rupees per hectare area. In that way, the horticulture may enhance the economic and other social values of communities, particularly when such plants are grown for commercial purposes (Abegunde 2012, Forestry Sub-sector Studies-Briefing 2005).

8.21 Future Perspectives of Fern Horticulture

Future perspectives may be looked within the benefits of fern horticulture, as the horticulture practice results in increased production, yield, and returns, utilization of lands, export of products, employment to the rural people, and availability of fern raw materials as medicines, foods, bio-fertilizers, and potential bio-molecules. There were many assertions on the future perspectives of horticulture (Chandran et al. 2009), but more precious attempt (Ide et al. 1993) enlightened us about the past, present, and future tactic for fern horticulture. Increased fern production may offer extensive opportunity for societal and sustainable economic development. Use of fern horticulture for development of various products would involve economic input as well as output, besides providing fern-based commodity to the society. The best example can be the USA, where the horticulture-based industry has increased the total income receipt at the rate of 10% per year (Johnson and Christensen 1995). Lewis (1977) reported that in low-income housing localities in America, the gardening programs have resulted in cleaned and painted buildings, cleaned streets, reduced destruction, and satisfactory environments confirming that fern horticulture signifies human perspectives, too. A variety of ferns, viz., "dhekiya," "ostrich fern," and "ekpatiya," provide nutritious foods (Aderkas 1984; Singh and Johari 2015); such ferns may be recommended as dietary and nutritional supplement. *Huperzia serrata* resulted Huperzine A and Huperzimine compounds, which were potential bio-molecules in the treatment of diseases (Ma et al. 2005; Yu et al. 2014); such horticultural produce is a major source of biologically active compounds (Mahajan et al. 2017). A variety of ferns, viz., *Asplenium trichomanes* grow in cold, the *Selaginella bryopteris* in desiccation, and the *Ceratopteris thalictroides* in rains.

Such ferns synthesize the proteins of specific composition in a variety of habitat and can be utilized as the genetic resource for development of high economic transgenic plants. *Tectaria macrodonta* is reported to exhibit Tma12 protein that is insecticidal to whitefly at sublethal doses (Shukla et al. 2016). Ferns could also be a useful organism in transcriptome sequencing for assembly, annotation, and marker discovery, as they belong to lower plants (Singh et al. 2015). *Azolla* species are efficient for mineralized organic nitrogen production and capable of reducing weed competition and resistant to water polluted with herbicides (Bocchi and Malgioglio 2010; Brouwer et al. 2017). *Pteris vittata* is known for its arsenic hyper-accumulator potentiality (Kalve et al. 2011; Cesaro et al. 2015; Chen et al. 2017), but a number of ferns could be screened out for metal detoxification potentiality and could be used to inhibit entry of the pollutant causing toxicity in the food chain system. In a wide spectrum, fern horticulture portrays sustainability issues from each of the economic, environmental, and social perspective, as it can have both positive and negative environmental impacts (Hallett et al. 2016). In that way, fern horticulture may result in increased fern production, product development, opportunity of employment, and high returns to provide societal commodity and elevate the economic status and livelihood of unprivileged rural people.

8.22 Conclusion

India exhibits diverse climatic condition and agroecological region and soil that provide favorable opportunity for fern horticulture. Fern horticulture may considerably contribute in total horticultural produce of the country. Fern horticulture may play a significant role in improving the Indian economy by improving the income of the farmers, women, and unprivileged and rural people. Studies have suggested that the *Adiantum capillus-veneris* and *Huperzia serrata* have distinct glycosides, flavonoids, terpenoids, alkaloids, and other primary as well as secondary metabolites (Ma et al. 2005; Swamy et al. 2006; Singh et al. 2010; Yu et al. 2014, 2017) and could be used as potential raw materials for development of drugs. *Azolla* has potential properties to enhance the agricultural productivity (Bocchi and Malgioglio 2010, Brouwer et al. 2017), whereas *Pteris vittata* and a few other ferns are hyper-accumulator of heavy metals (Singh et al. 2010; Kalve et al. 2011). Such ferns could be utilized for the management of agriculture and environmental pollution (Cesaro et al. 2015, Chen et al. 2017). New fronds and rhizomes of *Diplazium esculentum* and *Cyathea spinulosa* are highly nutritious and could be used as dietary supplement or food. *Nephrolepis tuberosa* and *Thelypteris dentata* are on demand in Indian florist's community. After large-scale production, these ferns could be provided to the florists for preparation of bouquets and beautification and decoration of the events places. Some of the ferns adapted to cold, desiccation, and rains have insecticidal proteins and synthesize the proteins of specific composition. Such ferns could be used as genetic resources for development of high economic transgenic plants (Shukla et al. 2016). Eventually, they could be used in transcriptome sequencing for assembly, annotation, and marker discovery (Singh et al. 2015), as ferns belong to cryptogams. For the reason of diverse uses, ferns could be utilized to make billion dollars turn over globally. Unfortunately, the raw materials are met from the forests because they are not under horticulture practices. Estimates on horticultural produce in India suggested that there is still an urgent need to fortify the production rate of edible, useful vegetables and ornamentals, including ferns, to ascertain the nutrition security and supply for better livelihood and economic development. Fern horticulture is also needed to bring the profitable land in use, as about 240 million acres of cultivable wasteland in India can be brought to use for cultivation (Joshi et al. 2013). Thus fern horticulture practice is more precious because it could provide resource for production of fern-based products. Since the fern horticultural practices are labor intensive, it could create a lot of employment opportunities for the farmers, women, and unprivileged and rural people to strengthen their economic condition and livelihood.

Acknowledgments Authors are thankful to the Director of CSIR-National Botanical Research Institute, Lucknow, for providing necessary facilities. One of the authors (APS) especially thank Prof. Eelena M. F. Gonzalez, the editor of the book, for her bighearted support in discounting the deadline, in lack of which, the chapter could not have appeared in its present form. Thanks to the Council of Scientific & Industrial Research (CSIR), New Delhi, for endowing financial assistance.

References

Abegunde AA (2012) Urban horticulture and community economic development of Lagging regions. In: Maldonado AIL (ed) Horticulture. InTech, China, pp 133–144

Adam CR, Bamford KM, Early MP (2012) Principles of horticulture. Routledge, Abingdon, Oxon, OX14 4RN, pp 1–379

Aderkas PV (1984) Economic history of Ostrich Fern, *Matteuccia struthiopteris*, the edible fiddlehead. Econ Bot 38(1):14–23

Bennici A (2008) Origin and early evolution of land plants problems and considerations. Commun Integr Biol 1(2):212–218

Berruezo F, de Souza FSJ, Picca PI, Nemirovsky SI, Martı'nez Tosar L, Rivero M et al (2017) Sequencing of small RNAs of the fern *Pleopeltis minima* (Polypodiaceae) offers insight into the evolution of the microrna repertoire in land plants. PLoS One 12(5):1–22

Bhattacharjee SK (1999) Post-harvest management of cut flowers, cut foliage and post production management of potted plants. J Ornamental Hortic Plants New Series 2:32–39

Bocchi S, Malgioglio A (2010) Azolla-Anabaena as a biofertilizer for rice paddy fields in the Po valley, a temperate rice area in Northern Italy. Int J Agro 2010:1–5

Brouwer P, Bräutigam A, Buijs VA, Tazelaar AOE, Van der Werf A, Schlüter U, Reichart GJ, Bolger A, Usadel B, Weber APM, Schluepmann H (2017) Metabolic adaptation, a specialized leaf organ structure and vascular responses to diurnal N_2 fixation by *Nostoc azollae* sustain the astonishing productivity of *Azolla* ferns without nitrogen fertilizer. Front Plant Sci 8:442

Bunnett RB (2002) Interactive Geography 4. SNP Pan Pacific Publishing. pp 98

Cardelus CL, Colwell RK, Watkins JE (2006) Vascular epiphyte distribution patterns: explaining the midelevation richness peak. J Ecol 94(1):144–156

Cesaro P, Cattaneo C, Bona E, Berta G, Cavaletto M (2015) The arsenic hyper accumulating *Pteris vittata* expresses two arsenate reductases. Sci Rep 5:14525. https://doi.org/10.1038/srep14525

Chandra H, Rinkoo AV, Verma JK, Verma S, Kapoor R, Sharma RK (2013) Supply chain management with cost-containment & financial-sustainability in a tertiary care hospital. J Health Hum Serv Adm Summer 36(1):3–23

Chandran P, Ray SK, Durge SL, Raja P, Nimkar AM, Bhattacharyya T, Pal DK (2009) Scope of horticultural land-use system in enhancing carbon sequestration in ferruginous soils of the semi-arid tropics. Curr Sci 97(7):1039–1046

Chen Y, Hua CY, Jia MR, Fu JW, Liu X, Han YH, Liu Y, Rathinasabapathi B, Cao Y, Ma LQ (2017) Heterologous expression of *Pteris vittata* Arsenite antiporter PvACR3;1 reduces arsenic accumulation in plant shoots. Environ Sci Technol 51(18):10387–10395

Ching M (2016) Disparity and conflict: a case of the Gorkhas in Darjeeling. Int J English Lang Literatures Humanities 4(11):1–17

Dalvi NV, Salvi BR, Chavan SA, Kandalkar MP (2010) High density planting in mango cv. Alphonso J Hort Sci 5(2):117–119

Dellecker WM (1997) Sleeve and method for packaging cut foliage. United States Patent (Patent no. 5,695,058)

Dzwonko Z, Kornas J (1994) Patterns of Species Richness and Distribution of Pteridophytes in Rwanda (Central Africa): A Numerical Approach. J Biogeogr 21(5):491

Fernandez H, Revilla MA (2003) *In vitro* culture of ornamental ferns. Plant Cell Tissue Organ Cult 73:1–13

Forest Practices Authority, Tasmania (2012) Tree fern management plan for the sustainable harvesting, transporting or Trading of *Dicksonia antarctica* in Tasmania. Environment Protection and Biodiversity Conservation Act 1999

Forestry Sub-Sector Studies-Briefing (2005) Forest ferns and foliage Trade, Water and Forestry. Republic of South Africa, pp 1–2

de Gasper AL, Eisenlohr PV, Salino A (2015) Climate-related variables and geographic distance affect fern species composition across a vegetation gradient in a shrinking hotspot. Plant Ecology Diversity 8(1):25–35

Hallett S, Hoagland L, Toner E (2016) Urban agriculture: environmental, economic and social perspectives. In: Janick J (ed) Horticultural reviews. Wiley, Hoboken, pp 65–120

Horticulture Statistics at a Glance (2015) Horticulture statistics division, Department of Agriculture, Cooperation & Farmers Welfare, Ministry of Agriculture & Farmers Welfare Government of India, Oxford University Press, YMCA Library Building, Jai Singh Road, New Delhi, India, pp 1–463

Huang YW, Chiang TY, Chiou WL (2008) Isolation and characterization of microsatellite loci of *Lycopodium fordii* Bak. (Lycopodiaceae, Pteridophyta). Conserv Genet 9(3):775–777

Hvoslef-Eide AK (1992) Effects of pre-storage conditions on storage of in-vitro cultures of *Nephrolepis exaltata* (L.) Schott and *Cordyline fruticosa* (L.) A. Chev. Plant Cell Tissue Organ Cult 28:167–174

Ide JM, Jermy C, Paul AM (1993) Fern horticulture: past present and future perspectives. Intercept Ltd., Andover, UK, pp 1–522

Islam MN (2011) Sustainable market linkages for rural entrepreneur products: strategies and challenges. Indian. J Manag Sci 1(1):36–42

Johnson DC, Christensen RL (1995) The green industry today – some issues and future prospects. J Agr Bus 13(1):63–76

Jones D (1987) Encyclopaedia of ferns: an introduction to ferns, their structure, biology, economic importance, cultivation and propagation. Lothian Publishing Company Pvt. Ltd., Melbourne, pp (viii–xvii) 1–433

Jones MM, Szyska B, Kessler M (2011) Microhabitat partitioning promotes plant diversity in a tropical montane forest. Glob Ecol Biogeogr 20:558–569

Jones MM, Ferrier S, Condit R, Manion G, Aguilar S, Pérez R, Zotz G (2013) Strong congruence in tree and fern community turnover in response to soils and climate in central Panama. J Ecol 101(2):506–516

Joshi R, Agarwal ML, Nandwani D (2013) Importance and scope of horticultural crops in India and Mdash; a commercial approach retrieved from https://www.researchgate.net/publication/267342485

Kalve S, Sarangi BK, Pandey RA, Chakrabarti T (2011) Arsenic and chromium hyperaccumulation by an ecotype of *Pteris vittata* – prospective for phytoextraction from contaminated water and soil. Curr Sci 100:888–894

Kawano T (2015) Pteridophytes as active components in gardening, agricultural and horticultural ecosystems in Japan. Adv Hort Sci 29(1):41–47

Kumar N (1997) Introduction to horticulture. Rajalakshmi Publications, Vepamoodu Junction, Nagercoil, pp 1547–1550

Lee (2014) Flowers in season now: 6 Fern-Filled wedding bouquets. https://www.brides.com/story/ferns-wedding-flowers-bouquets

Lewis CA (1977) Human perspectives in horticulture. In: Children, nature and the urban environment: proceedings of a symposium-fair; Gen. Tech. Rep. NE-30. Upper Darby, PA: US Department of Agriculture, Forest service, North-eastern Forest experiment station, pp 187–190

Ma X, Tan C, Zhu D, Gang DR (2005) Is there a better source of huperzine A than *Huperzia serrata*? Huperzine-A content of Huperziaceae species in China. J Agric Food Chem 53(5):1393–1398

Mahajan PV, Caleb OJ, Gil MI, Izumi H, Colelli G, Watkins CB, Zude M (2017) Quality and safety of fresh horticultural commodities: recent advances and future perspectives. Food Packag Shelf Life 14(A):2–11

Matsuura T, Sugimura K, Miyamoto A, Tanaka H, Tanaka N (2014) Spatial characteristics of edible wild fern harvesting in mountainous villages in North-eastern Japan using GPS tracks. Forests 5:269–286

Mickel JT (2003) Ferns for American gardens. Timber Press Oregon, USA, pp 1–370

Milton SJ (1987) Effects of harvesting on four species of forest ferns in South Africa. Biol Conserv 41(2:133–146

Moran RC (2008) Diversity, biogeography, and floristics. In: Ranker TA, Haufler CH (eds) Biology and evolution of ferns and lycophytes. Cambridge University Press, Cambridge, pp 367–394

Mustacisa MM (2016) Diversity, uses and economic value of ferns: an instrument for epistemological perception. Int J Environ Sci Educ 11(18):13111–13146

Nondorf LS, Dooley AM, Palmieri M, Swatzell J (2003) The effects of pH, temperature, light intensity, light quality, and moisture levels on spore germination in *Cheilanthes feei* of Southeast Missouri. Am Fern J 93:56–69

Nowak J, Rudnicki RM (1990) Postharvest handling and storage of cut flowers, florist greens and potted plants. Chapman and Hall, London, p 209

Page CN (1979) The diversity of ferns: an ecological perspective. In: Dyer AF (ed) The experimental biology of ferns. Academic Press, London

Plastic in Food Packaging (2003) Packaging aspects of Fruits and Vegetables, pp 116–132

Ravichandra NG (2014) Horticulture and its role in the national economies. In: Horticultural nematology. Springer, New Delhi

Reid MS (2009) Handling of cut flowers for export. In: Proflora bulletin. University of California, Davis

Rye T (2000) Creative wedding florals you can make. North Light Books

Shukla AK, Upadhyay SK, Mishra M, Saurabh S, Singh R, Singh H, Thakur N, Rai P, Pandey P, Hans AL, Srivastava S, Rajapure V, Yadav SK, Singh MK, Kumar J, Chandrashekhar K, Verma PC, Singh AP, Nair KN, Bhadauria S, Wahajuddin M, Singh S, Sharma S, Omkar S, Upadhyay RS, Ranade SA, Tuli R, Singh PK (2016) Expression of an insecticidal fern protein in cotton protects against whitefly. Nat Biotechnol 34:1046–1051

Singh AP, Johari D (2015) Ethnobotany and social context of pteridophytes used by tharu tribes in terai regions of Uttar Pradesh, India. J Econ Taxon Bot 39(3–4):512–526

Singh AP, Khare PB (2011) Status of ethno-pteridology in India. Appl Bot Abstr 31:332–361

Singh P, Singh K, Kumar R (2003) Study on refrigerated storage of *Nephrolepis* fronds. J Fruit Ornamental Plant Res 11:121–126

Singh AP, Rawat VK, Behera SK, Khare PB (2010) Perspectives of pteridophytes biodiversity: a source of economy elevation. In: Uttar Pradesh State Biodiversity Board (ed). Proceedings of National Conference on Biodiversity, Development and Poverty alleviation, pp 46–49

Singh AP, Johari D, Singh A, Behera SK, Khare PB (2012) Studies on regeneration of gametophytes and mass multiplication of *Anemia rotundifolia* Schrad. (PTERIDOPHYTE). Int J Curr Res 4(12):235–240

Singh AP, Johari D, Singh A, Khare PB (2013) Reproductive biology of a rare fern species-*Adiantum peruvianum* (Adiantaceae: Pteridophyta). Phytomorphology 63(1&2):105–114

Singh H, Rai KM, Upadhyay SK, Pant P, Verma PC, Singh AP, Singh PK (2015) Transcriptome sequencing of a thalloid bryophyte; *Dumortiera hirsuta* (Sw) Nees: assembly, annotation and marker discovery. Sci Rep. https://doi.org/10.1038/srep15350

Singh AP, Johari D, Khare PB (2017) Studies on ontogeny and reproductive behaviour of *Lepisorus nudus* (Hook.) Ching (Polypodiaceae). J Plant Res 130(2):281–290

Soni D, Ranjan S, Singh R, Khare PB, Pathre UV, Shirke PA (2012) Photosynthetic characteristics and the response of stomata to environmental determinants and ABA in *Selaginella bryopteris*, a resurrection spike moss species. Plant Sci 191–192:43–52

Srivastava K (2007) Ethnobotanical studies of some important ferns. Ethnobotanical Leaflets 11:164–172

Stamps RH, Chase AR (1984) Fungal inoculation, fungicide treatments and storage affect postharvest decay and vase life of leatherleaf fern fronds. Hort Sci 19(2):292–293

Swamy R, Kunert O, Schühly W, Bucar F, Ferreira D, Rani V, Kumar B, Appa Rao A (2006) Structurally unique biflavonoids from *Selaginella chrysocaulos* and *Selaginella bryopteris*. Chem Biodivers 3:405–414

Talukdar AD, Tarafdar RG, Choudhury MD, Nath D, Choudhury S (2011) A review on pteridophyte antioxidants and their potential role in discovery of new drugs. Assam Univ J Sci Technol: Biol Environ Sci 7(1):151–155

The Biological Diversity Act 2002 Ministry of Law and Justice (Legislative department), Government of India

Walliser J (2016) Grow fiddle-head ferns for market. http://www.hobbyfarms.com/grow-fiddlehead-ferns-for-market/

Wiggins S, Keats S (2013) Leaping & learning: linking smallholders to markets in Africa. Agriculture for Impact, Imperial College, London, pp 1–120

Yhome K, Chaturvedy RR (2017) Emerging trans-regional corridors: perspectives from south and Southeast Asia. In: Yhome K, Chaturvedy RR (eds) Global policy GP-ORF series emerging trans-regional corridors: perspectives from South and Southeast Asia. Mohit Enterprises Printers, New Delhi, pp 1–7

Yu CM, Calhoun LA, Konder RM, Grant AS (2014) Huperzimine, a novel *Lycopodium* alkaloid from *Huperzia serrata*. Can J Chem 92(5):406–410

Yu S, Yan H, Zhang L, Shan M, Chen P, Ding A, Li SFY (2017) A review on the phytochemistry, pharmacology, and pharmacokinetics of Amentoflavone, a naturally-occurring Biflavonoid. Molecules 22(299):1–23

Part II
Evolution, Biodiversity and Conservation of Ferns

Chapter 9
Evolution and Classification of Ferns and Lycophytes

Emily B. Sessa

9.1 Overview of Fern and Lycophyte Evolution

Ferns and lycophytes are among the oldest lineages of plants on Earth, and the living members of these lineages – along with the extant seed plants – are descendants of the first plants that evolved vascular tissue. These vascular plants, which possess xylem and phloem tissues specifically adapted for transport of water and nutrients, are collectively referred to as tracheophytes (Fig. 9.1). The tracheophytes today include approximately 310,000 named species of angiosperms, gymnosperms, lycophytes, and ferns (the latter are sometimes referred to as monilophytes, but "ferns" is used here for simplicity) (Christenhusz and Byng 2016) and are the basis of most global food webs as well as human civilizations and economies. Ferns and lycophytes represent a relatively small proportion of all tracheophytes, with ca. 12,000 species (PPG 1 2016), but they are critical members of Earth's modern floras and are disproportionately represented in some extinct and extant habitats (e.g., tropical rainforest canopies (Cardelus et al. 2006; Watkins and Cardelús 2012) and Carboniferous coal forests (Gastaldo 1987)).

The evolution of ferns and lycophytes as distinct lineages "begins", we might say, between 472 and 416 million years ago, in the Ordovician or Silurian. This is when most phylogenetic studies and analyses of fossil plants have dated the divergence that separated the bryophytes from the ancestor of modern tracheophytes, to which ferns and lycophytes belong (Clarke et al. 2011; Kenrick et al. 2012; Magallón et al. 2013; Gerrienne et al. 2016; Lenton et al. 2016; Plackett and Coates 2016). The earliest tracheophytes would have closely resembled bryophytes, and the fossil record reveals that numerous early tracheophyte lineages ultimately went extinct. As Plackett and Coates put it, the fossil record from this time period is "littered with now-extinct plant groups" (Plackett and Coates 2016). Gradually, key traits related

E. B. Sessa (✉)
Department of Biology, University of Florida, Gainesville, FL, USA
e-mail: emilysessa@ufl.edu

H. Fernández (ed.), *Current Advances in Fern Research*,
https://doi.org/10.1007/978-3-319-75103-0_9

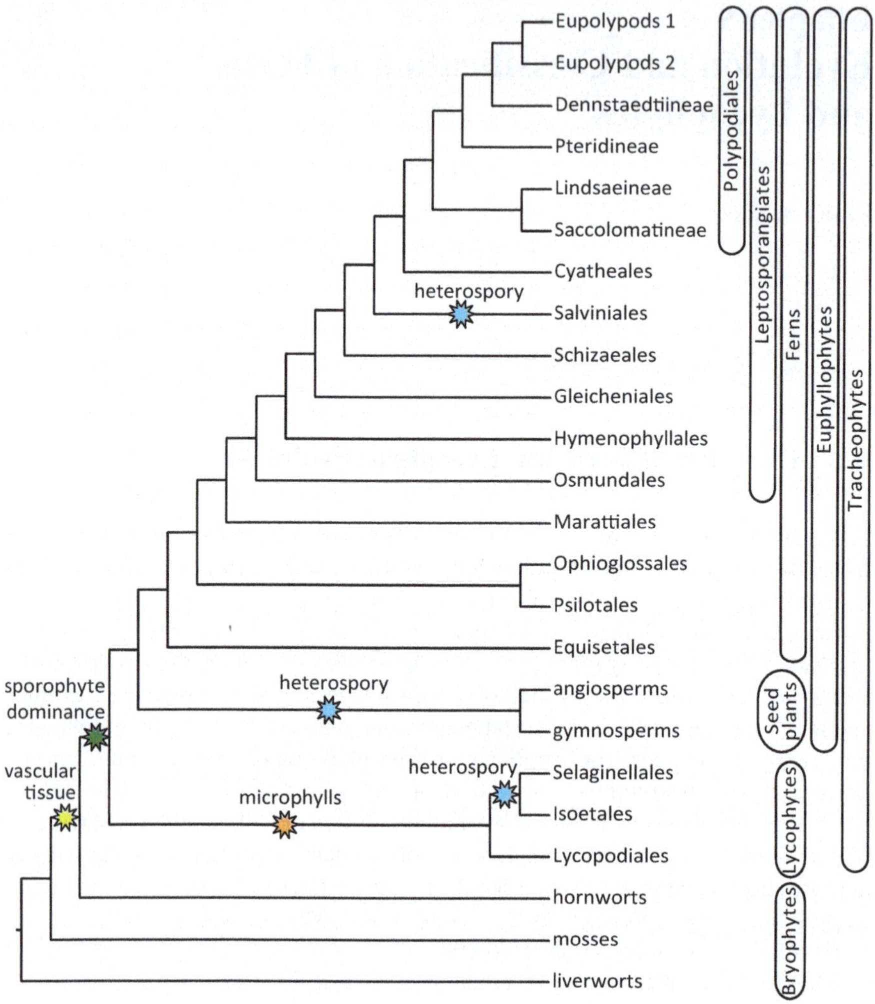

Fig. 9.1 Phylogeny of major groups of land plants. The topology is based on Clarke et al. (2011), Magallón et al. (2013), and Wickett et al. (2014). The transition from a gametophyte-dominant to a sporophyte-dominant life cycle at the root of tracheophytes is indicated, as are three transitions from homospory to heterospory. Note that the relationships of bryophytes to one another and to tracheophytes remain difficult to resolve. These relationships are depicted as reported by Magallón et al. (2013) and others, but see Wickett et al. (2014) for a discussion of the extent to which various resolutions of these relationships are supported by available data

to movement of water and nutrients (xylem and phloem tissues) evolved, as did other features associated with modern tracheophytes, such as distinct roots, stems, and leaves (i.e., megaphylls – which may have evolved multiple times in tracheophytes (Vasco et al. 2013, 2016)). One of these lineages managed to survive the ancient winnowing of early tracheophytes to become the ancestor of all modern vascular plants, including ferns and lycophytes.

9.2 Fern and Lycophyte Phylogenetics and Classification

9.2.1 Early Divergences in Vascular Plants: Ferns and "Fern Allies"

The term "fern allies" was long used to refer to a group of plants of dubious affinity but which were suspected to be closely related to ferns, including the horsetails (*Equisetum*), whisk ferns (*Psilotum* and *Tmesipteris*), and lycophytes (*Isoetes*, *Selaginella*, and genera of Lycopodiaceae) (Moran 2004). The confusion surrounding their relationships stems largely from their morphological features (or lack thereof, in the case of the whisk ferns). Despite the fact that they all possess vascular tissue, the hallmark of the tracheophytes, Linnaeus classified several members of these groups with the mosses because of their superficial resemblance to bryophytes, and their persistent common names reflect this (e.g., spike mosses (*Selaginella*) and club mosses (*Lycopodium*)). He also placed the whisk ferns with the mosses, but grouped *Equisetum* and *Isoetes* with the ferns (Linnaeus 1753; Moran 2004).

We now recognize that the "fern allies" in fact belong to two separate groups, neither of them mosses, but one of them the ferns. The other group is the lycophytes, which includes Selaginellaceae, Isoetaceae, and Lycopodiaceae, the first two sister to one another and together sister to Lycopodiaceae (Fig. 9.1). These three groups represent the extant lycophytes, and the earliest recognizable lycophyte fossils date to the Silurian, between 445 and 420 million years ago (Hueber 1992; Kenrick et al. 1997; Clarke et al. 2011; Magallón et al. 2013; Gerrienne et al. 2016; Plackett and Coates 2016). The lycophytes and ferns were recognized as distinct groups, but still classified together as Pteridophyta in the mid-1800s (Moran 2004), when botanists discovered that all members of these former "ferns and fern allies" share a feature of their life cycle that is unique among land plants: the two generations, the gametophyte and sporophyte, are ecologically and nutritionally independent of one another (Klekowski and Baker 1966; Haufler et al. 2016). This "separation of generations" sets the two lineages of the Pteridophyta apart from both the bryophytes and seed plants (Bryophyta and Spermatophyta, in the ranking system of the time), and "pteridophytes" has persisted as a useful term for distinguishing ferns and lycophytes from other land plants, though we now recognize that it describes a paraphyletic assemblage. A shift in the dominant phase of the life cycle is also a distinguishing feature of tracheophytes as a whole: in bryophytes, the gametophyte generation is considered dominant, as it is the independent and long-lived stage, while the sporophyte is short-lived and nutritionally dependent on the gametophyte. The relative status of the stages is reversed in the tracheophytes, including ferns and lycophytes, with the gametophytes typically short-lived relative to the sporophytes. It is only in ferns and lycophytes that the two stages are independent; however, during seed plant evolution, the gametophyte has become reduced to only a few cells that are entirely dependent on the sporophyte (Haufler et al. 2016).

A key divergence within tracheophytes splits the "pteridophytes" in half and separates the lycophytes from the remaining tracheophytes, which are collectively called the euphyllophytes (to acknowledge their possession of "true" leaves, i.e., megaphylls). This divergence between lycophytes and euphyllophytes is estimated to have occurred around 424–416 mya based on molecular phylogenies (Clarke et al. 2011; Magallón et al. 2013), and the earliest euphyllophyte fossils date to the Lower Devonian, ca. 416–398 mya (Gensel 1984; Magallón et al. 2013; Plackett and Coates 2016). Within euphyllophytes, ferns are sister to seed plants, a divergence that is estimated to have occurred 410–388 mya (Clarke et al. 2011; Magallón et al. 2013), and ferns are first represented in the fossil record by *Ibyka*, an equisetophyte from the Middle Devonian (392–385 mya) (Skog and Banks 1973; Kenrick et al. 1997; Magallón et al. 2013; Plackett and Coates 2016). Phylogenetic advances in the late 1990s and early 2000s confirmed that *Equisetum* and the other former "fern allies" *Psilotum* and *Tmesipteris* are strongly supported as members of a monophyletic fern lineage that is sister to the seed plants, with lycophytes sister to the ferns and seed plants together (Pryer et al. 1995; 2001a; 2004; Hasebe et al. 1995; Nickrent et al. 2000; Schneider et al. 2004a; Wikström and Pryer 2005; Schuettpelz et al. 2006; Smith et al. 2006; Qiu et al. 2006).

9.2.2 Relationships and Classification of Extant Lycophytes

Lycophytes are represented today by approximately 1,300 extant species that belong to three lineages, each an order composed of a single family in the most recent classification of ferns and lycophytes (PPG 1 2016) (Fig. 9.1). The Selaginellales (spike mosses) and Isoetales (quillworts) are sister to one another, and together they are sister to the Lycopodiales (club mosses). The lycophytes as a whole are characterized by possession of microphylls, leaves that are often small in size, with a single unbranched vein and no leaf gaps where the stele splits to plumb the leaves (Tomescu 2009).

A major distinction between the groups of extant lycophytes is whether they are homosporous (possessing a single type of spore) or heterosporous (with separate male and female spores). Homospory is assumed to be the ancestral condition of land plants (Bateman and DiMichele 1994), and transitions to heterospory have occurred three times during land plant evolution (Fig. 9.1). The lycophytes and ferns are the only lineages that include both homosporous and heterosporous members; the third transition occurred in the ancestor of the seed plants, which are entirely heterosporous. In the lycophytes, heterospory is inferred to have evolved in the shared ancestor of extant Selaginellales and Isoetales, while the Lycopodiales are homosporous. Selaginellales and Isoetales have one genus and 700 species and one genus and 250 species, respectively, while Lycopodiales includes 16 genera and an estimated 388 species (PPG 1 2016).

Among the lycophytes, *Selaginella* in particular has received extensive taxonomic and systematic attention, with much discussion focusing on whether its 700 species belong in one genus or should be recognized as several genera (Jermy 1986; Korall et al. 1999; Korall and Kenrick 2002, 2004; Arrigo et al. 2013; Zhou and Zhang 2015; Zhou et al. 2016) (see Zhou et al. 2016 for a historical overview of work on the genus). The most recent, phylogenetically informed classification for *Selaginella* takes the former approach and recognizes one genus with seven subgenera that are strongly supported both by phylogenetic analyses and morphological characters (Weststrand and Korall 2016a, b). Studies of the sister lineage to *Selaginella*, *Isoetes*, have likewise supported a single genus with several well-supported subclades (Hickey 1986; Taylor and Hickey 1992; Hoot and Taylor 2001; Rydin et al. 2002; Schuettpelz and Hoot 2006). Resolution of relationships within *Isoetes* is complicated by conservatism both in morphological and molecular characters and the lack of a close out-group (Hoot and Taylor 2001; Rydin et al. 2002; Hoot et al. 2006; Schuettpelz and Hoot 2006), but ongoing research suggests that the group has a complex history of relationships and biogeographic movements that deserves further study (Larsén and Rydin 2016).

The homosporous lycophytes belong to the family Lycopodiaceae, which includes three subfamilies (Lycopodielloideae, Lycopodioideae, and Huperzioideae) (Øllgaard 2015). This family has undergone extensive taxonomic upheaval, and its species have been assigned to anywhere from 2 to 16 genera over the years (Holub 1983; Wagner and Beitel 1992; Christenhusz et al. 2011; Field and Bostock 2013; Øllgaard 2015; Field et al. 2016). The PPG system recognizes 16 genera (PPG 1 2016), three of which belong to Huperzioideae. A deep split separates this lineage from the Lycopodielloideae plus Lycopodioideae (Wikström and Kenrick 2001; Ji et al. 2008; Field et al. 2016), which include four and nine genera, respectively (PPG 1 2016). The members of Lycopodiaceae as a whole are morphologically and ecologically diverse, with the greatest range of life forms and occupying the widest range of habitats of any lycophyte lineage (Field et al. 2016).

9.2.3 Relationships Among Major Clades of Extant Ferns

Equisetum Is Sister to Extant Ferns

Within ferns, the first divergence among extant taxa is now generally (though not universally) accepted as separating *Equisetum* (the horsetails) from all other species (Fig. 9.1), but determining the phylogenetic placement of *Equisetum* has been one of the most recalcitrant issues in fern systematics of the last several decades. *Equisetum* was most consistently recovered as sister to the marattioids in early molecular analyses of one or a handful of plastid, mitochondrial, and nuclear markers (Pryer et al. 2001a; Des Marais et al. 2003; Pryer et al. 2004; Wikström and Pryer 2005; Qiu et al. 2006, 2007; Schuettpelz and Pryer 2008), but it has also been resolved as sister to the Ophioglossales plus Psilotales (Grewe et al. 2013), the leptosporangiates (Lehtonen 2011), or the leptosporangiates plus marattioids (Rothwell and Nixon 2006; Schuettpelz et al. 2006), though in all cases with only low or moderate statistical support. Morphological data have added another possible resolution, with *Equisetum* sister to Psilotales in some analyses of extant taxa (Pryer et al. 1995; Schneider et al. 2009). However, most recent phylogenetic and phylogenomic studies support the placement of *Equisetum* as the sister group to all other extant ferns, based on analyses of plastid (Wolf 1997; Rai and Graham 2010; Kuo et al. 2011; Testo and Sundue 2016), mitochondrial (Vangerow et al. 1999; Knie et al. 2015), and low-copy nuclear sequence data (Wickett et al. 2014; Rothfels et al. 2015). While modern *Equisetum* is recognized as belonging to an ancient lineage (Rothwell 1999; Rothwell and Nixon 2006; Stanich et al. 2009), its 15 extant species represent a relatively recent radiation (Des Marais et al. 2003; Guillon 2004, 2007). The ancestor of this group is thought to have undergone a polyploidization event in the late Cretaceous that lead to the high chromosome numbers seen in the extant species (n=108) (Vanneste et al. 2015).

Remaining Eusporangiate Relationships

After the divergence of *Equisetum*, the next two divergence events within ferns separate the Ophioglossales plus Psilotales from all remaining species and then the Marattiales from the leptosporangiates (Fig. 9.1). These four groups (*Equisetum*, Ophioglossales, Psilotales, Marattiales), while not a clade, share a feature that differentiates them from the leptosporangiates: the production of eusporangia (as opposed to leptosporangia). This distinction refers to the development and structure of the sporangia, which produce spores via meiosis. A eusporangium arises from several epidermal initial cells and produces a large and variable number of spores, while a leptosporangium arises from a single initial cell and produces a fixed number of spores (64) during typical meiosis in a diploid sporophyte (Kumar 2001). The leptosporangium also has a unique catapult-like mechanism for spore dispersal via a specialized row of cells called the annulus (Schuettpelz and Pryer 2008; Noblin et al. 2012; Llorens et al. 2016). All leptosporangiate ferns produce a leptosporangium, which is inferred to have evolved in their common ancestor and which is the primary and eponymous synapomorphy for the group. The four other non-leptosporangiate fern lineages produce eusporangia.

The sister group relationship between Ophioglossales (moonworts) and Psilotales (whisk ferns) (Fig. 9.1) was established early on and is supported by both molecular analyses and morphological characters (Pryer et al. 1995; Manhart 1995; Vangerow et al. 1999; Nickrent et al. 2000; Pryer et al. 2001a, 2004; Wikström and Pryer 2005; Schuettpelz et al. 2006; Rothwell and Nixon 2006; Qiu et al. 2006, 2007; Kuo et al. 2011; Grewe et al. 2013; Knie et al. 2015; Testo and Sundue 2016). Both orders include one family each (Psilotaceae and Ophioglossaceae), with two genera and 17 species in the whisk ferns and four subfamilies, ten genera, and 112 species in the moonworts (PPG 1 2016). Although the affinities of Marattiales have been more challenging to untangle (some analyses infer a sister relationship with *Equisetum* (see previous section), others with Ophioglossales plus Psilotales (Wickett et al. 2014)), most recent studies place Marattiales sister to the remaining leptosporangiates with moderate to strong support (Rai and Graham 2010; Kuo et al. 2011; Knie et al. 2015; Rothfels et al. 2015; Testo and Sundue 2016). The marattioids have long been thought to form a natural group (Murdock 2008a; b), and the order includes one family (Marattiaceae) with six genera and an estimated 111 species (PPG 1 2016).

Leptosporangiates

The monophyly of the leptosporangiates has been moderately to strongly supported since the earliest molecular phylogenetic studies of ferns (Table 9.1), and possession of the leptosporangium (see previous section) has long been recognized as an unambiguous synapomorphy of the group (Moran 2004). Paleobotanical studies have identified three major radiations of leptosporangiates, the first of which began in the early Carboniferous (Lovis 1977; Rothwell 1987; Stockey and Rothwell 2006; Rothwell and Stockey 2008). This first radiation generated a set of families that had all gone extinct by the end of the Paleozoic (Rothwell 1987, 1999; Stockey and Rothwell 2006; Rothwell and Stockey 2008), but the second and third radiations produced all of the extant orders and families of leptosporangiates, beginning in the early Permian (Collinson 2001; Skog 2001; Rothwell and Stockey 2008; Schuettpelz and Pryer 2009).

The second leptosporangiate radiation produced the orders Osmundales, Hymenophyllales, Gleicheniales, Schizeales, Salviniales, and Cyatheales (Collinson 2001; Skog 2001; Rothwell and Stockey 2008), which are successively sister to the remaining leptosporangiates (Table 9.1, Fig. 9.1). Osmundales includes one family, Osmundaceae, which is strongly supported as both monophyletic and as the sister lineage to all other leptosporangiates, based on both molecular and morphological (including paleobotanical) data (Table 9.1, Rothwell and Stockey 2008). Osmundaceae fossils represent the oldest known specimens that are assignable to an extant leptosporangiate clade (Delevoryas et al. 1964; Tidwell and Ash 1994; Phipps et al. 1998), and today the group includes six genera and 18 species (Yatabe et al. 1999; Metzgar et al. 2008; PPG 1 2016).

The filmy ferns (Hymenophyllales) include one family (Hymenophyllaceae) with two subfamilies, which correspond to the trichomanoid and hymenophylloid clades (Pryer et al. 2001b; Ebihara et al. 2006). The two groups are each monophyletic (Pryer et al. 2001a; Dubuisson et al. 2003; Ebihara et al. 2007) and differ substantially from one another in their rates of molecular evolution (Schuettpelz and Pryer 2006). While the family and subfamilies are each strongly supported as monophyletic, relationships within the two subfamilies (especially the trichomanoids) have been less well supported, and intrafamilial classification has been controversial, with anywhere from 2 to 47 genera suggested by various authors (reviewed in Ebihara et al. 2006; Dubuisson et al. 2003; Hennequin et al. 2003, 2006; Ebihara et al. 2006, 2007). The PPG classification follows the system of Ebihara et al. (2006), recognizing nine genera with an estimated 434 species (PPG 1 2016).

The relationship between Hymenophyllales and Gleicheniales has also been difficult to resolve and remains somewhat uncertain. Most analyses have recovered the

Table 9.1 Overview of studies investigating deep relationships of ferns. Whether the results of each study are congruent with the overall consensus of relationships (shown in Fig. 9.1) is indicated, along with whether a given study supports (✓) monophyly of each group or not (–)

		Monophyly of:						
	Overall (as in Fig. 9.1)	Osmundales	Hymenophyllales	Gleicheniales	Schizeales	Salviniales	Cyatheales	Polypods
Hasebe et al. (1994)	✓	✓	✓	✓	✓	✓	✓	✓
Hasebe et al. (1995)	✓	✓	✓	✓	✓	✓	✓	✓
Pryer et al. (1995)	§	†	†	–	–	✓	✓	–
Pryer et al. (2001a)	✓	†	†	✓	†	✓	✓	✓
Pryer et al. (2004)	§§	✓	✓	✓	✓	✓	✓	✓
Schneider et al. (2004a)	✓	†	†	†	†	✓	✓	✓
Wikström and Pryer (2005)	✓	†	†	✓	†	*	✓	✓
Schuettpelz et al. (2006)	✓	✓	✓	✓	✓	✓	✓	✓
Schuettpelz and Pryer (2007)	✓	✓	✓	✓	✓	✓	✓	✓
Schuettpelz and Pryer (2009)	✓	✓	✓	✓	✓	✓	✓	✓
Rai and Graham (2010)	✓	✓	✓	✓	✓	✓	✓	✓
Lehtonen (2011)	✓	✓	✓	✓	✓	✓	✓	✓
Kuo et al. (2011)	§§	✓	✓	§§§	✓	✓	✓	✓
Rothfels et al. (2015)	§§	✓	✓	§§§	✓	✓	✓	✓
Knie et al. (2015)	§§	✓	✓	✓	✓	✓	✓	✓
Testo and Sundue (2016)	✓	✓	✓	✓	✓	✓	✓	✓

§Lack of monophyly of several groups leads to differences compared to Fig. 9.1
§§Gleicheniales and Hymenophyllales differ from Fig. 9.1
§§§Gleicheniales is not monophyletic
†Sampling insufficient to test monophyly
*Monophyletic in some but not all analyses

topology shown in Fig. 9.1, but only two (Rai and Graham 2010; Lehtonen 2011) have done so with strong support. Other analyses have recovered a sister group relationship between the two lineages (Pryer et al. 2004) or reversed their positions relative to the remaining leptosporangiates (i.e., recovering the topology (Osmundales (Gleicheniales (Hymenophyllales, remaining leptosporangiates))) as opposed to the topology shown in Fig. 9.1) (Knie et al. 2015). While most studies have recovered a monophyletic Gleicheniales (Table 9.1), Rothfels et al. (2015) and Kuo et al. (2011) did not, and further study on the group is clearly required. As currently circumscribed (PPG 1 2016), Gleicheniales includes three families (Matoniaceae, Dipteridaceae, and Gleicheniaceae), with ten genera and 172 species in total. In the studies that recover a monophyletic Gleicheniales with sampling from all three families, Dipteridaceae and Matoniaceae form a clade that is sister to Gleicheniaceae (Pryer et al. 2004; Schuettpelz et al. 2006; Schuettpelz and Pryer 2007, 2009; Lehtonen 2011; Testo and Sundue 2016).

The three remaining orders that are successively sister to the polypod ferns (Schizaeales, Salviniales, and Cyatheales) have each been almost universally recovered as monophyletic with strong support (Table 9.1). The Schizaeales comprises three families, four genera, and 190 species (Wikström et al. 2002; PPG 1 2016). Lygodiaceae is sister to Schizaeaceae plus Anemiaceae, and all three have been the subjects of family-level phylogenetic study (e.g., Skog et al. 2002; Wikström et al. 2002; Madeira et al. 2008; Labiak et al. 2015a). The monogeneric Lygodiaceae includes the vining ferns of the genus *Lygodium*, which are characterized by indeterminate growth (Mueller 1982, 1983) and include several documented pest species (Pemberton and Ferriter 1998; Volin et al. 2004; Madeira et al. 2008).

The Salviniales, or water ferns, are characterized by heterospory and by their aquatic habitat. Two major clades of water ferns are recognized, and these correspond to the two families of the order, Salviniaceae and Marsileaceae (Pryer et al. 2004; PPG 1 2016). These families are clearly distinguished by habit and morphology: members of the Salviniaceae are free-floating, while members of Marsileaceae are rooted below the water surface (Nagalingum et al. 2008). The two groups were initially thought to be independently derived but have been strongly supported as sister lineages in almost all major molecular studies (Table 9.1) and by morphological analyses of fossil taxa (Rothwell and Stockey 1994). Marsileaceae is the larger of the two families, with three genera (*Marsilea*, *Pilularia*, *Regnellidium*) and 61 species (Johnson 1986; Pryer 1999; Nagalingum et al. 2007, 2008; Whitten et al. 2012; PPG 1 2016), while Salviniaceae has two genera (*Salvinia*, *Azolla*) and 21 species (Reid et al. 2006; Metzgar et al. 2007; Nagalingum et al. 2008; Pereira et al. 2011; PPG 1 2016). *Azolla* has gained prominence in recent years as a candidate for whole genome sequencing (Qiu and Yu 2003; Li and Pryer 2014; Sessa et al. 2014) and for its symbiotic relationship with the nitrogen-fixing bacterium *Nostoc*, which has enabled its use as a green fertilizer in Southeast Asia for millennia (Pabby et al. 2004; Perkins and Peters 2006; Brouwer et al. 2014; Li and Pryer 2014).

The tree fern order, Cyatheales, is sister to the remaining leptosporangiates, which constitute the polypod ferns (Fig. 9.1; next section). The Cyatheales is now recognized as including eight families (with 13 genera and 713 species) that form two subclades: Thyrsopteridaceae, Loxsomataceae, Culcitaceae, and Plagiogyriaceae belong to one clade, while Cibotiaceae, Metaxyaceae, Dicksoniaceae, and Cyatheaceae make up the other (Korall et al. 2006; Schuettpelz and Pryer 2007; PPG 1 2016). The latter clade includes the species traditionally recognized as "tree ferns," which possess distinct trunks and belong to the families Cibotiaceae, Dicksoniaceae, and Cyatheaceae (Korall et al. 2006, 2007; Korall and Pryer 2014). Within this group of four families, Metaxyaceae (whose species have creeping rhizomes rather than erect trunks like the other three (Smith et al. 2001, Cárdenas et al. 2016)) is sister to Dicksoniaceae, and together they are sister to Cibotiaceae, with Cyatheaceae sister to the group of three: (Cyatheaceae, (Cibotiaceae, (Metaxyaceae, Dicksoniaceae))) (Korall et al. 2006; Schuettpelz and Pryer 2007). The other group of four families includes a number of morphologically disparate species and genera that were historically thought to be allied with other families that are now known to be distantly related to this order (e.g., *Plagiogyria* and monotypic *Loxsomopsis* and *Loxoma*) (Pryer et al. 2004; Korall et al. 2006). Within this group, Culcitaceae and Plagiogyriaceae are sisters, then Loxsomataceae, and then Thyrsopteridaceae: (Thy rsopteridaceae, (Loxsomataceae, (Culcitaceae, Plagiogyriaceae))) (Korall et al. 2006). Although there is no morphological synapomorphy known for the entire tree fern order, Cyatheales is strongly supported as monophyletic in all analyses of deep fern phylogeny (Table 9.1).

Polypods

The third leptosporangiate radiation began in the early Cretaceous (Rothwell and Stockey 2006, 2008) and corresponds to the polypod ferns, the order Polypodiales (Smith et al. 2006; PPG 1 2016). The polypods are strongly supported as monophyletic in studies of deep fern phylogeny (Table 9.1) and share a morphological synapomorphy: a vertical annulus that forms a ring of cells interrupted by the sporangial stalk (Pryer et al. 1995).

Within the polypod clade are six suborders. Two of these correspond to the eupolypods I (Polypodiineae) and II (Aspleniineae), which are sister groups (see next section). The remaining four are the Dennstaedtiineae, Pteridineae, Lindsaeineae, and Saccolomatineae (PPG 1 2016), and relationships within and between these four and the eupolypods have been challenging to resolve. The Saccolomatineae is equivalent to the family Saccolomataceae of Smith et al. (2006) and includes one genus and 18 species, though additional study is needed of this group in particular (Smith et al. 2006; PPG 1 2016); at least one study has suggested that it may be polyphyletic (Perrie et al. 2015), but this is not supported by other analyses (e.g., Testo and Sundue 2016) and may be the result of error (PPG 1 2016). Several studies have recovered Saccolomataceae as sister to the Lindsaeineae suborder (Fig. 9.1) (Schuettpelz and Pryer 2009; Lehtonen 2011; Testo and Sundue 2016), which includes the families Cystodiaceae, Lonchitidaceae, and Lindsaeaceae, but Lehtonen et al. (2012) and Kuo et al. (2011) both found Saccolomataceae to be sister instead to a group including the pteridoids, dennstaedtioids, and eupolypods, and Rai and Graham (2010) found it to be sister to these three plus the Lindsaeineae. These conflicts underscore the need for further study of Saccolomataceae and its phylogenetic affinities. Within the Lindsaeineae, the families differ substantially in size: Lindsaeaceae includes seven genera and 234 species, while Lonchitidaceae and Cystodiaceae each have one genus, with two and one species, respectively (Lehtonen et al. 2010, 2012; PPG 1 2016).

The suborders Dennstaedtiineae and Pteridineae are equivalent to the families Dennstaedtiaceae and Pteridaceae, and the relationships between these two and the eupolypods remain uncertain. The monophyly of all three groups receives consistent strong support, but all three possible relationships have been recovered: Pteridaceae sister to Dennstaedtiaceae plus the eupolypods (Fig. 9.1) (Lehtonen 2011; Knie et al. 2015; Rothfels et al. 2015), Dennstaedtiaceae sister to Pteridaceae plus the eupolypods (Schuettpelz and Pryer 2007, 2009; Kuo et al. 2011; Testo and Sundue 2016), and the eupolypods sister to Dennstaedtiaceae plus Pteridaceae (Rai and Graham 2010). In almost all studies, support is low for at least one of the branches that describe these relationships. Pteridaceae is one of the largest fern families, with five subfamilies, 53 genera, and just over 1200 species, and it includes roughly 10% of extant fern diversity (Schuettpelz and Pryer 2008; PPG 1 2016). The subfamilies correspond to five primary clades: the ceratopteridoids (Schuettpelz et al. 2007), cryptogrammoids (Metzgar et al. 2013), pteridoids (Zhang et al. 2015), adiantoids (Rothfels and Schuettpelz 2014; Schuettpelz et al. 2016), and cheilanthoids (Schuettpelz et al. 2007). Both the family as a whole and numerous individual subgroups and genera have been the subjects of systematic study (e.g., Sanchez-Baracaldo 2004; Kirkpatrick 2007; Prado et al. 2007; Schuettpelz et al. 2007, 2015; Beck et al. 2010; Sigel et al. 2011; Link-Pérez et al. 2011; Li et al. 2012; Grusz and Windham 2013; Chao et al. 2014; Zhang et al. 2015). Dennstaedtiaceae comprises 10 genera and 265 species (Perrie et al. 2015; PPG 1 2016) and includes the cosmopolitan genus *Pteridium aquilinum* (bracken fern), which was one of the first globally distributed plant species for which a worldwide phylogenetic analysis was undertaken (Der et al. 2009).

Eupolypods I and II

The eupolypods as a whole and the two subclades, eupolypods I and II, have traditionally been recognized as natural groups and receive strong phylogenetic support in molecular analyses (Schneider et al. 2004a; Schuettpelz and Pryer 2007, 2009; Rai and Graham 2010; Kuo et al. 2011; Lehtonen 2011; Rothfels et al. 2015; Testo and Sundue 2016). All three clades (eupolypods and the subclades) are supported morphologically by stelar anatomy (the number and arrangement of vascular bundles present in the stipe). All eupolypods include at least two large vascular bundles, while many non-eupolypod ferns have only one (Pittermann et al. 2015). Within eupolypods, vascular anatomy differs between the subclades: members of eupolypods I have two major bundles with several smaller strands usually present, while members of eupolypods II typically have just two bundles (with one exception:

Blechnaceae, a eupolypods II member whose anatomy resembles that of eupolypods I) (Pittermann et al. 2015).

Eupolypods I corresponds to the suborder Polypodiineae and includes nine families, 108 genera, and an estimated 4208 species (Smith et al. 2006, PPG 1 2016). The two largest fern families belong to this clade: Dryopteridaceae, with three subfamilies, 26 genera, and 2115 species, and Polypodiaceae, with six subfamilies, 65 genera, and 1652 species (PPG 1 2016). The earliest divergences among extant members of this suborder have been challenging to resolve, but several recent studies agree that the first two divergence events separate the monotypic family Didymochlaenaceae and then Hypodematiaceae from the remaining members of eupolypods I (Kuo et al. 2011; Zhang and Zhang 2015; Testo and Sundue 2016). Most studies are consistent with regard to the relationships of the remaining families, with Dryopteridaceae diverging next (Liu et al. 2016), followed by Lomariopsidaceae plus Nephrolepidaceae either sister to one another or successively sister to the group of (Tectariaceae (Oleandraceae (Davalliaceae, Polypodiaceae) (Zhang et al. 2016). Nephrolepidaceae and Lomariopsidaceae remain the primary points of phylogenetic controversy in eupolypods I, as recent studies have recovered two potential relationships: the two as sister groups (Schuettpelz and Pryer 2007, 2009; Hennequin et al. 2010; Zhang and Zhang 2015; Zhang et al. 2016) or Nephrolepidaceae sister to the remaining four-family clade (Tectariaceae through Polypodiaceae) (Kuo et al. 2011; Lehtonen 2011; Liu et al. 2013; Testo and Sundue 2016). Within the two largest families of eupolypods I, a number of groups and genera have been the subjects of study in recent years, including *Elaphoglossum* (Rouhan et al. 2004; Vasco et al. 2009, 2015; Lóriga et al. 2014; Matos and Mickel 2014), *Dryopteris* (Sessa et al. 2012; Zhang and Zhang 2012; Zhang et al. 2012), and the polystichoid (Little and Barrington 2003; Li et al. 2008; Le Péchon et al. 2016a, 2016b), lastreopsid (Labiak et al. 2014, 2015b), bolbitidoid (Moran et al. 2010), and polybotryoid (Moran and Labiak 2015) clades in Dryopteridaceae, and in Polypodiaceae the grammitid (Ranker et al. 2004; Schneider et al. 2004b; Labiak et al. 2010; Sundue et al. 2010, 2014; Sundue 2012; Bauret et al. 2017), loxogrammoid (Kreier and Schneider 2006), polygrammoid (Schneider et al. 2014c), microsoroid (Kreier et al. 2008), and drynarioid (Schneider et al. 2008) groups, and genera including *Pleopeltis* (Otto et al. 2009) and *Pecluma* (Assis et al. 2016), among many others.

Eupolypods II is equivalent to the suborder Aspleniineae and includes 11 families, 72 genera, and an estimated 2775 species (Smith et al. 2006; PPG 1 2016). The clade as a whole is well supported by molecular data, but it represents an ancient rapid radiation and as such is characterized internally by short backbone internodes, rate heterogeneity, and rooting uncertainty (Rothfels et al. 2012a). These characteristics have contributed to ambiguity with respect to interfamilial relationships within eupolypods II, but recent work has clarified the topology of the group considerably (Rothfels et al. 2012a, b; Mynssen et al. 2016). Most studies recover Cystopteridaceae as sister to the rest of the suborder, with the remaining ten families grouping into sister clades of five families each (Schuettpelz and Pryer 2007; Kuo et al. 2011; Rothfels et al. 2012a, b, 2013, 2015; Mynssen et al. 2016). One of these groups has the topology (Rhachidosoraceae (Diplaziopsidaceae (Desmophlebiaceae (Hemidictyaceae, Aspleniaceae)))) and the other (Theylpteridaceae (Woodsiaceae (Athyriaceae (Onocleaceae, Blechnaceae)))) (Li et al. 2011; Rothfels et al. 2012a, b; Mynssen et al. 2016). Thelypteridaceae is the largest of the eupolypods II families, with two subfamilies, 30 genera, and an estimated 1034 species (PPG 1 2016). It has been one of the most taxonomically complex fern families, with many or even all of its species historically included in a single massive genus *Thelypteris* that has since been broken up (Smith and Cranfill 2002; He and Zhang 2012; Almeida et al. 2016). Many other eupolypods II members have also been the focus of study in recent years, including *Rhachidosorus* and *Diplaziopsis* (Li et al. 2011), *Asplenium* and *Hymenasplenium* (Schneider et al. 2004a), *Woodsia* (Shao et al. 2015), *Diplazium* (Wei et al. 2013, 2015), and members of Blechnaceae (de Gasper et al. 2016, 2017).

9.3 Conclusion

Tremendous advances in fern and lycophyte phylogenetics and classification have been made in recent years, and the PPG 1 (2016) classification system exemplifies the dedication and community-driven spirit that has fueled much of this research. While several large-scale and myriad smaller problems of phylogeny, classification, and taxonomy in these two major clades remain to be resolved, current and future generations of fern and lycophyte biologists are up to the challenge. It is clear that there has never been a more exciting time to be a pteridologist.

References

Almeida TE, Hennequin S, Schneider H, Smith AR, Batista JAN, Ramalho AJ, Proite K, Salino A (2016) Towards a phylogenetic generic classification of Thelypteridaceae: additional sampling suggests alterations of neotropical taxa and further study of paleotropical genera. Mol Phylogenet Evol 94:688–700

Arrigo N, Therrien J, Anderson CL, Windham MD, Haufler CH, Barker MS (2013) A total evidence approach to understanding phylogenetic relationships and ecological diversity in *Selaginella* subg. *Tetragonostachys*. Am J Bot 100:1672–1682

Assis FC, Almeida TE, Russell SJ, Schneider H, Salino A (2016) Molecular phylogeny and recircumscription of the fern genus *Pecluma* (Polypodiaceae–Polypodiopsida). Phytotaxa 247:235

Bateman RM, DiMichele WA (1994) Heterospory: the most iterative key innovation in the evolutionary history of the plant kingdom. Biol Rev 69:345–417

Bauret L, Gaudeul M, Sundue MA, Parris BS, Ranker TA, Rakotondrainibe F, Hennequin S, Ranaivo J, Selosse M-A, Rouhan G (2017) Madagascar sheds new light on the molecular systematics and biogeography of grammitid ferns: new unexpected lineages and numerous long-distance dispersal events. Mol Phylogenet Evol 111:1–17

Beck JB, Windham MD, Yatskievych G, Pryer KM (2010) A diploids-first approach to species delimitation and interpreting polyploid evolution in the fern genus *Astrolepis* (Pteridaceae). Syst Bot 35:223–234

Brouwer P, Bräutigam A, Külahoglu C, Tazelaar AOE, Kurz S, Nierop KGJ, van der Werf A, Weber APM, Schluepmann H (2014) *Azolla* domestication towards a biobased economy? New Phytol 202:1069–1082

Cardelus CL, Colwell RK, Watkins JE (2006) Vascular epiphyte distribution patterns: explaining the mid-elevation richness peak. J Ecol 94:144–156

Cárdenas GG, Tuomisto H, Lehtonen S (2016) Newly discovered diversity in the tropical fern genus *Metaxya* based on morphology and molecular phylogenetic analyses. Kew Bull 71:5

Chao Y-S, Rouhan G, Amoroso VB, Chiou W-L (2014) Molecular phylogeny and biogeography of the fern genus *Pteris* (Pteridaceae). Ann Bot 114:109–124

Christenhusz MJM, Byng JW (2016) The number of known plants species in the world and its annual increase. Phytotaxa 261:201–217

Christenhusz MJM, Zhang XC, Schneider H (2011) A linear sequence of extant families and genera of lycophytes and ferns. Phytotaxa 19:7

Clarke JT, Warnock RCM, Donoghue PCJ (2011) Establishing a time-scale for plant evolution. New Phytol 192:266–301

Collinson ME (2001) Cainozoic ferns and their distribution. Brittonia 53:173–235

Delevoryas T, Hueber F, Banks H, Beck C, Eggert DA, Arnold CA, White RA, Tryon RM, Wagner WH (1964) The origin and evolution of ferns. Mem Torrey Bot Club 21:3–95

Der JP, Thomson JA, Stratford JK, Wolf PG (2009) Global chloroplast phylogeny and biogeography of Bracken (*Pteridium*; Dennstaedtiaceae). Am J Bot 96:1041–1049

Des Marais DL, Smith AR, Britton DM, Pryer KM (2003) Phylogenetic relationships and evolution of extant horsetails, *Equisetum*, based on chloroplast DNA sequence data (*rbcL* and *trnL-F*). Int J Plant Sci 164:737–751

Dubuisson JY, Hennequin S, Douzery EJP, Cranfill R, Smith AR, Pryer KM (2003) *rbcL* phylogeny of the fern genus *Trichomanes* (Hymenophyllaceae), with special reference to Neotropical taxa. Int J Plant Sci 164:753–761

Ebihara A, Dubuisson J-Y, Iwatsuki K, Hennequin S, Ito M (2006) A taxonomic revision of Hymenophyllaceae. Blumea – Biodivers Evol Biogeogr Plants 51:221–280

Ebihara A, Iwatsuki K, Ito M, Hennequin S, Dubuisson J-Y (2007) A global molecular phylogeny of the fern genus *Trichomanes* (Hymenophyllaceae) with special reference to stem anatomy. Bot J Linn Soc 155:1–27

Field AR, Bostock PD (2013) New and existing combinations in Palaeotropical *Phlegmariurus* (Lycopodiaceae) and lectotypification of the type species *Phlegmariurus phlegmaria* (L.) T.Sen & U.Sen. PhytoKeys 20:33–51

Field AR, Testo W, Bostock PD, Holtum JAM, Waycott M (2016) Molecular phylogenetics and the morphology of the Lycopodiaceae subfamily Huperzioideae supports three genera: *Huperzia*, *Phlegmariurus* and *Phylloglossum*. Mol Phylogenet Evol 94:635–657

de Gasper AL, Dittrich VADO, Smith AR, Salino A (2016) A classification for Blechnaceae (Polypodiales: Polypodiopsida): new genera, resurrected names, and combinations. *Phytotaxa* 275:191

de Gasper AL, Almeida TE, Dittrich VAO, Smith AR, Salino A (2017) Molecular phylogeny of the fern family Blechnaceae (Polypodiales) with a revised genus-level treatment. Cladistics Int J Willi Hennig Soc 33:429–446

Gastaldo RA (1987) Confirmation of Carboniferous clastic swamp communities. *Nature* 326:869–871

Gensel PG (1984) Plant life in the devonian. Praeger, New York

Gerrienne P, Servais T, Vecoli M (2016) Plant evolution and terrestrialization during Palaeozoic times—The phylogenetic context. Rev Palaeobot Palynol 227:4–18

Grewe F, Guo W, Gubbels EA, Hansen AK, Mower JP (2013) Complete plastid genomes from *Ophioglossum californicum, Psilotum nudum,* and *Equisetum hyemale* reveal an ancestral land plant genome structure and resolve the position of Equisetales among monilophytes. BMC Evol Biol 13:8

Grusz AL, Windham MD (2013) Toward a monophyletic *Cheilanthes*: the resurrection and recircumscription of *Myriopteris* (Pteridaceae). *PhytoKeys*:49–64

Guillon J-M (2004) Phylogeny of horsetails (*Equisetum*) based on the chloroplast *rps4* gene and adjacent noncoding sequences. Syst Bot 29:251–259

Guillon J-M (2007) Molecular phylogeny of horsetails (*Equisetum*) including chloroplast *atpB* sequences. J Plant Res 120:569–574

Hasebe M, Omori T, Nakazawa M, Sano T, Kato M, Iwatsuki K (1994) rbcL gene sequences provide evidence for the evolutionary lineages of leptosporangiate ferns. Proc Natl Acad Sci U S A 91:5730–5734

Hasebe M, Wolf PG, Pryer KM, Ueda K, Ito M, Sano R, Gastony GJ, Yokoyama J, Manhart JR, Murakami N, Crane EH, Haufler CH, Hauk WD (1995) Fern phylogeny based on rbcL nucleotide sequences. Am Fern J 85:134

Haufler CH, Pryer KM, Schuettpelz E, Sessa EB, Farrar DR, Moran R, Schneller JJ, Watkins JE, Windham MD (2016) Sex and the single gametophyte: revising the homosporous vascular plant life cycle in light of contemporary research. *Bioscience* 66:928–937

He L-J, Zhang X-C (2012) Exploring generic delimitation within the fern family Thelypteridaceae. Mol Phylogenet Evol 65:757–764

Hennequin S, Ebihara A, Ito M, Iwatsuki K, Dubuisson J-Y (2003) Molecular systematics of the fern genus *Hymenophyllum* s.l. (Hymenophyllaceae) based on chloroplastic coding and noncoding regions. Mol Phylogenet Evol 27:283–301

Hennequin S, Ebihara A, Ito M, Iwatsuki K, Dubuisson JY (2006) Phylogenetic systematics and evolution of the genus *Hymenophyllum* (Hymenophyllaceae: Pteridophyta). Fern Gaz 17:247–257

Hennequin S, Hovenkamp P, Christenhusz MJM, Schneider H (2010) Phylogenetics and biogeography of *Nephrolepis* – a tale of old settlers and young tramps. Bot J Linn Soc 164:113–127

Hickey RJ (1986) The early evolutionary and morphological diversity of *Isoetes*, with descriptions of two new neotropical species. Syst Bot 11:309

Holub J (1983) Validation of generic names in Lycopodiaceae: with a description of a new genus *Pseudolycopodiella*. Folia Geobotanica et Phytotaxonomica 18:439–442

Hoot SB, Taylor WC (2001) The utility of nuclear ITS, a LEAFY homolog intron, and chloroplast *atpB-rbcL* space region data in phylogenetic analyses and species delimitation in *Isoetes*. Am Fern J 91:166–177

Hoot SB, Taylor WC, Napier NS (2006) Phylogeny and biogeography of *Isoëtes* (Isoëtaceae) based on nuclear and chloroplast DNA sequence data. Syst Bot 31:449–460

Hueber FM (1992) Thoughts on the early lycopsids and zosterophylls. Ann Mo Bot Gard 79:474

Jermy AC (1986) Subgeneric names in *Selaginella*. Fern Gaz 13:117–118

Ji SG, Huo KK, Wang J, Pan SL (2008) A molecular phylogenetic study of Huperziaceae based on chloroplast rbcL and psbA-trnH sequences. J Syst Evol 46:213–219

Johnson DM (1986) Systematics of the new world species of *Marsilea* (Marsileaceae). Syst Bot Monogr 11:1

Kenrick P, Crane PR, Kenrick P, Crane P (1997) The origin and early evolution of plants on land. *Nature* 389:33–39

Kenrick P, Wellman CH, Schneider H, Edgecombe GD (2012) A timeline for terrestrialization: consequences for the carbon cycle in the Palaeozoic. Philos Trans R Soc Lond B Biol Sci 367:519–536

Kirkpatrick REB (2007) Investigating the monophyly of *Pellaea* (Pteridaceae) in the context of a phylogenetic analysis of cheilanthoid ferns. Syst Bot 32:504–518

Klekowski EJ, Baker HG (1966) Evolutionary significance of polyploidy in the pteridophyta. *Science* 153:305–307

Knie N, Fischer S, Grewe F, Polsakiewicz M, Knoop V (2015) Horsetails are the sister group to all other monilophytes and Marattiales are sister to leptosporangiate ferns. Mol Phylogenet Evol 90:140–149

Korall P, Kenrick P (2002) Phylogenetic relationships in Selaginellaceae based on RBCL sequences. Am J Bot 89:506–517

Korall P, Kenrick P (2004) The phylogenetic history of Selaginellaceae based on DNA sequences from the plastid and nucleus: extreme substitution rates and rate heterogeneity. Mol Phylogenet Evol 31:852–864

Korall P, Pryer KM (2014) Global biogeography of scaly tree ferns (Cyatheaceae): evidence for Gondwanan vicariance and limited transoceanic dispersal. J Biogeogr 41:402–413

Korall P, Kenrick P, Therrien JP (1999) Phylogeny of selaginellaceae: evaluation of generic/subgeneric relationships based on *rbc*L gene sequences. Int J Plant Sci 160:585–594

Korall P, Pryer KM, Metzgar JS, Schneider H, Conant DS (2006) Tree ferns: monophyletic groups and their relationships as revealed by four protein-coding plastid loci. Mol Phylogenet Evol 39:830–845

Korall P, Conant DS, Metzgar JS, Schneider H, Pryer KM (2007) A molecular phylogeny of scaly tree ferns (Cyatheaceae). Am J Bot 94:873–886

Kreier H-P, Schneider H (2006) Reinstatement of *Loxogramme dictyopteris*, based on phylogenetic evidence, for the New Zealand endemic fern, *Anarthropteris lanceolata* (Polypodiaceae, Polypodiidae). Aust Syst Bot 19:309

Kreier H-P, Zhang X-C, Muth H, Schneider H (2008) The microsoroid ferns: Inferring the relationships of a highly diverse lineage of Paleotropical epiphytic ferns (Polypodiaceae, Polypodiopsida). Mol Phylogenet Evol 48:1155–1167

Kumar K (2001) Reproductive biology of pteridophytes. In: Johri BM, Srivastava PS (eds) Reproductive biology of plants. Springer, New York, pp 175–214

Kuo L-Y, Li F-W, Chiou W-L, Wang C-N (2011) First insights into fern matK phylogeny. Mol Phylogenet Evol 59:556–566

Labiak PH, Sundue M, Rouhan G (2010) Molecular phylogeny, character evolution, and biogeography of the grammitid fern genus *Lellingeria* (Polypodiaceae). Am J Bot 97:1354–1364

Labiak PH, Sundue M, Rouhan G, Hanks JG, Mickel JT, Moran RC (2014) Phylogeny and historical biogeography of the lastreopsid ferns (Dryopteridaceae). Am J Bot 101:1207–1228

Labiak PH, Mickel JT, Hanks JG (2015a) Molecular phylogeny and character evolution of Anemiaceae (Schizaeales). *Taxon* 64:1141–1158

Labiak PH, Sundue M, Rouhan G, Moran RC (2015b) New combinations in *Lastreopsis* and *Parapolystichum* (Dryopteridaceae). *Brittonia* 67:79–86

Larsén E, Rydin C (2016) Disentangling the phylogeny of *Isoetes* (Isoetales), using nuclear and plastid data. Int J Plant Sci 177:157–174

Le Péchon T, He H, Zhang L, Zhou X-M, Gao X-F, Zhang L-B (2016a) Using a multilocus phylogeny to test morphology-based classifications of *Polystichum* (Dryopteridaceae), one of the largest fern genera. BMC Evol Biol 16:55

Le Péchon T, Zhang L, He H, Zhou X-M, Bytebier B, Gao X-F, Zhang L-B (2016b) A well-sampled phylogenetic analysis of the polystichoid ferns (Dryopteridaceae) suggests a complex biogeographical history involving both boreotropical migrations and recent transoceanic dispersals. Mol Phylogenet Evol 98:324–336

Lehtonen S (2011) Towards resolving the complete fern tree of life. *Plos One* 6:e24851

Lehtonen S, Tuomisto H, Rouhan G, Christenhusz MJM (2010) Phylogenetics and classification of the pantropical fern family Lindsaeaceae. Bot J Linn Soc 163:305–359

Lehtonen S, Wahlberg N, Christenhusz MJM (2012) Diversification of lindsaeoid ferns and phylogenetic uncertainty of early polypod relationships. Bot J Linn Soc Linn Soc Lond 170:489–503

Lenton TM, Dahl TW, Daines SJ, Mills BJW, Ozaki K, Saltzman MR, Porada P (2016) Earliest land plants created modern levels of atmospheric oxygen. Proc Natl Acad Sci U S A 113:9704–9709

Li F-W, Pryer KM (2014) Crowdfunding the *Azolla* fern genome project: a grassroots approach. *GigaScience* 3:16

Li C-X, Lu S-G, Barrington DS (2008) Phylogeny of Chinese *Polystichum* (Dryopteridaceae) based on chloroplast DNA sequence data (trnL-F and rps4-trnS). J Plant Res 121:19–26

Li C, Lu S, Sun X, Yang Q (2011) Phylogenetic positions of the enigmatic asiatic fern genera *Diplaziopsis* and *Rhachidosorus* from analyses of four plastid genes. Am Fern J 101:142–155

Li F-W, Pryer KM, Windham MD (2012) *Gaga*, a new fern genus segregated from *Cheilanthes* (Pteridaceae). Syst Bot 37:845–860

Link-Pérez MA, Watson LE, Hickey RJ (2011) Redefinition of *Adiantopsis* Fée (Pteridaceae): systematics, diversification, and biogeography. *Taxon* 60:1255–1268

Linnaeus C (1753) Species plantarum. Laurentius Salvius, Stockholm

Little DP, Barrington DS (2003) Major evolutionary events in the origin and diversification of the fern genus *Polystichum* (Dryopteridaceae). Am J Bot 90:508–514

Liu H-M, Jiang R-H, Guo J, Hovenkamp P, Perrie LR, Shepherd L, Hennequin S, Schneider H (2013) Towards a phylogenetic classification of the climbing fern genus *Arthropteris*. *Taxon* 62:688–700

Liu H-M, Zhang X-C, Wang M-P, Shang H, Zhou S-L, Yan Y-H, Wei X-P, Xu W-B, Schneider H (2016) Phylogenetic placement of the enigmatic fern genus *Trichoneuron informs* on the infra-familial relationship of Dryopteridaceae. Plant Syst Evol 302:319–332

Llorens C, Argentina M, Rojas N, Westbrook J, Dumais J, Noblin X (2016) The fern cavitation catapult: mechanism and design principles. J R Soc Interface 13:20150930

Lóriga J, Vasco A, Regalado L, Heinrichs J, Moran RC (2014) Phylogeny and classification of the Cuban species of *Elaphoglossum* (Dryopteridaceae), with description of *Elaphoglossum* sect. *Wrightiana* sect. nov. Plant Syst Evol 300:937–951

Lovis JD (1977) Evolutionary patterns and processes in ferns. Adv Bot Res 4:229–440

Madeira PT, Pemberton RW, Center TD (2008) A molecular phylogeny of the genus *Lygodium* (Schizaeaceae) with special reference to the biological control and host range testing of *Lygodium microphyllum*. Biol Control 45:308–318

Magallón S, Hilu KW, Quandt D (2013) Land plant evolutionary timeline: gene effects are secondary to fossil constraints in relaxed clock estimation of age and substitution rates. Am J Bot 100:556–573

Manhart JR (1995) Chloroplast 16S rDNA sequences and phylogenetic relationships of fern allies and ferns. Am Fern J 85:182

Matos FB, Mickel JT (2014) The Brazilian species of *Elaphoglossum* section *Polytrichia* (Dryopteridaceae). *Brittonia* 66:371–395

Metzgar JS, Schneider H, Pryer KM (2007) Phylogeny and divergence time estimates for the fern genus *Azolla* (Salviniaceae). Int J Plant Sci 168:1045–1053

Metzgar JS, Skog JE, Zimmer EA, Pryer KM (2008) The paraphyly of *Osmunda* is confirmed by phylogenetic analyses of seven plastid loci. Syst Bot 33:31–36

Metzgar JS, Alverson ER, Chen S, Vaganov AV, Ickert-Bond SM (2013) Diversification and reticulation in the circumboreal fern genus *Cryptogramma*. Mol Phylogenet Evol 67:589–599

Moran RC (2004) A natural history of ferns. Timber Press, Inc., Portland

Moran RC, Labiak PH (2015) Phylogeny of the polybotryoid fern clade (Dryopteridaceae). Int J Plant Sci 176:880–891

Moran RC, Labiak PH, Sundue M (2010) Phylogeny and character evolution of the bolbitidoid ferns (Dryopteridaceae). Int J Plant Sci 171:547–559

Mueller RJ (1982) Shoot morphology of the climbing fern *Lygodium* (Schizaeaceae): general organography, leaf initiation, and branching. *Bot Gaz* 143:319–330

Mueller RJ (1983) Indeterminate growth and ramification of the climbing leaves of *Lygodium japonicum* (Schizaeaceae). Am J Bot 70:682

Murdock AG (2008a) A taxonomic revision of the eusporangiate fern family Marattiaceae, with description of a new genus Ptisana. *Taxon* 57:737–755

Murdock AG (2008b) Phylogeny of marattioid ferns (Marattiaceae): inferring a root in the absence of a closely related outgroup. Am J Bot 95:626–641

Mynssen CM, Vasco A, Moran RC, Sylvestre LS, Rouhan G (2016) Desmophlebiaceae and *Desmophlebium*: a new family and genus of Eupolypod II ferns. *Taxon* 65:19–34

Nagalingum NS, Schneider H, Pryer KM (2007) Molecular phylogenetic relationships and morphological evolution in the heterosporous fern genus *Marsilea*. Syst Bot 32:16–25

Nagalingum NS, Nowak MD, Pryer KM (2008) Assessing phylogenetic relationships in extant heterosporous ferns (Salviniales), with a focus on *Pilularia* and *Salvinia*. Bot J Linn Soc 157:673–685

Nickrent DL, Parkinson CL, Palmer JD, Duff RJ (2000) Multigene phylogeny of land plants with special reference to bryophytes and the earliest land plants. Mol Biol Evol 17:1885–1895

Noblin X, Rojas NO, Westbrook J, Llorens C, Argentina M, Dumais J (2012) The fern sporangium: a unique catapult. Science 335:1322

Øllgaard B (2015) Six new species and some nomenclatural changes in neotropical Lycopodiaceae. Nord J Bot 33:186–196

Otto EM, Janssen T, Kreier H-P, Schneider H (2009) New insights into the phylogeny of *Pleopeltis* and related Neotropical genera (Polypodiaceae, Polypodiopsida). Mol Phylogenet Evol 53:190–201

Pabby A, Prasanna R, Singh PK (2004) Biological significance of *Azolla* and its utilization in agriculture. Proc Indian Natl Sci Acad B70:299–333

Pemberton RW, Ferriter AP (1998) Old world climbing fern (*Lygodium microphyllum*), a dangerous invasive weed in Florida. Am Fern J 88:165

Pereira AL, Martins M, Oliveira MM, Carrapiço F (2011) Morphological and genetic diversity of the family Azollaceae inferred from vegetative characters and RAPD markers. Plant Syst Evol 297:213–226

Perkins SK, Peters GA (2006) The *Azolla-Anabaena* symbiosis: endophyte continuity in the *Azolla* life-cycle is facilitated by epidermal trichomes. New Phytol 123:53–64

Perrie LR, Shepherd LD, Brownsey PJ (2015) An expanded phylogeny of the Dennstaedtiaceae ferns: *Oenotrichia* falls within a non-monophyletic *Dennstaedtia*, and *Saccoloma* is polyphyletic. Aust Syst Bot 28:256

Phipps C, Taylor T, Taylor E, Cúneo R, Boucher L, Yao X (1998) *Osmunda* (Osmundaceae) from the Triassic of Antarctica: an example of evolutionary stasis. Am J Bot 85:888

Pittermann J, Watkins JE, Cary KL, Schuettpelz E, Brodersen C, Smith AR, Baer A (2015) The structure and function of xylem in seed-free vascular plants: an evolutionary perspective. In: Hacke U (ed) Functional and ecological xylem anatomy. Springer International Publishing, Cham, pp 1–37

Plackett ARG, Coates JC (2016) Life's a beach – the colonization of the terrestrial environment. New Phytol 212:831–835

PPG 1 (2016) A community-derived classification for extant lycophytes and ferns. J Syst Evol 54:563–603

Prado J, Del Nero Rodrigues C, Salatino A, Salatino ML (2007) Phylogenetic relationships among Pteridaceae, including Brazilian species, inferred from rbcL sequences. Taxon 56:355–368

Pryer KM (1999) Phylogeny of marsileaceous ferns and relationships of the fossil *Hydropteris pinnata* reconsidered. Int J Plant Sci 160:931–954

Pryer KM, Smith AR, Skog JE, Present A-W (1995) Phylogenetic relationships of extant ferns based on evidence from morphology and rbcL sequences. Am Fern J 85:205–282

Pryer KM, Schneider H, Smith AR, Cranfill R, Wolf PG, Hunt JS, Sipes SD (2001a) Horsetails and ferns are a monophyletic group and the closest living relatives to seed plants. *Nature* 409:618–622

Pryer KM, Smith AR, Hunt JS, Dubuisson JY (2001b) rbcL data reveal two monophyletic groups of filmy ferns (Hymenophyllaceae). Am J Bot 88:1118–1130

Pryer KM, Schuettpelz E, Wolf PG, Schneider H, Smith AR, Cranfill R (2004) Phylogeny and evolution of ferns (monilophytes) with a focus on the early leptosporangiate divergences. Am J Bot 91:1582–1598

Qiu YL, Yu J (2003) Azolla – a model organism for plant genomic studies. Genomics Proteomics Bioinformatics 1:15–25

Qiu Y-L, Li L, Wang B, Chen Z, Knoop V, Groth-Malonek M, Dombrovska O, Lee J, Kent L, Rest J, Estabrook GF, Hendry TA, Taylor DW, Testa CM, Ambros M, Crandall-Stotler B, Duff RJ, Stech M, Frey W, Quandt D, Davis CC (2006) The deepest divergences in land plants inferred from phylogenomic evidence. Proc Natl Acad Sci U S A 103:15511–15516

Qiu Y, Li L, Wang B, Chen Z, Dombrovska O, Lee J, Kent L, Li R, Jobson RW, Hendry TA, Taylor DW, Testa CM, Ambros M (2007) A nonflowering land plant phylogeny inferred from nucleotide sequences of seven chloroplast, mitochondrial, and nuclear genes. Int J Plant Sci 168:691–708

Rai HS, Graham SW (2010) Utility of a large, multigene plastid data set in inferring higher-order relationships in ferns and relatives (monilophytes). Am J Bot 97:1444–1456

Ranker TA, Smith AR, Parris BS, Geiger JMO, Haufler CH, Straub SCK, Schneider H (2004) Phylogeny and evolution of grammitid ferns: a case of rampant morphological homoplasy. *Taxon* 53:415–428

Reid JD, Plunkett GM, Peters GA (2006) Phylogenetic relationships in the heterosporous fern genus *Azolla* (Azollaceae) based on DNA sequence data from three noncoding regions. Int J Plant Sci 167:529–538

Rothfels CJ, Schuettpelz E (2014) Accelerated rate of molecular evolution for vittarioid ferns is strong and not driven by selection. Syst Biol 63:31–54

Rothfels CJ, Larsson A, Kuo L-Y, Korall P, Chiou W-L, Pryer KM (2012a) Overcoming deep roots, fast rates, and short internodes to resolve the ancient rapid radiation of eupolypod II ferns. Syst Biol 61:490–509

Rothfels CJ, Sundue MA, Kuo L-Y, Larsson A, Kato M, Schuettpelz E, Pryer KM, Rothfels C, Sundue M, Kuo L-Y, Larsson A, Kato M, Schuettpelz E, Pryer K (2012b) A revised family-level classification for eupolypod II ferns (Polypodiidae: Polypodiales). *Taxon* 61:515–533

Rothfels CJ, Windham MD, Pryer KM (2013) A plastid phylogeny of the cosmopolitan fern family cystopteridaceae (Polypodiopsida). Syst Bot 38:295–306

Rothfels CJ, Li F-W, Sigel EM, Huiet L, Larsson A, Burge DO, Ruhsam M, Deyholos M, Soltis DE, Stewart CN, Shaw SW, Pokorny L, Chen T, dePamphilis C, DeGironimo L, Chen L, Wei X, Sun X, Korall P, Stevenson DW, Pryer KM (2015) The evolutionary history of ferns inferred from 25 low-copy nuclear genes. Am J Bot 102:1089–1107

Rothwell GW (1987) Complex paleozoic filicales in the evolutionary radiation of ferns. Am J Bot 74:458

Rothwell GW (1999) Fossils and ferns in the resolution of land plant phylogeny. Bot Rev 65:188–218

Rothwell GW, Nixon KC (2006) How does the inclusion of fossil data change our conclusions about the phylogenetic history of euphyllophytes? Int J Plant Sci 167:737–749

Rothwell GW, Stockey RA (1994) The role of *Hydropteris pinnata* gen. et. sp. nov. in reconstructing the cladistics of heterosporous ferns. Am J Bot 81:479

Rothwell GW, Stockey RA (2006) Combining characters of pteridaceae and tree ferns: *Pterisorus radiata* gen. et sp. nov., a permineralized lower Cretaceous filicalean with radial sori. Int J Plant Sci 167:695–701

Rothwell GW, Stockey RA (2008) Phylogeny and evolution of ferns: a paleontological perspective. In: Ranker TA, Haufler CH (eds) Biology and evolution of ferns and lycophytes. Cambridge University Press, Cambridge, pp 332–366

Rouhan G, Dubuisson J-Y, Rakotondrainibe F, Motley TJ, Mickel JT, Labat J-N, Moran RC (2004) Molecular phylogeny of the fern genus *Elaphoglossum* (Elaphoglossaceae) based on chloroplast non-coding DNA sequences: contributions of species from the Indian Ocean area. Mol Phylogenet Evol 33:745–763

Rydin C, Wikström N, Wikstrom N (2002) Phylogeny of *Isoëtes* (Lycopsida): resolving basal relationships using rbcL sequences. *Taxon* 51:83

Sanchez-Baracaldo P (2004) Phylogenetic relationships of the subfamily Taenitidoideae, Pteridaceae. Am Fern J 94:126–142

Schneider H, Russell SJ, Cox CJ, Bakker F, Henderson S, Rumsey F, Barrett J, Gibby M, Vogel JC (2004a) Chloroplast phylogeny of asplenioid ferns based on *rbcL* and *trnL-F* Spacer Sequences (Polypodiidae, Aspleniaceae) and its implications for biogeography. Syst Bot 29:260–274

Schneider H, Schuettpelz E, Pryer KM, Cranfill R, Magallón S, Lupia R (2004b) Ferns diversified in the shadow of angiosperms. *Nature* 428:553–557

Schneider H, Smith AR, Cranfill R, Hildebrand TJ, Haufler CH, Ranker TA (2004c) Unraveling the phylogeny of polygrammoid ferns (Polypodiaceae and Grammitidaceae): exploring aspects of the diversification of epiphytic plants. Mol Phylogenet Evol 31:1041–1063

Schneider H, Kreier HP, Hovenkamp P, Janssen T (2008) Phylogenetic relationships of the fern genus *Christiopteris* shed new light onto the classification and biogeography of drynarioid ferns. Bot J Linn Soc 157:645–656

Schneider H, Smith AR, Pryer KM (2009) Is morphology really at odds with molecules in estimating fern phylogeny? Syst Bot 34:455–475

Schuettpelz E, Hoot SB (2006) Inferring the root of *Isoëtes*: exploring alternatives in the absence of an acceptable outgroup. Syst Bot 31:258–270

Schuettpelz E, Pryer KM (2006) Reconciling extreme branch length differences: decoupling time and rate through the evolutionary history of filmy ferns. Syst Biol 55:485–502

Schuettpelz E, Pryer KM (2007) Fern phylogeny inferred from 400 leptosporangiate species and three plastid genes. *Taxon* 56:1037

Schuettpelz E, Pryer KM (2008) Fern phylogeny. In: Ranker TA, Haufler CH (eds) Biology and evolution of ferns and lycophytes. Cambridge University Press, Cambridge, pp 395–416

Schuettpelz E, Pryer KM (2009) Evidence for a Cenozoic radiation of ferns in an angiosperm-dominated canopy. Proc Natl Acad Sci U S A 106:11200–11205

Schuettpelz E, Korall P, Pryer KM (2006) Plastid atpA data provide improved support for deep relationships among ferns. *Taxon* 55:897–906

Schuettpelz E, Schneider H, Huiet L, Windham MD, Pryer KM (2007) A molecular phylogeny of the fern family Pteridaceae: assessing overall relationships and the affinities of previously unsampled genera. Mol Phylogenet Evol 44:1172–1185

Schuettpelz E, Pryer KM, Windham MD (2015) A unified approach to taxonomic delimitation in the fern genus *Pentagramma* (Pteridaceae). Syst Bot 40:629–644

Schuettpelz E, Chen C-W, Kessler M, Pinson JB, Johnson G, Davila A, Cochran AT, Huiet L, Pryer KM (2016) A revised generic classification of vittarioid ferns (Pteridaceae) based on molecular, micromorphological, and geographic data. Taxon 65:708–722

Sessa EB, Zimmer EA, Givnish TJ (2012) Phylogeny, divergence times, and historical biogeography of New World *Dryopteris* (Dryopteridaceae). Am J Bot 99:730–750

Sessa EB, Banks JA, Barker MS, Der JP, Duffy AM, Graham SW, Hasebe M, Langdale J, Li F-W, Marchant DB, Pryer KM, Rothfels CJ, Roux SJ, Salmi ML, Sigel EM, Soltis DE, Soltis PS, Stevenson DW, Wolf PG (2014) Between two fern genomes. *GigaScience* 3:15

Shao Y, Wei R, Zhang X, Xiang Q (2015) Molecular phylogeny of the cliff ferns (Woodsiaceae: Polypodiales) with a proposed infrageneric classification. *Plos One* 10:e0136318

Sigel EM, Windham MD, Huiet L, Yatskievych G, Pryer KM (2011) Species relationships and farina evolution in the cheilanthoid fern genus *Argyrochosma* (Pteridaceae). Syst Bot 36:554–564

Skog JE (2001) Biogeography of Mesozoic leptosporangiate ferns related to extant ferns. *Brittonia* 53:236–269

Skog JE, Banks HP (1973) *Ibyka amphikoma*, gen. et sp. n., a new protoarticulate precursor from the late Middle Devonian of New York State. Am J Bot 60:366–380

Skog JE, Zimmer EA, Mickel JT (2002) Additional support for two subgenera of *Anemia* (Schizaeaceae) from data for the chloroplast intergenic spacer region trnL-F and morphology. Am Fern J 92:119

Smith AR, Cranfill RB (2002) Intrafamilial relationships of the Thelypteroid Ferns (Thelypteridaceae). Am Fern J 92:131

Smith AR, Tuomisto H, Pryer KM, Hunt JS, Wolf PG (2001) *Metaxya lanosa*, a second species in the genus and fern family Metaxyaceae. Syst Bot 26:480–486

Smith AR, Pryer KM, Schuettpelz E, Korall P, Schneider H, Wolf PG (2006) A classification for extant ferns. *Taxon* 55:705–731

Stanich NA, Rothwell GW, Stockey RA (2009) Phylogenetic diversification of *Equisetum* (Equisetales) as inferred from lower cretaceous species of British Columbia, Canada. Am J Bot 96:1289–1299

Stockey RA, Rothwell GW (2006) Introduction: evolution of modern ferns. Int J Plant Sci 167:613–614

Sundue M (2012) *Galactodenia*, a new genus of grammitid ferns segregated from *Terpsichore* (Polypodiaceae). Subterranean Biol 37

Sundue MA, Islam MB, Ranker TA (2010) Systematics of grammitid ferns (Polypodiaceae): using morphology and plastid sequence data to resolve the circumscriptions of *Melpomene* and the polyphyletic genera *Lellingeria* and *Terpsichore*. Syst Bot 35:701–715

Sundue MA, Parris BS, Ranker TA, Smith AR, Fujimoto EL, Zamora-Crosby D, Morden CW, Chiou W-L, Chen C-W, Rouhan G, Hirai RY, Prado J (2014) Global phylogeny and biogeography of grammitid ferns (Polypodiaceae). Mol Phylogenet Evol 81:195–206

Taylor WC, Hickey RJ (1992) Habitat, evolution, and speciation in *Isoetes*. Ann Mo Bot Gard 79:613

Testo W, Sundue M (2016) A 4000-species dataset provides new insight into the evolution of ferns. Mol Phylogenet Evol 105:200–211

Tidwell WD, Ash SR (1994) A review of selected triassic to early Cretaceous ferns. J Plant Res 107:417–442

Tomescu AMF (2009) Megaphylls, microphylls and the evolution of leaf development. Trends Plant Sci 14:5–12

Vangerow S, Teerkorn T, Knoop V (1999) Phylogenetic information in the mitochondrial nad5 gene of pteridophytes: RNA editing and intron sequences. Plant Biol 1:235–243

Vanneste K, Sterck L, Myburg AA, Van de Peer Y, Mizrachi E (2015) Horsetails are ancient polyploids: evidence from *Equisetum giganteum*. Plant Cell 27:1567–1578

Vasco A, Moran RC, Rouhan G (2009) Circumscription and phylogeny of the *Elaphoglossum ciliatum* group (E. sect. *Lepidoglossa*, Dryopteridaceae) based on cpDNA sequences. *Taxon* 58:825–834

Vasco A, Moran RC, Ambrose BA (2013) The evolution, morphology, and development of fern leaves. Front Plant Sci 4:345

Vasco A, Lóriga J, Rouhan G, Ambrose BA, Moran RC (2015) Divided leaves in the genus *Elaphoglossum* (Dryopteridaceae): a phylogeny of *Elaphoglossum* section *Squamipedia*. Syst Bot 40:46–55

Vasco A, Smalls TL, Graham SW, Cooper ED, Wong GK-S, Stevenson DW, Moran RC, Ambrose BA (2016) Challenging the paradigms of leaf evolution: class III HD-Zips in ferns and lycophytes. New Phytol 212:745–758

Volin JC, Lott MS, Muss JD, Owen D (2004) Predicting rapid invasion of the florida everglades by old world climbing fern (*Lygodium microphyllum*). Divers Distrib 10:439–446

Wagner WH, Beitel JM (1992) Generic classification of modern North American Lycopodiaceae. Ann Mo Bot Gard 79:676

Watkins JE, Cardelús CL (2012) Ferns in an angiosperm world: cretaceous radiation into the epiphytic niche and diversification on the forest floor. Int J Plant Sci 173:695–710

Wei R, Schneider H, Zhang XC (2013) Toward a new circumscription of the twinsorus-fern genus *Diplazium* (Athyriaceae): a molecular phylogeny with morphological implications and infrageneric taxonomy. Taxon 62:441–457

Wei R, Xiang Q, Schneider H, Sundue MA, Kessler M, Kamau PW, Hidayat A, Zhang X (2015) Eurasian origin, boreotropical migration and transoceanic dispersal in the pantropical fern genus *Diplazium* (Athyriaceae). J Biogeogr 42:1809–1819

Weststrand S, Korall P (2016a) Phylogeny of Selaginellaceae: there is value in morphology after all! Am J Bot 103:2136–2159

Weststrand S, Korall P (2016b) A subgeneric classification of *Selaginella* (Selaginellaceae). Am J Bot 103:2160–2169

Whitten WM, Jacono CC, Nagalingum NS (2012) An expanded plastid phylogeny of *Marsilea* with emphasis on North American species. Am Fern J 102:114–135

Wickett NJ, Mirarab S, Nguyen N, Warnow T, Carpenter E, Matasci N, Ayyampalayam S, Barker MS, Burleigh JG, Gitzendanner MA, Ruhfel BR, Wafula E, Der JP, Graham SW, Mathews S, Melkonian M, Soltis DE, Soltis PS, Miles NW, Rothfels CJ, Leebens-Mack J (2014) Phylotranscriptomic analysis of the origin and early diversification of land plants. Proc Natl Acad Sci U S A 111:E4859–E4868

Wikström N, Kenrick P (2001) Evolution of Lycopodiaceae (Lycopsida): estimating divergence times from rbcL gene sequences by use of nonparametric rate smoothing. Mol Phylogenet Evol 19:177–186

Wikström N, Pryer KM (2005) Incongruence between primary sequence data and the distribution of a mitochondrial atp1 group II intron among ferns and horsetails. Mol Phylogenet Evol 36:484–493

Wikström N, Kenrick P, Vogel JC (2002) Schizaeaceae: a phylogenetic approach. Rev Palaeobot Palynol 119:35–50

Wolf PG (1997) Evaluation of atpB nucleotide sequences for phylogenetic studies of ferns and other pteridophytes. Am J Bot 84:1429–1440

Yatabe Y, Nishida H, Murakami N (1999) Phylogeny of osmundaceae inferred from rbcL nucleotide sequences and comparison to the fossil evidences. J Plant Res 112:397–404

Zhang LB, Zhang L (2012) The inclusion of *Acrophorus, Diacalpe, Nothoperanema,* and *Peranema* in *Dryopteris*: the molecular phylogeny, systematics, and nomenclature of *Dryopteris* subg. *Nothoperanema* (Dryopteridaceae). *Taxon* 61:1199–1216

Zhang L-B, Zhang L (2015) Didymochlaenaceae: a new fern family of eupolypods I (Polypodiales). *Taxon* 64:27–38

Zhang L-B, Zhang L, Dong S-Y, Sessa EB, Gao X-F, Ebihara A (2012) Molecular circumscription and major evolutionary lineages of the fern genus *Dryopteris* (Dryopteridaceae). BMC Evol Biol 12:180

Zhang L, Rothfels CJ, Ebihara A, Schuettpelz E, Le Péchon T, Kamau P, He H, Zhou X-M, Prado J, Field A, Yatskievych G, Gao X-F, Zhang L-B (2015) A global plastid phylogeny of the brake fern genus *Pteris* (Pteridaceae) and related genera in the Pteridoideae. Cladistics Int J Willi Hennig Soc 31:406–423

Zhang L, Schuettpelz E, Rothfels CJ, Zhou X-M, Gao X-F, Zhang L-B (2016) Circumscription and phylogeny of the fern family Tectariaceae based on plastid and nuclear markers, with the description of two new genera: *Draconopteris* and *Malaifilix* (Tectariaceae). *Taxon* 65:723–738

Zhou X-M, Zhang L-B (2015) A classification of *Selaginella* (Selaginellaceae) based on molecular (chloroplast and nuclear), macromorphological, and spore features. *Taxon* 64:1117–1140

Zhou X, Rothfels CJ, Zhang L, He Z, Le Péchon T, He H, Lu NT, Knapp R, Lorence D, He X, Gao X, Zhang L (2016) A large-scale phylogeny of the lycophyte genus *Selaginella* (Selaginellaceae: Lycopodiopsida) based on plastid and nuclear loci. Cladistics Int J Willi Hennig Soc 32:360–389

Chapter 10
Reciprocal Illumination and Fossils Provide Important Perspectives in Plant Evo-devo: Examples from Auxin in Seed-Free Plants

Kelly K. S. Matsunaga and Alexandru M. F. Tomescu

10.1 Introduction

Understanding the evolutionary origins of modern plant form is a central goal of plant evolutionary biology. Data relevant to understanding morphological evolution come from varied fields of study including paleontology, physiology, comparative anatomy and morphology, and developmental genetics. Only integration of all these data can provide a holistic understanding of plant evolution that spans plant phylogeny and deep evolutionary time. Studies on development are particularly important for understanding macroevolutionary patterns because all mature structures are formed through tightly regulated developmental processes, and these processes can differ depending on the homology of structures and the phylogenetic affinities of taxa. Development can therefore be used as an additional source of data, along with morphological and molecular data, for understanding evolution. When applied broadly to living and extinct representatives of a group, anatomical, physiological, and molecular aspects of development can thus provide novel insights on important aspects of plant evolution, including the origins of different growth forms, body plans, and major morphological innovations such as leaves, roots, and seeds.

Hormones such as auxins, cytokinins, and gibberellins are key regulators of a vast array of developmental processes (e.g., Durbak et al. 2012; Lacombe and Achard 2016). Auxins are particularly interesting because they have numerous documented roles in nearly every major aspect of plant development (Benjamins and

K. K. S. Matsunaga
Department of Earth and Environmental Sciences, University of Michigan, Ann Arbor, MI, USA

A. M. F. Tomescu (✉)
Department of Biological Sciences, Humboldt State University, Arcata, CA, USA
e-mail: mihai@humboldt.edu

H. Fernández (ed.), *Current Advances in Fern Research*,
https://doi.org/10.1007/978-3-319-75103-0_10

Scheres 2008). Indole-3 acetic acid (IAA) is the predominant form of auxin in the plant body (Finet and Jaillais 2012). Once auxin enters a cell either through diffusion or active transport by influx carriers, they can only exit the cell via auxin efflux carriers in the cell membrane known as PIN proteins (Jones 1998). Polarization of PIN protein position within the cell establishes a directional movement of auxin through the cell; at the tissue scale this directional movement is referred to as polar auxin transport (Benková et al. 2003; Friml et al. 2003).

From the perspective of understanding major morphological transitions of embryophytes, polar auxin transport is especially intriguing in several respects: (1) auxin is involved in numerous aspects of embryonic and postembryonic development, and auxin transport networks have an integral role in coordinating growth and development on a whole-plant scale (Cooke et al. 2004; Benjamins and Scheres 2008; Leyser 2011); (2) on a smaller scale, auxin accumulation and polar auxin transport have important functions in initiation and development of leaves, roots, and vascular tissues (Reinhardt 2005; Blilou et al. 2005; Miyashima et al. 2013); and (3) the molecular mechanisms underlying polar auxin transport appear to be conserved across vascular plants (Sanders et al. 2011; Bennett 2015). A recent synthesis of PIN protein evolution indicates that PIN proteins have undergone relatively little structural change throughout plant phylogeny, suggesting that changes in the PIN proteins themselves are unlikely to have directly driven morphological evolution. However, it remains unclear whether changes in the regulatory networks controlling the timing, position, and duration of auxin signaling can explain some of the major morphological transitions of embryophytes (Bennett 2015).

To date, most of the available data on auxin action come from studies of seed plants, particularly angiosperms. However, because significant evolutionary time separates angiosperms from the earliest vascular plants, these data may not be entirely applicable to seed-free plants. Understanding differences in the roles of auxin in development between angiosperms and seed-free lineages is therefore critical for developing and testing hypotheses on morphological evolution. Here, we summarize current knowledge of the roles of auxin, particularly polar auxin transport, in the development of seed-free tracheophytes. This survey encompasses data from extant and fossil plants, and we suggest hypotheses to be tested and directions for future research.

10.2 Auxin in the Seed Plant Sporophyte

Polar auxin transport through the plant body is essential for establishing the overall axial organization of the plant and the patterning of vascular tissues (Berleth 2001; Scarpella and Meijer 2004; Sawchuk and Scarpella 2013; Lucas et al. 2013; Harrison 2016; Campbell and Turner 2017). Our detailed knowledge of the drivers and effects of polar auxin transport in the plant sporophyte come almost exclusively from *Arabidopsis*. This means that as soon as we leave the angiosperm (and certainly the seed plant) realm, we are walking on terra incognita in terms of almost everything except for the most basic notions on the presence and sometimes the direction of polar auxin transport. Without trying to present a comprehensive review of polar auxin transport in angiosperm development, we review here its major aspects to provide context for a survey of polar auxin transport in seed-free plants. In doing this, we deliberately left out the interactions of auxin with genes and with other hormones, because there is little comparative data available on these aspects for the seed-free plants.

10.2.1 Embryogenesis

Auxin polarization is present as early as the asymmetric division of the zygote, which generates auxin-transporting and auxin-responsive daughter cells (Friml et al. 2003). This generates a steady auxin activity gradient responsible for the establishment and maintenance of polarity in the embryo (Friml et al. 2003; Aida et al. 2004). Initially, polar transport of auxin produced at the base of young embryos is directed toward the apical cell of the proembryo (Friml et al. 2003; Robert et al. 2013). The apical auxin response maximum thus generated contributes to the specification of apical embryonic structures, leading to localized apical auxin biosynthesis and subsequent reversal of auxin transport polarity within the embryo (Aida et al. 2004). The resulting basipetal (toward the base) polar auxin transport in the embryo specifies the root pole (radicle) and contributes positional information for establishment of the root stem cell niche (Aida et al. 2004; Moller and Weijers 2009; Robert et al. 2013).

Interestingly, looking at spruce (*Picea*) embryos, Larsson et al. (2008) showed that polar auxin transport is essential for the correct patterning of apical and basal parts of the embryos. They also demonstrated that auxin transport inhibitors generate phenotypes comparable with those of auxin response and transport mutants in *Arabidopsis*, suggesting that the role of polar auxin transport in embryogenesis is conserved between angiosperms and gymnosperms.

10.2.2 Shoot Apical Meristem

At the shoot apical meristem, auxin gradients are necessary for leaf initiation and phyllotactic patterning. The sites of new leaf primordia are specified by localized auxin maxima in the tunica (Reinhardt et al. 2000, 2003; Reinhardt 2005). The maxima are generated by the patterns of auxin transport in this superficial layer and are positioned according to constraints imposed by pre-existing primordia (Reinhardt 2005). Phyllotactic patterns result from the interplay between (1) meristem size, (2) spatial perturbations in the field of auxin gradients in superficial meristem layers induced by the positioning of pre-existing primordia that act as sinks (initially) or sources (subsequently) (Reinhardt 2005; Smith et al. 2006), and (3) differential modulation of the elastic properties of cell walls across the same superficial layers of the meristem (Kierzkowski et al. 2012). The auxin of the leaf primordium maxima is channeled basipetally (Reinhardt et al. 2003; Benková et al. 2003), specifying in the process the trajectory of procambial strands that will form the leaf traces (Benková et al. 2003), and feeds into the stream of auxin that is transported basipetally through xylem parenchyma of the pre-existing stem below (Smith et al. 2006; Leyser 2011; Bennett et al. 2014).

Leaves are the predominant source of auxin in the shoot. After the early growth stages when they act as auxin sinks, leaf primordia switch to being sources of auxin. Auxin flows basipetally in the leaf, from maxima positioned initially at the leaf tip and subsequently at additional locations along the leaf margin (Scarpella et al. 2006; Barkoulas et al. 2007; Bennett et al. 2014_1). These maxima generate auxin gradients that are responsible for auxin flow canalization and, ultimately, leaf venation (Scarpella et al. 2006; Sawchuk et al. 2007). They are also involved in leaf margin serration (Barkoulas et al. 2007; Zhou et al. 2013; Kasprzewska et al. 2015; Wang et al. 2016_1) and the development of compound leaves (Barkoulas et al. 2008; Runions et al. 2017).

10.2.3 Root Apical Meristem

In roots, patterns of polar auxin transport conform with, and are a direct continuation of, the pattern established early in the embryo and defined by an auxin maximum located at the growing tip (Sabatini et al. 1999; Aida et al. 2004). First documented by studies of the root branching, polar auxin transport in the growing root conforms to the "fountain model" (Benková et al. 2003), with a "reflux loop" (Blilou et al. 2005). Auxin moves acropetally (toward the apex) through the center of the root (procambium), maintaining an auxin maximum at the root tip (Benková et al. 2003; Blilou et al. 2005). This auxin maximum promotes stem cell specification (Aida et al. 2004) and is, thus, responsible, for meristem maintenance (Aida et al. 2004; Blilou et al. 2005; Benjamins and Scheres 2008). Auxin is transported away from the apical maximum, basipetally, through the peripheral layers of the root. Away from the root tip, the peripheral auxin flux is redirected in a radial centripetal direction, to close the loop and rejoin the central acropetal stream of auxin that transits through the procambium. This reflux loop localizes meristem and cell expansion zones in the proximal meristem and regulates final cell size (Blilou et al. 2005). At the root apex, the role of the auxin maximum recapitulates the role of auxin maxima in the establishment of the embryonic root pole and the of lateral root primordia, by specifying the stem cell niche, with the mediation of *PLT* genes, and promoting cell division (Benková et al. 2003; Aida et al. 2004; Beveridge et al. 2007; Overvoorde et al. 2010).

10.2.4 Embryo to Mature Sporophyte

Canalized polar flow of auxin is essential for overall elongation and branching of the plant body, as well as for the patterning of vascular tissues, in terms of both specification and differentiation (Tuominen et al. 1997; Berleth et al. 2000; Björklund et al. 2007; De Rybel et al. 2013; Lucas et al. 2013; Miyashima et al. 2013; Fabregas et al. 2015; Harrison 2016). In all these, pathways of high auxin concentration specify the position and trajectories of vascular initials (procambium, vascular cambium). This role of auxin is expressed as early as the division and specification of the first preprocambial cells in the globular embryo (De Rybel et al. 2013; Wendrich and Weijers 2013) and is maintained throughout a plant's lifespan, during both primary and secondary growth (Sachs 1969, 1981; Sachs and Cohen 1982; Hejnowicz and Kurczyńska 1987; Lev-Yadun and Aloni 1990; Aloni 1995; Berleth 2001; Dengler 2006).

The longitudinal polarity (apical-basal axis) of the embryo and the associated apical-to-basal polarization of auxin transport are maintained and amplified continuously throughout plant development and are responsible for the patterns of polarity in auxin transport in the mature sporophyte. Current understanding of auxin homeostasis in relation to plant development suggests that the initiation and orientation of growth axes during embryonic and postembryonic development result from the combined activity of local auxin sources and polar auxin transport (Aida et al. 2004; Robert et al. 2013; Wabnik et al. 2013; Turchi et al. 2015). The result of these is a general pattern of basipetal polar auxin transport in stems and acropetal auxin flow in roots, maintained throughout the lifespan of the mature sporophyte (Sachs 1991; Berleth and Sachs 2001; Sawchuk and Scarpella 2013). Later, as the plant transitions to secondary growth, the basipetal polar auxin transport of the shoot is canalized through the cambium, where it is responsible for the patterning of secondary tissues (Sachs and Cohen 1982; Björklund et al. 2007). These roles of polar auxin transport in shoot patterning are conserved across seed plants (Sanders and Langdale 2013).

Additionally, Benková et al. (2003) pointed out that local accumulations of auxin and redirection of auxin flow play a fundamental role during processes of organ formation, whether those are roots or leaves. In these, the role of auxin is central in both specification of the site of primordium initiation, by accumulation of an auxin maximum, and in the establishment of a new growth axis, by activation of polar auxin transport as a result of the auxin gradient (generated by the primordium-specifying auxin maximum).

10.3 Polar Auxin Transport in Extant and Extinct Seed-Free Plants

Mechanisms for auxin synthesis have been documented in all embryophytes, as well as charophycean algae, and PIN genes have been identified in nearly all embryophyte lineages (Finet and Jaillais 2012; Bennett 2015). These suggest that the basic molecular machinery for the production and regulation of auxin evolved early in the history of embryophytes and was potentially inherited from charophyte ancestors. In angiosperms, the roles of auxin in vascular tissue development and organogenesis are well documented. However, polar auxin transport and the roles of auxin in development and the organization of sporophytes have been documented to a considerably lesser extent in the seed-free tracheophytes (Cooke et al. 2002). Below we review current data on polar auxin transport in seed-free plants.

10.3.1 Embryogenesis

All seed plants have bipolar body plans, defined by two major poles of growth in mature sporophytes: the shoot and root (Rothwell 1995; Tomescu 2011). These two poles of growth are derived directly from their cotyledonary embryos, which at maturity possess two dormant meristems that generate the shoot and rooting systems, respectively. In contrast, all seed-free plants have unipolar body plans, in which there is a single persistent pole of growth from which all organs originate in the mature sporophyte (Rothwell 1995). This unipolar body plan is derived from the non-cotyledonary embryos of seed-free plants. Although these embryos possess both a shoot and a root pole, growth at the root pole produces only an ephemeral embryonic root that is inactivated early in sporophyte development, and thus all subsequent sporophyte growth is derived from the shoot pole of the embryo (Tomescu 2011); the only potential exception occurs in the rhizomorphic lycophytes (Isoetales), in which the mature sporophyte exhibits bipolar growth derived from an early dichotomy of the unipolar embryo (Rothwell and Erwin 1985; Rothwell 1995).

Among seed plants, polar auxin transport is necessary for normal development of cotyledonary embryos, and treatment with auxin transport inhibitors disrupts embryo formation (Larsson et al. 2008; Liu et al. 1993). For seed-free plants the literature on auxin action in embryogeny is sparse, but experiments performed on embryos of the hydropteridalean (water fern) *Marsilea vestita* provide contrasting results (Poli 2005). Like in angiosperms, application of auxin biosynthesis inhibitors produced abnormal or abortive embryos. However, while application of auxin transport inhibitors (1-n-naphthylphthalamic acid [NPA] and 2,3,5-triiodobenzoic acid [TIBA]) resulted in some abnormal leaf morphology, the treatments did not disrupt embryogenesis. These results suggest that while auxin biosynthesis is necessary for normal embryo development, polar auxin transport may not play an integral role in these processes in seed-free plants (Poli 2005) and may be related to the structural differences between cotyledonary and non-cotyledonary embryos.

10.3.2 Polar Auxin Transport in the Mature Sporophyte

Among seed-free plants, patterns of polar auxin transport have been documented in axial organs of *Selaginella*, as well as several fossil plants: *Paralycopodites*, an arborescent lycophyte; *Archaeopteris*, which belongs to a group of seed-free lignophytes called progymnosperms; and *Arthropitys*, an arborescent sphenophyte whose closest modern relative is *Equisetum* (Rothwell and Lev-Yadun 2005; Rothwell et al. 2008; Sanders et al. 2011). In *Selaginella*, like in angiosperms, polar auxin transport is basipetal in shoots and is sensitive to the auxin transport inhibitor NPA (Wochok and Sussex 1973; Sanders and Langdale 2013). Polar auxin transport has not been investigated in roots of *Selaginella*, but it has been studied in the rhizophore – an organ unique to *Selaginella* and for which homology is uncertain. Rhizophores are leafless axial organs that develop exogenously from meristems positioned in branching angles of shoots ("angle meristems") and grow downward, producing roots at their tip on reaching the soil (Jernstedt et al. 1994). Studies of *Selaginella* rhizophores indicate that polar auxin transport is acropetal (Wochok and Sussex 1974) and that rhizophore development can be altered by auxin (Cusick 1954; Wochok and Sussex 1976; Jernstedt et al. 1994; Sanders and Langdale 2013). For instance, excised rhizophores cultured on media supplemented with auxin exhibit normal growth and morphology. When no auxin is present in the medium, rhizophores sometimes produce leaves and continue to grow as shoots (Wochok and Sussex 1976; Sanders and Langdale 2013). Similarly, at least in some species, severing the shoot apical meristems above a branching point causes the nearest angle meristems to develop as shoots rather than as rhizophores; if the severed apices are replaced with an auxin source, angle meristems develop as rhizophores (Cusick 1954).

Among fossil plants, polar auxin transport has been inferred based on the position of circular or distorted tracheids ("auxin swirls") in the secondary xylem of woody organs. In modern plants, tracheids elongate and differentiate along polar auxin gradients (Berleth et al. 2000; Aloni 2010), and disruptions in the flow of auxin can produce unusual tracheid geometries. For instance, in shoots circular vessels form above tissue wounds in herbaceous seedlings (Sachs and Cohen 1982) and above axillary buds and branches in woody stems (Hejnowicz and Kurczyńska 1987; Lev-Yadun and Aloni 1990; Lev-Yadun 1996). Importantly, the position of auxin swirls is consistent with auxin flow polarity – in shoots, auxin moves in a basipetal direction from the apex, and thus auxin swirls form on the apical side of disrupting structures (e.g., branches, leaf traces). In woody shoots of extinct *Paralycopodites*, *Archaeopteris*, and *Sphenophyllum*, auxin swirls are found on the apical side of diverging branches and leaf traces, consistent with basipetal polar auxin transport in shoots, similar to extant lignophytes (Rothwell and Lev-Yadun

2005; Rothwell et al. 2008; Decombeix et al. 2010; Sanders et al. 2011). Roots of *Archaeopteris* exhibit distorted tracheids around branch roots, on the side facing the shoot system rather than the root apex, consistent with acropetal polar auxin transport, like in seed plant roots (A.-L. Decombeix, personal communication, 2015). Similarly, in the rooting system of *Paralycopodites* auxin swirls are formed above rootlet vascular traces, on the side facing the shoot (Sanders et al. 2011). Interestingly, the rooting system of *Paralycopodites* represents a shoot homolog highly modified for rooting (Rothwell and Erwin 1985; Rothwell and Tomescu 2017), termed a rhizomorph. Thus, although rhizomorphs lack root homology, they exhibit auxin transport polarity characteristic of roots.

10.3.3 Leaf Development

Auxin has several roles in angiosperm leaf development, including in leaf initiation, vascular (vein) differentiation, and phyllotaxis. Moreover, developing leaves function as the primary source of auxin in shoot apical meristems (Scarpella et al. 2006; Barkoulas et al. 2007). Among seed-free plants, auxin action in leaves has been studied in lycophytes (*Selaginella*), Filicales, and Hydropteridales. With respect to auxin production, polar auxin transport, and vascular differentiation, the data from seed-free plants are consistent with observations of angiosperms. In *Adiantum* and *Osmunda* (Filicales), auxin is produced in leaf pinnules (Steeves and Briggs 1960; White 1971) and is involved in vascular differentiation in *Osmunda* leaves (Steeves and Briggs 1960). Within leaf rachises, basipetal polar auxin transport was demonstrated in *Regnellidium* (Hydropteridales) and *Osmunda* (Filicales), as well as in the young leaves of *Pteris* (Filicales) (Albaum 1938; Steeves and Briggs 1960; Walters and Osborne 1979).

Few studies have addressed auxin action in leaf initiation and phyllotaxis, but experiments on *Selaginella* indicate that the role of auxin in these processes may differ for lycophytes. Application of auxin transport inhibitors to shoot apical meristems of *Selaginella* resulted in abnormal phyllotaxis and vascular development and eventual inactivation of the meristem (Sanders and Langdale 2013). In contrast, in comparable experiments on tomato shoots, leaf production ceased, but the meristem remained active, continuing to grow as a leafless axis (Reinhardt et al. 2000).

10.3.4 Root Development

Data on root initiation and development come primarily from culturing experiments involving auxin treatments, either using seedlings, excised structures, or callus cultures. Some of these studies were not explicitly investigating the effects of auxin on root production and growth but nevertheless provide some evidence for growth responses to auxin in ferns. For instance, when beads of auxin were applied to cultured apices of *Matteuccia* (Filicales) to investigate changes in stem vasculature, adventitious roots were produced (Ma and Steeves 1992). Cultures of *Pteridium* roots supplemented with auxin showed reduced root elongation and increased lateral root production (Partanen and Partanen 1963). Roots were also induced from callus cultures of *Pteris* and *Nephrolepis* (Filicales) using auxin ratios comparable to those used for root induction from angiosperm callus (Bristow 1962; Byrne and Caponetti 1992).

Other studies were specifically aimed at investigating responses to auxin. In *Selaginella*, like in seed plants, auxin promoted root branching (Sanders and Langdale 2013) despite the fact that lycophyte roots branch apically – a mode of branching that is developmentally very different from root branching in euphyllophytes, which occurs via endogenous origin of branch roots from the pericycle (Esau 1965). In *Ceratopteris* (Filicales), auxin treatments inhibited root growth and, unlike in *Pteridium*, had no effect on branching (Hou et al. 2004). Similarly, auxin was shown to inhibit root growth in *Azolla* (Hydropteridales) (de Vries et al. 2016).

10.4 Similarities and Differences, Knowns and Unknowns

Summarizing the extent of current knowledge, although the literature on auxin in seed-free plant development is sparse as compared to seed plants, representatives from six out of ten major groups of seed-free plants have been studied (Table 10.1). Among extant taxa, polar auxin transport has been investigated in the shoot and rooting systems of *Selaginella* (Selaginellales), as well as in the leaves of *Regnellidium* (Hydropteridales) and *Osmunda* and *Pteris* (Filicales). The directionality of polar auxin transport has also been inferred for extinct woody taxa based on the position of auxin swirls in the secondary xylem of *Paralycopodites* (Isoetales), *Archaeopteris* (progymnosperm), and *Arthropitys* (sphenopsid). Roles for auxin in leaf and root development have further been demonstrated for filicalean (*Pteris, Nephrolepis, Ceratopteris, Pteridium, Matteuccia*) and hydropteridalean ferns (*Azolla*), as well as in *Selaginella*. Several general conclusions can be drawn from this survey of polar auxin transport in living and extinct seed-free plants.

- The basic directional patterns of polar auxin transport appear to be shared by all tracheophytes: basipetal transport in negatively gravitropic shoots and acropetal transport in positively gravitropic rooting systems, irrespective of the homology of those rooting systems (e.g., Wochok and Sussex 1974; Sanders et al. 2011; Sanders and Langdale 2013).
- PIN-mediated polar auxin transport regulation of branching is likely conserved across vascular plants (Harrison 2016).
- The role of auxin in vascular differentiation is conserved (e.g., Steeves and Briggs 1960; Rothwell and Lev-Yadun 2005; Rothwell et al. 2008; Lucas et al. 2013); at least in stems and leaves, vascular tissue development is similarly regulated by polar auxin transport in lycophytes and euphyllophytes (Sanders and Langdale 2013).

Table 10.1 Summary of literature on polar auxin transport (PAT) and auxin action in seed-free plant development

Taxon	Available data	Extant/ fossil	References
Lycopodiales	Not documented	–	
Selaginellales	Stem PAT Rhizophore PAT Leaf primordia & development	Extant	Wochok and Sussex (1973, 1974), Sanders and Langdale (2013)
Isoetales	Stem PAT Rhizomorph PAT	Fossil	Sanders et al. (2011)
Psilotum	Not documented	–	
Ophioglossales	Not documented	–	
Marattiales	Not documented	–	
Filicales	Leaf PAT Leaf development Root initiation and branching	Extant	Steeves and Briggs (1960), Bristow (1962), Partanen and Partanen (1963), White (1971), Walters and Osborne (1979), Byrne and Caponetti 1992, Ma and Steeves 1992, Hou et al. 2004
Hydropteridales	Leaf PAT Root branching Embryogeny	Extant	Walters and Osborne 1979, Poli 2005, de Vries et al. 2016
Sphenopsids	Stem PAT	Fossil	Rothwell et al. (2008)
Progymnosperms	Stem PAT Root PAT	Fossil	Rothwell and Lev-Yadun (2005), A.-L. Decombeix, "personal communication"

- Like in seed plants, leaves are a source of auxin in ferns, as suggested by basipetal polar auxin transport in the leaf rachises of *Osmunda* and *Regnellidium* (Steeves and Briggs 1960; Walters and Osborne 1979) and in young leaves of *Pteris* sporelings (Albaum 1938).
- Auxin-associated mechanisms that regulate apical meristematic growth in shoots are different between lycophytes and seed plants (Sanders and Langdale 2013). Specifically, the role of auxin in leaf initiation and meristem maintenance appears to be different in lycophytes and seed plants: whereas in *Selaginella* disruption of polar auxin transport leads to abnormal phyllotaxis, abnormal leaf vascular development, and loss of meristematic activity (Sanders and Langdale 2013), in tomato, the same treatment shuts down leaf production, but meristematic activity is maintained (Reinhardt et al. 2000).

- In ferns (e.g., *Pteris*, *Nephrolepis*), ratios of auxin and other hormones similar to those used for the same purpose in seed plants can induce root or shoot formation in callus culture (Bristow 1962; Byrne and Caponetti 1992).
- There is conflicting evidence for the role of auxin in promoting root initiation and root branching in seed-free plants. In some ferns (*Pteridium*, *Matteuccia*), auxin promotes lateral root production and adventitious root formation (Partanen and Partanen 1963; Ma and Steeves 1992). High auxin concentrations also induce root formation in *Pteris* and *Nephrolepis* callus (Bristow 1962; Byrne and Caponetti 1992), and auxin promotes root branching in *Selaginella* (Sanders and Langdale 2013). In contrast, auxin inhibits root growth in *Ceratopteris* and *Azolla*, leads to reduced root elongation in *Pteridium*, and also appears to have no effect on root branching of *Ceratopteris* (Partanen and Partanen 1963; Hou et al. 2004; de Vries et al. 2016).

Accordingly, auxin plays fundamental and broadly similar roles in tracheophyte development, but there remain significant gaps in our understanding of the variation that exists between and within major seed-free lineages. For instance, while in-depth studies of auxin transport have been conducted in *Selaginella*, comparable studies of other lycophytes are lacking. Consequently, it is unclear whether the differences in the response of *Selaginella* shoot apices to auxin inhibition, as compared to similar experiments in angiosperms, are characteristic of all lycophytes. Auxin production and transport have also never been investigated in *Psilotum*, which is the only extant tracheophyte exhibiting a simple body plan that lacks roots and leaves, like the early land plants (Stewart and Rothwell 1993; Tomescu 2011). Similarly, very little is known about auxin transport in *Equisetum*.

Deeper understanding of these developmental and physiological processes, and how such processes are reflected in anatomy, has important implications for understanding morphological evolution among plants. Owing to the central role of auxin in root and leaf development, auxin is particularly relevant to questions on the origins of these organs. This raises the possibility that changes in auxin signaling underlie major transitions in the body plans of land plants (Cooke et al. 2004; Finet and Jaillais 2012; Bennett 2015), such as the transition from the simple body plans of early tracheophytes to complex morphologies characterized by leaves and roots. Because roots and leaves evolved independently multiple times among tracheophytes (Boyce 2005a, 2005b, 2010; Langdale 2008; Sanders et al. 2009; Tomescu 2009; Harrison 2017), understanding how development differs between major lineages is imperative in refining the evolution of development and morphological evolution in plants.

The long and complex evolutionary history of plants, as well as the spectrum of morphological diversity documented in the fossil record, means that answers to these questions must come from reciprocal illumination, in which morpho-anatomical features of extant plants are used to understand the development and morphology of extinct plants, and where patterns observed in the fossil record form the basis of evolutionary hypotheses for explaining traits of morphology and anatomy in extant plants. This has been done extensively in vertebrate macroevolution (e.g., Zhu and Ahlberg 2004; Shubin 2008; Nakamura et al. 2016), but comparable examples for plants are exceptionally rare (Tomescu et al. 2017; Rothwell and Tomescu 2017).

For studying the evolution of development in plants, recognition of anatomical fingerprints of developmental processes that can be identified in fossils (Rothwell et al. 2014) is central to a reciprocal illumination approach. Auxin is particularly interesting in the context of developmental fingerprints because (1) auxin has a demonstrated role in patterning vascular tissues (e.g., Sachs 1969, 1981; Aloni 1995; Mattsson et al. 1999; Berleth et al. 2000; Dengler 2006; Dettmer et al. 2009; Sawchuk and Scarpella 2013; Lucas et al. 2013) and (2) because tracheary elements of the xylem have some of the highest preservation potential of all plant cells. To illustrate this point, we present two examples in which a reciprocal illumination approach utilizing developmental fingerprints of polar auxin transport can be used to better understand the evolution of leaves and rooting systems among lycophytes and to formulate testable hypotheses for future studies.

10.5 Is Auxin Involved in the Evolution of Lycophyte Leaves?

Lycophyte leaves, sometimes referred to as microphylls (Tomescu 2009), are morphologically simple and typically contain a single vein. Although there is abundant evidence that lycophyte leaves evolved independently from those of euphyllophytes (Kenrick and Crane 1997; Boyce and Knoll 2002; Friedman et al. 2004; Langdale 2008; Boyce 2010; Harrison 2017), their evolutionary origins are unresolved (Kenrick and Crane 1997; Tomescu 2009). Two prevailing hypotheses posit that lycophyte leaves evolved through modification of pre-existing structures (e.g., Stewart 1964; Stewart and Rothwell 1993; Kenrick and Crane 1997). (1) The enation hypothesis (Bower 1935; Banks 1968; Gensel 1975) proposes that leaves arose through the vascularization of enations – lateral flap-like appendages lacking vasculature and regular phyllotaxis, which are seen in many extinct basal tracheophytes, as well as in extant *Psilotum*. (2) The sterilization hypothesis (Kenrick and Crane 1997) proposes that leaves evolved through the sterilization of sporangia. [A third hypothesis – associated with Zimmermann's (1938, 1952) telome theory – proposed lycophyte leaf evolution by reduction of more complex, branched structures; Stewart and Rothwell (1993)]. There is some evidence supporting each of the two prevailing hypotheses, from developmental genetics and the fossil record.

Studies on comparative gene expression reveal that Class III HD-Zip genes are expressed in both leaf primordia and sporangium primordia in *Selaginella*. This has been interpreted as support for the origin of lycophyte leaves through the sterilization of sporangia (Vasco et al. 2016), with an implicit assumption of homology of leaves and sporangia. However, it is also possible that the shared expression of HD-Zip III genes in sporangia and leaf primordia is not directly relevant to leaf-sporangium homology (Rothwell and Tomescu 2017). HD-Zip III genes also have a role in the vascular tissue development in all tracheophytes (Floyd and Bowman 2006, 2010; Prigge and Clark 2006), and this role could have evolved after duplication of the ancestral HD-Zip III, which regulated sporangium development. Therefore, it is possible that HD-Zip III expression patterns in lycophyte leaves have more to do with vascular tissue identity and the regulation of radial (and adaxial-abaxial) polarity in vascular tissues, than with leaf identity and homology (Floyd and Bowman 2010).

In support of the enation hypothesis, the basal lycophyte *Asteroxylon* possesses leaflike appendages exhibiting regular phyllotaxis, but which have a vascular strand that terminates at the base of the leaf and does not extend into the lamina (Lyon 1964). The partial vascularization of these appendages in *Asteroxylon* is thought to reflect an intermediate stage in the evolution of leaves (Banks 1968; Gensel 1975). However, *Asteroxylon* is coeval with and even younger than lycophytes possessing fully vascularized leaves, such as *Sengelia* (Matsunaga and Tomescu 2016), which raises the question of whether *Asteroxylon* is truly an evolutionary intermediate or simply possessed an unusual morphology.

Clues to the nature of lateral appendages in *Asteroxylon* may come from studies of auxin in extant seed-free plants. Experiments on *Coleus* shoot apices demonstrated that excising leaf primordia, thereby terminating the basipetal flux of auxin from the leaf, results in arrested development of the leaf trace (Wangermann 1967). Similarly, treatment of *Selaginella* shoot apices with auxin transport inhibitors results in abnormal phyllotaxis and incomplete development of leaf traces (Sanders and Langdale 2013). In this context, the morphology of *Asteroxylon* could be addressed in terms of auxin physiology. We can hypothesize that the incomplete vascularization of lateral appendages in *Asteroxylon* is the result of a weak or transient flux of auxin from the primordia of these appendages. In extant plants, vascular tissues of leaf traces mature acropetally, from the stem stele toward the leaf base and toward the leaf tip (opposite the transport direction of auxin) (Esau 1965; Dengler 2001). If this were also the case in *Asteroxylon*, a transient auxin flux that ceases before differentiation of vascular tissues along the entire length of lateral appendages could result in a vascular trace that is differentiated into conducting tissues only over part of its entire potential length, at the base.

A closer look at well-preserved *Asteroxylon* specimens from the Rhynie chert has shown that the vascular trace departing from the stele toward the base of the lateral appendage continues into the appendage as a strand of narrower, elongated cells (Hueber 1992; Edwards 1994) – cell patterning consistent with procambial identity and, thus, polar auxin transport. These strands support a developmental model in which basipetal polar auxin transport ceases prior to differentiation of tracheary elements from procambium inside lateral appendages, resulting in the differentiation of those procambial cells as a strand of elongated parenchyma rather than as vascular tissue. This developmental hypothesis could be tested using leaf excision and auxin inhibition experiments in lycopodialean lycophytes (e.g., *Lycopodium*, *Huperzia*), which are morphologically much more similar to *Asteroxylon* than *Selaginella*. Developmental transitions between enations and leaves could potentially be further explored in *Psilotum*, which, while not a lycophyte, is possibly the only extant seed-free plant bearing enations (Stewart and Rothwell 1993; Tomescu 2011).

Presently, there is not enough data supporting either of the hypotheses for leaf evolution in lycophytes. However, we note that the two morphological characters separating leaves from enations are regular arrangement (phyllotaxis) and vascularization, both of which are regulated at least in part by auxin. In proposing the sterilization hypothesis, Kenrick and Crane (1997) pointed out that the enations of basal lycophytes lacked regular phyllotactic patterns, which rendered the enation hypothesis an unsatisfying explanation for the origin of leaves among lycophytes. While this is indeed true, experiments on tomato plants suggest that regular phyllotaxis and formation of leaf vasculature are developmentally related, at least in seed plants – laser ablation of the incipient midvein, which caused an accumulation of auxin in the leaf primordium, resulted in irregular positioning of subsequent leaf primordia (Deb et al. 2015). If this developmental relationship between phyllotaxis and the formation of leaf vasculature is borne out in future studies, we can speculate that regular phyllotaxis and leaf vascularization may be evolutionarily coupled and the two characters might be expected to emerge concurrently in evolutionary time without a morphological intermediate. However, these relationships would need to be explicitly tested for lycophytes, since numerous studies hint at subtle differences in lycophyte and euphyllophyte development. For instance, a study of *Lycopodium* by Gola et al. (2007) found evidence for uncoupling between leaf phyllotaxis and stem vasculature. This suggests limited developmental interaction between stem and leaf vasculature, unlike such correlations documented in the euphyllophytes (e.g., Wardlaw 1944, 1946; Meicenheimer 1986; Ma and Steeves 1992; Kwiatkowska 1992).

10.6 Polar Auxin Transport and Lycophyte Rooting Structures

The lycophytes (Lycophytina Kenrick and Crane 1997) comprise six major lineages. Three of these are represented in modern floras by clubmosses (Lycopodiales), *Selaginella* (Selaginellales), and *Isoetes* (Isoetales), but the majority of lycophyte diversity is extinct and known only from their long and extensive fossil record, which stretches at least 400 million years, into the Early Devonian (Gensel and Andrews 1984; Gensel 2008). Extinct lineages include Zosterophyllopsida (Kenrick and Crane 1997), a monophyletic group of stem lycophytes (hereafter zosterophylls); Drepanophycales, which combine characters seen in zosterophylls with characters of crown group lycophytes (Gensel and Andrews 1984; Matsunaga and Tomescu 2017); the poorly understood Protolepidodendrales, a clade of heterosporous ligulate lycophytes; and the rhizomorphic lycophytes (Lepidodendrales,

Pleuromeiales, and Isoetales) – a derived clade containing many arborescent taxa (Pigg and Rothwell 1983; Pigg 2001).

The body plans of living and extinct lycophytes encompass a significant amount of morphological diversity, particularly with respect to rooting systems. Starting with the most basal lineages, zosterophylls had simple body plans lacking true leaves and roots. Instead, they possessed photosynthetic stems and in some cases specialized axes with inferred downward growth for rooting (Walton 1964; Gensel et al. 2001; Hao et al. 2007). Drepanophycalean lycophytes exhibit the earliest examples of roots in the fossil record (Schweitzer and Giesen 1980; Rayner 1984; Li and Edwards 1995; Matsunaga and Tomescu 2016, 2017). Among them, *Drepanophycus* bore fine dichotomously branching roots on rhizomatous portions of the shoot system (Schweitzer 1980; Rayner 1984; Li et al. 2000). Similarly, rhizomatous stems of *Asteroxylon* produced downward-growing branching axes, which are interpreted as having exogenous origin (Kidston and Lang 1920), in contrast to true roots, which originate endogenously (Raven and Edwards 2001; Kenrick and Strullu-Derrien 2014). Another drepanophycalean, *Sengelia*, exhibits a more complex rooting system, consisting of downward-growing stems bearing lateral branching structures interpreted as roots (Matsunaga and Tomescu 2016, 2017).

Two different types of rooting structures are seen among the extant Lycopodiales. Nearly all species produce adventitious roots in a mode similar to that of most seed-free euphyllophytes (e.g., *Lycopodium*, *Huperzia*; Gifford and Foster 1989). However, *Phylloglossum* has a body plan comprised of a single reduced shoot and a bulbous, downward-growing stem called a tuber that functions in nutrient storage and perennation. Roots are borne laterally at the junction between the shoot and tuber and have been described as arising exogenously (Bower 1885). In Selaginellales (*Selaginella*), roots are borne at the apex of rhizophores, which are themselves rooting organs of uncertain homology (Jernstedt et al. 1994).

Living and extinct rhizomorphic lycophytes exhibit a body plan wherein the rooting system is comprised of an organ termed a "rhizomorph." Studies of comparative anatomy and the fossil record have demonstrated the rhizomorph as a shoot homolog, with appendages termed "rootlets" corresponding to highly modified leaves (Rothwell and Erwin 1985; Rothwell and Tomescu 2017). Rhizomorphs are either cormose, forming a bulbous or lobed base of the plant, with numerous radiating rootlets seen in taxa like *Isoetes*, *Pleuromeia*, and *Chaloneria* (Stewart and Rothwell 1993), or are elongated and often highly branched as in Lepidodendrales (e.g., Frankenberg and Eggert 1969). In some cormose lycophytes, the rootlets lack the rootlike morphology of other members of the clade, and the corm instead bears laminar leaves that grow straight down into the sediment (e.g., Cúneo and Andreis 1983; Jasper and Guerra-Sommer 1999), further illustrating their shoot homology. To date the rooting structures of protolepidodendraleans are unknown.

When considering the rooting structures of lycophytes, one salient pattern emerges despite their seemingly wide morphological disparity – in every major lineage we can recognize rooting organs derived from stems or undifferentiated axes. These include the rooting axes of some zosterophylls, the root-bearing stems of *Sengelia* and the rootlike axes of *Asteroxylon* (Drepanophycales), the tuber of

Phylloglossum, the *Selaginella* rhizophore (the homology of which is unresolved), and the isoetalean rhizomorph. Importantly, all these organs exhibit either positive gravitropism or downward growth orientation indicative of positive gravitropic responses (Kidston and Lang 1920; Bower 1885; Edwards 1994; Matsunaga and Tomescu 2017; Matsunaga et al. 2017). This is very different from what we see in euphyllophytes, among which rooting systems are almost universally comprised only of roots; when stems do form part of the rooting system, they are typically rhizomatous and lack positive gravitropism. As discussed above, studies of extant plants demonstrate that auxin has important roles in tropic responses. Moreover, polar auxin transport is acropetal in roots of all plants for which it has been studied. This raises the question of whether aspects of auxin physiology that characterize euphyllophyte roots are shared by the diverse types of rooting structures of lycophytes.

To date patterns of auxin transport have been studied in only two lycophytes: *Selaginella* (Wochok and Sussex 1973, 1974; Sanders and Langdale 2013) and *Paralycopodites* (Sanders et al. 2011), an extinct arborescent lepidodendralean. Despite such sparse taxonomic sampling, these data on lycophytes are intriguing. In *Selaginella* rhizophores and in the rhizomorph of *Paralycopodites*, polar auxin transport is acropetal like in the roots of seed plants. Rhizomorphs have been demonstrated as shoot homologs (Rothwell and Erwin 1985), and while the homology of the rhizophore is unresolved, their exogenous development and capacity to produce leaves under certain experimental conditions (Cusick 1954; Wochok and Sussex 1976; Sanders and Langdale 2013) argue strongly against their interpretation as roots. The downward growth of rhizophores and the presence of acropetal polar auxin transport are the primary features that have been used to argue for root homology (Wochok and Sussex 1974). However, in the context of the auxin transport pattern documented in the *Paralycopodites* rhizomorph, and alongside it, the acropetal polar auxin transport of the *Selaginella* rhizophore suggests that rather than being a defining feature of roots, acropetal polar auxin transport is related to a more general growth habit characterized by downward growth and rooting function. Under this hypothesis we would also expect the rooting structures of other lycophytes to exhibit acropetal polar auxin transport, regardless of their homology. While this is relatively easy to ascertain experimentally for extant taxa like *Phylloglossum* and *Isoetes*, testing this hypothesis for fossil lycophytes presents a significant challenge. This is where anatomical fingerprints of development are essential for understanding the evolution of development in deep time.

10.7 Anatomical Fingerprints of Polar Auxin Transport

We have already briefly mentioned one example of an anatomical fingerprint for polar auxin transport, which has been applied to studies of plant development in the fossil record: auxin swirls in the secondary xylem of woody plants. Auxin swirls form in the secondary xylem of stems and roots in positions upstream (along the polar flow of auxin) of obstructions such as branches, leaf traces, and root traces (Lev-Yadun and Aloni 1990). Consequently, when looking at fossilized wood, if the direction of the apex is known and structures like branches or vascular traces are present in the secondary xylem, the polarity of auxin transport can be inferred. Auxin swirls have been documented in fossilized secondary xylem of seed plants, as well as arborescent lycophytes, sphenopsids, and progymnosperms, in which secondary growth evolved independently, thus demonstrating shared mechanisms in the developmental regulation of secondary growth in shoots and rooting systems (Rothwell and Lev-Yadun 2005; Rothwell et al. 2008). This anatomical fingerprint was used to infer the direction of polar auxin transport in the rhizomorph of the lepidodendralean lycophyte *Paralycopodites* (Sanders et al. 2011). However, because auxin swirls only form in secondary xylem, they cannot be used to infer the polarity of auxin transport in herbaceous basal lycophytes. Testing this hypothesis requires different anatomical fingerprints.

In a recent anatomical study of *Selaginella* branching (Matsunaga et al. 2017), we observed an anatomical feature that may serve as a fingerprint for auxin flow redirection at branching points and which is relevant to herbaceous basal lycophytes that produce rooting axes through branching of the aboveground axial system (e.g., some Drepanophycales and zosterophylls). Because tracheids elongate and mature along auxin gradients (Sachs 1981; Sachs and Cohen 1982; Lev-Yadun and Aloni 1990), their orientation can be used as a marker of the pathways of polar auxin transport. In *Selaginella*, the shape and position of tracheids reveals changes in the polarity of auxin flow from basipetal in shoots to acropetal in the rhizophore. Observations in cleared *Selaginella* shoots show that in shoots without rhizophores, at branching points tracheids from each branch converge smoothly and run parallel to one another below the bifurcation. In contrast, at branching points with attached rhizophores, some tracheids curve from each branch into the rhizophore base, consistent with an auxin flux that enters the rhizophore from the shoot. This pattern was observed in all nine *Selaginella* species studied, regardless of stele type (Matsunaga et al. 2017). Importantly, it was also observed in incipient and young rhizophores that had not yet produced roots, indicating that these tracheids did not form later in development as a consequence of water transport and distribution into the shoot system after the establishment of roots.

Curved tracheids like those documented in *Selaginella* could be searched for in fossil specimens that exhibit exogenously produced rooting structures, potentially providing evidence for developmental processes related to polar auxin transport in basal lycophytes. However, the utility of this anatomical feature for understanding development in early vascular plants is yet to be tested and requires exceptional preservation of anatomy. Nevertheless, both examples of anatomical signatures of polar auxin transport discussed here, auxin swirls and curved tracheids, demonstrate that some aspects of auxin action are recorded in anatomy. This is certainly promising, but more data are sorely needed on auxin physiology in lycophytes and other seed-free plants.

10.8 Conclusions and Future Directions

In this chapter we reviewed current knowledge of polar auxin transport in seed-free plant development and suggested areas where this knowledge can be used to better understand development in the fossil record. We exemplify a reciprocal illumination approach to address questions on morphological evolution, in proposing developmental models that provide testable hypotheses for understanding leaf evolution and rooting systems among lycophytes. However, despite significant advances in methods for studying development and physiology over the last few decades, our understanding of seed-free plant development lags far behind that of seed plants, and there remains a long road ahead toward a comprehensive understanding of auxin action across all tracheophyte lineages. In 2002, Cooke and collaborators undertook an exhaustive review of the literature on auxin in plant development. Now, more than 15 years later, considering the exciting and promising advances in our understanding of auxin in seed plant development, it is striking how little comparable progress has been made in seed-free plants – only a handful of studies, since, have gathered new data relevant to these questions (e.g., Hou et al. 2004; Sanders and Langdale 2013; de Vries et al. 2016).

Further progress toward understanding the role of auxin in the evolution of development and morphology in plants must come from continued study of seed-free lineages. Many major aspects of morphological evolution and homology across tracheophyte body plans are still unresolved (Rothwell and Tomescu 2017), and we are still far from a detailed understanding of the genetic regulation of development in all vascular plant lineages (Tomescu 2011). However, even just understanding the different roles of polar auxin transport throughout development, from the embryo to the mature sporophyte, in all the major tracheophyte lineages would help us make great strides toward these goals. For this to happen, work on seed-free plants must continue to be conducted and funded, and non-angiosperm model systems need to be developed for understanding developmental genetics. Incremental but necessary progress can also be made through basic research aimed at replicating results of older studies of auxin physiology, with treatments designed to tease apart different aspects of auxin action (e.g., controlling for auxin diffusion or testing responses to auxin inhibition), or by empirically confirming in each major lineage the presence of general patterns of polar auxin transport documented from seed plants.

While there is still a lot we do not know about developmental mechanisms in seed-free plants, a perspective centered on the deep evolutionary history and fossil record of plants has important potential for progress toward better understanding the evolution of plant form. This is not a novel idea – two great plant morphologists of the twentieth century, Frederick Orpen Bower and Wilson Nichols Stewart, long advocated such a view (Bower 1935; Stewart 1964). In this context, anatomical fingerprints of developmental processes are essential both for testing current evolutionary hypotheses and for observing patterns in living and extinct taxa that lead to the formation of novel hypotheses. Discovery and application of anatomical fingerprints requires not only an understanding of molecular and physiological processes but also of how these processes are reflected phenotypically in terms of anatomy and morphology. Because the evolutionary history of plants stretches hundreds of millions of years and includes numerous extinct lineages, inclusion of fossils in studies of phylogeny and evo-devo is imperative for obtaining increased resolution. For some evolutionary questions, particularly those pertaining to deep nodes in plant phylogeny, the problem of missing taxa cannot be overcome by sampling only extant representatives (Rothwell and Nixon 2006). Progress in many aspects of plant evolutionary biology must therefore come from studies that incorporate fossils, and doing so requires the continued study of plant anatomy and morphology, whether it is for building rigorous phylogenetic matrices or for understanding the biology of extinct organisms. The approaches and perspectives advocated here, as well as others that integrate data from the fossil record with analytical and methodological innovations (e.g., Bateman et al. 2006; Smith et al. 2009; Escapa and Catalano 2013; Rudall et al. 2013; Bomfleur et al. 2014, 2017; Hetherington and Dolan 2016; Hetherington et al. 2016; Fujinami et al. 2017; Wilson et al. 2017), provide an exciting frontier in plant evolutionary biology and move us closer toward understanding the processes that have generated the immense diversity of plant form through time.

References

Aida M, Beis D, Heidstra R et al (2004) The *PLETHORA* genes mediate patterning of the *Arabidopsis* stem cell niche. Cell 119:109–120

Albaum HG (1938) Inhibition due to growth hormones in fern prothallia and sporophytes. Am J Bot 25:124–133

Aloni R (1995) The induction of vascular tissues by auxin and cytokinin. In: Davies PJ (ed) Plant hormones: physiology, biochemistry, and molecular biology, 2nd edn. Kluwer Academic Publishers, Dordrecht, pp 531–546

Aloni R (2010) The induction of vascular tissues by auxin. In: Davies PJ (ed) Plant hormones. Springer, Dordrecht, pp 485–506

Banks HP (1968) The early history of land plants. In: Drake ET (ed) Evolution and environment. Yale University Press, New Haven, pp 73–107

Barkoulas M, Galinha C, Grigg SP et al (2007) From genes to shape: regulatory interactions in leaf development. Curr Opin Plant Biol 10:660–666

Barkoulas M, Hay A, Kougioumoutzi E et al (2008) A developmental framework for dissected leaf formation in the *Arabidopsis* relative *Cardamine hirsuta*. Nat Genet 40:1136–1141

Bateman RM, Hilton J, Rudall PJ (2006) Morphological and molecular phylogenetic context of the angiosperms: contrasting the 'top-down' and 'bottom-up' approaches used to infer the likely characteristics of the first flowers. J Exp Bot 57:3471–3503

Benjamins R, Scheres B (2008) Auxin: the looping star in plant development. Annu Rev Plant Biol 59:443–465

Benková E, Michniewicz M, Sauer M et al (2003) Local, efflux-dependent auxin gradients as a common module for plant organ formation. Cell 115:591–602

Bennett T (2015) PIN proteins and the evolution of plant development. Trends Plant Sci 20:498–507

Bennett T, Hines G, Leyser O (2014) Canalization: what the flux? Trends Genet 30:41–48

Berleth T (2001) Top-down and inside-out: directionality of signaling in vascular and embryo development. J Plant Growth Regul 20:14–21

Berleth T, Sachs T (2001) Plant morphogenesis: long-distance coordination and local patterning. Curr Opin Plant Biol 4:57–62
Berleth T, Mattsson J, Hardtke CS (2000) Vascular continuity and auxin signals. Trends Plant Sci 5:387–393
Beveridge CA, Mathesius U, Rose RJ et al (2007) Common regulatory themes in meristem development and whole-plant homeostasis. Curr Opin Plant Biol 10:44–51
Björklund S, Antti H, Uddestrand I et al (2007) Cross-talk between gibberellin and auxin in development of *Populus* wood: gibberellin stimulates polar auxin transport and has a common transcriptome with auxin. Plant J 52:499–511
Blilou I, Xu J, Wildwater M et al (2005) The PIN auxin efflux facilitator network controls growth and patterning in *Arabidopsis* roots. Nature 433:39–44
Bomfleur B, McLoughlin S, Vajda V (2014) Fossilized nuclei and chromosomes reveal 180 million years of genomic stasis in royal ferns. Science 343:1376–1377
Bomfleur B, Grimm GW, McLoughlin S (2017) The fossil Osmundales (Royal Ferns) – a phylogenetic network analysis, revised taxonomy, and evolutionary classification of anatomically preserved trunks and rhizomes. PeerJ 5:e3433
Bower FO (1885) On the development and morphology of *Phylloglossum drummondii*. Phil Trans R Soc Lond 176:665–678
Bower FO (1935) Primitive land plants, also known as the Archegoniatae. Macmillan, London
Boyce CK (2005a) Patterns of segregation and convergence in the evolution of fern and seed plant leaf morphologies. Paleobiology 31:117–140
Boyce CK (2005b) The evolutionary history of roots and leaves. In: Holbrook NM, Zwieniecki MA (eds) Vascular transport in plants. Elsevier, Amsterdam, pp 479–499
Boyce CK (2010) The evolution of plant development in a paleontological context. Curr Opin Plant Biol 13:102–107
Boyce CK, Knoll AH (2002) Evolution of developmental potential and the multiple independent origins of leaves in Paleozoic vascular plants. Paleobiology 28:70–100
Bristow MJ (1962) The controlled in vitro differentiation of callus derived from a fern, *Pteris cretica* L., into gametophytic or sporophytic tissues. Dev Biol 4:361–375
Byrne TE, Caponetti JD (1992) Morphogenesis in three cultivars of Boston fern. III. Callus production and plantlet differentiation from cell suspensions. Am Fern J 82:12–22
Campbell L, Turner S (2017) Regulation of vascular cell division. J Exp Bot 68:27–43
Cooke TJ, Poli D, Sztein AE et al (2002) Evolutionary patterns in auxin action. Plant Mol Biol 49:319–338
Cooke TJ, Poli D, Cohen JD (2004) Did auxin play a crucial role in the evolution of novel body plans during the late Silurian – early Devonian radiation of land plants? In: Hemsley AR, Poole I (eds) The evolution of plant physiology. From whole plants to ecosystems. Elsevier, Amsterdam, pp 85–107
Cúneo NR, Andreis RR (1983) Estudio de un bosque de licofitas en la Formación Nueva Lubecka, Pérmico de Chubut, Argentina. Implicancias paleoclimáticas y paleogeográficas. Ameghiniana 20:132–140
Cusick F (1954) Experimental and analytical studies of pteridophytes. XXV. Morphogenesis in *Selaginella willdenowii* Baker. II. Angle-meristems and angle-shoots. Ann Bot 18:171–181
De Rybel B, Moller B, Yoshida S et al (2013) A bHLH complex controls embryonic vascular tissue establishment and indeterminate growth in *Arabidopsis*. Dev Cell 24:426–437
de Vries J, Fischer AM, Roettger M et al (2016) Cytokinin-induced promotion of root meristem size in the fern *Azolla* supports a shoot-like origin of euphyllophyte roots. New Phytol 209:705–720
Deb Y, Marti D, Frenz M et al (2015) Phyllotaxis involves auxin drainage through leaf primordia. Development 142:1992–2001
Decombeix A-L, Taylor EL, Taylor TN (2010) Epicormic shoots in a Permian gymnosperm from Antarctica. Int J Plant Sci 171:772–782
Dengler NG (2001) Regulation of vascular development. J Plant Growth Regul 20:1–13

Dengler NG (2006) The shoot apical meristem and development of vascular architecture. Can J Bot 84:1660–1671

Dettmer J, Elo A, Helariutta Y (2009) Hormone interactions during vascular development. Plant Mol Biol 69:347–360

Durbak A, Yao H, McSteen P (2012) Hormone signaling in plant development. Curr Opin Plant Biol 15:92–96

Edwards D (1994) Toward an understanding of pattern and process in the growth of early vascular plants. In: Ingram DS, Hudson A (eds) Shape and form in plants and fungi. Academic Press & Linnean Society, London, pp 39–59

Esau K (1965) Plant anatomy, 2nd edn. Wiley & Sons, New York

Escapa IH, Catalano SA (2013) Phylogenetic analysis of Araucariaceae: integrating molecules, morphology, and fossils. Int J Plant Sci 174:1153–1170

Fabregas N, Formosa-Jordan P, Confraria A et al (2015) Auxin influx carriers control vascular patterning and xylem differentiation in *Arabidopsis thaliana*. PLoS Genet 11:e1005183

Finet C, Jaillais Y (2012) AUXOLOGY: when auxin meets plant evo-devo. Dev Biol 369:19–31

Floyd SK, Bowman JL (2006) Distinct developmental mechanisms reflect the independent origins of leaves in vascular plants. Curr Biol 16:1911–1917

Floyd SK, Bowman JL (2010) Gene expression patterns in seed plant shoot meristems and leaves: homoplasy or homology? J Plant Res 123:43–55

Frankenberg JM, Eggert DA (1969) Petrified *Stigmaria* from North America: part I. *Stigmaria ficoides*, the underground portions of Lepidodendraceae. Palaeontographica B 128:1–47

Friedman WE, Moore RC, Purugganan MD (2004) The evolution of plant development. Am J Bot 91:1726–1741

Friml J, Vieten A, Sauer M et al (2003) Efflux-dependent auxin gradients establish the apical-basal axis of *Arabidopsis*. Nature 426:147–153

Fujinami R, Yamada T, Nakajima A et al (2017) Root apical meristem diversity in extant lycophytes and implications for root origins. New Phytol 215:1210–1220

Gensel PG (1975) A new species of *Sawdonia* with notes on the origin of microphylls and lateral sporangia. Am J Bot 136:50–62

Gensel PG (2008) The earliest land plants. Annu Rev Ecol Evol Syst 39:459–477

Gensel PG, Andrews HN (1984) Plant life in the Devonian. Praeger, New York

Gensel PG, Kotyk ME, Basinger JF (2001) Morphology of above- and belowground structures in early Devonian (Pragian–Emsian) plants. In: Gensel PG, Edwards D (eds) Plants invade the land. Columbia University Press, New York, pp 83–102

Gifford EM, Foster AS (1989) Morphology and evolution of vascular plants, 3rd edn. Freeman, New York

Gola EM, Jernstedt JA, Zagorska-Marek B (2007) Vascular architecture in shoots of early divergent vascular plants, *Lycopodium clavatum* and *Lycopodium annotinum*. New Phytol 174:774–786

Hao S, Xue J, Liu Z et al (2007) *Zosterophyllum* Penhallow around the Silurian-Devonian boundary of northeastern Yunnan, China. Int J Plant Sci 168:477–489

Harrison CJ (2016) Auxin transport in the evolution of branching forms. New Phytol 215:545. https://doi.org/10.1111/nph.14333

Harrison CJ (2017) Development and genetics in the evolution of land plant body plans. Philos Trans R Soc B 372:20150490

Hejnowicz Z, Kurczyńska EU (1987) Occurrence of circular vessels above axillary buds in stems of woody plants. Acta Soc Bot Pol 56:415–419

Hetherington AJ, Dolan L (2016) The evolution of lycopsid rooting structures: conservatism and disparity. New Phytol 215:538–544

Hetherington AJ, Dubrovski JG, Dolan L (2016) Unique cellular organisation in the oldest root meristem. Curr Biol 26:1629–1633

Hou G, Hill JP, Blancaflor EB (2004) Developmental anatomy and auxin response of lateral root formation in *Ceratopteris richardii*. J Exp Bot 55:685–693

Hueber FM (1992) Thoughts on the early lycopsids and zosterophylls. Ann Mo Bot Gard 79:474–499
Jasper A, Guerra-Sommer M (1999) Licófitas arborescentes in situ como elementos importantes na definição de modelos deposicionais (Formação Rio Bonito - Bacia do Paraná - Brasil). Pesq Geoci 26:49–58
Jernstedt JA, Cutter EG, Lu P (1994) Independence of organogenesis and cell pattern in developing angle shoots of *Selaginella martensii*. Ann Bot 74:343–355
Jones AM (1998) Auxin transport: down and out and up again. Science 282:2201–2202
Kasprzewska A, Carter R, Swarup R et al (2015) Auxin influx importers modulate serration along the leaf margin. Plant J 83:705–718
Kenrick P, Crane PR (1997) The origin and early diversification of land plants. Smithsonian Institution Press, Washington
Kenrick P, Strullu-Derrien C (2014) The origin and early evolution of roots. Plant Physiol 166:570–580
Kidston R, Lang WH (1920) On old red sandstone plants showing structure, from the Rhynie Chert Bed, Aberdeenshire. Part III. *Asteroxylon mackiei*, Kidston and Lang. Trans R Soc Edinb 52:643–680
Kierzkowski D, Nakayama N, Routier-Kierzkowska A-L et al (2012) Elastic domains regulate growth and organogenesis in the plant shoot apical meristem. Science 335:1096–1099
Kwiatkowska D (1992) The relationships between the primary vascular system and phyllotactic patterns of *Anagallis arvensis* (Primulaceae). Am J Bot 79:904–913
Lacombe B, Achard P (2016) Long-distance transport of phytohormones through the plant vascular system. Curr Opin Plant Biol 34:1–8
Langdale JA (2008) Evolution of developmental mechanisms in plants. Curr Opin Genet Dev 18:368–373
Larsson E, Sitbon F, Ljung K et al (2008) Inhibited polar auxin transport results in aberrant embryo development in Norway spruce. New Phytol 177:356–366
Lev-Yadun S (1996) Circular vessels in the secondary xylem of *Arabidopsis thaliana* (Brassicaceae). IAWA J 17:31–35
Lev-Yadun S, Aloni R (1990) Vascular differentiation in branch junctions of trees: circular patterns and functional significance. Trees 4:49–54
Leyser O (2011) Auxin, self-organisation, and the colonial nature of plants. Curr Biol 21:R331–R337
Li C-S, Edwards D (1995) A re-investigation of Halle's *Drepanophycus spinaeformis* Goepp. from the Lower Devonian of Yunnan Province, southern China. Bot J Linn Soc 118:163–192
Li C-S, Hueber FM, Hotton CL (2000) A neotype for *Drepanophycus spinaeformis* Göppert 1852. Can J Bot 78:889–902
Liu C-m, Xu Z-h, Chua N-h (1993) Auxin polar transport is essential for the establishment of bilateral symmetry during early plant embryogenesis. Plant Cell 5:621–630
Lucas WJ, Groover A, Lichtenberger R et al (2013) The plant vascular system: evolution, development and functions. J Integr Plant Biol 55:294–388
Lyon AG (1964) Probable fertile region of *Asteroxylon mackiei* K. and L. Nature 203:1082–1083
Ma Y, Steeves TA (1992) Auxin effects on vascular differentiation in ostrich fern. Ann Bot 70:277–282
Matsunaga KKS, Tomescu AMF (2016) Root evolution at the base of the lycophyte clade: insights from an early Devonian lycophyte. Ann Bot 117:585–598
Matsunaga KKS, Tomescu AMF (2017) An organismal concept for *Sengelia radicans* gen. et sp. nov. – morphology and natural history of an early Devonian lycophyte. Ann Bot 119:1097–1113
Matsunaga KKS, Cullen NP, Tomescu AMF (2017) Vascularization of the *Selaginella* rhizophore: anatomical fingerprints of polar auxin transport with implications for the deep fossil record. New Phytol 216:419–428
Mattsson J, Sung ZR, Berleth T (1999) Responses of plant vascular systems to auxin transport inhibition. Development 126:2979–2991

Meicenheimer RD (1986) Role of parenchyma in *Linum usitatissimum* leaf trace patterns. Am J Bot 73:1649–1664
Miyashima S, Sebastian J, Lee J-Y et al (2013) Stem cell function during plant vascular development. EMBO J 32:178–193
Moller B, Weijers D (2009) Auxin control of embryo patterning. Cold Spring Harb Perspect Biol 1:a001545
Nakamura T, Gehrke AR, Lemberg J et al (2016) Digits and fin rays share common developmental histories. Nature 537:225–228
Overvoorde P, Fukaki H, Beeckman T (2010) Auxin control of root development. Cold Spring Harb Perspect Biol 2:a001537
Partanen JN, Partanen CR (1963) Observations on the culture of roots of the bracken fern. Can J Bot 41:1657–1661
Pigg KB (2001) Isoetalean lycopsid evolution: from the Devonian to the present. Am Fern J 91:99–114
Pigg KB, Rothwell GW (1983) Megagametophyte development in the Chaloneriaceae fam. Nov., permineralized Paleozoic Isoetales (Lycopsida). Bot Gaz 144:295–302
Poli D (2005) The role of auxin on the evolution of embryo development and axis formation in land plants. PhD Dissertation, University of Maryland
Prigge MJ, Clark SE (2006) Evolution of the class III HD-Zip gene family in land plants. Evol Dev 8:350–361
Raven JA, Edwards D (2001) Roots: evolutionary origins and biogeochemical significance. J Exp Bot 52:381–401
Rayner RJ (1984) New finds of *Drepanophycus spinaeformis* Goppert from the lower Devonian of Scotland. Trans R Soc Edinb Earth Sci 75:353–363
Reinhardt D (2005) Phyllotaxis – a new chapter in an old story about beauty and magic numbers. Curr Opin Plant Biol 8:487–493
Reinhardt D, Mandel T, Kuhlemeier C (2000) Auxin regulates the initiation and radial position of plant lateral organs. Plant Cell 12:507–518
Reinhardt D, Pesce E-R, Stieger P et al (2003) Regulation of phyllotaxis by polar auxin transport. Nature 426:255–260
Robert HS, Grones P, Stepanova AN et al (2013) Local auxin sources orient the apical-basal axis in *Arabidopsis* embryos. Curr Biol 23:2506–2512
Rothwell GW (1995) The fossil history of branching: implications for the phylogeny of land plants. In: Hoch PC, Stephenson AG (eds) Experimental and molecular approaches to plant biosystematics. Missouri Botanical Garden, St. Louis, pp 71–86
Rothwell GW, Erwin DM (1985) The rhizomorph apex of *Paurodendron*: implications for homologies among the rooting organs of Lycopsida. Am J Bot 72:86–98
Rothwell GW, Lev-Yadun S (2005) Evidence of polar auxin flow in 375 million-year-old fossil wood. Am J Bot 92:903–906
Rothwell GW, Nixon KC (2006) How does the inclusion of fossil data change our conclusions about the phylogenetic history of euphyllophytes? Int J Plant Sci 167:737–749
Rothwell GW, Tomescu AMF (2017) Structural fingerprints of development at the intersection of evolutionary developmental biology and the fossil record. In: Nuño de la Rosa L, Müller GB (eds) Evolutionary developmental biology – a reference guide. Springer International Publishing, Cham (Switzerland) https://doi.org/10.1007/978-3-319-33038-9_169-1
Rothwell GW, Sanders H, Wyatt SE et al (2008) A fossil record for growth regulation: the role of auxin in wood evolution. Ann Mo Bot Gard 95:121–134
Rothwell GW, Wyatt SE, Tomescu AMF (2014) Plant evolution at the interface of paleontology and developmental biology: an organism-centered paradigm. Am J Bot 101:899–913
Rudall PJ, Hilton J, Bateman RM (2013) Several developmental and morphogenetic factors govern the evolution of stomatal patterning in land plants. New Phytol 200:598–614
Runions A, Tsiantis M, Prusinkiewicz P (2017) A common developmental program can produce diverse leaf shapes. New Phytol 216:401–418

Sabatini S, Beis D, Wolkenfelt H et al (1999) An auxin-dependent distal organizer of pattern and polarity in the *Arabidopsis* root. Cell 99:463–472
Sachs T (1969) Polarity and the induction of organized vascular tissues. Ann Bot 33:263–275
Sachs T (1981) The control of the patterned differentiation of vascular tissues. Adv Bot Res 9:151–262
Sachs T (1991) Pattern formation in plant tissues. Cambridge University Press, Cambridge
Sachs T, Cohen D (1982) Circular vessels and the control of vascular differentiation in plants. Differentiation 21:22–26
Sanders HL, Langdale JA (2013) Conserved transport mechanisms but distinct auxin responses govern shoot patterning in *Selaginella kraussiana*. New Phytol 198:419–428
Sanders H, Rothwell GW, Wyatt SE (2009) Key morphological alterations in the evolution of leaves. Int J Plant Sci 170:860–868
Sanders H, Rothwell GW, Wyatt SE (2011) Parallel evolution of auxin regulation in rooting systems. Pl Syst Evol 291:221–225
Sawchuk MG, Scarpella E (2013) Polarity, continuity, and alignment in plant vascular strands. J Integr Plant Biol 55:824–834
Sawchuk MG, Head P, Donner TJ et al (2007) Time-lapse imaging of *Arabidopsis* leaf development shows dynamic patterns of procambium formation. New Phytol 176:560–571
Scarpella E, Meijer AH (2004) Pattern formation in the vascular system of monocot and dicot plant species. New Phytol 164:209–242
Scarpella E, Marcos D, Friml J et al (2006) Control of leaf vascular patterning by polar auxin transport. Genes Dev 20:1015–1027
Schweitzer H-J, Giesen P (1980) Uber *Taeniophyton inopinatum*, *Protolycopodites devonicus* und *Cladoxylon scoparium* aus dem Mitteldevon von Wuppertal. Palaeontographica B 173:1–25
Schweitzer H-J (1980) Über Drepanophycus spinaeformis Göppert. Bonner Paläobotan Mitt (Selbstverl d Paläobotan Abt d Inst für Paläontologie d Univ Bonn) 7:1–29
Shubin NH (2008) Your inner fish. Vintage Books, New York
Smith RP, Guyomarc'h S, Mandel T et al (2006) A plausible model of phyllotaxis. Proc Natl Acad Sci U S A 103:1301–1306
Smith SY, Collinson ME, Rudall PJ et al (2009) Virtual taphonomy using synchrotron tomographic microscopy reveals cryptic features and internal structure of modern and fossil plants. Proc Natl Acad Sci U S A 106:12013–12018
Steeves TA, Briggs WR (1960) Morphogenetic studies on *Osmunda cinnamomea* L. The auxin relationships of expanding fronds. J Exp Bot 11:45–67
Stewart WN (1964) An upward look in plant morphology. Phytomorphology 14:120–134
Stewart WN, Rothwell GW (1993) Paleobotany and the evolution of plants, 2nd edn. Cambridge University Press, Cambridge
Tomescu AMF (2009) Megaphylls, microphylls and the evolution of leaf development. Trends Plant Sci 14:5–12
Tomescu AMF (2011) The sporophytes of seed-free vascular plants – major vegetative developmental features and molecular genetic pathways. In: Fernandez H, Kumar A, Revilla MA (eds) Working with ferns. Issues and applications. Springer, New York, pp 67–94
Tomescu AMF, Escapa IH, Rothwell GW et al (2017) Developmental programmes in the evolution of *Equisetum* reproductive morphology: a hierarchical modularity hypothesis. Ann Bot 119:489–505
Tuominen H, Puech L, Fink S et al (1997) A radial concentration gradient of indole-3-acetic acid is related to secondary xylem development in hybrid aspen. Plant Physiol 115:577–585
Turchi L, Baima S, Morelli G et al (2015) Interplay of HD-ZIP II and III transcription factors in auxin-regulated plant development. J Exp Bot 66:5043–5053
Vasco A, Smalls TL, Graham SW et al (2016) Challenging the paradigms of leaf evolution: class III HD-Zips in ferns and lycophytes. New Phytol 212:745–758
Wabnik K, Robert HS, Smith RS et al (2013) Modeling framework for the establishment of the apical–basal embryonic axis in plants. Curr Biol 23:2513–2518

Walters J, Osborne DJ (1979) Ethylene and auxin-induced cell growth in relation to auxin transport and metabolism and ethylene production in the semi-aquatic plant, *Regnellidium diphyllum*. Planta 146:309–317

Walton J (1964) On the morphology of *Zosterophyllum* and some other early Devonian plants. Phytomorphology 14:155–160

Wang Q, Hasson A, Rossmann S et al (2016) *Divide et impera*: boundaries shape the plant body and initiate new meristems. New Phytol 209:485–498

Wangermann E (1967) The effect of the leaf on differentiation of primary xylem in the internode of *Coleus blumei* Benth. New Phytol 66:747–754

Wardlaw CW (1944) Experimental and analytical studies of pteridophytes. IV. Stelar morphology: experimental observations on the relation between leaf development and stelar morphology in species of *Dryopteris* and *Onoclea*. Ann Bot 8:387–399

Wardlaw CW (1946) Experimental and analytical studies of pteridophytes. VII. Stelar morphology: the effect of defoliation on the stele of *Osmunda* and *Todea*. Ann Bot 9:97–107

Wendrich JR, Weijers D (2013) The *Arabidopsis* embryo as a miniature morphogenesis model. New Phytol 199:14–25

White RA (1971) Experimental and developmental studies of the fern sporophyte. Bot Rev 37:509–540

Wilson JP, Montanez IP, White JD et al (2017) Dynamic Carboniferous tropical forests: new views of plant function and potential for physiological forcing of climate. New Phytol 215:1333. https://doi.org/10.1111/nph.14700

Wochok ZS, Sussex IM (1973) Morphogenesis in *Selaginella*. Auxin transport in the stem. Plant Physiol 51:646–650

Wochok ZS, Sussex IM (1974) Morphogenesis in *Selaginella*. II. Auxin transport in the root (rhizophore). Plant Physiol 53:738–741

Wochok ZS, Sussex IM (1976) Redetermination of cultured root tips to leafy shoots in *Selaginella willdenowii*. Plant Sci Lett 6:185–192

Zhou C, Han L, Fu C et al (2013) The trans-acting short interfering RNA3 pathway and NO APICAL MERISTEM antagonistically regulate leaf margin development and lateral organ separation, as revealed by analysis of an *argonaute7/lobed leaflet1* mutant in *Medicago truncatula*. Plant Cell 25:4845–4862

Zhu M, Ahlberg PE (2004) The origin of the internal nostril of tetrapods. Nature 432:94–97

Zimmermann W (1938) Die Telometheorie. Biologe 7:385–391

Zimmermann W (1952) Main results of the "Telome theory". Palaeobotanist 1:456–470

Chapter 11
Fern Conservation: Spore, Gametophyte, and Sporophyte Ex Situ Storage, In Vitro Culture, and Cryopreservation

Daniel Ballesteros and Valerie C. Pence

11.1 Introduction

Ferns (monilophytes) and lycophytes are estimated to be about 4% (c. 15,000 species) of all Earth's plants (Chapman 2009). Within this richness are species with diverse ecological roles, economic importance, and medicinal, food, or biotechnological value (De Long and Prange 2008; Fernández and Revilla 2003; Fernández et al. 2010; Liu et al. 2012; Mannan et al. 2008; Matsuura et al. 2014; Mehltreter et al. 2010; Stamps 2006). However, a large number of fern and lycophyte species are threatened, due to habitat destruction, over-collecting, invasive species, and climate change, and plans to secure their in situ and ex situ conservation are urgently needed (Arcand and Ranker 2008; Ibars and Estrelles 2012). Because of the alternation of life forms, ferns offer several options for ex situ conservation. This chapter will review the current status of fern conservation, and, using a recent case study on long-term cryo-survival, evaluate the current state of fern conservation technology, and outline the potential of current methods and areas for improving future efforts.

D. Ballesteros (✉)
Center for Conservation and Research of Endangered Wildlife, Cincinnati Zoo & Botanical Garden, Cincinnati, OH, USA

Comparative Seed Biology Group, Comparative Plant and Fungal Biology Department, Royal Botanic Gardens Kew, Richmond, UK
e-mail: d.ballesteros@kew.org

V. C. Pence
Center for Conservation and Research of Endangered Wildlife, Cincinnati Zoo & Botanical Garden, Cincinnati, OH, USA

H. Fernández (ed.), *Current Advances in Fern Research*,
https://doi.org/10.1007/978-3-319-75103-0_11

11.2 Current Status of Fern Conservation

11.2.1 Living Collections

Ferns have been collected and displayed for centuries, particularly in botanic gardens or private collections (e.g. https://ebps.org.uk/ferns/growing/where-to-see-ferns/). There are notable living collections of fern sporophytes, such as at the Royal Botanic Gardens, Edinburgh in the UK (http://www.rbge.org.uk/the-gardens/edinburgh/the-glasshouses/ferns-and-fossils), the Royal Botanic Gardens Victoria in Australia (https://www.rbg.vic.gov.au/visit-melbourne/attractions/plant-collections/fern-gully), or the Lankester Botanical Garden of the University of Costa Rica (http://www.jbl.ucr.ac.cr/helechos). Traditionally, these have been primarily for display, although more recently there are research and conservation missions developing around the collections (e.g. Page et al. 1992). Most species of ferns have not yet been evaluated for their conservation status (IUCN, 2017), but, for example, China alone has 182 species on its Red List, with 79 of those being endemic (Dong et al. 2017). Thus, there may be hundreds of threatened fern species globally. Because the space and labour to conserve living collections are substantial, living collections alone will likely not be able to conserve the wide genetic diversity of the fern species that are threatened. However, banking spores and other alternative propagules can provide a cost-effective supplement to what can be preserved in living collections (Li and Pritchard, 2009).

11.2.2 Conserving Fern Spores

Spore storage is an efficient and effective method for the ex situ conservation of ferns and lycophytes. Their desiccation tolerance (Chap. 19) and small size offer a simple, compact, and economical ex situ conservation tool for most of the biodiversity of ferns and lycophytes (Ashcroft and Sheffield 2000; Ballesteros 2010; Ballesteros et al. 2011, 2012, 2017; Dyer 1994; Ibars and Estrelles 2012; Mikula et al. 2015; Pence, 2008a, b). In addition, spores are demonstrated to be a ready source of germplasm to aid reestablishment of waning fern populations (Pennisi 2010). Ex situ spore collections were initially created ca. 30 years ago (Dyer 1994), but ex situ spore collections are currently rare if we compare them with the large network of seed banks worldwide (about 1300 throughout the world as of 2006, Rajasekharan 2015). To our knowledge (Alexander et al. 2008; Dyer 1994; Ibars et al. 2011; Pence 2008a several pers. comm. to authors), there are less than ten ex situ spore collections across the world, which represent the majority of the current collections for fern spore ex situ conservation (Table 11.1). In addition, there are spores that are stored for exchange programmes in fern societies (e.g. the British Pteridological Society and the American Fern Society), although the conditions used for these are focused on short-term storage and are not optimal for long-term conservation (Pérez-García and Reyes-Jaramillo 1993).

Table 11.1 Active fern spore banks in the world. Information gathered in 2016 and 2017 after a survey that involved 20 researchers from 12 different countries along America (USA, Mexico, Brazil), Europe (UK, Spain, Italy, Poland), Asia (Taiwan, China), Oceania (Australia, New Zealand), and Africa (South Africa)

Region	Country	Germplasm bank and institution	Temperature of storage	Moisture condition of spores	Start of fern spore storage	Number of taxa stored	Reference
Europe	UK	Royal Botanic Gardens, Edinburgh	1–5 °C and −20 °C	Dried at ambient temperature and RH	Early 1990s, new management from 2006	1300 accessions representing 438 species	Dyer (1994); Alexander et al. (2008); Barber S pers. comm.
		Millennium seed Bank at the Royal Botanic Gardens, Kew	−20 °C and LN	Dried at 15 °C and 15% RH	2000 (2013 for LN)	~90 species at −20 °C, ~20 species in LN	Dickie J and Davies R, pers. comm.
	Spain	Germplasm Bank of the Botanical Garden of Valencia University	4, −20, and −80 °C and LN	Dried at ambient temperature and RH or over silica gel	2000	127 samples representing 42 species	Ibars et al. (2011); Estrelles E pers. comm.
	Italy	Germplasm Bank of Tuscia University (Viterbo)	Mainly −25 °C (diverse temperatures for research)	Dried at 15 °C and 15% RH	2006	130 accessions representing 56 species	Magrini S pers. comm.
America	USA	CREW's CryoBioBank, Cincinnati Zoo and Botanical Garden	Mainly LN (4, −20, and −80 °C for research)	Dried at ambient temperature and RH or over silica gel	1992	8 at −20 °C, ~29 in LN	Pence (2004, 2008a, b), author's contribution
		USDA-ARS-NCGRP, Fort Collins, Colorado	Mainly LN (4, −20, and −80 °C for research)	Dried at ca. 20 °C and ~20–40% RH	2005	~103 accessions representing ~30 species	Walters C and Hill L pers. comm.
		National Tropical Botanical Garden in Kalāheo, Hawaii	5 °C (new collections will be stored at −80 °C)	Unknown desiccation (new collections will be dried to 42% RH at 20 °C)	1990, new management from 2016	~116 accessions representing ~49 taxa	Wolkis D pers. comm.
Asia	Taiwan (ROC)	Taiwan Forestry Research Institute	3 °C	26% RH	1998	5180 accessions representing 807 species	Huang YM pers. comm.

LN liquid nitrogen temperatures (−150 to −196 °C)

The scarcity of ex situ spore collections likely represents the fact that (1) ferns have been less studied and are less understood than seed plants and (2) there has been less funding and institutional focus on their conservation when compared with domesticated or wild seed plants. Because of this, there has not been the development of standard methods for fern spore storage as there has been for seed storage, with the development of international best practice standards over the past several decades (e.g. FAO Genebank standards such as FAO 2014). There are only a few publications, most in Spanish and in grey literature, describing procedures for routinely collecting and storing fern spores in germplasm banks (e.g. Bacchetta et al. 2008; Ibars and Estrelles 2015; Ibars et al. 2011; Pence 2008b). A result of this lack of standardized procedures for fern spore storage is the variety of conditions used in the different ex situ spore collections surveyed around the world (Table 11.1). This chapter is not intended to provide a final standard for ex situ spore collections, but, based on the experience of two decades of fern spore storage, aims to contribute to the development of such protocols for fern spore ex situ conservation and to encourage the establishment of fern spore collections within existing germplasm banks.

11.2.3 Non-spore Options for Fern Conservation

While storing spores is the most efficient method for ex situ fern conservation, there are also situations where storing tissues from gametophytes or sporophytes can be useful conservation tools. If only a few spores are available, they can be multiplied by germinating them into gametophytes and sporophytes, and either of these life forms can be propagated vegetatively, to provide more material for research, restoration, and preservation. There are also fern species for which the only life form known is that of the gametophyte (Farrar 1967, 2016), and thus, no spores are available for storage.

Gametophytes can be maintained as living collections, but because of their diminutive size, they are not as often used for display as sporophytes. Since they are small and easily manipulated, they have often been grown aseptically from spores or gemmae and propagated in vitro (e.g. Kamachi and Noguchi 2012; Marimuthu and Manickam 2011; Raine and Sheffield 1997). Gametophytes are well adapted for in vitro growth, as they consist of a small, photosynthetic thallus, one cell thick, which is highly regenerative and can reproduce clonally by fragmentation or budding (Maeda and Ito 1981; Ong and Ng 1998). In vitro cultures of gametophytes can also serve as a source of tissues for cryopreservation (Barnicoat et al. 2011; Makowski et al. 2015, 2016; Mikula et al. 2009, 2010; Pence 2000a; Wilkinson 2002). Because of the highly regenerative nature of gametophyte tissue and its relatively undifferentiated structure, even if some of the tissues are damaged during cryopreservation, only a small surviving fragment is needed to regenerate the culture (Mikula et al. 2010). In vitro and cryopreservation methods for maintaining gametophytes have been reported for a number of fern species (e.g. Mikula et al. 2010; Pence 2000a), but only a few of these have been endangered ferns (Barnicoat et al. 2011; Pence 2015; Raine and Sheffield 1997; Sara and Manickam 2007).

Sporophytes are also grown in vitro for research and horticulture, as well as for conservation (Fernández and Revilla 2003; From 2010; Yu et al. 2017), and in vitro propagation methods for the sporophyte life form are very similar to those used for many species of seed plants (Dolinsek and Camloh 1997; Higuchi et al. 1987; Khan et al. 2008; Rogers and Banister 1992). Such cultures may be initiated through the aseptic germination of spores or from pieces of the sporophyte, sterilized for initiation into culture (Dolinsek and Camloh 1997; Higuchi et al. 1987; Khan et al. 2008; Pence 2014). Many fern species have been propagated in vitro for horticulture (Dolinsek and Camloh 1997; Higuchi et al. 1987; Khan et al. 2008; Rogers and Banister 1992) as a way of providing uniform quality for commercial use. Similarly, a number of rare species of ferns have been propagated in vitro to conserve the species ex situ, as well as to provide material for restoration (Aguraiuja 2010a, b; Ashmore et al. 2011; Houser et al. 2016; Martin et al. 2006; Pence 2014, 2015; Sara and Manickam 2007). However, although there are a few institutions maintaining rare ferns in vitro, including the Lyon Arboretum's Hawaiian Rare Plant Program, the Center for Conservation and Research at Omaha's Henry Doorly Zoo, and the Center for Conservation and Research of Endangered Wildlife (CREW) at the Cincinnati Zoo and Botanical Garden, the use of in vitro collections of ferns for conservation has not been widely explored.

Long-term conservation of in vitro fern sporophyte cultures by cryopreservation is possible using growing, meristematic tissues, such as shoot tips or green globular bodies (Pence 2001, 2014, 2015). Cryopreservation of such tissues offers the opportunity to maintain stock lines for future use. It also has potential for conserving pteridophyte species for which spores are difficult to collect or produce in low numbers (Raine and Sheffield 1997). However, as with in vitro collections, there are only a few reports of cryopreserved sporophyte tissues of rare fern taxa (Pence 2014, 2015).

11.3 Case Study: Long-Term Storage of Ferns in CREW's Frozen Garden

CREW initiated its Frozen Garden of its CryoBioBank® in the late 1980s (Pritchard et al. 2017). This collection includes propagules that have been stored in liquid nitrogen (LN) (−196 °C) and some also at other temperatures (from about 20 to −20 °C). The Frozen Garden was initiated with the aim of supporting the long-term conservation of native and endangered plant species of the USA but also with the aim of providing resources to contribute to the understanding of long-term storage of plant germplasm in LN. About 25 years after the collection was started, the Institute of Museum and Library Services (IMLS) of the USA funded a complete evaluation of the viability and genetic integrity of CREW's collection. This evaluation has provided results that can be used to inform current practice for ex situ conservation of ferns.

11.3.1 Spore Banking

Collection and Storage of Fern Spores

Fern spores stored in the Frozen Garden were collected from several sources over a period of 14 years (from 1992 to 2006). When fronds were collected by CREW's staff, they were usually air-dried for 5–10 days under ambient conditions in the laboratory in paper envelopes. Subsequently, spores that fell on the paper were collected, dried as indicated below, and cryopreserved. However, if insufficient spores were observed on the paper, sori were scraped, and spores were collected along with sporangia and frond tissues. Some fern spores were sent directly by mail to CREW from several sources (e.g. Hawaii, Alaska, or China), which also included some frond and sporangia tissues. These samples, with spores and non-spore tissues, were dried and cryopreserved together due to the small amount of spores in the samples.

Spores (and other tissues if present) were dried at room temperature and RH < 40% for 2–7 days before storage, using silica gel as a desiccant (RH 20 ± 5%) if room RH was higher than 40%. Samples of the spores were then transferred to 2 ml polypropylene cryovials and immersed directly into LN. For some species, replicates were placed in a mechanical freezer (−20 ± 2 °C), a refrigerator (4 ± 2 °C), or left on the benchtop in the laboratory (21 ± 3 °C) for comparison with the samples stored in LN.

Germination of Fern Spores for Viability Assays

After storage of 9–21 years (e.g. Table 11.2), vials were removed from LN and placed on the benchtop to warm at ambient temperature for 20–60 min. Viability assays were afterward performed by germinating the fern spores in 15 x 150 mm petri dishes on mineral culture medium solidified with agar and prepared with the fungicide nystatin (100 U·ml −1) (Ballesteros et al. 2011, 2012, 2017). Petri dishes were then sealed with Parafilm (Bemis NA, Neenah, WI, USA) and put in a growth chamber set at 20 ± 2 °C with a 16/8 h light/dark cycle (Pence 2000b). Spores were observed using a dissecting microscope at 40× magnification, and germination was considered complete when the outer wall of the spore had ruptured and the rhizoid or the first chlorophyllic cell emerged (Ballesteros et al. 2011, 2012, 2017). Germination was measured regularly up to 30 days in order to calculate total germination and the rate of germination (Ballesteros et al. 2011, 2012, 2017). Then, the normal developmental transition from one-dimensional growth to two-dimensional growth was assessed in gametophytes after 60 days in culture (Ashcroft and Sheffield 2000; Ballesteros et al. 2011, 2012), a sufficient time to observe this transition.

Summary of Results

The original fern spore collection consisted of 162 samples representing 56 species of ferns and 1 species of lycophyte. We tested 93 samples and germination occurred in 45 of the 56 species of ferns tested. Lab notes from the time of banking suggested that the species that did not germinate were likely dead before storage. These were primarily species with green spores (e.g. *Matteuccia* sp., *Equisetum* sp., and *Onoclea* sp.) for which no germination was observed initially and species with non-green spores for which only a few spores were banked along with sporangia and frond tissues.

The percent germination was variable and depended on the species (Table 11.2). Germination of spores did not seem to be lower for the older species, and spores of most species stored for ≥20 years showed germination >60% (Table 11.2). When germination was measured in replicate samples (cryovials) of a particular species, no large differences were found between them, confirming the repeatability of the results obtained. In most cases, >80% of the spores that germinated developed laminar gametophytes (Table 11.2), except for some accessions for which germination was very low or plates were highly contaminated. Contamination was mostly by fungi growing out of sori and frond tissues.

In a few species, the spores were stored at several temperatures (21, 5, −20, and −196 °C) for 10 or 21 years. In these species, spores stored in LN maintained higher rates of germination than spores stored at −20 or 5 °C (Ballesteros et al. in prep.). Actually, for most of these species, germination at 5 or −20 °C was very low (3–10 times lower than germination of spores stored in LN) or nil for both storage times. There was no remaining viability in spores stored at room temperature (21 °C).

Production and Acclimatization of Sporophytes

Gametophytes in the laminar phase from spore cultures were moved to soil by scrapping the surface of the agar medium, washing the gametophytes into a small beaker, and subsequently distributing the gametophytes onto a 1:1 mix of soilless potting mix: pine mulch fines in small plastic boxes (Phytatrays™, Sigma-Aldrich) with lids. When growing on soil, most of the gametophytes developed into the heart-shaped stage after a few weeks. They were sprayed with purified water occasionally to aid in fertilization, and sporophytes usually appeared spontaneously after 3–12 months, depending on the species.

Table 11.2 Spore germination and gametophyte and sporophyte development of selected species stored in liquid nitrogen in the Frozen Garden up to 21 years

Species	Storage time (years)	Germination	Laminar gametophytes	Sporophyte growth in soil[a]
Asplenium verecundum	9	***	***	Y
Cibotium menziesii	10	**	N/A	Y
Asplenium kaulfusii	11	**	***	Y
Acrostichum danaeifolium	11	**	***	Y
Athyrium microphyllum	11	*	*	Y
Cibotium chamissoi	11	***	***	N
Diplazium sandwichianum	11	**	***	Y
Tectaria gaudichaudii	11	***	***	Y
Asplenium fragile	11	***	*	Y
Asplenium sp.	12	***	***	Y
Asplenium platyneuron	12	***	***	Y
Polystichum acrostichoides	12	***	***	Y
Blechnum occidentale	13	***	***	N
Cyathea microdonta	14	**	***	Y
Macrothelypteris torresiana	14	*	***	N/A
Tectaria trifoliata	14	***	***	N
Thelypteris dentata	14	**	***	Y
Osmunda regalis	17	*	***	N/A
Polystichum aleutichum	19	**	*	N/A
Adiantum fragile	19	**	***	Y
Cyclopelta semicordata	19	**	*	N/A
Pityrogramma calomelanos	20	***	***	Y
Cyrtomium falcatum	21	***	***	Y
Polystichum tsus-simense	21	*	***	Y
Pteris vittata	21	***	***	Y

Key: * (1–30%), ** (31–60%), *** (61–100%), N/A (data non-available), [a]at least one sporophyte grew in the period 2014–2017

Sporophytes were acclimatized first by replacing the solid lid with a lid with several holes, then by cracking the lid open, and finally by removing the lids but moving the boxes to a tray under a larger dome (Fig. 11.1g). After several weeks, vents in the domes were gradually opened to allow further acclimatization to the ambient greenhouse conditions. In a few cases, sporophytes appeared spontaneously from gametophytes while they were still on nutrient medium. These sporophytes were similarly moved to soil boxes and acclimatized. Sporophytes that were moved to the field (see section below) were also acclimatized from greenhouse conditions to field conditions (see below). Table 11.2 shows the species for which we obtained sporophytes in the greenhouse during the first months of growth using the methods described above.

11.3.2 Gametophyte Banking

Growth and Storage of Gametophytes

In vitro cultures of gametophytes of six species of fern were banked in 1995, including *Adiantum tenerum*, *A. trapeziforme*, *Cibotium glaucum*, *Davallia fejeensis*, *Drynaria quercifolia*, and *Phlebodium aureum.* Cultures were initiated in the early 1990s from the aseptic in vitro germination of spores. Spores were germinated on a half-strength Murashige and Skoog (1962) medium and the resulting gametophytes maintained on the same medium (Pence 2000a). Both the encapsulation dehydration method (Fabre and Dereuddre 1990) and drying without encapsulation, with and without a 7-day preculture on abscisic acid (ABA), were tested for their effectiveness in preserving viability of the gametophyte tissue through a 1 h exposure to LN (Pence 2000a). Encapsulation dehydration with preculture on ABA provided the best results, and this protocol was subsequently used for preparing tissues for long-term banking of samples. Dried beads were stored in 2 ml cryovials and rapidly cooled by immersion in LN. Vials were stored submerged in LN. A total of 21 samples from the six species were banked.

Testing the Viability of Gametophytes

Gametophytes were recovered after 20 years of storage with the same methods used to test prestorage viability after 1 h exposure to LN. Tissues were grown on 60 x 15 mm petri plates with half-strength MS medium. Tissues were incubated as for spores (above), although at 26 °C, rather than 20 °C. Prestorage viability was scored as the percent of tissue containing alginate beads showing some tissue growth.

Fig. 11.1 Aspect of immature (**a**) and mature (**b**) sporangia in *Dryopteris* sp. Aspect of mature (**c**-left) sporangia and sporangia after spore dehiscence (**c**-right) in *Osmunda regalis*. Mature spores dehisced on paper for their collection (**d**). Gametophytes of *Adiantum tenerum*, cryopreserved using the encapsulation dehydration procedure, 2 months after resuming growth in vitro

Summary of Results

Eight samples (cryovials) were tested representing all six species, two with replicates banked on different dates. After 20 years of storage in LN, tissues from all 6 species resumed growth in at least one replicate (Pence, unpublished). Survivals ranged from 0% to 90% per vial or 39% to 90% per species, while pre-banking survivals ranged from 7% to 92%. The percent regrowth of a sample was often higher than the original viability recorded for the sample, likely due to the small sample size and the variability of the samples, and thus it was not possible to determine if any deterioration of samples had occurred over time in storage. However, the sample sizes proved large enough to provide material for recovery, and the gametophytic fragments resumed growth into viable cultures (Fig. 11.1e).

11.3.3 Sporophyte Banking

Growth and Storage of Sporophytes

Shoot cultures of sporophytes of the lycophyte, *Selaginella uncinata*, had been initiated from surface-sterilized shoots that were grown on half-strength MS medium as for gametophyte cultures. Tips, approximately 1 mm in length, were dissected aseptically from the in vitro-grown shoots, treated for 7 days on half-strength MS medium with and without 10 μM ABA, and cryoprotected using the encapsulation dehydration procedure (Pence 2001). Tissues were banked in 1995.

Summary of Results

The survival of samples exposed to LN for 1 h at the time of banking was higher for tissues treated with ABA than for tissues precultured without ABA (Fig. 11.2) (Pence 2001).

After 3.5 years in LN, the shoot tips without ABA preculture showed no survival, while there was still survival, although reduced, in the ABA-treated tissues. However, when samples were removed after 20 years, there was no viability in either the ABA-treated or untreated samples. These results suggest that *S. uncinata* responded to the encapsulation dehydration procedure with some tolerance to desiccation, but that response was enhanced when the tissues were precultured on a

Fig. 11.1 (continued) after 20 years of storage in liquid nitrogen, bar = 0.5 cm (**e**). Spores of *Adiantum melanoleucum* germinated on 0.8% agar (left) and half-strength MS medium with 0.25% gellan gum (right), 11 weeks after surface sterilization and plating, bar = 2 cm (**f**). Sporophytes of *Polystichum tsus-simense* growing on soil under a dome, spores were stored in LN for 21 years (**g**). *Thelypteris patens* outplanted back into their native habitat in Florida (Possley, 2014) to help increase the native populations (F)

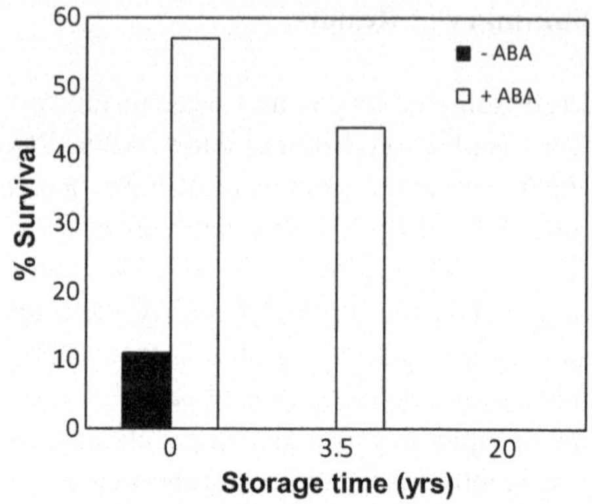

Fig. 11.2 Percent survival of shoot tips of *S. uncinata*, with and without ABA treatment, over time in cryostorage. Data for 0 and 3.5 years taken from Pence (2001)

medium with ABA. In addition, the ABA preculture appeared to enhance the ability to survive longer in LN storage than tissues without such treatment. These results are significant, in that they demonstrate that even in LN, viability may decline. Such declines have also been observed in short-lived seeds banked in LN (Ballesteros and Pence 2017), and decreases in longevity appear to be related to stresses on the tissues before banking. The results with *S. uncinata* also demonstrate that factors which improve the ability of tissues to survive the stress of short-term exposures to LN at the time of banking will likely improve the longevity of the tissue during storage, as well.

11.4 Practical Considerations for Conserving Ferns

11.4.1 Spore Banking

Fern Spore Maturity and Quality During Collection

Ideally, spores that are stored should be mature, of good quality, and free from other frond tissues. This is important, for example, to avoid the storage of dead or immature spores (Huang et al. 2014), to avoid the storage of spores that have been aged on the frond after dispersal and likely will show poor longevity during storage (Ballesteros et al. 2011; Li and Shi 2015), and also to avoid fungal and bacterial contamination during fern spore germination and gametophyte growth. Fern spores can be sterilized before storage (Agrawal et al. 1993; Beri and Bir 1993; Constantino et al. 2000; Dyer 1979; Fernández et al. 2010; Ford 1992; Pence 2000b; Simabukuro et al. 1998; Whittier and Pintaud 1999); however, this process can damage fern spores and reduce their germination capacity and gametophyte development (Camloh 1999; Simabukuro et al. 1998). Spores collected properly and sown without frond residues are less likely to become contaminated and more likely to produce normal gametophytes (Aragón and Pangua 2004; Colli Aurea and Perez Sonia 1999; Quintanilla et al. 2002; Whittier 1973).

In order to perform an optimal fern spore harvest, it is very important to plan the collection when the sporangia in the fertile leaves or strobili are mature (Bacchetta et al. 2008; Ibars and Estrelles, 2015; Ibars et al. 2011). A good knowledge of the phenology of the fern species and a few field trips to check its maturity will help to optimize the collection date, as is recommended for fern spores, as well as for seeds and pollen (e.g. Bacchetta et al. 2008; FAO 2014). The identification of mature sporangia can vary among species, but generally they will be found on the youngest healthy leaves presenting closed sporangia that are bright in colour (e.g. Figs 11.1b, c). The sporangia should not be white (nor green for nongreen spores, Fig. 11.1a). Sporangia of green spores generally turn dark green when mature (Fig. 11.1c – right), and sporangia of nongreen spores generally turn brown, yellow, or orange, depending on species (Fig. 11.1b). For species with sori covered by indusia, the colour of this structure (turning from green to yellow or brownish) helps to identify the mature sporangia (Bacchetta et al. 2008; Ibars and Estrelles 2015, Fig. 11.1b).

When fertile fronds with mature sporangia are collected, they can be briefly washed to eliminate impurities on the fronds. They can be stored and transported to the laboratory in paper envelopes or, if the transport is done within the same day of collection, in plastic bags (Bacchetta et al. 2008; Ibars and Estrelles 2015; Ibars et al. 2011). Then, in the laboratory, fronds are usually air-dried for a few days under ambient conditions in paper envelopes or sheets, so the spores fall on the surface of the paper (Fig. 11.1d). When fronds have been transported in paper envelopes, these can be used for spores' collection. Some fern species produce fertile leaves for which spore collection is not as direct as the one indicated above. For example, *Matteuccia* and *Onoclea* sp. fertile leaves must be submerged in water for a number of hours, and later the fertile leaves can be dried to release the spores. Without the first hydration step, the fertile leaves do not open (Templeman et al. 1987; Towill and Ikuma 1973).

One of the challenges of dealing with wild species, including ferns, is that it may not always be possible to collect an ideal sample, particularly if the plants are in a location that cannot be visited repeatedly to time the collections to the optimum level of spore maturity. In cases where the sporangia have opened, one may collect any remaining spores from open sporangia (Fig. 11.1c – left) by scraping the frond over paper. However, this can increase the risk of contamination during germination and the high possibility of getting spores of low viability or spores that are negatively affected by LN storage (Li and Shi 2015). If a collection must be made from open sporangia, spores might be germinated in vitro, to grow the gametophytes and then proceed either to gametophyte cryopreservation (described in Sects. 11.3.2 and 11.4.2) or to the growth of sporophytes for spore collection under more controlled conditions (e.g. greenhouse, botanical garden). If spores must be taken while they are still immature, germination may be lower than for mature spores (Huang et al. 2014). However, LN exposure may increase germination (Mikula et al. 2009) or germination rate, as it can with some mature spores (De Brum and Randi 2006; Rogge et al. 2000). Finally, whether the spores are collected at optimal maturity or not, there is the possibility of spores of one or more different species being included in the collection. Such spores may travel by wind to the leaves and be mixed in with the collection. Thus, it is important to monitor the gametophytes and sporophytes produced for the presence of species other than the target species.

Optimal Moisture Content for Fern Spore Storage

Spores in the Frozen Garden were stored dry (open dried at RH about 40% or over silica gel). This desiccation did not seem to affect fern spore viability, whether they were green or nongreen (Pence 2000b). Several authors have recently indicated that both green and nongreen spores are tolerant to desiccation and have greater longevity when stored dry at RHs between 10% and 25% (Ballesteros and Walters 2007a; Ballesteros et al. 2017; Li and Shi 2015; Mikula et al. 2015; Walters et al. 2005). Based on this optimal RH, and taking into account the changes in RH with temperature (and hence to avoid overdrying the spores during LN storage, e.g. Ballesteros and Walters 2011; Vertucci et al. 1994), it is recommended that after spores are collected, they are to be dried for 2–5 days at 20 °C and RHs between 20 and 40% and subsequently stored at the temperature of LN.

Optimal Temperatures for Fern Spore Storage

The results from long-term storage at CREW strongly support an optimal temperature of liquid nitrogen (−196 °C) for long-term storage of fern spores, when compared with −20 °C (e.g. Ballesteros et al. 2011, 2012). Storing in the vapour phase of liquid nitrogen (from −130 to −180 °C) or mechanical freezers providing −80 °C may also be as effective (e.g. Ballesteros et al. 2011, 2012). However, dry storage at −20 °C may be damaging for some species (reviewed in Ballesteros 2010). Poor longevity of spores at −20 °C is probably due to the lipid content and TAG composition of some nongreen spores (Ballesteros and Walters, 2007b), although this may need to be evaluated species by species. For example, *Pteris vittata* spores stored for about 20 years in both LN and −20 °C showed good viability under both conditions (Ballesteros et al. in prep). Cycles of freezing/defrost (Ballesteros et al. 2012), as would happen if all spores of an accession were stored in one container and it was moved in and out of the freezer each time spores were required, should be avoided at any freezing temperatures. Refrigeration of spores (3–7 °C) or storage of spores at room temperature is not recommended for the long-term conservation of fern spores.

11.4.2 Gametophytes

Gametophytes are highly adaptable and easily manipulated propagules for ex situ conservation, either in vitro or in cryopreservation.

In Vitro Culture of Gametophytes

Most in vitro cultures of gametophytes have been generated by the aseptic germination of spores, and thus, the considerations described above (Sect. 11.4.1) concerning spore stage, sterilization, and purity are equally important when initiating gametophyte cultures. Media used for in vitro gametophyte growth are usually low in nutrients, and the most commonly used include plain agar or agar or gellan gum with added nutrients, often either Knop's medium or half-strength Murashige and Skoog (MS) medium (e.g. Marimuthu and Manickam 2011; Mikula et al. 2009; Pence 2000a). For handling during surface sterilization, spores are often enclosed in a paper packet, such as one made by folding filter paper, and the packet is sterilized along with the spores, which are then blotted onto the medium (Pence 2000a). Alternatively, the spores can be suspended in the sterilant and then collected by centrifugation or filtering (de Brum and Randi 2006; Somer et al. 2009). An alternative method for gametophyte culture initiation has been reported for *Trichomanes speciosum*, in which gemmae were isolated and similarly surface sterilized and cultured (Raine and Sheffield 1997).

The rate of spore germination into gametophytes will vary with the species from days to months, and both germination and growth are influenced by the medium used. In this lab, it has been observed that spores sown on plain agar result in many small individual gametophytes, while spores sown on 0.25% gellan gum containing half-strength MS medium result in fewer spores germinating, but the gametophytes are larger and begin to propagate vegetatively (Fig. 11.1f). Aseptic germination has been reported for many species of leptosporangiate ferns, as well as for several eusporangiate species, including *Equisetum* spp. (Guillon and Fievet 2003; Kuriyama et al. 1990; Srinivasan and Kaufman 1978) and several Marattiaceae species (Barnicoat et al. 2011). Thus, it appears that gametophyte cultures of many fern species can be readily generated and maintained as in vitro collections, if spores are available. However, while there have been some examples of rare species being conserved as gametophytes in vitro (Ashmore et al. 2011; Marimuthu and Manickam 2011; Pence 2015; Raine and Sheffield 1997), thus far, long-term maintenance of in vitro collections has not been applied extensively to fern ex situ conservation.

Cryopreservation of Gametophytes

Conservation of in vitro-grown gametophyte tissues has more commonly been done through cryopreservation. Fern gametophyte cultures have been shown to be highly adaptable to cryopreservation, particularly with the encapsulation dehydration procedure (Fabre and Dereuddre 1990). This has been the most widely used cryoprotection method and is the one that successfully maintained viability of gametophytes for 20 years in LN (Pence, unpublished). In this procedure, gametophytes or gametophyte fragments are encapsulated in a sphere of alginate gel and cryoprotected using a combination of osmotic dehydration with concentrated sucrose and drying, before rapid cooling by immersion in LN. This method has worked well for most species reported, although a few have not survived the treatment (Barnicoat et al. 2011). Other cryoprotectant procedures have been tested, particularly encapsulation vitrification, in which chemical cryoprotectants are used, to remove water and protect the tissues (Hirai and Sakai 1999). While some survival of fern gametophytes through this method has been reported, it has generally not been as high as with the encapsulation dehydration method (Makowski et al. 2016; Mikula et al. 2010). The use of vitrification without encapsulation has resulted in extreme plasmolysis of the tissues and no survival after LN exposure. However, dehydration without encapsulation has resulted in some survival, especially when the tissues were pretreated with abscisic acid (ABA) (Pence 2000a). Treatment with ABA may mimic natural signals within the plant that provide desiccation tolerance to the tissues, as has been widely documented in moss gametophytes (Proctor et al. 2007).

Actively growing gametophytes appear to survive cryopreservation better than mature, non-growing tissues (Mikula et al. 2009). Whole gametophytes of some species are too large for encapsulation, but they can either be reduced in size by growth on paclobutrazol (Barnicoat et al. 2011) or cut into smaller fragments for encapsulation (Pence 2015). Whether whole or fragment, tissues from gametophytes often undergo damage during cryoprotection and cryopreservation (Mikula et al. 2009), but the high regenerative capacity of the tissues can reconstitute a culture from only a small portion of the tissue and possibly from single cells (Maeda and Ito 1981).

The application of encapsulation dehydration should be to a much wider range of pteridophyte taxa. Gametophyte cryopreservation has been reported primarily for taxa within the leptosporangiate ferns. Gametophytes of one eusporangiate species, *Macroglossum smithii*, have shown some survival after LN exposure using encapsulation dehydration, although two other species from the same family (Marattiaceae), *Marattia purpurascens* and *M. werneri*, did not show any survival (Barnicoat et al. 2011). To our knowledge, there have been no reports of gametophyte cryopreservation in other eusporangiate genera, such as *Psilotum*, *Equisetum*, or *Ophioglossum*, or in gametophytes of lycophyte genera, such as *Lycopodium* and *Selaginella*. In addition, methods are needed for species known only from the gametophyte stage (e.g. Farrar 1967) that would allow routine initiation of gametophyte cultures by direct surface sterilization of gametophytes. Finally, more species should be cryopreserved for long-term studies to more clearly understand the survival of these tissues over the course of decades.

While there are many areas for further research, the recovery of gametophyte growth after 20 years in LN, as well as reports from other labs of survival through short-term LN exposure, indicates that methods are available for the long-term ex situ conservation of gametophytes from likely a large number of fern species. However, these methods have been applied to an only a few rare fern species (Barnicoat et al. 2011; Mikula et al. 2009; Pence 2015; Wilkinson 2002) and could be more widely utilized.

11.4.3 Sporophytes

Like gametophytes, the sporophytes of ferns and lycophytes can be maintained as in vitro collections, or the in vitro-grown tissues can be cryopreserved for long-term ex situ conservation.

In Vitro Culture of Sporophytes

Most in vitro culture lines of sporophytes are initiated from in vitro-grown gametophytes, which, in turn, are initiated from aseptically germinated spores (see Sect. 11.4.2). When this is possible, it may take a few weeks to several years for the gametophytes to generate sporophytes (Pence 2015). However, additional methods for initiating aseptic sporophyte cultures have been reported, including regeneration from rhizomes, runner tips, leaves, leaf callus, and shoot tips, as well as bud scales and other tissues from in vitro-grown sporophytes (Bertrand et al. 1999; Camloh and Ambrozic-Dolinsek 2010; Higuchi et al. 1987; Jha et al. 2013; Li et al. 2017; Pence 2001; Shukla and Khare 2012; Winarto and Teixeira da Silva 2012), depending on the species and its growth habit. In vitro cultures have been reported for several rare species (Ashmore et al. 2011; Pence 2014, 2015; Taha et al. 2011; Yu et al. 2017), and the potential exists for utilizing such cultures for ex situ conservation as in vitro collections and for using them for propagating plants for restoration projects (Aguraiuja 2010a, b; Possley 2014).

Cryopreservation of Sporophytes

Once established, fern sporophyte cultures can provide materials for cryopreservation in a manner analogous to that of shoot tip cryopreservation in angiosperms (Reed 2008). Species with growing tips can provide shoot tips, and this approach has been successful for *Selaginella uncinata* and *Trichomanes punctatum* var. *floridanum* (Pence 2001; Pence 2015). For other species, such as those that grow from a basal meristem, it may be possible to manipulate in vitro culture conditions to generate the meristematic, green globular bodies (GGBs) (Bertrand et al. 1999; Fernández et al. 1996; Higuchi and Amaki 1989; Higuchi et al. 1987). GGBs of *Asplenium scolopendrium* var. *americanum* have been successfully cryopreserved (Pence 2015), and further work is needed to demonstrate this approach in other species.

The long-term survival of fern sporophyte tissues should also be explored further. Shoot tips of *Selaginella uncinata* precultured with ABA survived 3.5 years of cryostorage, while those without ABA preculture did not. However, even the ABA precultured tissues did not survive 20 years in LN. Thus, questions of longevity, as well as the effects of preculture treatments and of the cryoprotection methods on longevity in cryostorage, must be explored further. In the same study of long-term viability at CREW, shoot tips of several angiosperm species showed good survival after 15–20 years in LN (Pence et al. 2017). As with gametophytes, sporophyte tissue banking has been explored less than spore banking for ferns, although sporophytes of some rare ferns have been cryopreserved in CREW's Frozen Garden for long-term storage. While there may be fewer situations that require sporophyte banking, it can be important for some species, particularly in the context of conservation, when plants are few and difficult to access and/or produce few spores.

11.5 Ex Situ Conservation of Ferns: Case Study – Rare Ferns of South Florida

In addition to providing in vitro collections and tissues for cryopreservation, in vitro lines of rare fern species can be used to propagate plants for restoration. In collaboration with the Fairchild Tropical Botanic Garden (Miami, USA), CREW has propagated ferns in vitro of several species that are endangered in Florida, including *Thelypteris patens*, *Adiantum melanoleucum*, and *Odontosoria clavata*. These species were received as spores and germinated in vitro and the gametophytes propagated. After several months, sporophytes were produced, which also could be propagated in vitro. Plants were removed from culture and acclimatized in soil for 2–3 months and then sent to FTBG, where they were grown further into larger plants. Two of the species have been outplanted back into their native habitat to help increase the native populations (Possley 2014) (Fig. 11.1h).

11.6 Conclusions

Several explants from the life cycle of ferns and lycophytes can be used as a ready source of germplasm for ex situ conservation. Storing spores is the most efficient method, but there are situations where storing gametophytes or sporophytes can be extremely useful, as well. There is a body of work that has shown that current methods available for the storage of spores, gametophytes, and sporophytes in liquid nitrogen are relatively straightforward and similar to protocols employed for seeds and non-seed tissues of flowering plants. Our studies have shown that spores and gametophyte cryostorage can provide viable materials after at least 20 years for the implementation of restoration programmes of endangered fern species, and the potential exists for sporophyte tissues as well. However, given the increasing number of fern and lycophyte species of conservation concern, these ex situ conservation methods have been underutilized. There is a need for the publication of guidelines and protocols to encourage increased attention to these groups and to facilitate best practices in future banking efforts. This would allow current seed and plant tissue banks to utilize their expertise and infrastructure to expand their focus to include ferns and lycophytes and to provide guidance to new efforts, thereby helping to secure the diversity of these species into the future.

Acknowledgements We would like to thank Callihan A and Lindsey K for their contribution to the propagation and acclimation of diverse fern species at CREW and Thomas A and Stewart K for the work on the initial banking of spores and gametophytes. Also special thanks to Barber S, Pritchard H, Dickie J, Davies R (UK), Estrelles E (Spain), Mikula A (Poland), Magrini S (Italy), Walters C, Hill L, Wolkis D (USA), Reyes-Jaramillo I (Mexico), Randi M (Brazil), Sershen (South Africa), Huang YM (Taiwan), Cicuzza D, Bin W, Yang X-Y (China), Schnell J, McGill C (New Zealand), and Offord C (Australia) for their contributions in the development of Table 11.1 and Suggii N and From M for information on in vitro sporophyte collections. Fern spore, gametophyte, and sporophyte collections used in this work would not be possible without the help and the donations of multiple collectors and researchers from diverse institutions, including Bannister S, Yoshinaga A, Kawakami K, and staff of the Krohn Conservatory.

References

Agrawal DC, Pawar SS, Mascarenhas AF (1993) Cryopreservation of spores of *Cyathea spinulosa* wall. Ex. Hook. F., an endangered tree fern. J Plant Physiol 142:124–126

Aguraiuja R (2010a) Reinforcement of the populations of the critically endangered endemic fern, *Diellia pallid,* Kaua'I, Hawwaiian Islands, U.S.A. In: Soorae PS (ed) Global re-introduction perspectives: additional case-studies from around the globe. IUCN/SSC Re-introduction Specialist Group. Abu Dhabi, UAE, pp 319–323

Aguraiuja R (2010b) Conservation introduction of a locally extinct fern species in Estonia during 1998–2008. In: Soorae PS (ed) Global re-introduction perspectives: additional case-studies from around the globe. IUCN/SSC Re-introduction Specialist Group. Abu Dhabi, UAE, pp 345–348

Alexander A, Dudas G, Herren G, Kolben G, Kuhara Y, Murdoch L, Slaven J, Paultre D, Rae N, Rayner J, Vileitaite J, Chamberlain M (2008) Testing the viability of spores held in the fern spore bank at the Royal Botanic Garden, Edinburgh. Newsletter of the Botanical Society of Scotland 91:29–31

Aragón CF, Pangua E (2004) Spore viability under different storage conditions in four rupicolous *Asplenium* L. taxa. Am Fern J 941:28–38

Arcand NN, Ranker TA (2008) Conservation biology. In: Ranker TA, Haufler CH (eds) Biology and evolution of ferns and lycophytes. Cambridge University Press, Cambridge, pp 257–283

Ashcroft CJ, Sheffield E (2000) The effect of spore density on germination and development in *Pteridium*, monitored using a novel culture technique. Am Fern J 90:91–99

Ashmore SE, Hamilton KN, Offord CA (2011) Conservation technologies for safeguarding and restoring threatened flora: case studies from Eastern Australia. In Vitro Cell Dev Biol Plant 47:99–109

Bacchetta G, Bueno Sánchez A, Fenu G, Jiménez-Alfaro B, Mattana E, Piotto B, Virevaire M (eds) (2008) Conservación *ex situ* de plantas silvestres. Principado de Asturias/La Caixa, Principado de Asturias, Spain

Ballesteros D (2010) Conservation of fern spores. In: Kumar A, Fernández H, Revilla-Bahillo A (eds) Working with ferns. Issues and applications. Springer, New York, pp 165–172

Ballesteros D, Pence VC (2017) Survival and death of seeds during liquid nitrogen storage: a case study on seeds with short lifespans. CryoLetters 38(4):278–289

Ballesteros D, Walters C (2007a) Water properties in fern spores: sorption characteristics relating to water affinity, glassy states and storage stability. J Exp Bot 58:1185–1196

Ballesteros D, Walters C (2007b) Calorimetric properties of water and triacylglycerols in fern spores relating to storage at cryogenic temperatures. Cryobiology 55:1–9
Ballesteros D, Walters C (2011) Detailed characterization of mechanical properties and molecular mobility within dry seed glasses: relevance to the physiology of dry biological systems. The Plant J 68:607–619
Ballesteros D, Estrelles E, Walters C, Ibars AM (2011) Effect of storage temperature on green spore longevity for the ferns *Equisetum ramosissimum* and *Osmunda regalis*. CryoLetters 32(2):89–98
Ballesteros D, Estrelles E, Walters C, Ibars AM (2012) Effects of temperature and desiccation on *ex situ* conservation of non-green fern spores. Am J Bot 99:721–729
Ballesteros D, Hill LM, Walters C (2017) Variation of desiccation tolerance and longevity in fern spores. J Plant Physiol 211:53–62
Barnicoat H, Cripps R, Kendon J, Sarasan V (2011) Conservation *in vitro* of rare and threatened ferns—case studies of biodiversity hotspot and island species. In Vitro Cell Dev Biol Plant 47:37–45
Beri A, Bir SS (1993) Germination of stored spores of *Pteris vittata* L. Am Fern J 833:73–78
Bertrand AM, Albuerne MA, Fernández H, González A, Sánchez-Tamés R (1999) *In vitro* organogenesis of *Polypodium cambricum*. Plant Cell Tissue Organ Cult 57:65–69
Camloh M (1999) Spore age and sterilization affects germination and early gametophyte development of *Platycerium bifurcatum*. Am Fern J 89:124–132
Camloh M, Ambrozic-Dolinsek J (2010) *In vitro* regeneration systems of *Platycerium*. In: Fernández H, Kumar A, Angeles-Revilla A (eds) Working with ferns: issues and applications. Springer Science & Business Media, New York, pp 111–125
Chapman AD (2009) Numbers of living species in Australia and the world. Australian Biodiversity Information Services, Toowoomba, Australia
Colli Aurea MT, Perez Sonia CJG (1999) The effect of red light on the germination of a brazilian pteridophyte. Brazil Arch Biol Technol 422:169–174
Constantino S, Santamaria LM, Hodson E (2000) Storage and *in vitro* germination of tree fern spores. Bot Garden Microprop News 24:58–60
De Brum FMR, Randi AM (2006) Germination of spores and growth of gametophytes and sporophytes of *Rumohra adiantiformis* (Forst.) Ching (Dryopteridaceae) after spore cryogenic storage. Rev Bras Bot 29:489–495
De Long JM, Prange RK (2008) Fiddlehead fronds: nutrient rich delicacy. Chronica Hort 48:12–15
Dolinsek JA, Camloh M (1997) Gametophytic and sporophytic regeneration from bud scales of the fern *Platycerium bifurcatum* (Cav.) C.Chr. *in vitro*. Ann Bot 80:23–28
Dong S, Zuo Z, Yan Y, Xiang J (2017) Red list assessment of lycophytes and ferns in China. Biodivers Sci 25(7):765–773
Dyer AF (1979) The culture of fern gametophytes for experimental investigation. In: Dyer AF (ed) The experimental biology of ferns. Academic Press, London
Dyer AF (1994) Natural soil spore banks—can they be used to retrieve lost ferns? Biodivers Conserv 3:160–175
Fabre J, Dereuddre J (1990) Encapsulation-dehydration: a new approach to cryopreservation of *Solanum* shoot tips. CryoLetters 11:413–426
FAO (2014) Genebank standards for plant genetic resources for food and agriculture, Rev. edn, Rome. Food and Agriculture Organization of the United Nations, Rome
Farrar DR (1967) Gametophytes of four tropical fern genera reproducing independently of their sporophytes in the southern Appalachians. Science 155:1266–1267
Farrar DR (2016) *Vittaria appalachiana* continues to provide insight into the biology of ferns: a commentary on two studies recently published in American Journal of Botany. Am J Bot 103:593–595
Fernández H, Revilla MA (2003) *In vitro* culture of ornamental ferns. Plant Cell Tissue Organ Cult 73:1–13

Fernández H, Bertrand A, Sánchez-Tamés R (1996) Micropropagation and phase change in *Blechnum spicant* and *Pteris ensiformis*. Plant Cell Tissue Organ Cult 44:261–265
Fernández H, Kumar A, Revilla A (eds) (2010) Working with ferns: issues and applications. Springer Science & Business Media, New York
Ford MV (1992) Growing ferns from spores in sterile culture. In: Ide JM, Jermy AC, Paul AM (eds) Fern horticulture: past, present and future perspectives. Intercept, Andover, pp 295–297
From MM (2010) Conservation status and re-introduction of Bermuda's Governor Laffan fern—an endangered endemic. In: Sourae PS (ed) Global Re-introduction Perspectives—2010: additional Case Studies from around the Globe. IUCN/SSC Re-introduction Specialist Group, Abu Dhabi, UAE, pp 316–318
Guillon JM, Fievet D (2003) Environmental sex determination in response to light and biased sex ratios in *Equisetum* gametophytes. J Ecol 91:49–57
Higuchi H, Amaki W (1989) Effects of 6-benzylaminopurine on the organogenesis of *Asplenium nidus* L. through *in vitro* propagation. Sci Hortic 37:351–359
Higuchi H, Amaki W, Suzuki S (1987) *In vitro* propagation of *Nephrolepis cordifolia* Presl. Sci Hortic 32:105–113
Hirai D, Sakai A (1999) Cryopreservation of *in vitro*-grown axillary shoot-tip meristems of min (*Mentha spicata* L.) by encapsulation vitrification. Plant Cell Rep 19:150–155
Houser DC, From M, Landry M, Copeland A, Kellar R (2016) Systematics of *Diplazium laffanianum* (Athyriaceae), a fern species endemic to Bermuda. Am Fern J 106:206–222
Huang Y-M, Lee P-H, Fu C-H, Chang Y-H (2014) The relation between spore maturity and viability of tree ferns. J Natl Taiwan Mus 67(2):45–58
Ibars AM, Estrelles E (2012) Recent developments in *ex situ* and *in situ* conservation of ferns. Fern Gaz 19(3):67–86
Ibars AM, Estrelles E (2015) Protocolo de propagación de helechos. Botanic as PPECTS 1 1:1–5
Ibars AM, Gómez Serrano MA, Mayoral O, Estrelles E (2011) Prioridades para la conservación en el ámbito de los helechos en Castilla-La Mancha. In: Hernandez JE, Herranz JM (eds) Protección de la diversidad vegetal y de los recursos fitogenéticos de Castilla-La Mancha. Instituto de Estudios Albacetenses- Botánico de Castilla-La Mancha. Serie I Estudio, Albacete, Spain, pp 191–213
IUCN (2017) Table 3a: Status category summary by major taxonomic group (plants). In: The IUCN red list of threatened species, version 2017–2. Last updated 14 Sept 2017, http://cms-docs.s3.amazonaws.com/summarystats/2017-2_Summary_Stats_Page_Documents/2017_2_RL_Stats_Table_3b.pdf. Accessed 29 Sept 2017
Jha TB, Mukherjee S, Basak A, Adhikari J (2013) *In vitro* morphogenesis in *Selaginella microphylla* (Kunth.) Spring. Plant Biotechnol Rep 7:239–245
Kamachi H, Noguchi M (2012) Negative gravitropism in dark-grown gametophytes of the fern *Ceratopteris richardii*. Am Fern J 102:147–153
Khan S, Raziq M, Kayani HA (2008) *In vitro* propagation of bird's nest fern (*Asplenium nidus*) from spores. Pakistan J Bot 40:91–97
Kuriyama A, Sugawara Y, Matsushima H, Takeuchi M (1990) Production of sporophytic structures from gametophytes by cytokinin in Equisetum Arvense. Naturwissenschaften 77(1):31–32
Li D, Pritchard HW (2009) The science and economics of *ex situ* plant conservation. Trends Plant Sci 14:614–621
Li Y, Shi L (2015) Effect of maturity level and desiccation process on liquid nitrogen storage of green spores of *Osmunda japonica*. Plant Cell Tissue Organan Cult 120:531–538
Li X, Han J-D, Fang Y-H, Bai S-N, Rao G-Y (2017) Expression analyses of embryogenesis-associated genes during somatic embryogenesis of *Adiantum capillus-veneris* L. *in vitro*: new insights into the evolution of reproductive organs in land plants. Front Plant Sci 8:1–12
Liu Y, Wujisguleng W, Long C (2012) Food uses of ferns in China: a review. Acta Soc Bot Pol 81(4):263–270
Maeda M, Ito M (1981) Isolation of protoplasts from fern prothallia and their regeneration to gametophytes. Bot Mag Tokyo 94:35–40

Makowski D, Rybczyński JJ, Mikula A, Klimaszewska K (2015) A simple way to overcome the recalcitrance of the water fern *Ceratopteris thalictroides* (L.) Brongn. to cryopreservation. Acta Soc Bot Pol 84:385–388

Makowski D, Tomiczak K, Rybczyński JJ, Mikula A (2016) Integration of tissue culture and cryopreservation methods for propagation and conservation of the fern *Osmunda regalis* L. Acta Physiol Plant 38:1–12

Mannan M, Maridass M, Victor B (2008) A review on the potential uses of ferns. Ethnobotanical Leaflets 12:281–285

Marimuthu J, Manickam VS (2011) *Ex situ* conservation of two threatened ferns of the Western Ghats through *in vitro* spore culture. J Threat Taxa 3:1919–1928

Martin KP, Sini S, Zhang CL, Slater A, Madhusoodanan PV (2006) Efficient induction of apospory and apogamy *in vitro* in silver fern (*Pityrogramma calomelanos* L.) Plant Cell Rep 25:1300–1307

Matsuura T, Sugimura K, Miyamoto A, Tanaka H, Tanaka N (2014) Spatial characteristics of edible wild fern harvesting in mountainous villages in northeastern Japan using GPS tracks. Forests 5:269–286

Mehltreter K, Walker LR, Sharpe JM (eds) (2010) Fern ecology. Cambridge University Press, Cambridge

Mikula A, Jata K, Rybczynksi JJ (2009) Cryopreservation strategies for *Cyathea australis* (R.Br.) Domin. CryoLetters 30:429–439

Mikula A, Makowski D, Walters C, Rybczynski JJ (2010) Exploration of cryo-methods to preserve tree and herbaceous fern gametophytes. In: Fernández H, Kumar A, Angeles-Revilla A (eds) Working with ferns: issues and applications. Springer Science & Business Media, New York, pp 173–192

Mikula A, Tomiczak K, Makowski D, Niedzielski M, Rybczynski JJ (2015) The effect of moisture content and temperature on spore aging in *Osmunda regalis*. Acta Physiol Plant 37:229–240

Murashige T, Skoog F (1962) A revised medium for rapid growth and bio assays with tobacco tissue cultures. Physiol Plant 15:473–497

Ong BL, Ng ML (1998) Regeneration of drought-stressed gametophytes of the epiphytic fern, *Pyrrosia piloselloides* (L.) price. Plant Cell Rep 18:225–228

Page CN, Dyer AF, Lindsay S, Mann DG (1992) Conservation of Pteridophytes: the ex situ approach. In: Ide JM, Jermy AC, Paul AM (eds) Fern horticulture: past, present and future perspectives. Intercept, Andover, pp 269–278

Pence VC (2000a) Cryopreservation of *in vitro* grown fern gametophytes. Am Fern J 90:16–23

Pence VC (2000b) Survival of chlorophyllous and nonchlorophyllous fern spores through exposure to liquid nitrogen. Am Fern J 90:119–126

Pence VC (2001) Cryopreservation of shoot tips of *Selaginella uncinata*. Am Fern J 91:37–40

Pence VC (2004) Ex situ conservation methods for bryophytes and pteridophytes. In: Guerrant Jr. EO, Havens K, Maunder M (eds) Ex situ plant conservation. Supporting species survival in the wild. Island Press,Washington, DC, pp 206–227

Pence VC (2008a) *Ex situ* conservation of ferns and lycophytes – approaches and techniques. In: Ranker TA, Haufler CH (eds) Biology and evolution of ferns and lycophytes. Cambridge University Press, Cambridge, pp 284–300

Pence VC (2008b) Cryopreservation of bryophytes and ferns. In: Reed BM (ed) Plant cryopreservation: a practical guide. Springer, New York, pp 117–140

Pence VC (2014) *In vitro* propagation and cryopreservation of the endangered filmy fern, *Trichomanes punctatum* subsp. *floridanum* (Hymenophyllaceae). Fern Gaz 19:307–317

Pence VC (2015) Propagation and cryopreservation of *Asplenium scolopendrium* var. *americanum*, the American hart's-tongue fern. Am Fern J 105:211–225

Pence VC, Philpott M, Culley TM, Plair B, Yorke SR, Lindsey K, Vanhove A-C, Ballesteros D (2017) Survival and genetic stability of shoot tips of *Hedeoma todsenii* R.S. Irving after long-term cryostorage. In Vitro Cellular and Developmental Biology Plant 53(4):328–338

Pennisi E (2010) Tending the global garden. Science 329:1274–1277

Pérez-García B, Reyes-Jaramillo I (1993) Helechos: propagación y conservación. Ciencias 30:10–17
Possley J (2014) Fern conservation in a biodiversity hotspot. The Palmetto 31(2):1–9
Pritchard HW, Nadarajan J, Ballesteros D, Thammasiri K, Prasongsom S, Malik SK, Chaudhury R, Kim H-H, Lin L, Li W-Q, Yang X-Y, Popova E (2017) Cryobiotechnology of tropical seeds – scale, scope and hope. Acta Hortic 1167:37–47
Proctor MCF, Oliver MJ, Wood AJ, Alpert P, Stark LR, Cleavitt NL, Mishler BD (2007) Desiccation-tolerance in bryophytes: a review. Bryologist 110:595–621
Quintanilla LG, Amigo J, Pangua E, Pajaron S (2002) Effect of storage method on spore viability in five globally threatened fern species. Ann Bot 904:461–467
Raine ACA, Sheffield E (1997) Establishment and maintenance of aseptic culture of *Trichomanes speciosum* gametophytes from gemmae. Am Fern J 87:87–92
Rajasekharan PE (2015) Gene banking for *ex situ* conservation of plant genetic resources. In: Bahadur B, Rajam MV, Sahijram L, Krishnamurthy KV (eds) Plant biology and biotechnology Vol II: Plant genomics and biotechnology. Springer, New Delhi, pp 445–459
Reed BM (ed) (2008) Plant cryopreservation: a practical guide. Springer, New York
Rogers SMD, Banister S (1992) Micropropagation of *Notholaena*. Hortscience 27:1224–1225
Rogge GD, Viana AM, Randi AM (2000) Cryopreservation of spores of *Dicksonia sellowiana*: an endangered tree fern indigenous to South and Central America. CryoLetters 21:223–230
Sara SC, Manickam VS (2007) *In vitro* developmental ontogeny and life cycle of a rare fern species – *Thelypteris confluens* (Thunb.) Morton. Indian J Biotechnol 6:372–380
Shukla SP, Khare PB (2012) *In vitro* shoot regeneration via caulogenesis in fern, *Pteris vittata* L. J Environ Biol 33:683–687
Simabukuro EA, Dyer AF, Felippe GM (1998) The effect of sterilization and storage conditions on the viability of the spores of *Cyathea delgadii*. Am Fern J 88:72–80
Somer M, Arbesu R, Menéndez V, Revilla MA, Fernández H (2009) Sporophyte induction studies in ferns *in vitro*. Euphytica 171:203–210
Srinivasan JC, Kaufman PB (1978). Sex expression in Equisetum scirpoides in vitro. Phytomorphology 28:331–5
Stamps RH (2006) Irrigation and nutrient management practices for commercial leatherleaf fern production in Florida. University of Florida, IFAS, Florida
Taha RM, Haron NW, Wafa SN (2011) Morphological and tissue culture studies of *Platycerium coronarium*, a rare ornamental fern species from Malaysia. Am Fern J 101:241–251
Templeman TS, DeMaggio AE, Stetler DA (1987) Biochemistry of fern spore germination: globulin storage proteins in *Matteuccia struthiopteris* L. Plant Physiol 85:343–349
Towill LR, Ikuma H (1973) Photocontrol of the germination of *Onoclea* spores. Plant Physiol 51:973–978
Vertucci CW, Roos EE, Crane J (1994) Theoretical basis of protocols for seed storage III. Optimum moisture contents for pea seeds stored at different temperatures. Ann Bot 74(5):531–540
Walters C, Hill LM, Wheeler LJ (2005) Dying while dry: kinetics and mechanisms of deterioration in desiccated organisms. Integr Comp Biol 45:751–758
Whittier DP (1973) The effect of light and other factors on spore germination in *Botrychium dissectum*. Can J Bot 51:1791–1794
Whittier DP, Pintaud JC (1999) Spore germination and early gametophyte development in Stromatopteris. Am Fern J 89:142–146
Wilkinson T (2002) *In vitro* techniques for the conservation of *Hymenophyllum tunbrigense* (L.) Sm. Fern Gaz 16:458
Winarto B, Teixeira da Silva JA (2012) Improved micropropagation protocol for leatherleaf fern (*Rumohra adiantiformis*) using rhizomes as donor explants. Sci Hortic 140:74–80
Yu R, Zhang G, Li H, Cao H, Mo X, Gui M, Zhou X, Jiang Y, Li S, Wang J (2017) *In vitro* propagation of the endangered tree fern *Cibotium barometz* through formation of green globular bodies. Plant Cell Tissue Organ Cult 128:369–379

Chapter 12
Azolla and Bougainville's Voyage Around the World

Francisco Carrapiço

12.1 Introduction

The French expedition that was led by Louis Antoine de Bougainville (1729–1811) and that sailed around the world from 1766 to 1769 is well known for its contribution to science, namely, geography, astronomy, zoology and botany, in particular the discovery of *Bougainvillea* in Brazil. Unknown to many, however, is the fact that it was during this voyage that the first *Azolla* samples for science purposes were collected and included in Commerson's herbarium. This year (2017), this botanical event commemorates 250 years. The geographic origin of these *Azolla* samples collected by Philibert Commerson (1727–1773) and Jeanne Baret (1740–1807) was and still is a controversial matter, given that the information published by Jean-Baptiste Lamarck (1744–1829) in 1783 indicates that they were collected at the Strait of Magellan (Carrapiço 2010a; Lamarck 1783). In addition, the origin of *Azolla*'s name is often a source of confusion, based on two contradictory ideas. In conclusion, much of what is known about this fern's history and biology needs to be updated and complemented with new information. It is our goal in this chapter to shed new light on the untold history and biology of this important aquatic fern.

F. Carrapiço (✉)
Centro de Ecologia, Evolução e Alterações Ambientais e Centro de Filosofia das Ciências, Departamento de Biologia Vegetal, Faculdade de Ciências, Universidade de Lisboa, Lisboa, Portugal
e-mail: fcarrapico@fc.ul.pt

H. Fernández (ed.), *Current Advances in Fern Research*,
https://doi.org/10.1007/978-3-319-75103-0_12

12.2 A New Look on the Taxonomy and Biology of *Azolla*

Azolla is a small-leafed floating or semiaquatic heterosporous fern presenting overlapping scale-like bilobed leaves covering a slender and branched rhizome that floats horizontally on the freshwater surface of tropical and temperate environments (Carrapiço 2017; Svenson 1944). The roots of this fern are simple and emerging on the ramification points of the ventral side of the stem. The genus *Azolla* belongs to the family Azollaceae (Schneller 1990; Wettstein 1903) and includes seven species divided into two sections: *Rhizosperma* (*A. pinnata* and *A. nilotica*) and *Azolla* (*A. filiculoides*, *A. caroliniana*, *A. mexicana*, *A. microphylla* and *A. rubra*) (Adams et al. 2013; Pereira et al. 2011). However, the systematics of the family Azollaceae, namely, the section *Azolla*, is an open issue still controversial and confusing. In 2004, Evrard and Van Hove carried a review of the taxonomy of the New World *Azolla* species (section *Azolla*) reintroducing the species *A. cristata*, which was introduced in 1847 by the German botanist Georg Heinrich Mettenius (1823–1866), based on the previous classification made by Georg Friedrich Kaulfuss (1786–1830) in 1824 (Evrard and Van Hove 2004; Mettenius 1847; Svenson 1944). Evrard and Van Hove based their work on three morphological characters: leaf papillae, megaspore perine surface and number of glochidia septa. They rearranged the *Azolla* cluster in two species: *A. filiculoides* (grouping *A. filiculoides*, *A. caroliniana* Willd. and *A. microphylla* Kaulf.) and *A. cristata* (including *A. caroliniana* auct. non-Willd., *A. microphylla* auct. non-Kaulf. and *A. mexicana* Presl.) (Evrard and Van Hove 2004; Pereira et al. 2011). However, one of the characters used, the "number of glochidia septa", varies among specimens of *A. caroliniana* and *A. filiculoides*, and for that reason, it is not a good taxonomic character (Pereira et al. 2011). Furthermore, our research carried out on the species of the sections *Rhizosperma* and *Azolla* obtained from the IRRI germplasm collection and based on morphological and molecular data, namely, RAPD markers, does not support this proposed classification, meaning that the results indicate the existing of a two-section ranking, the *A. pinnata* and *A. nilotica* in section *Rhizosperma* and five distinct species on section *Azolla*: *A. mexicana*, *A. microphylla*, *A. caroliniana*, *A. filiculoides* and *A. rubra* (Pereira et al. 2011).

Each bilobed leaf of *Azolla* has an immersed, thick, greenish or reddish and photosynthetic dorsal lobe and a very thin (one cell thick), immersed hyaline ventral lobe. The chlorophyllous dorsal lobe contains an ellipsoid cavity, measuring 0.15 × 0.30 mm, which opens to the exterior through a pore (Caiola and Forni 1999; Carrapiço 2002, 2010a, b; Veys et al. 1999). This extracellular compartment contains a permanent and specific endosymbiotic prokaryote community, a heterocyst-forming filamentous nitrogen fixing cyanobacterium, usually assigned to *Anabaena azollae*, as well as several genera of bacteria, such as *Arthrobacter*, *Corynebacterium* and *Agrobacterium* (Carrapiço 1991, 2017). These microsymbionts are specific of this association and live immobilized in a mucilaginous fibrillary network, which fills the peripheral area of the cavity. The mucilaginous material of the leaf cavity is delimited by two envelopes, an external and an internal one, leaving the centre empty and probably filled with gas or liquid (Carrapiço 2010a, b). The presence of the prokaryote community, which lives in permanent symbiosis with *Azolla*, plays a relevant role in the plant's survival skills when faced with new conditions (Carrapiço 2010b). This is related to the fact that this community constitutes a source of evolutionary innovation, where symbionts are the means through which a rapid renovation and adaptation of the entire symbiotic system occurs, as a response to new environmental challenges and with consequences at the metabolical, physiological, and even genetic levels (Carrapiço 2010b).

Azolla is able to maintain an inoculum of the prokaryote community (cyanobacteria and bacteria) between plant generations (Carrapiço 2010a, 2017; Ran et al. 2010), and sporocarps act as the transfer vehicles during sexual reproduction (Ran et al. 2010). The majority of these studies refer to the cyanobionts, but we believe that the same process can also be applied to the bactobionts (Carrapiço 1991, 2010a, 2017). This process relies on the capacity of the cyanobiont to differentiate motile non-heterocystous hormogonial filaments, which are attracted to and enter the megasporocarp through a narrow pore (Ran et al. 2010). The mechanism by which the cyanobacteria cells become associated with the developing sporocarps involves branched epidermal trichomes, which arise on the ventral surface of the apical meristem and probably includes chemical mediators (Adams et al. 2013; Carrapiço 2017; Perkins and Peters 1993). On entering the megasporocarp, the hormogonia differentiate into akinetes in the indusium chamber above the megaspore, remaining dormant until the plant germinates (Adams et al. 2013), initiating a new generation of the life cycle of *Azolla*.

Azolla has been traditionally used as green manure for rice cultivation and animal feed not only in China and in Vietnam, in the past few centuries (Lumpkin and Plucknett 1980; Moore 1969), but also in Africa (Carrapiço 2010a) and in Central and South America (Mariano Montaño 2017, personal communication). In addition to agricultural purposes, there are other areas where the *Azolla-Anabaena* association has been applied. The use of *Azolla* as biofilter and bioremediation tool for contaminated waters has been developed in the last three decades with success (Costa et al. 1999; Costa et al. 2009; Forni et al. 2001; Sela et al. 1989). Recently, new areas of research have emerged related to this fern. The use of this symbiotic system as a model for biological, genomic, phytochemical and physiological studies (De Vries et al. 2016; Qiu and Yu 2003), namely, evolutionary developmental biology (De Vries and De Vries, in Chap. 2 of this book), and as an epistemological tool to understand the post-neo-Darwinian evolutive process (Carrapiço 2015) is currently taking place. In biotechnological research, including biological based life-support systems (BLSS), *Azolla* can be incorporated in bioregenerative space devices for cosmos exploration (Carrapiço 2010a).

The *Azolla* leaf cavity behaves as both the physiological and dynamic interface unit of this symbiotic association, where the recognition of molecular functions happens and the main metabolic and energetic flows occur (Caiola and Forni 1999; Carrapiço 2010b). Two morphologically distinct classes of epidermal trichomes (simple hairs and multicellular branched hairs), present in the leaf cavity, are involved in the transfer and uptake of metabolites from the fern to the prokaryote colony and from the endosymbionts to the host. In the vacuoles of the apical cells of the simple hairs, a mixture of several chemical compounds, namely, lipids, polysaccharides, polyphenols and alkaloids or alkaloid-like compounds, was detected (Carrapiço 2017; Carrapiço and Tavares 1989; Pereira and Carrapiço 2007). These chemical compounds may play a role in the communication between the host and the symbionts, in the selection of microorganisms that are not useful to the fern, in the control of the endosymbionts in the cavity and in the establishment and maintenance of the symbiosis (Carrapiço 2017). In this sense, this cavity can be considered to be a natural microcosm (Carrapiço 2002), a special micro-ecosystem which reveals a self-organization and an ecological defined structure, allowing the prokaryotes speciation or to form ecotypes adapted to specific environmental conditions (Carrapiço 2017).

Although traditionally considered as a lower vascular plant, *Azolla* exhibits symbiotic characteristics more evolved than the other vascular plant-cyanobacterial symbioses, such as *Cycas* and *Gunnera* (Adams et al. 2013; Carrapiço 2006). There appears to be no direct correspondence between the fern's evolutionary phylogeny and the complexity of the symbiosis (Carrapiço 2010a). This unique symbiosis is sustained throughout the fern's life cycle, where the cyanobiont and bacteria are always present and transferred from one generation to the next, either in the dorsal leaf cavities or in the megasporocarps, indicating the obligatory nature of the symbiosis (Carrapiço 2010a). It means that *Azolla* is not infected de novo, suggesting a specific phylogenetic co-evolution of the symbionts and the host, and can be considered as a natural photobioreactor with millions of years of evolution (Shi and Hall 1988). It also represents a good example of hereditary symbiosis, showing that the complexity of the relationship between the host and the symbionts can be recognized as a new level of biological organization in evolution. All this data leads to the idea that the *Azolla-Anabaena*-bacteria association works in a synchronized way and can be considered to be a polygenomic entity, in which the different genomes operate together in a complementary and synergistic way for the whole. It also suggests that we are in the presence of a superorganism or symbiome/holobiont in both biological and ecological terms (Carrapiço 2010a, 2015, 2017) in which new metabolic and organic capabilities are acquired and developed by the partners, which establish a new level of organization that goes beyond the individual capabilities of any individual partner. A good example of it can be found in the nitrogen metabolism shared by the host and its partners in this symbiotic system. The atmospheric nitrogen fixed by the cyanobacterium through the heterocysts is converted into ammonia and released into the leaf cavity. It has been shown that intracellular ammonia pools of symbiotically associated *Azolla* are five times bigger than those of endophyte-free *Azolla* (Braun-Howland and Nierzwicki-Bauer 1990). The activities of ammonia assimilating enzymes in the isolated trichomes of the dorsal leaf cavity were much higher than those in *Azolla* leaves, while these activities in the *Anabaena* filaments were repressed to very low levels. The host accounts for at least 90% and 80% of the total glutamine synthetase and NADH-dependent glutamate dehydrogenase activities, respectively. These results suggest that hair cells play an important role in the assimilation of the nitrogen fixed and released into the cavity by the cyanobiont. The nitrogen is transferred to the fern, after having been acquired during the development of the symbiotic process (Carrapiço 2010b; Uheda 1990). This data indicates that the synergies associated with symbiosis had and continue to have a leading role in the morphological, reproductive, physiological and metabolical complexification of the organisms in evolution (Corning 1983, 2005). In fact, the fossil records also suggest that the relationship between the fern and the cyanobiont was established in the mid-Cretaceous, which implies that the two organisms have been co-evolving for at least 80 million years (Jonathan Bujak 2017, personal communication; Bujak 2007; Bujak and Bujak 2014).

Data from the 2004 Arctic Coring Expedition (ACEX) cores, drilled in the central Arctic Ocean near the North Pole, show the presence of fossil *Azolla* in Eocene sediments (~48,5 Ma) (Brinkhuis et al. 2006). The plant's remains occur as laminations, reflecting seasonal or longer cycles, and they have been observed in more than 50 Arctic wells from northern Alaska, the Canadian Beaufort and the Chukchi Sea (Bujak 2007). The presence of repeated *Azolla* laminations in the central Arctic Ocean also indicates that the *Azolla* plants grew in situ on freshwater layers that repeatedly developed on the surface of the Arctic Ocean, rather than being transported from freshwater bodies on the surrounding land. Brinkhuis et al. (2006) and Bujak (2007) also suggested that the enormous quantities of *Azolla* inhabiting the Eocene Arctic Ocean for almost a million years may have triggered the initial shift from the Mesozoic greenhouse world towards our present icehouse state. According to their model, CO_2 absorption by the fern resulted in an abrupt reduction in this atmospheric gas with critical consequences in the climatic change and implications for the global biogeochemical cycles. This period is referred to as the *Azolla* event (Brinkhuis et al. 2006; Brouwer 2017; Bujak 2007; Carrapiço 2010a).

12.3 The Bougainville's Voyage: A Journey Through Utopia

In 1766, Louis-Antoine de Bougainville begins the political and scientific voyage around the world under the auspicious of the French King Louis XV (Taillemite 2006). This expedition involved two ships, the frigate *Boudeuse*, commanded by Bougainville and Nicolas-Pierre Duclos-Guyot (1722–1794) second in command, and the storeship *Étoile*, commanded by François Chenard de la Giraudais (1727–1776) (Carr 1968; Dunmore 2002; Williams 2013). There were 214 persons on board the *Boudeuse* and 116 on board the *Étoile* (Ridley 2010). Philibert Commerson and Jeanne Baret were travelling aboard the latter, him as royal botanist, naturalist and physician and her as his assistant or valet with knowledge in medicinal plants as it was suggested by Ridley (2010). Knapp (2011), however, considers Ridley's idea that Jeanne Baret was an "herb-woman" to be mere a speculation. Commerson was also a follower of the Genevan philosopher Jean-Jacques Rousseau (1712–1778) and disciple of the Swedish naturalist Carl Linnaeus (1707–1778), who in 1754 asked him to carry out research on marine plants, fishes and shells of the Mediterranean (Williams 2013).

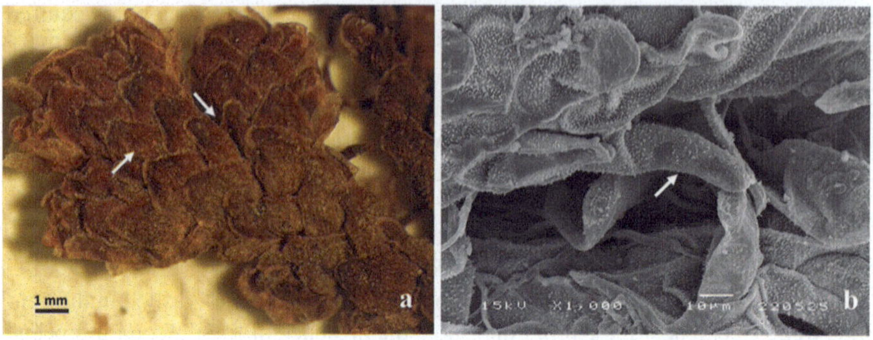

Fig. 12.1 *Azolla* samples from Commerson's herbarium, MNHN, Paris. Sporophyte of *Azolla* observed with stereo microscope and presenting the typical overlapping scale-like leaves (arrows) (**a**). Detail of unicellular papilla of dorsal leaf lobe (arrow) observed in SEM and characteristic of *A. filiculoides* (**b**)

The first ship to sail from France (Nantes) was the *Boudeuse* on 15 November 1766. However, but due to a storm in the Bay of Biscay, the ship had to return to France, to Brest, where she was repaired and returned to the sea on 6 December. From there, she sailed to South of America en route to the Falklands (Malvinas in Spanish or Malouines in French) (Bougainville 1771; Monnier et al. 1993). The *Étoile* was not ready to sail at that time and only left France (Rochefort) on 1 February 1767 (Monnier et al. 1993). The first mission of Bougainville was to deliver the Falklands to Spain, which happened on 1 April 1767 (Monnier et al. 1993; Taillemite 2006). After that, the *Boudeuse* sailed to Rio de Janeiro to meet the *Étoile*, which has arrived to this city on 13 June. A month later, on 15 July, both ships sailed to the Rio de la Plata, having arrived on 31 July to the city of Montevideo (Monnier et al. 1993; Taillemite 2006). They stayed there until November, visiting Buenos Aires too. During this period, several field trips were carried out, in which many biological samples were collected by Commerson and Baret (Williams 2013), including the first *Azolla* samples for scientific purposes. Some of this material prepared for herbarium was sporulated, which enabled us to identify its species using scanning electron microscopy (SEM) of the megaspore perine surface and also the leaf papillae. As it was expected, the *Azolla* species was *A. filiculoides* (Figs. 12.1 and 12.2).

We must mention that these field trips in the region of La Plata were really challenging for Commerson, who was hindered by a bad leg (Williams 2013), meaning that the success of the botanizing work was in great part due to Baret. In fact, François Vivez (1744–1828), surgeon of the *Étoile*, referring to Jeanne Baret's work in his journal, reinforces this idea saying that "she worked like a black. During our period of call at the River Plate, she went to collect plants in the plain, in the mountains two or three leagues away carrying a musket, a game-bag, food supply and paper for the plants" (Taillemite 2006; Williams 2013).

On 5 December 1767, the two ships began crossing the Strait of Magellan, which gives access to the Pacific Ocean. The crossing took almost 2 months, more precisely

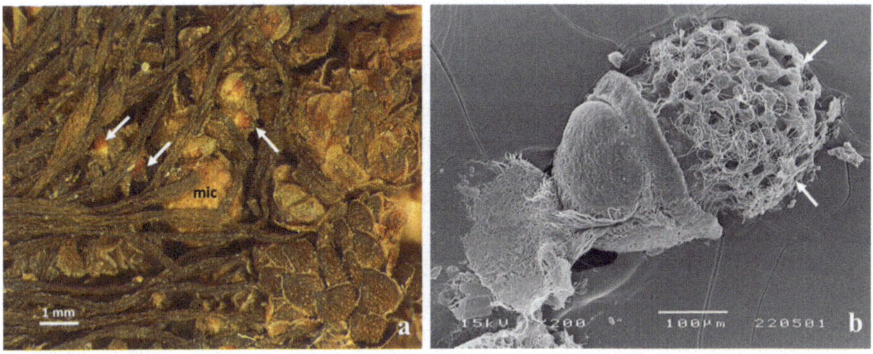

Fig. 12.2 *Azolla* samples from Commerson's herbarium, MNHN, Paris. Some of this material was sporulated. Megasporocarps (arrows) and microsporocarp (mic) observed with stereo microscope (**a**). Megasporocarp with a visible megaspore perine surface ornamentation (arrows) observed in SEM and characteristic of *A. filiculoides* (**b**)

52 days (Taillemite 2006) with a stormy and cold weather in unpleasant environmental conditions (Williams 2013). It is interesting to mention that the Portuguese navigator and explorer Fernão de Magalhães (1480–1521), at the service of the Spanish Kingdom, took 27 days to cross the Strait in 1520 but with much better weather conditions (Taillemite 2006). During the crossing, Baret and Commerson did several trips inland to collect biological specimens (Dunmore 2002; Olivier 1909; Williams 2013). François Vivez referring to Jeanne Baret in this period said "she spent entire days in the forest with snow, rain and ice, to seek plants, or along the seashore for shells. It remains for me to say in her praise that she generally surprised everyone by the work she did" (Dunmore 2002; Taillemite 2006).

After sailing through the Strait of Magellan, the two ships entered into the Pacific Ocean en route to the unknown Tahiti, where they arrived on 2 April 1768 (Taillemite 2006). This island was named "Nouvelle Cythère" by the Bougainville's French expedition, without knowing that in June of the previous year the British naval officer and explorer Samuel Wallis (1728–1795), commander of the *Dolphin*, declared Tahiti to be an English Crown territory on behalf of King George III (Dunmore 2002; Ridley 2010; Williams 2013). In fact, Wallis only returned to England in May 1768 which explains why Bougainville had no knowledge of this discovery (Dunmore 2002). The name "Nouvelle Cythère" was "inspired from the Greek island where Aphrodite, the goddess of love and beauty, was said to have been born" (Dunmore 2002). The island was also claimed as French territory (Williams 2013) and is currently part of the French Polynesia (French Overseas Country) following its annexation in 1880, when it was proclaimed a colony of France.

Many of the officers of the Bougainville's expedition were Rousseau's followers, including Bougainville himself as well as Commerson. For this reason, they had high expectations in relation to the inhabitants of the New Cythera island and the evolution of their society. The theory of the "noble savage", according to which man in its natural state was born essentially good, without guilty, was widely accepted in the eighteenth century and served as a key philosophical reference for these men. In earlier primitive societies, mankind would have been happier and less exploitative than in modern societies where corruption and oppression established the pattern for development, exploitation and poverty (Dunmore 2002; Williams 2013). According to personal journals of several members of the expedition, Tahiti seemed to correspond to a utopian and romantic world, where happiness and love were the main rules in society (Dunmore 2002; Ridley 2010; Williams 2013). The island's society, however, was not as romantic and peaceful as they thought (Williams 2013).

It was in these conditions that an incident happened with Jeanne Baret (while disguised as a man) when she was collecting samples onshore in Tahiti. It was Bougainville who described it in his journal. When Commerson and Baret were doing herborization work, a number of native men surround her yelling "Ayenne" (girl in local language) and attempted to carry her off, revealing that the Commerson's assistant was a woman (Williams 2013). Jeanne Baret travelled much of the trip disguised as a man until she was discovered on April 1768 (Dunmore 2002; Ridley 2010; Schiebinger 2003; Williams 2013). Although she has denied that Commerson knew her real identity, we know that they had lived together after Commerson's wife, Antoinette Vivante Beau, passed away on 19 April 1762, following the birth of a son 3 days earlier (Ridley 2010). In fact, Jeanne Baret was hired as housekeeper in 1764 when she was 5 months pregnant. When she was declaring her pregnancy (as required by law) on 22 August 1764, in Digoin outside Lyon, she refused to tell the name of the child's father (Ridley 2010; Schiebinger 2003, 2004). In September, they both moved to Paris, where in January 1765, a baby boy, called Jean-Pierre Baret, was born. The baby only lived a few months (Monnier et al. 1993; Ridley 2010; Schiebinger 2003, 2004).

After leaving the paradisiacal Tahiti, the French expedition visited several Pacific islands before arriving to Port Louis in the Isle de France (Mauritius) on November 1768. This island had become a French territory in 1715 and would remain so until 1810, when Great Britain took control of the island (Dunmore 2002). In 1768, the intendant was Pierre Poivre (1719–1786), a naturalist and friend of Commerson, who invited him to stay on the island to carry out botanical work (Dunmore 2002; Williams 2013). Having accepted the invitation, Commerson and Baret would conduct research, over the following 4 years, on the natural resources of Isle de France, Bourbon (Réunion) and Madagascar. They focused in particular on plants with pharmacological interest, as well as on spice plants whose trade was controlled by the Dutch. For this natural work, they were helped by two young illustrators provided by Pierre Poivre: Paul Philippe Sanguin de Jossigny (1750–1827) and Poivre's godson, Pierre Sonnerat (1748–1814) (Williams 2013).

In the meantime, Commerson's health had been deteriorating. In a letter dated 19 October 1772 and sent from the Isle of France to Paris, to his friend the astronomer Jérôme Lalande (1732–1807), he refers to these health problems in a prophetic way saying that "after an attack of rheumatic gout, which kept me in bed for nearly three months, I thought I was convalescent, when, in addition, dysentery attacked me; up to the present it was been incurable, and it was brought me to the very edge of the grave. My strength is almost utterly exhausted, and I am already more than half worn out. If the country air and diet of rice and fish do not cure me of this attack, you may well, as you once said (prophetically, no doubt), begin to work on the history of my martyrology". (Olivier 1909). He tried to obtain some relief from these health problems by moving to the East coast of the Isle of France, but on 13 March 1773 at 11:45 p.m., Commerson died at Le Flacq, at a friend's house, only a few weeks after his last botanical excursion and before the news that he had been elected member of the Académie des Sciences could reach him on 21 March 1773 (Dunmore 2002; Olivier 1909; Williams 2013). He was buried the next day in the Flacq cemetery, although his grave was never identified (Dunmore 2002).

12.4 The Winds of Liberty and the Emotions of the Natural World

As we have mentioned previously, Philibert Commerson and Jeanne Baret travelled in the same cabin on the *Étoile*. It was a very risky and delicate decision, but they were in a lovely relation not only with science (Knapp 2011; Ridley 2010; Schiebinger 2003, 2004). The sanitary conditions in both ships were very precarious, namely, inadequate for women (Carr 1968), which reveals a very brave and courageous attitude on the part of Jeanne Baret who decided to travel on the *Étoile* with Commerson and with an all-male crew. When the voyage took place, it was forbidden by royal naval ordinance for officers and sailors to have women on board, which led her to disguise herself as a man (Schiebinger 2003, 2004). The importance of her work to herbalize and to discover new botanical species has not been emphasized by many authors, but it was thanks to her work that many of the biological specimens have seen the daylight and reached France on May 1774 after the Commerson's death at Flacq in the Mauritius Island (Isle de France) on 13 March 1773 at the age of 45 (Monnier et al. 1993). Despite Philibert Commerson being a very good and determined botanist with a large expertise, it is well known that he was not a very organized person and was often ill or physically diminished during the voyage, namely, due to an injured leg made by a coach accident in Niort prior the journey and later gout crisis, which jeopardized part of his goals as naturalist (Bour 2015; Dunmore 2002; Williams 2013). However, Jeanne Baret, 13 years younger than Commerson, provided him with support and assistance during field trips, namely, by collecting and preparing biological specimens. As Bougainville referred in his 1771 historical book *Voyage autour du monde par la frégate du roi la Boudeuse et la flûte l'Étoile; en 1766,1767,1768&1769*, Jeanne Baret was "an expert botanist", and also Commerson "praised her for the countless plants she collected, the herbaria she constructed, the many collections of insects and shells she curated" (Schiebinger 2003). As a tribute to Baret's work, Commerson named one of the collected plants as *Baretia bonnafidia*, a flowering plant from Madagascar known for its "doubtful sexual characteristics" (Schiebinger 2004). However, Commerson never published it, and posteriorly this plant was renamed and included in the genus *Turraea* belonging to the family Meliaceae (Knapp 2011; Ridley 2010; Schiebinger 2004). It was only in 2012 that a definitive tribute to Jeanne Baret was achieved in recognition to the importance of her botanical work in science through the paper of Tepe et al. (2012), which named a new species of *Solanum* (*S. baretiae*) in her honour.

A considerable part of the biological material first collected in Brazil and La Plata failed to arrive to France (Olivier 1909; Williams 2013), leading Commerson to vow never to separate himself from his collections (Olivier 1909). After Commerson's death, several bureaucratic problems emerged between the Isle de France administration and the Commerson's biological material executors, which complicated the transfer of the collections to the Jardin du Roi in Paris (renamed Jardin des Plantes in 1793, after the 1789 revolution) (Williams 2013). Furthermore, many of the collections and manuscripts were not organized. Since it was decided to expedite the transfer of the material to France, it was not possible for the executors to catalogue it appropriately (Williams 2013). In November 1773, the boxes with the biological collections and his manuscripts were sent to the Jardin du Roi in Paris through the ship the *Victoire* which arrived at Lorient on May 1774 (Taillemite 2006). The number of boxes sent is still a matter for discussion. According to Christinat (1996), it seems there were 32 boxes, although the number indicated by the inventory letter sent by the Intendant Jacques Maillart-Dumesle (1730–1782) from Port Louis was 34. It was Paul de Jossigny, who has worked with Commerson as illustrator in the Isle de France, that travelled and escorted this material to Paris and delivered it to Georges-Louis Leclerc de Buffon (1707–1788) and Bernard de Jussieu (1699–1777) at the Cabinet d'Histoire Naturelle du Roi, including the drawings of the specimens collected by Commerson (Bour 2015; Williams 2013).

Using the new excellent database of the Muséum National d'Histoire Naturelle (MNHN) in Paris, we have found 3986 specimens of vascular plants collected by Philibert Commerson and, although not referred, also by Jeanne Baret in several locations during the Bougainville's voyage and also during his stay at Mauritius, Madagascar and Réunion (Le Bras et al. 2017). These plants sent to Paris were in most cases new for science. This material and other biological samples collected by Commerson and Baret were almost "cannibalized" by the scholarly world, namely, by the established French scientific community, though often without proper attributions (Williams 2013). I believe that a good example of it can be found in the well-known *Bougainvillea*, a flowering plant from Brazil and named by Commerson in honour of Bougainville in 1767. This plant, although with the name of *Buginvillaea*, and other 36 with their attributions were described by Antoine Laurent de Jussieu in 1789 in the *Genera Plantarum* (Jussieu 1789; Williams 2013) few days before the storming of the Bastille and the beginning of the French Revolution. In 1799, the German botanist Carl Ludwig Willdenow (1765–1812) published in the *Species Plantarum* the description of the species *Buginvillaea spectabilis* Willd. (currently *Bougainvillea spectabilis* Willd.) (Willdenow 1799) without any reference to Commerson. The same seems to have happened to the *Azolla* specimens collected by Philibert Commerson and Jeanne Baret, which were studied and published on by several scientists without reference to the original herborization work and research. Nevertheless, it is important to mention that Commerson never published his observations, as he has died before having the possibility to deepen his research. According to Georges Cuvier (1769–1832), "it is astounding that one man should have been able to do so much in so short of time and in a tropical climate" (Olivier 1909; Williams 2013). The collaborative partnership between Commerson and Baret was key in completing this task successfully (Knapp 2011).

What happened to Jeanne Baret after Commerson's death? We know that she became the owner of a small tavern at Port Louis (Isle de France), because she was fined for selling alcohol during the mass period in December 1773 (Dunmore 2002; Ridley 2010). After this reference, her name reappears in the official records in connection with her marriage to Jean Duberna (or Dubernat), a French officer, on 17 May 1774 in Port Louis (Dunmore 2002; Knapp 2011; Ridley 2010). The couple returned to France in early to mid 1775, settling down in Saint-Aulaye in Dordogne, some 80 km north of Bordeaux (Dunmore 2002). In 1785, she was granted a pension of 200 livres a year by the Ministry of Marine, as well as the official and public recognition of being an "extraordinary woman" for assisting Commerson's work during the voyage, for sharing with him the dangers of the mission and for showing great courage in her duty (Dunmore 2002). Although not mentioned, it is suspected that Bougainville was behind this decision (Dunmore 2002). On 5 August 1807, she died at home in the small village of Saint-Aulaye at the age of 67 (Dunmore 2002), leaving everything she possessed to Commerson's heirs, namely, his son Archambault (1762–1834) (Olivier 1909). She was the first woman to do the circumnavigation of the world (Ridley 2010).

12.5 Who Discovered *Azolla*?

The first reference and crude illustration of this plant in the literature was made by the French priest and naturalist Louis Feuillée (1660–1732), based on a Peruvian specimen. Published in 1725 in the *Journal des Observations Physiques, Mathematiques et Botaniques* under the name of *Muscus squamosas aquaticus elegantissimus*, the reference also included that the plant was used for improving the production of chicken eggs (Carrapiço 2010a, b, 2017; Evrard and Van Hove 2004; Feuillée 1725). The first samples for herbarium purposes of this aquatic pteridophyte were collected in 1767 by Philibert Commerson and Jeanne Baret in Montevideo and Buenos Aires region. After Commerson's death, this material was freely studied by several naturalists, among them were Marc Antoine Louis Claret de La Tourrette (1729–1793), Jean-Baptiste Lamarck, Antoine Laurent de Jussieu (1748–1836), Carl Ludwig Willdenow, Augustin Pyramus de Candolle (1778–1841) and Antoine Laurent Apollinaire Fée (1789–1874). Although Lamarck describes in 1783, in the *Encyclopédie Méthodique*, the species *Azolla filiculoides* referring that the plant was brought from Magellan (Carrapiço 2010a; Lamarck 1783), we believe this was an error. This assumption is based on the information in Bougainville's journal, which makes reference to these places, dates and existing samples at the

Fig. 12.3 Sheets from Commerson's herbarium featuring *Azolla* samples stemming from several botanical institutions: Muséum National d'Histoire Naturelle de Paris (**a**), Conservatoire et Jardin botaniques de la Ville de Genève (**b**), Jardin botanique de Lyon (**c**), Institut de Botanique de l'Université de Montpellier (**d**)

Herbarium of the Jardin botanique de Lyon, the Herbarium of the Muséum National d'Histoire Naturelle de Paris, the Herbarium of the Institut de Botanique de l'Université de Montpellier and the Herbarium of the Conservatoire et Jardin botaniques de la Ville de Genève (Fig. 12.3). An interesting note can be found in the Antoine Laurent de Jussieu's herbarium sheet existing at the Muséum National d'Histoire Naturelle of Paris (MNHN-P-P00668168) and probably written by the hand of this botanist, indicating that "M. Lamarck l'a annoncé du detroit de Magellan, par erreur" and referring to these samples from Buenos Aires and Montevideo and were included in Commerson herbarium (Fig. 12.4). It is also important to mention that the environmental conditions of the Strait of Magellan were not adequate to the biology, development and ecology of *A. filiculoides* (Roberto Rodríguez-Rios 2010, personal communication).

The *Azolla* ecological and biological description made by Lamarck was only possible if he had the opportunity to observe the live plant. This was, however, not the case, as *Azoll*a had arrived in dried form on a herbarium sheet, and this fern did not exist in Europe at the time (Lamarck 1783; Marsh 1914). In fact, *Azolla* was only introduced in Europe in 1872 in several botanic gardens, from which it escaped into ditches and ponds (Marsh 1914). The solution to this Lamarck's enigma can probably be found in the information included by Philibert Commerson with the collected *Azolla* and its ecology, which was observed by him and by Jeanne Baret in loco (Carrapiço 2010a). The published description made by Lamarck indicates that *Azolla* "C'est une petite plante aquatique, qui paroît flotter à la surface des eaux à la manière des Lenticules, (Lemna) avec lesquelles elle semble avoir beaucoup de

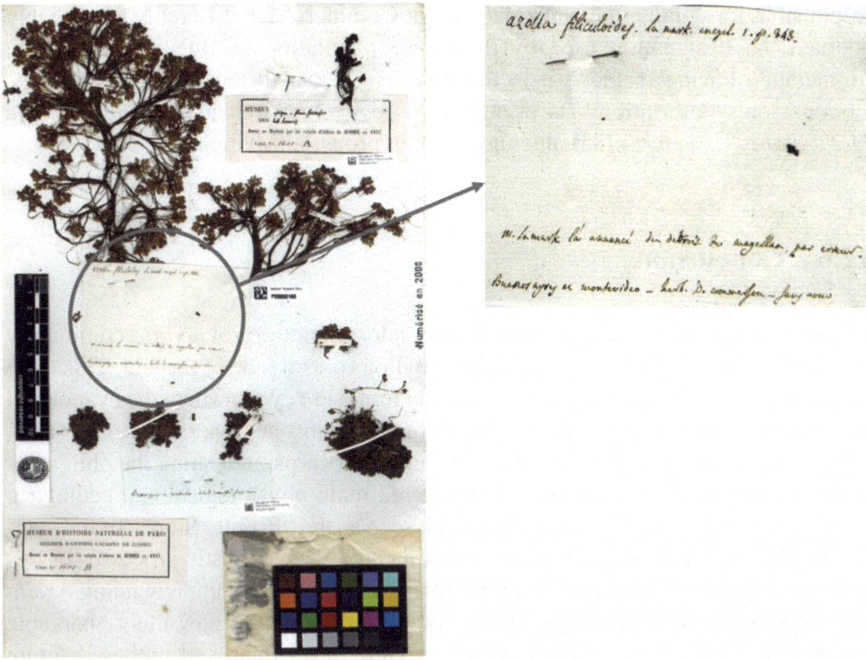

Fig. 12.4 Antoine Laurent de Jussieu's herbarium sheet with a note referring to Lamarck's error, Muséum National d'Histoire Naturelle of Paris (MNHN-P-P00668168). Probably handwritten by Jussieu, indicating that "M. Lamarck l'a annoncé du detroit de Magellan, par erreur"

rapports, et qui a néanmoins l'aspect d'une très petite fougère" and "cette plante a eté rapportée de Magellan par M. de Commerson" (Carrapiço 2010a; Lamarck 1783). This error made by Lamarck related to the *Azolla* location had consequences for the fern systematics. As an example, Carl Ludwig Willdenow studied the *Azolla* samples from the Commerson herbarium and published in 1810 in *Species Plantarum* that there was a new species of this fern, named *Azolla magellanica* Willd. (Willdenow 1810) (meaning *Azolla* from the Strait of Magellan). This species was in fact *A. filiculoides*. It was also in this publication that *Azolla caroliniana* was described for the first time based on material derived from Richard (Louis Claude Richard, 1754–1821) in Paris, and probably collected by Michaux (André Michaux, 1746–1802) in the Southeastern United States (Svenson 1944), referring to the geographic location of the plant collected.

Regarding the origin of the name *Azoll*a, we know that Lamarck described *A. filiculoides* without referring to it. Several authors have suggested that the genus is a result of two words of Greek origin: *azo*, to dry, and *ollyo*, to kill, alluding to death from drought (Gledhill 2008; Lumpkin and Plucknett 1980; Moore 1969; Stearn 1992). This interpretation seems contrived, and we believe that the name may correspond to a native word for the plant (Dyce 1988, edited and expanded by John Edgington in 2016; Carrapiço 2010a). We do not know for certain who was responsible for coining the name *Azolla*, but Commerson and Baret have probably adopted the designation used by the native population in Buenos Aires or in Montevideo having included it in the description sent to the Jardin des Plantes. Commerson's description was eventually used by Lamarck in the *Encyclopédie Méthodique* (Lamarck 1783) although without proper attributions.

12.6 Conclusion

Azolla is an aquatic fern that presents an endosymbiotic prokaryotic community (cyanobacteria and bacteria) in the chlorophyllous dorsal lobe of the leaf. This symbiotic association is sustained throughout the fern's life cycle, where the cyanobiont and bacteria are always present and transferred from one generation to the next, either in the dorsal leaf cavities or in the megasporocarps, indicating the obligatory nature of the symbiosis. Given all the referred main physiological and ecological characteristics, the *Azolla* leaf cavity can be seen as the "Rosetta Stone" of the symbiosis, allowing us to understand the dynamic and the synergistic processes of this symbiotic system (Carrapiço 2017). It represents the key to comprehending *Azolla* as a superorganism (Carrapiço 2010a) and to understanding how this remarkable and unique fern can be "domesticated" in view of an improved biobased future economy (Brouwer et al. 2014).

Azolla did not exist in Europe when the Bougainville's voyage around the world took place in the eighteenth century, from 1766 to 1769. It was in Montevideo and Buenos Aires, and not in the Strait of Magellan, that this plant was collected for the first time for science purposes and prepared for herbarium by Philibert Commerson and Jeanne Baret. Following Commerson's death in the Isle de France (Mauritius) in 1773, his herbarium and manuscripts, including all biological specimens collected by himself and Baret, were shipped to the Jardin du Roi in Paris and distributed among several scientists affiliated to a number of scientific institutions. In 1783, Lamarck describes in the *Encyclopédie Méthodique* the species *Azolla filiculoides* referring that the plant was brought from Magellan by Philibert Commerson (Carrapiço 2010a; Lamarck 1783). This reference, however, was an error related to the geographic location of the biological material collected by Commerson.

Acknowledgements The author is grateful and expresses his sincere thanks to Caroline Loup from the Herbarium of the Institut de Botanique de l'Université de Montpellier (MPU); Frédèric Danet from the Herbarium of the Jardin botanique de Lyon (LYJB); Christine Habashi and Isabella Valette from the Herbarium of the Conservatoire et Jardin botaniques (CJB) de la Ville de Genève; and Cécile Aupic, Gérard Aymonin and Germinal Rouhan from the Herbarium of the Muséum National d'Histoire Naturelle de Paris (MNHN) for the facilities to study the historical *Azolla* samples existing in their institutions and also for the information sent on the issue. A special thanks to Jeannine Monnier for the information on Philibert Commerson's life and work; to Jonathan Bujak (The *Azolla* Foundation) for the discussion and comments on the *Azolla* fossil data; to Roberto Rodríguez-Rios, Universidad de Concepción, Chile, for the discussion on the ecological conditions on the Strait of Magellan related to *Azolla*; to Mariano Montaño from SCIAPLI,

Ecuador, regarding the use of *Azolla* in South America; to Pedro Costa of the Faculty of Sciences and Technology of the New University of Lisbon on some of the light microscope observations; to Telmo Nunes for the support in the SEM facilities and observations at the Centre for Ecology, Evolution and Environmental Changes, Department of Plant Biology, Faculty of Sciences, University of Lisbon; to Ana Luisa Pereira from the University of Coimbra on the research developed on the biology of *Azolla* during her PhD thesis. And at last but not least, a particular thanks to Maria Helena Costa for the comments and suggestions on the text and to Helena Farrand-Carrapiço for the English revision of the manuscript.

References

Adams DG, Bergman B, Nierzwicki-Bauer SA, Duggan PS, Rai AN, Schussler A (2013) In: Rosenberg E et al (eds) The prokaryotes – prokaryotic biology and symbiotic associations. Springer, Berlin/Heiderberg, pp 359–400

Bougainville LA (1771) Voyage autour du monde par la frégate du roi la Boudeuse et la flûte l'Étoile; en 1766, 1767, 1768 &1769. Saillant & Nyon, Paris

Bour R (2015) Paul Philipe Sanguin de Jossigny (1750–1827), artiste de Philibert Commerson. Les dessins de reptiles de Madagascar, de Rodrigues et des Seychelles. Zoosystema 37(3):415–448

Braun-Howland EB, Nierzwicki-Bauer SA (1990) *Azolla-Anabaena* symbiosis: biochemistry, physiology, ultrastructure, and molecular biology. In: Rai AN (ed) CRC handbook of symbiotic cyanobacteria. CRC Press, Boca Raton, pp 65–117

Brinkhuis et al (2006) Episodic fresh surface waters in the Eocene Arctic Ocean. Nature 441:606–609

Brouwer P (2017) Turning the aquatic weed *Azolla* into a sustainable crop. PhD Dissertation, University of Utrecht, The Netherlands

Brouwer P, Brautigam A, Kulahoglu C, Tazelaar AOE, Kurz S, Nierop KGJ, van der Werf A, Weber APM, Schluepmann H (2014) *Azolla* domestication towards a biobased economy? New Phytol 202:1069–1082

Bujak J (2007) The *Azolla* story. Climate change and Arctic hydrocarbons. GEO ExPro September 4(4):66–72

Bujak A, Bujak J (2014) The Arctic *Azolla* event. Geoscientist 24:10–15

Caiola MG, Forni C (1999) The hard life of prokaryotes in the leaf cavities of Azolla. In: Seckbach J (ed) Enigmatic microorganisms and life in extreme environments. Kluwer Academic Press, Dordrecht, pp 629–639

Carr A (1968) L'Expédition de Bougainville et l'hygiène navale de son temps. Journal de la Société des Océanistes. Tome 24:63–75. doi: https://doi.org/10.3406/jso.1968.2232

Carrapiço F (1991) Are bacteria the third partner of the *Azolla-Anabaena* symbiosis? Plant Soil 137:157–160

Carrapiço F (2002) The *Azolla-Anabaena*-bacteria system as a natural microcosm. Proc SPIE 4495:261–265

Carrapiço F (2006) Is the *Azolla-Anabaena* symbiosis a co-evolution case? In: Sitnykov A (ed) General Botany: traditions and perspectives. Materials of the international conference, dedicated to 200th anniversary of the Kazan Botanical School (January 23–27, 2006). Part I Kazan University, 2006, pp 193–195

Carrapiço F (2010a) *Azolla* as a superorganism. Its implication in symbiotic studies. In: Seckbach J, Grube M (eds) Symbioses and stress: joints ventures in biology, cellular origin, life in extreme habitats and astrobiology 17. Springer Science, Dordrecht, pp 225–241

Carrapiço F (2010b) How symbiogenic is evolution? Theory Biosci 129(2–3):135–139

Carrapiço F (2015) Can we understand evolution without symbiogenesis? In: Gontier N (ed) Reticulate evolution, vol 3. Springer International Publishing, Cham, pp 81–105

Carrapiço F (2017) The *Azolla-Anabaena*-bacteria association: a case of symbiotic abduction? In: Muggia L, Seckbach J, Grube M (eds) Algae and cyanobacteria symbioses. World Scientific Publishing Company, pp 329–345, New Jersey, ISBN 978-1-78634-057-3

Carrapiço F, Tavares R (1989) New data on the *Azolla–Anabaena* symbiosis. I. Morphological and histochemical aspects. In: Skinner FA, Boddey RM, Fendrik I (eds) Nitrogen fixation with non-legumes. Kluwer Academic Press, Dordrecht, pp 89–94

Christinat C (1996) Une femme globe-trotter avec Bougainville: Jeanne Baret (1740–1807). Revue Française d'Histoire d'Outre-Mer Tome 83(310):83–95

Corning PA (1983) The synergism hypothesis: a theory of progressive evolution. MacGraw-Hill, New York

Corning PA (2005) Holistic darwinism. Synergy, cybernetics, and the bioeconomics of evolution. The University of Chicago Press, Chicago

Costa ML, Santos MCR, Carrapiço F (1999) Biomass characterization of *Azolla filiculoides* grown in natural ecosystems and wastewater. Hydrobiologia 415:323–327

Costa ML, Santos MCR, Carrapiço F, Pereira AL (2009) *Azolla-Anabaena*'s behaviour in urban wastewater and artificial media – influenced of combined nitrogen. Warer Research 43:3743–3750

De Vries J, Fischer AM, Roettger M, Rommel S, Schluepmann H, Brautigam A, Carlsbecker A, Gould SB (2016) Cytokinin-induced promotion of root meristem size in the fern *Azolla* supports a shoot-like origin of euphyllophyte roots. New Phytol 209:705–720. https://doi.org/10.1111/nph.13630

Dunmore J (2002) Monsieur Baret. First woman around the world. 1766–1768. Heritage Press, Auckland

Dyce JW (1988) Fern names and their meanings. A glossary for the fern grower and collector. British Pteridological Society, London. 2nd edition, 2016. Edited and expanded by John Edgington

Evrard C, Van Hove C (2004) Taxonomy of the American *Azolla* species (Azollaceae): a critical review. Syst Geogr Pl 74:301–318

Feuillée L (1725) Journal des observations physiques, mathematiques et botaniques, faites par ordre du roi sur les côtes orientales de l'Amerique Meridionale, & aux Indes Occidentales et dans un autre voyage faite par le même ordre à la nouvelle Espagne, & aux Illes de l'Amerique. Jean Mariette, Paris

Forni C, Chen J, Tancioni L, Caiola MG (2001) Evaluation of fern *Azolla* for growth, nitrogen and phosphorus removal from wastewater. Water Res 35(6):1592–1598

Gledhill D (2008) The names of plants, 4th edn. Cambridge University Press, Cambridge

Jussieu AL (1789) Genera plantarum: secundum ordines naturales disposita, juxta methodum in Horto regio parisiensi exaratam, anno M.DCC.LXXIV. Apud viduam Herissant et Theophilum Barrois, Paris

Knapp S (2011) The plantswoman who dressed as a boy. Nature 40:36–37

Lamarck J-B (1783) *Azolla filiculoides*. Encyclopédie Méthodique: Botanique. Tome 1: 343, Paris

Le Bras G, Pignal M, Jeanson ML, Muller S, Aupic C, Carré B, Flament G, Gaudeul M, Gonçalves C, Invernón VR, Jabbour F, Lerat E, Lowry PP, Offroy B, Pimparé EP, Poncy O, Rouhan G, Haevermans T (2017) Data descriptor: the French Muséum national d'histoire naturelle vascular plant herbarium collection dataset. Scientific Data 4:170016. https://doi.org/10.1038/sdata.2017.16

Lumpkin TA, Plucknett DL (1980) Azolla: botany, physiology, and use as a green manure. Econ Bot 34(2):111–153

Marsh AS (1914) History of the occurrence of *Azolla* in the British Isles and in Europe generally. Proc Cambridge Phil Soc 17:383–386

Mettenius GH (1847) Uber *Azolla*. Linnea 20:259–282

Monnier J, Lavondes A, Jolinon J-C, Elouard P (1993) Philibert Commerson. Le découvreur du bougainvillier. In: Association Saint-Guignefort (ed). Châtillon-sur-Chalaronne, France

Moore AW (1969) *Azolla*: biology and agronomic significance. Bot Rev 35:17–34

Olivier SP (1909) The life of Philibert Commerson, D.M. naturaliste du roi. An old-world story of French travel and science in the days of Linnaeus. GF Scott Elliot (ed) John Murray, London
Pereira AL, Carrapiço F (2007) Histochemistry of simple hairs from the foliar cavities of *Azolla filiculoides*. Plant Biosys 141:323–328
Pereira AL, Martins M, Oliveira MM, Carrapiço F (2011) Morphological and genetic diversity of the family Azollaceae inferred from vegetative characters and RAPD markers. Plant Syst Evol 297:213–222. https://doi.org/10.1007/s00606-011-0509-0
Perkins SK, Peters GA (1993) The *Azolla–Anabaena* symbiosis: endophyte continuity in the *Azolla* life–cycle is facilitated by epidermal trichomes. I. Partitioning of the endophytic *Anabaena* into developing sporocarps. New Phytol 123:53–64
Qiu YL, Yu J (2003) *Azolla* – a model organism for plant genomic studies. Geno Prot Bioinfo 1(1):15–25
Ran LA, Larsson J, Vigil-Stenman T, Nylander JAA, Ininbergs K, Zheng WW, Lapidus A, Lowry S, Haselkorn R, Bergman B (2010) Genome erosion in a nitrogen-fixing vertically transmitted endosymbiotic multicellular cyanobacterium. PLoS One 5(7):e11486
Ridley G (2010) The discovery of Jeanne Baret. A story of science, the high seas, and the first woman to circumnavigate the globe. Crown Publishers, New York
Schiebinger L (2003) Jeanne Baret: the first woman to circumnavigate the globe. Endeavour 27(1):22–25
Schiebinger L (2004) Plants and empire. Colonial bioprospecting in the atlantic world. Harvard University Press, Cambridge
Schneller JJ (1990) Azollaceae. In: Kramer KU, Green PS (eds) The families and genera of vascular plants. Springer-Verlag, Berlin, pp 57–60
Sela M, Garty J, Tel-Or E (1989) The accumulation and the effect of heavy metals on the water fern *Azolla filiculoides*. New Phytol 112:7–12
Shi DJ, Hall DO (1988) *Azolla* and immobilized cyanobacteria (blue-green algae): from traditional agriculture to biotechnology. Plants Today 1:5–12
Stearn WT (1992) Stearn's dictionary of plants names for gardeners. Cassel, London
Svenson HK (1944) The new world species of *Azolla*. Am Fern J 34:69–84
Taillemite E (2006) Bougainville et ses compagnons autour du monde. 1766–1769. Imprimerie Nationale Editions, Paris. ISBN 2-7427-6271-x
Tepe EJ, Ridley G, Bohs L (2012) A new species of Solanum named for Jeanne Baret, an overlooked contributor to the history of botany. PhytoKeys 8:37–47
Uheda E (1990) Isolation of hair cells from *Azolla filiculoides var. japonica* leaves. Plant Cell Physiol 27(7):1255–1261
Veys P, Waterkeyn L, Lejeune A, Van Hove C (1999) The pore of the leaf cavity of *Azolla*: morphology, cytochemistry and possible functions. Symbiosis 27:33–37
Wettstein R (1903) Azollaceae Wettst. Handb. Syst Bot 2:77
Willdenow CL (1799) *Buginvillaea spectabilis* Willd. Species Plantarum. Editio Quarta 2(1):348
Willdenow CL (1810) *Azolla magellanica* Willd. Species Plantarum. Editio Quarta. Tome 5:541.
Williams G (2013) Naturalists at sea. Scientific travellers from Dampier to Darwin. Yale University Press, London

Part III
Ferns as Genetic and Metabolic Resources

Chapter 13
The Power of Gametophyte Transformation

Linh Thuy Bui, Holly Long, Erin E. Irish, Angela R. Cordle, and Chi-Lien Cheng

Abbreviations

dsRNA	Double-stranded RNA
gRNA	Guide RNA
GUS	β-Glucuronidase
HR	Homologous recombination
RNAi	RNA interference
siRNA	Small interfering RNA
T_0	Primary transformant of gametophyte or sporophyte
T_1	Generation produced by selfing T_0
TALEN-mediated genome editing	Transcription activator-like effector nuclease-mediated genome editing

13.1 Introduction

Transgenesis is an essential tool to study gene function, and a simple, efficient DNA-delivery system is a criterion for any model organism. Ferns, the most diverse group of non-seed vascular plants, are sisters to the seed plants (Raubeson and Jansen 1992; Stevenson and Loconte 1996; Kenrick and Crane 1997; Pryer et al.

L. T. Bui
Department of Biology, The University of Iowa, Iowa City, IA, USA

Department of Biology, Indiana University, Bloomington, IN, USA

H. Long · E. E. Irish · A. R. Cordle · C. -L. Cheng (✉)
Department of Biology, The University of Iowa, Iowa City, IA, USA
e-mail: chi-lien-cheng@uiowa.edu

H. Fernández (ed.), *Current Advances in Fern Research*,
https://doi.org/10.1007/978-3-319-75103-0_13

2001). The phylogenic position of ferns provides an excellent opportunity for investigating fundamental questions in land plant evolution such as those concerning alternation of generations, invention of seeds, and meristem organization. The homosporous fern *Ceratopteris richardii* has been championed as a model organism for over 20 years (Hickok et al. 1995; Chatterjee and Roux 2000; Leroux et al. 2013) but only recently have transgenesis methods been developed successfully (Muthukumar et al. 2013; Plackett et al. 2014; Bui et al. 2015) in the fern. In this chapter, we will review methods of transgenesis developed in *C. richardii* and other non-seed land plant models and then focus on gametophyte transformation and its deployment for the study of asexual alternation of generations in the fern.

13.2 Transgenesis in the Moss Model

The moss *Physcomitrella patens* is a well-developed land plant model. A rare feature of *P. patens*, and possibly mosses in general (Trouiller et al. 2007), among multicellular organisms is its preference for using homologous recombination to repair DNA double-strand breaks discovered during the development of gene-targeting procedures (Schaefer and Zyrd 1997; Strepp et al. 1998) and later verified (Kamisugi et al. 2006). This feature has been exploited further to develop highly efficient targeted gene deletion and replacement procedures (Hohe et al. 2004). Although highly efficient, gene replacement in *P. patens* using linear DNA relying on the natural capacity of homologous recombination repair proves to be difficult, often with complex rearrangements at the target site (Wendeler et al. 2015). Recently, CRISPR-Cas9-mediated gene targeting has been developed in *P. patens*. With a single DNA double-strand break created at the target site, both single-copy integration and precision were increased compared with the conventional method (Collonnier et al. 2017). Multiplex gene editing has also been achieved using CRISPR-Cas9-mediated gene targeting in *P. patens* (Lopez-Obando et al. 2016). In addition, an RNAi-based method has been developed for gene suppression (Bezanilla et al. 2003, 2005) and used successfully in knocking down the expression of single (Nakaoka et al. 2012) or multiple (Vidali et al. 2009; Burkart et al. 2015) genes. In this system, the RNAi construct is integrated through homologous recombination into a locus that shows no phenotypic consequences when disrupted.

The dominant generation of *P. patens* is the gametophyte generation; haploid cells are preferable recipients in transgenesis. Ever since the first demonstration of successful stable transformation of *P. patens* (Schaefer et al. 1991), protonemata, the juvenile phase of gametophytes, have been used as the source of protoplasts, and PEG-mediated DNA transfer has been the commonly used DNA delivery system (Hohe et al. 2004, and see any recent publications such as Horst et al. (2016), Li et al. (2017)). Another method for DNA delivery is particle bombardment, which has been shown to stably integrate DNA in both random and targeted fashions, depending upon the vectors designed, into the moss genome (Cho et al. 1999;

Šmídková et al. 2010), but heritability has not been tested. An advantage of this system is that intact protonemata can be used, bypassing protoplast preparation and subsequent regeneration.

13.3 Transgenesis in the Liverwort Model

The liverwort *Marchantia polymorpha* is another well-developed model organism in the nonvascular lineages of land plants (Ishizaki et al. 2016). Like *P. patens*, the gametophyte of *M. polymorpha* is the dominant, long-lived generation. The haploid *M. polymorpha* generation is dioicous, conferred by heteromorphic sex chromosomes, X or Y. A *M. polymorpha* spore will inherit either an X or Y chromosome and upon germination develops into a thallus with archegonia or antheridia, respectively. In addition to reproducing sexually, a mature thallus can multiply through vegetative propagation by forming clustered gemmae on the dorsal side of the thallus. Each gemma will develop into a free-living, individual thallus. Both spore-derived and gemma-derived immature thalli have been used, without the need for making protoplasts, as the plant source for transgenesis experiments. The ability to propagate thalli vegetatively through gemmae provides a resource of plant material with pure genetic background for transformation and, after transformation, allows clonal maintenance of transgenic lines indefinitely. In addition, because each gemma originates from a single cell, vegetative propagation can avoid potential chimerism of a transgenic thallus by going through just one round of gemma formation (Ishizaki et al. 2016).

Particle bombardment using either spore-derived (Chiyoda et al. 2008) or gemma-derived (Takenaka et al. 2000; Kajikawa et al. 2008) young thalli was successful in generating random insertions for gene function analyses. *Agrobacterium*-mediated transformation of a non-seed land plant was first made successful in *M. polymorpha* in 2008 (Ishizaki et al. 2008) and has become the method of choice for DNA delivery in that species. Both gemma-derived and spore-derived young thalli were used for transformation, and the latter grown in liquid media was used to generate a large number of T-DNA tagged lines for mutant screening (Ishizaki et al., 2008; Ishizaki et al. 2013; Honkanen et al. 2016).

Gene targeting has been made possible by using two approaches, natural homologous recombination and genome editing to knock out genes. At first, a positive/negative double-selection procedure previously used in rice gene targeting was used in *M. polymorpha*; HR products reached 2% of the surviving transformants (Ishizaki et al. 2013), and knockout mutants were successfully produced (Shimakawa et al. 2017). CRISPR-Cas9-mediated genome editing was used successfully targeting an auxin response gene (Sugano et al. 2014). In this system, an endogenous U6 promoter was used to drive the gRNA. In another design, strategies of selecting high Cas9 expressers and shorter gRNA proved to increase the efficiency and avoid off-target effects (Osakabe et al. 2016). TALEN-mediated genome editing was also used to generate targeted mutations in *M. polymorpha* (Kopischke et al. 2017).

13.4 Transient and Stable Transformation in the Fern Model

Unlike moss and liverwort species whose dominant generation is the gametophyte, the dominant generation in ferns, as in seed plants, is the sporophyte. In contrast to seed plants, where gametophytes are retained by and depend upon the sporophytes for growth and development, the spores of ferns are shed from the sporophytes and develop into free-living gametophytes. This life cycle, as that of the moss and liverwort, offers an opportunity to use gametophytes as the target for transgenesis.

Transient suppression of gene expression by RNAi was made possible by direct uptake of dsRNA into germinating spores of *Marsilea vestita* (Klink and Wolniak 2001) and *C. richardii* (Stout et al. 2003) and by particle bombardment of DNA constructs into gametophytes of *Adiantum capillus-veneris* (Kawai-Toyooka et al. 2004) and *C. richardii* (Rutherford et al. 2004). In an elegant system, young hermaphroditic gametophytes of *C. richardii* were co-bombarded with an RNAi construct using the partial sequence of a *C. richardii* gene encoding protoporphyrin IX magnesium chelatase (CrChlI) under the control of P_{35S} plus a construct of P_{35S}-driven *GUS* reporter. The CrChlI protein is required for chlorophyll biosynthesis; without it, the gametophyte becomes colorless. RNAi silencing in plants such as *Arabidopsis* is systemic due to amplification of siRNA by RNA-dependent RNA polymerases (see a review by Willmann et al. 2011). Presumably, similar mechanisms act in the fern, as all the cells, including those formed after bombardment, were colorless. In contrast, GUS expression was restricted to cells that received the DNA constructs. The chlorophyll synthesis suppression was carried into the sporophyte after gametogenesis and fertilization, but the degree of this epigenetic inheritance was variable. Because the white sporophytes eventually died, further genetic inheritance of the suppression could not be followed. A similar system was successfully applied to study the function of an arsenite transporter in the arsenic-tolerant fern *Pteris vittata* by RNAi suppression (Indriolo et al. 2010). More recently, *Agrobacterium*-mediated transient transformation in *C. richardii* was also demonstrated (Bui et al. 2015).

Transgenesis of ferns was first demonstrated in *P. vittata* and *C. thalictroides* (Muthukumar et al. 2013). In this report, spores were chosen as the plant material for *Agrobacterium*-mediated transformation, while spores, 5-day-old germinating spores, and 15-day-old gametophytes were used in the particle bombardment transformation. The results for *Agrobacterium*-mediated transformation are encouraging because GUS reporter expression persisted to the T_2 sporophytes of two independent transgenic lines. However, this system lacks an effective method to select for transformants, which makes it difficult to determine the transformation efficiency and to evaluate the efficacy of the method. As for the particle-bombarded tissues, GUS expression was detected in the whole young sporophytes when spores and 5-day-old germinating spores were used, and, as expected, patterns of GUS expression suggested chimerism when 15-day-old gametophytes were bombarded. The stability of the transgene is not known as these bombarded, GUS-expressing gametophytes were not followed through into the sporophyte generation.

A tractable transgenesis system using particle bombardment was established in *C. richardii* and *C. thalictroides* (Plackett et al. 2014) and an updated protocol soon followed (Plackett et al. 2015). This protocol uses sporophytes, not gametophytes, as the recipient for transgenesis. Young sporophytes developed from cultured gametophytes are induced to produce callus tissue, excised, subcultured, and then bombarded with a DNA construct carrying P_{35S} driving a *GUS* reporter and P_{NOS} driving *HPH* for hygromycin selection. The bombarded calli are selected and T_0 sporophytes regenerated on hygromycin before being transplanted to soil. This well-documented protocol produces T_0 sporophytes with high efficiency, and the transgenes are stably integrated with the majority persisting into the T_1 sporophytes. The necessity of callus induction followed by sporophyte regeneration is a major drawback of this transgenesis method. Regeneration from callus often results in chimeric T_0 sporophytes, necessitating a delay on direct genefunction evaluation until the next generation. In addition, the inherent problem with particle bombardment of multiple insertions of the transgene was also true in this case.

13.5 *Agrobacterium*-Mediated Transformation of *C. richardii* Gametophytes

The preferred method to generate transgenic plants is the simple, floral-dip *in-planta* method first developed in *Arabidopsis* (Bechtold et al. 1993). The procedure involves dipping a bolting *Arabidopsis* plant into an *Agrobacterium* culture carrying the desired constructs, allowing the plant to set seed, then selecting for transformed seeds. Typically, 0.5% to 3% of seeds are transformed. Importantly, in this system, the target of the *Agrobacteria* is predominantly the female gametophyte (Ye et al. 1999; Bechtold et al. 2000; Desfeux et al. 2000). Two important lessons can be learned from this now classical method of transformation: one is to avoid tissue culture if at all possible, and the other is the susceptibility of female gametophytes to *Agrobacterium* infection. With those points in mind and encouraged by the successful and efficient transformation of the non-seed plant *M. polymorpha* by *Agrobacteria* (Ishizaki et al. 2008) and by a patent application published online for fern spore transformation (the work was later published by Muthukumar et al. 2013), we set out to develop an *Agrobacterium*-mediated gametophyte transformation system for *C. richardii*.

From our previous studies of asexual alternation of generations by induced apogamy and apospory in *C. richardii*, we learned that although the fern is readily coaxed into asexual developmental pathways, both sporophytes and gametophytes also have the propensity to regenerate (Cordle et al. 2007; Bui et al. 2012). With the knowledge of the gametophyte's capacity to regenerate, we were further convinced that gametophytes are ideal for transformation. After failed attempts at transforming intact gametophytes with *Agrobacteria*, we decided to partially digest the cell wall of gametophytes before co-cultivation to facilitate direct con-

Fig. 13.1 Stable transformation of *C. richardii* gametophytes. Two weeks after co-incubation with *Agrobacteria*, transformed gametophytic cells remained green and survived on selective media (**b**, arrows), while the untransformed gametophytes failed to grow (**a**). A single (**c**) or multiple (**d**, orange arrows) regenerated hermaphroditic gametophytes may arise on one gametophyte. A male gametophyte developed from transformed cells of the hermaphroditic gametophytes (**d**, white arrow). The young T_1 sporophytes growing in the soil after 1 week (**e**) or 2 months (**f**). Bars in (**a**), (**b**), and (**d**) = 1 mm; bar in (**c**) = 0.5 mm; and bars in (**e**) and (**f**) = 2 cm (Adapted from Bui et al. 2015)

tact of *Agrobacteria* with fern cells. This strategy avoided the care needed in handling protoplasts and allowed transformed gametophytes to quickly regenerate. This protocol is simple to perform, is reproducible, and has an efficiency of around 5% (number of transformed gametophytes divided by total hermaphrodites used in co-cultivation). The regenerated T_0 hermaphrodites are moved to individual fertilization wells for phenotype observation and for selfing or crossing to another line. After the T_1 sporophytes have produced a few roots, they are transplanted to soil (Fig. 13.1). It takes approximately 12–15 weeks between co-cultivation and transplanting to the soil. Segregation analysis of the T_1 spores from individual lines showed that 50% of the lines contain single-site insertions. Detailed procedures and results are described in Bui et al. (2015) where a slightly different protocol is also described for transient transformation (Fig. 13.2), which can be useful for a quick test of whether a promoter is expressed in the gametophytes, among other applications.

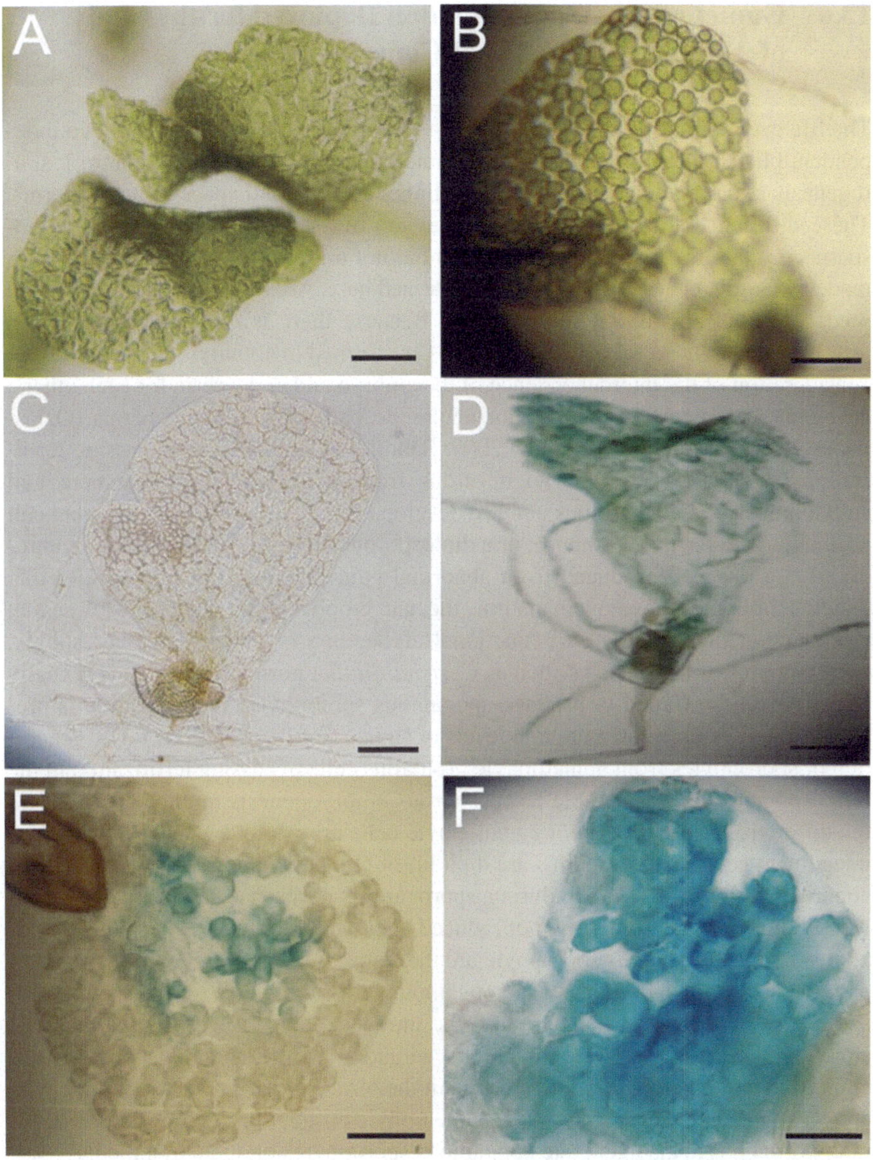

Fig. 13.2 Enzyme treatments facilitate transient transformation. A 12-day-old gametophytes treated with 1.5% (w/v) cellulase (**a**) or with 1.5% (w/v) cellulase and 0.5% (w/v) macerozyme (**b**) for 2 h. Histological GUS analysis of transiently transformed gametophytes (**d**, **e**, and **f**) treated with 1.5% (w/v) cellulase and 0.5% (w/v) macerozyme as in (**b**), or control (no enzyme treatment, **c**). Bar = 0.5 mm (Adapted from Bui et al. 2015)

13.6 Gametophyte Transformation Deployed for the Study of Asexual Alternation of Generations

The life cycle of the fern partitions sporogenesis and gametogenesis into two independent plants, the sporophyte and the gametophyte, respectively. In contrast, sporogenesis and gametogenesis are tightly linked in time and space in angiosperms. These differences make the direct comparison of asexual reproduction in angiosperms to that of ferns somewhat difficult but not impossible. The comparison was made by Cordle et al. (2011) and is summarized here. Asexual reproduction through seeds is called apomixis (Nogler 1984). Because there is no seed production in ferns, we will concentrate on how embryos are formed in apomixis and set aside the consideration of the interdependence of the embryo and endosperm for each other's development. There are three major apomictic pathways in angiosperms (Ozias-Akins 2006; Tucker and Koltunow 2009; Schmidt et al. 2015), all bypassing fertilization: an aposporous embryo develops from a nucellus or integument cell autonomously, an adventitious embryo develops from a nucellus or integument cell adjacent to the fertilized embryo, or a diplosporous embryo develops from an unreduced egg-like cell produced from abnormal gametogenesis. In fern species that undergo obligate apogamy in nature, the gametophytes are incapable of sexual reproduction due to a lack of functional archegonia or antheridia. Interestingly, gametophytes of fern species such as *C. richardii* that normally reproduce sexually in nature can be forced to produce apogamous sporophytes if fertilization is prevented and they are grown on high levels of sugar (Cordle et al. 2007). The generation of sporophytes from gametophyte prothallus cells bypassing fertilization in the fern (apogamy) is comparable to aposporous embryogenesis in angiosperms with the difference that *C. richardii* gametophyte cells are haploid whereas nucellus or integument cells in angiosperms are diploid. In obligately apogamous fern species, abnormal mitosis or meiosis during sporogenesis gives rise to restitution nuclei whose chromosome number is not reduced. Thus the ploidy levels of the gametophyte and the apogamous sporophyte are the same. It is therefore tempting to compare the abnormal sporogenesis in apogamous ferns to the diplosporous embryo of apomixes, although in the latter case, the unreduced egg-like cell is the result of abnormal gametogenesis. A comparable situation to the apomictic adventitious embryo in angiosperms cannot be found in the fern. Rather, in our experience with apogamy induction in *C. richardii*, a gametophyte producing zygotic sporophytes is rarely observed to also generate apogamous sporophytes on induction media.

One of the most exciting recent findings about apomixis is the discovery that the *BABY BOOM (BBM)* gene resides in the apospory-specific genomic region (ASGR) of *Pennisetum squamulatum* (Conner et al. 2015), a polyploidy grass whose mode of apomixis has been studied extensively. These researchers were able to insert a *P. squamulatum BBM* sequence identified in the ASGR under the control of its native promoter into the genome of a tetraploid, sexual relative, *P. glaucum*. Unfertilized, diploid eggs of the transgenic *P. glaucum* were able to develop into embryos. Although it is unclear how the ASGR *BBM* expression pattern compares with the

BBM gene residing outside of the ASGR, if indeed such a gene exists in the *P. squamulatum* genome, it is plausible to speculate that an ectopic and/or overexpressed ASGR *BBM* causes apomixis.

The *BBM* gene was identified as an early embryo-expressed gene from a *Brassica napus* microspore embryo culture where vegetative cells of immature pollen grains develop into embryos in vitro (Boutilier et al. 2002). Ectopic expression of *B. napus BBM* (*BnBBM*) or its soybean homolog results in spontaneous somatic embryogenesis *in planta* in *Arabidopsis* and *B. napus* (Boutilier et al. 2002; El Ouakfaoui et al. 2010). In addition to *in planta* somatic embryogenesis, ectopic expression of *BnBBM* in sweet pepper increases the regeneration rate in *Agrobacterium*-mediated transformation (Heidmann et al. 2011). In tobacco, ectopic expression of *Arabidopsis* or *B. napus BBM* enhanced the regenerative capacity and induced spontaneous callus, root, and shoot formation (Srinivasan et al. 2007). Interestingly, use of an inducible system for identifying BBM target genes uncovered many genes with no known roles in somatic embryogenesis but which were predominantly expressed in somatic embryos, suggesting multiple entries to somatic embryogenesis (Passarinho et al. 2008). This notion was further supported by the report of additive and in some cases synergistic effects of ectopic expression of *BBM* and *WUS2* in improving monocot transformation (Lowe et al. 2016).

The diverse effects of ectopic *BBM* expression in different species and cell types suggest an interaction between *BBM* targets and the existing program where the *BBM* gene is ectopically expressed, be it of an unfertilized egg of *P. glaucum*, the cells of an *Arabidopsis* cotyledon, or immature maize embryos. Will the BBM protein act within the gene network of gametophyte cells of *C. richardii*? The BBM protein belongs to the euANT lineage, a subfamily in the AP2/ERF family of transcription factors (Kim et al. 2006; Floyd and Bowman 2007). There are nine *euANT* genes in *Arabidopsis* including *AINTEGUMENTA* (*ANT*), *BBM*, and *AIL5* (Aoyama et al. 2012). Like BBM, ectopic expression of AIL5 produces *in planta* somatic embryos in *Arabidopsis* (Tsuwamoto et al. 2010). A single *euANT* gene, *CrANT*, the homolog of the *Arabidopsis ANT* gene, has been identified from a *C. richardii* transcriptome. Although uncertain without a sequenced genome, *C. richardii* may lack *BBM* and *AIL5*, as do *Selaginella moellendorffii* and *P. patens*, the two non-seed plants with sequenced and published genomes. Using the *Agrobacterium*-mediated gametophyte transformation system described above (13.5), *CrANT* and *BnBBM* were each overexpressed in *C. richardii* gametophytes under the control of P_{35S}. As observed in angiosperms, the transformation and/or regeneration efficiencies were increased by overexpression of *CrANT* and *BnBBM* in *C. richardii* gametophytes. More importantly, overexpression of either gene causes spontaneous apogamy at a rate higher than the sugar-induced apogamy. When *CrANT* expression was knocked down by RNAi, both the regeneration frequency and sugar-induced apogamy levels were decreased. The above work was published in 2017 (Bui et al.). More recently, the *Arabidopsis AIL5* was overexpressed in *C. richardii* with similar results, albeit with less potent effects, to that of the BBM overexpressing gametophytes (Fig. 13.3). There are multiple well-conserved domains among CrANT, BBM, and AIL5, but the two domains required for BBM to promote somatic

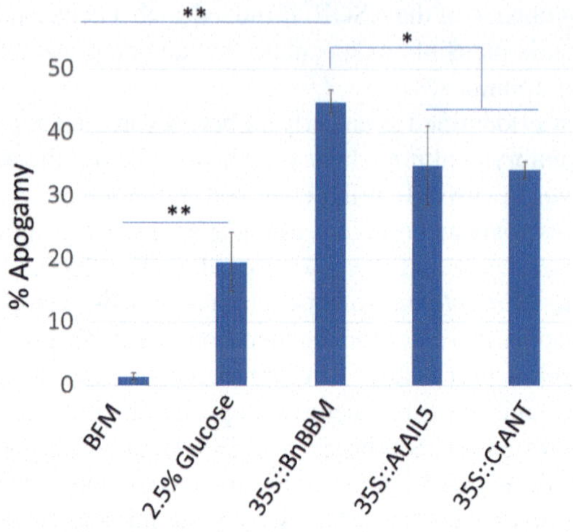

Fig. 13.3 Overexpression of *BnBBM*, *CrANT*, and *AtAIL5* in *C. richardii* resulted in spontaneous apogamy. The percentage of apogamy was calculated based on the number of gametophytes exhibiting at least one apogamous sporophyte divided by the total number of gametophytes that survived the selection. BFM, basal fern medium with no sugar supplements. 2.5% Glucose, basal fern medium supplemented with glucose. Vertical lines indicate standard deviation. Asterisks denote levels of significant difference between different groups at $p = 0.1$ (1 asterisk) or $p = 0.05$ (2 asterisks) calculated using Student's t-test using Excel Microsoft Office 2013. Unpublished results of AtAIL5 overexpression was incorporated into published results (Bui et al. 2017)

embryogenesis (El Ouakfaoui et al. 2010) are not present in CrANT (Bui et al. 2017), and only one is conserved in AIL5. A plausible interpretation of the results is that the interaction between a gene network and a conserved region (or regions) in the three genes that promotes apogamy in the fern no longer occurs to promote somatic embryogenesis in angiosperms.

13.7 Conclusion

A simple and reproducible *Agrobacterium*-mediated system to transform the gametophyte has been established in *C. richardii*. Direct transformation of gametophytes has made possible the study of candidate gene functions in apogamy, the asexual alternation from the gametophyte to the sporophyte generation. These studies connect apogamy in ferns to that of somatic embryogenesis and apomixis in angiosperms, providing a lead to trace the evolution of asexual reproduction in land plants.

References

Aoyama T, Hiwatashi Y, Shigyo M, Kofuji R, Kubo M, Ito M, Hasebe M (2012) AP2-type transcription factors determine stem cell identity in the moss *Physcomitrella patens*. Development 139:3120–3129

Bechtold N, Ellis J, Pelletier G (1993) In planta Agrobacterium-mediated gene transfer by infiltration of adult Arabidopsis thaliana plants. C.R. Acad Sci Sér 3 Sci Vie 316:1194–1199

Bechtold N, Jaudeau B, Jolivet S, Maba B, Vezon D, Voisin R, Pelletier G (2000) The maternal chromosome set is the target of the T-DNA in the in-planta transformation of *Arabidopsis thaliana*. Genetics 155:1875–1887

Bezanilla M, Pan A, Quatrano RS (2003) RNA interference in the moss *Physcomitrella patens*. Plant Physiol 133:470–474

Bezanilla M, Perroud PF, Pan A, Klueh P, Quatrano RS (2005) An RNAi system in *Physcomitrella patens* with an internal marker for silencing allows for rapid identification of loss of function phenotypes. Plant Biol 7:251–257

Boutilier K, Offringa R, Sharma VK, Kieft H, Ouellet T, Zhang L, Hattori J, Liu CM, van Lammeren AA, Miki BL (2002) Ectopic expression of *BABY BOOM* triggers a conversion from vegetative to embryonic growth. Plant Cell 14:1737–1749

Bui LT, Hurst A, Irish EE, Cheng CL (2012) The effects of sugars and ethylene on apospory and regeneration in *Ceratopteris richardii*. AJPS 3:953–961

Bui LT, Cordle AR, Irish EE, Cheng CL (2015) Transient and stable tranformation of *Ceratopteris richardii* gametophytes. BMC Res Notes 8:214–223

Bui LT, Pandzic D, Youngstrom CE, Wallace S, Irish EE, Szövényi P, Cheng CL (2017) A fern *AINTEGUMENTA* gene mirrors *BABY BOOM* in promoting apogamy in *Ceratopteris richardii*. Plant J 90:122–132

Burkart GM, Baskin TI, Bezanilla M (2015) A family of ROP proteins that suppresses actin dynamics, and is essential for polarized growth and cell adhesion. J Cell Sci 128:2553–2564

Chatterjee A, Roux SJ (2000) *Ceratopteris richardii*: a productive model for revealing secrets of signaling and development. J Plant Growth Regul 19:284–289

Chiyoda S, Ishizaki K, Kataoka H, Yamato KT, Kohchi T (2008) Direct transformation of the liverwort *Marchantia polymorpha L.* by particle bombardment using immature thalli developing from spores. Plant Cell Rep 27:1467–1473

Cho SH, Chung YS, Cho SK, Rim YW, Shin JS (1999) Particle bombardment mediated transformation and GFP expression in the moss *Physcomitrella patens*. Mol Cells 9:14–19

Collonnier C, Epert A, Mara K, Maclot F, Guyon-Debast A, Charlot F, White C, Schaefer DG, Nogué F (2017) CRISPR-Cas9-mediated efficient directed mutagenesis and RAD51-dependent and RAD51-independent gene targeting in the moss *Physcomitrella patens*. Plant Biotechnol J 15:122–131

Conner JA, Mookkan M, Huo H, Chae K, Ozias-Akins P (2015) A parthenogenesis gene of apomictic origin elicits embryo formation from unfertilized eggs in a sexual plant. Proc Natl Acad Sci U S A 112:11205–11210

Cordle AR, Irish EE, Cheng CL (2007) Apogamy induction in *Ceratopteris richardii*. Int J Plant Sci 168:361–369

Cordle AR, Bui LT, Irish EE, Cheng CL (2011) Laboratory-induced apogamy and apospory in *Ceratopteris richardii*. In: Kumar A, Fernández H, Revilla MA (eds) Working with ferns. Springer, New York, pp 25–36

Desfeux C, Clough SJ, Bent AF (2000) Female reproductive tissues are the primary target of Agrobacterium-mediated transformation by the Arabidopsis floral-dip method. Plant Physiol 123:895–904

El Ouakfaoui S, Schnell J, Abdeen A, Colville A, Labbé H, Han S, Baum B, Laberge S, Miki B (2010) Control of somatic embryogenesis and embryo development by AP2 transcription factors. Plant Mol Biol 74:313–326

Floyd SK, Bowman JL (2007) The ancestral developmental tool kit of land plants. Int J Plant Sci 168:1–35

Heidmann I, de Lange B, Lambalk J, Angenent GC, Boutilier K (2011) Efficient sweet pepper transformation mediated by the BABY BOOM transcription factor. Plant Cell Rep 30:1107–1115

Hickok LG, Warne TR, Fribourg RS (1995) The biology of the fern *Ceratopteris* and its use as a model system. Int J Plant Sci 156:332–345

Hohe A, Egener T, Lucht JM, Holtorf H, Reinhard C, Schween G, Reski R (2004) An improved and highly standardised transformation procedure allows efficient production of single and multiple targeted gene-knockouts in a moss, *Physcomitrella patens*. Curr Genet 44:339–347

Honkanen S, Jones VA, Morieri G, Champion C, Hetherington AJ, Kelly S, Proust H, Saint-Marcoux D, Prescott H, Dolan L (2016) The mechanism forming the cell surface of tip-growing rooting cells is conserved among land plants. Curr Biol 26:3238–3244

Horst NA, Katz A, Pereman I, Decker EL, Ohad N, Reski R (2016) A single homeobox gene triggers phase transition, embryogenesis and asexual reproduction. Nat Plants 2:15209. https://doi.org/10.1038/nplants.2015.209

Indriolo E, Na G, Ellis D, Salt DE, Banks JA (2010) A vacuolar arsenite transporter necessary for arsenic tolerance in the arsenic hyperaccumulating fern *Pteris vittata* is missing in flowering plants. Plant Cell 22:2045–2057

Ishizaki K, Chiyoda S, Yamato KT, Kohchi T (2008) Agrobacterium-mediated transformation of the haploid liverwort *Marchantia polymorpha L.*, an emerging model for plant biology. Plant Cell Physiol 49:1084–1091

Ishizaki K, Johzuka-Hisatomi Y, Ishida S, Iida S, Kohchi T (2013) Homologous recombination-mediated gene targeting in the liverwort *Marchantia polymorpha L.* Sci Rep 3:1532. https://doi.org/10.1038/srep01532

Ishizaki K, Nishihama R, Yamato KT, Kohchi T (2016) Molecular genetic tools and techniques for *Marchantia polymorpha* research. Plant Cell Physiol 57:262–270

Kajikawa M, Matsui K, Ochiai M, Tanaka Y, Kita Y, Ishimoto M, Kohzu Y, Shoji S, Yamato KT, Ohyama K, Fukuzawa H, Kohchi T (2008) Production of arachidonic and eicosapentaenoic acids in plants using bryophyte fatty acid Δ6-desaturase, Δ6-elongase, and Δ5-desaturase genes. Biosci Biotechnol Biochem 72:435–444

Kamisugi Y, Schlink K, Rensing SA, Schween G, von Stackelberg M, Cuming AC, Reski R, Cove DJ (2006) The mechanism of gene targeting in *Physcomitrella patens*: homologous recombination, concatenation and multiple integration. Nucleic Acids Res 34:6205–6214

Kawai-Toyooka H, Kuramoto C, Orui K, Motoyama K, Kikuchi K, Kanegae T, Wada M (2004) DNA interference: a simple and efficient gene-silencing system for high-throughput functional analysis in the fern Adiantum. Plant Cell Physiol 45:1648–1657

Kenrick P, Crane PR (1997) The origin and early evolution of plants on land. Nature 389:33–39

Kim S, Soltis PS, Wall K, Soltis DE (2006) Phylogeny and domain evolution in the APETALA2-like gene family. Mol Biol Evol 23:107–120

Klink VP, Wolniak SM (2001) Centrin is necessary for the formation of the motile apparatus in spermatids of Marsilea. Mol Biol Cell 12:761–776

Kopischke S, Schüßler E, Althoff F, Zachgo S (2017) TALEN-mediated genome-editing approaches in the liverwort *Marchantia polymorpha* yield high efficiencies for targeted mutagenesis. Plant Methods 13:20. https://doi.org/10.1186/s13007-017-0167-5

Leroux O, Eeckhout S, Viane RL, Popper ZA (2013) *Ceratopteris richardii* (C-Fern): a model for investigating adaptive modification of vascular plant cell walls. Front Plant Sci 4:367. https://doi.org/10.3389/fpls.2013.00367

Li C, Sako Y, Imai A, Nishiyama T, Thompson K, Kubo M, Hiwatashi Y, Kabeya Y, Karlson D, Wu SH, Ishikawa M, Murata T, Benfrey PN, Sato Y, Tamada Y, Hasebe M (2017) A Lin28 homologue reprograms differentiated cells to stem cells in the moss *Physcomitrella patens*. Nat Commun 8:14242. https://doi.org/10.1038/ncomms14242

Lopez-Obando M, Conn CE, Hoffmann B, Bythell-Douglas R, Nelson DC, Rameau C, Bonhomme S (2016) Structural modelling and transcriptional responses highlight a clade of PpKAI2-LIKE genes as candidate receptors for strigolactones in *Physcomitrella patens*. Planta 243:1441–1453

Lowe K et al (2016) Morphogenic regulators baby boom and wuschel improve monocot transformation. Plant Cell 28:1998–2015

Muthukumar B, Joyce BL, Elless MP, Stewart CN (2013) Stable transformation of ferns using spores as targets: *Pteris vittata* and *Ceratopteris thalictroides*. Plant Physiol 163:648–658

Nakaoka Y, Miki T, Fujioka R, Uehara R, Tomioka A, Obuse C, Kubo M, Hiwatashi Y, Goshima G (2012) An inducible RNA interference system in *Physcomitrella patens* reveals a dominant role of augmin in phragmoplast microtubule generation. Plant Cell 24:1478–1493

Nogler GA (1984) Gametophytic apomixis. In: Johri BM (ed) Embryology of angiosperms. Springer, Berlin/Heidelberg, pp 475–518

Osakabe Y, Watanabe T, Sugano SS, Ueta R, Ishihara R, Shinozaki K, Osakabe K (2016) Optimization of CRISPR/Cas9 genome editing to modify abiotic stress responses in plants. Sci Rep 6:26685. https://doi.org/10.1038/srep26685

Ozias-Akins P (2006) Apomixis: developmental characteristics and genetics. CRC Crit Rev Plant Sci 25:199–214

Passarinho P, Ketelaar T, Xing M, van Arkel J, Maliepaard C, Hendriks MW, Joosen R, Lammers M, Herdies L, den Boer B, van der Geest L, Boutilier K (2008) BABY BOOM target genes provide diverse entry points into cell proliferation and cell growth pathways. Plant Mol Biol 68:225–237

Plackett A, Huang L, Sanders HL, Langdale JA (2014) High-efficiency stable transformation of the model fern species *Ceratopteris richardii* via microparticle bombardment. Plant Physiol 165:3–14

Plackett AR, Rabbinowitsch EH, Langdale JA (2015) Protocol: genetic transformation of the fern *Ceratopteris richardii* through microparticle bombardment. Plant Methods 11:37. https://doi.org/10.1186/s13007-015-0080-8

Pryer KM, Schneider H, Smith AR, Cranfill R, Wolf PG, Hunt JS, Sipes SD (2001) Horsetails and ferns are a monophyletic group and the closest living relatives to seed plants. Nature 409:618–622

Raubeson LA, Jansen RK (1992) Chloroplast DNA evidence on the ancient evolutionary split in vascular land plants. Science 255:1697–1699

Rutherford G, Tanurdzic M, Hasebe M, Banks JA (2004) A systemic gene silencing method suitable for high throughput, reverse genetic analyses of gene function in fern gametophytes. BMC Plant Biol 4:6. https://doi.org/10.1186/1471-2229-4-6

Schaefer D, Zryd JP, Knight CD, Cove DJ (1991) Stable transformation of the moss *Physcomitrella patens*. Mol Gen Genet 226:418–424

Schaefer DG, Zrÿd JP (1997) Efficient gene targeting in the moss *Physcomitrella patens*. Plant J 11:1195–1206

Schmidt A, Schmid MW, Grossniklaus U (2015) Plant germline formation: common concepts and developmental flexibility in sexual and asexual reproduction. Development 142:229–241

Shimakawa G, Ishizaki K, Tsukamoto S, Tanaka M, Sejima T, Miyake C (2017) The liverwort, Marchantia, drives alternative electron flow using a flavodiiron protein to protect PSI. Plant Physiol 173:1636–1647

Šmídková M, Hola M, Angelis KJ (2010) Efficient biolistic transformation of the moss *Physcomitrella patens*. Biol Plant 54:777–780

Srinivasan C, Liu Z, Heidmann I, Supena ED, Fukuoka H, Joosen R, Lambalk J, Angenent G, Scorza R, Custers JB, Boutilier K (2007) Heterologous expression of the BABY BOOM AP2/ERF transcription factor enhances the regeneration capacity of tobacco (*Nicotiana tabacum L.)* Planta 225:341–351

Stevenson DW, Loconte H (1996) Ordinal and familial relationships of pteridophyte genera. In: Camus JM, Gibby M, Johns RJ (eds) Pteridology in perspective. Royal Botanic Gardens, Kew, pp 435–467

Stout SC, Clark GB, Archer-Evans S, Roux SJ (2003) Rapid and efficient suppression of gene expression in a single-cell model system, *Ceratopteris richardii*. Plant Physiol 131:1165–1168

Strepp R, Scholz S, Kruse S, Speth V, Reski R (1998) Plant nuclear gene knockout reveals a role in plastid division for the homolog of the bacterial cell division protein FtsZ, an ancestral tubulin. Proc Natl Acad Sci U S A 95:4368–4373

Sugano SS, Shirakawa M, Takagi J, Matsuda Y, Shimada T, Hara-Nishimura I, Kohchi T (2014) CRISPR/Cas9 mediated targeted mutagenesis in the liverwort *Marchantia polymorpha L.* Plant Cell Physiol 55:475–481

Takenaka M, Yamaoka S, Hanajiri T, Shimizu-Ueda Y, Yamato KT, Fukuzawa H, Ohyama K (2000) Direct transformation and plant regeneration of the haploid liverwort *Marchantia polymorpha* L. Transgenic Res 9:179–185

Trouiller B, Charlot F, Choinard S, Schaefer DG, Nogué F (2007) Comparison of gene targeting efficiencies in two mosses suggests that it is a conserved feature of Bryophyte transformation. Biotechnol Lett 29:1591–1598

Tsuwamoto R, Yokoi S, Takahata Y (2010) Arabidopsis EMBRYOMAKER encoding an AP2 domain transcription factor plays a key role in developmental change from vegetative to embryonic phase. Plant Mol Biol 73:481–492

Tucker MR, Koltunow AM (2009) Sexual and asexual (apomictic) seed development in flowering plants: molecular, morphological and evolutionary relationships. Funct Plant Biol 36:490–504

Vidali L, van Gisbergen PA, Guérin C, Franco P, Li M, Burkart GM, Augustine RC, Blanchoin L, Bezanilla M (2009) Rapid formin-mediated actin-filament elongation is essential for polarized plant cell growth. Proc Natl Acad Sci 106:13341–13346

Wendeler E, Zobell O, Chrost B, Reiss B (2015) Recombination products suggest the frequent occurrence of aberrant gene replacement in the moss *Physcomitrella patens*. Plant J 81:548–558

Willmann MR, Endres MW, Cook RT, Gregory BD (2011) The functions of RNA-dependent RNA polymerases in Arabidopsis. Arabidopsis Book 9:e0146. https://doi.org/10.1199/tab.0146

Ye GN, Stone D, Pang SZ, Creely W, Gonzalez K, Hinchee M (1999) Arabidopsis ovule is the target for Agrobacterium in planta vacuum infiltration transformation. Plant J 19:249–257

Chapter 14
Generation of Transgenic Spores of the Fern *Ceratopteris richardii* to Analyze Ca^{2+} Transport Dynamics During Gravity-Directed Polarization

Ashley E. Cannon, Mari L. Salmi, Araceli Cantero, and Stanley J. Roux

14.1 Introduction

Ceratopteris richardii is a homosporous, aquatic fern that is found in tropical and subtropical regions worldwide. Unlike seed-bearing model plants, such as *Arabidopsis*, ferns have an alternation of generations that is defined by two independent, autotrophic haploid and diploid generations (Hickok et al. 1995). *Ceratopteris richardii* is a particularly appealing fern system to study because it has a relatively short life cycle of approximately 4 months, is easily cultured and propagated, and can be used in genetic studies (Hickok et al. 1987). These valuable characteristics have led plant scientists to use this model system for over 30 years for conducting various physiological, biochemical, and genetic studies. To cite just a few examples of the diverse studies carried out, they include documentation of the role of a pheromone-induced gene in sex determination in *Ceratopteris* (Wen et al. 1999), the role of gravity in spore polarization (Edwards and Roux 1998), and the analysis of the *Ceratopteris* genome structure (Nakazato et al. 2006).

In addition to using the multicellular gametophyte and sporophyte, many studies have focused on the single-celled spore as a model, and this spore cell will be the focus of this chapter. As reviewed in this chapter, the *Ceratopteris* spore is a valuable cell to study because its development is well characterized and predictable (Chatterjee and Roux 2000), and it can be transformed for cellular and molecular studies (Plackett et al. 2014, 2015; Bui et al. 2015). Many of the studies using the

A. E. Cannon
BioDiscovery Institute, Department of Biology, The University of North Texas, Denton, TX, USA

M. L. Salmi · A. Cantero · S. J. Roux (✉)
Department of Molecular Biosciences, The University of Texas, Austin, TX, USA
e-mail: sroux@austin.utexas.edu

H. Fernández (ed.), *Current Advances in Fern Research*,
https://doi.org/10.1007/978-3-319-75103-0_14

Ceratopteris spore are focused on the developmental and signaling aspects of gravity-directed polarization, in which the transcellular transport of Ca^{2+} plays a key role in determining that the rhizoid will emerge and grow in a downward direction when the spore germinates. The focus of this chapter will be to review and update the data documenting the key role of Ca^{2+} transport changes in transducing the gravity stimulus into directed polarized growth and then to illustrate how RNA-seq analyses and genetic transformation studies still in progress have the potential to identify molecular components that transport Ca^{2+} and serve as Ca^{2+}-binding signaling agents.

The central role of Ca^{2+} transport in spore polarization first became evident from studies of the early steps leading to spore germination. In *Ceratopteris richardii*, spore germination is initiated by hydration and exposure to red light. Within hours of germination initiation, Ca^{2+} enters the spore through channels at the bottom and exits through pumps at the top, forming a 100-fold [Ca^{2+}] differential measured outside the cell between its top and bottom (Salmi et al. 2011; ul Haque et al. 2007). That this differential is regulated by gravity was demonstrated by Salmi et al. (2011) during parabolic flight experiments in which transient changes in the *g*-force were directly related to changes in the Ca^{2+} differential: the differential dropped to near zero in micro-*g* and doubled in magnitude at 2-*g*. Some of these results were independently confirmed in more recent hyper-*g* studies, in which the [Ca^{2+}] differential increased with increasing *g*-forces (Park et al. 2017).

Chemical and kinetic studies provided more insights into the role of Ca^{2+} transport changes in mediating the effects of gravity on cell polarization. If the entry of Ca^{2+} along the bottom of the spore is blocked by the Ca^{2+} channel antagonist nifedipine, the percentage of spores emerging downward is dramatically reduced (Chatterjee and Roux 2000). The efflux of Ca^{2+} at the top of the spore peaks between 7 h and 12 h after light induces the germination process. Polarization is irreversibly set by the force of gravity within 24–30 h after spore germination is initiated by light while the spore is still a single cell. After this time, all subsequent development is aligned with the direction of gravity perceived by the spore during this initial period.

As a result of gravity-directed polarity fixation, the nucleus migrates downward around 24 h after germination initiation. That this downward migration was directed by gravity was demonstrated by video microscopy of spores germinated in the Space Shuttle on orbit, which showed that in micro-*g*, the nuclei migrated in random directions (Roux et al. 2003). Both in space and on earth, this nuclear migration prepares the cell for an asymmetrical cell division at 48-h post-light exposure. Also both in space and on earth, the direction the nucleus migrates predicts the direction the rhizoid will emerge. On earth the direction of gravity perceived by the spore is displayed by the downward direction that the primary rhizoid grows after emerging around 72 h after germination is initiated (Edwards and Roux 1994). This highly predictable growth and developmental pattern allows scientists to visually observe the effects of disrupting cellular and molecular machinery during gravity-directed polarization or other developmental events of interest.

Previous studies focused on understanding gravity-directed polarization in *Ceratopteris* spores provided some insight into the cellular and molecular mechanisms governing this process. However, these studies were limited by a lack of gene sequence information and a reliable method of transformation. *Ceratopteris* spores were transiently transformed for the first time by incubating surface-sterilized spores with double-stranded RNA (dsRNA) (Stout et al. 2003).

Stout et al. (2003) used RNA isolated from spores 20 h after light exposure to generate a cDNA library, and dsRNA constructs were generated using these sequences. When dsRNA is produced by or introduced into cells, it causes the degradation of homologous transcripts. This technique is now routinely used to cause RNA suppression or silencing and is referred to as RNA interference (RNAi) (Watson et al. 2005).

Stout et al. (2003) successfully suppressed transcripts related to Ca^{2+} signaling in *Ceratopteris richardii* using this technique. The dsRNA constructs suppressed the steady-state mRNA levels of target genes during the entirety of the incubation period without affecting the expression of other genes assayed. The level and length of suppression were transient, but it successfully and selectively knocked down genes during developmental processes of interest, e.g., gravity-directed polarization. This was the first study to show that RNAi could be used to knock down gene expression in *Ceratopteris richardii*.

Since the initial RNAi experiment that transiently transformed *Ceratopteris* spores, scientists have tried using other techniques to stably transform ferns. Muthukumar et al. (2013) used *A. tumefaciens* to transform spores; however, the efficiency was so low, 0.03%, that this method was not a feasible option. Bui et al. (2015) also used *A. tumefaciens*-mediated transformation but increased the efficiency of transformation by using gametophytes that had been treated with cell wall-digesting enzymes. In this method, 12-day-old gametophytes were treated with 1.5% (w/v) cellulase and 0.5% (w/v) macerozyme. This treatment disrupted most of the cell walls, but the tissue remained intact. After breaking down the cell wall, gametophytes were co-incubated with *Agrobacterium* that contained a plasmid with the β-glucuronidase (GUS) reporter gene for 48 h (Bui et al. 2015). Using this method, 90% of the gametophytes showed GUS expression. This protocol worked to transiently transform *Ceratopteris richardii* gametophytes.

In addition, Bui et al. (2015) also developed a protocol to achieve stable transformation. Gametophytes of the same age were incubated in 1.5% (w/v) cellulase, the optimal treatment to ensure regeneration. After this treatment, cells were co-incubated with *Agrobacterium* containing a vector with a selection gene, hygromycin phosphotransferase (HPT) (Bui et al. 2015). A 35S promoter-driven GFP marker gene was used for phenotypic screens. This method led to 1.6%–2.6% stable transformation efficiency, with an average transformation-to-spore time period of 12–15 weeks (Bui et al. 2015).

In addition to *Agrobacterium*-mediated transformation techniques, a method using callus and microparticle bombardment was recently developed. In this method, Plackett et al. (2014, 2015) generated callus from the shoot apex of 14-day-old sporophytes. *Ceratopteris* calli were transformed with a plasmid that contained a GUS gene driven by a 35S promoter and an HPT selection gene, using microparticle bombardment. The transformation efficiency of this method was 71.58% ± 2.56%, a much higher level than previous methods.

By removing cytokinins from the culture medium, transformed calli were induced to develop into sporophytes in about 8 weeks. These sporophytes were transferred to soil, and T_1 spores were harvested 10–14 weeks later. Plackett et al. (2014, 2015) reported that on average, it took 4.5–5.5 months, from the time of bombardment to the time of T_1 spore harvest. Additionally, it took at least 6 weeks for *Ceratopteris richardii* spores to ripen and germinate at a high enough percentage to feasibly analyze hygromycin resistance. Although this method took a lot of time, the transformation efficiency was very high. It was the first method developed that could feasibly be used to insert genes or knockdown endogenous genes in order to analyze specific molecular aspects of gravity-directed development. Our lab utilized this method to generate *Ceratopteris* mutants that would allow us to test alternative hypotheses proposed to identify the key Ca^{2+} transporters and signaling agents by which the gravity stimulus is transduced into polarized development of spores. In the next several sections of this chapter, we will describe *Ceratopteris* transformation experiments in which we successfully targeted specific genes for suppression or overexpression in order to test hypotheses about the likely molecular components that initiate the Ca^{2+} transport changes and that bind Ca^{2+} to link the $\Delta[Ca^{2+}]_{cyt}$ to cell polarization. Although in each set of experiments the success of the transformation was verified, the tests of how these transformations altered the gravity response are still in progress, and the results of these tests will not be presented here.

The first genetic modification selected was to transform *Ceratopteris* cells with a FRET-based Ca^{2+} sensor, Yellow Cameleon 3.60. As described below, the success of this effort has generated a uniquely valuable tool that can be used to analyze intracellular Ca^{2+} dynamics and rapidly screen transformants in *Ceratopteris*, a primitive plant system. Most of the remainder of this chapter will describe the methods and technical hurdles that had to be overcome to achieve successful transformation and how this tool can now be used to identify and characterize the transporters and Ca^{2+}-activated signaling agents that are induced by gravity to direct the polarization of single-celled spores.

14.2 Screening Assays of Sporophytes and Gametophytes Transformed with Yellow Cameleon 3.60

The pioneering publications of Plackett et al. (2014, 2015) and of Bui et al. (2015) described very well the methods that could be used to achieve successful stable transformation of *Ceratopteris richardii*. Relatively minor modifications of the biolistic method of Plackett et al. (2014, 2015) were employed to transform *Ceratopteris* with the Yellow Cameleon gene 3.60. Once the multiple steps of those transformation methods were completed, it became necessary to confirm that the putative transformants were, indeed, stably transformed with the YC 3.60 gene. This was done by a combination of genotypic and phenotypic screening, as described in the following two sections.

14.2.1 Genotypic Screening of Transgenic Lines

Genomic DNA (gDNA) was isolated from fronds of T_0 sporophytes using the DNeasy Plant Mini Kit. Mutants were screened using a PCR with primers designed to amplify the YC 3.60 gene. After confirming the presence or absence of the YC 3.60 gene, the relative expression of YC 3.60 was analyzed using RT-PCR.

RNA was extracted from *Ceratopteris* sporophyte fronds using the Spectrum Plant Total RNA Kit. RNA was extracted from spores using a protocol previously published in Salmi et al. (2005). RNA isolated from spores or fronds was used for cDNA synthesis. The purity and concentration of RNA were determined using a NanoDrop UV-Vis spectrophotometer. The RNA integrity was checked by running a sample on a 1% agarose gel following the protocol of Aranda et al. (2012). The cDNA synthesis was done using the SuperScript First-Strand Synthesis System for RT-PCR. Primers specific to the YC 3.60 gene were used for a PCR using cDNA as a template. The *Ceratopteris richardii* actin and calmodulin genes were used as positive controls in semiquantitative RT-PCR.

14.2.2 Phenotypic Screening of Transgenic Lines

Imaging of YC 3.60 T_0 and T_1 roots of sporophytes was done using a Zeiss Axiovert 200 M Fluorescence Microscope with a 100 W Hg lamp variable intensity power supply using a 20×/0.8 NA objective. Fluorescence filter cubes for CFP and YFP were used to screen for the presence of YC 3.60 in mutant roots. Roots were cut from mature sporophyte plants, rinsed, and maintained in ½ strength MS, pH 6.3 until imaging. Each root was placed on a microscope slide with ½ strength MS, pH 6.3 and imaged using each filter cube. In order to account for autofluorescence, each mutant line was screened, and the optimal exposure time was recorded and used to screen wild-type plants with the same filter cubes. If the fluorescence intensity of the signal was significantly higher than wild type using the CFP and YFP filter cubes, the mutant was considered phenotypically confirmed. An example of a sporophyte root showing YC 3.60-specific fluorescence is shown in Fig. 14.1.

Imaging of *Ceratopteris* rhizoids that germinated from spores harvested from T_1-transformed sporophytes was done using an inverted Zeiss LSM 710 confocal laser scanning microscope using a 20X/0.8 NA objective. The excitation wavelengths used were 405 nm and 514 nm for CFP and YFP, respectively. Gametophytes were placed on microscope slides with ½ strength MS, pH 6.3. After measuring YC 3.60 rhizoid fluorescence, wild-type rhizoids were screened using the same exposure time

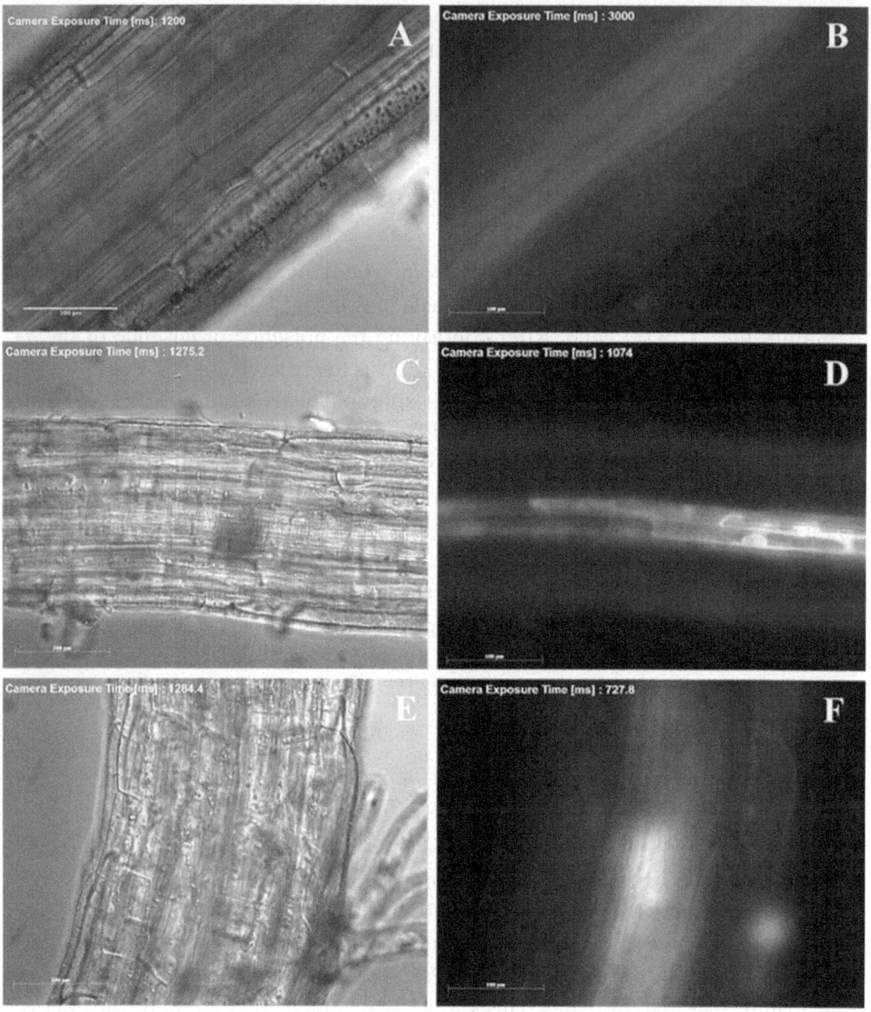

Fig. 14.1 Tissue-specific expression of Yellow Cameleon 3.60 in T_0 *Ceratopteris richardii* roots. Wild-type roots (**a** and **b**) showed a low level of autofluorescence in the yellow fluorescent protein (YFP) emission spectrum. Two independent lines (**c**, **d**, **e**, and **f**) showed varying levels of YFP expression, but in every line screened, the signal was isolated to the vasculature. Images **a**, **c**, and **e** are the bright field images associated with the YFP fluorescence images shown in **b**, **d**, and **f**. All of the scale bars = 100 μm (Cannon 2016)

in order to look for differences in intensity of fluorescent signals. An example of a rhizoid that is showing YC 3.60-specific fluorescence is shown in Fig. 14.2.

Additional phenotypic screening of the gametophyte generation was also done by fluorescence light microscopy of T_2 spores harvested from transformed sporophytes. Imaging of YC 3.60 T_2 spores was done using a Leica SP8 microscope with a Coherent Chameleon Ti:Sapphire mode-locked laser, a 63x/1.3 NA glycerol

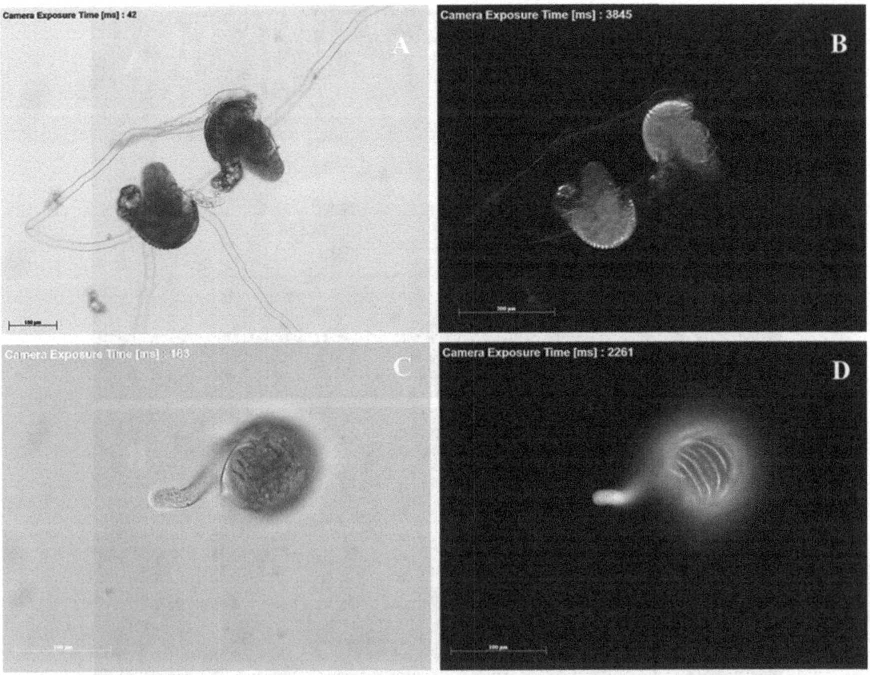

Fig. 14.2 *Ceratopteris richardii* T_1 gametophytes showed YC 3.60 expression in rhizoids. Wild-type gametophytes (**a** and **b**) showed a low level of autofluorescence in rhizoids in the yellow fluorescent protein emission spectrum. The spore coat in wild-type (**a** and **b**) and YC 3.60 (**c** and **d**) gametophytes showed a very strong autofluorescent signal. The rhizoids in YC 3.60 gametophytes (**c** and **d**) showed a relatively strong YFP fluorescent signal. All of the scale bars = 100 μm (Cannon 2016)

objective, and a hybrid (HyD) non-descanned detector with a CFP-YFP beam splitter to detect the fluorescence. Fluorescence lifetime imaging microscopy (FLIM) was performed using a PicoHarp 300 router and analyzed using the SymPhoTime software package. Two methods, acceptor photobleaching and FLIM, were used in order to determine if the YC 3.60 protein was present in the spores. A video was constructed using images from a single time point that were assembled using Imaris x64 v8.0.1. An image of the intracellular region of *Ceratopteris* spores filled with intracellular, autofluorescent granules is shown in Fig. 14.3.

14.2.3 Insights Gained from Transforming Ceratopteris *with Yellow Chameleon 3.60*

By using the protocol developed by Plackett et al. (2014, 2015), we were able to achieve a similar high level of transformation. When compared to previous methods used, the transformation efficiency of 79% achieved by the biolistic-callus method

Fig. 14.3 Multiphoton fluorescence microscopy was used to visualize the intracellular region of *Ceratopteris* spores at a high resolution. (**a**) The *Ceratopteris* spore coat has a strong autofluorescent signal. (**b**) Ceratopteris richardii spores are full of autofluorescent granules with a fluorescence lifetime similar to the autofluorescent spore coat (Cannon 2016)

used to transform *Ceratopteris* with YC 3.60 is much higher than what was achieved using *Agrobacterium*-mediated methods. This method successfully generated the first set of stable lines of transformed *Ceratopteris richardii* sporophytes that express a Ca^{2+} reporter gene that can be used both to rapidly determine if a fern has been transformed and to monitor intracellular Ca^{2+} dynamics during development. Transformation experiments also confirmed the efficacy of the method, which can now be used for multiple other gene transformations, as described below. Due to the novelty of this work, discoveries about the molecular characteristics and challenges associated with *Ceratopteris* transformation were made. These lessons can be used to optimize transformation in the future.

One discovery made during these experiments was a high level of autofluorescence in the spore, gametophytic, and sporophytic tissues screened during phenotypic assays. Autofluorescence is the fluorescence emission produced from endogenous molecules as a result of excitation from UV-Vis radiation. The major challenge associated with autofluorescence is the ability to distinguish the intrinsic signal from fluorescent markers that have been introduced by exogenous application or genetically encoded after transformation.

The most common autofluorescent molecule in green plants is chlorophyll, which emits light at a wavelength of 600–700 nm in the red/far-red region of the spectrum. Another common cause of autofluorescence in plants is secondary metabolites such phenolics, alkaloids, and terpenoids (Talamond et al. 2015). Secondary metabolites autofluoresce in the blue region (430–450 nm) and some of the green region (520–530 nm) of the spectrum. Plant secondary metabolites help defend the plant against herbivores and pathogens but are also involved in plant growth and development. The high level of autofluorescence in the CFP emission range could be attributed to these secondary metabolites.

In general, the autofluorescence in the YFP emission range was very low. Because of the low level of autofluorescence in this region of the spectrum, all phenotypic assays were done using the YFP filter cube that collects light between 520 nm and 550 nm. Even so, however, every tissue screened showed some level of autofluorescence. In order to overcome this issue, consistent controls were needed in order to subtract the average autofluorescent signal of wild-type sporophytes from fluorescent mutants.

Since the YC 3.60 protein is of limited use in autofluorescent tissues, alternative Ca^{2+}-specific sensors may be used in future transformations. Recently, plant scientists have started using a new Ca^{2+} sensor called GFP-aequorin (G5A). This is a chimeric protein composed of a green fluorescent protein (GFP) and aequorin, a bioluminescent protein. Like the yellow cameleon protein, the transfer of energy in this protein is Ca^{2+} dependent, so it can be used to monitor Ca^{2+} dynamics (Xiong et al. 2014). It is the transfer of energy from GFP to aequorin, referred to as bioluminescent resonance energy transfer (BRET), which defines the mechanism of Ca^{2+} detection. Since luminescence is the emission of light from a substance that does not require radiation, it is detected with a cooled charge-coupled device (cooled CCD) camera. This method eliminates problems associated with autofluorescence and has been successfully used to image Ca^{2+} dynamics in *Arabidopsis* leaves, a very autofluorescent tissue (Xiong et al. 2014).

In addition to GE5, another Ca^{2+} sensor that may be considered for future experiments is R-GECO-1. R-GECO-1 is an intensity-based reporter that has recently been used in plant tissues. When Ca^{2+} binds to the CaM domain of R-GECO-1, it causes CaM to fold around the M13 domain. This change alters the chemical environment of the chromophore leading to an increase in fluorescence intensity (Keinath et al. 2015). The GECO family of reporters has been optimized, and the newest version has an improved dynamic range, stability, and sensitivity. In addition, the GECO family of Ca^{2+} reporters has been developed in multiple colors. A GECO reporter could be used if the emission wavelength was outside of the autofluorescent range of *Ceratopteris* tissues. In conclusion, the issues associated with autofluorescence in *Ceratopteris* could be overcome by using the GE5 or a GECO Ca^{2+} reporter. This would eliminate or decrease the problems associated with autofluorescent interference.

Beyond revealing the autofluorescence properties of *Ceratopteris* tissues, this YC 3.60 transformation process also led to a better understanding of the tissue specificity of the 35S promoter in this fern system. The Cauliflower Mosaic Virus (CaMV) 35S promoter is the most commonly used general purpose, constitutive promoter in plant transformation. The expression profile of this promoter has been analyzed in multiple systems where it shows constitutive expression with varying levels of tissue specificity (Odell et al. 1985; Battraw and Hall 1990; Yang and Christou 1990; Benfey et al. 1990). When the expression profile of the 35S promoter was analyzed in cotton, longitudinal sections through the root showed the highest level of fluorescence at the tip of the root and a moderate level of expression in the vasculature (Sunilkumar et al. 2002). When cotton pollen was analyzed, both wild-type and 35S::GFP pollen showed a moderate level of autofluorescence. However, some mutant pollen showed higher levels of fluorescence, indicating that this pollen was expressing GFP (Sunilkumar et al. 2002).

Due to the similarity in their structure, pollen is the best flowering plant tissue for comparison to spores. Pollen and spores are both surrounded by a thick, protective coat. The pollen tube and spore rhizoid are also both polarly growing single cells. Scientists have studied the expression of the 35S promoter in pollen of multiple flowering plants. The expression of the 35S promoter in tobacco pollen is low, and it is not detectable in *Arabidopsis* pollen (Wilkinson et al. 1997). However, Petunia pollen expressed the 35S promoter (Benfey and Chua 1989). These conflicting reports support the conclusion that the 35S promoter has tissue-specific expression patterns that are specific to the plant in which it is expressed.

One solution to solve the problems associated with the 35S promoter in fern spores would be to change the promoter in future *Ceratopteris richardii* transformations. An alternative promoter that has been used in multiple plant systems is the *Arabidopsis* ubiquitin-10 (UBQ10) gene promoter (Norris et al. 1993). This promoter has a higher level of stability in transiently transformed plants when compared to the 35S promoter (Grefen et al. 2010; Krebs et al. 2012). Additionally, in a direct side-by-side comparison, the UBQ10 promoter leads to a higher number of plant lines with suitable expression levels when compared to the 35S promoter (Behera et al. 2015). An alternative to a generic promoter is a fern-specific promoter or a spore-specific promoter. Studies looking into this option are limited by the lack of gene sequence information available for *Ceratopteris richardii*. Overall, the tissue specificity and stability of expression issues could be overcome by changing the promoter from 35S to UBQ10 or a fern-specific promoter.

Although there were multiple issues associated with transforming *Ceratopteris richardii* with 35S::YC 3.60, one major success of these experiments was the discovery of a fluorescence microscopy method that can be used to visualize the intracellular region of spores at a high level of resolution. The technique used, multiphoton fluorescence microscopy, has been used to visualize dynamic processes in multiple living systems (Hoover and Squier 2013). Multiphoton microscopes are chosen over other fluorescent microscope systems due to their inherent optical sectioning method that restricts excitation of a fluorophore to the focal plane of the objective. This prevents photobleaching outside of the field of view. In addition, multiphoton microscopes can be used to visualize and penetrate highly scattering tissue because they can utilize longer wavelengths than standard confocal microscope systems.

The unique attributes of the multiphoton microscope were necessary in order to visualize the inside of *Ceratopteris* spores. The serrated edges, high autofluorescent signal, and density of the spore coat made it very difficult to penetrate using a standard confocal microscope. The multiphoton system overcame the scattering and penetration issue associated with the spore coat. In order to visualize the entirety of the cell, a glycerol objective with a large numerical aperture, 1.3, was used. The near-infrared wavelengths along with a large numerical aperture objective allowed us to visualize the autofluorescent intracellular components of an optically turbid *Ceratopteris* spore for the first time (Fig. 14.3).

In these experiments, *Ceratopteris richardii* was stably transformed with a Ca^{2+} sensor that allowed for the first time an analysis of both Ca^{2+} dynamics during development and a rapid detection of transformed ferns. Because of the novelty of this project, many issues came up during the process that would have been hard to predict. However, establishing a reliable method of transformation and detection of transformation success will allow scientists to efficiently proceed through this process in the future. Additionally, using the current YC 3.60 lines, scientists will be able to analyze intracellular Ca^{2+} dynamics during different physiological events that cause changes in Ca^{2+} flux, e.g., mechanostimulation. These Ca^{2+} signatures can be compared to other systems and lead to a better understanding of the evolutionary aspects of Ca^{2+} signaling by comparing pteridophytes to gymnosperms and angiosperms.

14.3 Successful Transformations Altering Expression of *Ceratopteris* Genes and Their Potential to Help Identify Ca^{2+} Transporters and Signaling Steps Activated by Gravity

As discussed in the Introduction of this chapter, a key purpose of developing methods to stably transform *Ceratopteris* was so that genetic approaches could be used to identify key transporters and signaling agents that are activated by gravity to directionally polarize the spore cell. In the following sections, we will describe the specific gene transformations that have been carried out thus far and discuss the alternative hypotheses about mechanisms of gravity signal transduction these transformations are likely to help resolve. Although the experiments to test how these transformations alter the gravity response of spores are still in progress, the following texts well illustrate the potential value of the transformations achieved to clarify the mechanisms by which gravity directs cell polarization in *Ceratopteris*. Protein sequences for the proteins encoded by the three different genes transformed (CrMscS, AnnCr2, psNTP9) and their alignments with related proteins are available at https://docs.google.com/document/d/17zyakvK8_m9aDS8RbPOwGjzFJV607F3_jV1WSDlFa_I/edit?usp=sharing.

14.3.1 Mechanosensitive Channels

As noted in the Introduction, one of the earliest gravity-directed changes in spores is that Ca^{2+} enters the spore through channels at the bottom and then exits through pumps at the top, resulting in a Ca^{2+} differential between the top and bottom of the cell (Salmi et al. 2011; ul Haque et al. 2007). The fact that chemically blocking the entry of Ca^{2+} along the bottom of the spore, but not inhibiting the activity of Ca^{2+}-ATPAse pumps, disrupts the gravity response favored the conclusion that Ca^{2+} entry through channels along the bottom of the spore was the more important transport change that induces the subsequent gravity-directed polarization changes. So what kinds of Ca^{2+} channels would be likely to be activated by gravity?

In both animal and plant cells, mechanosensitive channels have been proposed to be among the earliest molecular responders to the gravity stimulus. According to this hypothesis, some settling mass such as a statolith or protoplasm can exert a force on the bottom membrane of the cell to induce the opening of mechanosensitive cation channels. This would allow Ca^{2+} to enter from the extracellular matrix into the cytoplasm of cells. To test whether such channels are critical to initiate the gravity response in *Ceratopteris* spores, three different transcripts encoding members of the MscS (mechanosensitive channel, small conductance) family of MS channels in *Ceratopteris* were identified by an RNA-seq analysis of genes being expressed at the time cell polarization is fixed by gravity. We obtained the full-length sequence of one of these genes (*CrMscS1*) and, in collaboration with the laboratory of Paul Blount at the University of Texas Southwestern Medical Center, obtained evidence in support of the conclusion that it actually has Msc activity by showing that it could partially rescue a bacterial knockout of Msc. We used a modified version of the method of Bui et al. (2015) to stably transform *Ceratopteris* with a short hairpin construct of *CrMscS1*. This construct generated an RNAi product that resulted in stably suppressing the expression of this MscS gene in mature sporophytes of *Ceratopteris*.

Spores from these sporophytes are being generated now, and they will be used to test whether suppression of *CrMscS1* has any effect on the ability of gravity to polarize the cell. Mechanosensitive channels can be localized both on internal membranes and on the plasma membrane, and localization of the CrMscS1 protein in spores has not yet been confirmed. It is possible that only those mechanosensitive channels that are localized on the plasma membrane would mediate gravity-induced entry of Ca^{2+} at the bottom pole of the cell, so companion experiments are being carried out to learn whether *CrMscS1* or any of the other MscS-like genes in spores encode a channel that functions primarily on the plasma membrane. These localization experiments paired with gene suppression experiments will clarify whether MscS-like channels are key molecular transducers of the gravity response in single-spore cells.

14.3.2 Annexins

Annexins belong to a multigene family of multifunctional Ca^{2+}-dependent membrane proteins found in animal and plant cells (Clark et al. 2012). They play an important regulatory role in Ca^{2+}-mediated signaling (Laohavisit and Davies 2011), including polar development of plant cells and Ca^{2+} influx (Wang et al. 2010; Tang et al. 2014; Zhu et al. 2014). With specific reference to gravity responses, annexins appear to be targets of Ca^{2+} action because the gravity stimulus alters their distribution in pea plumules, as judged by immunocytochemical analyses (Clark et al. 2000). One way annexins could mediate gravity responses is by their documented function in the directional secretion of materials to the plasma membrane and the wall (Konopka-Postupolska et al. 2011). Such localized secretion would be required to establish and maintain polarity in plant cells (Belanger and Quatrano 2000), and also to mediate the response phase of root and shoot gravitropism, because this phase requires that wall polysaccharides and membrane lipids be delivered asymmetrically to the expanding cell periphery.

Both in *Arabidopsis* and in *Ceratopteris*, an early step in gravity-induced signaling is the asymmetrical opening of Ca^{2+} channels (Cannon 2016). Thus, another potential connection between the function of annexins and gravity responses is the ability of annexins to facilitate the passive Ca^{2+} transport into cells (Laohavisit and Davies 2009), a function demonstrated genetically in *Arabidopsis* roots (Laohavisit et al. 2012). To genetically test the potential role of annexin in mediating the gravity-induced entry of Ca^{2+} along the bottom of spore cells, stable transformants of an annexin gene in *Ceratopteris* have been successfully engineered.

In *Ceratopteris* two full-length annexin transcripts, *AnnCr1* and *AnnCr2*, have been identified. Because *AnnCr2* is more abundantly expressed during the first 24 h of spore germination, when gravity is fixing the polarity of the cell and the transcellular Ca^{2+} current peaks, this annexin was selected for the transformations. Using both the methods of Plackett et al. (2014, 2015) and Bui et al. (2015), we have successfully generated two different lines of transgenic *Ceratopteris* sporophytes stably altered in their expression of AnnCr2, one constitutively expressing a spore annexin and one suppressed in annexin expression. Analysis of the gravity response of spores harvested from these sporophytes will clarify whether AnnCr2 plays a significant role in mediating either the gravity-induced entry of Ca^{2+} into spores or the subsequent Ca^{2+}-mediated signaling steps needed for gravity to polarize these cells.

14.3.3 Apyrase

Apyrases are nucleoside triphosphate diphosphohydrolases (NTPDases) that hydrolyze nucleoside tri- and diphosphates but do not hydrolyze nucleoside monophosphates. Some apyrases are ectoapyrases, whose function is to limit the concentration of extracellular nucleotides like extracellular ATP and ADP (eATP, eADP). These nucleotides can function outside the cell membrane as cell regulators when they are released from the cytoplasm by cells as they grow, when they are wounded, or when they are mechanically stimulated (Roux and Steinebrunner 2007). Experiments have shown that certain ectoapyrases play a significant role in plant growth and development (Deng et al. 2015; Wu et al. 2007; Lim et al. 2014; Liu et al. 2012), perhaps by controlling the rate of transport of the growth hormone, auxin, which can rise or fall depending on the eATP (Tang et al. 2003; Liu et al. 2012).

In *Arabidopsis*, inhibiting apyrase activity using antibodies and chemical inhibitors caused the ATP in pollen growth media to increase (Wu et al. 2007). This experiment supported the hypothesis that APYrase 1 (APY1) and APY2 in *Arabidopsis* regulate eATP like their homologs in animal systems. Additional evidence supporting this hypothesis came from experiments showing that the genetic suppression of *APY1* increases eATP (Lim et al. 2014) and mimics the effects of increased eATP in modulating root skewing in *Arabidopsis* seedlings (Yang et al. 2015).

In animal systems, there are two main classes of ATP receptors that belong to the P2 superfamily, P2X and P2Y. The P2X receptors are transmitter-gated ion channels that allow Ca^{2+} ions to enter cells when ATP is bound. Although the P2Y receptors are not channels, the binding of ATP to them also leads to an increase in $[Ca^{2+}]_{cyt}$ through the release of intracellular stores of Ca^{2+}. In both cases, Ca^{2+} acts as a secondary messenger directing molecular events that lead to physiological changes. Recently, Choi et al. (2014) identified the first eATP receptor in plants. The receptor, DORN1 (DOes not Respond to Nucleotides 1), is a lectin receptor kinase I.9 (At5g60300). When this receptor is mutated, the ATP-induced increase in $[Ca^{2+}]_{cyt}$ is absent (Choi et al. 2014).

A role for apyrases and eATP in the gravity-directed polarization of *Ceratopteris* spores would be consistent with previously mentioned studies showing that Ca^{2+} channels in plant cells can be induced to open by extracellular nucleotides and that the concentration of these nucleotides is regulated by ectoapyrases (Clark and Roux 2011). Because a transcript encoding an apyrase-like enzyme is expressed in *Ceratopteris* spores during the period of polarity fixation, we hypothesized that extracellular nucleotides and apyrases might play a role in gravity-directed early growth and development in this system (Edwards and Roux 1998). Data in Bushart et al. (2013) and in Cannon (2016) show that spores release ATP as they germinate and grow and that applied nucleotides and a purinoceptor antagonist that blocks the effects of eATP suppress gravity-directed polarization. Collectively, these observations are consistent with the hypothesis that extracellular nucleotides could influence Ca^{2+} transport in *Ceratopteris* spores as they do in *Arabidopsis* (Demidchik et al. 2009, 2011).

Since previous experiments demonstrated that extracellular nucleotides could play a role in gravity-directed polarization of *Ceratopteris* spores, and an apyrase-like enzyme is expressed during this process, the pea (*Pisum sativum*) apyrase (psNTP9), a documented ectoapyrase (Thomas et al. 1999), was used for *Ceratopteris* transformation. Using the methods previously described by Plackett et al. (2014, 2015), we successfully generated one stably transformed line of transgenic

Ceratopteris sporophytes overexpressing psNTP9. Analyzing the gravity response of the spores produced by this sporophyte is in progress. These experiments will help determine the role of eATP in gravity-directed polarization. Based on data from previous studies, the constitutive expression of an ectoapyrase could decrease the concentration of eATP leading to a decrease in gravity-directed polarization. In addition to looking at the role of apyrase in the gravity response, it would also be valuable to determine if apyrase modulates eATP in *Ceratopteris* by measuring the concentration of eATP in the transgenic spores. Overall, transgenic spores overexpressing apyrase will help us determine the role of extracellular nucleotides and apyrase in the early growth and development of *Ceratopteris* spores, especially gravity-directed polarization.

14.3.4 Brief Note on a Nonviable Transformation

In addition to the successful transformations described above, we also attempted to ectopically express a Ca^{2+}-ATPase pump, whose identity, sequence, and central role in the gravity-induced transcellular Ca^{2+} transport were well documented by Bushart et al. (2013). In contrast to the multiple successful transformations described above, three separate attempts to generate transformants stably expressing the Ca^{2+}-ATPase gene, using the method of Plackett et al. (2014, 2015), all failed to yield any T_0 sporophytes with this transgene. This result could be an indication that some genes cannot be transformed into *Ceratopteris* callus using the biolistic method, but it could also indicate that enhanced expression of the Ca^{2+}-ATPase gene is incompatible with the growth of cells in the callus culture.

14.3.5 Platform for Testing Effects of Transformations on Ca^{2+} Transport Dynamics

Initial indirect tests of whether any of the individual transformations described above altered the gravity response of the spore could be done by assessing the directionality of rhizoid emergence, which is typically more than 85% downward in wild-type spores. However, more rigorous direct tests of whether any of the transgenic mutations altered gravity-induced changes in Ca^{2+} transport would require a device that could measure differential Ca^{2+} ion concentrations outside of individual cells. Self-referencing Ca^{2+}-selective electrodes, such as used by Chatterjee et al. (2000), or a lab-on-a-chip device, such as used by Salmi et al. (2011), would be apt platforms that could assess even subtle Ca^{2+} transport changes that resulted from the introduced transgene. An even more valuable device for these measurements would be the recently described autonomous lab on a chip that successfully measured gradual changes in gravity-induced transcellular Ca^{2+} polarization induced by gradual changes in the *g*-force applied to the spores (Park et al. 2017). This device, called a BioCD, has 32-spore wells laid out on a compact disc CD in four different rings at different distances from the center axis of rotation, so that when the disc is rotated, increasing *g*-forces are experienced by spores the further they are from the center. Because this device can be operated autonomously, it could be used in a spacecraft flying in low earth orbit, i.e., a microgravity environment, to assess the threshold of *g*-force needed to initiate the Ca^{2+} differential both in wild-type and in transgenic spores. Such an experiment would be a novel and valuable follow-up to earlier experiments examining the response of spores to the microgravity environment (Roux et al. 2003; Salmi et al. 2011).

14.4 Conclusion

This chapter had a twofold focus: an updated review of data demonstrating the key role of Ca^{2+} transport changes in transducing the gravity signal into directed polarization of *Ceratopteris* spore cells and documentation of multiple successful transformations of *Ceratopteris*. Along with the transformation descriptions, we discussed how these transformants can be used to test hypotheses on how the gravity stimulus directs cell polarization of spores. The review also provided a website to access the full-length sequences for CrMscS, Ann2Cr, and CrAPY.

Tests of how/if the transformations alter the gravity response in transgenic spores are now in progress. Because current hypotheses implicate roles for MscS, annexin, apyrase, and $\Delta[Ca^{2+}]_{cyt}$ in gravity-induced responses of the spores, both positive and negative outcomes of these tests will be unique contributions that will advance an understanding of molecular mechanisms affected by gravity in spore cells. Certainly, also, the value of *Ceratopteris* spores as a model system for understanding gravity responses in single cells and for understanding cell polarization has been greatly enhanced by the transformation methods described by Plackett et al. (2014, 2015) and by Bui et al. (2015) and verified here as readily applicable to multiple different gene transformations.

Acknowledgments We acknowledge the critically valuable assistance of Dr. Benjamin Smith at the University of Oklahoma who did the FLIM imaging and video assembly and Dr. Greg Clark at the University of Texas at Austin in facilitating all the research described here, which was supported by grants NNX13AM54G and NNX15AB85A from NASA awarded to S. Roux and G. Clark.

References

Aranda PS, LaJoie DM, Jorcyk CL (2012) Bleach gel: a simple agarose gel for analyzing RNA quality. Electrophoresis 33(2):366–369. https://doi.org/10.1002/elps.201100335

Battraw MJ, Hall TC (1990) Histochemical analysis of CaMV 35S promoter-beta-glucuronidase gene expression in transgenic rice plants. Plant Mol Biol 15(4):527–538

Behera S, Wang N, Zhang C, Schmitz-Thom I, Strohkamp S, Schultke S, Hashimoto K, Xiong L, Kudla J (2015) Analyses of Ca^{2+} dynamics using a ubiquitin-10 promoter-driven Yellow Cameleon 3.6 indicator reveal reliable transgene expression and differences in cytoplasmic Ca^{2+} responses in *Arabidopsis* and rice (*Oryza sativa*) roots. New Phytol 206(2):751–760. https://doi.org/10.1111/nph.13250

Belanger KD, Quatrano RS (2000) Polarity: the role of localized secretion. Curr Opin Plant Biol 3(1):67–72. https://doi.org/10.1016/s1369-5266(99)00043-6

Benfey PN, Chua NH (1989) Regulated genes in transgenic plants. Science (New York, NY) 244(4901):174–181. https://doi.org/10.1126/science.244.4901.174

Benfey PN, Ren L, Chua NH (1990) Tissue-specific expression from CaMV 35S enhancer subdomains in early stages of plant development. EMBO J 9(6):1677–1684

Bui LT, Cordle AR, Irish EE, Cheng CL (2015) Transient and stable transformation of *Ceratopteris richardii* gametophytes. BMC Res Notes 8:214. https://doi.org/10.1186/s13104-015-1193-x

Bushart TJ, Cannon AE, Ul Haque A, San Miguel P, Mostajeran K, Clark GB, Porterfield DM, Roux SJ (2013) RNA-seq analysis identifies potential modulators of gravity response in spores of Ceratopteris (Parkeriaceae): evidence for modulation by calcium pumps and apyrase activity. Am J Bot 100(1):161–174. https://doi.org/10.3732/ajb.1200292

Cannon AE (2016) Investigation of the role of extracellular nucleotide gradients in plant gravity responses. Dissertation, University of Texas, Austin

Chatterjee A, Porterfield DM, Smith PS, Roux SJ (2000) Gravity-directed calcium current in germinating spores of *Ceratopteris richardii*. Planta 210:607–610

Chatterjee A, Roux S (2000) Ceratopteris richardii: a productive model for revealing secrets of signaling and development. J Plant Growth Regul 19:284–289. https://doi.org/10.1007/s003440000032

Choi J, Tanaka K, Cao Y, Qi Y, Qiu J, Liang Y, Lee SY, Stacey G (2014) Identification of a plant receptor for extracellular ATP. Science (New York, NY) 343(6168):290–294. https://doi.org/10.1126/science.343.6168.290

Clark G, Roux SJ (2011) Apyrases, extracellular ATP and the regulation of growth. Curr Opin Plant Biol 14(6):700–706. https://doi.org/10.1016/j.pbi.2011.07.013

Clark GB, Morgan RO, Fernandez MP, Roux SJ (2012) Evolutionary adaptation of plant annexins has diversified their molecular structures, interactions and functional roles. New Phytol 196(3):695–712. https://doi.org/10.1111/j.1469-8137.2012.04308.x

Clark GB, Rafati DS, Bolton RJ, Dauwalder M, Roux SJ (2000) Redistribution of annexin in gravistimulated pea plumules. Plant Physiol Biochem 38(12):937–947. https://doi.org/10.1016/s0981-9428(00)01206-7

Demidchik V, Shang Z, Shin R, Colaco R, Laohavisit A, Shabala S, Davies JM (2011) Receptor-like activity evoked by extracellular ADP in Arabidopsis root epidermal plasma membrane. Plant Physiol 156(3):1375–1385. https://doi.org/10.1104/pp.111.174722

Demidchik V, Shang Z, Shin R, Thompson E, Rubio L, Laohavisit A, Mortimer JC, Chivasa S, Slabas AR, Glover BJ, Schachtman DP, Shabala SN, Davies JM (2009) Plant extracellular ATP signalling by plasma membrane NADPH oxidase and Ca2+ channels. Plant J 58(6):903–913. https://doi.org/10.1111/j.1365-313X.2009.03830.x

Deng S, Sun J, Zhao R, Ding M, Zhang Y, Sun Y, Wang W, Tan Y, Liu D, Ma X, Hou P, Wang M, Lu C, Shen X, Chen S (2015) Populus Euphratica APYRASE2 enhances cold tolerance by modulating vesicular trafficking and extracellular ATP in Arabidopsis plants. Plant Physiol 169(1):530-+. https://doi.org/10.1104/pp.15.00581

Edwards E, Roux S (1998) Influence of gravity and light on the developmental polarity of Ceratopteris Richardii fern spores. Planta 205(4):553–560

Edwards ES, Roux SJ (1994) Limited period of graviresponsiveness in germinating spores of Ceratopteris Richardii. Planta 195(1):150–152. https://doi.org/10.1007/bf00206304

Grefen C, Donald N, Hashimoto K, Kudla J, Schumacher K, Blatt MR (2010) A ubiquitin-10 promoter-based vector set for fluorescent protein tagging facilitates temporal stability and native protein distribution in transient and stable expression studies. The Plant J cell Mol Biol 64(2):355–365. https://doi.org/10.1111/j.1365-313X.2010.04322.x

Hickok L, Warne T, Fribourg R (1995) The biology of the fern *Ceratopteris* and its use as a model system. Int J Plant Sci 156:332–345

Hickok LG, Warne TR, Slocum MK (1987) *Ceratopteris richardii*: applications for experimental plant biology. Am J Bot 74:1304–1316

Hoover EE, Squier JA (2013) Advances in multiphoton microscopy technology. Nat Photon 7(2):93–101

Keinath NF, Waadt R, Brugman R, Schroeder JI, Grossmann G, Schumacher K, Krebs M (2015) Live cell imaging with R-GECO1 sheds light on flg22- and chitin-induced transient $[Ca^{2+}]_{cyt}$ patterns in *Arabidopsis*. Mol Plant 8(8):1188–1200. https://doi.org/10.1016/j.molp.2015.05.006

Konopka-Postupolska D, Clark G, Hofmann A (2011) Structure, function and membrane interactions of plant annexins: an update. Plant Sci 181(3):230–241. https://doi.org/10.1016/j.plantsci.2011.05.013

Krebs M, Held K, Binder A, Hashimoto K, Den Herder G, Parniske M, Kudla J, Schumacher K (2012) FRET-based genetically encoded sensors allow high-resolution live cell imaging of Ca^{2+} dynamics. The Plant J cell Mol Biol 69(1):181–192. https://doi.org/10.1111/j.1365-313X.2011.04780.x

Laohavisit A, Davies JM (2009) Multifunctional annexins. Plant Sci 177(6):532–539. https://doi.org/10.1016/j.plantsci.2009.09.008

Laohavisit A, Davies JM (2011) Annexins. New Phytol 189(1):40–53. https://doi.org/10.1111/j.1469-8137.2010.03533.x

Laohavisit A, Shang Z, Rubio L, Cuin TA, Very AA, Wang A, Mortimer JC, Macpherson N, Coxon KM, Battey NH, Brownlee C, Park OK, Sentenac H, Shabala S, Webb AA, Davies JM (2012) *Arabidopsis* annexin1 mediates the radical-activated plasma membrane Ca^{2+}- and K^{+}-permeable conductance in root cells. Plant Cell 24(4):1522–1533. https://doi.org/10.1105/tpc.112.097881

Lim MH, Wu J, Yao J, Gallardo IF, Dugger JW, Webb LJ, Huang J, Salmi ML, Song J, Clark G, Roux SJ (2014) Apyrase suppression raises extracellular ATP levels and induces gene expression and Cell Wall changes characteristic of stress responses. Plant Physiol 164(4):2054–2067. https://doi.org/10.1104/pp.113.233429

Liu X, Wu J, Clark G, Lundy S, Lim M, Arnold D, Chan J, Tang W, Muday GK, Gardner G, Roux SJ (2012) Role for apyrases in polar auxin transport in *Arabidopsis*. Plant Physiol 160(4):1985–1995. https://doi.org/10.1104/pp.112.202887

Muthukumar B, Joyce BL, Elless MP, Stewart CN (2013) Stable transformation of ferns using spores as targets: *Pteris vittata* and *Ceratopteris thalictroides*. Plant Physiol 163(2):648–658. https://doi.org/10.1104/pp.113.224675

Nakazato T, Jung M-K, Housworth EA, Rieseberg LH, Gastony GJ (2006) Genetic map-based analysis of genome structure in the homosporous fern *Ceratopteris richardii*. Genetics 173(3):1585–1597

Norris SR, Meyer SE, Callis J (1993) The intron of *Arabidopsis thaliana* polyubiquitin genes is conserved in location and is a quantitative determinant of chimeric gene-expression. Plant Mol Biol 21:895–906

Odell JT, Nagy F, Chua NH (1985) Identification of DNA sequences required for activity of the cauliflower mosaic virus 35S promoter. Nature 313(6005):810–812

Park J, Salmi ML, Wan Salim WW, Rademacher A, Wickizer B, Schooley A, Benton J, Cantero A, Argote PF, Ren M, Zhang M, Porterfield DM, Ricco AJ, Roux SJ, Rickus JL (2017) An autonomous lab on a chip for space flight calibration of gravity-induced transcellular calcium polarization in single-cell fern spores. Lab Chip 17(6):1095–1103. https://doi.org/10.1039/c6lc01370h

Plackett ARG, Huang LD, Sanders HL, Langdale JA (2014) High-efficiency stable transformation of the model Fern species *Ceratopteris richardii* via microparticle bombardment. Plant Physiol 165(1):3–14. https://doi.org/10.1104/pp.113.231357

Plackett ARG, Rabbinowitsch EH, Langdale JA (2015) Protocol: genetic transformation of the fern *Ceratopteris richardii* through microparticle bombardment. Plant Methods 11. https://doi.org/10.1186/s13007-015-0080-8

Roux SJ, Chatterjee A, Hillier S, Cannon T (2003) Early development of fern gametophytes in microgravity. Adv Space Res Off J Committee Space Res (COSPAR) 31(1):215–220
Roux SJ, Steinebrunner I (2007) Extracellular ATP: an unexpected role as a signaler in plants. Trends Plant Sci 12(11):522–527. https://doi.org/10.1016/j.tplants.2007.09.003
Salmi ML, Bushart TJ, Stout SC, Roux SJ (2005) Profile and analysis of gene expression changes during early development in germinating spores of *Ceratopteris richardii*. Plant Physiol 138(3):1734–1745. https://doi.org/10.1104/pp.105.062851
Salmi ML, ul Haque A, Bushart TJ, Stout SC, Roux SJ, Porterfield DM (2011) Changes in gravity rapidly alter the magnitude and direction of a cellular calcium current. Planta 233(5):911–920. https://doi.org/10.1007/s00425-010-1343-2
Stout SC, Clark GB, Archer-Evans S, Roux SJ (2003) Rapid and efficient suppression of gene expression in a single-cell model system, *Ceratopteris richardii*. Plant Physiol 131(3):1165–1168. https://doi.org/10.1104/pp.016949
Sunilkumar G, Mohr L, Lopata-Finch E, Emani C, Rathore KS (2002) Developmental and tissue-specific expression of CaMV 35S promoter in cotton as revealed by GFP. Plant Mol Biol 50(3):463–474
Talamond P, Verdeil JL, Conejero G (2015) Secondary metabolite localization by autofluorescence in living plant cells. Molecules (Basel, Switzerland) 20(3):5024–5037. https://doi.org/10.3390/molecules20035024
Tang W, Brady SR, Sun Y, Muday GK, Roux SJ (2003) Extracellular ATP inhibits root gravitropism at concentrations that inhibit polar auxin transport. Plant Physiol 131(1):147
Tang WX, He YH, Tu LL, Wang MJ, Li Y, Ruan YL, Zhang XL (2014) Down-regulating annexin gene GhAnn2 inhibits cotton fiber elongation and decreases Ca^{2+} influx at the cell apex. Plant Mol Biol 85(6):613–625. https://doi.org/10.1007/s11103-014-0208-7
Thomas C, Sun Y, Naus K, Lloyd A, Roux S (1999) Apyrase functions in plant phosphate nutrition and mobilizes phosphate from extracellular ATP. Plant Physiol 119(2):543–552
ul Haque A, Rokkam M, De Carlo AR, Wereley ST, Roux SJ, Irazoqui PP, Porterfield DM (2007) A MEMS fabricated cell electrophysiology biochip for in silico calcium measurements. Sens Actuators B Chem 123(1):391–399. https://doi.org/10.1016/j.snb.2006.08.043
Wang LK, Niu XW, Lv YH, Zhang TZ, Guo WZ (2010) Molecular cloning and localization of a novel cotton annexin gene expressed preferentially during fiber development. Mol Biol Rep 37(7):3327–3334. https://doi.org/10.1007/s11033-009-9919-2
Watson JM, Fusaro AF, Wang MB, Waterhouse PM (2005) RNA silencing platforms in plants. FEBS Lett 579(26):5982–5987. https://doi.org/10.1016/j.febslet.2005.08.014
Wen CK, Smith R, Banks JA (1999) ANI1. A sex pheromone-induced gene in *Ceratopteris* gametophytes and its possible role in sex determination. Plant Cell 11(7):1307–1318
Wilkinson JE, Twell D, Lindsey K (1997) Activities of CaMV 35S and nos promoters in pollen: implications for field release of transgenic plants. J Exp Bot 48(2):265–275. https://doi.org/10.1093/jxb/48.2.265
Wu J, Steinebrunner I, Sun Y, Butterfield T, Torres J, Arnold D, Gonzalez A, Jacob F, Reichler S, Roux SJ (2007) Apyrases (nucleoside triphosphate-diphosphohydrolases) play a key role in growth control in *Arabidopsis*. Plant Physiol 144(2):961–975. https://doi.org/10.1104/pp.107.097568
Xiong TC, Ronzier E, Sanchez F, Corratge-Faillie C, Mazars C, Thibaud JB (2014) Imaging long distance propagating calcium signals in intact plant leaves with the BRET-based GFP-aequorin reporter. Front Plant Sci 5:43. https://doi.org/10.3389/fpls.2014.00043
Yang N-S, Christou P (1990) Cell type specific expression of a CaMV 35S-GUS gene in transgenic soybean plants. Dev Genet 11(4):289–293. https://doi.org/10.1002/dvg.1020110407
Yang XY, Wang BC, Farris B, Clark G, Roux SJ (2015) Modulation of root skewing in *Arabidopsis* by apyrases and extracellular ATP. Plant Cell Physiol 56(11):2197–2206. https://doi.org/10.1093/pcp/pcv134
Zhu JE, Wu XR, Yuan SJ, Qian D, Nan Q, An LZ, Xiang Y (2014) Annexin5 plays a vital role in Arabidopsis pollen development via Ca^{2+}-dependent membrane trafficking. PLoS One 9(7):15. https://doi.org/10.1371/journal.pone.0102407

Chapter 15
Secondary Metabolites of Ferns

Janos Vetter

15.1 Introduction

A number of essential metabolic processes and metabolites are common to all plants, including photosynthesis, respiration, and protein and nucleic acid metabolism. These processes and their components are very similar (and in some cases, identical) across different systematic groups. They are necessary for the primary metabolism of plants and are the subjects of plant physiology and biochemistry. Differences may be noted among certain systematic or ecological groups of plants, such as differences in photosynthesis among C_3, C_4, and Crassulacean acid metabolism plants.

Secondary metabolism is the sum of specific biochemical processes. These reactions (and the participating molecules) are specific for a certain group, such as a family, genus, or species. The secondary metabolites from a given group are synthesized in similar or identical ways; furthermore, the precursor molecules (the main substrates for biosynthesis) are the same. Differences may be noted in the reactions of biosynthetic pathways (e.g., the participation of other enzymes, different factors of reactions). Large and diverse groups of molecules are formed by the secondary metabolites. These substances are synthesized from certain intermediary or final products of primary metabolism, but their functions can be very different. Their biological roles include comprehensive protection against microbes or different animals; they also are attractants and may play a role in plant-plant or plant-animal relations.

A plant's secondary metabolites may be grouped in a variety of logical ways. For example, one can consider the chemical structure of molecules and group alkaloids based on their "skeleton" type: tropane alkaloids, isoquinoline alkaloids, etc. Alternatively, the biological or physiological effects can be used to group the

J. Vetter (✉)
Department of Botany, University of Veterinary Sciences, Budapest, Hungary
e-mail: Vetter.Janos@univet.hu

H. Fernández (ed.), *Current Advances in Fern Research*,
https://doi.org/10.1007/978-3-319-75103-0_15

metabolites in a pharmacological sense. Thirdly, the principle of biogenetics can be used in a grouping based on the modes and precursors of their biosynthesis. In the latter approach, the plant's secondary metabolites would have five main groups: saccharides, phenolics, polyketides, terpenoids, and nitrogen-containing metabolites (mainly alkaloids). The following is a brief general overview of these groups:

1. *Saccharides.* Saccharoid compounds originate directly from sugars or from similar derivatives. This group includes simple (or more complicated) sugar molecules, uronic acids, amino sugars, and glycosides. According to the non-sugar parts (aglycons), the main subgroups are phenol-, N-, tio-, and cyanogenic glycosides.
2. *Phenolics.* This group of metabolites is very large and important. A common biosynthesis process uses the shikimic acid pathway; its precursors originate from universal glycolysis (as a C_3 molecule) and pentose phosphate cycles (as a C_4 molecule). The produced shikimic acid is the starting point for the biosynthesis of different phenolics, including cinnamoic derivatives, coumarins, tannins, and flavan derivatives (including isoflavonoids and phytoestrogens).
3. *Polyketides.* This smallest group of secondary metabolites includes β-polyketone derivatives, composed of specific C_2 units $[(CH_2\text{-}CO)_n]$. Their synthesis is attached to universal fatty acids by Ac-CoA participation.
4. *Terpenoids.* These terpene-like compounds are composed of C_5-units. The primary precursors are Ac-CoA molecules (as acetyl donors) from reactions of respiration; three Ac-CoA molecules produce mevalonic acid (a C_6 molecule). Through different enzyme reactions, the mevalonic pathway produces hemiterpenes (C_5), monoterpenes (C_{10}), sesquiterpenes (C_{15}), diterpenes (C_{20}), triterpenes (C_{30}), tetraterpenes (C_{40}), and politerpenes (in which the number of C-atoms is greater than 40).
5. *Nitrogen-containing metabolites.* These compounds are synthesized from some universal molecules (e.g., amino acids, nucleic acid basis molecules) in one or more steps. The most important reactions are decarboxylation and transamination. The most efficient molecules are the alkaloids; the number of described alkaloids today exceeds 14,000. Alkaloid biosynthesis starts from a given amino acid; its main purpose is the production of N-heterocycles.

The number and diversity of plant secondary metabolites has created a reservoir of potentially defensive chemical compounds. Some of these compounds occur in the wide spectrum of plant taxa, whereas other molecules are only found in certain taxa (e.g., in species, genera, or families). The protecting compounds are stored in different but separate compartments (mainly in vacuoles); thus, the probability of self-intoxication is very low or nonexistent (Mithöfer and Boland 2012). Defense reactions, based on secondary plant metabolites, have an inducible characteristic. The starting "signal" is the attack (entering) of herbivorous animals. The exact mode of action of these defense reactions is partly unknown; in general, the effects of secondary metabolites include disrupting animal cell membranes, influencing transport processes, and inhibiting various metabolic processes of animals (Mithöfer and Boland 2012).

This chapter summarizes the chemical, biochemical, and botanical aspects of the secondary metabolites of ferns. The discovery of these new molecules through isolation and structural/physiological investigations illuminates the chemical and biological properties of the secondary metabolites of ferns.

15.2 Secondary Metabolites of Ferns

15.2.1 Saccharides

The most characteristic active substances are the cyanogenic glycosides, the glycosides of α-hydroxynitrils (Fig. 15.1). These molecules are enzymatically hydrolyzed (by β-glucosidase and by hydroxynitrile lyase enzymes) and release HCN molecules. The cyanogenic glycosides are located in the vacuoles of cells. The β-glucosidase enzymes are located in the apoplastic space, are bound to cell walls, or are found in vesicles or chloroplasts. Liberation of HCN (cyanogenesis) begins with the mechanical destruction of a living plant structure.

Cyanogenic glycosides occur mainly in angiosperms and gymnosperms; however, some species of Pteridophyta contain certain cyanogenic glycosides. A Brazilian study examined the occurrence of these molecules in 19 fern species on a monthly basis (Santos et al. 2005). Cyanogenic glycosides (based on picric acid tests) were continuously recorded in the species *Pteridium aquilinum var. arachnoideum* and *Microgramma vacciniifolia* only. Prunasin is the most important cyanogenic glycoside in ferns (Fig. 15.2). The different varieties of *Pteridium aquilinum* (fam.: Dennsteadtiaceae) were found to contain different quantities of prunasin. In the early phase of the plant, the prunasin concentration in crosiers varied between 1.8 and 107.7 mg/g dry matter (DM) (Alonso-Amelot and Oliveros-Bastidas 2005). Jiang et al. (2013) reported the presence of vicianine (a glycoside with an aromatic substituent/phenol group and a vicianose disaccharide sugar component) in dried

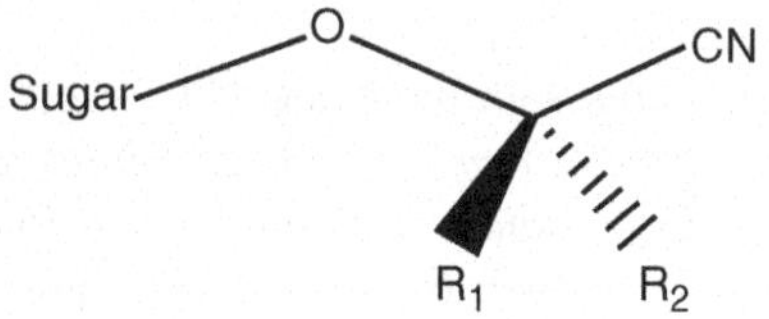

Fig. 15.1 General chemical structure of cyanogenic glycosides

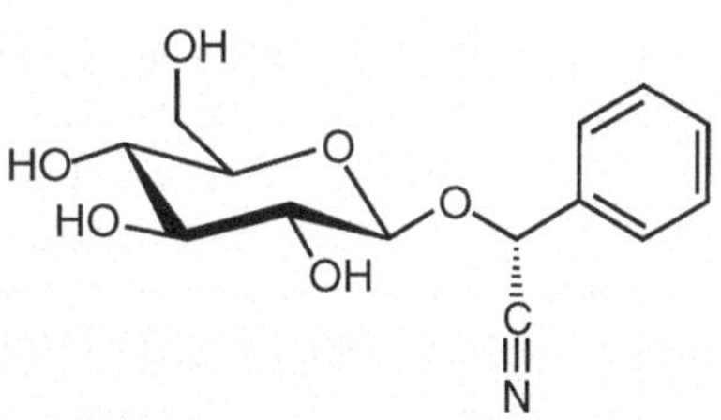

Fig. 15.2 Structure of prunasin cyanogenic glycoside

Glaphyropteridopsis erubescens (fam. Thelypteridaceae) in greater quantity. New data from Africa (Oloyede et al. 2014) indicated the presence of cyanide in leaflets of *Nephrolepis exaltata*, although the content (0.12–0.53 mg cyanide per 100 g FM= Fresh Matter) was much lower (by orders of magnitude) than the content of cyanogenic glycosides found in *Pteridium* taxa. Cyanogenic glycosides play a role in plant protection, acting as a feeding deterrent against insects and herbivore animals (da Costa et al. 2012). The release of HCN, which is one of the most effective plant toxins, inhibits cellular respiration by blocking cytochrome c oxidase in mitochondria.

15.2.2 Phenolics

To determine the total content of phenolics in ferns, Folin-Ciocalteous reagent is frequently used. The total level of phenolics is generally calculated using the gallic acid standard, with the concentration given in gallic acid equivalent units (GAE). Different methodological details (e.g., substance, mode and character of extraction, reagent, evaluation) can affect the chemical data remarkably.

Investigations have analyzed and compared the total phenolic content (TPC) for some common fern species using identical methods (Lai and Lim 2011; Gupta et al. 2014). TPC values for some Indian ferns (Gupta et al. 2014) fluctuated between 4.33 mg GAE/g FM and 77.4 mg GAE/g FM for *Lindsaea odorata* and *Aleuritopteris flava*, respectively. However, the majority of fern species had TPC values between 5 and 22 mg GAE/g FM, with the arithmetical mean (n = 8) of TPC data being 19.1 mg GAE/g FM. Investigations by Lai and Lim (2011) established different groups of ferns:

1. Ferns with the highest TPC content (21–32 mg GAE/g FM): *Cyathea latebrosa, Cibotium barometz, Drynaria quercifolia, Dicranopteris linearis,* and *Blechnum orientale*
2. Ferns with high TPC levels (10–19 mg GAE/g FM): *Adiantum raddianum* and *Pteris vittata)*
3. Ferns with moderate TPC levels (5–9.9 mg GAE/g FM): *Acrostichum aureum, Nephrolepis biserrata, Diplazium esculentum, Pityrogramma calomelanos, Lygopodium circinnatum,* and *Pyrrosia nummularifolia*
4. Ferns with the lowest TPC quantities (3.20 and 2.00 mg GAE/g FM, respectively): *Pteris venulosa* and *Microsorium punctatum*

To compare TPC values among different species, it seems logical to perform the evaluation according to their systematic position, such as at the level of genera or families. The two *Pteris* species (*P. vittata* and *P. venulosa*) had different TPC contents. The species from the *Pteridaceae* family (*Adiantum radianum, Pityrogramma calomelanos, Pteris venulosa, P. vittata, Acrostichum aureum*) had TPC contents between 4.2 and 180 mg GAE/g FM, with the contents seeming to be independent of the family. The three Polypodiaceae species showed a wide distribution, with phenolic contents between 2.3 and 25 mg GAE/g FM.

Table 15.1 Phenolic contents of different fern families

Fern family	Number of data	Phenolic content mg/100 g DW	± SD	Coefficient of variation (%)	Median	Minimum value	Maximum value
Dryopteridaceae	8	1493	513	35	1355	820	2350
Blechnaceae	4	2596	577	23	2575	2095	3140
Pteridaceae	10	1976	2126	108	1335	320	7742
Polypodiaceae	10	1485	806	55	1660	200	2540

Based on data of Alonso-Amelot et al. 2004, 2007; Gupta et al. 2014; Lai and Lim 2011

Based on the data from previous studies (Lai and Lim 2011; Gupta et al. 2014; Alonso-Amelot et al. 2004, 2007), the total phenolic content of ferns was grouped by family. Table 15.1 lists the calculated biometrical parameters for the number of available data, arithmetical means of phenolic contents, standard deviations, coefficients of variation, and median, minimum, and maximum values. The average phenolic contents (means) for three families have a coefficient of variation of 23–55%; however, this value is very high (108%) for the Pteridaceae family. A simple comparison of arithmetical means indicates remarkable differences between the families, with the greatest value for the Blechnaceae family. However, significant, confirmative differences were not found between the phenolic contents of the given fern families (see Table 15.1).

An accepted fact in plant physiology and ecology is that plant phenolics are induced in response to stress factors in the environment. These molecules are thought to protect against damage from ultraviolet-B radiation (Lai and Lim 2011). Ecological factors can induce significantly higher levels of TPC, such as the increased ultraviolet radiation in the ecosystems of higher-elevation mountains. Species with high phenolic content can grow well in mountain habitats where sunlight intensity is higher, at altitudes up to 1500–1700 m. On the other hand, the species *Adiantum aureum* has only 9.45 mg GAE/g FM, probably because of its capability for shade tolerance (Lai and Lim 2011). It appears that the systematic position of a fern is not a strong regulating factor for TPC. However, certain environmental parameters (e.g., altitude, ultraviolet radiation, temperature, light conditions) can affect the actual level of phenolics (Alonso-Amelot et al. 2004, 2007).

Secondary metabolites can occur as free or bound molecules. The free molecules can directly produce one or more reactions, whereas the bound forms are responsible for transport and accumulation. The liberation of free forms is an essential precondition of later biological effects. The fern species *Ceterach officinarum, Asplenium trichomanes,* and *Adiantum radiantum* were analyzed in Serbia (Durdevic et al. 2007) for free and bound phenolics. The samples were pretreated by boiling in 2 N NaOH solution for 1 h. The amounts of total free phenolics exceeded their bound forms in *Ceterach officinarum,* with the leaves containing 11.7 mg/g DM of free phenolics and 9.13 mg/g DM of bound phenolics. The rhizomes had 2.81 mg/g DM of free phenolics and 1.96 mg/g DM of bound phenolics. *Asplenium trichomanes* had the same characteristics in leaves; however, in rhizomes, the quantity of bound phenolics was higher than in the leaves.

Han and co-workers (2015) summarized the phytochemical investigations of *Dryopteris* species. Due to the important biological (anthelmintic) effects of *Dryopteris* species, the phytochemistry of these ferns has been studied since the nineteenth century. The number of their identified chemical compounds is high (approximately 260 until 2015); 15 compounds of phenolics were published (Han et al. 2015). The main sources of these molecules were the species *Dryopteris sublaeta, D. fragrans, D. ryotoana,* and *D. erythrosora.* Five derivatives belong to the free phenolics, including some metoxy derivatives (Fig. 15.3). These molecules have one or three aromatic rings, with two or four metoxy substituents. The majority of phenolics are phenol-glycosides, such as the small arbutin (Fig. 15.4) or the fragranoside G and fragranoside F (Fig. 15.5) (Han et al. 2015). The isolated phenol glycosides of *Dryopteris* genus have one or two aromatic rings.

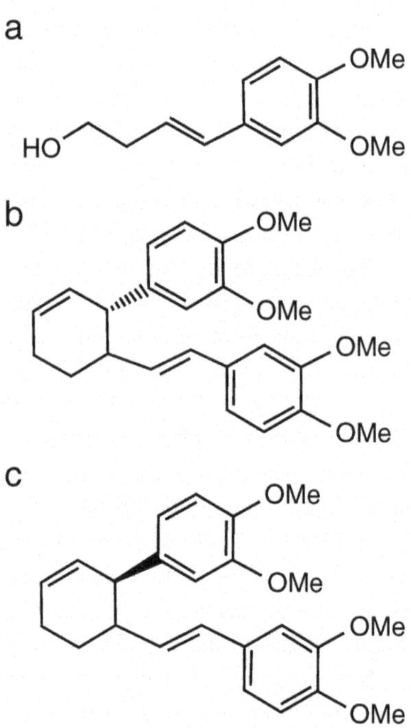

Fig. 15.3 (**a**) Structure of (E)-4-(3,4-Dimethoxyphenyl) but-3-en-1-ol methoxyderivative isolated from *Dryopteris fragrans* (**b**) Structure of *trans*-3-(3,4-Dimethoxyphenyl)-4-[(E)-3,4-dimethoxystyryl] cyclohex-1-ene (isolated from *Dryopteris fragrans*) (**c**) Structure of *cis*-3-(3,4-Dimethoxyphenyl)-4-[(E)-3,4-dimethoxystyryl] cyclohex-1-ene (isolated from *Dryopteris fragrans*)

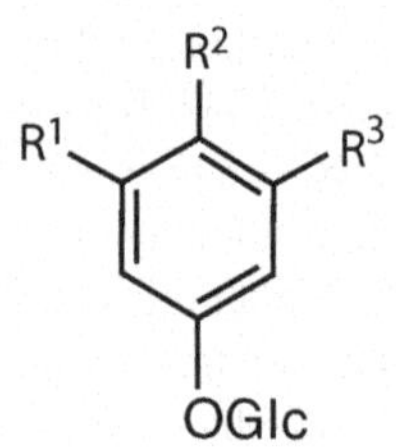

Fig. 15.4 Structure of the arbutin phenol-glycoside (R^1, R^2 = H; R^3 = OH; Glc = Glucose)

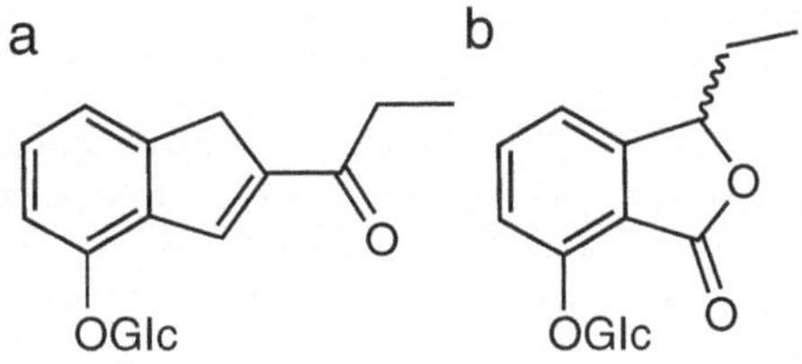

Fig. 15.5 Chemical structure of Fragranoside G (A) and Fragranoside F (B)

The isolation and quantification of phenolics with high-performance liquid chromatography were conducted for *Pteris ensiformis* (Pteridaceae) by Chen and co-workers (2007). Their group measured the quantity of eight phenolics, three of which were isolated first. The authors used whole fern plants and produced an aqueous extract. The highest concentration was found for a caffeic derivative (9.99 mg/g DM), whereas the lowest content was found for 5-O-caffeoylquinic acid (0.86 mg/g DM); the other contents were somewhere between these high and low values. In a test of antioxidant capacity, the two di-O-caffeoylquinic acid derivatives had the highest 1, 1-diphenyl-2-picrylhydrazyl (DPPH) scavenging potential and the strongest 6-hydroxy-2,5,7,8-tetramethylchroman-2 carboxylic acid (Trolox) antioxidant capacity. The data from Chen et al. (2007) indicate one essential step in the biological roles of phenolics: the fundamental antioxidant potential of ferns.

Tannins

Tannins are great macromolecules consisting of polyphenols. Their main groups are hydrolysable tannins composed of gallic acid monomers and condensed tannins composed of flavone monomers. The former group of tannin molecules has lower molecular masses (500–3000 Da), whereas the latter group has higher molecular masses (1000–20,000 Da). Tannins occur in all groups of higher plants (in Pteridophyta, gymnosperm, and angiosperm plants). Their main biological function is protection against phytophagous insects. They also can decrease the nutrient uptake of animals: high tannin levels diminish the palatability of forages, the intensity of digestion is reduced, and certain rejection of food is induced. The most problematic fern species (*Pteridium aquilinum*, bracken fern) has approximately 120 mg/g of condensed tannin content (da Costa et al. 2012). Tannins and tannin derivatives probably have no direct role in bracken toxicity, but they can cause irritation of the gastrointestinal tract in bracken-fed animals. The progressive accumulation of phenolics (including condensed tannins as high molecular phenolics) was recorded with frond expansion for *Pteridium* species (Alonso-Amelot et al. 2004). Tannin levels were correlated positively with the increasing elevation of mountains in *P. caudatum* and *P. arachnoideum*.

Flavonoids

Flavonoids are a large group of polyphenols (more than 4000 compounds are known) with high and mainly beneficial biological activity. Their name originates from the Latin *flava* (meaning yellow). The flavonoids have a C_{15}-carbon skeleton (two phenyl rings and a heterocyclic ring in the middle: C_6-C_3-C_6 structure). The flavone backbone has a 2-phenyl-1,4-benzopyrone structure.

The majority of investigations include the fractionation of flavonoids, isolation of constituents, and the determination of their chemical structure. However, the number of publications about the quantification of individual components or of total flavonoid concentrations is very low. The qualitative aspects of fern flavonoids are better known than are the quantitative characteristics. Total flavonoid levels were measured in the leaves, stems, rachis, and roots of *Dryopteris erythrosora* (Zhang et al. 2012). Their contents fluctuate from 2.1 to 8.26%.

Flavonoid occurrence was analyzed in different samples of *Dryopteris filix-mas* and *D. dilatata* from different locations, years, and fractions by Vogler et al. (2012). Data were obtained from thin-layer chromatography and high-performance liquid chromatography for both ferns. Very characteristic distributions were found: the flavonoids kaempferol-rutinoside and rutin occurred exclusively in *D. filix-mas*, whereas the caffeoyltartronate and kaempferol-3-O-β-D-glucopyranosido-7-O-α-L-rhamnopyranoside molecules were detected in *D. dilatata*. These data may indicate a species-dependent flavonoid distribution. A review by Han et al. (2015) contains a long list of flavonoids isolated from the *Dryopteris* genus, with 43 flavonoids obtained from five *Dryopteris* species (*D. erythrosora, D. fragrans, D. villarii, D. crassirhizoma* and *D. sublaeta*) based on data from 15 publications.

A comparative analysis of flavonoids was carried out in China by Wang et al. (2017). The total flavonoid content of 49 fern species ranged from 3.3 mg/g (*Selaginella uncinata*) to 191.7 mg/g (*Woodwardia japonica*). Samples of this analysis were gathered from different habitats of Tianmu Mountain, which has characteristically various ecological conditions in altitude, soil condition, and light intensity. Fern samples originating from habitats with less sunlight had higher flavonoid contents than those growing in strong sunlight. Fern species from higher altitudes (between 500 and 1000 m) had higher total flavonoid content than those growing at lower altitudes (under 500 m). However, these associations were not confirmed by statistical methods.

Protoflavones are specific, relatively oxidized flavonoids with an unusual non-aromatic B-ring (Fig. 15.6) and a hydroxyl group at C-1′ (Hunyadi et al. 2014). Their main method of biosynthesis is largely unknown, but the probable first step is the oxidation of the 4′-hydroxyflavon precursor (e.g., apigenin, genkwanin). The next phase of synthesis is the reduction of the 4-oxogroup and/or saturation of one or both double bonds in the B-ring (Lin et al. 2007). The biological effects of protoapigenone and its analogues have been studied extensively. These molecules have strong cytotoxic activity against various cancer cell lines (e.g., lung, liver, ovarian, breast), as discussed in the review by Hunyadi et al. (2014).

Fig. 15.6 Chemical structure of the protoflavones (Hunyadi et al. 2014)

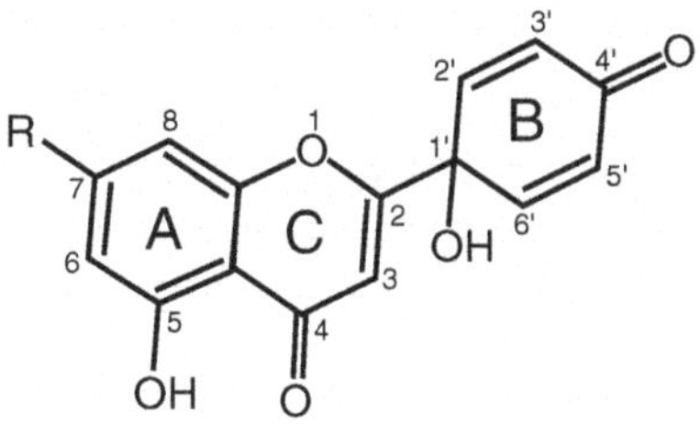

Coumarins

Coumarin and its derivatives can cause specific problems in blood clotting. In the plant kingdom—mainly in certain angiosperm plant groups—some coumarin derivatives occur; they can later produce bishydroxicoumarin molecules that cause inhibition of the blood clotting process. In the fern species *Polypodium hartatum* (Polypodiaceae), three such derivatives (coumaric acid glycosides) were isolated: trans-melilotoside; trans-coumaric acid-3-*O*-β-D-glucopyranoside, and trans-coumaric acid-4-*O*-D-glycopyranoside (Yao et al. 2016).

Antioxidant Activity of Ferns

The normal biochemical processes of life are oxygen and oxidation dependent. However, reactive oxygen radicals may be overproduced under the effects of various pathological and environmental, thus produce oxidative stress. Oxidative status has a determining role in the induction and development of many diseases, such as cardiovascular problems, cancer, atherosclerosis, rheumatic damage, or the different processes of aging. Antioxidant agents or effects can help to overcome or at least improve the problem. There are synthetic antioxidants for this use, but they can have side effects and unexpected consequences. Several components of plants have demonstrated antioxidant capacity, such as β-carotene, ascorbic acid, α-tocopherol, phenolics (including flavonoids), and others. Ferns and fern allies have considerable antioxidant potential, mainly via their phenolics. The antioxidant activity (potential) of a plant or plant extract can be estimated by in vitro tests. These reactions are based on the reduction capacity of a model compound in a regulated chemical process. One of the most frequent methods is the measurement of the reduction degree for the free radical 1,1-dyphenil-2-picrylhydrazil (DPPH).

Different samples of *Pteris vittata* were compared for certain phytochemical properties (Singh et al. 2015). The total phenolic and flavonoid contents showed remarkable variations; the means were 0.718 mg ± 0.53 GAE/g and 0.455 mg ± 0.37 QE/g(Quercetin Equivalent/g) for phenolics and flavonoids, respectively. The measured tendency of data was unexpected, while all analyzed plants belonged to one species (*Pteris vittata*). Unfortunately, the ecological conditions of habitat were not described in the cited publication. The great differences reported for total

Table 15.2 DPPH radical scavenging activity in different fern families

Fern family	Number of data(n)	DPPH scavenging activity IC_{50} (mg/mM),mean	± SD	Coefficient of variation(%)	Median	Minimum value	Maximum value
Dryopteridaceae	11	0.0022	0.0191	869	0.018	0.0059	0.071
Blechnaceae	4	0.0736	0.076	104	0.073	0.072	0.14
Pteridaceae	11	0.810	0.833	103	0.54	0.087	2.70
Polypodiaceae	11	0.407	0.081	20	0.054	0.014	2.51
Aspleniaceae	4	4.86	8.28	171	1.08	0.071	17.21

Based on data of Ding et al. 2008; Gupta et al. 2014; Lai et al. 2009; Lai and Lim 2011; Valizadeh et al. 2015

phenolic and flavonoid levels may be consequences of various ecological or other stress factors.

An important aspect of biochemical-physiological studies is the investigation of the antioxidant power from a given plant's isolated compounds. Eight such molecules were isolated from *Pteris ensiformis* (Chen et al. 2007). The contents of these derivatives fluctuated between 0.86 mg/g DM and 9.99 mg/g DM for 7-*O*-caffeoylhydroxymaltol 3-*O*-β-D-glucopyranoside and 5-*O*-caffeoyl-quinic acid, respectively. All of these compounds produced considerable radical scavenging activities based on DPPH and Trolox equivalent antioxidant capacities. Caffeic acid, its four derivatives, and hispidin glycosides had stronger DPPH scavenging activities than "standard" α-tocopherol or its equivalent. Among the isolated and studied molecules, di-O-caffeoylquinic acids were the strongest antioxidant components. Other work from the same team (Chen et al. 2008) compared the antioxidant activities of 12 compounds isolated from *Davallia solida* rhizomes. The water extract of rhizomes had high phenolic content (210.8 ± 4.6 mg catechin equivalent/g DW) and a strong DPPH scavenging activity (IC_{50} = 0.0159 ±1.21 mg/g DW). The majority of these isolated molecules were catecholic polyphenols.

The connection between antioxidant activities and systematic appurtenance seems to be a logical relationship, but its validation is a goal of researchers. Some data on the antioxidant activity of different fern species are available in the literature. The antioxidant activities based on in vitro DPPH scavenging activities were gathered from recent publications (Lai and Lim 2011; Lai et al. 2009; Gupta et al. 2014; Ding et al. 2008; Chang et al. 2007) and were recalculated in the same units for comparison. The transformed data of different species have a high variability; the basic conclusion may be that the found antioxidant "power" is practically independent from the systematic position of a species. In the next step of the evaluation, data were organized based on the appurtenance to fern families. We analyzed the families when the number of species was at least four; thus, five fern families (Aspleniaceae, Blechnaceae, Dryopteridaceae, Polypodiaceae and Pteridaceae) were included in the statistical analysis. The main parameters (arithmetical mean, standard deviation, coefficient of variation, median, and minimum and maximum values) are given on Table 15.2.

The standard deviations of the means for the antioxidant capacities of fern families are markedly different. The coefficient of variation was 27% for Polypodiaceae ($n = 11$), 103% for Pteridaceae ($n = 11$), and 869% for Dryopteridaceae ($n = 11$). The highest antioxidant potentials (i.e., the lowest numbers) were produced in families Dryopteridaceae and Blechnaceae. A comparison of DPPH radical scavenging activities based on data from four research teams showed remarkable numerical differences. Statistical verification, however, was impossible by an analysis of variance method because of disparities between data groups.

There are two theoretical possibilities for the chemical-biochemical factors and substances that regulate the actual antioxidant activity:

1. Specific molecules with antioxidant characteristics (phenolics, especially certain flavonoids)
2. Enzyme molecules capable ofr antioxidative reactions (catalase, superoxide dismutase and others)

Phenolics and Antioxidant Activity

A meta-analysis was performed to examine the possible connection between phenolic content and antioxidant activity based on five publications (Chang et al. 2007; Ding et al. 2008; Gupta et al. 2014; Lai et al. 2009; Lai and Lim 2011). Data pairs of total phenolic content and free radical scavenging capacity (measured by DPPH radicals) were analyzed. We performed linear regression analyses separately for the different data groups. The calculated statistical parameters are summarized in Table 15.3.

Significant linear connections were demonstrated for 5 samples of Study 1 (at 0.05 level), for 15 samples of Study 2 (at 0.05, as well as 0.01 and 0.001 significance levels), and for 6 samples of Study 5 at the 0.05 level. Interestingly, aqueous extraction produced a significant connection, but methanol extraction did not (Chang et al. 2007). The connections between phenolics and free radical scavenging activities were not verified statistically for data from Studies 3, 5, and 6, as well as part of Study 1. A probable explanation for this inconsistency is complex: although there were small differences in methodologies, the in vitro measurable free radical scavenging activity is a complicated property that depends not only on phenolics but also on other antioxidant molecules (e.g., polysaccharides).

The pool of antioxidant substances in ferns seems to be great and complex, including more than just phenolics in the broad sense. Water-soluble polysaccharides were fractionated and analyzed from *Pteridium aquilinum* (Xu et al. 2009). The average molecular mass was 458 kDa. Monosaccharide components included glucose (58%), galactose (18.7%), rhamnose (10.2%), and arabinose (6.1%). The main conclusion of this study was that fern polysaccharides can be components of the antioxidative system. Another study (Wang and Wu 2013) produced an

Table 15.3 Linear connections between phenolics and antioxidant activity for different fern groups: a meta-analysis of data

No. of study	Number of analyzed species (n)	Equation for the linear connection of phenolics (X) and antioxidant activity (Y)	r	Level of significance 0.05	0.01	0.001	Origin of basic data
1	5	$Y = 1.99–0.000975X$	−0.890	+	∅	∅	Lai et al. (2009)
2	15	$Y = 1.66–0.00064X$	−0.780	+	+	+	Lai and Lim (2011)
3	8	$Y = 9.09–0.00123X$	−0.442	∅	∅	∅	Gupta et al. (2014)
4	6 Extract in methanol	$Y = 0.349–0.0000134X$	−0.800	∅	∅	∅	Chang et al. (2007)
5	6 Extract in water	$Y = 0.423–0.00000213X$	−0.838	+	∅	∅	Chang et al. (2007)
6	31	$Y = 0.069–0.00000127X$	−0.200	∅	∅	∅	Ding et al. (2008)

+, significant connection; ∅, lack of significance

oligosaccharide fraction by chemical hydrolysis from *P. aquilinum*. The water-soluble oligosaccharide fraction had high sugar content and remarkable hydroxyl radical scavenging activity.

A study by De Long et al. (2011) on the chemical composition of *Matteuccia struthiopteris* demonstrated the chemical diversity of antioxidant potential. Considerable concentrations were measured from ascorbic acid (3.0 μmol/g DM), α- and β-tocopherol (314 and 80.8 μg/g DM, respectively), α- and β–carotenes (438 and 122 μg/g DM, respectively), and xantophylls (violaxanthin 225 μg/g DM; zeaxanthin 127 μg/g DM; lutein 238 μg/g DM). These compounds, under high phenolics levels, provide an important reservoir of antioxidant molecules.

15.2.3 Terpenes and Terpenoids

Terpenes and terpenoids are the oldest and most widespread group of secondary metabolites. Terpenoids are modified terpenes (i.e., their derivatives), in which methyl groups are eliminated/changed or oxygen atoms are added. The number of isolated terpenoids is very high, with more than 23,000 structures having been identified in the plant kingdom in 2007 (Chen et al. 2007).

Monoterpenes

Leaves of the bracken fern (*Pteridium aquilinum*) contain different terpenoids as well as other monoterpenes. These molecules have mainly volatile characteristics; consequently, this plant can emit certain monoterpene molecules. Terpenoid emission was the subject of field measurements in *Pteridium aquilinum* (Mandronich et al. 2012). Emitted monoterpene levels ranged from 0.002 to 13 μg C/g DW per hour in understory plants; open air plants emitted monoterpenes at a rate of 0.005 to 2.21 μg C/g DW per hour. Regulation of this monoterpene emission is not clear; there were no significant connections between terpene emissions and temperature or photosynthetically active radiation. The documented data of such emissions provide evidence for the effects on plant metabolism, ecosystems, and the chemical composition of the atmosphere (Mandronich et al. 2012).

Sesquiterpenes

From a toxicological point of view, the most important sesquiterpene derivative is ptaquiloside (Fig. 15.7). The occurrence, structure, toxicology, and other characteristics have been discussed in detail by different authors, including our own group (Rasmussen et al. 2013, 2015; Rizgar 2016; Somvanshi et al. 2006; Vetter 2009, 2011 etc.); therefore, a very short summary is presented in this chapter. Ptaquiloside is a colorless compound ($C_{20}H_{30}O_8$, molecular weight is 398.45) of amorphous character with good water solubility. Ptaquiloside belongs to the norsesquiterpene glycosides of illudane type. It has been found not only in *Pteridium aquilinum*, but also in *Pteris cretica*, *Cheilanthus myriophyla*, *Pityrogramma calomelanos* (all from the Pteridaceae family), *Dennstaedtia scabra*, *Histiopteris incisa* (from the family Dennstaedtiaceae), and *Cibotium barometz* (from the family Cibotiaceae) species (Somvanshi et al. 2006). The occurrence of ptaquiloside has been reported (Pathania et al. 2012) in *Onychium cryptogrammoides*, *O. tenuifrons* (both from Pteridaceae), *Hypodematium crenatum* (from Hypodematiaceae), and *Dryopteris cochleata* (from Dryopteridaceae).

Fig. 15.7 Ptaquiloside (a norsesquiterpene glycoside of illudane type)

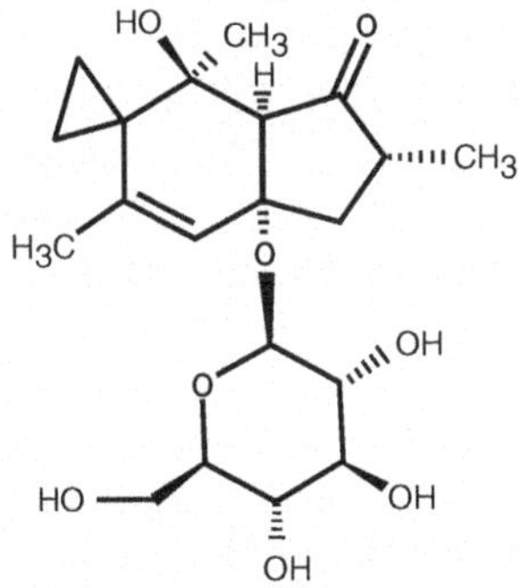

Fig. 15.8 Chemical structure of 1-indanone

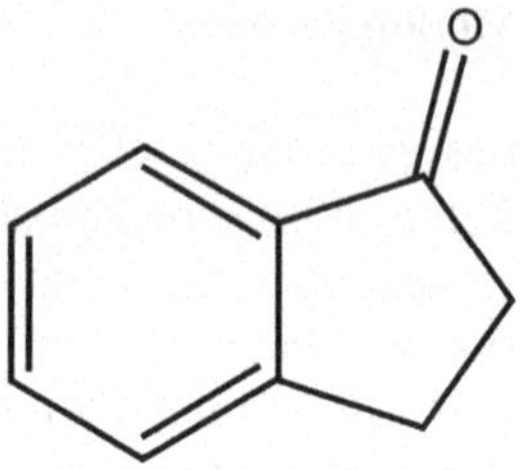

The highest ptaquiloside content can be measured in young plants (in crosiers) at the beginning of the growing season. All organs of bracken ferns (the shoot and root systems, including rhizomes) were shown to have ptaquiloside molecules. Rasmussen et al. (2015) confirmed the hypothesis about ptaquiloside transport into rhizomes at the end of the vegetation period. Other data from this team indicate that the maximum ptaquiloside content occurs in May, with a strong decrease during maturation of fronds (Rasmussen et al. 2013). An important publication from Rasmussen et al. (2003) demonstrated the detection of ptaquiloside molecules in soil around the bracken plant. This fact and its certain consequences (e.g., the possibility of ptaquiloside contamination in drinking water via ptaquiloside secretion into the soil) prompted new investigations into the stability and hydrolysis of ptaquiloside under different environmental conditions, including temperature, pH, and other factors. The essential conclusion of these studies is that ptaquiloside molecules are stable for a few months under common soil conditions at low temperatures (5 °C) at pH values of 4.0–6.5. The degradation of molecules is rapid under pH 4 or above pH 6.5. Other factors affecting this degradation (hydrolysis) include a microbiological attack. Microbes in upper soil layers can cause such degradation (Rasmussen et al. 2015). Ptaquiloside molecules showed lability at room temperature, in both acidic and alkaline solutions. Their hydrolysis produces 1-indanon compounds (pterosin O and pterosin B) and glucose liberation.

Pterosins

Pterosins are a large group of sesquiterpenes (mainly from the species of the Dennstaedtiaceae and Pteridaceae families) with 1-indanon skeletons (Fig. 15.8). Pterosin B was first isolated in 1970 from *P. aquilinum* var. *latiusculum* (Hikino et al. 1970). The distribution of pterosins was summarized and published by Chen et al. (2015). They identified approximately 31 pterosins from approximately 30 taxa (27 species, 1 subspecies, and 2 varieties from families Adiantaceae, Dennstaedtiaceae, Dicsoniaceae, Pteridaceae and Polypodiaceae). The genera *Microlepis* and *Pteridium* each had five species from which one or more pterosins were isolated and elucidated. The species *Hypolepis punctata* had 11, whereas *Pteridium aquilinum* var. latiusculum had 10 pterosins. Three new pterosin derivatives (pterosin A, Z, and I) were isolated and described from species *Ceratopteris thalictroides* (Pteridaceae), *Hypolepis punctata* (Dennstaedtiaceae), and *Pteridium revolutum* (Dennstaedtiaceae) (Chen et al. 2015). Pterosins with unusual characteristics (e.g., dimer structures) were isolated and elucidated from *Pteris multifida*, including bimutipterosin A and B (Liu et al. 2011).

The biological effects of pterosins are wide ranging. They have resistance against certain insects. Pterosins B and O can inhibit *Bacillus subtilis* bacteria (Rizgar 2016). Further investigations reported anti-diabetic and anti-obesity effects for these molecules. Certain extracted or synthesized molecules (pterosins A, C, D, G, I, L, N, and X and pteroside Z) showed biological activity against diabetes, for both type 1 (insulin-dependent) and type 2 (insulin-independent) diabetes. Their anti-diabetic effects may stem from the stimulation of glucose consumption (uptake) and improved insulin sensitivity (Hsu et al. 2013).

Diterpenoids

Diterpenoids of plants are secondary metabolites with strong biological activity. Examples include the various heart glycosides (digitalis glycosides) and others. The number of known diterpene glycosides in ferns is not too high; likewise, the number of related studies and publications is low. Wang et al. (2011) isolated six new molecules from the bioactive *ent*-kaurene diterpenoids, so the total number of known diterpenoids of ferns (together with the earlier described molecules) is 12. New molecules were obtained from *Pteris semipinnata* (Pteridaceae): ent-15-oxokauran-19-oic acid derivatives, known as pterisolic acids A–F. Another medicinal fern (*Pteris dispar*) was the object of studies in China (Gou et al. 2011). *Pteris dispar* is used widely in folk medicine against tumors and inflammations. The isolated new diterpene derivative ($C_{20}H_{28}O_6$)—a trihydroxylated product of *ent*-kaur-16-en-19-oic acid—showed cytotoxic properties against KB cells during in vitro tests (IC_{50} value: 59.8 μmol/L).

Triterpenes

Triterpene molecules have six isoprene units and are C_{30} substances. More than 200 triterpene molecules have been reported. The structures of fern triterpenoid derivatives are distinct from the triterpenoids of higher plant taxa. The most frequently occurring fern triterpenoids are diplopterol, diploptene, and fernene molecules (Fig. 15.9). The most essential chemical difference in fern triterpenoids is the lack of oxygen functionality at the C-3 atom (Shinozaki et al. 2008). Fern triterpenoids are formed by direct cyclization of squalen (which is different from higher plant groups), so the specific squalen cyclase enzymes occur in ferns.

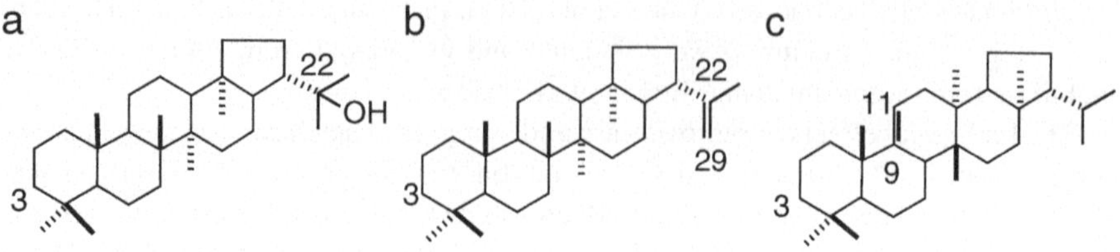

Fig. 15.9 Characteristic triterpenes from ferns (**a**) Diplopterol (22-hydroxyhopane) (**b**) Diploptene (hop-22(29)-ene) (**c**) Fernene (fern-9(11)-ene

Phytoecdysteroids

Phytoecdysteroids are specific secondary metabolites of ferns that are triterpene derivatives. They are analogues of the molting hormones of insects. Their distribution, biosynthesis, and functions were summarized by Lafon et al. (2011). The number of isolated and elucidated molecules is more than 40, and most of them have 27 C atoms. Phytoecdysteroids occur only in some fern families, and their distribution in ferns is not equal. For example, their concentration in *Pteridium aquilinum* is greatest in the rhizomes, followed by the roots and fronds. However, in *Microsorum membranifolium*, it is more concentrated in the fronds than the rhizomes; in *M. scolopendria*, it is more concentrated in the rhizomes than the fronds (Lafon et al. 2011). Fluctuations in ecdysteroid contents during vegetation season is characteristic for *Pteridium vulgare* rhizome, in which it was reported to be 0.4% DM in January but tenfold lower (0.04% DM) in October (Messeguer et al. 1998). New phytoecdysteroid molecules were isolated from *Polypodium vulgare* by Hungarian researchers (Simon et al. 2011). The names and molecular formulas of these compounds include 5-hydroxiecdysone ($C_{27}H_{45}O_7$), 20-deoxyshidasterone ($C_{27}H_{43}O_5$), and polypodine B 2-β-D-glucoside ($C_{33}H_{55}O_{13}$).

15.2.4 Nitrogen-Containing Secondary Metabolites; Alkaloids

Alkaloids are the most important members of nitrogen-containing secondary metabolite group. The plant kingdom is rich in alkaloids and their estimated numbers are high, although not in ferns. The fern species *Hypodematium squamuloso-pilosum* was analyzed by Zhou et al. (1998), who identified two β-carboline alkaloids: 1-acetyl-β-carboline and its hydroxyl derivative 1-acetyl-8-hydroxy-β-carboline (Fig. 15.10). Both molecules showed positive Dragendorff reactions. The β-carboline alkaloids contain a pyridine ring and are attached to an indole skeleton.

A phytochemical screening may be the first step in the elucidation of these substances. Screening reactions are used frequently in phytochemistry for medicinal plants, their candidate plants, or the production of a first proximate analysis. The documented occurrence of alkaloids in ferns is a rare phenomenon. Thus, it seemed to be logical to carry out a meta-analysis of alkaloid occurrence from screenings

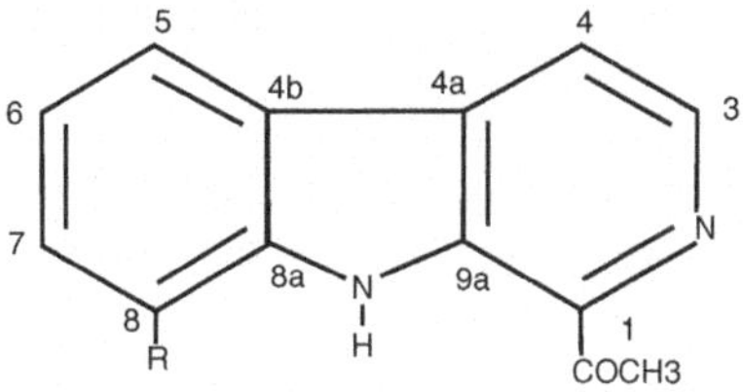

Fig. 15.10 Chemical structure of β–carboline alkaloids (isolated from *Hypodematium squamuloso-pilosum*)

based on recent publications. The interpretation of data in publications was simple: positive alkaloid tests clearly indicate their occurrence.

We summarized the results of 11 such studies on fern species (Table 15.4) and grouped all data according to family. Various extracts were produced in different investigations (extracts in water, ethanol, methanol, petroleum ether, chloroform, acetone, and benzene). The reaction of Mayer was used mostly for alkaloid detection. In a study by Patil et al. (2013), aqueous extracts were produced and reactions of four kinds were carried out: tests of Mayer, Wagner, Dragendorff, and reaction with picric acid.

Positive reactions were documented for some fern species in certain extracts, including alcohol (ethanol or methanol). The strongest reactions were experienced by *Pyrrosia lanceolata* (Polypodiaceae) in the frond and rhizome (Ruby and Sara 2014). Four extracts showed positive reactions, but negative tests were reported for the other analyzed species of this family (*Drynaria quercifolia;* Polypodiaceae) (Rajesh et al. 2014). Four positive reactions (with the exception of aqueous extract) were found for *Blechnum orientale* (Devi et al. 2016). An interesting question is the value of different alkaloid reactions. Patil and his team (2013) used four such tests and observed interesting differences. In *Angiopteris erecta* (Marattiaceae), the aqueous extract showed positive reactions in Mayer's and Wagner's tests, but negative reactions were detected with Dragendorff reagent and picric acid.

An essential question of a comparative evaluation is the role of the systematic position of the taxa. Alkaloid occurrence is characteristic for the families Dryopteridaceae, Pteridaceae, Dennstaedtiaceae, Athyraceae, and Cyatheaceae (Table 15.4). These fern families—except Cyatheaceae—belong to the order Polypodiales. In a given family and genus, however, there are alkaloid-containing and alkaloid-deficient species. For example, *Cyathea gigantea*, *C. nilgirensis,* and *C. crinita* produced positive reactions (Janakiraman and Johnson 2015), but *C. brunoniana* had no alkaloids (Talukdar et al. 2010). *C. gigantea* produced a positive test in one of the studies (Janakiraman and Johnson 2015). In a second study, its test was negative both in the stem and in the leaf (Talukdar et al. 2010). A fundamental problem is the type of extractant used in chemical determinations. Extraction in water is successful for certain species only, whereas alcohol (ethanol, methanol) seems to be effective. Only one or two extractants were used in some publications, for other solvents, Table 15.4 indicates that no data were available. There are differences between the alkaloid content of different plant organs (shoot system, root system); however, the available data are limited and mostly consider the entire plant.

Table 15.4 Occurrence of alkaloids in fern species detected in different extracts by various test reactions

Family species	Type of reaction	Extract in water	Extract in ethanol/ methanol	Extract in petrol-ether	Extract in chloro-form	Extract in acetone
Dryopteridaceae fam.						
Dryopteris filix-mas	a b c d	+ + - -	n.d.	n.d.	n.d.	n.d.
Dryopteris cochleata	a	–	+	–	n.d.	n.d.
Dryopteris nigropaleacea		–	+	n.d.	n.d.	+
Polystichum bakerianum		–	+	n.d.	n.d.	+
P. prescottianum		–	+	n.d.	n.d.	–
Pteridaceae fam.						
Adiantum lunulatum	a b c d	- + - +	n.d.	n.d.	n.d.	n.d.
Adiantum incisum	a b c d	- - - -	n.d.	n.d.	n.d.	n.d.
Adiantum venustum		–	+	n.d.	n.d.	+
Actinopteris radiata	a	+	+	–	n.d.	n.d.
Pityrogramma calomelanos	a	–	+	–	n.d.	n.d.
Onychium cryptogrammoides		–	+	n.d.	n.d.	+
Pteris confusa		n.d.	+	n.d.	n.d.	n.d.
Pteris vittata		n.d.	+	n.d.	n.d.	n.d.
Pteris argyrea		n.d.	+	n.d.	n.d.	n.d.
Pteris biaurita		n.d.	+	n.d.	n.d.	n.d.
Pteris multiaurita		n.d.	+	n.d.	n.d.	n.d.
Cyatheaceae fam.						
Cyathea nilgirensis		–	+	–	–	+
Cyathea gigantea		–	+	–	–	+
Cyathea crinita		–	+	–	–	-
Cyathea gigantea	a	n.d.	–	–	–	–
Cyathea gigantea	a	–	–	–	–	n.d.
Cyathea brunoniana	a	–	–	–	–	n.d.
Dennstaedtiaceae fam.						

(continued)

Table 15.4 (continued)

Family species	Type of reaction	Extract in water	Extract in ethanol/ methanol	Extract in petrol-ether	Extract in chloro-form	Extract in acetone
Pteridium aquilinum	a	n.d.	+	n.d.	n.d.	n.d.
Pteridium aquilinum		–	+	n.d.	n.d.	+
Pteridium aquilinum		n.d.	+	n.d.	n.d.	n.d.
Athyraceae fam.						
Athyrium atkinsonii		–	+	n.d.	n.d.	+
Athyrium mackinnonii		–	+	n.d.	n.d.	+
Athyrium wallichianum		–	+	n.d.	n.d.	+
Deparia acuta		–	+	n.d.	n.d.	+
Deparia allantodiodes		–	+	n.d.	n.d	+
Diplazium maximum		–	+	n.d.	n.d.	–
Polypodiaceae fam.						
Drynaria quercifolia	a	–	–	–	n.d.	n.d.
Pyrrosia lanceolata		+	++	++	+	n.d.
Pyrrosia lanceolata		+	++	++	+++	n.d.
Aspleniaceae fam.						
Asplenium trichomanes		–	+	n.d.	n.d.	+
Salviniaceae fam.						
Salvinia auriculata	a	n.d.	–	n.d.	n.d.	+
Blechnaceae fam.						
Blechnum orientale		–	+	+	+	+
Marattiaceae fam.						
Angiopteris erecta	a b c d	+ + - -				

a, Mayer reaction; b, Wagner reaction; c, Dragendorff reaction; d, reaction with picric acid; +, positive test; –, negative test); n.d., no data

Origin of basic data: Patil et al. 2013; Rajesh et al. 2014; Mir et al. 2013; Gracelin et al. 2013; Janakiraman and Johnson 2015; Talukdar et al. 2010; Kardong et al. 2013; Awe and Amobi 2015; Ruby and Sara 2014; Devi et al. 2015; Devi et al. 2016

The interpretation of certain results presented in Table 15.4 is not easy. These data do not provide reliable and unambiguous information on the alkaloid content of ferns. Two items need to be resolved for a clear discussion of this problem:

1. Isolation, elucidation, and clear demonstration of alkaloids in ferns
2. Production of comparable, quantitative data for ferns with positive test reactions

15.3 Conclusions

The development and use of new chemical methods for the isolation, elucidation, and quantification of secondary metabolites may produce new data for phytochemistry, biochemistry, and various disciplines of applied botany and phytotherapy. With the currently available data, the main conclusions of this chapter are as follows:

1. Great diversity has been found in the secondary metabolites of ferns. Recent investigations can extend our knowledge on actual molecule groups.
2. The number of elucidated phenolics and terpenoids is high, but amount of isolated saccharides is low. Furthermore, the occurrence of alkaloids seems to be very limited. Recent data on alkaloids are based on positive alkaloid tests, but these findings have not yet been confirmed by isolation, elucidation, and quantification of the actual molecules.
3. The biological effects of secondary metabolites in ferns are very heterogeneous. Studies on this topic have proliferated in recent years. The exact chemical composition of ferns can substantiate or confirm the common use of a given species. Therefore, these investigations have great practical importance because many fern species have been used in folk medicine of different countries.
4. The most important chemical fact may be the occurrence of various secondary metabolites with antioxidant characteristics. Phenolics (including flavonoids and their derivatives) and other molecules (oligo- and polysaccharides) may have such properties. The antioxidant capacity of certain ferns depends on the concentration of some secondary metabolites. These connections are known and have been documented for various higher plant groups; for ferns, however, the evidence is currently incomplete.
5. The elucidation of new bioactive molecules from ferns can increase the range of substances that are suitable for medicinal and veterinary use. Recognition of their biosynthesis and biological effects can result in the creation of new products. The future industrial synthesis of these molecules may be a new research field.
6. Some secondary metabolites of ferns have toxicological importance because of their poisonous characteristics for both animals and humans. Certain sesquiterpenes and sesquiterpene glycosides (namely, ptaquiloside) can cause acute and chronic toxicity via direct (consumption) and indirect (milk secreted by ingested animals) methods. Various diseases (e.g., bovine enzootic hematuria, acute hemorrhagic disease, upper alimentary carcinoma for cattle, bright blindness for sheep, carcinogenic problems for humans) indicate the toxicological relevance of some fern's products and thus the importance of these metabolites.
7. The ferns can be the objects of animal attacks, from worms and herbivores. The plant's defense mechanisms may be attributed to the occurrence and effects of certain secondary metabolites. Ferns containing cyanogenic glycoside will produce free HCN molecules to work against the biological attackers. Ferns with tannin molecules can deter or inhibit some phytophagous insects. The filicin

molecule (*Dryopteris filix-mas*) has anthelmintic properties, whereas phytoecdysones can cause damage to the developmental processes of insects. Mono- and sesquiterpenes are produced as volatile substances after animal damage. These materials have positive effects on an herbivore's enemies; certain sesqui- and diterpenes can act as phytoalexins. Thus, some fern species may be safe alternatives to synthetic polluting pesticides.

8. Further investigations are required on the secondary metabolites of ferns. The isolation, detection, and characterization of new compounds of ferns may lead to practical medicinal, biotechnological, and industrial benefits.

References

Alonso-Amelot A, Oliveros-Bastidas A (2005) Kinetics of the natural evolution of hydrogen cyanide in plant of neotropical *Pteridium arachnoideum* and its ecological significance. J Chem Ecol 31:315–331

Alonso-Amelot ME, Oliveros A, Calcagno-Pisarelli MP (2004) Phenolics and condensed tannins in relation to altitude in neotropical *Pteridium* spp. a field study in the Venezuelan Andes. Biochem Syst Ecol 32:969–981

Alonso-Amelot ME, Oliveros-Bastidas A, Calcagno-Pisarelli MP (2007) Phenolics and condensed tannins of high altitude *Pteridium arachnoideum* in relation to sunlight exposure, elevation, and rain regime. Biochem Syst Ecol 35:1–10

Awe S, Amobi OO (2015) Antibacterial, phytochemical and proximate analysis of *Pteridium aquilinum*. Int J Res Pharm Biosci 2:1–7

Chang HC, Huang GJ, Agrawal DC, Kuo CH, Wu CR, Tsay HS (2007) Antioxidant activities and polyphenol contents of six folk medicinal ferns used as "Gusuibu". Bot Stud 48:397–406

Chen YH, Chang FR, Lin YJ, Wang L, Chen JF, Wu YC, Wu MJ (2007) Identification of phenolic antioxidants from a Sword Brake fern (*Pteris ensiformis* Burm.) Food Chem 105:48–56

Chen YH, Chang FR, Lin YJ, Hsieh PW, Wu MJ, Wu YC (2008) Identification of antioxidants from rhizome of *Davallia solida*. Food Chem 107:684–691

Chen CH, Chiu FY, Lin Y, Huang WJ, Hsieh PS, Hsu FL (2015) Chemical constituents analysis and antidiabetic activity validation of four fern species from Taiwan. Int J Mol Sci 16:2497–2516

Da Costa RMG, Bastos MMSM, Oliveira PA, Lopes C (2012) Bracken-associated human and animal health hazards: chemical, biological and pathological evidence. J Hazard Mater 203-204:1–12

DeLong JM, Hodges DM, Prange RK, Forney CF, Toivenon PMA, Bishop MC, Elliot L, Jordan MA (2011) The unique fatty acid and antioxidant composition of ostrich fern (*Matteucia struthiopteris*) fiddleheads. Can J Plant Sci 91:919–930

Devi PS, Rukmini K, Himabindu N, Savithramma N (2015) Antibacterial acitivity and phytochemical screening of *Salvinia auriculata* Aubl. from Tirumala Hills Tirupati. Int J Pharm Sci Rev Res 30:35–38

Devi RK, Vasantha S, Panneerselvam A, Rajesh NV, Jeyatbilakan N (2016) Phytochemical constituents and *in vitro* trematocidal activity of *Blechnum orientale* Linn. Against *Gastrothylax crumenifer*. Ann Phytomed 5:127–134

Ding ZT, Fang YS, Tai ZG, Yang MH, Xu YQ, Li F, Cao QE (2008) Phenolic contents and radical scavenging capacity of 31 species of ferns. Fitoterapia 79:581–583

Durdevic L, Mitrovic M, Pavlovic P, Bojovic S, Jaric S, Oberan L, Gajic G, Kostic O (2007) Total phenolics and phenolic acids content in leaves, rhizomes and rhizosphere soil under *Ceterach officinarum* D.C., *Asplenium trichomanes* L., and *A. adiantum-nigrum* L. in The Gorge of Sicevo (Serbia). Ekológia 26:164–173

Gou ZP, Liang NC, Hou J, Li L, Tian Y, Wang CY, Cui X, Zhang BJ, Ma XC (2011) Two new diterpene and sesquiterpene from *Pteris dispar*. Chin Chem Lett 22:1451–1453

Gracelin DH, de Britto AJ, Kumar PBJR (2013) Qualitative and quantitative analysis of phytochemicals in five *Pteris* species. Int J Pharm Pharm Sci 5:105–107

Gupta SK, Ghosal M, Biswas R, Saha BK (2014) Evaluation of *in vitro* antioxidant activity of methanolic axtracts of some ferns from Mawsynram of Meghalaya, India. Int J Curr Sci 12:E 87–E 89

Han X, Li Z, Li CY, Jia WN, Wang HT, Wang CH (2015) Phytochemical constituents and biological activities of plants from the genus *Dryopteris*. Chem Biodivers 12:1131–1162

Hikino H, Takahashi T, Arihara S, Takemoto T (1970) Structure of pteroside B glucoside of *Pteridium aquilinum* var. latiusculum. Chem Pharm Bull 18:1488–1489

Hsu FL, Huang CF, Chen YW, Yen YP, Wu CT, Uang BJ, Yang RS, Liu SH (2013) Antidiabetic effects of Pterosin A, a small molecular weight natural product on diabetic mouse models. Diabetes 62:628–638

Hunyadi A, Martins A, Danko B, Chang FR, Wu YC (2014) Protoflavones: a class of unusual flavonoids as promising novel anticancer agents. Phytochem Rev 13:69–77

Janakiraman N, Johnson M (2015) Inter specific variation studies on *Cyathea* species using phytochemical and fluorescence analysis. Res Rev: J Pharmacol Toxicol Stud 3:25–31

Jiang J, Tian L, Wang L, Liu Y, Chen Y (2013) Phenolic compounds from the fern *Glaphyropteridopsis erubescens* (Hook.) Ching. Biochem Syst Ecol 50:136–138

Kardong D, Upadhyaya S, Saikia LR (2013) Screening of phytochemical, antioxidant and antibacterial activity of crude extract of *Pteridium aquilinum* Kuhn. J Pharm Res 6:179–182

Lafon R, Ho R, Raharivelomanana P, Dinan L (2011) Ecdysteroids in ferns: distribution, diversity, biosynthesis and functions. In: Fernandez H, Kumar A, Revilla MA (eds) Working with ferns, issues and applications. Springer, New York/London, pp 305–319

Lai HY, Lim YY (2011) Antioxidant properties of some Malaysian Ferns. In: 3rd International Conference on Chemical, Biological and Environmental Engineering 20:8–12

Lai HY, Lim YY, Tan SP (2009) Antioxidative, Tyrosinase inhibiting and antibacterial activities of leaf extracts from medicinal ferns. Biosci Biotechnol Biochem 73:1362–1366

Lin AS, Chang FR, Yen HF, Björkeborn H, Norlen P, Wu YC (2007) Novel flavonoids of *Thelypteris torresiana*. Chem Pharm Bull 55:635–637

Liu J, Shu J, Zhang R, Zhang W (2011) Two new pterosin dimers from *Pteris multifida* Poir. Fitoterapia 82:1181–1184

Mandronich MB, Greenberg JP, Wessman CA, Guenther AB (2012) Monoterpene emissions from an understory species, *Pteridium aquilinum*. Atmos Environ 54:308–312

Messeguer J, Melé E, Reixach N, Irure-Santilari J, Casas J (1998) *Polypodium vulgare* L. (wood fern): in vitro cultures and the production of phytoecdysteroids. In: Bajaj YPS (ed) Biotechnology in Agriculture and Forestry 41; Medicinal and aromatic Plants X. Springer, Berlin/Heidelberg, pp 333–348

Mir SA, Mishra AK, Reshi ZA, Sharma MP (2013) Preliminary phytochemical screening of some Pteridophytes from district Shopia (J & K). Int J Pharm Pharm Sci 5:632–637

Mithöfer A, Boland W (2012) Plant defense against Herbivores: chemical aspects. Annu Rev Plant Biol 63:431–450

Oloyede FA, Airemionkhale M, Olatunji TY (2014) Chemical, nutritional and anti-nutritional compositions of *Nephrolepis exaltata* (L.) Schott. (Nephrolepidaceae). FUTA J Res Sci 2:252–257

Pathania S, Kumar P, Singh S, Khatoon S, Rawat AKS, Punetha N, Jensen DJ, Lauren DR, Somvanshi R (2012) Detection of ptaquiloside and quercetin in certain Indian ferns. Curr Sci 102:1683–1692

Patil DT, Gurav KD, Kadam AS, Thite SV, Thoke RB, Kore AB (2013) Qualitative analysis of secondary metabolites from some Filicales members. Int J Res Phar Chemistry 3:300–302

Rajesh KD, Vasantha S, Rajesh NV, Panneerselvam A (2014) Qualitative and quantitative phytochemical analysis in four Pteridophytes. Int J Pharm Sci Rev Res 27:408–412

Rasmussen LH, Jensen LS, Hansen HCB (2003) Distribution of the carcinogenic terpene ptaquiloside in bracken fronds, rhizomes (*Pteridium aquilinum*) and litter in Denmark. J Chem Ecol 29:771–778
Rasmussen LH, Schmidt B, Sheffield E (2013) Ptaquiloside in bracken spores from Britain. Chemosphere 90:2539–2541
Rasmussen LH, Donelly E, Strobel BW, Holm PE, Hansen HCB (2015) Land management of bracken needs to account for bracken carcinogens. A case study for Britain. J Environ Manag 151:258–266
Rizgar HM. 2016 Pterosins and pterosides in bracken (*Pteridium aquilinum* (L.) Kuhn) Ph.D. Thesis Bangor University Bangor, pp 1–340, UK
Ruby GD, Sara CS (2014) Preliminary phytochemical analysis of *Pyrrosia lanceolata* (L.) Farwell. Int Res J Pharm 5:879–883
Santos MG, Carvalho CEM, Kelecom A, da Costa Ribeiro MLR, de Freitas CVC, da Costa LM, de Fernandes LVG (2005) Cianogenese em esporofitos de pteridófitas avaliada pelo teste do ácido pícrico. Acta Bot Bras 19:783–788
Shinozaki J, Shibuya M, Masuda K, Ebizuka Y (2008) Squalene cyclase and oxidosqualene cyclase from a fern. FEBS Lett 582:310–318
Simon A, Ványolós A, Béni Z, Dékány M, Tóth G, Báthori M (2011) Ecdysteroids from *Polypodium vulgare* L. Steroids 76:1419–1424
Singh M, Malhotra K, Singh C (2015) Phytochemical evaluation and antioxidant activity of different samples of *Pteris vittata* in Doon Valley, Uttarakhand Region. Int J Pure App Biosci 3:296–304
Somvanshi R, Lauren DR, Smith BL, Dawra BL, Sharma OP, Sharma VK, Singh AK, Gangwar NK (2006) Estimation of the fern toxin, ptaquiloside, in certain Indian ferns other than bracken. Curr Sci 91:1547–1552
Talukdar AD, Choudhury MD, Chakraborty M, Dutta BK (2010) Phytochemical screening and TLC profiling of plant extracts of *Cyathea gigantea* (Wall. ex. Hook.) Haltt. and *Cyathea brunoniana* Wall. ex. Hook. (Cl. & Bak.) Assam University Sci Technol: Biol Environ Sci 5:70–74
Valizadeh H, Sonboli A, Kordi FM, Dehghan H, Bahadori MB (2015) Cytotoxicity, antioxidant activity and phenolic content of eight fern species from North Iran. Pharm Sci 21:18–24
Vetter J (2009) A biological hazard of our age: bracken fern [*Pteridium aquilinum* (L.) Kuhn]- a review. Acta Vet Hung 57:183–196
Vetter J (2011) Toxicological and medicinal aspects of the most frequent fern species, *Pteridium aquilinum* (L.) Kuhn. In: Fernandez H, Kumar A, Revilla MA (eds) Working with ferns, issues and applications. Springer, New York/London, p 36375
Vogler G, Donath O, Saukel J, Rauch AE, Kählig H, Krenn L (2012) Polar phenolic compounds in *Dryopteris filix-mas* and *Dryopteris dilatata*. Verh Zool-Bot Ges Österreich 148(149):279–289
Wang H, Wu S (2013) Preparation and antioxidant activity of *Pteridium aquilinum*-derived oligosaccharide. Int J Biol Macromol 61:33–35
Wang F, Li YJ, Ren FC, Wei GZ, Liu JK (2011) Pterisolic acids A-F, new *ent*-kaurene diterpenoids from the fern *Pteris semipinnata*. Chem Pharm Bull 59:484–487
Wang X, Wang M, Cao J, Wu Y, Xiao J, Wang Q (2017) Analysis of flavonoids and antioxidants in extracts of ferns from Tianmu Mountain in Zhejiand Province (China). Ind Crop Prod 97:137–145
Xu W, Zhang F, Luo Y, Ma L, Kou X, Huang K (2009) Antioxidant activity of a water-soluble polysaccharide purified from *Pteridium aquilinum*. Carbohydr Res 344:217–222
Yao H, Duan J, Zhang C, Li Y, Liu C (2016) Coumaric acid glycosides from the Chinese fern *Polypodium hastatum*. Chem Nat Compd 52:669–671
Zhang M, Cao J, Dai XI, Chen XF, Wang QX (2012) Flavonoid contents and free radical scavenging activity of extracts from leaves, stems, rachis and roots of *Dryopteris erythrosora*. J Pharm Res 11:991–997
Zhou TS, Cai W, Wang ZT, Che CT, Zhou RH, Xu GJ, Xu LS (1998) β-carboline alkaloids from *Hypodematium squamuloso-pilosum*. Phytochemistry 49:1807–1809

Chapter 16
Current Trends in Pteridophyte Extracts: From Plant to Nanoparticles

Liliana Cristina Soare and Nicoleta Anca Şuţan

16.1 Introduction

The diversity of pteridophytes is estimated at 10,000–15,000 species (Akeroyd and Synge 1992; Roos 1996), with a distribution of species that increases from the Poles towards the equator (Moran 2004). Areas of high diversity are found in the Andes (with approximately 2500 species of ferns and lycophytes), New Guinea (approximately 2000 species), Borneo (1200 species), Ecuador (1250 species), Costa Rica (1165 species) (Moran 2008), China (2600 species) (Lu 2007), the Philippines (1000 species), Malaysia (550 species), Thailand (700 species), India (600 species) (Lu 2007; Schneider et al. 2004), etc.

Such areas of high pteridophyte biodiversity stimulate the use of a great number of species for human consumption or in traditional medicine. Thus, in countries such as Malaysia, the Philippines, Hawaii, Australia, Nepal, India etc., ferns are used for consumption (Liu et al. 2012). China is home to "more than 20% of the global pteridophyte flora" (Lu 2007). For over three millennia, 52 edible fern species have been used for consumption in China, and other 144 species may be used for this purpose. Their dry and salted leaves, the starch or the tea made from these leaves are commercial products in China (Liu et al. 2012).

The use of pteridophytes in the multimillennial traditional medicine of many people has been the foundation of modern therapeutical research. Crude extracts, standardized extracts or purified substances have been tested to discover new properties with therapeutic potential.

The current research methods, techniques and equipment allow us to obtain results which help validate ethnopharmacological treatments. The identification of secondary fern metabolites, specific to this group, explains their bioactivity. Modern research studies have shown that *Blechnum orientale* (the Centipede fern), an edible

L. C. Soare (✉) · N. A. Şuţan
Department of Natural Sciences, University of Piteşti, Piteşti, Romania

H. Fernández (ed.), *Current Advances in Fern Research*,
https://doi.org/10.1007/978-3-319-75103-0_16

species used in the Asian traditional medicine for various skin diseases, stomach pain, gall bladder disorders, etc., may be used in the treatment of colon cancer and that of multidrug-resistant bacterial infections (methicillin resistant *Staphylococcus aureus*, MRSA) (Lai et al. 2010) as well as in the treatment of diabetic ulcer wounds (Lai et al. 2016).

New technologies for identifying pteridophyte species of medicinal importance have been discovered and developed. Thus, the classical identification of medicinal fern species based on their morphological characters has been replaced by "DNA barcoding" (Ma et al. 2010).

Some of the current problems identified in the medical field may be solved by using natural products obtained from ferns, products that possess a wide range of biological properties (antibacterial, antifungal, antiviral, antioxidant, anti-inflammatory, antitumor, hypoglycemic, neuroprotective, etc.). One of the problems that poses a major threat to the population is the rapid increase in the resistance of microorganisms to synthetic antibiotics, including those that were recently discovered. Superbacteria, some of them resistant to more than 20 antibiotics, have appeared in recent years and added to the range of multidrug-resistant bacteria. In 2013, The Centers for Disease Control and Prevention (CDC; https://www.cdc.gov/) published a study on the "Current Antibiotic Resistance Threats in the United States, by microorganism", which mentions three groups of microorganisms:

1. Microorganisms with a threat level of urgent (*Clostridium difficile*, Carbapenem-resistant Enterobacteriaceae, drug-resistant *Neisseria gonorrhoeae*)
2. Microorganisms with a threat level of serious (multidrug-resistant Acinetobacter, drug-resistant Campylobacter, fluconazole-resistant *Candida*, extended spectrum β-lactamase producing Enterobacteriaceae (ESBLs), vancomycin-resistant *Enterococcus* VRE, multidrug-resistant *Pseudomonas aeruginosa*, drug-resistant non-typhoidal *Salmonella*, drug-resistant *Salmonella typhi*, drug-resistant *Shigella*, methicillin-resistant *Staphylococcus aureus* MRSA, drug-resistant *Streptococcus pneumoniae*, drug-resistant tuberculosis)
3. Microorganisms with a threat level of concerning (vancomycin-resistant *Staphylococcus aureus* VRSA, erythromycin-resistant Group A *Streptococcus*, clindamycin-resistant Group B *Streptococcus*)

16.2 Crude Extracts

The impressive number of research studies on fern use in traditional medicine has inspired researchers to create databases that provide information on "medicinal uses, chemical constituents as well as protein/enzyme sequences" of some species (Thakar et al. 2015). The information stored in such databases can help us not only to discover new medicinal drugs but also to conserve pteridophytes (Thakar et al. 2015), in case the species used for therapeutic purposes are rare, vulnerable, etc. and they would consequently require protection.

Extracts were obtained, and phytochemical characterization along with bioactivity testing was performed for different spontaneously occurring species, such as *Stenochlaena palustris* (swampy land, Kampar, Malaysia) (Chai et al. 2012), for species cultivated in botanical gardens, for example, *Blechnum orientale* (Putrajaya Botanical Garden, Kuala Lumpur, Malaezia) (Lai et al. 2010), *Azolla microphylla* (Centre for Conservation and Utilization of Blue Green Algae, Indian Agricultural Research Institute, New Delhi) (Abraham and Kaushik 2015), *Psilotum nudum*, *Nephrolepis biserrata* and *N. cordifolia* (fern house of National Botanical Institute, Lucknow, India) (Rani et al. 2010), or for in vitro plants, for example, the gametophyte of *Asplenium trichomanes*, *A. trichomanes-ramosum* and *Polypodium vulgare* (Deliu et al. 2013), the gametophyte and sporophyte of *P. vulgare* or the gametophyte of *Cystopteris fragilis* (Soare et al. 2012a).

Plant parts (leaves, rhizomes, roots, spores or gametophytes) or the whole plant were used to obtain extracts. The extracts from the whole plant of *Dicranopteris linearis* (Rajesh et al. 2016) were tested to establish their phytochemical composition and antioxidant activity. Cao et al. (2013) used the whole plant of *Dryopteris erythrosora* to obtain extracts, which were tested for their antioxidant, anticarcinogenic and acetylcholinesterase inhibitory properties. The analysis also comprised the characterization of the flavonoids encountered in these extracts. The leaves of *Blechnum orientale* (Lai et al. 2010), *Asplenium adiantum-nigrum* and *A. trichomanes* (Hammami et al. 2016), *A. nidus* (Nath et al. 2013), *Acrostichum aureum*, *Asplenium nidus*, *Blechnum orientale*, *Cibotium barometz*, *Dicranopteris linearis* var. *linearis* (Lai et al. 2009), *Asplenium scolopendrium* (Şuţan et al. 2016), etc. were used to obtain extracts, which were subsequently tested for bioactivity. The leaves of the edible species *Stenochlaena palustris*, at different ontogenetic stages (young/mature sterile frond, young/mature fertile frond), were used by Chai et al. (2012) to obtain extracts, which were characterized in terms of polyphenol content and antioxidant activity. They proved to be a healthy source of food. Fern leaves are a renewable resource, which may be used even when the species require protection.

The antioxidant activity and the polyphenol content were tested for the extracts obtained from rhizomes of *Drynaria fortunei*, *Pseudodrynaria coronans*, *Davallia divaricata*, *D. mariesii*, *D. solida* and *Humata griffithiana*, species that compose the Chinese traditional formula "Gusuibu" (Chang et al. 2007). Rhizomes of *Polypodium vulgare* (Souri et al. 2008), *Drynaria quercifolia* (Kandhasamy et al. 2008), *Polypodium interjectum*, *Polystichum woronowii*, *P. aculeatum*, *Asplenium scolopendrium*, *A. adiantum-nigrum*, *Dryopteris affinis*, *Pteris cretica* and *Athyrium filix-femina* were also used to obtain and characterize extracts (Valizadeh et al. 2015); some of the extracts were supplemented with silver nanoparticles (Şuţan et al. 2016).

Spores, which are produced in large amounts by ferns, have constituted the biological material for some extracts, whose allergenic, carcinogenic and cytogenetic effects were tested. In high biodiversity areas, pteridophyte spores are natural atmospheric pollutants (Bunnag et al. 1989). These are found in large amounts in the closed spaces of botanical gardens or in closed private spaces (Simán et al. 2000).

They may affect human health due to their allergenic or carcinogenic action. Thus, the spores of the species *Acrostichum aureum* (Bunnag et al. 1989), *Lycopodium* (Devi et al. 1989), *Asplenium nidus*, *Dicranopteris curranii*, *D. linearis*, *Nephrolepis auriculata*, *Pteridium aquilinum* or *Stenochlaena palustris* (Chew et al. 2000) are allergenic, while those of the species *Anemia phyllitidis*, *Dicksonia antarctica*, *Pteridium aquilinum*, *Pteris vittata* and *Sadleria pallida* (Simán et al. 2000) are known for their carcinogenic action. Furthermore, the extracts from spores of *Athyrium filix-femina* were evaluated for their potential in the photosynthesis of silver nanoparticles (AgNPs) and for their cytogenetic effects (see Subchap. 16.7).

The physical condition of the plant material used for obtaining extracts can influence their composition and bioactivity. For example, certain flavonoids can be degraded when using fresh plant material (Marston and Hostettmann 2006). The raw plant material necessary for the extracts was fresh, frozen or dried and pulverized, as follows: fresh leaves of *Athyrium filix-femina*, *Dryopteris affinis*, *D. filix-mas* (Soare et al. 2012b) and *Asplenium nidus* (Nath et al. 2013); fresh rhizomes of *Drynaria quercifolia* (Kandhasamy et al. 2008); dried leaves of *Asplenium adiantum-nigrum*, *A. trichomanes* (Hammami et al. 2016) and *Stenochlaena palustris* (Chai et al. 2012); dried leaves and rhizomes of *Polypodium interjectum*, *Polystichum woronowii*, *P. aculeatum*, *Asplenium scolopendrium*, *A. adiantum-nigrum*, *Dryopteris affinis*, *Pteris cretica* and *Athyrium filix-femina* (Bahadori et al. 2015); dried rhizomes of *Drynaria fortunei*, *Pseudodrynaria coronans*, *Davallia divaricata*, *D. mariesii*, *D. solida* and *Humata griffithiana* (Chang et al. 2007); and frozen rhizomes and leaves (−18 °C) of *Asplenium scolopendrium* (Șuțan et al. 2016).

In some cases, the extracts were obtained from plant material pulverized in liquid nitrogen. Thus, in order to determine the polyphenol content, and the antioxidant, antibacterial and tyrosinase-inhibiting activity, fresh leaves of some medicinal plants were pulverized in liquid nitrogen. The extraction was performed using methanol, in a rotary orbital shaker (Lai et al. 2009).

Specialist literature mentions extracts obtained with the help of one or more solvents. Diverse categories of solvents selectively extract the components from the plant material, which may show different bioactivity. The solvents that are frequently used to obtain crude extracts include water, ethanol, methanol, dichloromethane, petroleum ether, hexane, benzene, chloroform and acetone (Chai et al. 2012, 2015; Chang et al. 2007; Kandhasamy et al. 2008; Lai et al. 2009; Rajesh et al. 2016; Roudsari et al. 2012; Souri et al. 2008; Valizadeh et al. 2015; Xie et al. 2015).

The evaluation of the bioactivity of crude extracts was either directly performed on these extracts or after their concentration and fractionation. For example, crude ethanol extracts were obtained from the dried and pulverized leaves of the species *Psilotum nudum*, *Nephrolepis biserrata* and *N. cordifolia*. The extracts were concentrated in a rotary orbital shaker, freeze-dried and afterwards fractionated with hexane, chloroform, ethanol and water (Rani et al. 2010). Each of the obtained fractions was tested for antibacterial and antifungal properties.

16.3 Antimicrobial Activity of New Extracts and Formulations

Pteridophytes contain a wide range of bioactive compounds, from the terpenoid, phenolic, flavonoid and alkaloid classes (Ho et al. 2011). Alcaraz et al. (2000), Cushnie and Lamb (2005) and Yusuf (1994) mentioned the presence of secretory epidermal tissues on fern leaves and tissues that produce alkaloids, flavonoids, glycosides and phenolic compounds, which are substances with antimicrobial properties. Flavaspidic acids PB and AB, isolated from the methanol extract obtained from the rhizome of *Dryopteris crassirhizoma*, are responsible for the powerful antimicrobial activity against *Staphylococcus aureus* MRSA, *Streptococcus mutans* and *Bacillus subtilis* (Lee et al. 2009). Eicosenes and heptadecanes have an antibacterial and antifungal activity (Abraham and Kaushik 2015; Dalli et al. 2007). Recent research showed that some phenolic compounds inhibit bacterial virulence factors (T3SS-type III secretion systems) present in many Gram-negative pathogenic bacteria (Khokhani et al. 2013). It was demonstrated that fern extracts have antimicrobial effects on some Gram-negative and Gram-positive pathogenic bacteria, including multidrug-resistant bacteria and on certain species of fungi.

Numerous research studies have emphasized the link between the antimicrobial activity of the extracts and the solvents used to obtain them. Thus, Pal (2012) observed that the most effective antimicrobial activity against *E. coli* was exhibited by acetone extracts obtained from leaves of *Athyrium filix-femina* (inhibition zone/IZ = 18 mm). Ethanol extracts produced an inhibition zone of 15 mm and methanol extracts of 12 mm; the weakest activity was observed in the case of aqueous extracts (IZ = 7 mm). Aqueous extracts from leaves of *Psilotum nudum*, *Nephrolepis biserrata* and *N. cordifolia* showed a more powerful antimicrobial activity against nine bacterial strains (Rani et al. 2010) than the chloroform, hexane or ethanol extracts. The extracts from the rhizome of *Drynaria quercifolia* obtained using ethanol, methanol, petroleum ether, hexane, benzene or chloroform exhibited a different antibacterial activity, dependent on the solvent. Thus, the diameter of the inhibition zone produced by the extracts against *Escherichia coli* was 10 mm for the ethanol, methanol and benzene extracts and 8 mm for the chloroform extract, while the extracts for which petroleum ether and hexane had been used showed no antimicrobial activity (Kandhasamy et al. 2008).

The part of the plant used to obtain extracts influences their antimicrobial activity. Thus, for the methanol extracts from leaves of spontaneously occurring plants of *Polypodium vulgare* and from gametophytes with sporelings (generated in vitro), the antibacterial effects were different. The extract from leaves exhibited an inhibition zone of 12 mm against a bacterial strain isolated from the soil, while the extract from the gametophyte with sporelings created an inhibition zone of 8 mm (Soare et al. 2012a).

For multidrug-resistant bacteria, new antibacterial products can be developed, starting from the pteridophyte extracts that have been tested. Aqueous, methanol and chloroform extracts from the leaves of *Lygodium flexuosum* presented an important antimicrobial activity against multidrug-resistant bacteria such as methicillin-resistant *Staphylococcus aureus* (MRSA) and vancomycin-resistant *Enterococcus faecalis* (VRE), as well as against other clinically isolated bacterial strains (Nayak et al. 2013). The antibiotic activity of the extract, which, in some cases, was more effective than that of the antibiotic used as a control sample, indicates the fact that the plant is a source that must be taken into consideration and further investigated for this property.

The antibacterial effects of pteridophyte extracts were evaluated for their effectiveness against phytopathogenic bacteria. The fraction obtained with anhydrous ethyl acetate from the methanol extract of *Azolla microphylla* inhibited the growth of *Xanthomonas oryzae* (Abraham and Kaushik 2015), one of the most important rice pathogens (Fun et al. 2017). The ethyl acetate fraction contains a mixture of eicosenes and heptadecanes, compounds that were identified in the extract of *Pteris biaurita*, an extract that was tested against the fungi *Curvularia lunata*, *Fomes lamaoensis*, *Poria hypobrumea*, *Fusarium oxysporum* and against the bacterium *Bacillus pumilus* (Dalli et al. 2007). The antifungal activity of pteridophyte extracts was also emphasized for the following species: *Saccharomyces cerevisiae*, *Aspergillus niger* (Soare et al. 2012b), *Candida albicans* (Amoroso et al. 2014), *Microsporum gypseum*, *Trichophyton mentagrophytes* and *T. rubrum* (Rani et al. 2010), etc.

In human or animal organisms, the natural compounds ingested are metabolized producing certain biological responses. As such, the results reached for a crude extract, for whole fruits or vegetables may be different from those reached for compounds isolated from that particular extract or for parts of fruits or vegetables (Lila and Raskin 2005; Liu 2003).

A series of interactions, known as endointeractions, occurs among the existing components of a plant extract (Lila and Raskin 2005); they produce the biological response. The combination of a plant extract with another extract or with a synthetic medical drug produces exointeractions (Lila and Raskin 2005), which modify the biological response. These phytochemical interactions can explain the state of health of people from a certain geographical area, known for a certain food behaviour (Lila and Raskin 2005).

Antimicrobial activity of pteridophyte extracts may be modified through combinations with other extracts or other synthetic medical drugs, in new formulations with therapeutic potential. The interactions that take place in such combinations can be cummulative, if the effect obtained for the new formulation sums up the separate effects. A particular interest is shown to those associations that produce synergetic effects, significantly higher than the sum of the separate effects. It is possible for the phytochemical interactions that take place in the new formulations to produce negative effects (antagonistic effects). In such cases, the activity of the active compounds is inhibited (Lila 2009). In some other cases, the combination enhances bioavailability, solubility and absorption of bioactive components (Eder and Mehnert 1998; Gartner et al. 1997).

The research on the synergetic effect of the combination antibiotic plant product has become a key instrument in phytomedicine (Santos et al. 2015). The combination between curcumin and some synthetic antibiotics produced synergetic antibacterial effects against *Staphylococcus aureus* (Teow and Ali 2015). The generation of nanoparticles of the solid lipid (SLNs) type in formulations that contain ampicillin and curcumim determined a decrease in the MIC (minimum inhibitory concentration) and MBC (minimum bactericidal concentration) as compared to the formulation that contains only ampicillin and curcumin, a result that also suggests a decrease in the risk of developing resistance to antibiotics (Alihosseini et al. 2016). Green tea polyphenols, combined with oxacillin, produce a decrease in the MIC of the antibiotic in the fight against *Staphylococcus aureus* (MRSA) (Yoshida et al. 2005).

Formulations that contain kaempferol isolated from *Impatiens balsamina* and erythromycin or clindamycin showed synergetic antibacterial effects against *Propionibacterium acnes* (Lim et al. 2007). Different flavonoids and carotenoids in combination with antibiotics also led to synergetic antibacterial effects against *Pseudomonas aeruginosa* (Santos et al. 2016). The methanol extract of *Punica granatum*, combined with the antibiotics chloramphenicol, gentamicin, ampicillin, tetracycline and oxacillin, has a synergetic effect against methicillin-resistant *Staphylococcus aureus* (MRSA) and methicillin-sensitive *S. aureus* (MSSA) (Braga et al. 2005).

Pteridophyte extracts have been considered in the evaluation of the possible enhancing antibacterial effects produced in combination with antibiotics. New combinations between methanol extracts from leaves of *Dryopteris filix-mas* (D) and the antibiotics piperacillin (Pip) and tetracycline (Te), respectively, produced larger inhibition zones than those corresponding to the antibiotic or the extract when used independently. The new combinations were created by impregnating standard antibiotic discs with the extract. The formulation D + Te produced better results against *Staphylococcus aureus* and *Enterobacter cloacae* and the formulation D + Pip determined larger inhibition zones against *S. aureus*, *Enterobacter cloacae* and a bacterial strain isolated from the soil (Soare et al. 2012c). Similar results were reported by Parihar et al. (2006) for the aqueous and alcoholic extracts from leaves, rhizomes and roots of *Athyrium pectinatum*. Thus, a mixture of equal parts of extract and antibiotic may show, in some cases, greater effectiveness than the antibiotic alone.

16.4 Synthesis of "Green" Nanoparticles Using Pteridophyte Extracts

The therapeutic use of pteridophytes has had a spectacular evolution, starting from their use in the traditional medicine of different peoples to the current stage in which pteridophytes are used in the form of nanoparticles.

The nanotechnology industry was a global business evaluated at about 4 million dollars about 10 years ago and 1 billion dollars in 2015 (Roco 2005). At present, hundreds of products containing biomaterials have applicability and are used on a large scale in the fields of electronics, optics, in the process of food packaging, in medicine and the cosmetic industry, in wastewater treatment technology or in environmental remediation processes (Aitken et al. 2006; Nowack and Bucheli 2007; Handy et al. 2008a; Klain et al. 2008). A few thousand other nanomaterials are being researched at the moment.

Nanoparticles are the smallest increments of the physical world; they are of similar scale to the biomolecules that govern the existence of life. The DNA molecule has a diameter of approximately 2 nm, while the C_{60} diameter is approximately 5 nm. These structures, as a result of their nanometric dimensions (variable between 1 and 100 nm), have some modified characteristics (resistance, conductivity, catalytic reactivity, magnetic or optical properties), leading to a collateral interpretation of the common laws of science (Handy et al. 2008b). The nanoparticle behaviour and its potential side or favourable effects on the living systems do not depend only on the dosage but also on their surface, size and shape. These effects are also based on their physical and chemical properties. For example, the solubility of otherwise insoluble substances may increase dramatically when the size of their particles is smaller than 100 nm. Furthermore, nanoparticles have a larger surface compared to the particles with a similar mass, which leads to a higher surface reactivity (Auffan et al. 2009). Surfactants and other additives can modify the surface characteristics of nanoparticles and can prevent their aggregation (European Comission 2006). The similarity between natural biomolecules and designed nanomaterials may be translated into an enormous potential of the latter to interfere in biological processes. These interactions might affect the cell membrane behaviour and biochemical processes or even the genetic code itself (Klaine et al. 2012).

In the last decade, the elaboration of one-step methods for the synthesis of metal nanoparticles has been stimulated by their multiple practical applications. Even though the synthesis of metal nanoparticles may be induced through various physicochemical methods, the biogenic reduction of metal ions is a rapid process, which can be performed at room temperature and pressure, with increased intensity. At the same time, nanoparticle biosynthesis, as a cost-effective and nontoxic method, safe to the environment, is preferable when the finite product is to be used in biomedical and pharmaceutical applications (Mittal et al. 2013). Nanoparticles offer the advantage that they can be encapsulated and released in a controlled and targeted manner.

Diverse biological systems, such as plants, algae, bacteria, yeasts and fungi, were proved to be excellent bioreactors for the in vivo synthesis of nanoparticles, but the use of plant extracts, combined with particular acids or some metal salts, such as silver (Ag), gold (Au), copper (Cu), platinum (Pt), zinc (Zn), cesium (Cs), titanium (Ti), indium (In) salts and many others, offer the advantage of an effective control over nanoparticle biosynthesis and purification.

Specialist literature indicates a relatively recent date for the first nanoparticle synthesis using fern extracts. Kang et al. (2008), using Pteridophyta, reported the synthesis of silver nanoparticles (AgNPs) by means of "green chemistry"; these are the most popular types of nanoparticles encountered in biosynthesis methods. Later on, extracts from *Actiniopteris radiata* were used to reduce silver ions to nanoparticles with a diameter ranging from 20.8 nm to 30.6 nm, while AgNPs measuring between 35 and 65 nm were synthesized in extracts from *Adiantum capillus-veneris.* Silver nanoparticles, whose synthesis was mediated by the extract from *Nephrolepis exaltata*, without supplementing the reaction medium with capping and stabilizing agents, were characterized as polydispersed, with a large amplitude of their mean size between 22 and 44 nm (Raut et al. 2010 cited by Chrislyn et al. 2016).

Analyses performed by means of transmission electronic microscope (TEM) and selected area electron diffraction (SAED) showed that gold and silver nanoparticles photosynthesized through the reduction of tetrachloroauric acid and $AgNO_3$ in the extract of *Adiantum philippense*, respectively, were monocrystalline and anisotropic and had variable sizes, between 10 and 18 nm (Sant et al. 2013). It was appreciated that the reduced sizes of these NPs, approximately 11 nm for AuNPs and about 13 nm for AgNPs, were due to the alkaline pH value and the photosynthetic environment. The same method was applied to the Au ions which were reduced to nanoparticles with an average size of 8.3 nm in the extract from *Azolla microphylla.* The shape of Ag and Au nanoparticles, whose synthesis was mediated by fern extracts, was predominantly spherical, but the HRTEM analysis with higher magnification indicated that they were spherical, triangular, hexagonal and rod shaped (Kunjiappan et al. 2015).

16.5 The Role of Metabolites from the Pteridophyte Extracts in Nanoparticle Synthesis

For nanoparticle biosynthesis, fern extract is mixed with a metal salt at room temperature. The reaction is completed in a few minutes to a few hours. The molecular basis for the biosynthesis of AgNPs is not known, but it has been speculated that the organic matrix favours the bonds between Ag and proteins, through the aminoacid fragments that serve as nucleation sites.

Most often metal nanoparticles have been produced by condensation, a principle that was proposed by Turkevich et al. (1951). According to Balaji et al. (2008), proteins/enzymes are responsible for reducing metal ions when plant extracts are used in nanoparticle synthesis. Polysaccharides, polyols, heterocyclic compounds soluble in water and, generally, phytochemical compounds with reductive or antioxidant properties are recognized as the main agents responsible for reducing metal compounds and for stabilizing nanoparticles, respectively (Huang et al. 2007; Geethalakshmi and Sarada 2010; Prathna et al. 2010; Ghosh et al. 2011, 2012a, b; Park et al. 2011).

Phytochemical screening of fern extracts showed their content of proteins and secondary metabolites, such as phenols, alkaloids, tannins, flavonoids, carbohydrates, saponins, glycosides, steroids and terpenoids (Pan et al. 2011; Panneerselvam et al. 2016; Xavier et al. 2016), to which the reductive and stabilizing properties are attributed.

For example, the FTIR (Fourier-transform infrared spectroscopy) analysis of the extracts from *A. capillus-veneris* and *Cyathea nilgerensis* demonstrated phenols, when combined with AgNPs, prevent their agglomeration (Samidoss et al. 2013; Johnson et al. 2017). For the extracts from *Pteris tripartita*, flavonoids were mentioned to have an extremely important role in reducing the Ag ions. The implication of phenols, as well as that of aliphatic amines, alcohols, amides and carbonyl groups from biomolecules, was also suggested in the process of stabilization of silver nanoparticles in the extracts from *Dicranopteris linearis* (Rajaganesh et al. 2016; Baskaran et al. 2016).

The reductive capacity of metal ions was tested for different types of extracts, such as the fresh aqueous extracts from dried plants of *Actiniopteris radiata* (Sant et al. 2013) or from plants/green leaves of *Adiantum capillus-veneris* (Santhoshkumar and Nagarajan 2014), *A. philippense* (Sant et al. 2013) and *Azolla pinnata* (Korbekandi et al. 2014); boiled aqueous extracts from fresh leaves of *Cheilanthes farinosa* (Nalwade et al. 2013), *Nephrolepis exaltata* (Chrislyn et al. 2016), *Dicranopteris linearis* (Rajaganesh et al. 2016), *Pteris argyraea*, *P. confusa*, *P. biaurita* (Britto et al. 2014), *P. tripartita* (Baskaran et al. 2016) and *Nephrolepis exaltata* (Bhor et al. 2014); crude methanol extracts from dried plants of *Azolla microphylla* (Kunjiappan et al. 2015) or extracts concentrated by evaporation from dried plants of *Cyclosorus interruptus*, *Christella dentata* and *Nephrolepis cordifolia* (Xavier et al. 2016); and crude ethanol extracts from leaves or rhizomes of *Asplenium scolopendrium* (Şuţan et al. 2016) or *Dryopteris crassirhizoma* (Lee et al. 2016).

The antioxidant potential and the medicinal value of fern species have been the determining factors in selecting them for the synthesis of Ag and Au nanoparticles. At the same time, the quantitative and qualitative diversity of the bioactive ingredients from fern extracts may be an explanation for the morphological diversity of nanoparticles (Makarov et al. 2014).

The bioactive compounds found in the composition of the extracts, the solvent used for extraction, the concentration of the extract and that of the metal salt, the electrochemical potential of metal ions, the pH of the reaction mixture, the incubation period and the period of contact were often mentioned as influential factors in the nanoparticle production rate, as well as in the resulting characteristics.

The first indication of the biosynthesis of AgNPs is given by the change in the colour of the reaction mixture. For example, in only 20 s, the phytosynthesis of AgNPs induced by the extract from *Salvinia molesta* determined a colour change in the reaction mixture from green yellowish to reddish brown (Verma et al. 2016). The addition of the extract from *Adiantum philippense* to the solution of tetrachloroauric acid ($HAuCl_4.3H_2O$ 1 mM) determined a colour change in the mixture from greenish to dark blue (Sant et al. 2013). In the case of the extract from *Azolla microphylla*, through the biosynthesis of AuNPs, the reaction mixture turned dark pink (Kunjiappan et al. 2015).

The pH value of the reaction mixture has an extraordinary influence on the nanoparticle formation process. First, pH modification is correlated with quantitative and qualitative diversification of phytochemical compounds contained in the extract, as well as with the variation of their potential for binding and reducing metal cations and anions during nanoparticle synthesis, determining changes in shape, size and synthesis yield. For example, Chrislyn et al. (2016) showed that the synthesis speed of AgNPs in extracts from *Nephrolepis exaltata* was higher than in the alkaline medium. Similarly, the UV-Vis spectrum confirmed that pH = 11 was optimal for the synthesis of AuNPs. AgNPs were best synthesized when pH = 12, for the synthesis assisted by the extract from *Adiantum philippense* (Sant et al. 2013). However, the same authors pointed out that for a lower pH, the absorption capacity was lower and wider. These findings confirm the results according to which the alkaline pH stabilizes and prevents the agglomeration of synthesized nanoparticles.

The size of the phytosynthesized NPs varied depending on the reaction medium. Thus, AgNPs whose biosynthesis was mediated by the extract from *Cyathea nilgerensis* varied on average between 45 nm and 74 nm (Johnson et al. 2017). The smallest AgNPs with a diameter of 0.93 nm and a mean size of 6.5 nm were obtained in the biosynthesis reaction catalysed by the extract from *Azolla pinnata* (Korbekandi et al. 2014).

Kinetic studies revealed that temperature is a decisive factor in the process of nanoparticle biosynthesis. For example, the reaction rate and biosynthesis efficiency of anisotropic Au and Ag nanoparticles (spherical, triangular and decahedral) in extracts from *Diplazium esculentum* (Sinha and Ahmaruzzaman 2016) were favoured by the increase in temperature. It is estimated that the increase in temperature leads to the increase in the nucleation rate.

The concentration of the plant extract, as well as the concentration of metal salts, is a defining parameter for the reaction of nanoparticle phytosynthesis. Thus, Sant et al. (2013) showed that the undiluted aqueous extract from *Adiantum philippense* had induced an effective synthesis of the Au and Ag nanoparticles, while the diluted sample had displayed no significant nanoparticle synthesis. Moreover, mixing the plant extract with tetrachloroauric acid/silver nitrate in a higher concentration (from 1 mM to 10 mM) induced nanoparticle synthesis at higher rates. In the case of the tetrachloroauric acid, the increase in the biosynthesis rate was recorded up to a concentration of 5 mM; the increase in the concentration of metal ions resulted in a decrease of biosynthesis intensity. For silver nitrate the optimal concentration of metal ions was 9 mM.

In this context, it should be mentioned that the efficiency of metal nanoparticle synthesis depends on the electrochemical potential of the ions, so that the ability of plant extracts to reduce metal ions can be significantly higher in the case of ions with a higher electrochemical potential (Ag^+) compared to the ions with a lower electrochemical potential (e.g. $[Ag(S_2O_3)_2]^{3-}$) (Haverkamp and Marshall 2009, cited by Makarov et al. 2014).

16.6 Applications of Nanoparticles Phytosynthesized in Pteridophyte Extracts

Up to the present, numerous studies have promoted fern extracts as renewable sources, safe to the environment, which can be successfully used to reduce metal ions to bioparticles. In spite of the fact that nanoparticles biosynthesized in plant extracts are of real interest, till now, they have been tested in a relatively low number of in vitro or in vivo practical applications for their antioxidant, antimicrobial, anti-inflammatory, larvicidal and pupicidal activity. For example, the DPPH assay, ferrozine and phosphomolybdenum assay and $ABTS^+$ assay revealed a significant difference in the antioxidant activity of biosynthesized AgNPs when compared to the aqueous extract from *Pteris tripartita* (Baskaran et al. 2016). AuNPs biosynthesized in extracts from *Azolla microphylla* were shown to have a significant antioxidant capacity as well as a liver protective activity. They were promoted as potential protection and prevention agents against liver diseases (Kunjiappan et al. 2015).

Comprehensive studies on the in vitro antimicrobial activity of AgNPs biosynthesized using fern extracts demonstrated their powerful or moderate antimicrobial effect against certain pathogenic agents. It is considered that antimicrobial activity is dependent on the size and dosage of AgNPs and that their inhibitory activity is more effective against Gram-positive bacteria than against Gram-negative bacteria (Koteswaramma et al. 2017).

Silver nanoparticles whose synthesis was mediated by extracts from *Nephrolepis exaltata* exhibited an important antimicrobial activity as compared to those that were chemically synthesized, which presented either no antimicrobial activity or an antimicrobial activity comparable to that of standard antibacterial agents (Bhor et al. 2014; Chrislyn et al. 2016). AgNPs phytosynthesized in extracts from *Cheilanthes farinosa* induced an inhibition zone of 15 mm against *Staphylococcus aureus* and 20 mm against *Proteus morganii,* an activity similar to that manifested by tetracycline (Nalwade et al. 2013). In addition, Kalita et al. (2016) underlined the biocompatibility and functionalization of the AuNPs surface biosynthesized using aqueous extracts from *Adiantum philippense* with the broad-spectrum antibiotic beta lactam (amoxicillin). AuNPs combined with amoxicillin enhanced the in vivo and in vitro bactericidal activity against Gram-positive and Gram-negative bacteria.

The action of AgNPs and Ag^{2+} was reported to be similar, although nanoparticles proved to be more efficient at much lower concentrations than ions. The exact mechanism of their antibacterial activity has not been discovered yet, but some researchers claim that AgNPs attach to the cell membrane, disrupt its permeability and determine structural modifications of the cell membrane, eventually leading to the death of the cell (Krishnaraj et al. 2012).

Morones et al. (2005) and Baker et al. (2005) suggested that AgNPs may enter bacteria and fungi producing damage due to their interaction with phosphorus and sulphur electrons from compounds such as DNA and proteins.

Several studies have evaluated the mosquitocidal effect of nanoparticles whose synthesis was induced in pteridophyte extracts. Based on the findings resulted from the experiments focused on the ovicidal properties and especially on the oviposition deterrent effects of AgNPs synthesized using extracts from *Dicranopteris linearis*, Rajaganesh et al. (2016) recommended them particularly for the integrated control programs against arbovirus vectors.

Other studies emphasized the larvicidal and pupicidal potential of aqueous extracts from *Adiantum capillus-veneris* supplemented with AgNPs against the vector of malaria, *Anopheles stephensi* (Samidoss et al. 2013), or that of the extracts from *Cyclosorus interruptus*, *Christella dentata* and *Nephrolepis cordifolia* against the pest *Spodoptera litura* (Xavier et al. 2016). The same authors argued in favour of the use of extracts of *Nephrolepis exaltata* with AgNPs in the treatment of wastewater contaminated by *Escherichia coli.*

Gold and silver nanoparticles biosynthesized in extracts from *Diplazium esculentum* were tested for their catalytic properties in the remediation of two carcinogenic textile dyes (methyl violet 6B and rose bengal) and of the phenolic compound 4-nitrophenol. The depleted nanostructured materials and the intermediary products of the degradation process, along with the period of dye degradation by the nanostructured catalysts, recommend them as ideal candidates in wastewater treatment processes (Sinha and Ahmaruzzaman 2016).

The interest shown to nanoparticles and nanomaterials must be supported by empirical research (Bernhardt et al. 2010; Klaine et al. 2012) and scientific arguments that reduce fear and concern about potential side effects of nanotechnological products on human and natural ecosystems. The vast majority of the research conducted so far has tested the exposure of living systems to much higher nanoparticle concentrations (mg L^{-1}) than the evaluated natural risks (μg L^{-1}). Appreciating and defining the dosage-response curve in correlation with reality is essential, especially because most scientific results suggest that nanomaterials have a chronic action at low concentrations (Gottschalk et al. 2009). The determination of the correct dosage, validated by reality, must be followed by the creation of a scenario that includes the most efficient way to introduce nanoparticles in the experimental system and the evaluation of transportation, targets as well as the nanoparticle physicochemical changes during the experiment.

16.7 Evaluating Phytotoxicity and Cytogenetic Effects of the Ethanol Extracts from Spores of *Athyrium filix-femina* (L.) Roth

A considerable number of species, including bacteria (Kumar et al. 2011; Wang et al. 2011), algae (Manzo et al. 2013), plants (Makarov et al. 2014; Allafchian et al. 2016; Jeevanandam et al. 2017), invertebrates (Croteau et al. 2014) and vertebrates as well as fish (Ma et al. 2013) and rats (Magaye et al. 2014) have been the subject of many nanotoxicological studies.

Despite the fact that numerous fern species have been largely used as natural remedies for hundreds of years, at the moment there are relatively few data regarding the phytotoxicity and cytogenotoxicity of fern extracts. Using the cytokinesis-block micronucleus cytome assay, Glamočlija et al. (2014) remarked that the cytotoxic and genotoxic potential of the *Asplenium scolopendrium* extracts on in vitro human lymphocytes was very low and null, respectively.

The extracts from spores of *Anemia phyllitidis, Dicksonia antarctica, Pteridium aquilinum, Pteris vittata* and *Sadleria pallida* induced in vitro DNA damage (Simán et al. 2000). Chai et al. (2015) evaluated the cytotoxicity of aqueous extracts from *Christella arida, C. dentata, Cyclosorus interruptus, Microsorum punctatum, Nephrolepis acutifolia* and *Pleocnemia irregularis* on leukaemia cell line (K562), emphasizing their therapeutic effects. Lok et al. (2007) concluded that AgNPs do not have a direct effect on DNA or proteins.

In our research aimed to determine the cytogenetic effects of extracts obtained from spores of *Athyrium filix-femina* (L.) Roth, we used the test organism *Allium cepa* L. ($2n = 16$). The *Allium* test provides data that may be correlated with the results obtained for the prokaryotic systems (Fiskesjö 1985). The spores were collected from plants found in the Vâlsan Valley, Argeş, Romania, at the following location: N45°20′18″, E0.24°44′04.7″, at an altitude of 730 m. The ethanol extracts were prepared through spore mass suspension in ethanol (1:100 w/v). The mixture was kept for 48 h at room temperature. The extracts obtained (EEA) were filtered using filter paper (Whatman no. 1) and the filtrates were used for nanoparticle phytosynthesis and for evaluation of the cytogenotoxic potential using the *Allium* test.

The synthesis of AgNPs was induced by treating 100 ml of ethanol extract from spores of *A. filix-femina* with 100 ml of aqueous solution of $AgNO_3$ 1 mM and by incubating the mixture in the dark, at room temperature for 3 h.

A HITACHI SU8230 cold-field emission scanning electron microscope was used for scanning electron microscope analysis (SEM). Before subjected to analysis, plant extract solution was dropped and dried on scanning transmission electron microscopy (STEM) copper (Cu) grids. A secondary electrons detector (SE) was used for sample morphology and particle size measurements. For the energy dispersive X-ray spectroscopy analysis (EDS), several point scans were performed on relevant sample spots on each sample. Because of sample charging at higher voltage, quantitative EDS analysis could not be properly conducted. Still, qualitative EDS analysis revealed the presence of Ag. Additionally, for spore extract supplemented with AgNPs in a concentration of 25%, an EDS-mapping analysis was performed in order to observe Ag distribution on the analysed sample surface.

The cytogenetic effects of the extracts, before and after Ag nanoparticle phytosynthesis, were evaluated considering the modifications in mitotic index and the stages of mitotic division (prophase, metaphase, anaphase and telophase), as well as the frequency of chromosomal aberrations and nuclear anomalies induced in the meristematic root cells of *Allium cepa* L. (Şuţan et al. 2016). The evaluation of the cytogenetic effects consisted in a 96 h exposure of the bulbs, initially to the action

Fig. 16.1 Colour change in the extract from spores of *Athyrium filix-femina* (L.) Roth after adding $AgNO_3$ (left, before the addition of $AgNO_3$; right, after the formation of nanoparticles)

of distilled water (for 48 h) and of some varied concentrations of ethanol extracts from spores, without (EEA 5%, 15% and 25%) and with AgNPs (EEA + AgNPs 5%, 15% and 25%).

The microscopic preparations were analysed with an Olympus CX-31 microscope, using a ×400 objective. The microscopic analysis consisted in determining the number of cells found in different phases of mitosis and the frequency of chromosomal and nuclear aberrations reported for a number of 3000 cells for each experimental sample. The mitotic index was calculated as the percentage ratio of the number of cells in the mitosis process to the total number of analysed cells (Tedesco and Laughinghouse 2012). The percentage ratio of cells in prophase, metaphase, anaphase or telophase was determined based on the total number of cells in mitosis. The frequency of chromosomal aberrations and nuclear anomalies was determined by relating them to the adequate phase of the cell and mitotic cycle.

The experiments were performed in triplicate, and 3000 cells were analysed for each experimental sample. Statistical analysis of results was conducted using the IBM SPSS Statistics 20 program (2011). Statistical significance and significant differences between variables were determined using variance analysis (one way ANOVA) and Duncan's test for multiple comparison, respectively. The values $P \leq 0.05$ were considered statistically significant. Graphs and tables were compiled based on the mean values ± standard error resulted from several independent experiments.

The formation of Ag nanoparticles in extracts from spores of *A. filix-femina* was established by visual observation of colour change from light yellowish (EEA) to light brown (EEA + AgNPs), as shown in Fig. 16.1. SEM analysis confirmed nanoparticle phytosynthesis (Fig. 16.2). The presence of silicon, according to EDS results (Fig. 16.3), is not surprising, considering the fact that silicon was mentioned in the atomic composition of *Selaginella* spores (Tryon and Lugardon 1978) and that of the leaves of some fern species (Höhne and Richter 1981). Aluminium (Al) presence on EDS spectra is due to the sample holder interference, while the copper (Cu) signal is due to the STEM grid on which the plant extract solution was dropped and dried. Thus, the presence of Cu and Al on sample EDS spectra is not relevant

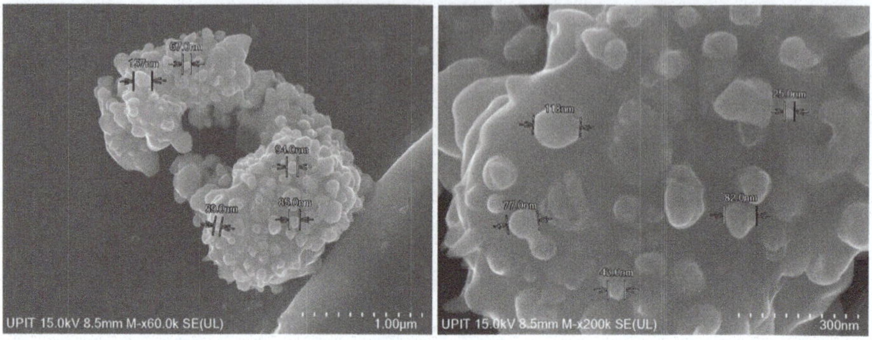

Fig. 16.2 SEM analysis of AgNPs phytosynthesized in extracts from spores of *Athyrium filix-femina* (L.) Roth

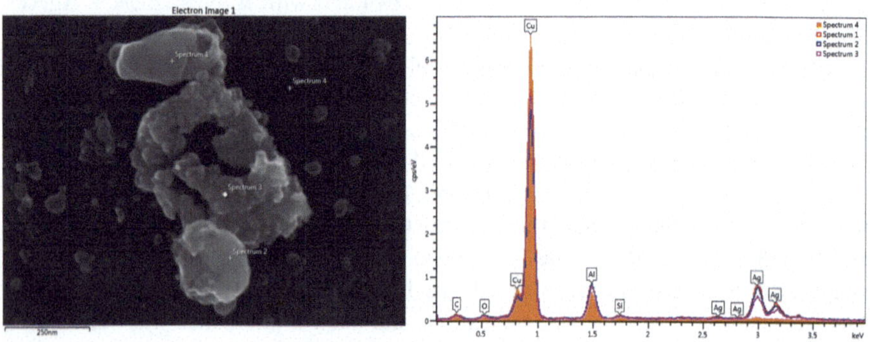

Fig. 16.3 EDS analysis of extracts from spores of *Athyrium filix-femina* (L.) Roth supplemented with AgNPs

for the chemical elemental composition of the sample and should be excluded from EDS analysis.

Root length was the macroscopic parameter taken into account in the evaluation of extract toxicity. For all experimental samples, root length was measured after 72 h, and the results were compared to the negative control sample. Analysis of results indicated growth inhibition in the roots of *A. cepa* exposed to different concentrations of extracts from spores of *A. filix-femina*, with and without AgNPs (Fig. 16.4).

The roots of *A. cepa* had an average length of 21.46 mm in the experimental control sample, while the roots exposed to treatments with extracts from spores reached an average of 16 mm, which was the length of the experimental sample (EEAs 5%). Statistical analysis of results showed a statistically significant correlation between root length and extract concentration. However, the results confirmed the stimulating effect of the extracts, compared to the samples defined by the corresponding ethanol concentrations for root growth. Improved root tolerance to

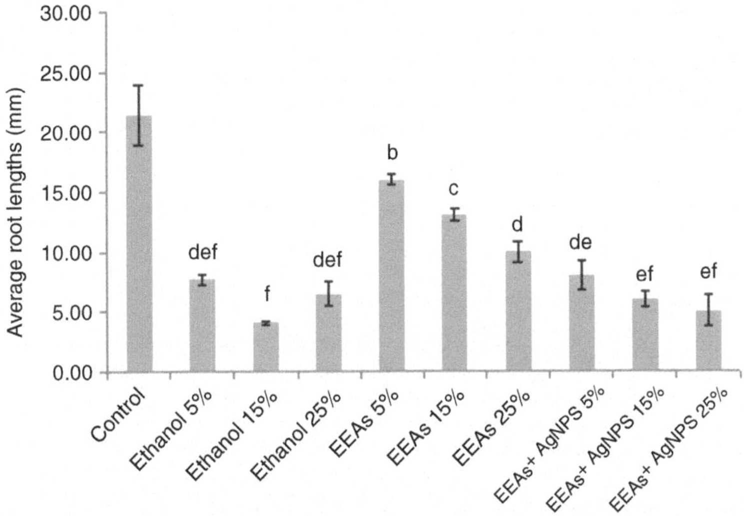

Fig. 16.4 The influence of the extract from spores of *Athyrium filix-femina* (L.) Roth on root length of *Allium cepa* L. before and after phytosynthesis of AgNPs (Data are shown as mean values ± SE of three replicates; a, b, c, d, e, f – interpretation of statistical significance and significant differences through Duncan's test, $p < 0.05$)

low water concentration can be attributed to the presence of silicon (Eneji et al. 2005; Hattori et al. 2005) identified in the spore extracts through EDS analysis.

According to Fiskesjö (1993), the growth of the roots is inhibited when they are exposed to the action of toxic substances, to an inappropriate pH value or to undissolved substances that limit their intake of nutrients. In our study, compared to the negative control sample and to the spore extracts, the incubation of the roots in extracts supplemented with AgNPs induced a significant growth delay, up to the maximum mean value of 8 mm (EEAs + AgNPs 5%), and minimum mean value of 5 mm (EEAs + AgNPs 25%), values that can be attributed to the action of AgNPs.

Cytogenetic Effects of Spore Extracts Analyses of mitotic index variation, distribution of the phases of mitotic division, types of chromosomal aberrations and their frequency were conducted to evaluate the cytogenotoxic effects of extracts from spores of *A. filix-femina*, before and after phytosynthesis of AgNPs.

The variation of the mitotic index was one of the dependent factors considered when analyzing variance and determining the significant differences between experiments. The results are presented in Fig. 16.5. Statistical interpretation of the microscopic results underlined a significant reduction in the frequency of cells in mitosis in the meristematic root tips incubated in extracts from spores of *A. filix-femina* compared to the negative control sample, for a confidence interval of 95%. The MI value determined for the negative control sample was 10.65%, significantly higher than the values presented by all the other experimental samples. Campos et al. (2008) explained the mitodepressive effects of the extracts of *Dicranopteris*

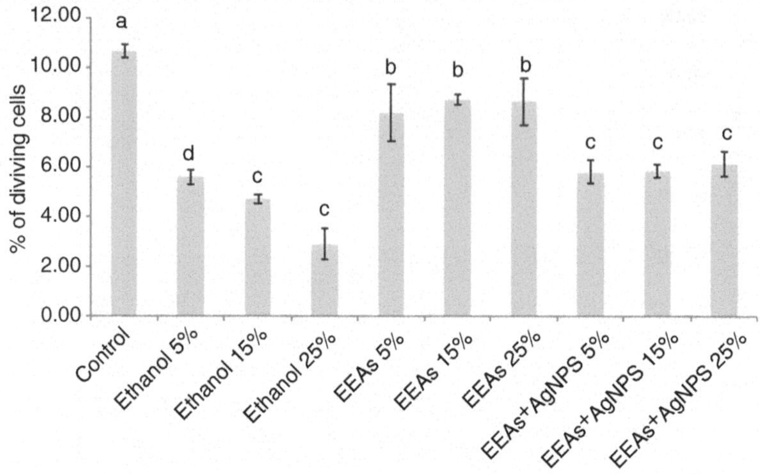

Fig. 16.5 Influence of the extract from spores of *Athyrium filix-femina* (L.) Roth on the mitotic index in meristematic root cells of *Allium cepa* L. before and after phytosynthesis of AgNPs (Data are shown as mean values ± SE of three replicates; a, b, c, d – interpretation of statistical significance and significant differences through Duncan's test, $p < 0.05$)

flexuosa and *Gleicheniella pectinata* through their allelopathic action on meristematic cells of *Lactuca sativa* and *Zea mays*.

The mitodepressive effects of spore extracts were independent of the extract concentration. In this context, we must highlight the stimulating effects of spore extracts compared to the corresponding concentrations of ethanol. At the same time, statistical analysis of results revealed that the mitoinhibitory effects of spore extracts supplemented with AgNPs were more powerful than those of extracts without nanoparticles, which were comparable with the mitodepressive effect of ethanol.

The Mitotic Phase Distribution The analysis of the results regarding the distribution of mitotic phases is presented in Fig. 16.6. Statistical interpretation of data emphasized a substantial prophase accumulation (98.83%) in the meristematic root tips incubated in EEAs 5%. The increase in the concentration of spore extracts to 15% and 25%, respectively, did not generate major changes when compared to the negative control sample. Ethanol treatments, as well as those with spore extracts enriched with AgNPs, induced an important accumulation of metaphases and telophases in the onion root tips.

The mitotic index is a solid parameter in cytogenotoxicity evaluation (Fiskesjö 1985; Fernandez et al. 2007). The reduction in the mitotic index can be explained by mitotic cell cycle arrest, either in one of the stages of the interphase, through inhibition of nuclear proteins or DNA synthesis, or in one of the stages of mitosis, especially during the metaphase. In our study, the important increase in the frequency of prophases after incubation of roots in spore extracts for 24 h suggested that a higher number had passed the G_2 phase. On the other hand, root treatment with spore

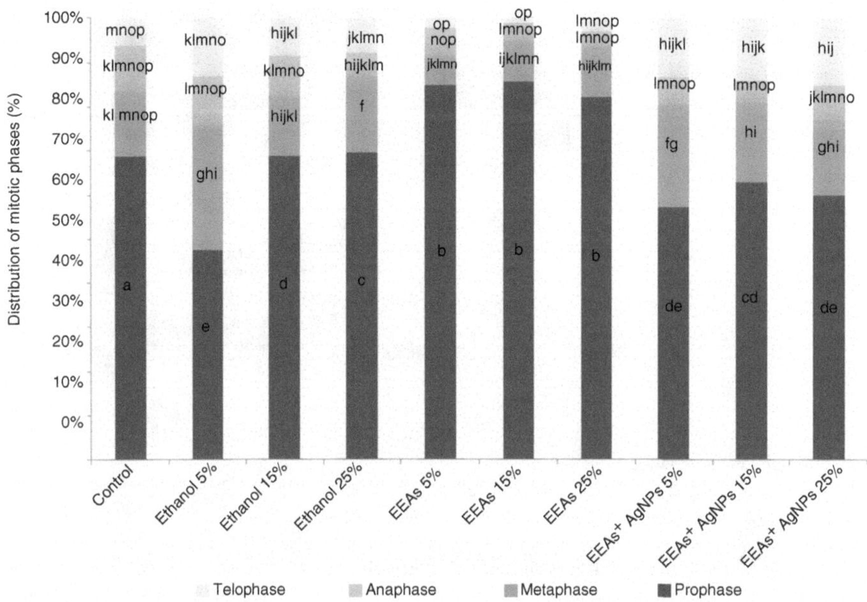

Fig. 16.6 Influence of the extract from spores of *Athyrium filix-femina* (L.) Roth, with and without AgNPs, on the distribution of mitotic phases in the meristematic root cells of *Allium cepa* L. (Data shown as mean values ± SE of three replicates; a, b, c, d, e, f, g, h, i, j, k, l, m, n, o, p – interpretation of statistical significance and significant differences through Duncan's test, $p < 0.05$)

extracts supplemented with AgNPs through biosynthesis induced the accumulation of metaphases, probably as a consequence of a blockage in the formation/functioning of the mitotic spindle. This explanation is supported by the presence of C-mitoses and polyploid cells, which were observed in the experimental samples EEAs + AgNPs. It is estimated that some disrupting agents could prevent the attachment of micronuclei to kinetochores, so that, following the signals generated by control points, the transition from the metaphase to the anaphase is blocked. However, prolonged treatments with disrupting agents help the mitotic checkpoint to adapt, allowing the cell to complete the mitotic cycle by forming tetraploid cells and cytokinesis (Jordan et al. 1991, cited by Blajeski et al. 2002). Genotoxic stress activates intracellular signaling molecules, which leads to cell growth arrest, DNA repair and/or apoptosis.

Chromosomal and Nuclear Aberrations Microscopic analysis of temporary squash preparations displayed different types of chromosomal and nuclear aberrations, such as laggards and vagrant chromosomes, sticky chromosomes, C-mitoses, binucleate cells, variation of chromosomes number or budding nuclei, which characterized the various experimental samples (Fig. 16.7). Thus, laggards and vagrant chromosomes were the most frequent chromosomal aberrations. Vagrant chromosomes were identified in all experimental samples, without major variations, while

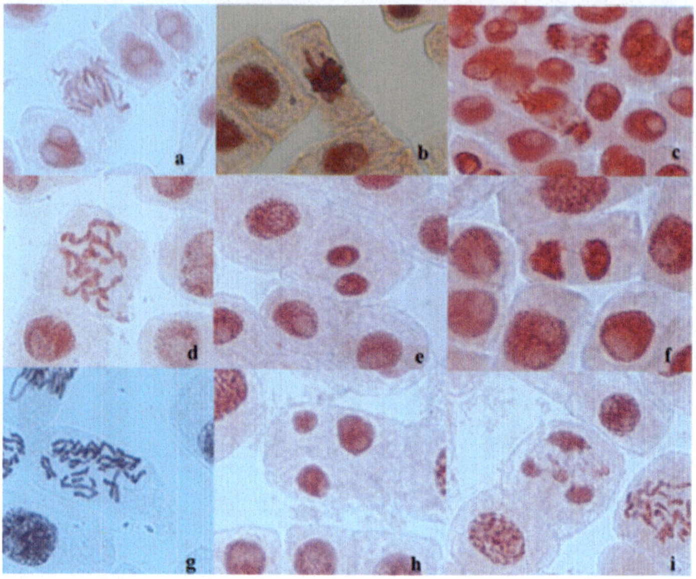

Fig. 16.7 Chromosomal aberrations and nuclear anomalies identified in meristematic root cells of *Allium cepa* L. that underwent treatment with spore extracts of *Athyrium filix-femina* (L.) Roth. ex Mert. (**a**) vagrant chromosomes, EEAs 5%; (**b**) stickiness, ethanol 15%; (**c**) anaphase bridges, EEAs + AgNPs 15%; (**d**) C-mitosis, EEAs + AgNPs 5%; (**e**) binucleate cell, EEAs 25%; (**f**) telophase bridge, EEAs + AgNPs 15%; (**g**) variation of chromosome number, EEAs + AgNPs 25%; (**h**) micronucleus, EEAs + AgNPs 25%; (**i**) nucleoplasmic bridges, EEAs + AgNPs 25%; (**a–i**): at a magnification 400×

the laggards had the highest frequency in the meristematic root cells incubated in ethanol 25% and a significantly lower frequency in the experimental samples EEAs + AgNPs, independent of their concentrations. Spore extracts from *A. filix-femina* induced a lower frequency formation of binuclear cells, while root incubation in spore extracts supplemented with AgNPs generated the formation of C-mitoses and their accumulation with a frequency ranging from 8.33% up to 19.44% (Table 16.1). These observations revealed the different influence of spore extracts on the mitotic apparatus, based on the presence or the absence of AgNPs.

With no correspondence between the aberration frequency and the extract concentration, the influence of bioactive compounds on cells in the division process is obvious. McFee and Tice (1990) appreciated that a series of factors, such as compound solubility, transport rate and biodistribution and the concentration at the target site, can modulate the time interval in which aberrations can be induced.

According to Rank (2003), the presence of vagrant chromosomes was an indicator of the toxic effect on the mitotic spindle. Furthermore, the formation of abnormal anaphases and the presence of C-mitoses were evidence of the side effects of AgNPs on the spindle. The detection of binucleate cells in meristematic cells of

Table 16.1 Chromosomal aberrations and nuclear anomalies observed in the meristematic root cells of *Allium cepa* L. incubated in ethanol extracts from spores of *Athyrium filix-femina* (L.) Roth, before and after phytosynthesis of AgNPs

Experimental variants	Chromosomal aberrations				Nuclear anomalies			
	Vagrants	Laggards	Stickiness	C-mitosis	Binucleate cell	Variation of chromosomes number	Nuclear buds	Others
Control	1.85 ± 1.85^{ab}	–	–	–		–	–	–
Ethanol 5%	10 ± 5.77^{ab}	–	–	10 ± 5.77^{ab}	–	–	–	–
Ethanol 15%	20 ± 11.55^{ab}	20.69 ± 14.98^{ab}	4.99 ± 2.54^{ab}	8 ± 0.04^{b}	–	–	–	0.04 ± 0.04^{b}
Ethanol 25%	25.5 ± 12.8^{ab}	30 ± 17.32^{a}	22.8 ± 4.28^{ab}	12 ± 0.3^{ab}	–	–	–	0.04 ± 0.04^{b}
EEAs 5%	6.61 ± 4.15^{ab}	–	–	–	0.76 ± 0.20^{b}	–	–	0.02 ± 0.02^{b}
EEAs 15%	8.79 ± 8.79^{ab}	21.42 ± 5.99^{ab}	2.38 ± 2.38^{ab}	–	0.18 ± 0.1^{b}	–	–	0.08 ± 0.08^{b}
EEAs 25%	10.18 ± 3.03^{ab}	13.88 ± 10.01^{ab}	8.33 ± 4.80^{ab}	–	0.24 ± 0.08^{b}	–	–	0.08 ± 0.04^{b}
EEAs + AgNPs 5%	2.22 ± 2.22^{ab}	0.66 ± 0.66^{b}	–	8.33 ± 8.33^{ab}	–	0.05 ± 0.05^{b}	–	–
EEAs + AgNPs 15%	9.86 ± 6.97^{ab}	0.08 ± 0.04^{b}	–	19.44 ± 7.09^{ab}	–	0.02 ± 0.02^{b}	1.12 ± 0.48^{b}	0.29 ± 0.03^{b}
EEAs + AgNPs 25%	22.81 ± 13.9^{ab}	1.63 ± 0.18^{b}	–	14.01 ± 1.6^{ab}	–	0.28 ± 0.15^{b}	2.04 ± 0.76^{ab}	–

Data shown as mean values ± SE of three replicates

[a, b] Interpretation of statistical significance and significant differences through Duncan's test, $p < 0.05$

A. cepa suggested that spore extracts inhibited cytokinesis (Kalcheva et al. 2009). Formation of binucleate cells was taken into account in genotoxicity analysis, due to the fact that the disruption of microtubules affects the consolidation of phragmoplasts in the telophase (Soliman 2001). Sticky metaphases appeared as a result of the toxic effect on chromatin and usually led to the death of the cell (Campos et al. 2008). In our study, the frequency of sticky chromosomes was specific to spore extracts without AgNPs. The variation in the number of chromosomes or chromosome sets is a tolerated and relatively common phenomenon in plants, due, at least partially, to the absence of p53 (Korthout et al. 2002).

The identification of C-mitoses and cells with a chromosome number higher than $2n = 16$ in the onion meristematic tips treated with spore extracts containing AgNPs may have resulted from the interference of some of their compounds with the assembly and disassembly process of tubulin molecules in the networks of microtubules (Cassimeris et al. 1988). The presence of C-mitoses, along with the variation in the chromosome number in the experimental samples EEAs + AgNPs, indicates the antitumour potential of AgNPs.

However, even though higher plants are fundamental test organisms for monitoring and testing genotoxins and the *Allium* test is marked by sensitivity and reproducibility (Fiskesjö 1985), it is still necessary to evaluate the genotoxic potential of each compound or bioactive chemical complex, as well as their capacity to damage the genetic material organized in the form of chromosomes. The evaluation has to be done through multiple tests that must certify the fidelity of the results (Repetto et al. 2001).

16.8 Conclusions

Phytosynthesis of Ag nanoparticles in ethanol extracts from spores of *Athyrium filix-femina* was demonstrated through SEMChai and EDS analyses. The sizes of AgNPs varied between 10 and 94 nm. In the meristematic tips of *Allium cepa*, mitosis was inhibited by ethanol extracts from spores. The more pronounced mitodepressive effect associated with the presence of C-mitoses was induced by spore extracts supplemented with AgNPs, which demonstrate potential antitumour effects. Depending on the absence or the presence of AgNPs, the experimental samples were defined by distinct chromosomal aberrations, as follows: sticky chromosomes and binucleate cells and C-mitoses and the variation in the chromosome number, respectively.

Acknowledgements The authors thank the following researchers for their contribution to the work: PhD Physicist Cătălin Ducu and PhD Physicist Denis Negrea (University of Pitești). The spores of *Athyrium filix-femina* were provided by *The Romanian Pteridological Society.*

References

Abraham G, Kaushik GK (2015) Antimicrobial activity and identification of potential antimicrobial compounds from aquatic pteridophyte, *Azolla microphylla* Kaulf. Indian J Exp Biol 53:232–235

Aitken RJ, Chaundhry MQ, Boxall ABA, Hull M (2006) Manufacture and use of nanomaterials: current status in the UK and global trends. Occup Med 56:300–306

Akeroyd J, Synge H (1992) Higher plant diversity. In: Groombridge B (ed) Global biodiversity status of the Earth's living resources, 1st edn. Springer, Dordrecht, pp 64–87

Alcaraz LE, Blanco SE, Puig ON, Tomas F, Ferretti FH (2000) Antibacterial activity of flavonoids against methicillin-resistant *Staphylococcus aureus* strains. J Theor Biol 205:231–240

Alihosseini F, Azarmi S, Ghaffari S, Haghighat S, Sorkhabadi RSM (2016) Synergic antibacterial effect of curcumin with ampicillin; free drug solutions in comparison with SLN dispersions. Adv Pharm Bull 6(3):461–465

Allafchian AR, Mirahmadi-Zare SZ, Jalali SHAH, Hashemi SS, Wahabi MR (2016) Green synthesis of silver nanoparticles using phlomis leaf extract and investigation of their antibacterial activity. J Nanostruct Chem 6(2):129–135

Amoroso VB, Antes DE, Buenavista DP, Coritico FP (2014) Antimicrobial, antipyretic, and anti-inflammatory activities of selected Philippine medicinal pteridophytes. Asian J Biodivers 5:18–40

Auffan M, Rose J, Wiesner MR, Bottero J (2009) Chemical stability of metallic nanoparticles: a parameter controlling their potential cellular toxicity *in vitro*. Environ Pollut 157:1127–1133. https://doi.org/10.1016/j.envpol.2008.10.002

Bahadori MB, Kordi FM, Ahmadi AA, Bahadori S, Valizadeh H (2015) Antibacterial evaluation and preliminary phytochemical screening of selected ferns from Iran. Res J Pharmacogn 2(2):53–59

Baker C, Pradhan A, Pakstis L, Pochan DJ, Shah SI (2005) Synthesis and antibacterial properties of silver nanoparticles. J Nanosci Nanotechnol 5(2):244–249

Balaji DS, Basavaraja RD, Mahesh DB, Belawadi KP, Abbaraju V (2008) Biosynthesis and stabilization of Au and Ag alloy nanoparticles by fungus *Fusarium semitectum*. Sci Technol Adv Mater 9:035012–035017

Baskaran X, Vigila AVG, Parimelazhagan T, Muralidhara-Rao D, Zhang S (2016) Biosynthesis, characterization, and evaluation of bioactivities of leaf extact-mediated biocompatible silver nanoparticles from an early tracheophyte, *Pteris tripartita* Sw. Int J Nanomedicine 11:5789–5806

Bernhardt ES, Benjamin PC, Hochella MF, Cardinale BJ, Nisbet RM, Richardson CJ, Yin L (2010) An ecological perspective on nanomaterial impacts in the environment. J Environ Qual 39:1–12

Bhor G, Maskare S, Hinge S, Singh L, Nalawade A (2014) Synthesis of silver nanoparticles by using leaflet extract of *Nephrolepis exaltata* L and evaluation of antibacterial activity against human and plant pathogenic bacteria. Asian J Pharm Technol Innov 2(7), p. 6.

Blajeski AL, Phan VA, Kottke TJ, Kaufmann SH (2002) G1 and G2 cell-cycle arrest following microtubule depolymerization in human breast cancer cells. J Clin Invest 110(1):91–99. https://doi.org/10.1172/JCI13275

Braga LC, Leite AA, Xavier KG, Takahashi JA, Bemquerer MP, Chartone-Souza E, Nascimento AM (2005) Synergic interaction between pomegranate extract and antibiotics against *Staphylococcus aureus*. Can J Microbiol 51(7):541–547

Britto JD, Gracelin DHS, Kumar PBJR (2014) Antibacterial activity of silver nanoparticles synthesis from a few medicinal ferns. Bull Environ Pharm Life Sci 3(3):224–227

Bunnag D, Dhorranintra B, Limsuvan S (1989) Ferns and their allergenic importance: skin and nasal provocation tests to fern spore extract in allergic and non-allergic patients. Ann Allergy 62(6):554–558

Campos JMS, Davide LC, Soares GLG, Viccini LF (2008) Mutagenic effects due to allelopathic action of fern (Gleicheniaceae) extracts. Allelopath J 22(1):143–151

Cao J, Xia X, Chen X, Xiao J, Wang Q (2013) Characterization of flavonoids from *Dryopteris erythrosora* and evaluation of their antioxidant, anticancer and acetylcholinesterase inhibition activities. Food Chem Toxicol 51:242–250

Cassimeris L, Pryor NK, Salmon ED (1988) Real-time observations of microtubules dynamic instability in living cells. J Cell Biol 107:2223–2231

Chai TT, Panirchellvum E, Ong H-C, Wong F-C (2012) Phenolic contents and antioxidant properties of *Stenochlaena palustris*, an edible medicinal fern. Bot Stud 53:439–446

Chai TT, Yeoh LY, Ismail NIM, Ong HC, Wong FC (2015) Cytotoxicity and antiglucosidase potential of six selected edible and medicinal ferns. Acta Pol Pharm Drug Res 72(2):297–401

Chang HC, Huang G-J, Agrawal D, Kuo C-L, Wu C-R, Tsay H-S (2007) Antioxidant activities and polyphenol contents of six folk medicinal ferns used as "Gusuibu". Bot Stud 48:397–406

Chew FT, Lim SH, Shang HS, Siti Dahlia M, Goh DYT, Lee BW (2000) Evaluation of the allergenicity of tropical pollen and airborne spores in Singapore. Allergy 55:340–347

Chrislyn G, Prachi S, Sangeeta S, Priya S (2016) Green synthesis and characterization of silver nanoparticles. Int J Adv Res 4(8):1563–1568. https://doi.org/10.21474/IJAR01/1362

Croteau MN, Misra SK, Lioma S, Valsami-Jones E (2014) Bioaccumulation and toxicity of CuO nanoparticles by a freshwater invertebrate after waterborne and diet borne exposures. Environ Sci Technol 48:10929–10937. https://doi.org/10.1021/es5018703

Cushnie TPT, Lamb AJ (2005) Antimicrobial activity of flavonoids. Int J Antimicrob Agents 26:343–356

Dalli AK, Saha G, Chakraborty U (2007) Characterization of antimicrobial compounds from a common fern, *Pteris biaurita*. Indian J Exp Biol 45(3):285–290

Deliu I, Bejan C, Vişoiu E, Soare LC (2013) The antimicrobial activity of some extracts of fern gametophytes. Curr Trends Nat Sci 2(2):10–13

Devi S, Yasmeen, Singh J, Shankar R (1989) Patch testing animals to allergenic fern spores. J Toxicol Cutan Ocul Toxicol 8(2):167–172

Eder M, Mehnert W (1998) Bedeutung planzlicher begleitstoffe in extrackten. Pharmazie 53:285–293

Eneji AE, Inanaga S, Muranaka S, Li J, Hattori T, Tsuji W (2005) Effect of calcium silicate on growth and dry matter yield of *Chloris gayana* and *Sorghum sudanense* under two soil water regimes. Grass Forage Sci 60:393–398

European Commission Scientific Committee on Emerging and Newly Identified Health Risks (2006) Modified opinion (after public consultation) on the appropriateness of existing methodologies to assess the potential risks associated with engineered and adventitious products of nanotechnologies, European Commission, Bruxelles, Belgium, p 3–14

Fernández OL, Mazzeo DEC, Marin-Morales MA (2007) Mechanism of micronuclei formation in polyploidization cells of *Allium cepa* exposed to trifluralin herbicide. Pestic Biochem Physiol 88:252–259

Fiskesjö G (1985) The *Allium* test as standard in environmental monitoring. Hereditas 102:99–112

Fiskesjö G (1993) *Allium* test I: a 2-3 day plant test for toxicity assessment by measuring the mean root growth of anions (*Allium cepa* L.) Environ Toxicol 8(4):461–470. https://doi.org/10.1002/tox2530080410

Fun S, Tian F, Li J, Hutchins W, Chen H, Yang F, Yuan X, Cui Z, Yang C-H, He C (2017) Identification of phenolic compounds that suppress the virulence of *Xanthomonas oryzae* on rice via the type III secretion system. Mol Plant Pathol 18(4):555–568

Gartner C, Etahl W, Sies H (1997) Lycopene is more bioavailable from tomato paste than from fresh tomatoes. Am J Clin Nutr 66:116–122

Geethalakshmi R, Sarada DVL (2010) Synthesis of plant mediated silver nanoparticles using *Trianthema decandra* extract and evaluation of their anti-microbial activities. Int J Eng Sci Technol 2(5):970–975

Ghosh S, Patil S, Ahire M et al (2011) Synthesis of gold nanoanisotrops using *Dioscorea bulbifera* tuber extract. J Nanomater 2011, Article ID 354793. https://doi.org/10.1155/2011/354793

Ghosh S, Patil S, Ahire M et al (2012a) Synthesis of silver nanoparticles using *Dioscorea bulbifera* tuber extract and evaluation of its synergistic potential in combination with antimicrobial agents. Int J Nanomedicine 7:483–496

Ghosh S, Patil S, Ahire M et al (2012b) *Gnidia glauca* flower extract mediated synthesis of gold nanoparticles and evaluation of its chemocatalytic potential. J Nanobiotechnol 10:17

Glamočlija U, Haverić S, Čakar J, Durmić A, Haverić A, Bajrović K (2014) Bioactivity and genotoxicity of centuries old remedy *Asplenium scolopendrium* L. Int J Pharm 4(2):38–41

Gottschalk F, Sonderer T, Scholz RW, Nowack B (2009) Modeled environmental concentrations of engineered nanomaterials (TiO, ZnO, Ag, CNT, 2 Fullerenes) for different regions. Environ Sci Technol 43:9216–9222. https://doi.org/10.1021/es9015553

Hammami S, Snène A, El Mokni R, Faidi K, Falconieri D, Dhaouadi H, Piras A, Mighri Z, Porcedda S (2016) Essential oil constituents and antioxidant activity of *Asplenium* ferns. J Chromatogr Sci 54(8):1341–1345. https://doi.org/10.1093/chromsci/bmw071

Handy RD, Owen R, Valsami-Jones E (2008a) The ecotoxicology of nanoparticles and nanomaterials: current satus, knowledge gaps, challenges, and future needs. Ecotoxicology 17:315–325. https://doi.org/10.1007/s10646-008-0206-0

Handy RD, Kamme F, Lead JR, Hasselov M, Owen R, Crane M (2008b) The ecotoxicology and chemistry of manufactured nanoparticles. Ecotoxicology 17:287–314. https://doi.org/10.1007/s10646-008-0199-8

Hattori T, Inanaga S, Araki H, Morita S, Luxuva M, Lux A (2005) Application of silicon enhanced drought tolerance in Sorghum bicolor. Physiol Plant 123:459–466

Haverkamp R, Marshall A (2009) The mechanism of metal nanoparticle formation in plants: limits on accumulation. J Nanopart Res 11(6):1453–1463

Ho R, Teai T, Bianchini J-P, Laffont R, Raharivelomanana P (2011) Ferns: from traditional uses to pharmaceutical development, chemical identification of active principles. In: Kumar A, Fernández H, Revilla MA (eds) Working with ferns: issues and applications. Springer, New York, pp 321–346

Höhne H, Richter B (1981) Untersuchungen über den Mineralstoff- und Stickstoffgehalt von Farnkrautern. Flora 171:1–10

Huang J, Li Q, Sun D, Lu Y, Su Y, Yang X, Wang H, Wang Y, Shao W, Hong NJ, Chen C (2007) Biosynthesis of silver and gold nanoparticles by novel sundried *Cinnamomum camphora* leaf. Nanotechnology 18:105104–105115

IBM Corp Released (2011) IBM SPSS statistics for Windows, version 200 Armonk. IBM Corp, New York

Jeevanandam J, Chan YS, Danquah MK (2017) Biosynthesis and characterization of MgO nanoparticles from plant extracts via induced molecular nucleation. New J Chem 41(7):2800–2814

Johnson M, Amutha S, Shibila T, Janakiraman N (2017) Green synthesis of silver nanoparticles using *Cyathea nilgirensis* Holttum and their cytotoxic and phytotoxic potentials. Part Sci Technol. https://doi.org/10.1080/02726351.2016.1278292

Jordan MA, Thrower D, Wilson L (1991) Mechanism of inhibition of cell proliferation by Vinca alkaloids. Cancer Res 51:2212–2222

Kalcheva VP, Dragoeva AP, Kalchev KN, Enchev DD (2009) Cytotoxic and genotoxic effects of Br-containing oxalosphole on *Allium cepa* L. root tip cells and mouse bone marrow cells. Genet Mol Biol 32(2):389–393

Kalita S, Kandimalla R, Sharma KK, Kataki AC, Deka M, Kotoky J (2016) Amoxicillin functionalized gold nanoparticles reverts MRSA resistance. Mater Sci Eng C 61:720–727

Kandhasamy M, Arunachalam KD, Thatheyus AJ (2008) *Drynaria quercifolia* (L.) J.Sm: a potential resource for antibacterial activity. Afr J Microbiol Res 2:202–205

Kang KC, Kim SS, Baik MH, Choi JW, Kwon SH (2008) Synthesis of silver nanoparticles by using the green chemical method. Appl Chem 12(2):281–284

Khokhani D, Zhang C, Li Y, Wang Q, Zeng Q, Yamazaki A, Hutchins W, Zhou SS, Chen X, Yang CH (2013) Discovery of plant phenolic compounds that act as type III secretion system inhibitors or inducers of the fire blight pathogen, *Erwinia amylovora*. Appl Environ Microbiol 79(18):5424–5436

Klain SJ, Alvarez PJJ, Batley GE, Fernandes TF, Handy RD, Lyon DY, Mahendra S, McLaughlin MJ, Lead JR (2008) Nanomaterials in the environment: behavior, fate, bioavailability, and effects. Environ Toxicol Chem 27:1825–1851

Klaine SJ, Koelmans AA, Horne N, Carley S, Handy RD, Kapustka L, Nowack B, Von der Kammer F (2012) Paradigms to assess the environmental impact of manufactured nanomaterials. Environ Toxicol Chem 31(1):3–14. https://doi.org/10.1002/etc.733

Korbekandi H, Chitsazi MR, Asghari G, Najafi RB, Badii A, Iravani S (2014) Green biosynthesis of silver nanoparticles using *Azolla pinnata* whole plant hydroalcoholic extract. Green Processes Synth 3(5):365–373. https://doi.org/10.1515/gps-2014-0042

Korthout HA, Casper MP, Kottenhagen MJ, Helmer Q, Wang M (2002) A tormentor in the quest for plant p-53-like proteins. FEBS Lett 526:53–57

Koteswaramma B, Kamakshamma J, Varalaksmi S (2017) Biological synthesis of silver nanoparticles from aqueous extract of *Actiniopteris radiata* and evaluation of their antimicrobial activity. Int J Pharm Biol Sci 8(1):121–125

Krishnaraj C, Ramachandran R, Mohan K, Kalaichelvan PT (2012) Optimization for rapid synthesis of silver nanoparticles and its effect on phytopathogenic fungi. Spectrochim Acta A 93:95–99

Kumar A, Pandey AK, Singh SS, Shanker R, Dhawan A (2011) A flow cytometric method to assess nanoparticle uptake in bacteria. Cytometry A 79A:707–712

Kunjiappan S, Bhattacharjee C, Chowdhury R (2015) Hepatoprotective and antioxidant effects of *Azolla microphylla* based gold nanoparticles against acetaminophen induced toxicity in a fresh water common carp fish (*Cyprinus carpio* L.) Nanomed J 2(2):88–110

Lai HY, Lim YY, Tan SP (2009) Antioxidative, tyrosinase inhibiting and antibacterial activities of leaf extracts from medicinal ferns. Biosci Biotechnol Biochem 73(6):1362–1366

Lai HY, Lim YY, Kim KH (2010) *Blechnum orientale* Linn a fern with potential as antioxidant, anticancer and antibacterial agent. BMC Complement Altern Med 10:15. http://wwwbiomedcentralcom/1472-6882/10/15

Lai JC, Lai HY, Rao NK, Ng SF (2016) Treatment for diabetic ulcer wounds using a fern tannin optimized hydrogel formulation with antibacterial and antioxidative properties. J Ethnopharmacol 189:277–289. https://doi.org/10.1016/jep.2016.05.032

Lee HB, Kim JC, Lee SM (2009) Antibacterial activity of two phloroglucinols, flavaspidic acids AB and PB, from *Dryopteris crassirhizoma*. Arch Pharm Res 32:655–659. https://doi.org/10.1007/s12272-009-1502-9

Lee JH, Lim JM, Velmurugan P, Park YJ, Park YJ, Bang KS, Oh BT (2016) Photobiologic-mediated fabrication of silver nanoparticles with antibacterial activity. J Photochem Photobiol B 162:93–99. https://doi.org/10.1016/j.jphotobiol.2016.06.029

Lila MA (2009) Interactions between flavonoids that benefit human health. In: Gould K, Davies K, Winefield C (eds) Anthocyanins biosynthesis, functions, and applications. Springer, New York, pp 305–324

Lila MA, Raskin I (2005) Health-related interactions of phytochemicals. J Food Sci 7:R20–R27

Lim YH, Kim IH, Seo JJ (2007) In vitro activity of kaempferol isolated from the *Impatiens balsamina* alone and in combination with erythromycin or clindamycin against *Propionibacterium acnes*. J Microbiol 45(5):473–477

Liu R (2003) Health benefits of fruit and vegetables are from additive and synergistic combinations of phytochemicals. Am J Clin Nutr 78:517S–520S

Liu Y, Wujisguleng W, Long C (2012) Food uses of ferns in China: a review. Acta Soc Bot Pol 81(4):263–270

Lok C, Ho C, Chen R, He Q, Tu W, Sun H, Tam PK, Chiu J, Chem C (2007) Silver nanoparticles: partial oxidation and antibacterial activities. J Biol Inorg Chem 12:527–534

Lu SG (2007) Pteridology. Higher Education Press, Beijing

Ma XY, Xie CX, Liu C, Song JY, Yao H, Luo K, Zhu YJ, Gao T, Pang XH, Qian J, Chen SL (2010) Species identification of medicinal pteridophytes by a DNA barcode marker, the chloroplast psbA-trnH intergenic region. Biol Pharm Bull 33(11):1919–1924

Ma H, Williams PL, Diamond SA (2013) Ecotoxicity of manufactured ZnO nanoparticles - a review. Environ Pollut 172:76–85. https://doi.org/10.1016/j.envpol.2012.08.011

Magaye RR, Wu A, Zhao J, Zou B, Shi H, Yu H, Liu K, Lin X, Xu J, Yang C (2014) Acute toxicity of nickel nanoparticles in rats after intravenous injection. Int J Nanomedicine 9(1):1393–1402

Makarov VV, Love AJ, Sinitsyna OV, Makarova SS, Yaminsky IV, Taliansky ME, Kalinina NO (2014) "Green" nanotechnologies: synthesis of metal nanoparticles using plants. Acta Nat 6(1):35–44

Manzo S, Miglietta ML, Rametta G, Buono S, Di Francia G (2013) Toxic effects of ZnO nanoparticles towards marine algae *Dunaliella tertiolecta*. Sci Total Environ 445-446:371–376. https://doi.org/10.1016/j.scitotenv.2012.12.051

Marston A, Hostettmann K (2006) Separation and quantification of flavonoids. In: Andersen ØM, Markham KR (eds) Flavonoids chemistry, biochemistry and applications. Taylor & Francis Group, Boca Raton, London, New York, pp 1–36

McFee AF, Tice RR (1990) Influence of treatment to sacrifice time and the presence of BrdUrd on chemically-induced aberration rates in mouse marrow cells. Mutat Res 241:95–108

Mittal AK, Chisti Y, Banerjee UC (2013) Synthesis of metallic nanoparticles using plant extracts. Biotechnol Adv 31:346–356. https://doi.org/10.1016/j.biotechadv.2013.01.003

Moran RC (2004) A natural history of ferns Portland. Timber Press, Portland

Moran RC (2008) Diversity, biogeography, and floristics. In: Ranker TA, Haufler CH (eds) Biology and evolution of ferns and Lycophytes. Cambridge University Press, New York, pp 367–394

Morones JR, Elechiguerra JL, Camacho A, Holt K, Kouri JB, Ramirez JT, Yacaman MJ (2005) The bactericidal effect of silver nanoparticles. Nanotechnology 16(10):2346–2353. https://doi.org/10.1088/0957-4484/16/10/059

Nalwade AR, Badhe MN, Pawale CB, Hinge SB (2013) Rapid biosynthesis of silver nanoparticles using fern leaflet extract and evaluation of their antibacterial activity. Int J Biol Technol 4(2):12–18

Nath K, Bhattacharya MK, Sen A, Kar S (2013) Antibacterial activity of frond extract of *Asplenium nidus* L., a threatened ethnomedicinal fern of North East India. International. J Pharmacogn Phytochem 28(2):1169–1172

Nayak N, Rath S, Mishra MP, Ghosh G, Padhy RN (2013) Antibacterial activity of the terrestrial fern *Lygodium flexuosum* (L.) Sw. against multidrug resistant enteric- and uro- pathogenic bacteria. J Acute Dis 2:270–276

Nowack B, Bucheli TD (2007) Occurrence, behavior and effect of nanoparticles in the environment. Environ Pollut 150(1):5–22

Pal SK (2012) Study of solvent extracts of some selected ferns for antimicrobial activity. Indian J Biol Sci 18:33–37

Pan C, Chen YG, Ma XY, Jiang JH, He F, Zhang Y (2011) Phytochemical constituents and pharmacological activities of plants from the genus *Adiantum*: a review. Trop J Pharm Res 10(5):681–692

Panneerselvam C, Murugan K, Roni M, Aziz AT, Suresh U, Rajaganesh R, Mathiyazhagan P, Subramaniam J, Devakumar D, Nicoletti M, Higuchi A, Alarfaj AA, Munusamy MA, Kumar S, Desneux N, Benelli G (2016) Fern-synthesized nanoparticles in the fight against malaria: LC/MS analysis of *Pteridium aquilinum* leaf extract and biosynthesis of silver nanoparticles with high mosquitocidal and antiplasmodial activity. J Parasitol Res 115(3):997–1013. https://doi.org/10.1007/s00436-015-4828-x

Parihar P, Parihar L, Bohra A (2006) Antibacterial activity of *Athyrium pectinatum* (Wall) Presl. Nat Prod Rad 5(4):262–265

Park Y, Hong YN, Weyers A, Kim YS, Linhardt RJ (2011) Polysaccharides and phytochemicals: a natural reservoir for the green synthesis of gold and silver nanoparticles. IET Nanobiotechnol 5(3):69–78. https://doi.org/10.1049/iet-nbt.2010.0033

Prathna TC, Mathew L, Chandrasekaran N, Raichur AM, Mukherjee A (2010) Biomimetic synthesis of nanoparticles: science, technology and applicability. In: Mukherjee A (ed) Biomimetics learning from nature. InTech. https://doi.org/10.5772/8776. https://www.intechopen.com/books/biomimetics-learning-from-nature/biomimetic-synthesis-of-nanoparticles-science-technology-amp-applicability

Rajaganesh R, Murugan K, Panneerselvam C, Jayashanthini S, Aziz AT, Roni M, Suresh U, Trivedi S, Rehman H, Nicoletti M, Benelli G, Higuchi A (2016) Fern-synthesized silver nanocrystals: towards a new class of mosquito oviposition deterrents? Res Vet Sci 109:40–51. https://doi.org/10.1016/j.rvsc.2016.09.012

Rajesh KD, Vasantha S, Panneerselvam A, Rajesh NV, Jeyathilakan N (2016) Phytochemical analysis, in vitro antioxidant potential and gas chromatography-mas spectrometry studies of *Dicranopteris linearis*. Asian J Pharm Clin Res 9(Suppl. 2):1–6

Rani D, Khare PB, Dantu PK (2010) *In vitro* antibacterial and antifungal properties of aqueous and non-aqueous frond extracts of *Psilotum nudum*, *Nephrolepis biserrata* and *Nephrolepis cordifolia*. Indian J Pharm Sci 72(6):818–822. https://doi.org/10.4103/0250-474X.84606

Rank J (2003) The method of *Allium* anaphase-telophase chromosome aberration assay. Ekologija (Vilnius) 1:38–42

Raut R, Kolekar N, Lakkakula J, Mendhulkar V, Kashid S (2010) Extracellular synthesis of silver nanoparticles using dried leaves of *Pongamia pinnata* (L.) Pierre. Nano-Micro Lett 2(2):106–113. https://doi.org/10.5101/nml.v2i2.p106-113

Repetto G, Jos A, Hayen MJ, Molero ML, del Peso A, Salguero M, del Castillo P, Rodriguez-Vicente MC, Repetto M (2001) A test battery for the ecotoxicological evaluation of pentachlorophenol. Toxicol In Vitro 15:503–509. https://doi.org/10.1016/S0887-2333(01)00055-8

Roco MC (2005) Environmentally responsible development of nanotechnology. Environ Sci Technol 39(5):106A–112A. https://doi.org/10.1021/es053199u

Roos M (1996) Mapping the world's pteridophyte diversity - systematics and floras. In: Camus JM, Gibby M, Johns RJ (eds) Pteridology in perspective. Kew: Royal Botanic Gardens, London, pp 29–42

Roudsari MT, Bahrami AR, Dehghani H, Iranshani M, Matin MM, Mahmoudi M (2012) Bracken-fern extract induce cell cycle arrest and apoptosis in certain cancer cell lines. Asian Pac J Cancer Prev 13(12):6047–6053

Samidoss CM, Murugn K, Mathath R, Sivaprizajothi S, Suganya NA (2013) Larvicidal potential of silver nanoparticles synthesized using *Adiantum capillus-veneris* against *Anopheles stephensi* (Diptera; Culicidae). Int J Curr Trop Med Health Res 1(1):9–18

Sant DG, Gujarathi TR, Harne SR, Ghosh S, Kitture R, Kale S, Chopade BA, Pardesi KR (2013) Adiantum philippense L. frond assisted rapid green synthesis of gold and silver nanoparticles. J Nanopart 2013:9. https://doi.org/10.1155/2013/182320

Santhoshkumar S, Nagarajan N (2014) AM fungal association in the rhizosphere soil of some pteridophytic plant species in Valparai Hills, Western Ghats of Tamilnadu, India. Int J Life Sci 2(3):201–206

Santos ATB, Silva Araujo TF, Silva LCN, Silva CB, Oliveira AFM, Araujo JM, Menezes Lima VL, Santos Correia MT (2015) Organic extract from *Indigofera suffruticosa* leaves have antimicrobial and synergic actions with erythromycin against *Staphylococcus aureus*. Front Microbiol 6:1–7. https://doi.org/10.3389/fmicb.2015.00013

Santos RD, Pimenta-Freire G, FJB F, Dias-Souza MV (2016) Interference of flavonoids and carotenoids on the antimicrobial activity of some drugs against clinical isolates of *Pseudomonas aeruginosa*. Int Food Res J 23(3):1268–1273

Schneider H, Schuettpelz E, Pryer KM, Cranfill R, Magallón S, Lupia R (2004) Ferns diversified in the shadow of angiosperms. Nature 428(6982):553–557. https://doi.org/10.1038/nature02361

Simán SE, Povey AC, Ward TH, Margison GP, Sheffield E (2000) Fern spore extracts can damage DNA. Br J Cancer 83(1):69–73

Sinha T, Ahmaruzzaman M (2016) Indigenous north eastern Indian fern mediated fabrication of spherical silver and anisotropic gold nano structured materials and their efficiency for the abatement of perilous organic compounds from waste eater – a green approach. RSC Adv 6:21076–21089. https://doi.org/10.1039/C5RA26124D

Soare LC, Ferdeş M, Deliu I, Gibea A (2012a) Studies regarding the antibacterial activity of some extracts of native pteridophytes. UPB Sci Bull Ser B 74(1):21–26

Soare LC, Ferdeş M, Stefanov S, Denkova Z, Nikolova R, Denev P, Bejan C, Păunescu A (2012b) Antioxidant activity, polyphenols content and antimicrobial activity of several native pteridophytes of Romania. Not Bot Hort Agrobot Cluj Napoca 40(1):53–57. https://doi.org/10.15835/nbha4016648

Soare LC, Deliu I, Iosub I, Dobrescu CM, Ferdeş M (2012c) New therapeutic formulations with an antibacterial effect, based on plant extracts. Current trends in. Nat Sci 1(2):79–82

Soliman MI (2001) Genotoxicity testing of neem plant (*Azadirachta indica* A. Juss) using the *Allium cepa* chromosome aberration assay. J Biol Sci 1(11):1021–1027

Souri E, Amin G, Farsam H, Jalalizadeh H, Barezi S (2008) Screening of thirteen medicinal plant extracts for antioxidant activity. Iran J Pharm Res 7(2):149–154

Şuţan AN, Fierăscu I, Fierăscu RC, Manolescu DŞ, Soare LC (2016) Comparative analytical characterization and in vitrocytogenotoxic activity evaluation of *Asplenium scolopendrium* L. leaves and rhizome extracts prior to and after Ag nanoparticles phytosynthesis. Ind Crop Prod 83:379–386. https://doi.org/10.1016/j.indcrop.2016.01.011

Tedesco SB, Laughinghouse HD IV (2012) Bioindicator of genotoxicity: the *Allium cepa* test. In: Srivastava JK (ed) Environmental contamination. InTech, Rijeka, pp 137–156

Teow SY, Ali SA (2015) Synergistic antibacterial activity of Curcumin with antibiotics against *Staphylococcus aureus*. Pak J Pharm Sci 28(6):2109–2114

Thakar SB, Ghorpade PN, Kale MV, Sonawane KD (2015) Fern Ethnomedicinal Plant Database: exploring fern ethnomedicinal plants knowledge for computational drug discovery. Curr Comput Aided Drug Des 11(3):266–271

Tryon AF, Lugardon B (1978) Wall structure and mineral content in *Selaginella* spores. Pollen Spores 20:315–340

Turkevich J, Stevenson PC, Hiller PCJ (1951) A study of the nucleation and growth processes in the synthesis of colloidal gold. Discuss Faraday Soc 11:55

Valizadeh H, Sonboli A, Kordi FM, Dehghan H, Bahadori MB (2015) Cytotoxicity, antioxidant activity and phenolic content of eight fern species from North of Iran. Pharm Sci 21:18–24. https://doi.org/10.15171/PS.2015.12

Verma DK, Hasan SH, Banik RM (2016) Photo-catalyzed and phyto-mediated rapid green synthesis of silver nanoparticles using herbal extract of *Salvinia molesta* and its antimicrobial efficacy. J Photochem Photobiol B 155:51–59. https://doi.org/10.1016/j.jphotobiol.2015.12.008

Wang S, Lawson R, Ray P, Yu H (2011) Toxic effects of gold nanoparticles on *Salmonella typhimurium* bacteria. Toxicol Ind Health 27(6):547–554. https://doi.org/10.1177/0748233710393395

Xavier GSA, Selvaraj P, Nida J (2016) Impact of phytoecdysone fractions of the ferns *Cyclosorous interruptus*, *Christella dentata* and *Nephrolepis cordifolia* on the biology of *Spodoptera litura* (Fab). J Biopest 9(2):125–134

Xie Y, Zheng Y, Dai X, Wang Q, Cao J, Xiao J (2015) Sesonal dynamics of total flavonoid contents and antioxidant activity of *Dryopteris erythrosora*. Food Chem 186:113–118

Yoshida T, Hatano T, Ito H (2005) High molecular weight plant polyphenols (tannins): prospective functions. In: Romeo JT (ed) Recent advances in phytochemistry, volume 39: chemical ecology and phytochemistry of forest ecosystems. Elsevier Inc., San Diego, CA, pp 163–190.

Yusuf UK (1994) Flavonoid glycosides in the fern *Blechnum orientale* Linn. Am Fern J 84:69–70

Part IV
Ferns and Environment

Chapter 17
Novel Genes of Hyperaccumulator Ferns in Arsenic Tolerance, Uptake, and Metabolism: Implications for Crop Improvement

Yanshan Chen, Yue Cao, Bala Rathinasabapathi, and Lena Ma

17.1 Arsenic-Hyperaccumulator *Pteris vittata* (Chinese Brake Fern)

17.1.1 Arsenic Toxicity

Arsenic (As) is a toxic and carcinogenic metalloid, which is ubiquitous in the environment. It ranks the 20th in abundance in the earth crust, with an average concentration of ~6 mg kg^{-1} in the upper earth crust (Matschullat 2000). Arsenic can be released into environment through both natural and anthropogenic sources, which elevates As levels in soil and water (Oremland and Stolz 2003; Jomova et al. 2011). Geochemical processes including rock weathering and volcanic eruption, as well as human activities like mining, metal smelting, coal combustion, the use of arsenical pesticides, and wood preservatives, all contribute to As release and cause As pollution in the environment (Jarup 2003; Zhu et al. 2014).

Arsenic in soils is mainly present in inorganic form including arsenate (AsV) and arsenite (AsIII). In aerobic soils, As is present predominantly as oxidized AsV,

Y. Chen (✉) · Y. Cao
State Key Laboratory of Pollution Control and Resource Reuse, School of the Environment, Nanjing University, Jiangsu, China
e-mail: chenyanshan@nju.edu.cn

B. Rathinasabapathi
Horticultural Sciences Department, University of Florida, Gainesville, FL, USA
e-mail: brath@ufl.edu

L. Ma
State Key Laboratory of Pollution Control and Resource Reuse, School of the Environment, Nanjing University, Jiangsu, China

Soil and Water Science Department, University of Florida, Gainesville, FL, USA
e-mail: lqma@ufl.edu

H. Fernández (ed.), *Current Advances in Fern Research*,
https://doi.org/10.1007/978-3-319-75103-0_17

while in anaerobic environments like flooded paddy soil, it exists mainly as reduced AsIII (Huang et al. 2011). Due to As methylation processes, small amount of organic As species may also exist in soil, including dimethylarsinic acid (DMA) and monomethylarsonic acid (MMA) (Huang et al. 2011). Arsenic toxicity in organisms varies with As species. Generally, AsV and AsIII are more toxic than methylated As species. For example, AsIII is ~100-fold more toxic than DMA to animal cells (Hirano 2004). As a phosphate analogue, AsV substitutes for phosphate at binding sites to incorporate into ATP and participate in phosphorylation reactions, thereby interfering with energy and phosphate metabolism (Meharg and Hartley-Whitaker 2002; Finnegan and Chen 2012). Different from AsV, AsIII has high affinity for sulfhydryl groups of the cysteine residues of proteins, thus affecting protein structure and enzyme function (Shen et al. 2013).

For humans, general population is exposed to As mainly via food intake and water consumption. Rice (*Oryza sativa* L.) is the staple food for half of the world's population and wildly cultivated in Asian countries. However, rice is an efficient accumulator of As, and some Asian countries are polluted by As, making rice consumption a major source of As exposure to local people (Sohn 2014; Chen et al. 2017a). Besides, in some areas like Bangladesh, As in drinking water is a significant source of As exposure (Argos et al. 2010). Arsenic exposure adversely impacts millions of people, increasing the risk of skin lesions and various cancers, greatly threatening human health (Stone 2008; Argos et al. 2010; Yang 2011).

17.1.2 Arsenic-Hyperaccumulator **Pteris vittata**

Similar to animals and humans, As is a nonessential element and toxic to plants. Arsenic exposure usually hampers plant growth and impairs plant reproductive capacity (Garg and Singla 2011; Finnegan and Chen 2012). At relatively high As concentrations, As interferes with critical metabolic processes, leading to plant death. In addition, it also impacts plant physiology, leading to generation of reactive oxygen species, increased lipid oxidation, and impaired membrane veracity and nutrient acquisition (Garg and Singla 2011). To survive under As stress, most plants possess mechanisms to restrict As uptake and retain much of the As in the roots, thereby protecting photosynthetic tissues (Ali et al. 2012; Finnegan and Chen 2012).

The first-known As-hyperaccumulating plant, *Pteris vittata* (Chinese brake fern), was reported by Ma et al. (2001). Different from other plants, *P. vittata* is extremely efficient in extracting As from soils and translocating it into the fronds, showing its extraordinary As detoxification ability (Ma et al. 2001). Though other plants can also take up and tolerate As from soils (Dekoe 1994; Meharg et al. 1994; Bech et al. 1997), they do not hyperaccumulate the metalloid. In contrast, As in *P. vittata* is found in much higher concentrations in the fronds than in the roots as well as in soils (Ma et al. 2001).

P. vittata possesses extremely high As tolerance and can live in soils containing as high as 1500 ppm As. Moreover, it can take up large amounts of As into its fronds, even up to >2.3% by dry weight, exhibiting a great potential to remediate As-contaminated soils (Ma et al. 2001). Following the discovery of *P. vittata*, several ferns belonging to the *Pteris* genus were found to hyperaccumulate As, including *Pteris cretica*, *Pteris longifolia*, *Pteris umbrosa*, *Pteris multifida*, *Pteris biaurita*, *Pteris quadriaurita*, *Pteris ryukyuensis*, *Pteris aspericaulis*, *Pteris fauriei*, *Pteris multifida*, and *Pteris oshimensis* (Zhao et al. 2002; Srivastava et al. 2006; Wang et al. 2006). Interestingly, not all species within the *Pteris* genus are As hyperaccumulators. For example, *Pteris straminea*, *Pteris tremula*, and *Pteris semipinnata* do not hyperaccumulate As (Meharg 2003; Wang et al. 2006). In addition, *Pityrogramma calomelanos*, a fern in the order of Pteridales, exhibits As-hyperaccumulating property (Visoottiviseth et al. 2002). Other ferns like *Actiniopteris radiata*, *Cheilanthes sinuata*, and *Polystichum acrostichoides* can also accumulate As and were identified as potential As accumulators, suggesting that As accumulation likely is an evolutionary advantage in ferns (Srivastava et al. 2010).

The discovery of As-hyperaccumulator *P. vittata* opens the door to an inexpensive and environmentally friendly technology to remediate As-contaminated soils (Kramer 2005). Besides its application in phytoremediation, *P. vittata* is also used as a model plant to investigate As metabolism in plants and As hyperaccumulation mechanisms (Gumaelius et al. 2004; Han et al. 2017a). Much research based on plant physiology, biochemistry, and molecular biology have been carried out to investigate its As uptake, translocation, speciation, distribution, and detoxification.

17.2 Arsenic Uptake, Translocation, and Detoxification in *P. vittata*

17.2.1 Arsenate Uptake Via Phosphate Transporters (Pht)

In non-hyperaccumulating plants, AsV uptake is inhibited by phosphate (P) supplement, and P starvation enhances AsV uptake in plants (Muchhal et al. 1996; Abedin et al. 2002). In addition, AsV treatment decreases P uptake by plants (Clark et al. 2000). The results showed that AsV and P compete in plant uptake, indicating AsV is taken up by phosphate transport system. Molecular biological studies confirmed that AsV is taken up by plants via P transporters (Pht) (Jia et al. 2011; Puckett et al. 2012; Wang et al. 2016).

P. vittata is highly efficient in taking up AsV (Wang et al. 2002). In *P. vittata*, AsV uptake is also inhibited by P, with increasing P supply decreasing As uptake, while P deficiency increases its As accumulation, indicating that AsV uptake may also occur via P transporters (Lessl and Ma 2013; Fu et al. 2017). Compared to non-hyperaccumulator, *P. vittata* exhibits enhanced AsV uptake, indicating that it may possess unique Pht. Recently, DiTusa et al. (2016) identified three *P. vittata* P transporter 1 (Pht1) genes, including *PvPht1;1*, *PvPht1;2*, and *PvPht1;3*. Since PvPht1;1 and PvPht1;2 share 98.5% identical amino acid residues and vary only by 8 amino acid residues, they represent probably the same P transporter, which shares 72% identity to PvPht1;3. Transcripts of both *PvPht1;1/2* and *PvPht1;3* are induced by P deficiency. However, in response to AsV, *PvPht1;1/2* transcript abundance decreases by 2-fold, whereas *PvPht1;3* transcript increases 1.5-fold. The results indicated similar P responses for *PvPht1;1/2* and *PvPht1;3* but with differential regulation by AsV (DiTusa et al. 2016).

Functional studies showed that PvPht1;1, PvPht1;2, and PvPht1;3 proteins all function as P transporters but with different transport properties. In yeast experiments, PvPht1;3 is more effective in complementing yeast mutant PAM2 (Δpho84Δpho89) defective in P uptake. Further studies showed that yeast cells expressing PvPht1;3 are sensitive to AsV by accumulating more AsV than empty vector control. Moreover, P uptake of PvPht1;3 is strongly inhibited by AsV, with the Km for P uptake by PvPht1;3 being 22.1 μM, and the inhibition constant (Ki) of AsV being 14.8 μM. The results indicated that PvPht1;3 has similar affinities for P and AsV, which are different from other Pht transporters that prefer P over AsV (DiTusa et al. 2016). Previous study showed that *P. vittata* had lower Km value for AsV as compared with a non-hyperaccumulating fern, suggesting a higher AsV affinity in *P. vittata* (Poynton et al. 2004). As a high-affinity AsV transporter, PvPht1;3 may be responsible for enhanced AsV uptake exhibited by *P. vittata* and play a critical role in efficient AsV uptake by *P. vittata* (DiTusa et al. 2016). However, to further understand AsV uptake in *P. vittata*, more Pht transporters should be identified and characterized.

17.2.2 Arsenite Uptake Via Aquaporin Channels

AsIII is not the predominant As species in aerobic soils where *P. vittata* grows (Xu et al. 2014; Han et al. 2016). However, small amount of AsIII may exist in the rhizosphere due to biochemical transformation by plant roots and AsV reduction by microbes (Moreno-Jiménez et al. 2012, 2013; Han et al. 2017b). AsIII is taken up by plant roots via aquaporin water channels (Major intrinsic proteins; MIPs), mainly a subfamily of the MIP called nodulin 26-like intrinsic proteins (NIPs) (Ma et al. 2008; Mitani-Ueno et al. 2011). In *Arabidopsis*, NIP1;1 and NIP3;1 play important roles in AsIII uptake and translocation (Kamiya et al. 2009; Xu et al. 2015). In rice, Lsi1 (OsNIP2;1) plays a critical role in mediating AsIII uptake and accumulation (Ma et al. 2008). Besides, NIP1;2, NIP5;1, and NIP7;1 in *Arabidopsis* and NIP1;1, NIP2;2, NIP3;1, and NIP3;2 in rice also show permeability to AsIII (Bienert et al. 2008; Isayenkov and Maathuis 2008; Kamiya et al. 2009).

MIP superfamily in higher plants can be classified into different subfamilies. Besides NIPs, plasma membrane intrinsic proteins (PIPs) and tonoplast intrinsic proteins (TIPs) are also important subfamilies (Quigley et al. 2002; Danielson and Johanson 2008). Some PIPs in rice, OsPIP2;4, OsPIP2;6, and OsPIP2;7, are also involved in AsIII transport (Mosa et al. 2012).

In *P. vittata*, aquaporin inhibitor $AgNO_3$ at 0.01 mM reduces AsIII uptake, indicating that AsIII uptake may be mediated by aquaporin transporters (Mathews et al. 2011). More recently, to select genes that mediate AsIII uptake in *P. vittata*, yeast mutant *Δfps1* was transformed with a cDNA library from *P. vittata* and an AsIII-sensitive yeast transformant expressing a new TIP gene was obtained (He et al. 2016). As a subfamily of MIP, TIPs are classified into six groups, including TIP1, TIP2, TIP3, TIP4, and TIP5 based on sequence homology (Chaumont et al. 2001; Quigley et al. 2002) and TIP6, which is specifically expressed in *Physcomitrella patens* (Danielson and Johanson 2008). The new TIP gene encoding a TIP4-like aquaporin (named PvTIP4;1) mediates AsIII uptake in yeast (He et al. 2016).

In plant experiments, *Arabidopsis* transgenic lines expressing PvTIP4;1 accumulate more As in vivo, so they are more sensitive to AsIII stress, indicating that PvTIP4;1 is capable of transporting AsIII in transgenic *Arabidopsis* (He et al. 2016). Moreover, *PvTIP4;1* is mainly expressed in *P. vittata* roots, with AsIII showing little effect on its transcriptional expression. In subcellular level, PvTIP4;1 localizes to the plasma membrane of *P. vittata*. The results strongly indicate that PvTIP4;1 may mediate AsIII uptake in *P. vittata* (He et al. 2016).

17.2.3 Arsenate Reduction Via ACR2

After being taken up by plant roots, AsV can be rapidly reduced to AsIII in root cells (Zhao et al. 2009). In eukaryote *Saccharomyces cerevisiae*, ACR2 (Arsenic Compounds Resistance 2) mediates the reduction of AsV to AsIII, which is the initial step of As detoxification (Bobrowicz et al. 1997; Mukhopadhyay and Rosen 1998). In plants, ACR2 arsenate reductases, which are similar to yeast ACR2, have also been identified. For example, reports showed that AtACR2, OsACR2.1, and OsACR2.2 are involved in AsV reduction in *Arabidopsis* and rice, respectively (Dhankher et al. 2006; Duan et al. 2007). In *Arabidopsis*, knockdown of AtACR2 using RNA interference leads to its sensitivity to high AsV concentrations than wild type, but not AsIII (Dhankher et al. 2006). However, other study also showed that knocking out ACR2 by T-DNA insertion does not affect As redox (Liu et al. 2012).

More recently, researchers found that a novel arsenate reductase, termed ARQ1 (Arsenate Reductase QTL1) or HAC1 (High Arsenic Content 1), which is different from canonical ACR2, plays critical roles in AsV reduction and AsV tolerance in *Arabidopsis* (Chao et al. 2014; Sánchez-Bermejo et al. 2014). In rice, OsHAC1;1/2 and OsHAC4 are also responsible for AsV reduction, reducing AsV to AsIII in the root cells to facilitate AsIII efflux into the external environment, thus decreasing As accumulation in plants (Shi et al. 2016; Xu et al. 2017).

In *P. vittata*, As species analysis showed that, after AsV exposure, As is mainly present as inorganic forms, with considerable amounts of AsIII in the roots but more AsIII than AsV in the fronds, showing AsV reduction occurs in *P. vittata* (Ma et al. 2001; Chen et al. 2016). Duan et al. (2005) showed that root extracts from *P. vittata* can reduce AsV to AsIII, but frond extracts cannot, indicating AsIII in the fronds may mainly come from AsV reduction in the roots including rhizomes. However, Bondada et al. (2004) showed that when AsV was foliar-applied to *P. vittata* frond surface, it occurred mainly as AsIII after being absorbed, indicating that efficient AsV reduction occurs in the fronds also (Tu et al. 2002).

Ellis et al. (2006) reported the isolation of PvACR2 gene from *P. vittata* gametophytes, which can suppress AsV sensitivity and As hyperaccumulation phenotypes of yeast lacking ScACR2. ACR2 belongs to the tyrosine phosphatases, which involves in cell division and cycle (CDC) (Bhattacharjee and Rosen 2007). Actually, ACR2 in *Arabidopsis* is initially identified as a dual-specificity CDC25 phosphatase (Landrieu et al. 2004). However, the phosphatase activity was found in AtACR2 and OsACR2, but not in PvACR2, making PvACR2 the first plant ACR2 that lacks phosphatase activity (Ellis et al. 2006; Duan et al. 2007). Besides, PvACR2 is also unique due to its unusual active site at a conserved motif that is essential for phosphatase and reductase activity. These unique properties of PvACR2, together with its constitutive expression, may play an important role in As hyperaccumulation by *P. vittata* (Ellis et al. 2006).

Besides ACR2, *P. vittata* triosephosphate isomerase (TPI) expression is capable of increasing AsV resistance in *E. coli* by directly or indirectly functioning as an arsenate reductase, suggesting that PvTPI may play a role related to AsV reduction in *P. vittata* (Rathinasabapathi et al. 2006). In addition, *P. vittata* may also express an AsV reductase named Pv2.5–8, which corresponds to a "predicted protein sequence of a putative AsV reductase" (Cesaro et al. 2015). Though functional ATQ/HAC arsenate reductases have been identified in *Arabidopsis* and rice (Chao et al. 2014; Sánchez-Bermejo et al. 2014; Shi et al. 2016; Xu et al. 2017), whether its homologue exists in *P. vittata* is still unclear.

17.2.4 Arsenic Translocation

Compared with non-hyperaccumulating ferns, *P. vittata* has a higher rate of AsV uptake and, more importantly, a much enhanced translocation of As from the roots to fronds (Huang et al. 2004; Poynton et al. 2004; Caille et al. 2005; Su et al. 2008; Fu et al. 2017). Following entry into the root cells, AsV can be reduced to AsIII. Analysis of As speciation in *P. vittata* showed that AsV is the dominant species in the roots, with AsIII accounting for 10–40%. In contrast, AsIII is the main As species in the fronds, accounting for 60–90% (Ma et al. 2001; Zhang et al. 2002; Poynton et al. 2004; Pickering et al. 2006; Chen et al. 2016). One explanation for much higher AsIII in the fronds than the roots is that some AsV reduction occurs in the fronds (Zhang et al. 2002; Kertulis et al. 2005; Mathews et al. 2011). However, it may also be due to the preferential translocation of AsIII from the roots to fronds, leaving more AsV in the roots (Su et al. 2008).

In fact, Duan et al. (2005) found that the activity of arsenate reductase was in the root extract of *P. vittata*, but not the frond extract, suggesting that As reduction occurs in the roots. Moreover, Su et al. (2008) showed that the As concentrations in xylem sap was 18–73 times higher than that in the growth solution, with AsIII being the predominant species, supporting preferential translocation of AsIII over AsV. However, both Tu et al. (2002) and Mathews et al. (2011) showed AsV reduction occurs in the fronds, therefore, more research is needed in this aspect.

17.2.5 Arsenic Efflux

Root Arsenic Efflux

Arsenite efflux into external environment is an important mechanism used by microbes to detoxify As. In both prokaryote *E. coli* and eukaryote *S. cerevisiae*, As detoxification involves AsIII extrusion from cytosol to the external medium (Ghosh et al. 1999; Rosen 1999). Arsenite efflux also occurs in plants, though it is unclear whether this is an As detoxification strategy. Xu et al. (2007) found that after taken up by plant roots, AsV can be rapidly reduced to AsIII, which was effluxed to the external medium. Actually, plants are capable of effluxing large amounts of AsIII (normally >50% of AsV uptake) to external environment in a short time after low AsV exposure (Xu et al. 2007; Shi et al. 2015; Chen et al. 2016). However, in *P. vittata*, there was little AsIII complexation and little AsIII efflux from the roots to external medium, which may partially explain As hyperaccumulation by *P. vittata* (Su et al. 2008). However, at high AsV exposure, substantial AsIII efflux was detected (Chen et al. 2016). Whether AsIII efflux is an As detoxification mechanism is unclear. But, AsIII efflux appears to be critical for plants to release As burden at high exposure.

Frond Arsenic Efflux

Besides the roots, the aerial organs such as the leaves can also absorb soluble heavy metals, including essential elements like Cu and Zn and toxic metals such as Cd and As (Lepp 1975; Handson 1984; Pandey et al. 2013). Similarly, *P. vittata* can also absorb foliar-applied As, with most of the As concentrations being in the fronds (Bondada et al. 2004).

P. vittata can not only absorb foliar-applied As but also extrude As from the fronds. This is consistent with that most of the As in the pinnae is distributed in the epidermal cells and multicellular trichomes (Lombi et al. 2002; Li et al. 2005). It has been also reported that the *P. vittata* is able to leach As from the fronds, especially when they are dry (Tu et al. 2003; Yan et al. 2009). Recently, Cantamessa et al. (2016) showed that *P. vittata* fronds can extrude crystals containing As after extremely high As exposure (334 mM AsV for 60 d). More recently, Ronzan et al. (2017) also showed that, after exposing to As at toxicity level, both saline secretions and discrete crystals containing As occurred on *P. vittata* fronds. These results are in agreement with reports that plants can extrude As, Cd, and Zn as crystals from the leaves through trichomes (Kupper et al. 2000; Gupta and Bhatnagar 2015). Furthermore, despite intracellular compartmentalization in *P. vittata*, Datta et al. (2017) showed evidence for a preponderance of exocellular As in *P. vittata* fronds, which is a potential mechanism of As detoxification. Further research is needed to understand the nature of As transport into the exocellular location.

17.2.6 Arsenic Detoxification and Hyperaccumulation

For non-hyperaccumulating plants, As tolerance can be achieved by suppressing high-affinity phosphate uptake system to decrease AsV uptake (Meharg and Macnair 1992; Meharg and Hartley-Whitaker 2002). Besides, AsV can be reduced to AsIII, which has high affinity to sulfhydryl (–SH) groups and can be complexed with peptide thiols glutathione (GSH) and phytochelatins (PCs) (Pal and Rai 2010). In many plants, As exposure induces enhanced synthesis and accumulation of PCs (Schmoger et al. 2000; Sneller et al. 2000). Either inhibition of PC synthesis by L-buthionine-sulfoximine, an inhibitor of γ-glutamylcysteine synthetase, or lacking functional PC synthase (PCS) leads to As sensitivity in plants, indicating the critical role of PCs in plant As tolerance (Schat et al. 2002). In contrast, overexpression of PCS significantly enhances As tolerance in transgenic plants (Liu et al. 2010; Pal and Rai 2010; Guo et al. 2008).

Though PCs play a crucial role in many plant species, they may play little role in As tolerance and detoxification in *P. vittata*. PC synthesis was induced upon AsV exposure in *P. vittata*, but the molar ratio of PC-SH to As was only 0.03–0.09 for the shoots and roots, suggesting only 1–3% of the As in *P. vittata* is chelated by PCs (Zhao et al. 2003). Moreover, the PCs in *P. vittata* contributed to little AsIII complexation with thiols in the roots, which is critical for its efficient AsIII translocation. In the xylem sap, AsIII is the predominant species, accounting for 93–98% of the total As, regardless AsV or AsIII being supplied (Su et al. 2008).

P. vittata is tolerant to As and can hyperaccumulate >1% As, showing little symptom of toxicity (Ma et al. 2001; Tu et al. 2002; Wang et al. 2002). Analyses of As distribution using energy dispersive X-ray microanalyses revealed that As is compartmentalized mainly in the upper and lower epidermal cells, probably in the vacuoles (Lombi et al. 2002). In gametophytes, X-ray absorption spectroscopy analyses showed that AsIII is compartmentalized within the cell vacuole, with As being excluded from cell walls, rhizoids, and reproductive areas (Pickering et al. 2006). Moreover, Yang et al. (2009) isolated cell wall, protoplast, and vacuoles from *P. vittata* callus and found that ~91% of total As is present in the vacuoles.

In 2010, two *P. vittata* genes, *PvACR3* and *PvACR3;1*, which encode proteins similar to yeast AsIII efflux transporter ACR3 (Arsenic Compounds Resistance 3), were isolated and characterized (Indriolo et al. 2010). PvACR3 is a functional AsIII antiporter when expressed in yeast, which mediates AsIII efflux from yeast cytoplasm to external environment. In *P. vittata* gametophyte, knocking down the expression of PvACR3 resulted in an As-sensitive phenotype, indicating that

PvACR3 plays a critical role in As detoxification (Indriolo et al. 2010). Moreover, protein analyses show that PvACR3 localizes to the vacuolar membrane in *P. vittata*, indicating that it may mediate AsIII transport into the vacuole for sequestration (Indriolo et al. 2010). However, when heterologously expressed in *Arabidopsis*, PvACR3 localizes to the plasma membrane (Chen et al. 2013; Wang et al. 2017) while PvACR3;1 localizes to the vacuolar membrane (Chen et al. 2017b).

Unlike ferns, ACR3 gene is lost in flowering plants. Even in plants like moss, lycophytes, ferns, and gymnosperms, only single-copy *ACR3* gene was found, whereas in *P. vittata*, two *ACR3* genes exist (Indriolo et al. 2010; Chen et al. 2017b). The duplication of *ACR3* in *P. vittata* and the loss of *ACR3* in angiosperms may explain *P. vittata*'s As tolerance and hyperaccumulation.

17.3 Novel Genes for Phytoremediation and Crop Improvement

Ferns resulting from a more recent diversification that is independent of angiosperm evolution represent an important, hitherto-unexplored gene diversity that can be utilized for improving plants (Rathinasabapathi 2006). For *P. vittata*, the crucial genes involved in As metabolism and other physiological process can benefit phytoremediation and crop production.

17.3.1 PvPCS1 Gene

Phytochelatins (PCs), with the general structure of $(\gamma\text{-Glu-Cys})_n\text{-Gly}$ ($n = 2\text{–}11$), play an important role in metal tolerance and homeostasis in plants (Cobbett 2000). In plants, PCs synthesis can be induced upon exposure to metals (Pal and Rai 2010; Ronzan et al. 2017). PCs are synthesized enzymatically from glutathione by phytochelatin synthase (PCS), which can be activated by metals. The in vitro activity of partially purified PCS is active only in the presence of metals, with the most efficient activator being Cd (Grill et al., 1989). When *P. vittata* is exposed to As and Cd, the presence of Cd increases As uptake but reduces As translocation to the fronds. This is probably because the co-exposure induced the synthesis As-PC complexes, indicating PCs may be a factor affecting As translocation (Ronzan et al. 2017).

Genes encoding PCS have been isolated from *Arabidopsis* (Vatamaniuk et al. 1999), wheat (Clemens et al. 1999), and *Brassica juncea* (Heiss et al. 2003). However, the PCS gene of *P. vittata*, *PvPCS1*, shows low identity in nucleotide sequence with most known plant *PCS* genes, which is intriguing. The predicted PvPCS1 protein contains 2 Cys-Cys motifs and 12 single Cys residues, with only 4 being conserved in known PCS (Dong et al. 2005). When expressed in *S. cerevisiae*, PvPCS1 increases Cd tolerance, showing that it is a functional PCS (Dong et al. 2005).

However, whether PvPCS1 increases As tolerance is unknown. Though little PCs are detected in *P. vittata*, PvPCS gene may provide important information to understand its genetic basis of As tolerance. Moreover, considering overexpression of PCS may enhance As tolerance in transgenic plants and affect their As accumulation (Liu et al. 2010; Pal and Rai 2010; Shri et al. 2014; Hayashi et al. 2017), thereby PvPCS may also help to regulate As accumulation in crops.

17.3.2 *PvGrx5* Gene

In 2008, a glutaredoxin (Grx) gene *PvGrx5* was isolated from a cDNA library of *P. vittata* fronds based on the ability of the cDNA to increase As tolerance in *E. coli* (Sundaram et al. 2008). Further investigation showed that expressing PvGrx5 lowers the intracellular AsIII levels in *E. coli* after AsV exposure, probably by modulating the aquaporin, making PvGrx5 the first plant Grx involving in As metabolism (Sundaram et al. 2008).

To further understand its function, PvGrx5 was constitutively expressed in transgenic *Arabidopsis*. The transgenic *Arabidopsis* is more tolerant to As than control lines based on germination, root elongation, and plant growth under As stress (Sundaram et al. 2009). Besides, PvGRX5 transgenic lines accumulate lower As than control lines after AsV exposure, partially due to more efficient AsV reduction (Sundaram et al. 2009). Due to its role in decreasing As levels in transgenic plants, PvGRX5 may be used in a novel biotechnological solution to decrease As accumulation in crops. Moreover, PvGrx5 also increases plant tolerance to high-temperature stress and reduces oxidative damage to proteins (Sundaram and Rathinasabapathi 2010), which may also benefit engineering oxidative stress tolerance in plants and improving crop growth.

17.3.3 Phytase Genes

Phosphorus (P) is one of the most limiting macronutrients, affecting crop growth and food production around the world (Cordell et al. 2009; Gilbert 2009). Plants take up inorganic phosphate to meet their P demand, but P often exists as insoluble form in soils due to soil sorption (Richardson et al. 2009). In addition, soil also contains large amount of organic P, which accounts for 30–80% of total soil P, predominantly as phytic acid or phytate (myo-inositol 1,2,3,4,5,6-hexakisphosphate). However, phytate is a stable compound resistant to biochemical degradation, rendering it unavailable for plant to utilize as a P source (Turner et al. 2002). Thus, identification of plants that are capable of utilizing phytate in soils can be of significance.

Lessl et al. (2013) tested six plants, including three angiosperms with known phytase activity (*Lactuca sativa*, *Trifolium subterraneum*, and *Allium schoenoprasum*) and three pteridophytes (*Pteris ensiformis*, *Thelypteris kunthii*, and *P. vittata*). They found that *P. vittata* is the only plant that can effectively utilize phytate. The novel phytase, which is resistant to AsV and heat shock, likely plays a critical role in phytate utilization in *P. vittata* (Lessl et al. 2013).

Phytase is the enzyme that can specifically cleave P from phytate (Brinch-Pedersen et al. 2002). Compared to phytase-specific activity in *P. vittata* roots, those in the root exudates are even higher, indicating that phytase in the root exudates may account for the P acquisition from the rhizosphere (Lessl et al. 2013). Moreover, the phytase of other plants may lose their activity due to sorption and precipitation reactions in soil, while *P. vittata* phytase is not deactivated by soils (Lessl et al. 2013). These results suggest *P. vittata* has a unique stable phytase that help it to directly obtain P from phytate in soils. Though the critical genes that encode *P. vittata* phytase are unclear, these genes may have potential use as a transgenic tool to reduce the need for P fertilizers and to increase crop yield in P-limiting soils.

17.4 ACR3 Genes

Both AsV and AsIII can be effluxed into external environment (Xu et al. 2007). Following AsV uptake, a small amount of AsV was released from the cells via efflux to external environment, which occurs when the plant is transferred into As-free solution. Compared to AsV efflux, AsIII efflux is more intriguing and play a more important role in plant As metabolism. This is probably because AsV can be rapidly reduced into AsIII, with AsIII efflux in plant roots being obvious and significant (Xu et al. 2007; Chen et al. 2016).

AsIII efflux by plant roots provides a potential strategy to decrease As accumulation in plants. However, the key transporters responsible for AsIII efflux in plant roots are largely unknown. AsIII enters plant roots via aquaporins (Ali et al. 2009). However, many aquaporins transport AsIII bidirectionally across membranes in yeast (Bienert et al. 2008; Ali et al. 2009; Xu et al. 2015; He et al. 2016). Thus, it is possible that aquaporins are involved in AsIII efflux in plants. This notion is proved by a functional study of OsNIP2;1, which mediates AsIII uptake and confers As accumulation in rice, but also mediates AsIII efflux, contributing to 15–20% of As efflux in rice (Zhao et al. 2010). However, manipulation of aquaporins to enhance AsIII efflux may be complicated, mainly because the aquaporin-mediated AsIII transport is a passive process, with the direction of flux depending on the concentration gradient (Zhao et al. 2010).

In yeast, AsIII is extruded into the external environment by the AsIII efflux transporter ACR3, which is a key As detoxification mechanism for yeast (Wysocki et al. 1997). Considering ACR3 mediates AsIII efflux in microbes, researchers also transformed the yeast ACR3 into plants to increase plant AsIII efflux. Ali et al. (2012) transformed ScACR3 into *Arabidopsis* and found that ScACR3 increased AsIII efflux at the cellular level but hardly affected As levels in plants. Duan et al. (2012) transformed ScACR3 into rice and found that heterologous expression of ScACR3 increases AsIII efflux by plant roots and decreases As concentrations in brown rice by 20%, which is of significance for food safety and crop improvement.

Though ACR3 is lost in flowering plants, it exists in *P. vittata* with duplication (Indriolo et al. 2010). It was reported that little AsIII efflux occurs in *P. vittata*, which partly explains its efficient As uptake and accumulation. However, at high As exposure, substantial AsIII efflux was observed in *P. vittata* (Chen et al. 2016). Whether ACR3s are involved in AsIII efflux in *P. vittata* remains unclear. However, Chen et al. (2013) showed that PvACR3 is capable of increasing AsIII efflux when heterologously expressed in *Arabidopsis*. The most interesting part is that PvACR3 is localized to the plasma membrane in *Arabidopsis* (Chen et al. 2013; Wang et al. 2017), which is different from its subcellular localization of vacuolar membrane in *P. vittata* (Indriolo et al. 2010).

In transgenic *Arabidopsis*, PvACR3 increases AsIII efflux, decreasing As accumulation by 90% in the roots compared to wild type after exposing to 5 μM AsIII (Chen et al. 2013). Due to its strong AsIII efflux, PvACR3 also decreases As accumulation in transgenic plant shoots after short-term As exposure (Chen et al. 2013). However, *PvACR3* transgenic plants accumulate more As in the shoots after long-term As exposure in soils, probably because PvACR3 conferred AsIII transport toward or into xylem in the roots and promoted As translocation, indicating PvACR3 transgenic approach can be used to engineer As hyperaccumulation plants for phytoremediation (Chen et al. 2013). More recent study also proved this notion, as they also showed that PvACR3 localizes to the plasma membrane and its heterologous expression increases root-to-shoot As translocation in *Arabidopsis* (Wang et al. 2017). However, the authors found no significant differences between WT and *PvACR3* transgenic plants in AsIII efflux, which is different from Chen et al. (2013). Moreover, they combined PvACR3 expression with HAC1 mutation, which led to even higher As accumulation in the shoots than that expressing PvACR3, indicating more efficient As hyperaccumulation can be engineered in plants by knocking out the HAC1 gene and expressing PvACR3 (Wang et al. 2017).

More recently, Chen et al. (2017b) cloned the *PvACR3;1* gene from *P. vittata* and tested its function in yeast and transgenic plants. In yeast, PvACR3;1 mediates AsIII efflux across yeast plasma membrane to decrease As accumulation in the cells, thus effectively suppressing As-sensitive phenotype of *Δacr3* yeast mutant. However, PvACR3;1 did not confer as strong AsIII efflux and As tolerance as PvACR3 in Δacr3, indicating that PvACR3;1 is less efficient than PvACR3 in mediating AsIII transport (Chen et al. 2017b). In *Arabidopsis* and tobacco, expression of PvACR3;1 increased As retention in the roots, thereby decreasing As accumulation in the shoots. Further analysis showed that PvACR3;1 localizes to vacuolar membrane and may mediate As sequestration into vacuoles in the roots, therefore reducing As translocation to the shoots (Chen et al. 2017b). This work also provides the basis for *PvACR3;1* transgenic strategy to decrease As accumulation in the shoots and edible parts of food crops, thus reducing As risks to humans.

Research in the last decade using the arsenic-hyperaccumulator *P. vittata* has uncovered multiple genes important in arsenic uptake, translocation, efflux, and metabolism. Many of these genes have been shown to function in transgenic plants.

Yet, arsenic-hyperaccumulating transgenic plants have not been realized. We can expect that in the near future, using existing transgenic technologies, we will be able to engineer multiple genes to create arsenic-hyperaccumulating transgenic plants targeted for phytoremediation of contaminated soil and water. Similarly some of the genes from *P. vittata* could be employed to improve abiotic stress tolerance or decrease As accumulation in crop plants.

Acknowledgments This work was supported by the National Natural Science Foundation of China (Grant No. 21707068 and 21637002), Jiangsu Provincial Natural Science Foundation of China (Grant No. BK20160649), and the National Key Research and Development Program of China (Grant No. 2016YFD0800801).

References

Abedin MJ, Feldmann J, Meharg AA (2002) Uptake kinetics of arsenic species in rice plants. Plant Physiol 128(3):1120–1128

Ali W, Isayenkov SV, Zhao FJ, Maathuis FJM (2009) Arsenite transport in plants. Cell Mol Life Sci 66(14):2329–2339

Ali W, Isner JC, Isayenkov SV, Liu W, Zhao FJ, Maathuis FJM (2012) Heterologous expression of the yeast arsenite efflux system ACR3 improves *Arabidopsis thaliana* tolerance to arsenic stress. New Phytol 194(3):716–723

Argos M, Kalra T, Rathouz PJ, Chen Y, Pierce B, Parvez F, Islam T, Ahmed A, Rakibuz-Zaman M, Hasan R, Sarwar G, Slavkovich V, van Geen A, Graziano J, Ahsan H (2010) Arsenic exposure from drinking water, and all-cause and chronic-disease mortalities in Bangladesh (HEALS): a prospective cohort study. Lancet 376(9737):252–258

Bech J, Poschenrieder C, Llugany M, Barcelo J, Tume P, Tobias FJ, Barranzuela JL, Vasquez ER (1997) Arsenic and heavy metal contamination of soil and vegetation around a copper mine in Northern Peru. Sci Total Environ 203(1):83–91

Bhattacharjee H, Rosen BP (2007) Arsenic metabolism in prokaryotic and eukaryotic microbes. In: Nies DH, Silver S (eds) Molecular microbiology of heavy metals. Springer, Berlin/Heidelberg, pp 371–406

Bienert GP, Thorsen M, Schussler MD, Nilsson HR, Wagner A, Tamas MJ, Jahn TP (2008) A subgroup of plant aquaporins facilitate the bi-directional diffusion of $As(OH)_3$ and $Sb(OH)_3$ across membranes. BMC Biol 6:26

Bobrowicz P, Wysocki R, Owsianik G, Goffeau A, Ulaszewski S (1997) Isolation of three contiguous genes, ACR1, ACR2 and ACR3, involved in resistance to arsenic compounds in the yeast *Saccharomyces cerevisiae*. Yeast 13(9):819–828

Bondada BR, Tu S, Ma LQ (2004) Absorption of foliar-applied arsenic by the arsenic hyperaccumulating fern (*Pteris vittata* L.) Sci Total Environ 332(1):61–70

Brinch-Pedersen H, Sorensen LD, Holm PB (2002) Engineering crop plants: getting a handle on phosphate. Trends Plant Sci 7(3):118–125

Caille N, Zhao FJ, McGrath SP (2005) Comparison of root absorption, translocation and tolerance of arsenic in the hyperaccumulator *Pteris vittata* and the nonhyperaccumulator *Pteris tremula*. New Phytol 165(3):755–761

Cantamessa S, D'Agostino G, Berta G (2016) Hydathode structure and localization in *Pteris vittata* fronds and evidence for their involvement in arsenic leaching. Plant Biosys Int J Dealing Asp Plant Biol 150(6):1208–1215

Cesaro P, Cattaneo C, Bona E, Berta G, Cavaletto M (2015) The arsenic hyperaccumulating *Pteris vittata* expresses two arsenate reductases. Sci Rep 5:14525

Chao DY, Chen Y, Chen J, Shi S, Chen Z, Wang C, Danku JM, Zhao FJ, Salt DE (2014) Genome-wide association mapping identifies a new arsenate reductase enzyme critical for limiting arsenic accumulation in plants. PLoS Biol 12(12):e1002009

Chaumont F, Barrieu F, Wojcik E, Chrispeels MJ, Jung R (2001) Aquaporins constitute a large and highly divergent protein family in maize. Plant Physiol 125(3):1206–1215

Chen Y, Xu W, Shen H, Yan H, He Z, Ma M (2013) Engineering arsenic tolerance and hyperaccumulation in plants for phytoremediation by a *PvACR3* transgenic approach. Environ Sci Technol 47(16):9355–9362

Chen Y, Fu J-W, Han Y-H, Rathinasabapathi B, Ma LQ (2016) High as exposure induced substantial arsenite efflux in As-hyperaccumulator Pteris vittata. Chemosphere 144:2189–2194

Chen Y, Han YH, Cao Y, Zhu YG, Rathinasabapathi B, Ma LQ (2017a) Arsenic transport in rice and biological solutions to reduce arsenic risk from rice. Front Plant Sci 8:268

Chen Y, Hua CY, Jia MR, Fu JW, Liu X, Han YH, Liu Y, Rathinasabapathi B, Cao Y, Ma LQ (2017b) Heterologous expression of *Pteris vittata* arsenite antiporter PvACR3;1 reduces arsenic accumulation in plant shoots. Environ Sci Technol 51(18):10387–10395

Clark GT, Dunlop J, Phung HT (2000) Phosphate absorption by *Arabidopsis thaliana*: interactions between phosphorus status and inhibition by arsenate. Aust J Plant Physiol 27(10):959–965

Clemens S, Kim EJ, Neumann D, Schroeder JI (1999) Tolerance to toxic metals by a gene family of phytochelatin synthases from plants and yeast. EMBO J 18(12):3325–3333

Cobbett CS (2000) Phytochelatins and their roles in heavy metal detoxification. Plant Physiol 123(3):825–832

Cordell D, Drangert J-O, White S (2009) The story of phosphorus: global food security and food for thought. Glob Environ Chang 19(2):292–305

Danielson JA, Johanson U (2008) Unexpected complexity of the aquaporin gene family in the moss *Physcomitrella patens*. BMC Plant Biol 8:45

Datta R, Das P, Tappero R, Punamiya P, Elzinga E, Sahi S, Feng H, Kiiskila J, Sarkar D (2017) Evidence for exocellular arsenic in fronds of *Pteris vittata*. Sci Rep 7(1):2839

Dekoe T (1994) *Agrostis castellana* and *Agrostis delicatula* on heavy metal and arsenic enriched sites in NE Portugal. Sci Total Environ 145(1–2):103–109

Dhankher OP, Rosen BP, McKinney EC, Meagher RB (2006) Hyperaccumulation of arsenic in the shoots of *Arabidopsis* silenced for arsenate reductase (ACR2). Proc Natl Acad Sci U S A 103(14):5413–5418

DiTusa SF, Fontenot EB, Wallace RW, Silvers MA, Steele TN, Elnagar AH, Dearman KM, Smith AP (2016) A member of the Phosphate transporter 1 (Pht1) family from the arsenic-hyperaccumulating fern *Pteris vittata* is a high-affinity arsenate transporter. New Phytol 209(2):762–772

Dong RB, Formentin E, Losseso C, Carimi F, Benedetti P, Terzi M, Lo Schiavo F (2005) Molecular cloning and characterization of a phytochelatin synthase gene, *PvPCS1*, from *Pteris vittata* L. J Ind Microbiol Biotechnol 32(11–12):527–533

Duan GL, Zhu YG, Tong YP, Cai C, Kneer R (2005) Characterization of arsenate reductase in the extract of roots and fronds of Chinese brake fern, an arsenic hyperaccumulator. Plant Physiol 138(1):461–469

Duan G-L, Zhou Y, Tong Y-P, Mukhopadhyay R, Rosen BP, Zhu Y-G (2007) A CDC25 homologue from rice functions as an arsenate reductase. New Phytol 174(2):311–321

Duan GL, Kamiya T, Ishikawa S, Arao T, Fujiwara T (2012) Expressing ScACR3 in rice enhanced arsenite efflux and reduced arsenic accumulation in rice grains. Plant Cell Physiol 53(1):154–163

Ellis DR, Gumaelius L, Indriolo E, Pickering IJ, Banks JA, Salt DE (2006) A novel arsenate reductase from the arsenic hyperaccumulating fern *Pteris vittata*. Plant Physiol 141(4):1544–1554

Finnegan PM, Chen W (2012) Arsenic toxicity: the effects on plant metabolism. Front Physiol 3:182

Fu JW, Liu X, Han YH, Mei H, Cao Y, de Oliveira LM, Liu Y, Rathinasabapathi B, Chen Y, Ma LQ (2017) Arsenic-hyperaccumulator *Pteris vittata* efficiently solubilized phosphate rock to sustain plant growth and as uptake. J Hazard Mater 330:68–75

Garg N, Singla P (2011) Arsenic toxicity in crop plants: physiological effects and tolerance mechanisms. Environ Chem Lett 9(3):303–321

Ghosh M, Shen J, Rosen BP (1999) Pathways of As(III) detoxification in *Saccharomyces cerevisiae*. Proc Natl Acad Sci U S A 96(9):5001–5006

Gilbert N (2009) Environment: the disappearing nutrient. Nature 461(7265):716–718

Grill E, Löffler S, Winnacker EL, Zenk MH (1989) Phytochelatins, the heavy-metal-binding peptides of plants, are synthesized from glutathione by a specific γ-glutamylcysteine dipeptidyl transpeptidase (phytochelatin synthase). Proc Natl Acad Sci U S A 86(18):6838–6842

Gumaelius L, Lahner B, Salt DE, Banks JA (2004) Arsenic hyperaccumulation in gametophytes of *Pteris vittata*. A new model system for analysis of arsenic hyperaccumulation. Plant Physiol 136(2):3198–3208

Guo JB, Dai XJ, Xu WZ, Ma M (2008) Overexpressing GSH1 and AsPCS1 simultaneously increases the tolerance and accumulation of cadmium and arsenic in *Arabidopsis thaliana*. Chemosphere 72(7):1020–1026

Gupta P, Bhatnagar AK (2015) Spatial distribution of arsenic in different leaf tissues and its effect on structure and development of stomata and trichomes in mung bean, *Vigna radiata* (L.) Wilczek. Environ Exp Bot 109:12–22

Han YH, Yang GM, Fu JW, Guan DX, Chen Y, Ma LQ (2016) Arsenic-induced plant growth of arsenic-hyperaccumulator *Pteris vittata*: impact of arsenic and phosphate rock. Chemosphere 149:366–372

Han Y-H, Liu X, Rathinasabapathi B, Li H-B, Chen Y, Ma LQ (2017a) Mechanisms of efficient as solubilization in soils and as accumulation by As-hyperaccumulator *Pteris vittata*. Environ Pollut 227:569–577

Han YH, Fu JW, Xiang P, Cao Y, Rathinasabapathi B, Chen Y, Ma LQ (2017b) Arsenic and phosphate rock impacted the abundance and diversity of bacterial arsenic oxidase and reductase genes in rhizosphere of As-hyperaccumulator *Pteris vittata*. J Hazard Mater 321:146–153

Handson PD (1984) Lead and arsenic levels in wines produced from vineyards where lead arsenate sprays are used for caterpillar control. J Sci Food Agric 35(2):215–218

Hayashi S, Kuramata M, Abe T, Takagi H, Ozawa K, Ishikawa S (2017) Phytochelatin synthase OsPCS1 plays a crucial role in reducing arsenic levels in rice grains. Plant J 91(5):840–848

He Z, Yan H, Chen Y, Shen H, Xu W, Zhang H, Shi L, Zhu YG, Ma M (2016) An aquaporin PvTIP4;1 from *Pteris vittata* may mediate arsenite uptake. New Phytol 209(2):746–761

Heiss S, Wachter A, Bogs J, Cobbett C, Rausch T (2003) Phytochelatin synthase (PCS) protein is induced in *Brassica juncea* leaves after prolonged Cd exposure. J Exp Bot 54(389):1833–1839

Hirano S (2004) The accumulation and toxicity of methylated arsenicals in endothelial cells: important roles of thiol compounds. Toxicol Appl Pharmacol 198(3):458–467

Huang JW, Poynton CY, Kochian LV, Elless MP (2004) Phytofiltration of arsenic from drinking water using arsenic-hyperaccumulating ferns. Environ Sci Technol 38(12):3412–3417

Huang J-H, Hu K-N, Decker B (2011) Organic arsenic in the soil environment: speciation, occurrence, transformation, and adsorption behavior. Water Air Soil Pollut 219(1–4):401–415

Indriolo E, Na G, Ellis D, Salt DE, Banks JA (2010) A vacuolar arsenite transporter necessary for arsenic tolerance in the arsenic hyperaccumulating fern *Pteris vittata* is missing in flowering plants. Plant Cell 22(6):2045–2057

Isayenkov SV, Maathuis FJ (2008) The *Arabidopsis thaliana* aquaglyceroporin AtNIP7;1 is a pathway for arsenite uptake. FEBS Lett 582(11):1625–1628

Jarup L (2003) Hazards of heavy metal contamination. Br Med Bull 68:167–182

Jia H, Ren H, Gu M, Zhao J, Sun S, Zhang X, Chen J, Wu P, Xu G (2011) The phosphate transporter gene OsPht1;8 is involved in phosphate homeostasis in rice. Plant Physiol 156(3):1164–1175

Jomova K, Jenisova Z, Feszterova M, Baros S, Liska J, Hudecova D, Rhodes CJ, Valko M (2011) Arsenic: toxicity, oxidative stress and human disease. J Appl Toxicol 31(2):95–107

Kamiya T, Tanaka M, Mitani N, Ma JF, Maeshima M, Fujiwara T (2009) NIP1;1, an aquaporin homolog, determines the arsenite sensitivity of *Arabidopsis thaliana*. J Biol Chem 284(4):2114–2120

Kertulis GM, Ma LQ, MacDonald GE, Chen R, Winefordner JD, Cai Y (2005) Arsenic speciation and transport in *Pteris vittata* L. and the effects on phosphorus in the xylem sap. Environ Exp Bot 54(3):239–247

Kramer U (2005) Phytoremediation: novel approaches to cleaning up polluted soils. Curr Opin Biotechnol 16(2):133–141

Kupper H, Lombi E, Zhao FJ, McGrath SP (2000) Cellular compartmentation of cadmium and zinc in relation to other elements in the hyperaccumulator *Arabidopsis halleri*. Planta 212(1):75–84

Landrieu I, Hassan S, Sauty M, Dewitte F, Wieruszeski J-M, Inzé D, Veylder LD, Lippens G (2004) Characterization of the *Arabidopsis thaliana* Arath; CDC25 dual-specificity tyrosine phosphatase. Biochem Biophys Res Commun 322(3):734–739

Lepp NW (1975) Potential of tree-ring analysis for monitoring heavy-metal pollution patterns. Environ Pollut 9(1):49–61

Lessl JT, Ma LQ (2013) Sparingly-soluble phosphate rock induced significant plant growth and arsenic uptake by *Pteris vittata* from three contaminated soils. Environ Sci Technol 47(10):5311–5318

Lessl JT, Ma LQ, Rathinasabapathi B, Guy C (2013) Novel phytase from *Pteris vittata* resistant to arsenate, high temperature, and soil deactivation. Environ Sci Technol 47(5):2204–2211

Li WX, Chen TB, Chen Y, Lei M (2005) Role of trichome of *Pteris vittata* L. in arsenic hyperaccumulation. Sci China C Life Sci 48(2):148–154

Liu WJ, Wood BA, Raab A, McGrath SP, Zhao FJ, Feldmann J (2010) Complexation of arsenite with phytochelatins reduces arsenite efflux and translocation from roots to shoots in *Arabidopsis*. Plant Physiol 152(4):2211–2221

Liu W, Schat H, Bliek M, Chen Y, McGrath SP, George G, Salt DE, Zhao FJ (2012) Knocking out ACR2 does not affect arsenic redox status in *Arabidopsis thaliana*: implications for as detoxification and accumulation in plants. PLoS One 7(8):e42408

Lombi E, Zhao FJ, Fuhrmann M, Ma LQ, McGrath SP (2002) Arsenic distribution and speciation in the fronds of the hyperaccumulator *Pteris vittata*. New Phytol 156(2):195–203

Ma LQ, Komar KM, Tu C, Zhang WH, Cai Y, Kennelley ED (2001) A fern that hyperaccumulates arsenic - a hardy, versatile, fast-growing plant helps to remove arsenic from contaminated soils. Nature 409(6820):579–579

Ma JF, Yamaji N, Mitani N, Xu XY, Su YH, McGrath SP, Zhao FJ (2008) Transporters of arsenite in rice and their role in arsenic accumulation in rice grain. Proc Natl Acad Sci U S A 105(29):9931–9935

Mathews S, Rathinasabapathi B, Ma LQ (2011) Uptake and translocation of arsenite by *Pteris vittata* L.: effects of glycerol, antimonite and silver. Environ Pollut 159(12):3490–3495

Matschullat J (2000) Arsenic in the geosphere - a review. Sci Total Environ 249(1–3):297–312

Meharg AA (2003) Variation in arsenic accumulation - hyperaccumulation in ferns and their allies. New Phytol 157(1):25–31

Meharg AA, Hartley-Whitaker J (2002) Arsenic uptake and metabolism in arsenic resistant and nonresistant plant species. New Phytol 154(1):29–43

Meharg AA, Macnair MR (1992) Suppression of the high-affinity phosphate-uptake system - a mechanism of arsenate tolerance in *Holcus lanatus* L. J Exp Bot 43(249):519–524

Meharg AA, Naylor J, Macnair MR (1994) Phosphorus nutrition of arsenate-tolerant and nontolerant phenotypes of velvetgrass. J Environ Qual 23(2):234–238

Mitani-Ueno N, Yamaji N, Zhao FJ, Ma JF (2011) The aromatic/arginine selectivity filter of NIP aquaporins plays a critical role in substrate selectivity for silicon, boron, and arsenic. J Exp Bot 62(12):4391–4398

Moreno-Jimenez E, Esteban E, Penalosa JM (2012) The fate of arsenic in soil-plant systems. Rev Environ Contam Toxicol 215:1–37

Moreno-Jimenez E, Six L, Williams PN, Smolders E (2013) Inorganic species of arsenic in soil solution determined by microcartridges and ferrihydrite-based diffusive gradient in thin films (DGT). Talanta 104:83–89

Mosa KA, Kumar K, Chhikara S, McDermott J, Liu Z, Musante C, White JC, Dhankher OP (2012) Members of rice plasma membrane intrinsic proteins subfamily are involved in arsenite permeability and tolerance in plants. Transgenic Res 21(6):1265–1277

Muchhal US, Pardo JM, Raghothama KG (1996) Phosphate transporters from the higher plant *Arabidopsis thaliana*. Proc Natl Acad Sci U S A 93(19):10519–10523

Mukhopadhyay R, Rosen BP (1998) *Saccharomyces cerevisiae* ACR2 gene encodes an arsenate reductase. FEMS Microbiol Lett 168(1):127–136

Oremland RS, Stolz JF (2003) The ecology of arsenic. Science 300(5621):939–944

Pal R, Rai JP (2010) Phytochelatins: peptides involved in heavy metal detoxification. Appl Biochem Biotechnol 160(3):945–963

Pandey N, Gupta B, Pathak GC (2013) Foliar application of Zn at flowering stage improves plant's performance, yield and yield attributes of black gram. Indian J Exp Biol 51(7):548–555

Pickering IJ, Gumaelius L, Harris HH, Prince RC, Hirsch G, Banks JA, Salt DE, George GN (2006) Localizing the biochemical transformations of arsenate in a hyperaccumulating fern. Environ Sci Technol 40(16):5010–5014

Poynton CY, Huang JWW, Blaylock MJ, Kochian LV, Elless MP (2004) Mechanisms of arsenic hyperaccumulation in Pteris species: root as influx and translocation. Planta 219(6):1080–1088

Puckett EE, Serapiglia MJ, DeLeon AM, Long S, Minocha R, Smart LB (2012) Differential expression of genes encoding phosphate transporters contributes to arsenic tolerance and accumulation in shrub willow (Salix spp.) Environ Exp Bot 75:248–257

Quigley F, Rosenberg JM, Shachar-Hill Y, Bohnert HJ (2002) From genome to function: the *Arabidopsis* aquaporins. Genome Biol 3(1):1–17

Rathinasabapathi B (2006) Ferns represent an untapped biodiversity for improving crops for environmental stress tolerance. New Phytol 172(3):385–390

Rathinasabapathi B, Wu S, Sundaram S, Rivoal J, Srivastava M, Ma LQ (2006) Arsenic resistance in *Pteris vittata* L.: identification of a cytosolic triosephosphate isomerase based on cDNA expression cloning in *Escherichia coli*. Plant Mol Biol 62(6):845–857

Richardson AE, Hocking PJ, Simpson RJ, George TS (2009) Plant mechanisms to optimise access to soil phosphorus. Crop Pasture Sci 60(2):124–143

Ronzan M, Zanella L, Fattorini L, Della Rovere F, Urgast D, Cantamessa S, Nigro A, Barbieri M, Sanità di Toppi L, Berta G, Feldmann J, Altamura MM, Falasca G (2017) The morphogenic responses and phytochelatin complexes induced by arsenic in *Pteris vittata* change in the presence of cadmium. Environ Exp Bot 133:176–187

Rosen BP (1999) Families of arsenic transporters. Trends Microbiol 7(5):207–212

Sánchez-Bermejo E, Castrillo G, del Llano B, Navarro C, Zarco-Fernández S, Martinez-Herrera DJ, Leo-del Puerto Y, Muñoz R, Cámara C, Paz-Ares J, Alonso-Blanco C, Leyva A (2014) Natural variation in arsenate tolerance identifies an arsenate reductase in *Arabidopsis thaliana*. Nat Commun 5:4617

Schat H, Llugany M, Vooijs R, Hartley-Whitaker J, Bleeker PM (2002) The role of phytochelatins in constitutive and adaptive heavy metal tolerances in hyperaccumulator and non-hyperaccumulator metallophytes. J Exp Bot 53(379):2381–2392

Schmoger ME, Oven M, Grill E (2000) Detoxification of arsenic by phytochelatins in plants. Plant Physiol 122(3):793–801

Shen S, Li X-F, Cullen WR, Weinfeld M, Le XC (2013) Arsenic binding to proteins. Chem Rev 113(10):7769–7792

Shi GL, Zhu S, Meng JR, Qian M, Yang N, Lou LQ, Cai QS (2015) Variation in arsenic accumulation and translocation among wheat cultivars: the relationship between arsenic accumulation, efflux by wheat roots and arsenate tolerance of wheat seedlings. J Hazard Mater 17:190–196

Shi S, Wang T, Chen Z, Tang Z, Wu Z, Salt DE, Chao D-Y, Zhao F-J (2016) OsHAC1;1 and OsHAC1;2 function as arsenate reductases and regulate arsenic accumulation. Plant Physiol 172(3):1708–1719

Shri M, Dave R, Diwedi S, Shukla D, Kesari R, Tripathi RD, Trivedi PK, Chakrabarty D (2014) Heterologous expression of *Ceratophyllum demersum* phytochelatin synthase, CdPCS1, in rice leads to lower arsenic accumulation in grain. Sci Rep 4(1):5784

Sneller FE, van Heerwaarden LM, Koevoets PL, Vooijs R, Schat H, Verkleij JA (2000) Derivatization of phytochelatins from *Silene vulgaris*, induced upon exposure to arsenate and cadmium: comparison of derivatization with Ellman's reagent and monobromobimane. J Agric Food Chem 48(9):4014–4019

Sohn E (2014) Contamination: the toxic side of rice. Nature 514(7524):S62–S63

Srivastava M, Ma LQ, Santos JAG (2006) Three new arsenic hyperaccumulating ferns. Sci Total Environ 364(1–3):24–31

Srivastava M, Santos J, Srivastava P, Ma LQ (2010) Comparison of arsenic accumulation in 18 fern species and four *Pteris vittata* accessions. Bioresour Technol 101(8):2691–2699

Stone R (2008) Arsenic and paddy rice: a neglected cancer risk? Science 321(5886):184–185

Su YH, McGrath SP, Zhu YG, Zhao FJ (2008) Highly efficient xylem transport of arsenite in the arsenic hyperaccumulator *Pteris vittata*. New Phytol 180(2):434–441

Sundaram S, Rathinasabapathi B (2010) Transgenic expression of fern *Pteris vittata* glutaredoxin PvGrx5 in *Arabidopsis thaliana* increases plant tolerance to high temperature stress and reduces oxidative damage to proteins. Planta 231(2):361–369

Sundaram S, Rathinasabapathi B, Ma LQ, Rosen BP (2008) An arsenate-activated glutaredoxin from the arsenic hyperaccumulator fern *Pteris vittata* L. regulates intracellular arsenite. J Biol Chem 283(10):6095–6101

Sundaram S, Wu S, Ma LQ, Rathinasabapathi B (2009) Expression of a *Pteris vittata* glutaredoxin PvGRX5 in transgenic *Arabidopsis thaliana* increases plant arsenic tolerance and decreases arsenic accumulation in the leaves. Plant Cell Environ 32(7):851–858

Tu C, Ma LQ, Bondada B (2002) Arsenic accumulation in the hyperaccumulator Chinese brake and its utilization potential for phytoremediation. J Environ Qual 31(5):1671–1675

Tu C, Ma LQ, Zhang W, Cai Y, Harris WG (2003) Arsenic species and leachability in the fronds of the hyperaccumulator Chinese brake (*Pteris vittata* L.) Environ Pollut 124(2):223–230

Turner BL, Paphazy MJ, Haygarth PM, McKelvie ID (2002) Inositol phosphates in the environment. Philos Trans R Soc Lond Ser B Biol Sci 357(1420):449–469

Vatamaniuk OK, Mari S, Lu YP, Rea PA (1999) AtPCS1, a phytochelatin synthase from *Arabidopsis*: isolation and in vitro reconstitution. Proc Natl Acad Sci U S A 96(12):7110–7115

Visoottiviseth P, Francesconi K, Sridokchan W (2002) The potential of Thai indigenous plant species for the phytoremediation of arsenic contaminated land. Environ Pollut 118(3):453–461

Wang JR, Zhao FJ, Meharg AA, Raab A, Feldmann J, McGrath SP (2002) Mechanisms of arsenic hyperaccumulation in *Pteris vittata*. Uptake kinetics, interactions with phosphate, and arsenic speciation. Plant Physiol 130(3):1552–1561

Wang HB, Ye ZH, Shu WS, Li WC, Wong MH, Lan CY (2006) Arsenic uptake and accumulation in fern species growing at arsenic-contaminated sites of southern China: field surveys. Int J Phytoremediation 8(1):1–11

Wang P, Zhang W, Mao C, Xu G, Zhao F-J (2016) The role of OsPT8 in arsenate uptake and varietal difference in arsenate tolerance in rice. J Exp Bot 67(21):6051–6059

Wang C, Na G, Bermejo ES, Chen Y, Banks JA, Salt DE, Zhao FJ (2017) Dissecting the components controlling root-to-shoot arsenic translocation in *Arabidopsis thaliana*. New Phytol 31(10):14761

Wysocki R, Bobrowicz P, Ulaszewski S (1997) The *Saccharomyces cerevisiae* ACR3 gene encodes a putative membrane protein involved in arsenite transport. J Biol Chem 272(48):30061–30066

Xu XY, McGrath SP, Zhao FJ (2007) Rapid reduction of arsenate in the medium mediated by plant roots. New Phytol 176(3):590–599

Xu JY, Li HB, Liang S, Luo J, Ma LQ (2014) Arsenic enhanced plant growth and altered rhizosphere characteristics of hyperaccumulator *Pteris vittata*. Environ Pollut 194:105–111

Xu W, Dai W, Yan H, Li S, Shen H, Chen Y, Xu H, Sun Y, He Z, Ma M (2015) *Arabidopsis* NIP3;1 plays an important role in arsenic uptake and root-to-shoot translocation under arsenite stress conditions. Mol Plant 8(5):722–733

Xu J, Shi S, Wang L, Tang Z, Lv T, Zhu X, Ding X, Wang Y, Zhao F-J, Wu Z (2017) OsHAC4 is critical for arsenate tolerance and regulates arsenic accumulation in rice. New Phytol 215(3):1090–1101

Yan X-L, Liao X-Y, T-b C (2009) Leaching potential of arsenic from *Pteris vittata* L. under field conditions. Sci Total Environ 408(2):425–430

Yang M (2011) A current global view of environmental and occupational cancers. J Environ Sci Health C Environ Carcinog Ecotoxicol Rev 29(3):223–249

Yang XX, Chen H, Dai XJ, Xu WZ, He ZY, Ma M (2009) Evidence of vacuolar compartmentalization of arsenic in the hyperaccumulator *Pteris vittata*. Chin Sci Bull 54(22):4229–4233

Zhang WH, Cai Y, Tu C, Ma LQ (2002) Arsenic speciation and distribution in an arsenic hyperaccumulating plant. Sci Total Environ 300(1–3):167–177

Zhao FJ, Dunham SJ, McGrath SP (2002) Arsenic hyperaccumulation by different fern species. New Phytol 156(1):27–31

Zhao FJ, Wang JR, Barker JHA, Schat H, Bleeker PM, McGrath SP (2003) The role of phytochelatins in arsenic tolerance in the hyperaccumulator *Pteris vittata*. New Phytol 159(2):403–410

Zhao FJ, Ma JF, Meharg AA, McGrath SP (2009) Arsenic uptake and metabolism in plants. New Phytol 181(4):777–794

Zhao FJ, Ago Y, Mitani N, Li RY, Su YH, Yamaji N, McGrath SP, Ma JF (2010) The role of the rice aquaporin Lsi1 in arsenite efflux from roots. New Phytol 186(2):392–399

Zhu YG, Yoshinaga M, Zhao FJ, Rosen BP (2014) Earth abides arsenic biotransformations. Annu Rev Earth Planet Sci 42:443–467

Chapter 18
Fern Phenology

Pei-Hsuan Lee, Yao-Moan Huang, and Wen-Liang Chiou

18.1 Introduction

Phenology, a relatively recent term (Charles Morren first introduced the word in French for phenology in 1853) derived from the Greek word *phaino* meaning to show or to appear, is the study of the timing of bioactivity (Baddour and Kontongomde 2009; Noormet 2009; Schwartz 2013). Lieth (1974) adopts a more detailed definition of phenology "the study of the timing of recurrent biological events, the causes of their timing regard to biotic and abiotic forces, and the interrelation among phases of the same or different species." Thus, plant phenology could be briefly described as a study of seasonal activity and its relationships with biotic/abiotic factors (van Schaik et al. 1993; Mehltreter 2008).

Phenological studies are basic to understanding the ecology, evolution, as well as biological processes of plants (van Schaik et al. 1993; Newstrom and Frankie 1994; Mehltreter 2008). This science is also applied to the agriculture and other sciences with a long history (Hudson and Keatley 2010; Schwartz 2013). The Australian aboriginals have developed a deep understanding of the interrelationships between the environment and its influence on fauna and flora for at least 50,000 years (Baddour and Kontongomde 2009). In China, phenological data have been recorded over the past 3000 years and could, to some extent, reflect past climate (Chen 2013). Hanami (cherry blooming viewing), a well-known and still ongoing phenological observation in Japan, can be traced back to the eighth century (Menzel 2013). The dates of grape harvest in Switzerland have been recorded more than 500 years (Meier et al. 2007). Recently, it has also been used to study the influence of climate change on plants' phenological characters (Sparks and Carey 1995; Menzel 2000; Keatley et al. 2002; Cleland et al. 2007; Doi and Takahashi 2008; Calinger et al. 2013; Li et al. 2013). However, most phenological studies focus on the flowering plants but seldom on ferns.

P. -H. Lee · Y. -M. Huang (✉) · W. -L. Chiou (✉)
Taiwan Forestry Research Institute, Taipei, Taiwan
e-mail: huangym@tfri.gov.tw

H. Fernández (ed.), *Current Advances in Fern Research*,
https://doi.org/10.1007/978-3-319-75103-0_18

Ferns are a lineage of vascular plants. They do not produce flowers and seeds but use spores to reproduce offspring. Leaves (may be called fronds) are the dominant organ of ferns. Fern leaves transfer energy and produce spores. They also contribute to the growth, survival, and regeneration of ferns (Mehltreter and Sharpe 2013). Thus, the developmental stages of leaves and spores are the foremost phases observed in fern phenology. Phenological studies of ferns have been gradually attracting researcher attention. Geldenhuys and van der Merwe (1988) studied the phenology to facilitate the leaf harvest of an ornamental fern, *Rumohra adiantiformis* (G. Forst.) Ching. The phenology of *Matteuccia struthiopteris* (L.) Tod., widely distributed in north temperate regions and used for locally as a vegetable and in the horticultural trade as a garden ornamental, was also studied to infer its optimal harvest season (von Aderkas and Green 1986). Phenological phases have also been used to support the separation of components of a species complex (e.g., the *Pteris cadieri* Christ complex (Chao et al. 2012)).

In this chapter, we collected a phenological dataset of 225 fern species from around the world to illustrate the distribution of studies during the past half century into leaf and spore production seasonality and the correlation with climate factors. Seasonality in leaf development, production rates, and lifespans of both fertile and sterile leaves are compared where such data are available. Future research priorities are also suggested.

18.2 Phenological Studies of Ferns in the World

Previous studies have documented fern phenology of at least 225 species, which is about 2% of the currently estimated 10,578 fern species (PPG I 2016), belonging to 73 genera and 26 families (Farrar 1976; Sato 1982; Page 1997; Liu et al. 2002; Mehltreter 2008; Lee et al. 2009b, 2016; Sharpe and Mehltreter 2010; and references therein). Most of those studies were conducted in northern hemisphere, especially in temperate regions, but relatively few in southern hemisphere or tropical regions, i.e., south of the Tropic of Cancer (Fig. 18.1).

Monomorphic ferns that have fertile and sterile leaves with similar morphologies are more often addressed in these studies than dimorphic ferns where the sterile and fertile leaves are distinctly different in shape. The phenology of terrestrial fern species has also been studied more often than other life forms. These biases may reflect the fact that monomorphic and terrestrial species are dominant among ferns (Watkins et al. 2016). Seasonal patterns in tree ferns have also received more attention, probably because their height raises them above the largely neglected herbaceous layer of the tropical forest (Table 18.1). Most of those phenological studies present some form of a "phenological calendar" (e.g., Sato 1982; Page 1997; Liu et al. 2002), showing time periods of phenological events and providing useful biological information.

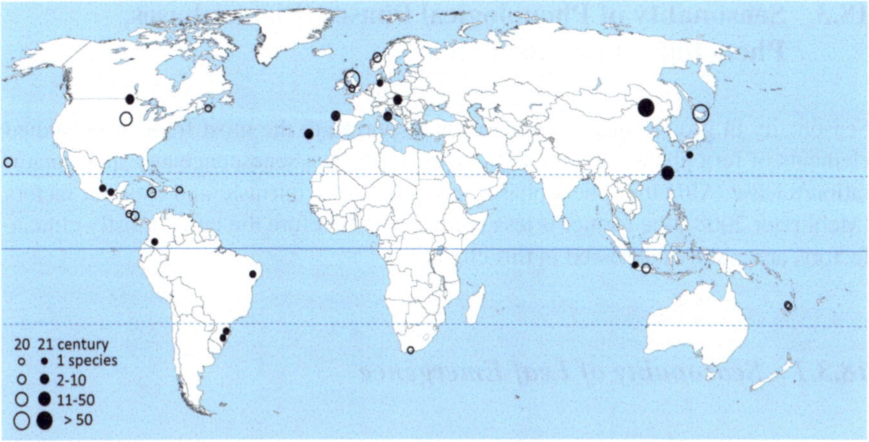

Fig. 18.1 The geographical distribution of published fern phenological research

Table 18.1 Summary of categories of ferns documented by previous phenological studies

Category	Number
Family	26
Genus	73
Species[a]	225
Fertile-sterile leaf dimorphy[b]	
Monomorphism	187 (83%)
Dimorphism	38 (17%)
Life form	
Terrestrial[c]	203 (90%)
Epiphytic	14 (6%)
Aquatic[d]	7 (3%)
Climbing	1 (1%)

[a]Including infraspecies
[b]Following Wagner and Wagner (1977); dimorphism includes hemidimorphism
[c]Including rupestral (21 species, 9%) and tree ferns (32 species, 14%)
[d]Including rheophytic and mangrove ferns

However, those reports did not include environmental parameters and were in general descriptive rather than quantitative. Without records of the numbers of plants/leaves observed, it is difficult to compare the results statistically with phenological data on ferns of different periods, regions, or species.

18.3 Seasonality of Phenological Phases (Phenophases, Phenological Events/Traits)

Seasonality in growth phases of leaves and spores are the most frequently studied elements of fern phenology, including leaf emergence/senescence and spore maturation/release. Although phenophases are affected by intrinsic and external factors (Mehltreter 2008), the former is less known, and therefore the latter, mostly climate factors, are mainly addressed in this chapter.

18.3.1 Seasonality of Leaf Emergence

Emergence of the fern leaf is signaled by the appearance of a coiled leaf (crosier), which is like a fiddlehead. It takes 1–2 months to uncoil and fully expand the leaf. During the expansion process, the leaf turns light green to dark green when it is fully uncoiled, and its area ceases to increase (full expansion). The leaf area is often estimated by its length and width (e.g., Lee et al. 2009a, b, 2016; Landi et al. 2014) or length only (de Paiva Farias et al. 2015).

Most fern species exhibit leaf-emergence seasonality, especially in temperate regions and in tropical regions with alternation of dry and wet seasons. In temperate regions, leaf emergence is inhibited by cold temperature, snow cover, or frozen soil (Odland 1995; Strandberg 2003; Ellwood 2012), as shown for as *Botrychium gallicomontanum* Farrar & Johnson-Groh, *B. mormo* W. H. Wagner (Johnson-Groh and Lee 2002), *Dryopteris affinis* (Lowe) Fras.-Jenk. ssp. *affinis*, *Polystichum aculeatum* (L.) Roth (Landi et al. 2014), *Osmunda claytoniana* L. (Lee et al. 2008), and *Osmundastrum cinnamomeum* (L.) C. Presl (Chiu et al. 2013). The leaves of those temperate fern species emerge mostly in spring. Sato (1982) recorded the phenology of 67 fern species in Hokkaido, a temperate region in Japan, and found that for most of them the leaves emerged in the spring, grew through summer, and senesced before the coming of winter. Many British and Ireland ferns also exhibit similar phenograms (Page 1997).

In tropical and subtropical regions with obvious dry and wet seasons, water is an important trigger for leaf growth. The leaves of the neotropical rheophyte *Thelypteris angustifolia* (Willd.) Proctor, for example, emerge mainly during the periods of higher rainfall and are lost during months with lower rainfall in a Puerto Rican rainforest that is not strongly seasonal with respect to precipitation (Sharpe 1997). *Oleandra pistillaris* (Sw) C.Chr., growing in western Java, decreases its leaf production when precipitation lessens, while production rates rise when precipitation increases although there is no obvious dry season (Takahashi and Mikami 2006). In a tropical monsoon forest of southern Taiwan, leaf emergence of 13 fern species occurs mainly during the wet season when the temperature is also higher (Lee et al. 2016). On the other hand, in many humid tropical and subtropical regions, temperature is more correlated with leaf emergence because water is sufficient throughout the year, as the cases in northern and northeastern Taiwan (Lee et al. 2009a, b). However, within a species, fertile and sterile leaves usually emerge during different seasons. If the emergent periods of both leaf types are combined in analysis, the emergence seasonality may become less obvious (e.g., Lee et al. 2009b, and see below sterile vs. fertile section). Because temperature and precipitation vary in different seasons, the correlations between temperature/precipitation and phenological events also correspond to the phenophase seasonality to some extent (Table 18.2).

Effects of day length on phenophases have also been reported (Schmitt and Windisch 2012; Neumann et al. 2014; Müller et al. 2016). Increase in day length often accompanies a rise in temperature. Leaves emerging in the period of longer day length and subsequently higher temperature gain benefit for growth and reproduction because photosynthesis and carbohydrate production are higher during that period (Braun 1984; Bauer et al. 1991; Watkins et al. 2016). However, leaves, especially fertile ones, emerging during the period of shorter day length and the resulting lower temperature, have also been observed (e.g., *Cibotium taiwanensis* Kuo (Chiou et al. 2001), *Plagiogyria dunnii* Copel., *P. adnata* (Blume) Bedd., *Pteris wallichiana* J.Agardh (Lee et al. 2009b), and *Osmundastrum cinnamomeum* (Chiu et al. 2013)). Fertile leaf production of *Osmundastrum cinnamomeum* (= *Osmunda cinnamomeum*) was found to be stimulated by short light exposure in a controlled experiment (Harvey and Caponetti 1972).

18.3.2 Seasonality of Leaf Senescence

Leaf senescence is the end process of leaf growth. This process results in nutrient relocation in plants and the mobilization of organic matter between plants and soil and affects the life cycle of plants (Gan and Amasino 1997; Zalamea and Gonzalez 2008). It is different from mortality which is counted in a specific time point. However, leaf senescence is often used to represent mortality in many reports and is adopted in this chapter. Several leaf appearances are used to indicate leaf senescence, e.g., leaf turning brown (Lee et al. 2009a, b, 2016; Landi et al. 2014), leaf turning dry, or remaining stipe only (Schmitt and Windisch 2012; Neumann et al. 2014).

Water is frequently closely related to leaf emergence as described above but seems not so obviously related to fern leaf senescence except for those deciduous ferns in the extremely dry season of tropical regions (Hietz 2010). In many cases, leaf senescence of most ferns is not significantly correlated with precipitation but with temperature. For example, in temperate or high elevation regions, leaves of summergreen ferns are deciduous, senescing, and dying in late fall or winter, while new leaves emerge in the next spring, e.g., *Osmunda claytoniana* (Lee et al. 2008), *Osmundastrum cinnamomeum* (Chiu et al. 2013), and *Botrychium* spp. (Johnson-Groh and Lee 2002), and for more examples, see Sato 1982. Most ferns in a subtropical forest of northeastern Taiwan, where abundant water is available all year, exhibit leaf senescence in warm summer demonstrating a significant correlation

Table 18.2 Correlations between leaf phenological events and temperature/precipitation

Species	Temperature		Precipitation		Location	References
	Emergence	Senescence	Emergence	senescence		
Acrostichum danaeifolium	0.66*	–	–	–	Mexico	Mehltreter and Palacios-Rios (2003)
Alsophila firma	0.56** (0.69**)[b]	0.80** (0.75**)[b]	0.65** (0.51*)[b]	0.69** (0.57*)[b]	Mexico	Mehltreter and Garcia-Franco (2008)
Cibotium taiwanense	0.06	–	0.15	–	N Taiwan	Chiou et al. (2001)
Cyathea atrovirens	0.69* 0.64* [f]	−0.15	−0.29 −0.39 [f]	0.32	Brazil	Schmitt and Windisch (2012)
Cyathea corcovadensis	0.63**	0.79**	0.37	0.11	Brazil	Neumann et al. (2014)
Cyathea hancockii	−0.51** [f] 0.39* [s]	0.29 0.13	−0.20 [f] 0.02 [s]	−0.18 −0.10	N Taiwan	Lee et al. (2009a)
Cyathea metteniana	−0.33[f] −0.12[s]	−0.24 0.00	−0.26[f] −0.20[s]	−0.08 −0.03	N Taiwan	Lee et al. (2009a)
Cyathea podophylla	−0.58**[f] 0.63**[s]	0.23 −0.07	−0.15[f] −0.03[s]	−0.01 −0.20	N Taiwan	Lee et al. (2009a)
Didymochlaena truncatula	0.08[f] −0.37[s]	0.43	0.11[f] 0.39[s]	−0.55*	NE Brazil	de Paiva Farias et al. (2015)
Dryopteris affinis ssp. *affinis*[a]	−3.87* (T_m) 3.71* (T_{d10})	−9.87* (T_m) 9.01* (T_{d10})	−0.67* (P_{mm})	0.48* (P_{mm}) −2.46* (P_d)	C Italy	Landi et al. (2014)
Lygodium venustum	0.37* (0.46**)[b]	– (0.56**)[b]	0.45* –	– (0.64**)[b]	Mexico	Mehltreter (2006)
Osmundastrum cinnamomeum	0.58*	–	−0.15	–	NE Taiwan	Chiu et al. (2013)
Polystichum aculeatum [a]	−1.07* (T_m)	−3.35* (T_m) 4.72* (T_{d10})	−0.65* (P_{mm}) 0.26* (P_d)	−1.02* (P_{mm}) 1.98* (P_d)	C Italy	Landi et al. (2014)
16 species[c,e]	−0.29* [f] 0.74** [s]	0.78** –	−0.36* [f] 0.09 [s]	0.08 –	NE Taiwan	Lee et al. (2009b)
13 species[d,e]	0.33**[f] 0.47*[s]	0.09 −0.10	0. 35**[f] 0.58*[s]	0.08 −0.15	S Taiwan	Lee et al. (2016)

Numbers without indication of fertility are averages of fertile (f) and sterile (s) leaves
*$0.01 < p < 0.05$; **$p < 0.01$; –, data not available
[a]Values calculated by GAMLSS and ranged >0 (positive correlation) or <0 (negative correlation). T_m mean monthly temperature, T_{d10} days > 10°C per month, P_{mm} monthly precipitation, P_d days of rainfall per month
[b]Values show the correlation between leaf phenological event and climatic data at 1 month before
[c]Sixteen species in NE Taiwan including *Acrorumohra hasseltii, Blechnum orientale, Cyathea podophylla, C. spinulosa, Dictyocline griffithii, D. griffithii* var. *wilfordii, Diplazium dilatatum,*

(continued)

Table 18.2 (continued)

Diplazium doederleinii, *Diplazium petri*, *Diplazium pullingeri*, *Histiopteris incisa*, *Plagiogyria adnata*, *Plagiogyria dunnii*, *Pleocnemia rufinervis*, *Pteris wallichiana*, and *Sphaerostephanos taiwanensis*

[d]Thirteen species in S Taiwan including *Arthropteris palisotii*, *Asplenium excisum*, *Colysis pothifolia*, *Colysis wrightii*, *Ctenitis subglandulosa*, *Ctenitopsis dissecta*, *Cyclosorus truncatus*, *Diplazium chinense*, *Diplazium dilatatum*, *Microlepia strigosa*, *Microsorum punctatum*, *Tectaria devexa*, and *Tectaria subtriphylla* var. *subtriphylla*

[e]Means of all species used in this Table. Some individual exceptions illustrated in the text of this chapter. See citation for detail

with temperature but not with precipitation (Lee et al. 2009b). However, in southern Taiwan where there are obvious dry and wet seasons, the correlation of leaf senescence with temperature or precipitation is not significant (Lee et al. 2016). Similar to the leaf emergence phase, the correlation of leaf senescence with temperature/precipitation somewhat reflects their seasonality as summarized in Table 18.2.

The timing of leaf senescence can also be affected by factors other than apparently favorable climate extremes. Leaves of *Alsophila firma* (Baker) D. S. Conant wither during the season of higher precipitation and temperature (Mehltreter and Garcia-Franco 2008). Insects also thrive in this season, and the leaf falling of *A. firma* is suggested as a mechanism to reduce herbivore damage (Lieberman and Lieberman 1984; Karban 2007; Mehltreter and Sharpe 2013). Leaf senescence of *Lygodium venustum* also occurred during the season of higher precipitation and temperature, but disturbance from mechanical damage or herbivores is inferred to be the main cause of senescence (Mehltreter 2006). Leaf senescence of some ferns, e.g., *Polystichum aculeatum* (L.) Roth ex Mert., varied from year to year, a pattern that was inferred to result from innate physiology (Landi et al. 2014). Declining plant vitality also may play an important role in leaf senescence (Thomas 2013).

18.3.3 Seasonality of Spore Maturation/Release

Spore maturation/release is an essential entry of subsequent spore germination and successful reproduction of ferns. Although a test of spore vigor in the laboratory is more reliable to determine the maturation of spores (Arosa et al. 2009; Huang et al. 2014), this indoor work is not very practical for observing the maturation status in the field. Instead, sporangium color is often used for the signal of spore maturation: light green indicates the inside spores are young, and dark (yellow, brown, or black) color indicates maturity. Degree of sporangium openness is recommended to represent spore release, i.e., ca. 5% of sori having open sporangia indicates the start of spore release; ca. 95% of sori having open sporangia indicate the end of spore release.

Table 18.3 The correlations of spore maturation and release dates with temperature and precipitation

Species	Temperature		Precipitation		Location	References
	Spore maturation	Spore release	Spore maturation	Spore release		
Cyathea atrovirens	0.82**	0.35	−0.41	−0.46	S Brazil	Schmitt and Windisch (2012)
C. hancockii	0.52**	−0.50**	0.03	−0.03	N Taiwan	Lee et al. (2009a)
C. metteniana	0.55**	0.06	0.11	0.24	N Taiwan	Lee et al. (2009a)
C. podophylla	0.39*	0.59**	−0.13	−0.02	N Taiwan	Lee et al. (2009a)
Dryopteris affinis ssp. *affinis*[a]	1.87**	−1.66**	0.17 (P_d)	0.67** (P_d)	C Italy	Landi et al. (2014)
Polystichum aculeatum[a]	1.49**	0.62*	–	−0.98** (P_d)	C Italy	Landi et al. (2014)
16 species[b,c]	0.70**	0.82**	0.15	0.12	NE Taiwan	Lee et al. (2009b)
13 species[b,c]	0.48**	0.46**	0.59**	0.60**	S Taiwan	Lee et al. (2016)

*$0.01 < p < 0.05$; **$p < 0.01$; –, data not available

[a]For these species values are calculated by GAMLSS. P_d days of rainfall per month

[b]Species are listed in Table 18.2

[c]Mean of all species used in this table. Some exceptions are discussed in the text of this chapter. See citation for detail

Spore maturation and release of most ferns mainly occur in a specific season. For many ferns, including those in tropical/subtropical/temperate regions, spores exhibit maturity in late spring to early autumn, and spores release 1–2 months later (e.g., Lee et al. 2008, 2009b, 2016; Landi et al. 2014). These two phenological events in those regions occurred in the relatively higher temperature season and are positively correlated with temperature (Table 18.3). Spores of many other ferns are also found to be mature in summer to early autumn in temperate regions, such as northeastern China (Liu et al. 2002), northern Japan (Sato 1982), and Britain and Ireland (Page 1997).

Precipitation is not significantly correlated with spore maturation for most ferns except those in southern Taiwan (Table 18.3). The wet season in southern Taiwan is also the warm season, and the higher temperature is inferred to be the crucial factor resulting in spore maturation there (Lee et al. 2016).

Spore liberation is affected by temperature/humidity. A temperature decrease and/or a humidity increase postponed the spore release of *Culcita macrocarpa* C.Presl and *Woodwardia radicans* (L.) Sm. (Arosa et al. 2009). The longer interval between spore maturation and release in subtropical than in tropical forests of Taiwan (1.6 vs. 1.0 month; $p < 0.01$, t-test) may also be attributed to lower temperature (average monthly temperature 18.2 vs. 25.4 °C) and higher humidity (annual precipitation 5000 vs. 2500 mm) (Lee et al. 2009b, 2016). However, the time intervals between maturation and release may vary from <1 month to >4 month, depending on species even in the same habitat in northeastern Taiwan (Lee et al. 2009b). Two other exceptions are *Dryopteris affinis* ssp. *affinis* and *Polystichum aculeatum*. The end of spore release of the former was positively significantly correlated with rainy days (more than 10 days/month), whereas it was negatively significantly correlated for the latter species (Landi et al. 2014).

Green spores occur in a wide variety of lineages of ferns, and such spores are capable of immediate germination and have a relatively short period of vitality in nature (Cousens 1988; Raghavan 1989; Ballesteros et al. 2011). Some green spores of ferns in temperate regions may mature and be released in late spring and germinate quickly, such as Osmundaceae (*Osmunda*, *Osmundastrum*) and most *Equisetum* species, while others mature in late autumn and remain in sporangia on contracted pinnae until release and germinate in the following spring, such as *Equisetum hymale* and Onocleaceae (*Matteuccia*, *Onoclea*) species (Hill and Wagner 1974; Sato 1982; Lee et al. 2008; Chiu et al. 2013 (but see exceptions for *Equisetum* in Liu et al. 2002). The green spores that are released and germinate in the spring have a whole growing season (i.e., the warm summer) during which gametophytes and sporophytes can develop (Sato 1982). However, little is known about the phenology of the fern species with green spores in the Hymenophyllaceae family or the grammitid ferns that are abundant in tropical and subtropical regions.

18.4 Sterile Vs. Fertile Leaves

Leaves produced by a fern plant may or may not bear sori and are termed fertile or sterile, respectively. The morphology of these two types of leaves may be very different or the leaves may look very similar. For dimorphic species, the leaves differ, and for monomorphic species they do not. Some species may develop fertile portion on a part of the leaf, and these species can be called hemidimorphic (Wagner and Wagner 1977). When the differences between the fertile and sterile leaves are extreme, a species may be called holodimorphic (e.g., Watkins et al. 2016). The phenological attributes of the production rates and lifespans of fertile and sterile leaves may also differ; however, in most phenological studies, these important differences may be ignored or only noted for obviously dimorphic species of ferns.

18.4.1 Seasonality

The phenology of fertile and sterile leaves could be similar or very different for both monomorphic and dimorphic fern species (Table 18.2). In forests of northern and northeastern Taiwan, fertile leaves of the emergence of most ferns occurs in early spring and is negatively significantly correlated with temperature, whereas the sterile leaf emergence mainly occurs later during the warmer season and is positively significantly correlated with temperature (Table 18.2, Lee et al. 2009a, b). In contrast, for about half of monomorphic ferns in a tropical monsoon forest of southern Taiwan, the emergence periods of fertile and sterile leaves did not show significant differences and positively correlated with temperature and precipitation (Table 18.2). The fertile leaves of *Acrostichum danaeifolium* Langsd. & Fisch. mostly emerged from April to August in Mexico, but the sterile leaves could emerge throughout the year (Mehltreter and Palacios- Rios 2003). Both fertile and sterile leaves of *Cyathea atrovirens* (Langsd. & Fisch.) Domin (Cyatheaceae) in secondary forests of southern Brazil synchronously emerged in spring and did not correlate with either temperature or day length (Schmitt and Windisch 2012).

Relatively few studies report the seasonal difference in leaf emergence between fertile and sterile leaves of temperate ferns. *Osmundastrum cinnamomeum* produce its fertile leaves 1 month before its sterile leaves in Taiwan (Chiu et al. 2013). Both of fertile and sterile leaves of *Osmunda claytoniana* start to emerge in May, but the former have all emerged within 1 month, while the latter continuously emerge till July (Lee et al. 2008). Sato (1982) has also reported that the fertile leaves of *Osmunda cinnamomeum* (=*Osmundastrum cinnamomeum*) and *O. japonica* in Japan emerge earlier than its sterile leaves.

18.4.2 Leaf Numbers for Fertile and Sterile Leaves

The ratio of the number of fertile leaves to sterile leaves produced by a fern during a specific time period (leaf cohort), together with their respective lifespans (see next section), may relate to the strategy of the growth and reproduction of a fern species. Studies reveal that numbers of fertile leaves produced were on average lower than sterile leaves for dimorphic ferns (Sharpe and Jernstedt 1990; Lee et al. 2008, 2009b; Chiu et al. 2013; Watkins et al. 2016). The ratios of fertile/sterile leaf numbers for dimorphic fern species in these studies averaged less than one, ranging from 0.16 to 0.73. Watkins et al. (2016) demonstrated an overall pattern of a reduction in fertile leaf number with an increase in the degree of dimorphism by six fern species: three temperate species (*Dryopteris marginalis* (L.) A. Gray (monomorphic), *Polystichum acrostichoides* (Michx.) Schott (hemidimorphic), and *Onoclea sensibilis* L. (dimorphic)) and three tropical species (*Adiantum latifolium* Lam. (monomorphic), *Thelypteris curta* (Christ) C.F. Reed (hemidimorphic), and *Lomariopsis vestita* E. Fourn. (dimorphic)). However, Table 18.4 shows that fertile/sterile leaf ratios for monomorphic ferns are variable, ranging from 0.08 to 12.67 and indicating that numbers of fertile leaves can be lower, close to, or higher than the numbers of sterile leaves of the same species. For two hemidimorphic fern species, *Acrostichum danaeifolium* and *Osmunda claytoniana*, it has been demonstrated that the production rate for fertile leaves is lower than for the sterile leaves (Mehltreter and Palacios-Rios 2003; Lee et al. 2008), but the ratio is not necessarily intermediate between monomorphic and dimorphic ferns. However, the fertile leaf lifespan

Table 18.4 Leaf numbers, lifespans, and ratios of fertile/sterile leaves

Species	Plants	Fn	Sn	Fn/ Sn	FL	SL	FL/ SL	Types	Locations
Bolbitis appendiculata[7]	25	155	213	0.73	7.5	19.3	0.39	D	SE Taiwan
Danaea wendlandii[9]	17	0.1[a]	1.6[a]	0.06	4	39.6	0.10	D	Costa Rica
Osmundastrum cinnamomeum[2]	80	123	755	0.16	1	6.3	0.16	D	NE Taiwan
Plagiogyria adnata[5]	5	58	101	0.57	4.7	30.3	0.16	D	NE Taiwan
Plagiogyria adnata[7]	25	37	190	0.19	8.2	25.4	0.32	D	SE Taiwan
Plagiogyria dunnii[5]	5	31	173	0.18	4.4	22.1	0.20	D	NE Taiwan
Thelypteris anustifolia[10]	>30	0.4[a]	4.2[a]	0.10	9.6[c]	11[c]	0.87	D	Puerto Rico
Acrostichum danaeifolium[8]	30	0.5 [a]	14.6 [a]	0.03	4.1 [a,b]	7.7 [a,b]	0.53	HD	Mexico
Osmunda claytoniana[3]	25	30	120	0.25	4.7	4.8	0.98	HD	C Taiwan
Acrorumohra hasseltii[5]	5	42	7	6.00	23.7	22.4	1.06	M	NE Taiwan
Asplenium excisum[6]	5	43	11	3.91	14.8	15.5	0.95	M	S Taiwan
Blechnum orientale[5]	5	67	19	3.53	25.9	27.5	0.94	M	NE Taiwan
Blechnum orientale[7]	25	195	88	2.22	23.3	22.1	1.05	M	SE Taiwan
Cibotium taiwanense[1]	19	24	39	0.62	26	15	1.73	M	N Taiwan
Colysis pothifolia[6]	5	14	42	0.33	11.8	15.5	0.76	M	S Taiwan
Colysis wrightii[6]	5	34	125	0.27	15.9	19.5	0.82	M	S Taiwan
Ctenitis subglandulosa[6]	5	31	43	0.72	17.5	9.6	1.82	M	S Taiwan
Ctenitopsis dissecta[6]	5	27	7	3.86	9.6	10.9	0.88	M	S Taiwan
Cyathea hancockii[4]	10	30	23	1.30	15.9	15	1.06	M	N Taiwan
Cyathea metteniana[4]	5	18	15	1.20	19.5	21.2	0.92	M	N Taiwan
Cyathea podophylla[4]	10	68	58	1.17	26.6	28.3	0.94	M	N Taiwan
Cyathea podophylla[5]	5	97	26	3.73	26.1	25.8	1.01	M	NE Taiwan
Cyathea podophylla[7]	25	157	32	4.91	27.9	26.5	1.05	M	SE Taiwan
Cyathea spinulosa[5]	5	111	211	0.53	6.6	8	0.83	M	NE Taiwan
Cyclosorus taiwanensis[5]	5	125	66	1.89	13.1	12.9	1.02	M	NE Taiwan
Cyclosorus taiwanensis[7]	25	376	179	2.10	12.7	13.8	0.92	M	SE Taiwan
Cyclosorus truncatus[6]	5	11	64	0.17	10.7	10.1	1.06	M	S Taiwan
Dictyocline griffithii wilfordii[5]	5	68	61	1.11	15.1	15	1.01	M	NE Taiwan
Didymochlaena truncatula[11]	23	3.27[a]	2.86[a]	1.14	10.2[b]	14.9 [b]	0.68	M	NE Brazil
Diplazium dilatatum[5]	5	93	13	7.15	18.3	19.8	0.92	M	NE Taiwan
Diplazium dilatatum[6]	5	59	65	0.91	15.6	15.4	1.01	M	S Taiwan
Diplazium dilatatum[7]	25	52	48	1.08	26.2	25.2	1.04	M	SE Taiwan
Diplazium doederleinii[5]	5	37	7	5.29	14.2	15.8	0.90	M	NE Taiwan
Diplazium petri[5]	5	43	14	3.07	17.5	20.1	0.87	M	NE Taiwan
Diplazium pullingeri[5]	5	76	6	12.67	16.8	16.7	1.01	M	NE Taiwan
Histiopteris incisa[5]	5	33	30	1.10	7.7	6.9	1.12	M	NE Taiwan
Microlepia strigosa[6]	5	9	115	0.08	17.1	11.9	1.44	M	S Taiwan

(continued)

Table 18.4 (continued)

Species	Plants	Fn	Sn	Fn/ Sn	FL	SL	FL/ SL	Types	Locations
Pleocnemia rufinervis[5]	5	16	28	0.57	17.2	18.1	0.95	M	NE Taiwan
Pteris wallichiana[5]	5	41	102	0.40	6.1	5.1	1.20	M	NE Taiwan
Pteris wallichiana[7]	25	85	246	0.35	5.2	4.2	1.24	M	SE Taiwan
Tectaria devexa[6]	5	39	37	1.05	10.1	11.7	0.86	M	S Taiwan

Sources: [1]Chiou et al. (2001), [2]Chiu et al. (2013), [3]Lee et al. (2008), [4]Lee et al. (2009a), [5]Lee et al. (2009b), [6]Lee et al. (2016), [7]Lee et al. (unpublished), [8]Mehltreter and Palacios-Rios (2003), [9]Sharpe and Jernstedt (1990), [10]Sharpe (1997), [11]de Paiva Farias et al. (2015)
Leaf lifespan counted from leaf emergence to senescence unless otherwise indicated
Fn number of fertile leaves, *Sn* number of sterile leaves, *FL* lifespan of fertile leaves (month), *SL* lifespan of sterile leaf (month), *D* dimorphic, *HD* hemidimorphic, *M* monomorphic
[a]Leaf number/plant
[b]Estimated by mean leaf number/leaf production rate
[c]Measured from leaf expansion to senescence

for *A. danaeifolium* is ca. half that of the sterile leaf lifespan whereas they are almost the same for *O. claytoniana* (Table 18.4, and see below) although fertile pinnae of both species wither soon after their spore release.

18.4.3 Leaf Lifespan (LLS)

Fern LLS of different species vary from 2.8 months (*Botrychium gallicomontanum*; Johnson-Groh and Lee 2002) to 37 months (*Asplenium wrightii* D.C. Eaton ex Hook.; Yoshida and Takasu 1993) and 4.9 years (*Terpsichore asplenifolia* (L.) A.R. Sm.; Monge González (2010) referred from Mehltreter and Sharpe 2013). In temperate regions, fern LLS are correlated to their growth habits, which can be classified into four categories, i.e., evergreen (14–26 months), summergreen (3–5 months), semievergreen (10 months), and wintergreen (11 months) in the temperate region (Sato 1982). Mehltreter and Sharpe (2013), based on the terminology of Lellinger (2002), combined the latter two types into wintergreen type and estimated the mean LLS of British and Ireland ferns (Page 1997) with approximately 16, 5, and 11 months for evergreen, summergreen, and wintergreen ferns, respectively. The fern LLS in both geographical regions are the longest for evergreen ferns, followed by those of wintergreen ferns and summergreen ferns.

Comparing the LLS of tropical ferns, Mehltreter and Sharpe (2013) report that sterile leaves of ferns of Puerto Rico has a mean LLS (26.5 months) longer than other four tropical regions, i.e., Hawaii, Mexico, Jamaica, and Taiwan with LLS of 15.0, 16.4, 17.3, and 17.5 months, respectively. However, this summary included temperate and tropical ferns and did not account for differences between LLS of sterile and fertile leaves or monomorphic and dimorphic species and therefore does not adequately address the potential differences LLS of these geographic datasets. For example, the LLS of fertile and sterile leaves of *Osmundastrum cinnamomeum* (a dimorphic temperate summergreen fern) in Taiwan is 1 and 6.3 months, respectively (Chiu et al. 2013), which averages into 3.7 months, an LLS estimate that is very misleading and also does not take into account the relative numbers of fertile and sterile leaves annually produced by the plant.

Dimorphic species produce a smaller number of fertile leaves, and those leaves have shorter LLS compared with sterile leaves of that species. The fertile/sterile LLS ratios of dimorphic ferns are generally lower than monomorphic ferns (0.10–0.39 (−0.87) vs. 0.76–1.82; Table 18.4). These lower ratios have been inferred to be resulted from the lower C/N ratios of fertile leaves (Chiu et al. 2013). Fertile leaves of dimorphic ferns have narrow blades and thus have less photosynthetic products and lower C/N ratios, which result in the shorter LLS (Karst and Lechowicz 2007). Compared with their sterile leaves, fertile leaves in dimorphic ferns have significantly higher respiration rates and energy investment for producing spores and results in their lower numbers and shorter lifespans (Watkins et al. 2016). The short LLS (4.1 and 7.7 months for fertile and sterile leaves, respectively) of *Acrostichum danaeifolium*, a mangrove hemidimorphic fern, may be attributed to the function of eliminating the accumulated salt (Sharpe 2010; Mehltreter and Sharpe 2013).

In monomorphic ferns, on the other hand, the trend of leaf number ratios is not consistent with lifespan ratios of fertile and sterile leaves. Table 18.4 shows that about 1/3 of monomorphic ferns exhibit shorter fertile LLS than their sterile LLS; however, only *Colysis pothfolia* (Buch.-Ham. ex D. Don) C. Presl and *Cyathea spinulosa* Wall. ex Hook. exhibit significant difference. Other monomorphic ferns show similar or longer fertile LLS compared to their sterile leaves, but only *Pteris wallichiana* J. Agardh exhibits significantly longer fertile LLS. Shorter fertile LLS with smaller fertile leaf number characterize dimorphic ferns, a correlation that is variable in monomorphic ferns. In dimorphic ferns, fertile leaves die soon after investing energy in reproduction by producing spores (Watkins et al. 2016), whereas in monomorphic ferns the fertile leaves continue to conduct photosynthesis after releasing spores. Therefore, fertile leaves in monomorphic ferns serve as a source of nourishment as well as reproduction.

18.4.4 Discrimination of LLS and Leaf Numbers for Species across Locations

Fertile and sterile LLS for the same species growing in different locations have been studied in a few cases. In Taiwan, LLS of six species, i.e., *Plagiogyria adnata* Blume Bedd., *Blechnum orientale* L., *Cyathea podophylla* (Hook.) Copel., *Cyclosorus taiwanensis* (C.Chr.) H.Ito, *Diplazium dilatatum* Blume, and *Pteris wallichiana* J.Agardh, were investigated at two sites: one in NE Taiwan and the other in SE Taiwan with *C. podophylla* monitored at a third site (N Taiwan) and *D. dilatatum* also at a third site (S Taiwan) as well (Table 18.4). The fertile and sterile LLS and their ratios of monomorphic species are very similar in the different locations, though there is a lower fertile LLS and higher FL/SL ratio in SE Taiwan for the dimorphic species (*P. adnata*).

For leaf numbers, there are fairly large differences in the ratio of fertile leaves to sterile leaves of the same species in different locations (Table 18.4), probably due to various nutrient, soil, and/or microhabitats. As for LLS of the same species in different locations, the fertile leaf numbers of the dimorphic species are lower than the sterile leaf number, but that is reversed for the monomorphic species with one exception: *P. wallichiana* has lower fertile leaf counts than sterile leaf counts. Ratios of the number of fertile and sterile leaves of *Cyathea podophylla* and *Diplazium dilatatum* are similar in two locations but very different in the third. The habitat conditions of the different locations and the habitat preference of different species might affect these results. Further studies of plant size in these two species are needed and may show size to be an explanation for this disparity. For example, the fertile leaf number has been found to be directly proportional to different measures of plant size in other studies: (1) the larger caudex diameter (*Cibotium barmetz* (L.) J.Sm., Lee et al. 2010), (2) the larger rosette diameter (*Asplenium nidus* L., Huang et al. 2008; *Osmundastrum cinnamomeum*, Kao et al. 2010), and (3) the longer leaf length (*Athyrium distentifolium* Tausch ex Opiz and *Thelypteris limbosperm* (All.) H.P. Fuchs, Odland 1995) within a population, the more fertile leaves.

18.5 Future Studies of Fern Phenology

Although the number of phenological studies of ferns has increased during the past half century, this field is still at a very young stage in terms of understanding the basic science and its applied usage for ferns. *Long-term research*, *broad-scale research*, *multi-disciplinary approach*, and *quantitative method* are recommended for future studies by Mehltreter (2008) and are goals to be taken seriously today and in the future. The distinction between fertile and sterile leaves is also an essential issue to be addressed when fern phenology is studied.

Most phenological investigations to date have been conducted in the field. Experiments in the laboratory and greenhouse or nursery are required to better understand the mechanism controlling phenological patterns that have been observed in the field.

Furthermore, fern gametophytes, as vital genetic transfer vectors, have been largely neglected, and since very few phenology studies of fern gametophytes have been done, they deserve more attention. One of the few exceptions was a study conducted by Galka and Szmeja (2013), who investigated the phenology of both sporophytes (including spore germination) and the development of gametophytes in *Salvinia natans* (L.) All. Working on gametophyte phenology is a big challenge because of the minute bodies and few characters of gametophytes making identification in the field difficult. A DNA barcode approach could facilitate solving this difficulty, e.g., Chen et al. (2013) and Ebihara et al. (2013).

To accurately determine the phenological patterns, the monitoring interval needs to be carefully considered. Most phenological investigations take monthly observation (e.g., Lee et al. 2009a, b, 2016; de Paiva Farias et al. 2015). Monthly phenological data are relatively convenient to compare with metrological data and calculate correlations between phenological traits and climate factors. However, for short LLS fern species, e.g., *Pteris wallichina*, 4–5 months; *Histiopteris incisa*, ca. 7 months; and *Osmundastrum cinnamomeum*, 1 month of fertile leaves (Table 18.4), weekly to bi-weekly investigation is recommended, at least at the beginning of leaf growth or/and spore maturation/release period.

For the long-term and broad scale studies, in addition to involving more researchers in field studies in additional regions of the world, accessing herbarium data can also be worthwhile sources of phenological data. There are more than 380 million specimens deposited in nearly 3000 herbaria of 176 countries (Thiers continuously updated). Those specimens can document character shifts across long time scales and broad geographical ranges. These historical records can also provide valuable information that can reveal phenological variation through time (Cousens 1988), especially compared with meteorological data to predict the trend of future phenological variation (e.g., Calinger et al. 2013; Li et al. 2013). Data extracted from the plant morphology on specimen sheets may not exactly correspond to data collected in the field (e.g., spore maturity dates) and will need careful examination to provide the proper adjustments (Sato 1982; Huang et al. 2014).

A network of observers is another efficient approach to accumulate worldwide phenological data, such as the International Phenological Gardens in Europe (IPG) and the Global Phenological Monitoring Programme (GPM) (Chmielewski et al. 2013) or the USA National Phenology Network (NPN) (2017) (https://www.usanpn.org/). However, projects in those networks have not included fern species although NPN is just beginning to assess fern protocols for observation (Joanne Sharpe, pers. comm.). Joining those networks or organizing a fern-phenology network would be very worthwhile. Some worldwide species, e.g., *Osmundastrum cinnamomeum* (dimorphic, temperate species), *Acrostichum danaeifolium* or/and *A. aureum* L. (hemidimorphic, tropical species), *Blechnum orientale*, *Pteris vitata* L. (both monomorphic, tropical, sunny species), *Pteridum aquilinum* (L.) Kuhn (monomorphic, sunny species, with various varieties from temperate to tropical regions), and *Diplazium dilatatum* (monomorphic, tropical, shady species), can be candidates of fern phenological studies of the network. Phenology of some invasive species and agriculture/horticulture species are also recommended for future studies. Phenological observation of a large diversity of life forms, including ferns, is absolutely important in assessing the potential impacts of climate change.

Acknowledgements We thank Dr. Joanne M. Sharpe for her useful comments and English editing. This research was partially supported by grant 100AS-8.2.1-FI-G9, 101AS-13.5.2-FI-G6, and 102AS-13.3.7-FI-G2 from the Taiwan Forestry Research Institute.

References

Arosa ML, Ramos JA, Quintanilla LG et al (2009) Spore maturation and release of two evergreen Macaronesian ferns, *Culcita macrocarpa* and *Woodwardia radicans*, along an altitudinal gradient. Am Fern J 99(4):1250–1140

Baddour O, Kontongomde H (2009) Guidelines for plant phenological observations. World Meteorological Organization WMO-TD No. 1484. Geneva

Ballesteros D, Estrelles E, Walters C et al (2011) Effect of storage temperature on green spore longevity for the ferns *Equisetum ramosissimum* and *Osmunda regalis*. CryoLetters 32(2):89–98

Bauer H, Gallmetzer C, Sato T (1991) Phenology and photosynthetic activity in sterile and fertile sporophytes of *Dryopteris filix-mas* (L.) Schott. Oecologia 86:159–162

Braun HJ (1984) The significance of the accessory tissues of the hydrosystem for osmotic water shifting as the second principle of water ascent, with some thoughts concerning the evolution of trees. IAWA Bull (N S) 5:275–294

Calinger KM, Queenborough S, Curtis PS (2013) Herbarium specimens reveal the footprint of climate change on flowering trends across north-central North America. Ecol Lett 16:1037–1044

Chao YS, Dong SY, Chiang YC et al (2012) Extreme multiple reticulate origins of the *Pteris cadieri* complex (Pteridaceae). Int J Mol Sci 13:4523–4544

Chen CW, Huang YM, Kuo LY et al (2013) *trnL-F* is a powerful marker for DNA identification of field vittarioid gametophytes (Pteridaceae). Ann Bot 111:663–673

Chen X (2013) East Asia. In: Schwartz MD (ed) Phenology: an integrative environmental science, 2nd edn. Springer, New York, pp 9–22

Chiou WL, Lin JC, Wang JY (2001) The phenology of *Cibotium taiwanense* (Dicksoniaceae). Taiwan. J For Sci 16:209–215

Chiu TY, Chiou WL, Huang YM (2013) Phenological differences between sterile and fertile fronds in a dimorphic fern, *Osmundastrum cinnamomeum* (L.) C.Presl (Osmundaceae). TW J Biodivers 15(4):311–322. [in Chinese with English abstract]

Chmielewski F-M, Heider S, Moryson S, Bruns E (2013) International phenological observation networks: concept of IPG and GPM. In: Schwartz MD (ed) Phenology: an integrative environmental science, 2nd edn. Springer, New York, pp 137–153

Cleland EE, Chuine I, Menzel A et al (2007) Shifting plant phenology in response to global change. Trends Ecol Evol 22:357–365

Cousens MI (1988) Reproductive strategies of Pteridophytes. In: Lovett-Doust J, Lovett-Doust L (eds) Plant reproductive ecology: patterns and strategies. Oxford University Press, Oxford, pp 307–328

de Paiva Farias R, da Costa LEN, da Silva IAA et al (2015) Phenological studies of selected leaf and plant traits of *Didymochlaena truncatula* (Dryopteridaceae) in a Brazilian submontane tropical rainforest. Nord J Bot 33:249–255

Doi H, Takahashi M (2008) Latitudinal patterns in the phenological responses of leaf colouring and leaf fall to climate change in Japan. Glob Ecol Biogeogr 17:556–561

Ebihara A, Yamaoka A, Mizukami N, Sakoda A, Nitta JH, Imaichi R (2013) A survey of the fern gametophyte flora of Japan: frequent independent occurrences of noncordiform gametophytes. Am J Bot 100(4):735–743

Ellwood ER (2012) Climate change and species phenology at three trophic levels. Doctor Dissertation, Boston University, Graduate School of Arts and Sciences

Farrar DR (1976) Spore retention and release from overwintering fern fronds. Am Fern J 66:49–52

Galka A, Szmeja J (2013) Phenology of the aquatic fern *Salvinia natans* (L.) All. in the Vistula Delta in the context of climate warming. Limnologica 43(2):100–105

Gan S, Amasino RM (1997) Making sense of senescence: molecular genetic regulation and manipulation of leaf senescence. Plant Physiol 113:313–319

Geldenhuys CJ, van der Merwe CJ (1988) Population structure and growth of the fern *Rumohra adiantiformis* in relation to frond harvesting in the southern Cape forests. S Afr J Bot 54(4):351–362

Harvey WH, Caponetti JD (1972) In vitro studies on the induction of sporogenous tissue on leaves of cinnamon fern. I. Environmental factors. Can J Bot 50:2673–2682

Hietz P (2010) Fern adaptations to xeric environments. In: Mehltreter K, Walker LR, Sharpe JM (eds) Fern ecology. Cambridge University Press, New York, pp 140–176

Hill RH, Wagner WH Jr (1974) Seasonality and spore type of the pteridophytes of Michigan. Mich Bot 13:40–44

Huang YM, Chou HM, Liu CJ et al (2008) The correlation of plant size and fertility of *Asplenium nidus* L. (Aspleniaceae; Pteridophyta). BioFormosa 43(2):77–83. [in Chinese with English summary]

Huang YM, Lee PH, Fu CH et al (2014) The relation between spore maturity and viability of tree ferns. J Natl Taiwan Mus 67(2):45–58. [in Chinese with English summary]

Hudson IL, Keatley MR (2010) Phenological research: methods for environmental and climate change analysis. Springer, New York

Johnson-Groh CL, Lee JM (2002) Phenology and demography of two species of *Botrychium* (Ophioglossaceae). Am J Bot 89(10):1624–1633

Kao YC, Chiou TY, Chu MK, Chiou WL, Huang YM (2010) Study of the fertility in *Osmunda cinnamomea* L. at Cao-Pi area. J Natl Taiwan Mus 63(4):15–25. [in Chinese with English abstract]

Karban R (2007) Deciduous leaf drop reduces insect herbivory. Oecologia 153:81–88

Karst AL, Lechowicz MJ (2007) Are correlations among foliar traits in ferns consistent with those in the seed plant? New Phytol 173:306–312

Keatley MR, Fletcher TD, Hudson IL et al (2002) Phenological studies in Australia: potential application in historical and future climate analysis. Int J Climatol 22:1769–1780

Landi M, Zoccola A, Bacaro G et al (2014) Phenology of *Dryopteris affinis* ssp. *affinis* and *Polystichum aculeatum*: modeling relationships to the climatic variables in a Mediterranean area. Plant Species Biol 29:129–130. https://doi.org/10.1111/1442-1984.12000

Lee PH, Huang YM, Chiou WL (2008) The phenology of *Osmunda claytoniana* L. in the Tataka area, Central Taiwan. Taiwan J For Sci 23:71–79

Lee PH, Chiou WL, Huang YM (2009a) Phenology of three *Cyathea* (Cyatheaceae) ferns in Northern Taiwan. Taiwan J For Sci 25(1):1–10

Lee PH, Lin TT, Chiou WL (2009b) Phenology of 16 species of ferns in a subtropical forest of northeastern Taiwan. J Plant Res 122:61–67

Lee PH, Chang YH, Huang YM (2010) Indicators of the maturity of *Cibotium barometz* (L.) J. Sm. (Dicksoniaceae, Pteridophyta). J Yuanpei Univ 17:43–52. [in Chinese with English abstract]

Lee PH, Chen SW, Chiou WL et al (2016) Phenology of 13 fern species in a tropical monsoon forest of southern Taiwan. Int J Plant Reprod Biol 8(1):87–97

Lellinger D (2002) A modern multi-lingual glossary for taxonomic pteridology. Pteridologia 3:1–263

Li Z, Wu N, Gao X et al (2013) Species-level phenological responses to 'global warming' as evidenced by herbarium collections in the Tibetan Autonomous Region. Biodivers Conserv 22:141–152

Lieberman D, Lieberman M (1984) The causes and consequences of synchronous flushing in a tropical dry forest. Biotropica 16:193–201

Lieth H (ed) (1974) Phenology and seasonality modelling. Springer, New York

Liu BD, Wang XN, Gao H (2002) An investigation on the spore mature period of pteridophyte in Heilongjiang Province of China. Bull Bot Res 22(4):422–427. [in Chinese with English abstract]

Mehltreter K (2006) Leaf phenology of the climbing fern *Lygodium venustum* in a semideciduous lowland forest on the Gulf of Mexico. Am Fern J 96(1):21–30

Mehltreter K (2008) Phenology and habitat specificity of tropical ferns. In: Tanker TA, Haufler CH (eds) Biology and evolution of ferns and lycophytes. Cambridge University Press, New York, pp 201–221

Mehltreter K, Garcia-Franco JG (2008) Leaf phenology and trunk growth of the deciduous tree fern *Alsophila firma* (Baker) D. S Conant in a Lower Montane Mexican Forest. Am Fern J 98(1):1–13

Mehltreter K, Palacios-Rios M (2003) Phenological studies of *Acrostichum danaefolium* (Pteridaceae, Pteridophyta) at a mangrove site on the Gulf of Mexico. J Trop Ecol 19:155–162

Mehltreter K, Sharpe JM (2013) Causes and consequences of the variability of leaf lifespan of ferns. Fern Gaz 19(6):193–202

Meier N, Rutishauser T, Pfister C et al (2007) Grape harvest dates as a proxy for Swiss April to August temperature reconstructions back to AD 1480. Geophys Res Lett 34(20):L20705. https://doi.org/10.1029/2007GL031381

Menzel A (2000) Trends in phenological phases in Europe between 1951 and 1996. Int J Biometeorol 44(2):76–81

Menzel A (2013) Europe. In: Schwartz MD (ed) Phenology: an integrative environmental science, 2nd edn. Springer, New York, pp 53–65

Monge González L (2010) Fenología de tres especies de helechos epifitos sobre dos grupos de hospederos en Las Cañadas, Huatusco, Veracruz. Bachelor thesis, Universidad Veracruzana, Facultad de Biologia, Xalapa

Müller A, Cunha S, Junges F et al (2016) Climate effects on the phenology of *Lindsaea lancea* (L.) Bedd. (Lindsaceae) in an Atlantic forest fragment in southern Brazil. Interciencia 41:34–39. [in portbrazil with English summary]

Neumann MK, Schneider PH, Schmitt JL (2014) Phenology, caudex growth and age estimation of *Cyathea corcovadensis* (Raddi) Domin (Cyatheaceae) in a subtropical forest in southern Brazil. Acta Bot Bras 28(2):274–280

Newstrom LE, Frankie GW (1994) A new classification for plant phenology based on flowering patterns in low tropical rain forest trees at La Selva, Costa Rica. Biotropica 26:141–159

Noormet A (2009) Phenology of ecosystem processes: applications in global change research. Springer, New York

Odland A (1995) Frond development and phenology of *Thelypteris limbosperma*, *Athyrium distentifolium*, and *Matteuccia struthiopteris* in Western Norway. Nord J Bot 15:225–236

Page CN (1997) The ferns of Britain and Ireland. Cambridge University press, Cambridge

PPG I (2016) A community-derived classification for extant lycophytes and ferns. J Syst Evol 54:561–603

Raghavan V (1989) Development biology of fern gametophyte. Cambridge University Press, New York

Sato T (1982) Phenology and wintering capacity of sporophytes and gametophytes of ferns native to northern Japan. Oecologia 55:53–61

Schmitt JL, Windisch PG (2012) Caudex growth and phenology of *Cyathea atrovirens* (Langsd. & Fisch.) Domin (Cyatheaceae) in secondary forest, southern Brazil. Braz J Biol 72(2):397–405

Schwartz MD (ed) (2013) Phenology: an integrative environmental science. Springer, New York

Sharpe JM (1997) Leaf growth and demography of the rheophytic fern *Thelypteris angustifolia* (Willdenow) Proctor in a Puerto Rican rainforest. Plant Ecol 130(2):203–212

Sharpe JM (2010) Responses of the mangrove fern *Acrostichum danaeifolium* Langsd. & Fisch. (Pteridaceae, Pteridophyta) to disturbances resulting from increased soil salinity and Hurricane Georges at the Jobos Bay National Estuarine Research Reserve, Puerto Rico. Wetl Ecol Manag 18(1):57–68

Sharpe JM, Jernstedt JA (1990) Leaf growth and phenology of the dimorphic herbaceous layer fern *Danaea wendlandii* (Marattiaceae) in a Costa Rican rain forest. Am J Bot 77(8):1040–1049

Sharpe JM, Mehltreter K (2010) Ecological insights from fern population dynamics. In: Mehltreter K, Walker LR, Sharpe JM (eds) Fern ecology. Cambridge University Press, New York, pp 61–110

Sparks TH, Carey PD (1995) The response of species to climate over two centuries: an analysis of the Marsham phenological records, 1736–1947. J Ecol 83:321–329

Strandberg JO (2003) Seasonal variations in production and development of leatherleaf fern leaves. Ann Appl Biol 143:235–243

Takahashi K, Mikami Y (2006) Effects of canopy cover and seasonal reduction in rainfall on leaf phenology and leaf traits of the fern *Oleandra pistillaris* in a tropical montane forest, Indonesia. J Trop Ecol 22:599–604

Thiers B (Continuously updated) Index herbariorum: a global directory of public herbaria and associated staff. New York Botanical Garden's virtual herbarium. http://sweetgum.nybg.org/science/ih/

Thomas H (2013) Senescence, ageing and death of the whole plant. New Phytol 197(3):696–711

USA National Phenology Network (NPN) (2017) https://www.usanpn.org/. Accessed 4 July 2017

van Schaik CP, Terborgh JW, Wright SK (1993) The phenology of tropical forests: adaptive significance and consequences for primary consumers. Annu Rev Ecol Syst 24:353–377

von Aderkas PV, Green PEJ (1986) Leaf development of the ostrich fern *Matteuccia struthiopteris* (L.) Todaro. Bot J Linn Soc 93(3):307–321

Wagner WH, Wagner FS (1977) Fertile-sterile leaf dimorphy in ferns. Gard Bull Singapore 30:251–267

Watkins JE Jr, Churchill AC, Holbrook NM (2016) A site for sori: ecophysiology of fertile–sterile leaf dimorphy in ferns. Am J Bot 103(5):1–11

Yoshida C, Takasu H (1993) Leaf life-span in some ferns of the Kii peninsula I. Acta Phytotax Geobot 44:59–66

Zalamea M, Gonzalez G (2008) Leaffall phenology in a subtropical wet forest in Puerto Rico: from species to community patterns. Biotropica 40(3):295–304

Chapter 19
Desiccation Tolerance in Ferns: From the Unicellular Spore to the Multi-tissular Sporophyte

Marina López-Pozo, Beatriz Fernández-Marín, Jose Ignacio García-Plazaola, and Daniel Ballesteros

19.1 Introduction

19.1.1 General Background About Plant Desiccation Tolerance

Plants are able to cope with environments in which the availability of liquid water within their tissues is severely restricted. Three main strategies are employed by plants to cope with this stress: escape, avoidance, and tolerance. In the first strategy, plants rapidly complete their life cycle, so that reproductive structures are developed before the plant dehydrates. The second one implies that the plant delays water loss and prioritizes the maintenance of cell turgor by imposing structural barriers to dehydration (cuticles, stomatal closure, enhanced boundary layer, etc.). For the third strategy, tolerance, the plant equilibrates its water content with that of the air by an organized restructuration of cell that prevents irreparable damage and allows resumption of normal metabolic activity after rewatering (Fernández-Marín et al. 2016).

Desiccation-tolerant (DT) plants can lose more than 90% of their relative water content and then resume normal functions after rehydration (Rascio and La Rocca 2005). Desiccation tolerance (DT) can be defined as the ability of some organisms to survive drying to below the absolute water content of around 0.08–0.10 g of water per g of dry mass ($H_2O\ g^{-1}$ DW) or lower, without suffering irreparable dam-

M. López-Pozo · B. Fernández-Marín · J. I. García-Plazaola
Department of Plant Biology and Ecology, University of the Basque Country (UPV/EHU), Vizcaya, Spain

D. Ballesteros (✉)
Comparative Seed Biology Group, Comparative Plant and Fungal Biology Department, Royal Botanic Gardens Kew, Richmond, UK
e-mail: d.ballesteros@kew.org

H. Fernández (ed.), *Current Advances in Fern Research*,
https://doi.org/10.1007/978-3-319-75103-0_19

age (Vertucci and Farrant 1995; Hong and Ellis 1996; Fernández-Marín et al. 2016). Nevertheless, one should be aware that any benchmark is relatively flexible after several factors have been considered. For example, these water content benchmarks can encompass drying at a wide range of relative humidity (RH) depending on the internal cell composition of the organisms (Ballesteros et al. 2017). All this considered, survival to the water contents achieved after equilibration with air at 50% RH at 20 °C is also used as reference for definition of a DT organism. This extent of drying can correspond to a drop in the water potential down to −100 MPa or even lower (Alpert 2006), although again these values are dependent on different cell composition. Finally, another proposed benchmark for DT is in the vicinity of the glass transition (Walters 2015). Among plants, DT is widespread in organs and structures related with dispersion and reproduction (Franchi et al. 2011; Gaff and Oliver 2013; Hoekstra 2005). Almost all photosynthetic species possess the capability to survive dehydration in their pollen, spores, or special organs such as rhizomes, seeds, etc. Indeed, desiccation sensitivity (DS) is the exception more than the rule in the seeds of angiosperm species. It is estimated that around 92% of seeds are tolerant to desiccation after complete maturation (orthodox seeds) (Wyse and Dickie 2017).

The presence of DT in photosynthetic tissues has a much more restricted distribution, being more common among species of algae, lichens, and bryophytes. It has been proposed that the initial development of DT in photosynthetic tissues was probably a crucial step in the colonization of land by early plants (Oliver et al. 2000). Within pteridophytes and angiosperms, this ability is rare (no gymnosperms are known to date) (Gaff 1980; Alpert 2000; Oliver et al. 2000). Two main groups of DT photosynthetic organisms can be distinguished according to the rate of water loss they can endure: "fully DT organisms" withstand very fast drying and base their tolerance mostly on constitutive mechanisms (i.e., most DT algae, lichen, and bryophytes) and "modified DT organisms" or "resurrection plants" refer to DT tracheophytes which rely mostly on inducible mechanisms and are thought to not be directly derived from primitive DT plants but from DS plants that had already lost DT in their photosynthetic tissues (i.e., the genetic information for DT was preserved in their genome although only expressed in reproductive organs, such as seeds) (Oliver and Bewley 1997).

19.1.2 General Mechanisms of Plant Desiccation Tolerance

During desiccation, but also in the dry state and throughout rehydration, DT cells must deal with two major threats: (1) mechanical stress due to changes in cell turgor and (2) chemical stress, which regards, on the one hand, the loss of native conformation of macromolecules and, on the other hand, an increase in oxidative stress that is particularly exacerbated in photosynthetic tissues (i.e., tissues containing chlorophyll) (Fernández-Marín et al. 2016). Coping with mechanical stress during dehydration implies three not mutually exclusive options (Rascio and La Rocca 2005):

(1) to keep cell volume (i.e., in some species water is replaced by low-molecular-weight osmoprotectants that allow preservation of cell volume), (2) to divide the vacuole into smaller vacuoles (or synthesize new small vacuoles – Georgieva et al. 2017) that increase mechanical stability of the cell, or (3) to produce flexible cell walls than can then follow bending of plasma membrane during desiccation without disruption of membrane wall interactions. The accumulation of specific types of proteins and small sugars and polyols seems to play indispensable roles in the maintenance of cellular and macromolecular structures upon desiccation.

Proteins and Amino Acids

The accumulation of high levels of late embryogenesis abundant (LEA) proteins is a common mechanism among DT plants also found in orthodox seeds. LEA proteins are highly hydrophilic and unstructured in aqueous solution but structured upon desiccation. Although their biochemical functions are still not completely understood, these proteins are thought to have important roles in the protection of nucleic acids, preservation of macromolecular structure (Close 1997), and glass formation (Farrant et al. 2015). LEA proteins are localized in the cytosol, the nucleoplasm, and the chloroplast and are synthesized in response to desiccation and abscisic acid (ABA) in vegetative tissues of DT plants (Bartels and Sunkar 2005). Although genes encoding LEA proteins are present also in DS plants, DT plants seem to accumulate higher amounts of LEA transcripts during dehydration (Juszczak and Bartels 2017).

The second important group of proteins are the early light-inducible proteins (ELIP) and ELIP-like proteins. These are thylakoidal pigment-binding proteins that have been found in DT angiosperms and bryophytes and that seem to play a key role in the protection and/or repair of the photosynthetic apparatus (Neale et al. 2000; Alamillo and Bartels 2001; Zeng et al. 2002).

The function of some amino acids as compatible solutes or as mobile nitrogen reserves for rehydrating tissues has also been suggested (revised in Dinakar and Bartels 2013).

Sugars

Nonreducing sugars have a very important role as stabilizers of membranes and proteins in the dry state (Crowe et al. 1992). Due to their propensity to form a glass through hydrogen-bonding interactions, they are thought to play a fundamental role in cytoplasm vitrification (see Sect. 19.1.2.3) (Hoekstra et al. 2001). Currently, it is acknowledged that besides sugars, other molecules such as proteins (including LEA proteins) play a pivotal role in intracellular glass formation (Buitink and Leprince 2008; Walters et al. 2010). Nonreducing sugars allow the removal of the closely associated water from proteins by satisfying the hydrogen-bonding requirement of polar groups preserving their functionally correct folding

(Hoekstra et al. 2001). Thus, sugars act as water substitutes on the surface of dried proteins. Soluble sugars seem to play important additional roles as a rapid source of energy during rehydration (Georgieva et al. 2017). The major sugar involved in these protective mechanisms is species-specific although sucrose is the most frequently found among DT plants (Farrant et al. 2007). Sucrose and raffinose accumulate during desiccation mainly in angiosperms (Scott 2000; Peters et al. 2007; Georgieva et al. 2017), whereas trehalose occurs predominantly in nonflowering plants (Liu et al. 2008; Pampurova and Van Dijck 2014). Constitutively high amounts of trehalose, sucrose, and glucose have been found in DT species of spike mosses (Yobi et al. 2013).

Slowing Down of Metabolism Under Desiccation: The Vitrification of Tissues and the Glassy State

Upon dehydration, the viscosity of the internal cellular compartments increases, and solutes are concentrated within the tissues. At RH ranges between 30 and 60%, the aqueous mixtures, such as cells, become more viscous and molecules become densely packed, reducing drastically molecular mobility up to a point that the cytoplasm resembles a solid (Buitink and Leprince 2008; Fernández-Marín et al. 2013; Leprince and Buitink 2015; Walters 2015; Ballesteros et al. 2017). Such types of solids are known as amorphous solids and have different thermodynamic properties than crystalline solids (Walters et al. 2010). The change from fluid to amorphous solid is known as vitrification (glass transition when the change is from solid to fluid), and the resulting amorphous solid is often called a glass. Glasses are known to decrease detrimental reactions, to increase the stability of enzymes, and to prevent conformational changes of proteins (Buitink and Leprince 2008). The formation of intracellular glasses is indispensable to survive the dry state, but glass formation itself is not a mechanism that initially confers the tolerance to desiccation during drying (Buitink and Leprince 2008; Walters et al. 2010).

Plant cellular glasses have been studied through different calorimetric and spectroscopic techniques (e.g., differential scanning calorimetry (DSC)) or electron paramagnetic resonance (EPR) and mainly in seeds because of its importance in longevity during storage (Walters et al. 2010). Recently, a different approach using dynamic mechanical thermal analysis (DMTA) has been used for the characterization of the mechanical properties and molecular mobility within dry seed glasses (Walters et al. 2010; Ballesteros and Walters 2011) and has provided first evidence for the absence of enzymatic reactions in dry photosynthetic tissues of bryophytes (Fernández-Marín et al. 2013). Still, many questions remain open about intracellular glasses and how (and for how long) they confer cell stability and longevity in dry plant tissues. This represents a gap of knowledge to be filled that could be of noticeable relevance to understanding DT and longevity in the dry state of plant germplasm.

Coping with Oxidative Stress

The chloroplast is the main source of reactive oxygen species (ROS) upon desiccation. ROS production can be exacerbated when desiccation occurs under strong irradiance. Consequently, the first barrier to prevent oxidative stress is the reduction of light energy reaching the photosynthetic apparatus. The presence of chlorophyll during desiccation represents a difference against DS plants in which damage upon desiccation includes the irreparable loss of photosynthetic pigments and thylakoid structure (Juszczak and Bartels 2017). During the loss of turgor, many DT species curl or fold their leaves exposing the abaxial surface which is frequently covered with hairs, scales, or increased content of anthocyanins that, in sum, efficiently reduce the interception of light by chlorophyll (Lebkuecher and Eickmeier 1991; Alpert 2000). The next step in photoprotection relies on the switching off of the photosynthetic apparatus and the efficient dissipation of light energy still absorbed by chlorophylls (revised in García-Plazaola et al. 2012, Fernández-Marín et al. 2016). Desiccation-induced deactivation of PSII seems to be an intrinsic protective mechanism of DT species since DS plants maintain activity even at very low water content, after photosynthetic electron transport has already been stopped (Nabe et al. 2007).

Antioxidant mechanisms play such a fundamental role in DT plants, that the survival after long storage in the desiccated state and after subsequent rehydration depends on their efficiency to counteract ROS production (Kranner et al. 2002). Despite the general increase in the antioxidant response upon desiccation (Yahubian et al. 2009), two molecules play outstanding roles in the DT strategy: the hydrophilic and ubiquitous glutathione (GSH_{tot}) and the lipophilic and chloroplast-located zeaxanthin (Z). The relevance of these two molecules is evidenced by their conserved response: their synthesis and accumulation are triggered by desiccation among species of phylogenetically diverse taxa, e.g., lichens, bryophytes, and angiosperms. The proportion of reduced glutathione to GSH_{tot} is related to the survival of DT organs, including seeds (Kranner et al. 2002; Nagel et al. 2015). Besides its role as an antioxidant, reduced glutathione is thought to interact with proteins and protect them from irreversible formation of disulfide bonds upon water loss (Kranner et al. 2008). Synthesis of Z is activated in DT plants by desiccation, even in the absence of incident irradiance (Fernández-Marín et al. 2009, 2010, 2011, 2013). Z can play up to three different protective roles in DT photosynthetic tissues: (1) it is involved in the thermal dissipation of energy, (2) it efficiently quenches triplet chlorophyll and singlet oxygen, and (3) it is thought to stabilize thylakoid membranes during desiccation and enhance the recovery of photochemical efficiency during rehydration (Alamillo and Bartels 2001; Deltoro et al. 1998; Gill and Tuteja 2010; Havaux et al. 2007; Kranner et al. 2002, 2008; Fernández-Marín et al. 2009, 2010, 2011, 2013).

19.1.3 Why Study Tolerance to Desiccation in Pteridophytes?

Ferns represent a highly valuable model group for the study of plant desiccation tolerance, as they have unique characteristics to study plant adaptations for the conquest of land and the mechanisms that allowed this colonization. Ferns have a strategic evolutionary position in the evolution to land, between bryophytes and seed plants, and their mechanisms of DT seem to be intermediate between those of "fully DT plants" (i.e., bryophytes) and those of flowering plants (Oliver et al. 2000). This intermediate position is, for example, expressed in terms of the vascular system and stomatal regulation, which present two different ways in the independence of plants from water (McElwain 2011; Cai et al. 2017). In this context, ferns have recently been used to demonstrate that ABA regulation of stomata evolved much earlier than the divergence of ferns and seed plants, as different fern species have shown ABA-induced closure of stomata, pointing to alterative and complex pathways for ABA-induced stomatal closure in plants (Cai et al. 2017; Hõrak et al. 2017). This characteristic makes ferns valuable models for the study of plant regulation of DT responses.

Another interesting characteristic of ferns is that they present three different life forms in their life cycle that can be used as diverse models: the unicellular spore (chlorophyllous or non-chlorophyllous), the simple multicellular gametophyte, and the complex multicellular and multi-tissular sporophyte. These life forms not only have different cellular and tissue complexity but also present different DT responses and mechanisms (described in the following sections). Sporophyte and gametophyte could resemble the behavior of both vascular and nonvascular plants, respectively. The lack of complexity, cuticle, and stomata in gametophyte tissues could allow an adaptation to water loss, due to the inability to control the exit and entry of water that is totally dependent on the ambient humidity. In turn, the sporophyte has full control over water loss because of the development of complex tissues, cuticle, and stomata.

In addition, spores offer a simple model to understand the mechanisms responsible of seed DT and longevity in the dry state and, in turn, what changes are necessary at the cellular level to lose this property. Both propagules, seeds and fern spores, differ in their structure and ontogeny, but they carry out the same function: they are survival units, suitable for dispersion, acting as population backup in soil banks, and often exposed to extreme environmental conditions but being able to survive them (Page 2002). Moreover, they are related in many physiological aspects, such as the induction or inhibition of germination by red or far-red light, the loss of water during maturation, or the role of gibberellins in germination (e.g., Hoekstra et al. 2001). Given the unicellular character of spores, the study of these is simpler than that of seeds (that contain a hypocotyl, cotyledon(s), embryo with plumule and radicle) where each part will act or have some mechanism of DT. Additionally, chlorophyllous spores can be understood as a miniature photosynthetic tissue whose compression would aid in the comprehension of more complex tissues.

19.2 Desiccation Tolerance in Fern Spores

The ability to survive extreme desiccation in vascular plants is generally expressed in reproductive cells, such as fern spores, pollen, and seeds (see Sect. 19.1.1). However, the extent of DT in fern spores, the stage when this tolerance is acquired, and the physical and biochemical mechanisms that confer tolerance to desiccation have been understudied in comparison to other plant propagules (e.g., seeds and pollen). This section will review the current knowledge in DT of fern spores and will discuss possible directions for future research.

19.2.1 Sporogenesis, Spore Dispersal, and Fern Spore Water Content

Ferns produce spores by meiosis in the sporangia in a process known as sporogenesis. During sporogenesis, after meiosis is completed, the fern spore goes through a maturation stage where plastids and storage reserves accumulate, including the precursors of proteins analogous to seed LEA proteins (Raghavan and Kamalay 1992; see Sects. 19.1.2 and 19.2.4). The characteristics of this phase suggest that the maturation stage of the fern spores could be similar to that occurring in DT seeds (Angelovici et al. 2010). However, most research published on fern spores has focused in describing the ultrastructural changes during the early stages of sporogenesis (e.g., Marengo 1979; Brown and Lemmon 2001a, 2001b), and little is known about the maturation phase of the spore and the changes in water content that may occur in the spores and if this is actually the stage where fern spores acquire any tolerance to desiccation. After sporogenesis, when the sporangia is mature and the environment is dry enough, evaporation of water from the annular cells of the sporangia induces sufficient stress to cause an abrupt dehiscence, which ends up in the release and dispersal of the spores (Noblin et al. 2012).

Fern spores are classified as green (chlorophyllous) or nongreen (nonchlorophyllous) depending on their color at maturity (Lloyd and Klekowski 1970). However, this traditional naming is not accurate since some chlorophyll-containing spores can visually appear as nongreen when the cover of the spore is not transparent (Sundue et al. 2011). Green spores usually have higher water contents than nongreen spores (Lloyd and Klekowski 1970). For example, according to water sorption isotherms calculated for several species at 25 °C, water contents of spores equilibrated at 50% RH vary from ca. 0.09 to ca. 0.04 g water/g dry mass for green spores and nongreen spores, respectively (Ballesteros and Walters 2007a; Ballesteros et al. 2017). The causes for the difference of water content between green and nongreen spores are not completely known, but it is likely that it is related to their chemical composition. For example, it is known in seeds that the moisture content at a particular RH varies according to the lipid composition (Vertucci and Roos 1990), and

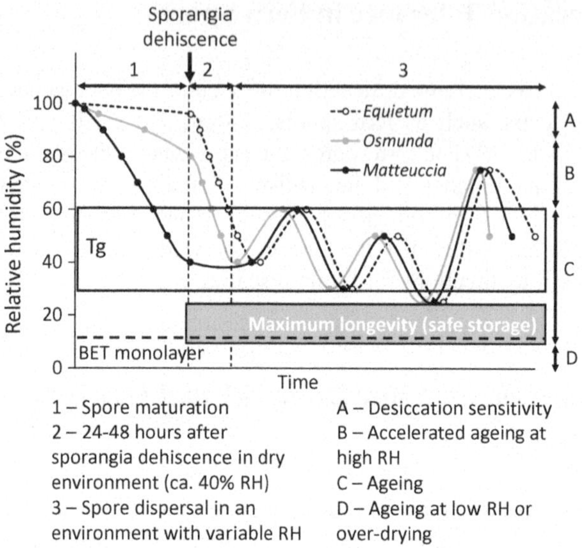

Fig. 19.1 The Moisture contents (in terms of equilibrium relative humidity, RH) of diverse green or chlorophyllous fern spores during spore maturation in the sporangia (1), during the initial hours after sporangia dehiscence (2), and during a hypothetical scenario of long-term dispersal with environmental RH changes (3). The moisture contents at which fern spore desiccation sensitivity would be expressed (A) or which are subjected to accelerated aging at high RH (B), aging (C), and aging at low RH or "overdrying" (D) are also indicated as suggested by Ballesteros et al. (2017)

nongreen spores have 10–40 times larger lipid contents than green spores (Ballesteros and Walters 2007b; Ballesteros et al. unpublished).

Measurements of water content for "freshly harvested" spores have usually been made in green spores collected 1 or 2 days after sporangia dehiscence or in nongreen spores collected about 1–5 days after sporangia dehiscence (e.g., Farrant et al. 2009, Li et al. 2010; Li and Shi 2014, 2015; Mikula et al. 2015; Lopez-Pozo et al. unpublished). Freshly harvested spores are viable, and water contents range from 0.11 to 0.14 g water/g fresh mass in green spores of *Osmunda* sp. and from 0.04 to 0.07 g water/g fresh mass in diverse nongreen spores or in the green spores of *Matteuccia struthiopteris*. These water contents correspond to those of spores equilibrated at 30–80% RH and 25 °C (Ballesteros and Walters 2007a; Ballesteros et al. 2017, Fig. 19.1). The exception is the spores of *Equisetum* sp. for which water contents above 0.50 g water/g dry mass (which correspond to the water contents of *Equisetum* sp. spores equilibrated at >96% RH at 25 °C, Fig. 19.1) have been measured in diverse species (Gabriel y Galan and Prada 2010; Lopez-Pozo et al. unpublished).

After sporangia dehiscence and spore dispersal, both green and nongreen spores can remain in the sporangia for several days or months, fall around the mother plant, or travel by air in very large distances (Farrar 1976; Parris 2001; Wolf et al. 2001; Page 2002; Geiger et al. 2007). In these scenarios, they may be exposed to (and tolerate) highly stressful conditions, including low and high RH (Fig. 19.1), freezing temperatures, fast winds at high altitudes, and UV exposure (Page 1979, 2002). After dispersal, it is likely that spores equilibrate their internal water content rapidly, in a few hours, depending on the ambient RH (Ballesteros and Walters 2007a). This capacity of fern spores to quickly modulate their internal water content to low levels and their tolerance to these low water contents seems to be key in their ability to disperse and tolerate the environmental stressful conditions that they may experience (Page 2002).

19.2.2 Variation of Desiccation Tolerance and Longevity Among Fern Spores

DT of nongreen spores has typically been described as similar to that of orthodox seeds and other DT organisms (Dyer 1979; Ballesteros 2010; Walters et al. 2005; Ballesteros et al. 2012). In contrast, there was a widely held belief that green spores had limited DT (Lloyd and Klekowski 1970; Pérez-Garcia et al. 1994; Gabriel y Galan and Prada 2010; Ibars and Estrelles 2012). This belief was based on conclusions of pioneer studies that suggested that the short longevity of green spores was related to their high water content, respiration rate, and catalase levels (Okada 1929; Hauke 1978) as well as to the lack of a spore wall protecting against drying (Wollersheim 1957). However, conflicting evidence suggests that green spores can tolerate substantial water loss, as they tolerate drying at low RHs (1–20%) to moisture contents <0.08 water/g fresh weight (Lebkuecher 1997; Pence 2000; Hoekstra 2005; Li and Shi 2014, 2015; Mikula et al. 2015). Recently, in agreement with these late reports, it has been shown that both green and nongreen fern spores possess a similar and remarkable level of DT (Ballesteros et al. 2017). Fern spores from diverse unrelated species (three green spore and two nongreen spore species) survived the immediate effects of extreme desiccation (i.e., initial drying to 1% RH). However, survival in the desiccated state was temporary, especially under the most severe desiccation treatments (RH <5%). Longest survival times were observed in spores placed at about 10–25% RH, independently of the water contents that the spores had at such conditions. This response is similar to what has been observed for other fern spores, orthodox seeds, DT pollen, and various DT organisms (e.g., Vertucci and Roos 1990; Buitink et al. 1998; Walters et al. 2005).

Though response to RH during storage is similar among fern species (Ballesteros et al. 2017), the kinetics of deterioration in the dry state vary considerably among species. Generally, green spores present shorter life-spans than nongreen spores, ranging from a few days to 1 year for green spores and from a few months to several decades in nongreen spores (Lloyd and Klekowski 1970). It is likely that the short longevity of green spores is related to their particular chemical and cellular composition (e.g., thin and transparent walls and the presence of chloroplasts in the dry cytoplasm). Analogously, the chlorophyllous seeds of *Salix nigra* deteriorate fast in the dry state due to photooxidative damage of the thylakoid membranes mediated by free radicals and ROS (Roqueiro et al. 2010). However, we may find nongreen spores of particular species with shorter longevity than the green spores of some species. For example, the non-green spores of *Culcita macrocarpa* show shorter longevity at the optimal RH for storage than the green spores of *Matteuccia struthiopteris* (Ballesteros et al. 2017), indicating that diverse physical and biochemical aspects of the fern spore may also be playing an important role in the differences in longevity among spores.

19.2.3 Structural and Biophysical Aspects of Desiccation Tolerance and Longevity in Fern Spores

Diverse definitions can be found to describe the extent of the desiccation stress that defines "desiccation-tolerant" organisms (see Sect. 19.1.1; Ballesteros et al. 2017). However, what seems to be common to all of the definitions is that DT organisms survive the compressive forces and cell volume reduction before glasses (see Sect. 19.1.2.3) are formed (Leprince and Buitink 2015; Walters 2015; Ballesteros et al. 2017). If green spores had limited DT as initially thought (see Sect. 19.2.2), one would expect that the most sensitive green spores, which are released at high water contents (e.g., *Equisetum* sp., see Sect. 19.2.1), would die before glasses are formed (A in Fig. 19.1). However, the cytoplasm of nongreen and green fern spores enter into a glassy state between 30 and 60% RH (Hoekstra 2005; Ballesteros et al. 2017), and all the species tested (including *Equisetum* sp.) survived this transition (Hoekstra 2005; Ballesteros et al. 2017).

Longest survival times for fern spores are observed when they are placed at about 10–25% RH, hence when they are in the glassy state (Walters et al. 2005; Ballesteros et al. 2017). At moisture contents above the RH that provides longest longevity, a timed response damage is expressed in fern spores, which is faster as the RH increases (Ballesteros et al. 2017; Fig. 19.1). This type of deterioration is classically referred to as aging (C in Fig. 19.1) and accelerated aging (B in Fig. 19.1) in the literature on seed longevity, and deterioration kinetics in fern spores are consistent with modeled behavior in seeds (Ellis and Roberts 1980; Ballesteros et al. 2017). Below the RH that provides longest longevity, desiccation damage occurs in DT cells, and this is also expressed as a time-dependent response (D in Fig. 19.1). Although in this case deterioration is faster as spores are moved to lower RH, it has been suggested that this deterioration could also be considered as aging (Ballesteros et al. 2017). However, as the aging kinetics at water contents above and below the RH that provides longest longevity have opposite responses to water content, it is suggested that the mechanisms of damage and protection involved in these conditions may be different (Ballesteros et al. 2017).

One of the explanations for the damage that occurs in fern spores at <10–25% RH is that they are dried below the BET monolayer (Ballesteros and Walters 2007a; Ballesteros et al. 2017). The BET monolayer is a parameter calculated from the

Brunauer–Emmett–Teller (BET) model which is used to explain the physical adsorption of water on the solid surface of biomolecules of the cell cytoplasm of diverse organisms, including fern spores (Ballesteros and Walters 2007a). The BET monolayer describes the water content at which all sites at the adsorbent surface are filled and generally has been indicated as a good marker of maximum stability during dry storage (Passot et al. 2012; Rahman 2009; Roos and Drusch 2015; Roussenova et al. 2010). It could be possible that the removal of the water that forms the BET monolayer destabilizes the physical structure of the cytoplasm, allowing also a greater attack of free radicals and ROS. These hypothetical situations could lead to death over the time as suggested for orthodox seeds (Mira et al. 2010; Walters et al. 2010). However, the mechanisms of fern spore deterioration in the dry state are still conjectural, and the complexity in structure and motion within the glass of fern spores conferring different responses to desiccation and longevity still need to be ascertained (Ballesteros et al. 2017).

19.2.4 Accumulation of Proteins

LEA Proteins

Studies carried out in *Onoclea sensibilis* determined that there is an accumulation of mRNAs related to survival, growth maturation, and germination during the sporogenesis (Raghavan and Kamalay 1992). These mRNAs detected in fern spores by Raghavan and Kamalay (1992) were related to members of group-3 LEA sequences described by Dure et al. (1989) from angiosperm embryos during seed maturation and desiccation. A similar process has also been shown during pollen development in angiosperms (Stinson et al. 1987). The sequences in *O. sensibilis* were spore-specific, and the accumulation was initiated only after meiosis and as long as the spores were mature (desiccated) but were absent in the gametophyte. For the case of seeds during the maturation phase, there is a decrease in cell water potential and a synthesis of LEA proteins that protect embryonic organs from this desiccation (Curry et al. 1991; Angelovici et al. 2010). Thus, it could be expected that spore maturation could suffer similar process and LEA proteins would play the same or an analogous role.

During spore germination, many other proteins are degraded in the cytoplasm and in thylakoid membranes, some of which are related to their maintenance. This is the case of 22 kDa protein of thylakoid membranes isolated from spores of *Osmunda japonica*. Before germination, high levels of this protein are found, but when germination starts the amount of protein decreases. Something similar was observed in spores of *Adiantum capillus-veneris* (non-chlorophyllous spores) (Minamikawa et al. 1984) whose spore proteins may be a homolog of ABA-inducible, desiccation-related, or LEA proteins. Several studies have reported degradation of proteins that could be related to DT in fern spores (Raghavan 1991; DeMaggio and Stetler 1980; Paless et al. 1984). Currently, the specific role of the 22 kDa protein in chloroplasts remains unknown (Inoue et al. 2000).

Other Proteins

Within gymnosperms and angiosperms seeds, vicilins and legumins are the two major classes of storage proteins, which have the function of nutrition once the seed germinates (Shutov et al. 2003). Shutov et al. (1998) concluded that these seed-storage proteins presented by phanerogamic species were involved in cellular processes of desiccation and rehydration and *Physarum polycephalum* presents similar proteins that allowed the acquisition of cellular desiccation tolerance (Lane et al. 1991). Thus, it may be possible that they are involved in desiccation tolerance in tissues. *Matteuccia struthiopteris* presents an intermediate vicilin-like protein (Kakhovskaya et al. 2003; Shutov et al. 1998), with features of both vicilins and legumins. These structural features resemble those of a non-spermatophyte (cryptogamic) "primitive" storage protein; thus, this protein probably represents an ancestor of the spermatophyte storage globulins (Shutov et al. 1998). The question exists whether this protein is related with storage or with desiccation tolerance in fern spores. On the one hand, it has been proposed that fern spores may also function as a storage tissue (for more details, see Schallau et al. 2008), but the idea that this type of protein descended from germin- and spherulin-like ancestors involved in various cellular desiccation processes (Shutov and Bäumlein 1999; Shutov et al. 1998; Wohlfarth et al. 1998) supports that it is related with DT and not with storage functions.

It seems that the role of these proteins is similar to LEA proteins (Baker et al. 1988). A similar role has been proposed for a variety of other glycine-rich proteins in mature seed embryos (Galau et al. 1987; Mundy and Chua 1988).

Osmotic Adjustment: Proline Accumulation in Cells Exposed to Stress Conditions

As occurs in tissues of higher plants exposed to osmotic stress, dry chlorophyllous spores of *Equisetum arvense* show high levels of arginine and proline. Also, spores of *O. japonica* present high contents of these amino acids, but they disappear in the gametophyte. Besides, it has been described that non-chlorophyllous spores of *Adiantum capillus-veneris* contain proline in the dry state (Minamikawa et al. 1984).

Protection of Fern Spores Against Oxidative Processes During Desiccation and Life in the Dry State

During the maturation phase of fern spores and their desiccation after dispersal the production and accumulation of ROS is likely. In the dry state, the repair of the oxidative damage caused by ROS may be limited by the formation of the glass and the reduced molecular mobility allowed in such state (see Sects. 19.1.2.3 and 19.2.3). Decreased or suppression of mitochondrial and chloroplastic activities during desiccation is the best way to reduce ROS generation, even at low water contents (Leprince et al. 2000). Antioxidative defenses play a major role once free radicals have already been formed. Two main antioxidant enzymes have been found in the desiccated green spores of *Equisetum arvense*: ascorbate peroxidase (APX) and dehydroascorbate reductase (DHAR). Both enzymes can be localized in the cytoplasm and chloroplast (Zhao et al. 2015) and take part in the glutathione-ascorbate cycle. Furthermore, a chloroplast drought-induced stress protein (CDSP32) has been found in these spores. This protein has two thioredoxin (TRX) modules, which are involved in the protection of the photosynthetic apparatus against oxidative damage. Hence, it could be suggested that these antioxidants play the main role in the ROS homeostasis of the chlorophyllous spores of *E. arvense* (Zhao et al. 2015). Other spores also exhibit enzymes of the glutathione-ascorbate cycle. In the dry green spores of *Osmunda cinnamomea*, all the enzymes of this cycle were found (Suo et al. 2015).

Photoprotection of Green Spores During Desiccation

When a chlorophyllous tissue starts to lose water, photosynthesis declines, but the total irradiance intercepted by the chlorophyll is essentially the same. In this context, the major threat is the formation of ROS by excited chlorophyll molecules and the consequent oxidative deterioration (Heber and Shuvalov 2005; Farrant et al. 2007). Thus, the simplest way to protect the photosynthetic tissue is to reduce the amount of light absorbed by chlorophyll molecules through several mechanisms as have been commented in Sect. 19.1. In green spores, *M. struthiopteris* has a brown coat that filters light and could act as the first barrier to reduce light absorbance; however, other spores have a transparent coat. In addition to a dark coat, quenching of chlorophyll could be another photoprotection mechanism in green spores. In this sense, Lebkuecher (1997) showed that the spores of *Equisetum hyemale* reduced the photochemical efficiency of PSII (Fv/Fm) during the first week of storage at 2% RH without major changes in spore viability and recovered their Fv/Fm after rehydration. This could be considered a type of photoprotection as observed in the DT algae *Trebouxia asymmetrica* (Wieners et al. 2012). The decrease in photochemical activity of PSII can be due to various mechanisms: (1) desiccation-induced detachment of the light-harvesting complex from PS II (LHCII) (Schreiber and Armond 1978), (2) transference excess of excitation energy from PS II to PS I associated with phosphorylation of LHC II (Williams and Allen 1987), and (3) thermal energy dissipation mediated by the VAZ cycle (Demmig et al. 1987; Gilmore et al. 1995). All of these mechanisms are reversible (deactivate) after, at least, 24 h of rehydration.

19.3 Desiccation Tolerance in Gametophytes

19.3.1 Presence and Extent of Desiccation Tolerance in Gametophytes

Most fern gametophytes are photosynthetic from the initial steps of development. Lacking a vascular structure, gametophytes are forced to be poikilohydric. However, their structure is not the fragile prothallus as described in the literature, and gametophytes of many species can establish in xeric places. Furthermore, gametophytes are frequently perennial and persist for years (Johnson et al. 2000), encountering successive periods of desiccation and rehydration during their long life-span. As a consequence of such poikilohydric commitment, it is considered that DT has to be much more frequent among gametophytes than in sporophytes (Pittermann et al. 2013; Watkins et al. 2007). In fact, the first experimental studies on DT in gametophytes date back to more than 100 years ago, when Pickett (1913) demonstrated that prothalli of the limestone-colonizing North American fern *Camptosorus rhizophyllus* were able to completely recover after desiccation. Since then, reports on DT in gametophytes have been scarce, describing its presence in several other species (Kappen 1965; Ong and Ng 1998; Kappen and Valladares 2007). Only recently more extensive surveys on tropical species (Pittermann et al. 2013; Watkins et al. 2007) have confirmed the intuitive belief that DT is probably widespread among fern gametophytes. These studies concluded that the degree to which gametophytes can withstand desiccation seems to be strongly linked to habitat preferences with species inhabiting the most xeric places also being the most DT. However, this general rule does not imply that gametophytes are necessarily DT, as indicated by the extreme sensitivity to desiccation of gametophytes from species growing in moist ecosystems such as the Mexican cloud forest (Riaño and Briones 2015) or European temperate environments (Kappen 1965). Examples of genera for which DT has been described in gametophytes can be found across a wide taxonomic diversity that includes most of the phylogenetic fern clades. Being a polyphyletic character, gametophyte habitat becomes the main factor determining DT. Specifically DT has been tested and confirmed in the following families and genera (Diamond et al. 2012; Kappen 1965; Ong and Ng 1998; Pickett 1913; Pittermann et al. 2013; Watkins et al. 2007): Aspleniaceae (*Asplenium*, *Ceterach*), Athyriaceae (*Diplazium*), Cibotiaceae (*Cibotium*), Davalliaceae (*Davallia*), Dennstaedtiaceae (*Dennstaedtia*), Hymenophyllaceae (*Trichomanes*), Lomariopsidaceae (*Nephrolepis*, *Cyclopeltis*), Polypodiaceae (*Campyloneurum*, *Drynaria*, *Pyrrosia*, *Phlebodium*, *Microgramma*, *Polypodium*), Pteridaceae (*Adiantum*, *Cheilanthes*, *Pteris*, *Pellaea*, *Pityrogramma*, *Vittaria*), and Thelypteridaceae (*Thelypteris*).

19.3.2 Mechanisms That Confer Tolerance to Desiccation in Gametophytes

DT is common in gametophytes of mosses and liverworts, and presumably the same mechanisms employed by bryophytes (see Sect. 19.1) can be used by ferns to withstand desiccation. However, there is an almost complete absence of physiological studies on DT in fern gametophytes. Among the few studied mechanisms, morphology seems to play a prominent role on slowing the rate of water loss, with the more complex structures retaining water for longer (Watkins et al. 2007). Regulation by ABA is also involved in some species, as demonstrated by the increase on survival rate after desiccation when gametophytes were cultured on medium containing ABA (Pence 2000). Interestingly, once desiccated, gametophytes can survive liquid nitrogen, allowing the development of protocols for cryopreservation (Pence 2000, Chap. 11). Furthermore, in the field, overwintering gametophytes can survive much lower temperatures than the respective sporophyte (Sato 1982).

19.3.3 Photoprotection of Gametophytes During Desiccation

From the initial steps of development, most fern gametophytes are photosynthetic but very little is known about their photosynthetic responses (Hagar and Freeberg 1980; Johnson et al. 2000; Watkins et al. 2007). Gametophytes, in comparison with their sporophytic counterparts, can be considered as shade plants, with a limited capacity for light acclimation and a pigment composition typical of low light grown plants (Fernández-Marín et al. 2012). For those species in which Fv/Fm has been measured (Johnson et al. 2000; Watkins et al. 2007, Fig. 19.2), control values are in

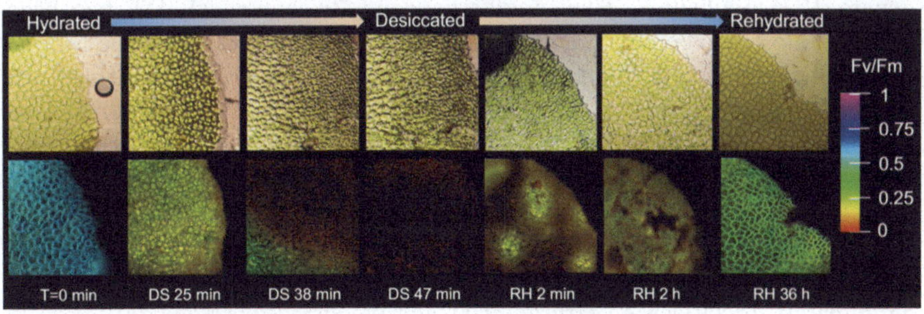

Fig. 19.2 Deactivation/activation of photochemical efficiency (measured as Fv/Fm) during a cycle of desiccation and rehydration in gametophytes of *Davallia canariensis*. Upper panels show images taken with the optical microscope and bottom panels the equivalent images of Fv/Fm captured with the imaging PAM microscopy fluorometer

the range 0.6–0.75, similar to gametophytes of mosses and liverworts, indicating that no major photoinhibition occurs under nonstress conditions. In addition, our lab has recently measured that some species, such as *Davallia canariensis*, are able to recover Fv/Fm after a dehydration-hydration cycle (Fig. 19.2, based in Lopez-Pozo et al. unpublished).

19.4 Desiccation Tolerance in Sporophytes

19.4.1 Presence and Extent of Desiccation Tolerance in the Leaves or Whole Plants in Some Species

In hydric terms, most fern sporophytes, with well-defined vascular tissue, cuticle, and stomata, can be considered as homoiohydric. Some of them are DT, but function as typical vascular plants when water is available. Others, such as *Mohria caffrorum*, produce two types of fronds: DT in the dry and DS in the rainy season (Farrant et al. 2009). The exception are the filmy ferns (family Hymenophyllaceae), whose leaves are formed by one layer of cells, lacking epidermis and stomata. This family, which comprises more than 600 species (Dubuisson et al. 2009), in some way has returned to the poikilohydric strategy exemplified by bryophytes (Proctor and Tuba 2002). In the absence of a specific evaluation for each species, it seems likely that most, if not all, members of this family are to some extent DT (Proctor 2003, 2012). Based on this assumption, Porembski (2011) estimated that the number of DT fern species should be between 200 and 1200, with the latter being more likely. Given that c.a. 12,000 fern and lycopod species have been described and accepted (Christenhusz and Byng 2016), the presence of vegetative DT in fern sporophytes could be as frequent as 10%, a proportion at least tenfold higher than in angiosperms and much higher than previous estimations of DT frequency around 1% for fern sporophytes (Oliver et al. 2000; Proctor and Pence 2002). Apart from Hymenophyllaceae (genera *Hymenophyllum*, *Hymenoglossum*, and *Trichomanes*), DT has been described mostly in epiphytic and rock-colonizing species including the following families and genera (Gaff 1977, 1987; Hietz 2010; Porembski 2011; Tuba and Lichtenthaler 2011): Aspleniaceae (*Asplenium*, *Ceterach*, *Pleurosorus*), Dryopteridaceae (*Dryopteris*, *Elaphoglossum*), Polypodiaceae (*Ctenopteris*, *Drynaria*, *Loxogramme*, *Melpomene*, *Pecluma*, *Polypodium*, *Phymatosorus*, *Platycerium*, *Pleopeltis*, *Pyrrosia*), Pteridaceae (*Actiniopteris*, *Adiantum*, *Cheilanthes*, *Doryopteris*, *Hemionitis*, *Notholaena*, *Pellaea*, *Paraceterach*, *Vittaria*), Schizaeaceae (*Anemia*, *Mohria*, *Schizaea*), Selaginellaceae (*Selaginella*), Tectariaceae (*Arthropteris*), and Woodsiaceae (*Woodsia*).

19.4.2 Mechanisms of Desiccation Tolerance in Fern Sporophytes

Mechanisms of DT in ferns are sometimes constitutively expressed, while others require a certain time for their activation, differentiating between constitutive DT (CDT) and inducible DT (IDT) (Stark et al. 2013). Most fern sporophytes behave as IDT, in the sense that, in general, a rapid desiccation causes more extensive damage than a slower one. Exceptions are filmy ferns that while lacking epidermis display a DT strategy (see Sect. 19.4.1), upon being exposed to rapid and continuous cycles of desiccation and rehydration. The importance of CDT mechanisms is evidenced in this group by the existence of a little variation (3–4%) on proteome upon desiccation (Garces et al. 2014). In species with IDT, ABA plays a fundamental role in the regulation of the expression of desiccation responsive genes (Wang et al. 2010), and the external application of this hormone increases DT as has been shown in *Polypodium virginianum* (Reynolds and Bewley 1993). To allow enough time for the activation of IDT, ferns employ mechanisms that slow down water loss; this is, for example, the role of the epidermal scales in *Pleopeltis polypodioides* that prevent rapid water loss from desiccating fronds (John and Hasenstein 2017).

An unavoidable consequence of water loss is cell and tissue shrinkage. To prevent mechanical damage, DT sporophytes have to be able to fold and unfold fronds in a plastic and organized way. The presence of arabinose-rich pectins and proteins makes cell walls more plastic (Moore et al. 2013). Besides, it has been described in *Polypodium polypodioides* that the rapid synthesis and degradation of dehydrins in response to desiccation are probably involved in the mechanical properties of cell walls (Layton et al. 2010).

Fern sporophytes also respond to dehydration by osmotic adjustment that in *Pleopeltis pleopeltifolia* is achieved by an increase in solute content (sugars) (Voytena et al. 2014), while in *Selaginella tamariscina*, sugars and proline contribute to the osmotic adjustment (Wang et al. 2010). A more detailed characterization of carbohydrate accumulation in *Mohria caffrorum* reported the presence of high amounts of sucrose, raffinose family oligosaccharides, and cyclitols (Farrant et al. 2009).

19.4.3 Role of Photoprotection Mechanisms in Desiccation Tolerance in Fern Sporophytes

Most fern species show a certain chlorophyll loss (Voytena et al. 2014; Wang et al. 2010; Tausz et al. 2001; Kavitha and Murugan 2016) that can be as high as 75% in the case of *Pleopeltis pleopeltifolia* (Voytena et al. 2014). Even in this kind of species that downregulate chlorophyll content, one of the main sources of stress in the desiccated state is the generation of oxidative damage by photoactivated chlorophyll. Thus, DT ferns need to activate a set of photoprotective responses to prevent damage to photosynthetic tissues during desiccation. The simplest and most effective way to prevent such photooxidative stress is to reduce photosynthetic surface exposed to illumination. In most DT fern sporophytes, this is simply achieved by the frond curling or shrinkage, as has been characterized in a wide representation of species such as *Selaginella lepidophylla* (Brighigna et al. 2002; Lebkuecher and Eickmeier 1991), *Hymenophyllum dentatum* and *Hymenoglossum cruentum* (Flores-Bavestrello et al. 2016), *Ceterach officinarum* (Fernández-Marín et al. 2009), *Pleopeltis pleopeltifolia* (Voytena et al. 2014), or *Polypodium polypodioides* (Helseth and Fischer 2005; Muslin and Homann 1992). In some of these species, such as *M. caffrorum* and *C. officinarum*, this strategy is reinforced with chlorophyll masking by highly reflective abaxial scales (Farrant et al. 2009).

As a second line of defense, in the desiccated state, fern photosynthetic apparatus enter into a highly dissipative condition with a maximum quenching capacity (Flores-Bavestrello et al. 2016), which prevents the formation of ROS by excited chlorophyll molecules. In fact, the desiccation-induced production of H_2O_2 has been demonstrated in *Adiantum latifolium* (Lubaina et al. 2016). The quenching of excited chlorophyll and ROS formation can be compensated by the enhanced synthesis of Z, a xanthophyll with important antioxidant activity that regulates the level of energy dissipation in the antenna (Demmig-Adams and Adams III 1996; Fernández-Marín et al. 2013). Interestingly it has been described in the DT fern *Ceterach officinarum* that the sole process of desiccation is able to trigger the synthesis of Z even in complete darkness (see Sect. 19.1.2.4) (Fernández-Marín et al. 2009). The synthesis or activities of other antioxidants are also upregulated during desiccation, as is the case of tocopherol in epiphytic ferns (Tausz et al. 2001) and superoxide dismutase, peroxidase, catalase, and glutathione reductase in *Selaginella tamariscina* (Wang et al. 2010).

19.5 Conclusions

DT is a trait that is relatively extended among ferns in all their life forms, being more common in the spores (all spores appear to be DT), followed by gametophytes (widespread but habitat dependent), and then sporophytes (10% of species may show vegetative DT). Gametophytes and sporophytes have mechanisms to cope with DT analogous to those present in poikilohydric and homoiohydric plants, respectively (except for Hymenophyllaceae sporophytes that show poikilohydric strategies). Fern spores appear to show comparable mechanism to seeds, which are likely developed during the maturation phase of the spore and rely in the formation of a glassy cytoplasm for long-term survival in the dry state. For all three life forms, protection against oxidative stress seems to be a common and important feature, and diverse strategies with some commonalities (e.g., accumulation of antioxidants, protection against photooxidation from the chloroplasts during desiccation and while in the dry state) are found depending on the life form. The intermediate position of ferns between bryophytes and spermatophytes in an evolutionary context, and the variation of responses to desiccation of the different life forms of ferns, can allow us to use them as diverse models to study plant adaptations to terrestrial environment, as well as the evolution of the mechanisms of DT in land plants.

Acknowledgments We kindly acknowledge Charlotte Seal for her review of the manuscript.

References

Alamillo JM, Bartels D (2001) Effects of desiccation on photosynthesis pigments and the ELIP-like dsp 22 protein complexes in the resurrection plant *Craterostigma plantagineum*. Plant Sci 160:1161–1170

Alpert P (2000) The discovery, scope, and puzzle of desiccation-tolerance in plants. Plant Ecol 151:5–17

Alpert P (2006) Constraints of tolerance: why are desiccation-tolerant organisms so small or rare? J Exp Biol 209:1575–1584

Angelovici R, Galili G, Fernie AR, Fait A (2010) Seed desiccation: a bridge between maturation and germination. Trends Plant Sci 15:211–218

Baker J, Steele C, Dure L III (1988) Sequence and characterization of 6 Lea proteins and their genes from cotton. Plant Mol Biol 11:277–291

Ballesteros D (2010) Conservation of fern spores. In: Kumar A, Fernández H, Revilla-Bahillo A (eds) Working with Ferns Issues and Applications. Springer, New York, pp 165–172

Ballesteros D, Walters C (2007a) Water properties in fern spores: sorption characteristics relating to water affinity, glassy states and storage stability. J Exp Bot 58:1185–1196

Ballesteros D, Walters C (2007b) Calorimetric properties of water and triacylglycerols in fern spores relating to storage at cryogenic temperatures. Cryobiology 55:1–9

Ballesteros D, Walters C (2011) Detailed characterization of mechanical properties and molecular mobility within dry seed glasses: relevance to the physiology of dry biological systems. Plant J 68:607–619

Ballesteros D, Estrelles E, Walters C, Ibars AM (2012) Effects of temperature and desiccation on ex situ conservation of non-green fern spores. Amer J Bot 99:1–9

Ballesteros D, Hill LM, Walters C (2017) Variation of desiccation tolerance and longevity in fern spores. J Plant Physiol 211:53–62

Bartels D, Sunkar R (2005) Drought and salt tolerance in plants. Crit Rev Plant Sci 24:23–58

Brighigna L, Bennici A, Tani C, Tani G (2002) Structural and ultrastructural characterization of *Selaginella lepidophylla*, a desiccation-tolerant plant, during the rehydration process. Flora 197:81–91

Brown RC, Lemmon BE (2001a) Sporogenesis in Eusporangiate Ferns: 1. Monoplastidic Meiosis in *Angiopteris* (Marattiales). J Plant Res 114:223–235

Brown RC, Lemmon BE (2001b) Sporogenesis in Eusporangiate Ferns: II. Polyplastidic Meiosis in *Ophioglossum* (Ophioglossales). J Plant Res 114:237–246

Buitink J, Leprince O (2008) Intracellular glasses and seed survival in the dry state. C R Biologies 331:788–795

Buitink J, Walters C, Hoekstra FA, Crane J (1998) Storage behavior of *Typha latifolia* pollen at low water contents: interpretation on the basis of water activity and glass concepts. Phys Plant 103:145–153

Cai S, Chen G, Wang Y, Huang Y, Marchant B, Wang Y, Yang Q, Dai F, Hills A, Franks PJ, Nevo E, Soltis D, Soltis P, Sessa E, Wolf PG, Xue D, Zhang G, Pogson BJ, Blatt MR, Chen Z-H (2017) Evolutionary Conservation of ABA signaling for stomatal closure in ferns. Plant Physiol. https://doi.org/10.1104/pp.16.01848

Christenhusz MJM, Byng JW (2016) The number of known plant species in the world and its annual increase. Phytotaxa 261:201–217

Close TJ (1997) Dehydrins: A commonalty in the response of plants to dehydration and low temperature. Physiol Plant 100:291–296

Crowe JH, Hoekstra FA, Crowe LM (1992) Anhydrobiosis. Annu Rev Physiol 54:579–599

Curry J, Morris CF, Walker-Simmons MK (1991) Sequence analysis of a cDNA encoding a group 3 LEA mRNA inducible by ABA or dehydration stress in wheat. Plant Mol Biol 16:1073–1076

Deltoro VI, Calatayud A, Gimeno C, Abadía A, Barreno E (1998) Changes in chlorophyll a fluorescence, photosynthetic CO2 assimilation and xanthophyll cycle interconversions during dehydration in desiccation-tolerant and intolerant liverworts. Planta 207:224–228

DeMaggio AE, Stetler DE (1980) Storage products in spores of *Onoclea sensibilis L.* Am J Bot 67:452–455

Demmig BK, Winter A, Krüger F, Czygan C (1987) Photoinhibition and zeaxanthin formation in leaves. Plant Physiol 84:218–224

Demmig-Adams B, Adams WW III (1996) The role of xanthophyll cycle carotenoids in the protection of photosynthesis. Trends Plant Sci 1:21–26

Diamond HL, Jones HR, Swatzell LJ (2012) The role of Aquaporins in water balance in *Cheilanthes lanosa* (Adiantaceae) Gametophytes. Am Fern J 102:11–31

Dinakar C, Bartels D (2013) Desiccation tolerance in resurrection plants: new insights from transcriptome, proteome, and metabolome analysis. Front Plant Sci 4:482

Dubuisson JY, Schneider H, Hennequin S (2009) Epiphytism in ferns: diversity and history. C R Biol 332:120–128

Dure L, Crouch M, Harada J, Ho TH, Mundy J, Quatrano R, Thomas T, Sung ZR (1989) Common amino acid sequence domains among the LEA proteins of higher plants. Plant Mol Biol 12:475–486

Dyer AF (1979) The experimental biology of ferns. Academic, London

Ellis RH, Roberts EH (1980) Improved equations for the prediction of seed longevity. Ann Bot 45:13–30

Farrant JM, Brandt W, Lindsey GG (2007) An overview of mechanisms of desiccation tolerance in selected angiosperm resurrection plants. Plant Stress 1:72–84

Farrant JM, Lehner A, Cooper K, Wiswedel S (2009) Desiccation tolerance in the vegetative tissues of the fern Mohria caffrorum is seasonally regulated. Plant J 57:65–79

Farrant JM, Cooper K, Hilgart A, Abdalla KO, Bentley J, Thomson JA, Dace HJW, Peton N, Mundree SG, Rafudeen MS (2015) A molecular physiological review of vegetative desiccation tolerance in the resurrection plant *Xerophyta viscosa* (Baker). Planta 242:407–426

Farrar DR (1976) Spore retention and release from overwintering fern fronds. Am Fern J 66(2):49–52

Fernández-Marín B, Balaguer L, Esteban R, Becerril JM, García-Plazaola JI (2009) Dark induction of the photoprotective xanthophyll cycle in response to dehydration. J Plant Physiol 166:1734–1744

Fernández-Marín B, Becerril JM, García-Plazaola JI (2010) Unravelling the roles of desiccation-induced xanthophyll cycle activity in darkness: a case study in *Lobaria pulmonaria*. Planta 231:1335–1342

Fernández-Marín B, Míguez F, Becerril JM, García-Plazaola JI (2011) Dehydration-mediated activation of the xanthophyll cycle in darkness: is it related to desiccation tolerance? Planta 243:579–588

Fernández-Marín B, Arroyo Alfaro SJ, Becerril JM, García-Plazaola JI (2012) Do fern gametophytes have the capacity for irradiance acclimation? Biol Plant 56:351–356

Fernández-Marín B, Kranner I, San Sebastián M, Artetxe U, Laza JM, Vilas JL, Pritchard HW, Nadajaran J, Míguez F, Becerril JM, García-Plazaola JI (2013) Evidence for the absence of enzymatic reactions in the glassy state. A case study of xanthophyll cycle pigments in the desiccation-tolerant moss *Syntrichia ruralis*. J Exp Bot 64:3013–3043

Fernández-Marín B, Holzinger A, García-Plazaola JI (2016) Photosynthetic strategies of desiccation-tolerant organism. In: Pessarakli M (ed) Handbook of photosynthesis, 3rd edn. Taylor & Francis Catalog #:K22682, Florida. ISBN: 978–1–4822-3073-4.

Flores-Bavestrello A, Krol M, Ivanov A, Hüner N, García-Plazaola JI, Corcuera LJ, Bravo LA (2016) Morphological, physiological, and structural changes during a desiccation-rehydration process in two *Hymenophyllaceae* (Pteridophyte) contrasting in their tree vertical distribution. J Plant Physiol 191:82–94

Franchi GG, Piotto B, Nepi M, Baskin CC, Baskin JM, Pacini E (2011) Pollen and seed desiccation tolerance in relation to degree of developmental arrest, dispersal, and survival. J Exp Bot 62:5267–5281

Gabriel y Galan JM, Prada C (2010) Pteridophyte spores viability. In: Kumar A, Fernández H, Revilla-Bahillo A (eds) Working with Ferns. Issues and applications. Springer, New York, pp 165–172

Gaff DF (1977) Desiccation tolerant vascular plants of southern Africa. Oecol 31:95–109

Gaff DF (1980) Protoplasmic tolerance of extreme water stress. In: Turner NC, Kramer PJ (eds) Adaptation of plants to water and high temperature stress. Wiley, New York, pp 207–230

Gaff DF (1987) Desiccation tolerant plants in South America. Oecol 74:133–136

Gaff DF, Oliver M (2013) The evolution of desiccation tolerance in angiosperm plants: a rare yet common phenomenon. Funct Plant Biol 40:315–328

Galau GA, Bijaisoradat N, Hughes DW (1987) Accumulation kinetics of cotton late embryogenesis-abundent (Lea) mRNAs and storage protein mRNAs: coordinate regulation during embryogenesis and role of abscisic acid. Dev Biol 123:198–212

Garces M, Claverol S, Alvear C, Rabert C, Bravo L (2014) Desiccation tolerance of Hymenophyllacea filmy ferns is mediated by constitutive and non-inducible cellular mechanisms. C R Biologies 337:235–243

García-Plazaola JI, Esteban R, Fernández-Marín B, Kranner I, Porcar-Castell A (2012) Thermal energy dissipation and xanthophyll cycles beyond the Arabidopsis model. Photosynth Res 113:89–103

Geiger JMO, Tom A, Ranker JM, Neale R, Klimas ST (2007) Molecular biogeography and origins of the Hawaiian fern flora. Brittonia 59(2):142–158

Georgieva K, Rapparini F, Bertzza G, Mihailova G, Sarvari E, Solti A, Keresztes A (2017) Alterations in the sugar metabolism and in the vacuolar system of mesophyll cells contribute to the desiccation tolerance of *Haberlea rhodopensis* ecotypes. Protoplasma 254:193–201

Gill SS, Tuteja N (2010) Reactive oxygen species and antioxidant machinery in abiotic stress tolerance in crop plants. Plant Physiol Biochem 48(12):909–930

Gilmore AM, Hazlett TL, Govindjee (1995) Xanthophyll cycle-dependent quenching of photosystem II chlorophyll a fluorescence: formation of a quenching complex with a short fluorescence lifetime. PNAS USA 92:2273–2277

Hagar WG, Freeberg JA (1980) Photosynthetic rates of sporophytes and gametophytes of the fern, Todea barbara. Plant Physiol 65:584–586

Hauke RL (1978) Taxonomic monograph of *Equisetum* subgenus *Equisetum*. Nova Hedwigia 30:385–455

Havaux M, Dall'Osto L, Bassi R (2007) Zeaxanthin has enhanced antioxidant capacity with respect to all other xanthophylls in Arabidopsis leaves and functions independent of binding to PSII antennae. Plant Physiol 145:1506–1520

Heber U, Shuvalov VA (2005) Photochemical reactions of chlorophyll in dehydrated photosystem II: two chlorophyll forms (680 and 700 nm). Photosynth Res 84:85–91

Helseth LE, Fischer TM (2005) Physical mechanisms of rehydration in *Polypodium polypodioides*, a resurrection plant. Physical Reviews 71:061903–061901

Hietz P (2010) Fern adaptations to xeric environments. In: Mehltreter K, Walker LR, Sharpe JM (eds) Fern ecology. Cambridge University Press, Cambridge, pp 140–170

Hoekstra FA (2005) Differential longevities in desiccated anhydrobiotic plant systems. Integr Comp Biol 45:725–733

Hoekstra FA, Golovina EA, Buitink J (2001) Mechanisms of plant desiccation tolerance. Trends Plant Sci 6:1360–1385

Hong TD, Ellis RH (1996) A protocol to determine seed storage behaviour (No.1). Bioversity International, Rome

Hõrak H, Kollist H, Merilo E (2017) Fern stomatal responses to ABA and CO2 depend on species and growth conditions. Plant Physiol. https://doi.org/10.1104/pp.17.00120

Ibars AM, Estrelles E (2012) Recent developments in ex situ and in situ conservation of ferns. Fern Gaz 19:67–86

Inoue H, Kamachi H, Yamaya D, Oguma M, Munenori N (2000) Characterization of a protease that acts specifically on the 22-kDa protein in thylakoid membranes from green spores of the fern *Osmunda japonica*. Physiol Plant 109:129–136

John SP, Hasenstein KH (2017) The role of peltate scales in desiccation tolerance of *Pleopeltis polypodioides*. Planta 245:207–220

Johnson GN, Rumsey FJ, Headley AD, Sheffield E (2000) Adaptations to extreme low light in the fern *Trichomanes speciosum*. New Phytol 148:423–431

Juszczak I, Bartels D (2017) LEA gene expression, RNA stability and pigment accumulation in three closely related Linderniaceae species differing in desiccation tolerance. Plant Sci 255:59–71

Kakhovskaya IA, Rudakova AS, Manteuffel R (2003) Legumin- like and vicilin-like proteins from spores of the fern *Matteuccia struthiopteris*. J Plant Physiol 160:583–588

Kappen L (1965) Untersuchungen über die Widerstandsfähigkeit der Gametophyten einheimischer Polypodiaceen gegenüber Frost, Hitze and Trockenheit. Flora 156:101–115

Kappen L, Valladares F (2007) Opportunistic growth and desiccation tolerance: the ecological success of poikilohydrous autotrophs. In: Pugnaire F, Valladares F (eds) Functional plant ecology. Taylor and Francis, New York, pp 8–48

Kavitha CH, Murugan K (2016) Desiccation- induced changes in lipid peroxidation superoxide level and cell membrane ATPase activity in forked fern *Dicranopteris linearis* (burm.f.) underw. Int J Adv Res 3:588–593

Kranner I, Beckett RP, Wornik S, Zorn M, Pfeifhofer HW (2002) Revival of a resurrection plant correlates with its antioxidant status. Plant J 31:13–24

Kranner I, Beckett R, Hochman A, Nash TH (2008) Desiccation tolerance in lichens: a review. Bryol 11:576–593

Lane BG, Bernier F, Dratewka-Kos E, Shafai R, Kennedy TD, Pyne C, Monro JR, Vaughan T, Walters D, Altomare F (1991) Homologies between members of the germin gene family in hexaploid wheat and similarities between these wheat germins and certain *Physarum spherulins*. J Biol Chem 266:10461–10469

Layton BE, Boyd MB, Tripepi MS, Bitonti BM, Dollahon MNR, Balsamo RA (2010) Dehydration-induced expression of a 31-kDa dehydrin in *Polypodium polypodioides* (Polypodiaceae) may enable large, reversible deformation of the cell walls. Am J Bot 97:535–544

Lebkuecher JG (1997) Desiccation-time limits of photosynthetic recovery in *Equisetum hyemale* (Equisetaceae) spores. Am J Bot 84:792–797

Lebkuecher JG, Eickmeier WG (1991) Reduced photoinhibition with stem curling in the resurrection plant *Selaginella lepidophylla*. Oecol 88:597–604

Leprince O, Buitink J (2015) Introduction to desiccation biology: from old borders to new frontiers. Planta 242:369–378

Leprince O, Harren FJM, Buitink J, Alberda M, Hoekstra FA (2000) Metabolic dysfunction and unabated respiration precede the loss of membrane integrity during dehydration of germinating radicles. Plant Physiol 122:597–608

Li Y, Shi L (2014) Effect of desiccation level and storage temperature on green spore viability of *Osmunda japonica*. Cryobiology 68(3):446–450

Li Y, Shi L (2015) Effect of maturity level and desiccation process on liquid nitrogen storage of green spores of *Osmunda japonica*. Plant Cell Tissue Organ Cult 120(2):531–538

Li Y, Zhang YL, Jiang CD, Wang T, Wang Q, Shi L (2010) Effect of storage temperature on spore viability and early gametophyte development of three vulnerable species of *Alsophila* (Cyatheaceae). Aust J Bot 58:89–96

Liu MS, Chien CT, Lin TP (2008) Constitutive components and induced gene expression are involved in the desiccation tolerance of *Selaginella tamariscina*. Plant Cell Physiol 49:653–663

Lloyd RM, Klekowski EJ (1970) Spore germination and viability in pteridophyta: evolutionary significance of chlorophyllous spores spore germination and viability in pteridophyta: evolutionary significance of chlorophyllous spores. Biotropica 2(2):129–137

Lubaina AS, Brijithlal ND, Murugan K (2016) Unravelling desiccation and rehydration tolerance mechanism in the fern *Adiantum latifolium*. Biosci Biotech Res Comm 9:672–679

Marengo N (1979) The fine structure of the pre-meiotic stages of sporogenesis in *Onoclea sensibilis*. Am FernJ 69:87–91

McElwain J (2011) Ferns: a xylem success story. New Phytol 192:307–310

Mikula A, Tomiczak K, Makowski D, Niedzielski M, Rybczyński JJ (2015) The effect of moisture content and temperature on spore aging in *Osmunda regalis*. Acta Physiol Plant 37:229

Minamikawa T, Koshiba T, Wada M (1984) Compositional changes in germinating spores of *Adiantum capillus-veneris* L. Bot Mag 97:313–332

Mira S, Gonzalez-Benito ME, Hill LM, Walters C (2010) Characterization of volatile production during storage of lettuce (*Lactuca sativa*) seed. J Exp Bot 61:3915–3924

Moore JP, Nguema-Ona EE, Vicre-Gibouin M, Sørensen I, Willats GTW, Driouich A, Farrant JM (2013) Arabinose-rich polymers as an evolutionary strategy to plasticize resurrection plant cell walls against desiccation. Planta 237:739–754

Mundy J, Chua NH (1988) Abscisic acid and water-stress induce the expression of a novel rice gene. EMBO J 7(8):2279–2286

Muslin EH, Homann PH (1992) Light as a hazard for the desiccation-resistant "resurrection" fern *Polypodium polypodioides* L. Plant Cell Environ 15:81–89

Nabe H, Funabiki R, Kashino Y, Koike H, Satoh K (2007) Responses to desiccation stress in bryophytes and important role of dithiothreitol-insensitive non-photochemical quenching against photoinhibition in dehydrated states. Plant Cell Physiol 48:1548–1557

Nagel M, Kranner I, Neumann K, Rolletschek H, Seal C, Colville L, Fernandez-Marin B, Borner A (2015) Genome-wide association mapping and biochemical markers reveal that seed ageing and longevity are intricately affected by genetic background and developmental and environmental conditions in barley. Plant Cell Environ 38:1011–1022

Neale AD, Blomstedt CK, Bronson P, Le TN, Guthridge K, Evans J, Gaff DF, Hamill JD (2000) The isolation of genes from the resurrection grass *Sporobolus stapfianus* which are induced during severe drought stress. Plant Cell Environ 23:265–277

Noblin X, Rojas NO, Westbrook J, Llorens C, Argentina M, Dumais J (2012) The fern sporangium: a unique catapult. Science 335:1322

Okada Y (1929) Notes on the germination of the spores of some Pteridophytes with special regard to their viability. Sci Rept Tdhoku Imp Univ Biol 4:127–182

Oliver MJ, Bewley JD (1997) Desiccation tolerance in plant tissues. A mechanistic overview. Hortic Rev 18:171–214

Oliver MJ, Tuba Z, Mishler BD (2000) The evolution of vegetative desiccation tolerance in land plants. Plant Ecol 151:85–100

Ong BL, Ng ML (1998) Regeneration of drought-stressed gametophytes of the epiphytic fern, *Pyrrosia pilosellodes* (L.) Price. Plant Cell Rep 18:225–228

Page CN (1979) Experimental aspects of fern ecology. In: Dyer AF (ed) The Experimental biology of Ferns. Academic Press, London, pp 551–589

Page CN (2002) Ecological strategies in fern evolution: a neopteridological overview. Rev Palaeobot Palynol 119:1–33

Paless G, Vida G, Nagy AH (1984) Phytochrome-induced control of protein synthesis during germination of fern spores. Acta Bot Hung 30:191–200

Pampurova S, Van Dijck P (2014) The desiccation tolerant secrets of *Selaginela lepidophylla:* what we have learned so far? Plant Physiol Biochem 80:285–290

Parris BS (2001) Circum-Antarctic continental distribution patterns in pteridophyte species. Brittonia 53:270–283

Passot S, Cenard S, Douania I, Tréléa IC, Fonseca F (2012) Critical water activity and amorphous state for optimal preservation of lyophilised lactic acid bacteria. Food Chem 132:1699–1705

Pence VC (2000) Survival of chlorophyllous and non-chlorophyllous fern Spores through exposure to liquid nitrogen. Am Fern J 90:119–126

Pérez-Garcia B, Orozco-Segovia A, Riba R (1994) The effects of white fluorescent light, far-red light, darkness, and moisture on spore germination of *Lygodium heterodoxum* (Schizaeaceae). Am J Bot 81:1367–1369

Peters S, Mundree SG, Thomson JA, Farrant JM, Keller F (2007) Protection mechanisms in the resurrection plant *Xerophyta viscosa* (Baker): both sucrose and raffinose family oligosaccharides (RFOs) accumulate in leaves in response to water deficit. J Exp Bot 58:1947–1956

Pickett FL (1913) Resistance of the prothallia of *Camptosorus rhizophyllus* to desiccation. Bul Torrey Botany Club 40:641–645

Pittermann J, Brodersen C, Watkins JE (2013) The physiological resilience of fern sporophytes and gametophytes: advances in water relations offer new insights into an old lineage. Front Plant Sci 4:285

Porembski S (2011) Evolution, diversity, and habitats of Poikilohydrous vascular plants. In: Lüttge U, Beck E, Bartels D (eds) Plant desiccation tolerance, Ecological studies, vol 215. Springer, Heidelberg, pp 139–154

Proctor MCF (2003) Comparative ecophysiological measurements on the light responses, water relations and desiccation tolerance of the filmy ferns *Hymenophyllum wilsonii* Hook. and *H. tunbrigense* (L.) Smith. Ann Bot 91:717–727

Proctor MCF (2012) Light and desiccation responses of some Hymenophyllaceae (filmy ferns) from Trinidad, Venezuela and New Zealand: poikilohydry in a light-limited but low evaporation ecological niche. Ann Bot 109:1019–1026

Proctor MCF, Pence VC (2002) Vegetative tissues: bryophytes, vascular resurrection plants and vegetative propagules. In: Black M, Pritchard HW (eds) Desiccation and survival in plants: drying without dying. CABI publishing, Wallingford, pp 207–237

Proctor MCF, Tuba Z (2002) Poikilohydry and homoihydry: antithesis or spectrum of possibilities? New Phytol 156:327–349

Raghavan V (1991) Gene activity during germination of spores of the fern, *Onoclea sensibilis*: RNA and protein synthesis and the role of stored mRNA. J Exp Bot 42:251–260

Raghavan V, Kamalay JC (1992) Expression of two cloned mRNA sequences during development and germination of spores of the sensitive fern, *Onoclea sensibilis* L. Planta 189:1–9

Rahman MS (2009) Food stability beyond water activity ad glass transitio: macro-micro region concept in the state diagram. Int J Food Prop 12:726–740

Rascio N, La Rocca N (2005) Resurrection plants: the puzzle of surviving extreme vegetative desiccation. Crit Rev Plant Sci 24:209–225

Reynolds TL, Bewley JD (1993) Abscisic acid enhances the ability of the desiccation-tolerant fern *Polypodium virginianum* to withstand drying. J Exp Bot 44:1771–1779

Riaño K, Briones O (2015) Sensitivity of three tree ferns during their first phase of life to the variation of solar radiation and water availability in a Mexican cloud forest. Am J Bot 102:1472–1481

Roos YH, Drusch S (2015) Phase Transitions in Foods, 2nd edn. Academic Press, Cambridge, MA

Roqueiro G, Facorro GB, Huarte MG, Rubín de Celis E, García F, Maldonado S, Maroder H (2010) Effects of photooxidation on membrane integrity in *Salix nigra* seeds. Ann Bot 105:1027–1036

Roussenova M, Murith M, Alam A, Ubbink J (2010) Plasticization,antiplasticization, and molecular packing in amorphous carbohydrate-glycerolmatrices. Biomacromolecules 11:3237–3247

Sato T (1982) Phenology and wintering capacity of sporophytes and gametophytes of ferns native to northern Japan. Oecol 55:53–61

Schallau A, Kakhovskaya I, Tewes A, Czihal A, Tiedemann J, Mohr M, Grosse I, Manteuffel R, Bäumlein H (2008) Phylogenetic footprints in fern spore- and seed-specific gene promoters. Plant J 53:414–424
Schreiber U, Armond PA (1978) Heat-induced changes of chlorophyll fluorescence in isolated chloroplasts and related heat damage at the pigment level. Biochim Biophys Acta 502:138–151
Scott P (2000) Resurrection plants and the secrets of eternal leaf. Ann Bot 85:159–166
Shutov AD, Bäumlein H (1999) Origin and evolution of seed storage globulins. In: Casey R, Shewry P (eds) Seed Proteins. Kluwer Academic Publishers, Dordrecht, pp 543–561
Shutov AD, Braun H, Chesnokov YV, Baümlein H (1998) A gene encoding a vicilin-like protein is specifically expressed in fern spores. Evolutionary pathway of seed storage globulins. Eur J Biochem 252:79–89
Shutov AD, Blattner FR Baümlein H, Müntz K (2003) Storage and mobilization as antagonistic functional constraints of seed storage globulin evolution. J Exp Bot 54:1645–1654
Stark LR, Greenwood JL, Brinda JC, Oliver MJ (2013) Physiological history may mask the inherent inducible desiccation tolerance strategy of the desert moss *Crossidium crassinerve*. Plant Biol 16:935–946
Stinson JR, Eisenberg AJ, Willing RP, Pe ME, Hanson DD, Mascarenhas JP (1987) Genes expressed in the male gametophyte of fowering plants and their isolation. Plant Physiol 83:442–447
Sundue MA, Vasco A, Moran RC (2011) Cryptochlorophyllous spores in ferns: nongreen spores that contain chlorophyll. Int J Plant Sci 172:1110–1119
Suo J, Zhao Q, Zhang Z, Chen S, Cao J, Liu G, Wei X, Wang T, Yang C, Dai S (2015) Cytological and proteomic analyses of *Osmunda cinnamomea* germinating spores reveal characteristics of fern spore germination and rhizoid tip growth. Mol Cell Proteomics 14:2510–2534
Tausz M, Hietz P, Briones O (2001) The significance of carotenoids and tocopherols in photoprotection of seven epiphytic fern species of a Mexican cloud forest. Aust J Plant Physiol 28:775–783
Tuba Z, Lichtenthaler HK (2011) Ecophysiology of homoiochlorophyllous and poikilochlorophyllous desiccation-tolerant plants and vegetations. In: Lüttge U, Beck E, Bartels D (eds) Plant desiccation tolerance. Ecological studies, vol 215. Springer, Heidelberg, pp 157–184
Vertucci CW, Farrant JM (1995) Acquisition and loss of desiccation tolerance. In: Kigel J, Galili G (eds) Seed development and germination. Marcel Dekker Presss Inc., New York, pp 237–272
Vertucci CW, Roos EE (1990) Theoretical basis of protocols for seed storage. Plant Physiol 94:1019–1023
Voytena APL, Minardi BD, Barufi JB, Santos M, Randi AM (2014) *Pleopeltis pleopeltifolia* (Polypodiopsida, Polypodiaceae), a poikilochlorophyllous desiccation-tolerant fern: anatomical, biochemical and physiological responses during water stress. Aust J Bot 62:647–656
Walters C (2015) Orthodoxy, recalcitrance and in-between: describing variation in seed storage characteristics using threshold responses to water loss. Planta 242:397–406
Walters C, Hill LM, Wheeler LJ (2005) Dying while dry: kinetics andmechanisms of deterioration in desiccated organisms. Integr Comp Biol 45:751–758
Walters C, Ballesteros D, Vertucci VA (2010) Structural mechanics of seed deterioration: Standing the test of time. Plant Sci 179:565–573
Wang X, Chen S, Zhang H, Shi L, Cao F, Guo L, Xie Y, Wang T, Yan X, Dai S (2010) Desiccation tolerance mechanism in resurrection fern-ally *Selaginella tamariscina* revealed by physiological and proteomic analysis. J Proteome Res 9:6561–6577
Watkins JE, Mack MC, Sinclair TR, Mulkey SS (2007) Ecological and evolutionary consequences of desiccation tolerance in tropical fern gametophytes. New Phytol 176:708–717
Wieners PC, Mudimu O, Bilger W (2012) Desiccation-induced non-radiative dissipation in isolated green lichen algae. Photosynth Res 113:239–247
Williams WP, Allen JF (1987) State 1/State 2 changes in higher plants and algae. Photosynth Res 13:19–45

Wohlfarth T, Braun H, Kirik V, Kölle K, Czihal A, Tewes A, Luerssen H, Misera S, Shutov A Baümlein H (1998) Regulation and evolution of seed globulin genes. J Plant Physiol 152:600–606

Wolf PG, Schneider H, Ranker TA (2001) Geographic distributions of homosporous ferns: does dispersal obscure evidence of vicariance? J Biogeogr 28(2):263–270

Wollersheim M (1957) Untersuchunger über die keimungsphysiologie der Sporen von *Equisetum arvense* und *Equisetum limosum*. Z Bot 45:145–159

Wyse SV, Dickie JB (2017) Predicting the global incidence of seed desiccation sensitivity. J Ecol 105:1082–1093

Yahubian G, Gozmanova M, Denev I, Toneva V, Minkov I (2009) Prompt response of superoxide dismutase and peroxidase to dehydration and rehydration of the resurrection plant *Haberlea rhodopensis*. Plant Growth Regul 57:49–56

Yobi A, Wone BWM, Xu W, Alexander DC, Guo L, Ryals JA, Olive MJ, Cushman JC (2013) Metabolomics profiling in Selaginella lepidophylla at various hydration states provides new insights into the mechanistic basis of desiccation tolerance. Mol Plant 6:369–385

Zeng O, Chen XB, Wood AJ (2002) Two early light-indudible protein (ELIP) cDNAs from the resurrection plant *Tortula ruralis* are differentially expressed in response to desiccation, rehydration, salinity, and high light. J Exp Bot 371:1197–1205

Zhao Q, Gao J, Suo J, Chen S, Wang T, Dai S (2015) Cytological and proteomic analyses of horsetail (*Equisetum arvense* L.) spore germination. Front Plant Sci 6:441

Chapter 20
New Insights on Atmospheric Fern Spore Dynamics

David Rodríguez de la Cruz, Estefanía Sánchez-Reyes, José Sánchez-Sánchez, and José Ángel Sánchez-Agudo

20.1 Introduction

Fern and fern allies, also known as pteridophytes, are a group of vascular plants with nearly 11,300 species (Ranker and Sundue 2015; Corlett 2016) which make them the second most abundant group among metaphytes (Prance 2001). From a phylogenetic perspective (Pryer et al. 2004), this group is paraphyletic, including lycophytes (clubmosses, spikemosses and quillworts) and monilophytes (ferns and horsetails). Both are considered to be an older group than flowering plants being the dominant plant elements in the landscape during the Carboniferous and Permian (Niklas et al. 1983), reaching its greatest diversification in the Cretaceous and Cenozoic (Testo and Sundue 2016). At present, the greatest diversity of these organisms is found in tropical areas (Kreft et al. 2010).

With respect to the rest of vascular plants, seed or flowering plants, there are some important differences to understand their biology (Vicent et al. 2014). Ferns and lycopods have a digenetic heteromorphous life cycle with a diploid dominant sporophyte producing haploid spores after meiosis. Some groups produce two types of spores, macro- and microspores (heterosporous, i.e. Salviniales or Isoetales), although most of them cause only one type of spore (homosporous). These spores, when germinated, will result in a free-living gametophyte that will produce sexual organs and after subsequent fertilization and embryonary development will result in a new sporophyte.

The number of spores produced by a sporophyte is variable and depends on several factors, mainly the species (Cousens 1988). Some studies estimate production

D. Rodríguez de la Cruz (✉) · E. Sánchez-Reyes · J. Sánchez-Sánchez · J. Á. Sánchez-Agudo
Department of Botany and Plant Physiology, University of Salamanca, Salamanca, Spain

Instituto Hispano-Luso de Investigaciones Agrarias (CIALE), University of Salamanca, Salamanca, Spain
e-mail: droc@usal.es

H. Fernández (ed.), *Current Advances in Fern Research*,
https://doi.org/10.1007/978-3-319-75103-0_20

of around half a million spores per year and individual in *Cyathea arborea* (L.) Sm. (Conant 1978), 50 million spores in *Dryopteris intermedia* (Muhl. ex Willd.) A. Gray, 90 million in *Osmunda claytoniana* L. (Mickel 1982) or up to 100 million spores in *Dryopteris filix-mas* (L.) Schott (Schneller 1975). The dispersion of these spores is mainly through wind, except in aquatic species (Sharpe et al. 2010), although there are some cases of spore dispersal by animals (Boch et al. 2016). However, the dispersal capacity of the spores is controversial, with some authors claiming that their potential capacity is high (Schneller and Liebst 2007) and others claiming that they are a few metres away from the sporophyte producer (Penrod and McCormick 1996). Recent studies support a limited wind dispersion capacity strongly influenced by the lipid and moisture content of the spores (Gómez-Noguez et al. 2016). In the case of spores dispersed by animals, there is also scattering at short distances, such as slugs passing through the digestive tract (Boch et al. 2013), or at long distances through bats (Sugita et al. 2013).

In this chapter we will focus on the airborne spores of ferns and lycopods, updating the advances and applications discussed by Rodríguez de la Cruz et al. (2010) and incorporating notes on spore morphology as well as a simple identification key for the different genera identified in the atmosphere.

20.2 Aerobiology: Concept, Methods and Applications

The term aerobiology was coined in 1930 by plant pathologist Fred C. Meier (1893–1938) to define the science that included the study of both living (fungal and fern spores, pollen grains, mites, small insects, etc.) and dead particles (dust and contaminants) present in the atmosphere (Gregory 1973). In 1975, Pathirane included the study of the release, retention, dispersion, deposition and incidence of spores, pollen and other airborne particles in the air, as an objective of this discipline.

Currently, it is defined as the science that deals with the study of animal or plant origin particles that are transported by the wind, as well as knowing the interrelationships that exist between these particles and the processes of production, release, transport and deposition (Spieksma 1992), since these will condition the incorporation of the different biological particles into the atmosphere, their permanence and their spatiotemporal variations.

Therefore, in order to know the atmospheric content of spores/pollen present in a specific place and time, firstly it is necessary to capture said particles. In this sense, there are different mechanisms on which the instruments developed for this purpose are based (Mandrioli et al. 1998; Ogden et al. 1974), and two major groups can be highlighted.

Gravimetric methods: based on passive or gravitational capture, since the particles are deposited by the action of gravity on a surface (slide) impregnated with an adhesive substance (usually glycero-gelatin) stained with fuchsin. The effectiveness of this methodology varies depending on particles size, wind speed and direction thus possible atmospheric turbulence. The main traps were developed by Durham (1946) and modified by Pla Dalmau (1958) and Tauber (1967). They are simple and economic methods, valid for any type of particles and which allow taking samples in places of complex access where there are no sources of electricity. However, they have the disadvantages of not being able to specify the hours in which the deposition was more intense; the inefficiency when capturing small biological particles, many of them considered strongly allergenic as some pollen or fungal spores; and the impossibility to quantify its content, for not being able to refer the sample to a known volume of air (Lacey et al. 1996).

Volumetric methods: by filtration of known volumes of air, can be divided into three subgroups.

- Impaction: based on the collision of air against a surface that intercepts and retains the particles it transports. The main samplers used are Rotorod (Perkins 1957) and Rotoslide (Ogden and Raynor 1967). It must be said that the efficiency of these devices depends on the wind speed, the morphological characters of the particles and the trap size (Mandrioli et al. 1998). Collection efficiency is very high, but as a consequence, they are not able to work continuously since the samples taken would become saturated with particles, which would make it impossible to study them at microscope.
- Suction: we will pay special attention to this group since it is the methodology used for all the European aeropalynology networks. In them, volumetric samplers known as spore traps (Hirst 1952) are used, which have an air inlet hole through which a volume of 10 l/m is aspirated (thanks to a vacuum pump) pretending to simulate the volume that a person breathes under normal conditions. Air particles impact on a tape coated with an adhesive. Furthermore, a wind vane is attached to the sampler head, which ensures that the orifice is always oriented into the wind in order to guarantee higher collection efficiency. There are two brands commercially available at present (Burkard® and Lanzoni®), which have the same advantages (simplicity and continuous recording) and disadvantages (variable results depending on wind speed and particle size). Although they were initially designed for the capture of spores, they are also perfectly valid for pollen.

The other suction sampler that is used is the so-called cascade impactor being the Andersen type (Davies 1971) the most used. In this case, a known volume of air passes through a set of stacked and perforated metal plates, which contain Petri dishes with agar. These perforations are progressively smaller as we approach the base, so that the smaller particles impact on the lower part, being analogous to the different levels and filters existing in the respiratory tract, from the nose to the pulmonary alveoli. Subsequently, these Petri plates are incubated in order to know the fungal spore types and bacteria present in each level.

- Filtration: they rely on the air passing through a porous surface in which the particles are retained depending on their size. The best known of these methods is the one designed by Cour in 1974, in which the filtering surface consists of five

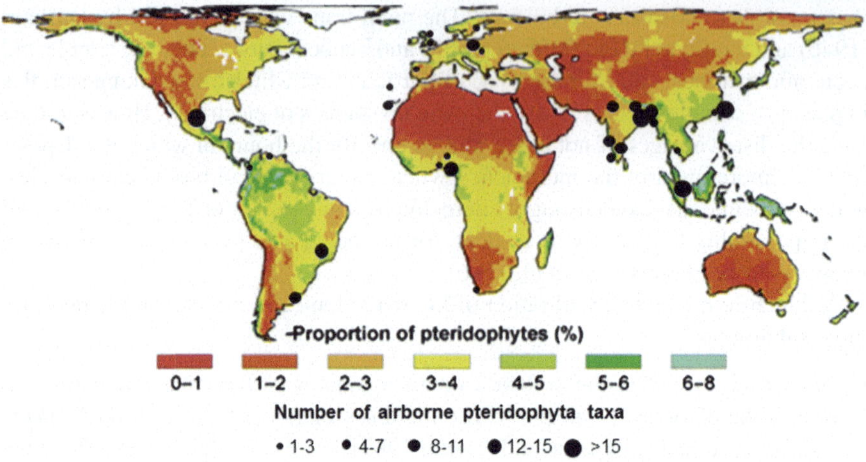

Fig. 20.1 The estimated diversity of pteridophytes and the number of aerovagant taxa in the atmosphere of different sampling points (Adapted from Kreft et al. (2010))

hydrophilic gauzes impregnated with silicone oil, supported by a metal frame and coupled to a weather vane and another horizontal motionless device, which is the one that collects the particles by gravitational sedimentation. In addition, wind speed can be measured by an anemometer attached at the level of the trap obtaining a subsequent volumetric transformation of the data. The main drawback of this sampler is that it does not allow knowing daily variations. In the same way, the analysis of the content of the gauzes is carried out through the acetolytic process (Erdtman 1960), which implies a complex and long manipulation, in addition to a loss of information. This is because fungal spores and certain types of pollen grains (especially those anemophilous) can be destroyed within the process, as well as the perispore layer (also denominated perine, Punt et al. 2007) of some fern spores (Large and Braggins 1990), not always acetolysis resistant (Devi 1980) which may be important for the identification of some Pteridophyta spores since it frequently has ornamentation.

Therefore, it could be said that the use of one or another methodology is conditioned to the specific characteristics that each one of them has, to the particles we want to capture and to the study we want to carry out.

20.3 New Observations on Airborne Fern and Lycopod Spore Dynamics

Pteridophyte airborne spores registered greater diversity in tropical and subtropical areas in line with the largest estimated number of species within this group of cormophytes (Kreft et al. 2010). Figure 20.1 shows the relationship between the

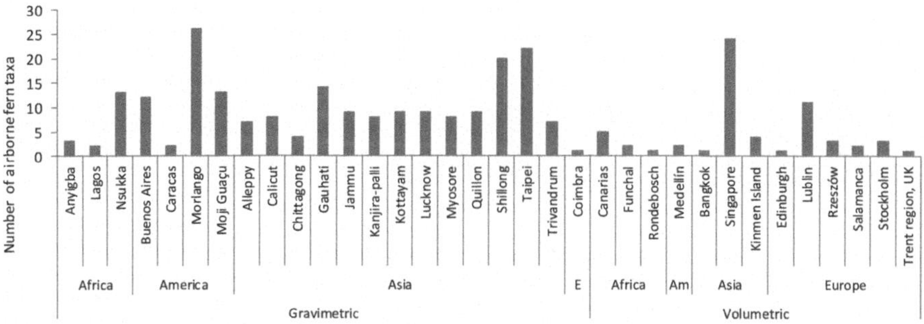

Fig. 20.2 Number of airborne pteridophyte taxa identified at different sampling points around the world. Am: America. E: Europe

estimated diversity of pteridophytes and the greater number of aerovagant taxa in the atmosphere of areas close to the tropics. The largest number of taxa was identified in Molango (Hidalgo State, México), near the Barranca de Metztitlán Biosphere Reserve (Gómez-Noguez et al. 2017), followed by two urban environments in southeast and central-east Asia (Chen and Chien 1986; Chen and Huang 1980; Huang et al. 2008; Ong et al. 2012; Yang et al. 2003), Singapore and Taipei, respectively. On the other hand, only a few taxa were cited in several urban and non-urban environments (Fig. 20.2) mainly in Europe (Leitão et al. 1996; Newson et al. 2000; Nilsson and Praglowski 1974; Rodríguez et al. 2011) but also in some parts of Asia or Africa (Berman and Hons 2013; Songnuan et al. 2015). However, it should be noted that many of these works did not have as their object the identification of ferns and lycopod spores and that they could therefore not reflect the actual atmospheric diversity of pteridophyte spores.

The number of scientific papers dealing with the atmospheric content of Pteridophyta spores was low either by volumetric or gravimetric methods, especially if we compare it to articles about atmospheric levels of pollen and fungal spores. Twenty-eight scientific articles were collected for this chapter, identifying spores of 81 taxa of different pteridophytes, 75 at genus level and 6 at family level (Table 20.1). The largest number of taxa was identified in studies conducted in Asian localities (51), followed by those registered in various parts of America (43). The works carried out in different parts of Africa and Europe revealed a lower diversity of atmospheric taxa (20 and 12, respectively). This diversification was also reflected in the number of specific taxa for the different studies grouped by continent, with Asia registering the highest number (25), followed by America (16). On the other hand, the surveys developed in Europe and Africa showed only three (*Athyrium*, *Cystopteris* and *Matteuccia*) and two specific taxa (*Lonchitis* and *Microgramma*) at continental level, respectively. In addition, these studies displayed five types of spores present in the different sites analysed at the four continents: *Adiantum*, *Asplenium*, *Lycopodium*, *Polypodium* and *Pteridium*.

Underlined references correspond to studies carried out by means of gravimetric methods.

Table 20.1 Identified airborne pteridophyte taxa following classification proposed by Smith et al. (2006) and Christenhusz et al. (2011)

Division Tracheophyta								
Subdivision	Class	Order	Family	Genera	Africa	America	Asia	Europe
Lycophytina	Lycopsida	Lycopodiales	Lycopodiaceae	*Lycopodiella*		25		
				Lycopodium	9	10	6,8,13	15,21
	Selaginellopsida	Selaginellales	Selaginellaceae	*Selaginella*	9		12,23,28	
Euphyllophytina	Psilotopsida	Ophioglossales	Ophioglossaceae			22		
				Botrychium		10		11
				Ophioglossum		18	12,23	
	Equisetopsida	Equisetales	Equisetaceae	*Equisetum*		10,18,22		11,15
	Polypodiopsida	Osmundales	Osmundaceae	*Osmunda*		25		
		Gleichniales	Gleicheniaceae	*Dicranopteris*			6,7,8,13,23	
				Diplopterygium			6	
		Schizales	Anemiaceae	*Anemia*			28	
			Lygodiaceae	*Lygodium*		10	6,23	
		Salviniales	Salviniaceae	*Azolla*		18		
		Cyatheales	Cyatheaceae	*Alsophila*		10	6,7,8,13	
				Cnemidaria		10		
				Cyathea		10,25	23,27	
			Dicksoniaceae	*Dicksonia*		10		
				Lophosoria		10		
		Polypodiales	Lindsaeaceae	*Lindsaea*		25	23	
				Lonchitis	20			
				Odontosoria		10		
				Tapeinidium			8	
			Dennstaedtiaceae	*Dennstaedtia*		10	6	
				Hypolepis	20	10	8	
				Microlepia			8,28	

Division Tracheophyta								
Subdivision	Class	Order	Family	Genera	Africa	America	Asia	Europe
				Pteridium	16	18	6,7,23	5,11,17,19,24
			Pteridaceae	*Acrostichum*	1		26,28	
				Adiantum	4,16	22	23,28	11
				Ceratopteris			23	
				Cheilanthes			28	
				Notholaena			28	
				Pellaea		12		
				Pityrogramma		25	23	
				Pteris	4,9,20		8,12,13,23,28	
				Taenitis			23	
				Vittaria			23	
			Aspleniaceae	*Asplenium*	16,20	2,10,25	8,23,28	11
			Thelypteridaceae		20		8,28	
				Cyclosorus	20		6,12,23,28	
				Thelypteris		25		
			Woodsiaceae		20	22	8,28	
				Athyrium				11
				Cystopteris				11
				Diplazium			6	
			Onocleaceae	*Matteuccia*				11
				Onocleopsis		10		
			Blechnaceae		20	18,22	28	
				Blechnum		10,18,25	23	11,24
				Stenochlaena			23,28	

(continued)

Table 20.1 (continued)

Division Tracheophyta								
Subdivision	Class	Order	Family	Genera	Africa	America	Asia	Europe
				Woodwardia		10		
			Dryopteridaceae				6	
				Bolbitis			28	
				Ctenitis		10		
				Cyclodium		25		
				Cyrtomium			28	
				Dryopteris	16,20		28	11,21
				Elaphoglossum	20	10		
				Lastreopsis		10		
				Lithostegia			28	
				Megalastrum		10		
				Phanerophlebia		10		
				Polystichum		10	28	
			Lomariopsidaceae	*Nephrolepis*	20	10,14,22	8,23,28	
			Tectariaceae	*Tectaria*			28	
			Davalliaceae	*Davallia*	16,20		8,23	
			Polypodiaceae		1	2,22		
				Campyloneurum		10,25		
				Colysis			6,28	
				Drynaria			23,28	
				Goniophlebium			23,28	
				Lemmaphyllum			6	
				Lepisorus			28	
				Microgramma	20			

Division Tracheophyta								
Subdivision	Class	Order	Family	Genera	Africa	America	Asia	Europe
				Microsorum			28	
				Pecluma		10,25		
				Phlebodium		10		
				Phymatosorus			23	
				Platycerium			23	
				Pleopeltis		10,25		
				Polypodium	3,16	14,18,25	8,28	11,15,21
				Pyrrosia			23,28	

[1] Ajikah et al. (2015), [2] Alzate Guarín et al. (2015), [3] Berman and Hons (2013), [4] Camacho (2015) [5] Caulton et al. (2000), [6] Chen (1984), [7] Chen and Chien (1986), [8] Chen and Huang (1980), [9] Christopher et al. (2013), [10] Gómez-Noguez et al. (2017), [11] Haratym et al. (2014), [12] Hossain and Pasha (2012), [13] Huang et al. (2008), [14] Hurtado and Riegler-Goihman (1986), [15] Kasprzyk (2004), [16] La Serna-Ramos and Domínguez-Santana (2003), [17] Leitão et al. (1996), [18] Majas and Romero (1992), [19] Newson et al. (2000), [20] Njokuocha (2006), [21] Nilsson and Praglowski (1974), [22] Noetinger et al. (1994), [23] Ong et al. (2012), [24] Rodríguez et al. (2011), [25] Simabukuro et al. (2000), [26] Songnuan et al. (2015), [27] Yang et al. (2003), [28] Yasmeen and Devi (1988)

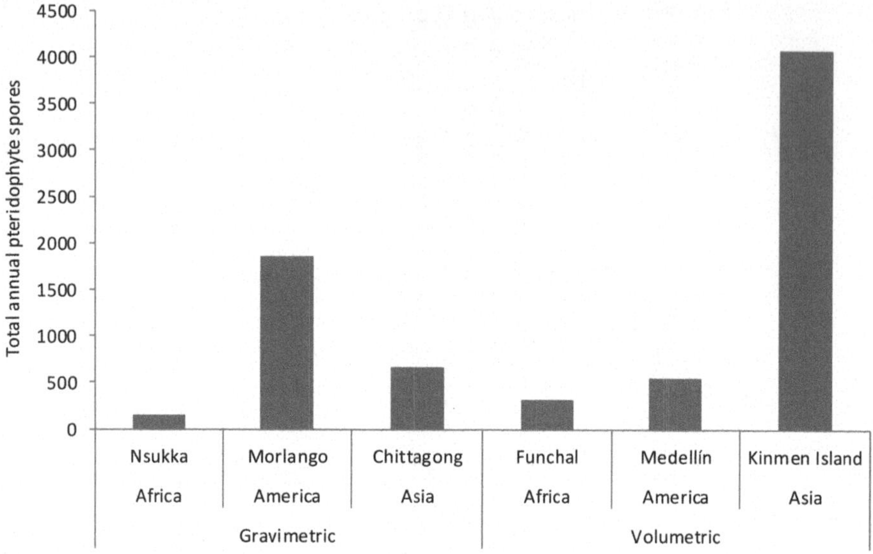

Fig. 20.3 Annual total records of some fern and lycopod spores obtained by means of gravimetric method or volumetric method

In general, the total annual number of fern and lycopod spores in the atmosphere did not exceed 1000 spores, although in some cases it exceeded these values broadly (Fig. 20.3), as in Taipei or Molango (Gómez-Noguez et al. 2017; Huang et al. 2008). In both cases, the sampling sites were located in areas close to more or less well-preserved humid forests or shrubs with a relevant presence of pteridophyte. This fact was able to explain these high annual levels, regardless of the sampling method used. Proximity to production sources together with probable medium-range transport could also explain certain daily concentrations (Favali et al. 2003; Rantio-Lehtimäki 1994). In urban areas remote from plant formations with remarkable populations of ferns, daily levels were low, as in some European cities (Haratym et al. 2014; Rodríguez de la Cruz et al. 2009), where concentrations of *Athyrium*, *Dryopteris* and *Pteridium* did not exceed 20 spores per cubic metre. On the other hand, also by means of volumetric collectors, daily airborne concentrations above 750 *Pteridium* spores/m^3 were observed in a rural environment between the end of August and the end of October, with a maximum value of 1800 spores/m^3 (Lacey and McCartney 1994). Other works carried out through gravimetric traps, facing forested hillsides, also displayed high daily punctual values for *Alsophila* (523 spores/cm^2), *Dicranopteris* (148 spores/cm^2) and *Pteridium* (82 spores/cm^2), as in Taipei during the month of July (Chen 1984).

In relation to the number of taxa identified in the different studies, a greater number of spore types appeared to be observed in the gravimetric surveys than those developed with different volumetric spore traps (Fig. 20.2). The average number of taxa identified in different places was 10 for the works developed with gravimetric methods and 5 with volumetric methods. This difference was greater in the sampling sites analysed throughout America, being practically negligible in the case of the surveys carried out in Asia. Therefore, it can be deduced that when conducting aerobiological studies with the aim to characterize the atmospheric spore content of pteridophytes from a qualitative point of view, the use of gravimetric collectors might be more suitable, although surveys developed by Levetin et al. (2000) with another group of cormophytes producing pollen grains indicated that both volumetric and gravimetric methods were valid for reflecting local anemophilous vegetation. It should not be overlooked, however, that certain types of spore may experience medium- or long-distance wind transport (Rodríguez de la Cruz et al. 2009) and that the production and subsequent presence of different spore types, as well as their atmospheric levels, depend on several factors (Caulton et al. 2000; Lee et al. 2009).

The influence of some meteorological factors on airborne levels of ferns spores has been scarcely analysed (Ong et al. 2012; Rodríguez de la Cruz et al. 2009). The results obtained through Spearman's nonparametric correlation coefficient indicated a positive relationship between spore concentrations with temperature and sunshine hours and negative with rainfall, relative humidity and wind speed, mainly for diurnal concentrations of *Pteridium*, *Dicranopteris* and *Stenochlaena*. Gómez-Nóguez et al. (2017) found also a negative effect of rainfall on the pteridophyte spore rain. However, there was a negative influence of temperature and sunshine hours on the atmospheric levels of *Asplenium*, *Dicranopteris* and *Stenochlaena*. These results showed that further studies on the influence of meteorological parameters on atmospheric spore concentrations are needed in order to better assess the relationship between atmospheric dynamics of fern spores and meteorology.

The seasonal distribution of pteridophyte spores varied between different sampling points studied by various papers, probably due to the phenology of each species and the environmental conditions of each site (Landi et al. 2014; Mehltreter and Palacios-Ríos 2003). Figure 20.4 shows the total monthly variation of pteridophyte spores in several localities located on various continents and with different collection methods. In several parts of the northern hemisphere, higher atmospheric levels of spores were observed between June and August (Haratym et al. 2014; Huang et al. 2008) while in other places they were concentrated between February and April (Njokuocha 2006) or even between February and June (Abu-Dieyeh and Ratrout 2012). As noted by Yasmeen and Devi (1988), in places with tropical forests, airborne levels of spores appeared to be more or less continuous throughout the year (Gómez-Noguez et al. 2017; Hossain and Pasha 2012). In other southern hemisphere locations, the largest number of spores was collected between October and November (Noetinger et al. 1994).

Some studies were able to differentiate between several types of spores over a year and consequently provided information on the variation in the atmospheric content of spores at these sites (Gómez-Noguez et al. 2017; Hossain and Pasha 2012; Huang et al. 2008; Noetinger et al. 1994). Thus, in Kinmen Island, East Asia, in a work carried out using the volumetric method, different spore types were observed throughout the year, with *Lycopodium* present only in the month of April, *Dicranopteris* between June and August, *Alsophila* between July and September and *Pteris* in two months, April and November (Fig. 20.5). In other studies,

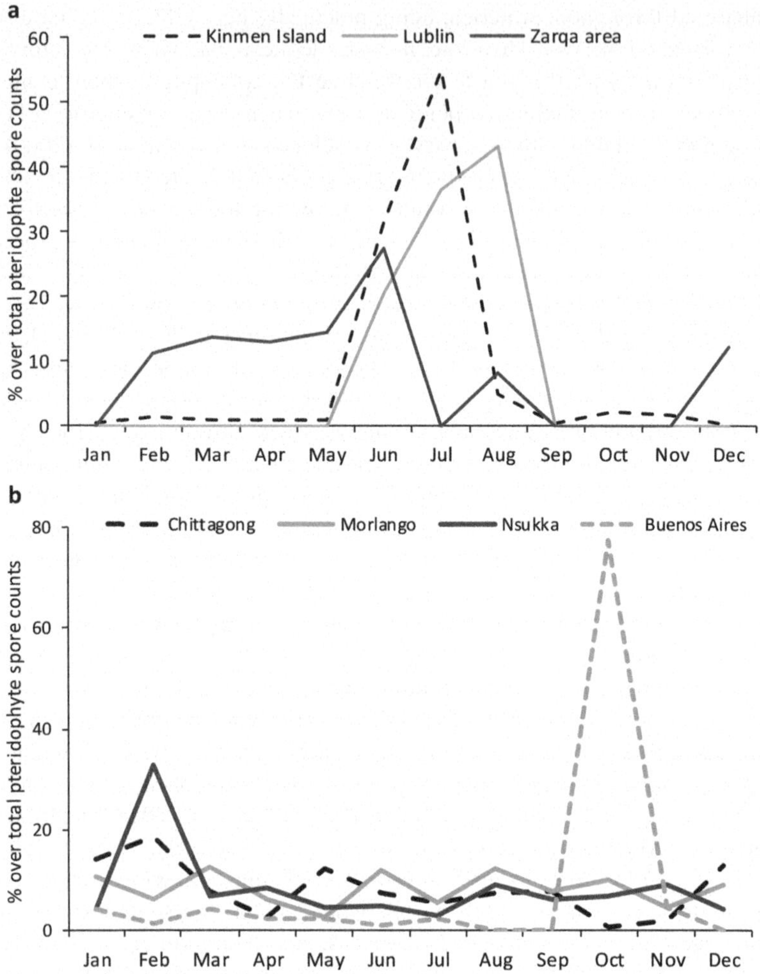

Fig. 20.4 Seasonal distribution of airborne pteridophyte spores by means of volumetric method (**a**) or gravimetric method (**b**)

developed using the gravimetric method, monthly distribution patterns were observed in the atmosphere for various genera, from a more or less continuous presence throughout the year in Molango, passing through more defined periods in Chittagong depending on the spore type analysed, to a specific monthly pattern as in Buenos Aires, with *Equisetum* only present in October and *Nephrolepis* in May and June (Fig. 20.5). This short presence of *Equisetum* spores in the atmosphere was also observed in Central Europe (Kasprzyk 2004) and coincided with this short production period, concentrated during its growing season (Duckett 1985).

Information on intra-daily variation of pteridophyte spore levels is even lower than that reported for annual and/or monthly values. A study developed in Singapore

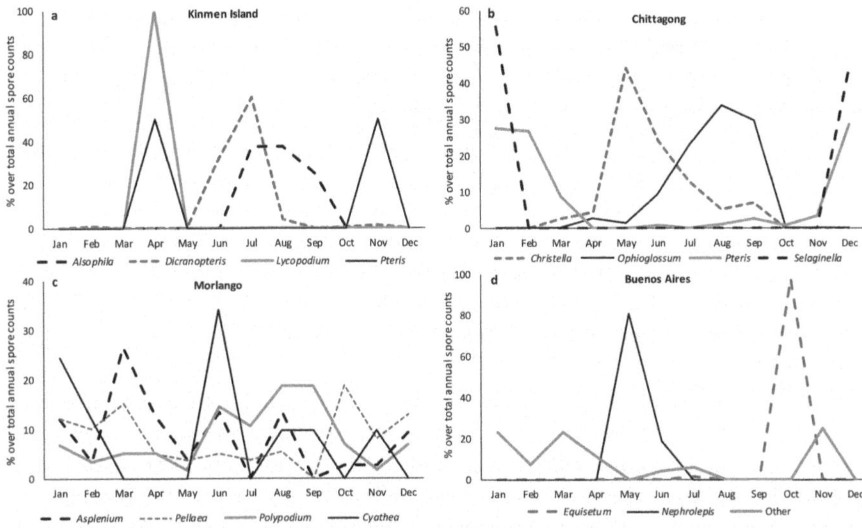

Fig. 20.5 Seasonal distribution of some pteridophyte taxa at various sampling points by means of volumetric method (**a**) or gravimetric method (**b–d**)

by volumetric methods displayed the highest concentrations of some airborne spores during the central hours of the day, between 10:00 and 18:00 h, except for *Asplenium nidus*, reaching higher levels between 08:00 and 14:00 h (Ong et al. 2012). The same paper indicated the highest hourly values for some species spores, with a time range between 14:00 and 16:00 (*Dicranopteris curranii* 15 spores/m^3, *Nephrolepis auriculata* 133 spores/m^3, *Pteridium aquilinum* 74 spores/m^3 and *Stenochlaena palustris* 30 spores/m^3), except for *Asplenium nidus* and *Dicranopteris linearis*, that registered their maximum values between 08:00 and 10:00 (30 spores/m^3) and between 16:00 and 18:00 (37 spores/m^3), respectively.

All this work highlights the need for further studies on the atmospheric content of pteridophyte spores, in order to increase the knowledge of these airborne particles, despite their low number in relation to pollen grains or fungal spores.

20.4 Applications and New Perspectives

Knowledge of the atmospheric dynamics of spores is related to other disciplines such as medicine and allergology to a greater extent. It is well known that pollen grains and spores (mainly fungal spores) produced by different organisms trigger allergic processes (D'Amato and Spieksma 1995; Hyde 1973). In recent years there has been an increase in the prevalence of these processes mainly due to pollution and climate change (Beggs 2010; Cakmak et al. 2012). The scientific literature showed little work on the allergenicity of different vegetative and reproductive parts of ferns and lycopods (Andersen and Paulsen 2016; Hausen and Schulz 1978; Rodríguez de la Cruz et al. 2009). The first references to pteridophyte spores allergy appeared in the mid-twentieth century, specifically for *Lycopodium* (Salén 1951), and it was not until the late 1980s that positive skin test reaction to extracts from ferns spores in a group of patients suffering from hay fever was reported (Geller-Bernstein et al. 1987). In later years, cases of allergy to spores of various taxa were displayed, such as *Acrostichum aureum* (Bunnag et al. 1989; Yasmeen and Devi 1988), *Asplenium nidus* (Chew et al. 2000), *Dicranopteris curranii* (Chew et al. 2000), *Lycopodium* (Devi et al. 1989), *Nephrolepis* (Chew et al. 2000; Kofler et al. 2000), *Pteridium aquilinum* (Chew et al. 2000) and *Stenochlaena palustris* (Chew et al. 2000).

Other human health problems, since the late twentieth and early twenty-first centuries, associated with the spores of various fern species were their carcinogenic potential (Potter and Baird 2000; Povey et al. 1996; Simán et al. 2000), mainly due to ptaquiloside and other genotoxic terpenoids (Tomšík 2014). Spores may be inhaled or ingested directly by humans or other livestock animals (Wilson et al. 1998) and also indirectly by consumption of products obtained from livestock holdings (Virgilio et al. 2015) or even receiving waters (O'Driscoll et al. 2016). However, this risk was reduced as Rasmussen et al. (2013) indicated, displaying the minor concentration of ptaquiloside in *Pteridium aquilinum* spores and the lower probability of aerial exposure to these spores. Conversely, several pteridophyte spores appeared to have beneficial effects for the treatment of some diseases, such as *Lycopodium clavatum* spores, from which lycopodine is derived causing apoptosis and cell cycle arrest in prostate cancer cells (Bishayee et al. 2013). For *L. clavatum* spores, ethnobotanical uses were also reported for the treatment of rheumatism, cramps and varices, as well as for the spores of *Angiopteris evecta* in the treatment of leprosy and other skin diseases or *Psilotum nudum* to stop childhood diarrhoea (Kumari et al. 2011). Furthermore, it should be noted that *L. clavatum* spores were also proposed for use by the pharmaceutical industry in the oral administration of peptides and proteins (Sudareva et al. 2017) and even for oral vaccination (Atwe et al. 2014).

In other disciplines related to aerobiology, forensic palynology and paleopalynology, ferns and lycopod spores could play an important role. Some murder cases could be solved thanks to the presence of pteridophyte spores, among other biological particles. Wiltshire (2009) reported a murder case in which the murderer failed to recall the place where he buried the body and could finally be found thanks to analysis of the palynomorphs of various objects, including *Polypodium* spores, which were very abundant at the burial site. In other police cases, the presence of *Lycopodium* spores in a sample of cocaine was very useful to guess its origin, considering the time when sporangia were mature (Stanley 1992). The analysis of different types of sediments, including pteridophyte spores, provided information for the reconstruction of paleoenvironments (Lindström et al. 2017) and even to assess the position of the vegetation belts on the adjacent continents (Hooghiemstra et al. 2006).

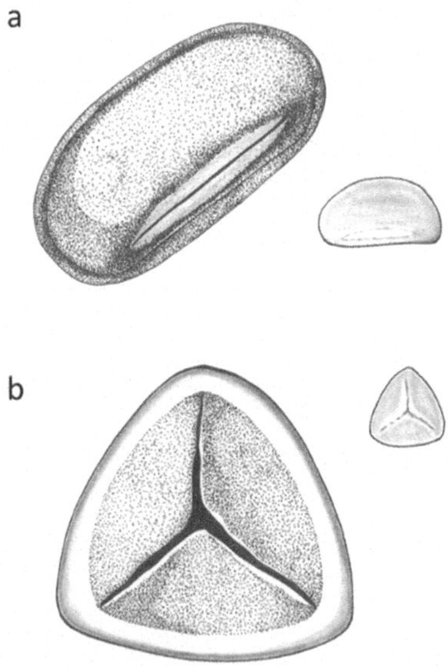

Fig. 20.6 Different types of pteridophyte spores in polar view: monolete spore (**a**) and trilete spore (**b**)

Ferns and lycopod airborne spores were occasionally found in studies from other disciplines such as entomology (Quamar and Bera 2016) or climatology (Adeonipekun and John 2011), causing the development of new specific traps for their analysis (Gómez-Noguez et al. 2014) and showing the importance of humidity in their dispersion through wind.

20.5 Some Notes on the Morphology of Pteridophyta Spores

This section discusses the most relevant aspects of the morphology of ferns and lycopods in order to better understand some important details for the identification of different pteridophyte spore types (Punt et al. 2007; Sáenz Laín 2004). The different types of shape and symmetry that spores may present cannot be understood without knowing their origin. In the sporangia of the different species, there are cells called sporocytes that will produce four haploid cells in tetrad form after meiosis. The part of the spore in this tetrad that is oriented toward the centre of the tetrad is called the proximal face, and that one which is on the opposite side, i.e. toward the outside of the tetrad, is the distal face. This tetrad eventually separates, leaving individual spores. At the time of germination, these spores can be opened by apertures or laesurae on the proximal face. In turn, the spores may have radial or bilateral symmetry.

Spores with radial symmetry have a triradiate laesurae formed by three arms converging at one point, the proximal pole: these are known as trilete spores (Fig. 20.6a). If observed in polar view, they may have a circular or more or less triangular contour, as a result of their general spherical or tetrahedral shape, respectively

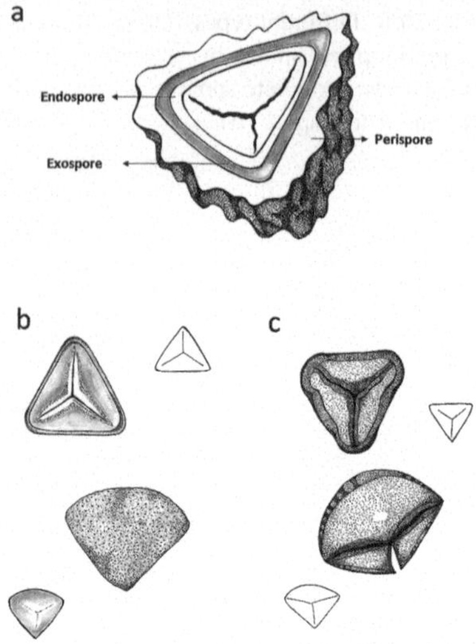

Fig. 20.7 Different pteridophyte spore wall forming layers (**a**) Trilete spore with triangular (**b**) and subtriangular (**c**) shape in polar (*above*) and equatorial (*below*) view

(Fig. 20.7b, c). Spores with bilateral symmetry have a linear laesura, and they are therefore monolete spores (Fig. 20.6b).

The shape of the spore in equatorial view can be elliptical, when elongated and with the poles of equal width; ovoid, when one of the two poles is narrower than the other (in some cases, also if they are both clearly narrower than in an elliptical spore); reniform (or bean-shaped) when in the equatorial zone there is a constriction; or spherical. There is another type of spore in the genus *Equisetum* that does not have laesurae, but it does have special structures, the elaters, with hygroscopic properties.

The size of the spores is usually between 25 and 60 micrometres (μm), although there are large spores, mainly in the case of some heterosporous taxa megaspores, such as *Azolla* or *Selaginella*.

The entire wall of a spore is called sporoderm and consists of a series of layers. The exospore is the main layer of a spore wall and acetolysis resistant. Perispore is the outer layer, is not always acetolysis resistant and lacks clear apertural differentiation. Endospore is the innermost layer of a spore wall, not acetolysis resistant, underlying the exospore and bordering the surface of the cytoplasm (Fig. 20.7a). Perispore and exospore are responsible for the surface relief, or topography, of the spores, variable between different taxa. Therefore, regardless of the layer in which the different ornamentations that make up different types of surface are arranged, spores can be found with smooth (psilate), perforate, reticulate, granulate and echinate, among other surfaces (Fig. 20.8).

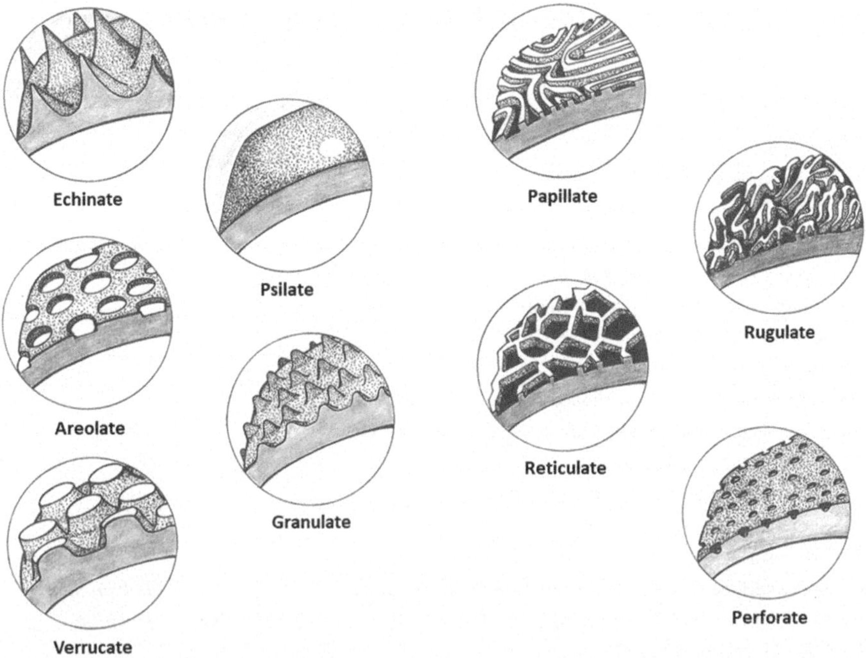

Fig. 20.8 Some types of surface in pteridophyte spores

20.6 Some Notes About the Key to Airborne Pteridophyte Spores

This last section (see Annex) presents a simple tool to identify airborne fern and lycopod spores, although it should be noted that this dichotomous key tends to simplify many aspects related to the morphology of spores and consequently has certain limitations to discriminate against certain taxa at the genus level. In this sense, the surface of the spore wall is used to avoid differentiating between perispore and exospore, since in many cases, and depending on various factors such as the sampling method used, it is difficult to distinguish them. Various works on pteridophyte spore morphology were used for its preparation (Contreras-Duarte et al. 2006; Gómez-Noguez et al. 2013; Makgomol 2006; Tyron and Lugardon 1991; Vijayakanth and Sathish 2016; Zenkteler 2012, among others). In order to better discriminate spores, we recommend checking the distribution of the different taxa in Annex 20.1, according to the origin of the samples under analysis.

As a conclusion to this chapter, there could be said that an increasing number of papers related to ferns and lycopods spores are being registered over the years and throughout the world. In that sense, seasonal behaviour is more or less established for various spore types/taxa in both hemispheres. However, more studies are required, in order to determine daily patterns as well as relations with meteorological parameters.

Annex 20.1 Key to Airborne Fern and Lycopod Spores

Spores inaperturate	*Equisetum*
Spores aperturate or scared	1
1a. Green spores (with chlorophyll)	1′
1b. Spores with a different colour	2
1’a. Spores with a single aperture (monolete) as a furrow	*Onocleopsis*
1’b. Spores with a trilete scar present	*Osmunda*
2a. Spores with a single aperture (monolete) as a furrow	3
2b. Spores with a trilete scar present	42
Monolete spores	
3a. Kidney-shaped (or bean-shaped) spores in equatorial view	4
3b. Ellipsoidal, ovoid or spherical spores in equatorial view	12
4a. Spore surface mainly verrucate	5
4b. Spore surface with another main ornamentation	10
5a. Spore surface only verrucate	6
5b. Spore surface verrucate or with other ornamentation	7
6a. Spore surface with perispore adhering to the exospore	*Polypodium*
6b. Spore surface with loose perispore wall, generally of a smaller size than the previous genus (<55 μm)	*Davallia*
7a. Spore surface verrucate or granulate	8
7b. Spore surface verrucate, echinate or papillate	9
8a. Spore surface verrucate or granulate	*Woodwardia*
8b. Spore surface verrucate or granulate, some species echinate	*Cystopteris*
9a. Spore surface verrucate and/or echinate	*Dryopteris*
9b. Spore surface verrucate, papillate or rugulate	*Thelypteris*
10a. Spore surface reticulate or cristate	*Cyclosorus*
10b. Spore surface rugulate or with other ornamentation	11
11a. Spore surface rugulate forming folds with a reticular appearance	*Elaphoglossum.*
11b. Spore surface rugulate, granulate or reticulate, rarely wedged	*Diplazium*
12a. Ellipsoidal to spherical spores	13
12b. Ellipsoidal to ovoid spores	26
13a. Spore surface mainly verrucate	14
13b. Spore surface with another main ornamentation	18
14a. Spore usually yellow or translucent	15
14b. Spore not yellow or translucent	16
15a. Translucent or light yellow spores	*Campyloneurum*
15b. Yellow spores	*Pecluma*
16a. Spore surface verrucate, sometimes areolate	*Phlebodium*
16b. Spore surface verrucate, rugulate, echinate or reticulate	17
17a. Spore surface verrucate or rugulate	*Microsorum*
17b. Spore surface verrucate, echinate or reticulate	*Asplenium*
18a. Spore surface mainly papillate	19
18b. Spore surface no papillate	20

19a. Spore surface finely papillate	*Tapeinidium*
19b. Spore surface papillate or cristate	*Matteuccia*
20a. Spore surface mainly foveolate	21
20b. Spore surface not foveolate	22
21a. Spore surface foveolate	*Lemmaphyllum*
21b. Spore surface foveolate, rugulate or even psilate	*Lepisorus*
22a. Spore surface echinate	*Ctenitis*
22b. Spore surface not echinate	23
23a. Spore surface psilate, cristate, sometimes with convoluted folds	*Lastreopsis*
23b. Spore surface psilate, rugulate or granulate	24
24a. Spore surface psilate with globules scattered	*Platycerium*
24b. Spore surface not always psilate	25
25a. Spore surface psilate or rugulate	*Phymatosorus*
25b. Spore surface psilate or granulate	*Odontosoria*
26a. Spore surface mainly verrucate	27
26b. Spore surface not verrucate	34
27a. Spore surface only verrucate	*Goniophlebium*
27b. Spore surface verrucate or sometimes with other types of ornamentation	28
28a. Spore surface verrucate with tubercles	*Nephrolepis*
28b. Spore surface without tubercles	29
29a. Spore surface verrucate or echinate	30
29b. Spore surface verrucate or other ornamentation not echinate	31
30a. Spore sometimes with a folded foveolate-echinate surface	*Tectaria*
30b. Spore surface without a folded foveolate-echinate surface	*Colysis*
31a. Spore surface verrucate or reticulate	*Microgramma*
31b. Spore surface verrucate or other ornamentation not reticulate	32
32a. Spore surface verrucate or granulate	*Pleopeltis*
32b. Spore surface verrucate, rugulate or cristate	33
33a. Spore surface verrucate or rugulate	*Stenochlaena*
33b. Spore surface verrucate or cristate, sometimes folded	*Bolbitis*
34a. Spore surface mainly papillate	35
34b. Spore surface not papillate	37
35a. Spore surface only papillate	*Vittaria*
35b. Spore surface papillate or reticulate	36
36a. Spore surface papillate or reticulate, sometimes areolate or echinate	*Polystichum*
36b. Spore surface papillate or reticulate, sometimes cristate	*Athyrium*
37a. Spore surface sometimes echinate	38
37b. Spore surface not echinate	40
38a. Spore surface echinate, some species granulate	*Drynaria*
38b. Spore surface echinate, cristate or tuberculate	39
39a. Spore surface echinate or cristate	*Megalastrum*
39b. Spore surface echinate or tuberculate	*Hypolepis*
40a. Spore surface psilate with prominent perforate wing-like folds	*Cyclodium*
40b. Spore surface not psilate	41
41a. Spore surface reticulate or vermiculate	*Phanerophlebia*

41b. Spore rugulate	*Blechnum*
Trilete spores	
42a. Spore surface with long pointed spines or granulate/verrucate sculpture (large megaspore >1000 μm)	*Selaginella*
42b. Spore surface without long pointed spines and/or other ornamentation	43
43a. Spore surface clearly reticulate	44
43b. Spore surface not reticulate	46
44a. Trilete scar with a thick border	*Ophioglossum*
44b. Trilete scar without a thick border	45
45a. Reticulum regular, sometimes with small spines	*Lycopodium*
45b. Reticulum irregular, occasionally near to rugulate	*Lycopodiella*
46a. Spore with tetrahedral/triangular shape	47
46b. Spore with rounded/globose or subtriangular shape	56
47a. Spore surface sometimes reticulate	48
47b. Spore surface not reticulate	50
48a. Spore surface reticulate, sometimes verrucate or echinate	*Pteris*
48b. Spore surface reticulate, granulate or psilate	49
49a. Spore surface reticulate or granulate or even psilate	*Lindsaea*
49b. Spore surface reticulate or granulate, never psilate	*Cyathea*
50a. Spore surface mainly rugulate	51
50b. Spore surface not rugulate	54
51a. Spore surface rugulate, sometimes folded, spongy in appearance	*Botrychium*
51b. Spore surface rugulate or psilate	52
52a. Spore surface low rugulate or psilate	*Diplopterygium*
52b. Spore surface only rugulate	53
53a. Spore surface only low rugulate	*Dicranopteris*
53b. Spore surface rugulate, sometimes cristate	*Pellaea*
54a. Spore surface granulate or verrucate	*Microlepia*
54b. Spore surface not granulate or verrucate	55
55a. Spore surface foveolate	*Cnemidaria*
55b. Spore surface papillate or even psilate	*Alsophila*
56a. Spore with rounded/globose shape	57
56b. Spore with a different shape (from rounded-subtriangular to triangular)	60
57a. Spore surface granulate	58
57b. Spore surface not granulate	59
58a. Spore surface simple granulate	*Lonchitis*
58b. Spore surface granulate (sometimes subtriangular shape)	*Odontosoria*
59a. Spore surface tuberculate	*Taenitis*
59b. Spore surface papillate/reticulate (large megaspore >200 μm with floats above and attached massula below containing microspores)	*Azolla*
60a. Spore with a rounded-subtriangular shape	61
60b. Spore with a subtriangular shape	68
61a. Spore surface with projecting rods by granulate material, more or less striate	*Ceratopteris*
61b. Spore surface without projecting rods	62

62a. Spore surface papillate	*Vittaria*
62b. Spore surface not papillate	63
63a. Spores surface granulate or reticulate	64
63b. Spore surface sometimes granulate, rarely reticulate	65
64a. Spore surface granulate or reticulate	*Dicksonia*
64b. Spore surface granulate or reticulate, some species cristate	*Cheilanthes*
65a. Spore surface granulate or verrucate, sometimes rugulate	66
65b. Spore surface granulate but never verrucate	67
66a. Spore surface granulate or verrucate	*Acrostichum*
66b. Spore surface finely granulate or verrucate, sometimes rugulate	*Adiantum*
67a. Spore surface granulate or cristate, sometimes reticulate	*Notholaena*
67b. Spore surface granulate, tuberculate, foveolate or even microreticulate	*Lophosoria*
68a. Spore surface granulate or baculate	*Anemia*
68b. Spore surface not granulate or baculate, mainly verrucate	69
69a. Spore surface verrucate or reticulate	*Dennstaedtia*
69b. Spore surface only verrucate	70
70a. Spore surface roughly verrucate	*Pteridium*
70b. Spore surface uniformly verrucate	*Lygodium*

References

Abu-Dieyeh MH, Ratrout YS (2012) Seasonal variation of airborne pollen grains in the atmosphere of Zarqa area, Jordan. Aerobiologia 28:527–539

Adeonipekun PA, John M (2011) Palynological investigation of haze dust in Ayetoro-Itele Ota, Southwest Nigeria. J Ecol Nat Environ 3(14):455–460

Ajikah L, Ogundipe OT, Bamgboye O (2015) Palynological survey of airborne pollen and spores in the University of Lagos, Akoka campus, Southwestern Nigeria. Ife Journal of Science 17(3):643–655

Alzate Guarín F, Quijano Abril MA, Álvarez A, Fonnegra R (2015) Atmospheric pollen and spore content in the urban area of the city of Medellin, Colombia. Hoehnea 42(1):9–19. http://www.scielo.br/scielo.php?script=sci_arttext&pid=S2236-89062015000100009

Andersen F, Paulsen E (2016) Allergic contact dermatitis caused by the Boston fern *Nephrolepis exaltata* 'Bostoniensis. Contact Derm 75:240–259

Atwe SU, Ma Y, Gill HS (2014) Pollen grains for oral vaccination. J Control Release 194:45–52

Beggs PJ (2010) Adaptation to impacts of climate change on aeroallergens and allergic respiratory diseases. Int J Environ Res Public Health 7(8):3006–3021

Berman D, Hons BA (2013) Regional-specific pollen and fungal spore allergens in South Africa. Curr Allergy Clin Im 26(4):196–209

Bishayee K, Chakraborty D, Ghosh S, Boujedaini N, Khuda-Bukhsh AR (2013) Lycopodine triggers apoptosis by modulating 5-lipoxygenase, and depolarizing mitochondrial membrane potential in androgen sensitive and refractory prostate cancer cells without modulating p53 activity: signaling cascade and drug-DNA interaction. Eur J Pharmacol 698:110121

Boch S, Berlinger M, Fischer M, Knop E, Nentwig W, Türke M, Prati D (2013) Fern and bryophyte endozoochory by slugs. Oecol 172:817–822

Boch S, Berlinger M, Prati D, Fischer M (2016) Is fern endozoochory widespread among fern-eating herbivores? Plant Ecol 27(1):13–20

Bunnag C, Dhorranintra B, Limsuvan S, Jareoncharsri P (1989) Ferns and their allergenic importance: skin and nasal provocation tests to fern spore extract in allergic and non-allergic patients. Ann Allergy 62(6):554–558

Cakmak S, Dales RE, Coates F (2012) Does air pollution increase the effect of aeroallergens on hospitalization for asthma? J Allergy Clin Immunol 129(1):228–231

Camacho IC (2015) Airborne pollen in Funchal city, (Madeira Island, Portugal) – First pollinic calendar and allergic risk assessment. Ann Agric Environ Med 22(4):608–613

Caulton E, Keddie S, Carmichael R, Sales J (2000) A ten year study of the incidence of spores of bracken, (*Pteridium aquilinum* (L.) Kuhn.) in an urban rooftop airstream in south east Scotland. Aerobiologia 16:29–33

Chen S-H (1984) Aeropalynological study of Nankang, Taipei (Taiwan). Taiwania 29:113–120

Chen S-H, Chien M-C (1986) Two-year investigation of the airborne pollen at Nankang, Taipei (Taiwan). Taiwania 31:33–40

Chen S-H, Huang T-C (1980) Aeropalynological study of Taipei Basin, Taiwan. Grana 19(2):147–155

Chew FT, Lim SH, Shang HS, Siti Dahlia MD, Goh DYT, Lee BW (2000) Evaluation of the allergenicity of tropical pollen and airborne spores in Singapore. Allergy 55:340–347

Christenhusz MJM, Zhang X-C, Schneider H (2011) A linear sequence of extant families and genera of lycophytes and ferns. Phytotaxa 19:7–54

Christopher EB, Ojone AS, Onojo IS (2013) A Study of Derived Savanna environment through airborne palynomorphs, Anyigba, Kogi State, Nigeria. Sch Acad J Biosci 1(6):313–317

Conant DS (1978) A radioisotope technique to measure spore dispersal of the tree fern *Cyathea arborea* SM. Pollen Spores 20:580–593

Contreras-Duarte AR, Bogotá-Ángel RG, Jiménez-Bulla LC (2006) Atlas de las esporas de Chipaque (Cundinamarca, Colombia). Cadalsia 28(2):327–357

Corlett R (2016) Plant diversity in a changing world: status, trends, and conservation needs. Plant Diver 38:10–16

Cour P (1974) Nouvelles techniques de détection des flux et des retombées polliniques: étude de la sedimentation des pollens et des spores á la surface du sol. Pollen Spores 16:103–141

Cousens MI (1988) Reproductive strategies of pteridophytes. In: Lovett-Doust J, Lovett-Doust L (eds) Plant reproductive ecology: patterns and strategies. Oxford University Press, New York, pp 307–328

D'Amato G, Spieksma FTM (1995) Aerobiologic and clinical aspects of mould allergy in Europe. Allergy 55(11):870–877

Davies RR (1971) Air sampling for fungi, pollens and bacteria. Academic Press, London

Devi S (1980) The concept of Perispore-an assessment. Grana 19:159–172

Devi S, Yasmeen Singh J, Shankar R (1989) Patch testing animals to allergenic Fern spores. Cutan Ocul Toxicol 8(2):167–172

Duckett JG (1985) Wild gametophytes of *Equisetum sylvaticum*. Am. Fern J. 75:120–127

Durham OC (1946) The volumetric incidence of atmospheric allergens, IV. A proposed standard method of gravity sampling counting and volumetric interpolation of results. J Allergy 17:79–86

Erdtman G (1960) The acetolysis method, a revised description. Sven Bot Tidskr 54:561–564

Favali MA, Gallo F, Maggi O, Mandrioli P, Pacini E, Pasquariello G, Piervittori R, Pietrini AM, Ranalli G, Ricci S, Roccardi A, Sorlini C (2003) Analysis of the biological aerosol. In: Mandrioli P, Caneva G, Sabbioni C (eds) Cultural heritage and aerobiology. Kluwer Academic Publishers, Dordrecht, pp 145–172

Geller-Bernstein C, Keynan N, Bejerano A, Shomer-Ilan A, Waisel Y (1987) Positive skin tests to fern spore extracts in atopic patients. Ann Allergy 58(2):125–127

Gómez-Noguez F, Pérez-García B, Mendoza-Ruiz A, Orozco-Segovia A (2013) Flora palinológica de los helechos y licofitas de Río Malila, Hidalgo, México. Bot Sci 91(2):135–154

Gómez-Noguez F, Pérez-García B, Mendoza-Ruiz A, Orozco-Segovia A (2014) A Pluviometric Fern spore, fungal spore, and pollen trap. Am Fern J 104(1):1–6

Gómez-Noguez F, Pérez-García B, Mehltreter K, Orozco-Segovia A, Rosas-Pérez I (2016) Spore mass and morphometry of some fern species. Flora 223:99–105

Gómez-Noguez F, Pérez-García B, Mendoza-Ruiz A, Orozco-Segovia A (2017) Fern and lycopod spores rain in a cloud forest of hidalgo, Mexico. Aerobiologia 33:23–35

Gregory PH (1973) The microbiology of the atmosphere. Leonard Hill, Plymouth, pp 39–42

Haratym W, Weryszko-Chmielewska E, Dmitruk M (2014) An analysis of the content of Pteridophyta spores in aeroplankton of Lublin (2013-2014). Acta Agrobot 67(3):21–28

Hausen BM, Schulz KH (1978) Occupational allergic contact dermatitis due to leatherleaf fern *Arachnioides adiantiformis* (Forst) Tindale. Br J Dermatol 98(3):325–329

Hirst JM (1952) An automatic volumetric spore trap. Ann Apl Biol 39(2):257–265

Hooghiemstra H, Lézine A-M, Leroy SAG, Dupont L, Marret F (2006) Late quaternary palynology in marine sediments: a synthesis of the understanding of pollen distribution patterns in the NW African setting. Quat Int 148:29–44

Hossain MS, Pasha MK (2012) Airborne fungal and pteridophytic spores in Chittagong University Campus, Chittagong. J. Asiat. Soc. Bangladesh, Sci 38(1):119–124

Huang T-C, Huang S-Y, Hsiao A, Chen S-H (2008) Aeropalynological study of Kinmen Island, Taiwan. Taiwania 53(4):369–382

Hurtado I, Riegler-Goihman, M (1986) Air-Sampling Studies in a Tropical Area. Grana 25(1):63–68

Hyde HA (1973) Atmospheric pollen and spores in relation to allergy. II. Clin Exp Allergy 3:109–126

Kasprzyk I (2004) Airborne pollen of entomophilous plants and spores of pteridophytes in Rzeszów and its environs (SE Poland). Aerobiologia 20:217–222

Kofler H, Hemmer W, Focke M, Jarisch R (2000) Fern allergy. Allergy 55:299–300

Kreft H, Jetz W, Mutke J, Barthlott W (2010) Contrasting environmental and regional effects on global pteridophyte and seed plant diversity. Ecography 33:408–419

Kumari P, Otaghvari AM, Govindapyari H, Bahuguna YM, Uniyal PL (2011) Some ethnomedicinally important Pteridophyte of India. Int J Med Arom Plants 1(1):18–22

Lacey ME, McCartney HA (1994) Measurement of airborne concentrations of spores of bracken (*Pteridium aquilinum*). Grana 33:91–93

Lacey J, Crook B, Janaku Bai A (1996) The detection of airborne allergens implicated in occupational asthma. In: Muillenberg ML, Burge HA (eds) Aerobiology. Boca Raton CRC Press, New York

Landi M, Zoccola A, Bacaro G, Angiolini C (2014) Phenology of *Dryopteris affinis* ssp. *affinis* and *Polystichum aculeatum*: modeling relationships to the climatic variables in a Mediterranean area. Plant Species Biol 29:129–137

Large MF, Braggins JE (1990) Effect of different treatments on the morphology and size of fern spores. Rev Paleobot Palynol 64(1):213–221

La Serna-Ramos IE, Domínguez-Santana MD (2003) Pólenes y esporas aerovagantes en Canarias. Servicio de Publicaciones de la Universidad de La Laguna, La Laguna

Lee PH, Lin TT, Chiou WL (2009) Phenology of 16 species of ferns in a subtropical forest of northeastern Taiwan. J Plant Res 122(1):61–67

Leitão MT, Santos MF, Sérgio C, Ormonde J, Carvalho GM (1996) Plantas criptogâmicas na atmosfera de Coimbra. Portugal Anales Jard Bot Madrid 54:30–36

Levetin E, Rogers CA, Hall SA (2000) Comparison of pollen sampling with a Burkard spore trap and a Tauber trap in a warm temperate climate. Grana 39:294–302

Lindström S, Erlström M, Piasecki S, Nielsen LH, Mathiesen A (2017) Palynology and terrestrial ecosystem change of the middle Triassic to lowermost Jurassic succession of the eastern Danish Basin. Rev Palaeobot Palynol 244:65–95

Majas FD, Romero EJ (1992) Aeropalynological research in the Northeast of Buenos Aires Province, Argentina. Grana 31(2):143–156
Makgomol K (2006) Morphology of Fern spores from Phu Phan National Park. Kasetsart J (Nat Sci) 40:116–122
Mandrioli P, Comtois P, Levizzani V (1998) Methods in aerobiology. Pitagora Editrice Bologna, Bologna
Mehltreter K, Palacios-Ríos M (2003) Phenological studies of *Acrostichum danaeifolium* (Pteridaceae, Pteridophyta) at a mangrove site on the Gulf of Mexico. J Trop Ecol 19(2):155–162
Mickel JT (1982) Fern spore? What for? Fiddlehead forum. News Am Fern Soc 9:5
Newson R, Strachan D, Corden J, Millington W (2000) Fungal and other spore counts as predictors of admissions for asthma in the Trent region. Occup Environ Med 57:786–792
Niklas KJ, Tiffney BH, Knoll AH (1983) Patterns in vascular land plant diversification. Nature 303:614–616
Nilsson S, Praglowski J (1974) Pollen and spore incident and phenology in the Stockholm area during 1972. Grana 14(2):78–84
Njokuocha RC (2006) Airborne pollen grains in Nsukka, Nigeria. Grana 45:73–80
Noetinger M, Romero EJ, Majas FD (1994) Airborne pollen and spores monitoring in Buenos Aires city: a preliminary report. Part II. Herbs, weeds (NAP) and spores. General discussion. Aerobiologia 10:129–139
O'Driscoll C, Ramwell C, Harhen B, Morrison L, Clauson-Kaas F, Hansen HCB, Campbell G, Sheahan J, Misstear B, Xiao L (2016) Ptaquiloside in Irish bracken ferns and receiving waters, with implications for land managers. Molecules 21(5):543
Ogden EC, Raynor GS (1967) A new sampler for airborne pollen emission in *Ambrosia*, *Phleum*, *Zea* and *Ricinus*. Amer J Bot 56(1):16–21
Ogden EC, Raynor GS, Hayes JV, Lewis DM, Haines JH (1974) Manual for sampling airborne pollen. Hapner Press, New York
Ong TC, Lim SH, Chen X, Dali SDM, Tan HTW, Lee BW, Chew FT (2012) Fern spore and pollen air spora profile of Singapore. Aerobiologia 28:135–151
Pathirane L (1975) Aerobiological literature in scientific periodicals. Grana 15:145–147
Penrod KA, McCormick LH (1996) Abundance of viable hay-scented fern spores germinated from hardwood forest soils at various distances from a source. Am Fern J 86:69–79
Perkins WA (1957) The rotorod sampler. Second Semiannual Report. CLM 186. Standford Univ, California
Pla Dalmau JM (1958) Aeropalinología gerundense. An Inst Est Gerundenses 12:63–88
Potter DM, Baird MS (2000) Carcinogenic effects of ptaquiloside in bracken fern and related compounds. Br J Cancer 83(7):1342–1348
Povey AC, Potter D, O'Connor PJ (1996) 32P-Postlabeling analysis of DNA adducts formed in the upper gastrointestinal tissue of mice fed bracken or bracken spores. Br J Cancer 74:1342–1348
Prance GT (2001) Discovering the plant world. Taxon 50(2):345–359
Pryer KM, Schuettpelz E, Wolf PG, Schneider H, Smith AR, Cranfill R (2004) Phylogeny and evolution of ferns (monilophytes) with a focus on the early leptosporangiate divergences. Amer J Bot 91:1582–1598
Punt W, Hoen PP, Blackmore S, Nilsson S, Le Thomas A (2007) Glossary of pollen and spore terminology. Rev Palaeobot Palynol 143:1–81
Quamar MF, Bera SF (2016) Pollen analysis of spider web samples from Korba District, Chhattisgarh (central India): an aerobiological aspect. Aerobiologia 32(4):645–655
Ranker TA, Sundue MA (2015) Why are there so few species of ferns? Trends Plant Sci 20:402–403
Rantio-Lehtimäki A (1994) Short, medium and long range transported airborne particles in viability and antigenic analyses. Aerobiologia 10:175–181
Rasmussen LH, Schmidt B, Sheffield E (2013) Ptaquiloside in bracken spores from Britain. Chemosphere 90:2539–2541
Rodríguez de la Cruz D, Sánchez-Reyes E, Sánchez-Sánchez J (2009) Effects of meteorological factors on airborne bracken (*Pteridium aquilinum* (L.) Kuhn.) spores in Salamanca (middle-west Spain). Int J Biometeorol 53:231–237

Rodríguez de la Cruz D, Sánchez-Reyes E, Sánchez-Sánchez J (2010) Aerobiology of Pteridophyta spores: preliminary results and applications. In: Kumar A, Fernández H, Revilla MA (eds) Working with ferns: issues and applications. Springer, New York, pp 271–281

Rodríguez de la Cruz D, Sánchez Reyes E, Martín Baz A, Sánchez Sánchez J (2011) Aerobiological survey in the biosphere reserve "sierras de Francia y Béjar" (MW Spain). In: Testillano PS, Pardo C, Risueño MC, López-Cepero JM (eds) Pollen biotechnology, diversity and function in a changing environment. Consejo Superior de Investigaciones Científicas (CSIC), Madrid, p 161

Sáenz Laín C (2004) Glosario de términos palinológicos. Lazaroa 25:93–112

Salén EB (1951) *Lycopodium* allergy. Acta Allergol 4:308–319

Schneller JJ (1975) Untersuchungen an einheimischen Farnen, insbesondere der *Dryopteris filix-mas* Gruppe 3. Teil Okologische Unters Ber Schweiz Bot Ges 85:110–159

Schneller J, Liebst B (2007) Patterns of variation in a common fern (*Athyrium filix-femina*; Woodsiaceae): population structure along and between altitudinal gradients. Am J Bot 94:965–971

Sharpe JM, Mehltreter K, Walker LR (2010) Ecological importance of ferns. In: Mehltreter K, Walker LR, Sharpe JM (eds) Fern ecology. Cambridge University Press, Cambridge, pp 1–21

Simán SE, Povey AC, Ward TH, Margison GP, Sheffield E (2000) Fern spore extracts can damage DNA. Br J Cancer 83(1):69–73

Simabukuro EA, Esteves LM, Felippe GM (2000) Fern spore rain collected at two different heights. Fern Gaz 16(3):147–166

Smith AR, Pryer KM, Schuettpelz E, Korall P, Schneider H, Wolf PG (2006) A classification for extant ferns. Taxon 55(3):705–731

Songnuan W, Bunnag C, Soontrapa K, Pacharn P, Wangthan U, Siriwattanakul U, Malainual N (2015) Airborne pollen survey in Bangkok, Thailand: a 35-year update. Asian Pac J Allergy Immunol 33:253–262

Spieksma FTM (1992) Allergological aerobiology. Aerobiologia 8:5–8

Stanley E (1992) Application in palynology to establish the province and travel history of illicit drugs. Microscope 40:149–152

Sudareva N, Suvorova O, Saprykina N, Vilesov A, Bel'tiukov P, Petunov S, Radilov A (2017) Two-level delivery systems for oral administration of peptides and proteins based on spore capsules of Lycopodium clavatum. J Mater Chem B 5:7711–7720

Sugita N, Ootsuki R, Fujita T, Murakami N, Ueda K (2013) Possible spore dispersal of a bird-nest fern *Asplenium setoi* by Bonin flying foxes *Pteropus pselaphon*. Mammal Study 38:225–229

Tauber H (1967) Investigations of the mode of pollen transfer in forested areas. Rev Palaeobot Palynol 3:277–286

Testo W, Sundue M (2016) A 4000-species dataset provides new insight into the evolution of ferns. Mol Phylogenet Evol 105:200–211

Tomšík P (2014) Ferns and Lycopods - a potential Treasury of anticancer agents but also a carcinogenic hazard. Phytother Res 28:798–810

Tryon AF, Lugardon B (1991) Spores of the Pteridophyta. Springer, New York

Vicent M, Gabriel y Galán JM, Ainoüche A (2014) Insight into fern evolution: a mechanistic approach to main concepts and study techniques. Bot Complut 38:7–24

Vijayakanth P, Sathish SS (2016) Studies on the spore morphology of pteridophytes from Kolli Hills, Eastern Ghats, Tamil Nadu, India. IJRBS 4(1):1–12

Virgilio A, Sinisi A, Russo V, Gerardo S, Santoro A, Galeone A, Taglialatela-Scafati O, Ropero F (2015) Ptaquiloside, the major carcinogen of bracken fern, in the pooled raw milk of healthy sheep and goats: an underestimated, global concern of food safety. J Agric Food Chem 63(19):4886–4892

Wilson D, Donaldson LJ, Sepai O (1998) Should be frightened by bracken? A review of the evidence. J Epidemiol Community Health 52:812–817

Wiltshire PEJ (2009) Forensic ecology, botany, and palynology: some aspects of their role in criminal investigation. In: Ritz K, Dawson L, Miller D (eds) Criminal and environmental soil forensics. Springer Science + Business Media, Dordrecht, pp 129–149
Yang Y-L, Huang T-C, Chen S-H (2003) Diurnal variations of airborne pollen and spores in Taipei City, Taiwan. Taiwania 48(3):168–179
Yasmeen JS, Devi S (1988) Pteridophyte aerospora of India. Grana 27:229–238
Zenkteler E (2012) Morphology and peculiar features of spores of fern species occurring in Poland. Acta Agrobot 65(2):3–10

Chapter 21
Ecological Significance of Brassinosteroids in Three Temperate Ferns

Aránzazu Gómez-Garay, Jose Maria Gabriel y Galán, Alberto Cabezuelo, Beatriz Pintos, Carmen Prada, and Luisa Martín

21.1 Introduction

Plant growth and development was thought to be regulated only by five groups of hormones: auxins, gibberellins, cytokinins, abscisic acid and ethylene. Nevertheless, there is another group of substances which is nowadays being considered as the sixth group of phytohormones: brassinosteroids (BRs). The structure of BRs is derived from the 5-cholestone skeleton, and these polyhydroxy steroid plant molecules have strong similarities to the steroid hormone of animals. Although numerous brassinosteroids have been identified (nearly 70 different compounds have been isolated from plants), only brassinolide and its precursor, castasterone, are active (Nomura et al. 2005; Nomura and Bishop 2006). Brassinolide was firstly isolated from *Brassica napus* L. pollen and castasterone from the insect galls of *Castanea sativa* Mill. Brassinosteroids have been identified from many angiosperm and gymnosperm families and also from bryophytes, pteridophytes and chlorophytes (Ross and Reid 2010).

Brassinosteroid function is linked with many plant processes (e.g. cell division and elongation, stress response, senescence and reproductive development). Brassinosteroids, in junction with auxins and gibberellins, form part of a key subset of plant hormones which are major determinants of plant growth and development (Ross and Reid 2010). Thus, brassinosteroids influence developmental processes like seed germination, rhizogenesis, maturation and senescence. These plant hormones not only promoted seed germination but also reversed the inhibitory effects of the abscisic acid (Vardhini and Rao 1997). Nevertheless, the research to discern the role of brassinosteroids in plants is focused in model species, and the results indicate that there are great differences in hormone physiology across the different model species. Most of these model species are angiosperms. Such physiological evidence is lacking from other plant groups (Ross and Reid 2010). Information on

A. Gómez-Garay · J. M. Gabriel y Galán (✉) · A. Cabezuelo · B. Pintos · C. Prada · L. Martín
Department of Plant Sciences I, Universidad Complutense, Madrid, Spain
e-mail: jmgabrie@ucm.es

H. Fernández (ed.), *Current Advances in Fern Research*,
https://doi.org/10.1007/978-3-319-75103-0_21

the occurrence of endogenous brassinosteroids in non-flowering plants provide important clues on physiological functions of these plant hormones. The biologically active castasterone was identified in most of the ferns studied by Yokota et al. (2017), but the other biologically active brassinosteroid, brassinolide, was not.

Brassinosteroids are present in plants at very low concentrations, and young growing tissues contain higher levels of this type of hormones than mature tissues (Yokota and Takahashi 1986; Clouse 2002). In this way, pollen and immature seeds are the richest sources (Takatsuto 1994). The levels of castasterone in fern leaves are as low as 25 pg g fr. wt^{-1} (Yokota et al. 2017). Up to now, the applied research on brassinosteroids was conducted mainly in agriculture (see Choudhary et al. 2012) and forestry (Li et al. 2005; Kunes et al. 2016). Furthermore, research focus mainly on seed germination (Steber and McCourt 2001; Leubner-Metzger 2001; Li et al. 2005; Kartal et al. 2009; Kuneš et al. 2016).

Germination is a crucial biological process for plants, as it directly influences population establishment and dynamics, thus setting the limits for plant distributions, species diversity and community composition (Baskin and Baskin 1998; Gabriel y Galán and Prada 2010). This is especially critical in ferns, plants that have a biological cycle with two independent generations. For the purposes of this work, we just want to note that spore germination is the first step in the establishment of a gametophytic population, from which the next sexual, sporophytic population is to be built (Shorina 2001). Factors affecting germination, of both exogenous and endogenous nature, could pose important consequences on populations and communities in nature (Gabriel y Galán and Prada 2010; Raghavan 1989; Sheffield 1996). The sporophytic populations can be increased by means of some biological capacities, shown by some fern species: a very quick biological cycle with highly effective and productive reproductions, a vegetative reproduction through proliferative organs or by avoiding sexual contacts through apomictic processes (Fernández et al. 1999; Mehltreter et al. 2010; Regalado et al. 2010).

Although the vast majority of ferns occur in the tropics, these plants can be also found in temperate areas, where they play an ecologically interesting role in some ecosystems (Mehltreter et al. 2010). Despite this importance, there is no previous information about the presence or the role of brassinosteroids in fern spore germination. The aim of this work is to assess the effect of brassinosteroids on spore germination of three fern species in order to elucidate their physiological behaviour. In addition, since the selected species present different ecological preferences, we also purport to make a first approach to the ecological meaning of brassinosteroids in temperate ferns.

21.2 Material and Methods

21.2.1 Plant Material

We selected three fern species with more or less extended presence in temperate areas of Europe (Fig. 21.1): *Pteridium aquilinum* (L.) Kuhn (Dennstaedtiaceae), *Polystichum lonchitis* (L.) Roth (Dryopteridaceae) and *Pteris vittata* L. (Pteridaceae).

Fig. 21.1 Selected fern species (**a**) *Polystichum lonchitis*, Cantabria, Los Urrieles, 10/2016, *Gabriel y Galán* (**b**) *Pteridium aquilinum*, Cantabria, Cosgaya, 10/2016, *Gabriel y Galán* (**c**) *Pteris vittata*, Porto, 07/2015, *Gabriel y Galán & Prada*

These three species present different ecological breath and biological capacities (Castroviejo et al. 1986; Salvo 1990): First, *Pteridium aquilinum* has a wide ecological tolerance but mainly occurs in open places, roadsides, pastures, etc. This plant hardly produces meiotic spores at least in our latitudes but is known for its high facility to extend by rhizome. Second, *Polystichum lonchitis* inhabits crevices of granitic rocks at high altitudes in the mountains and has been selected as a representative of a fern with typical cycle and populations limited by environmental conditions. Finally, *Pteris vittata* tends to occupy rocks and walls at low altitudes, even in moderately disturbed areas; this species is known for having a very quick and productive sexual cycle that allows the species to form large sexually formed sporophytic populations. However, it has long been known that apogamy can be induced in laboratory conditions in *P. aquilinum* (Witthier 1966) and *P. vittata* (Huang et al. 2011); whilst interesting, this reproductive aspect is not an essential part of our current investigation, even if we have cited it along the text.

The biological material came from new collections (*P. vittata* at Portugal, Porto, 07/2015; *P. aquilinum* and *P. lonchitis* at Spain, Cantabria, 10/2016). One voucher for each species has been deposited in the herbarium MACB (Faculty of Biology, Universidad Complutense, Madrid).

For each species, we sampled five different individuals. Fertile fronds were placed in sealed envelopes, and sporangia were allowed to dehisce for 24 h under a source of moderate heat. The spores from all five individuals were mixed to avoid possible individual differences in the response to brassinosteroid.

21.2.2 Brassinosteroid Treatments

Each treatment concentration was prepared separately, without dilution. A stock solution (2 10^{-6} M) was prepared by adding corresponding quantity of epibrassinolide (Sigma Aldrich) dissolved in DMSO in sterilized and distilled H_2O. From this stock solution, the corresponding quantity for each treatment was prepared. All the solutions were prepared fresh in a final volume of 10 ml.

For each species two different treatments were chosen: the control and the brassinosteroid treatment. We made a first pilot survey exposing the spores during 60′ to a concentration of 10^{-7} M epibrassinolide, selected as optimum exposure time and concentration for germination as reported in previous research (Yokota and Takahashi 1986; Clouse 2002), to assess positive effects in fern spore germination. Should any positive effect be detected, the range of exposure times and concentrations would be increased to assess 15, 30 and 60′ of exposure and 10^{-8}, 10^{-7} and 10^{-6} M of brassinosteroid.

The brassinosteroid-treated spores were immersed in a 2 ml Eppendorf tube into each solution (10^{-8}, 10^{-7} and 10^{-6} M of brassinosteroid) for 15, 30 or 60 min. Then each tube was centrifuged for 5 min at 5000 rpm. The supernatant was discarded, and 2 ml of sterilized and distilled H_2O was added to each tube. After suspension by vortexing, each tube was centrifuged again at 5000 rpm for 5 min. Supernatant was discharged, and 2 ml of sterilized and distilled H_2O was added to each tube. After suspension by vortexing, 600 μl was sowed in each Petri dish.

21.2.3 In Vitro Germination

Each experiment (i.e. species/concentration/exposure time) was sown in mineral agar and cultured in Petri plates in a germination chamber (Dyer 1979). The culture conditions were maintained constant, at more or less standardized conditions for fern spore culture (Dyer 1979; Gabriel y Galán and Prada 2010; Gomez-Garay et al. 2016): no-depletion mineral nutrition, culture density of about 20 spores/cm^2, temperature of 22 ± 1 °C, saturated humidity within the sealed plate, 16/8 h light photoperiod and constant light regimen from daylight fluorescent tubes at a photon irradiance of 30–45 μmol m^{-2} s^{-1}. Three replicates for each treatment were done.

Every 3 days, germination percentage was recorded from 100 spores randomly selected in each plate, until no further increase was detected in two consecutive observations. A spore was considered as germinated when a first rhizoid was evident, emerging from the spore wall (Gabriel y Galán and Prada 2010). Subsequent relevant critical events at different stages of gametophytes development (filament, spatulate, pre-cordate meristematic and cordate adult prothalli stages) were observed and photographed with a Nikon Labophot-2 compound light microscope and a Nikon Coolpix MDC camera. Gametophytic development is yet known for the three species here studied: for *P. aquilinum*, Banks (1999), Conway (1949), Korpelainen (1997) and Munroe and Sussex (1969); for *P. vittata*, Beri and Bir (1993), Martínez (2010) and Prada et al. (2008); and for *P. lonchitis*, Pangua et al. (2003). Therefore, these previous works allowed us to have references about the gametophytic developmental timing and patterns, with which to contrast the gametophytes of our experiments.

21.2.4 Statistical Data Treatment

Each concentration point was conducted in triplicate. A total of 300 spores were analysed in each treatment (100 in each replicate). Under a compound light microscope, germinated spores were counted from a pool of 100 spores randomly selected in each Petri dish, excluding those abortive or irregularly formed. Data were reported as mean ± standard error (SE). Data were analysed using a one-way analysis of variance and the Duncan's test. The Dunnett's test was used to calculate the minimum difference between the control and the treatment means detected as being statistically significant. Significant difference was defined as that with a p value < 0.05 in all statistical analyses. All the statistical analyses were implemented using the statistical package Statistica v. 9.

21.3 Results

21.3.1 Effects on Germination

The germination of the three species studied was obtained for both control and brassinosteroid treatments, following an expected pattern of species variation. We detected a positive effect of brassinosteroids on spore germination only in *P. lonchitis*, for which a wider range of concentrations and exposures were probed. However, other effects were detected on the other species.

The germination percentage of *P. aquilinum* spores was significantly affected by the exposure to epibrassinolide. The effect of brassinosteroid application on *P. aquilinum* spores was rather inhibitory (Fig. 21.2): the highest germination rate was detected for the spores germinated under control treatment, which reached almost 60%, whilst for 10^{-7} M BR treatment was about 32%.

In the case of the species *P. lonchitis*, there was a markedly positive effect of brassinosteroids on germination (Fig. 21.3). At the end of the experiment, spores

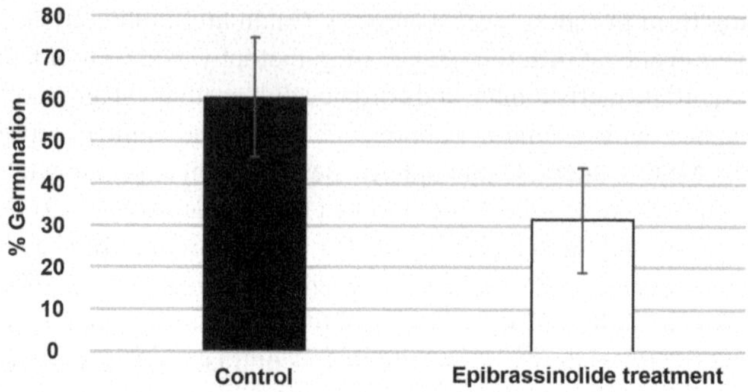

Fig. 21.2 Percentage of germination of *Pteridium aquilinum* spores for control and epibrassinolide treatment (60′ of BR 10^{-7} M), showing a difference in c. 30% of germination. Thus, brassinolide hormones produce a noteworthy inhibitory effect over germination in this species

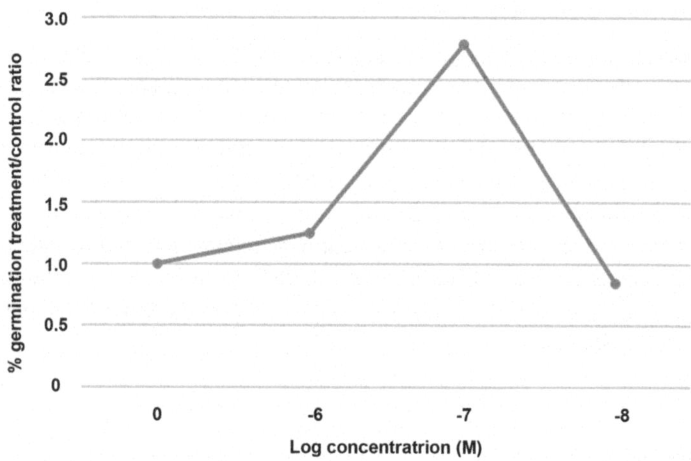

Fig. 21.3 Dose-response curve for the *Polystichum lonchitis* spore germination induced by 60′ of epibrassinolide exposure. A threshold effect concentration was estimated at 10^{-7} M

treated with the 10^{-7} M epibrassinolide treatment showed more than five times higher germination than the control cultures. For this species, we obtained a dose-response curve by determining the germination response of *P. lonchitis* spores after brassinosteroid application. A clear bell-shaped curve was identified with a threshold at 10^{-7} M, indicating that lower and higher amounts of hormones had an inhibitory effect over the germination in this species.

The germination of *P. lonchitis* showed also a correlation with the time of exposure to the hormone (Fig. 21.4). An exposure of 10^{-7} M BR 15 min had a statistically significant inhibitory effect, as the spores exposed germinated much less than the control ones (20% versus c. 100%). High exposure times (30′ and 60′) showed the contrary effect, resulting in much higher values than the control (c. three and five times higher, respectively).

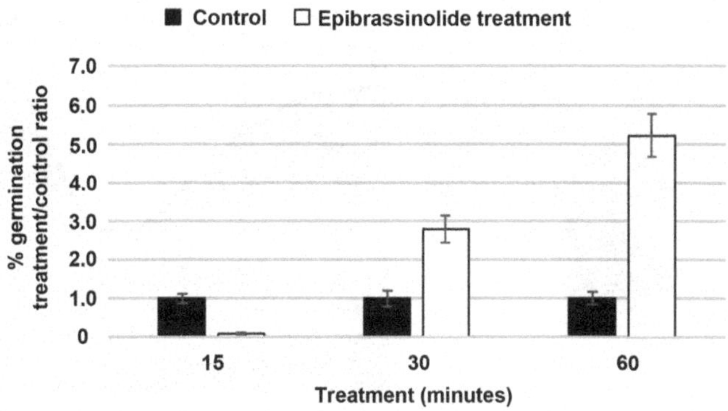

Fig. 21.4 Effects of brassinosteroids (BR 10^{-7} M) over germination of *P. lonchitis*, depending on exposure time of the spores to the hormone

Finally, the application of brassinosteroids did not affect the germination of *P. vittata* spores. The proportion of germinated spores in the control and brassinosteroid treatment remained almost equal in all counting dates, so germination rate of spores did not differ between the control and epibrassinolide treatment.

21.3.2 Effects on Gametophyte Development

Patterns of morphological development of the three species were less affected by the addition of epibrassinolide than germination.

For the species *P. vittata*, no significant deviation in timing and morphological features was detected between control gametophytes and those grown under hormone treatment (Fig. 21.5).

Pteridium aquilinum gametophytes grown from control and exposed spores also showed a more or less similar way and velocity of development (Fig. 21.6a–c). A rather large, albeit non-significant, number of gametophytes quickly developed sexual female organs, at the time that the rest remained sterile.

By far, gametophytes of *P. lonchitis* showed the higher variation in the timing and velocity of development in the different stages (Fig. 21.6d–f) induced by brassinosteroids. Firstly, spores from exposed cultures germinated progressively along the whole experimexntal period. As a result, as experiments progressed, gametophytes at very different stages of development coexisted, from simple filaments to bigger adult, cordate prothalli (Fig. 21.6e, f). Some exposed gametophytes reached a cordate, adult stage at the time that many control gametophytes were at a filamentous stage; i.e. brassinosteroids seemed to highly accelerate the gametophyte development. Besides this, brassinosteroid application induced some odd-branched gametophytes at early stages of the development (Fig. 21.6e).

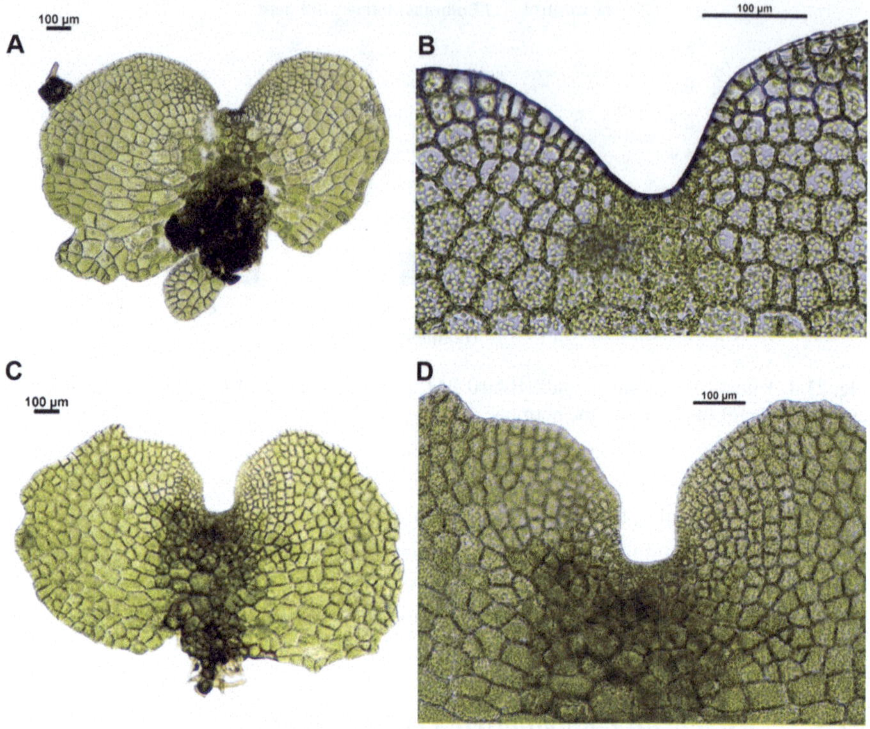

Fig. 21.5 Some gametophytes of *Pteris vittata* grown under different conditions, 30 days after sowing (**a, b**) controls (**c, d**) 60′ epibrassinolide 10^{-7} M treatment. Adult gametophytes (**a, c**) in a presexual stage developed as normal, winged cordate prothalli, whilst slight differences in general gametophyte size, depth of the apical notch and extension of the meristematic area (**b, d**) fall within the range of variation of the species and thus are not due to hormone application

21.4 Discussion

In this work, we contribute with the first essay devoted to relating the effect of brassinosteroids on the germination and development of fern spores. Our results have shown that this effect is species-dependent and display a set of different effects (Table 21.1): regarding germination, the observed effects range from a promotion of the germination in *P. lonchitis* to an inhibitory effect in *P. aquilinum* and near independence in *P. vittata*; regarding further gametophyte development, the effects range from the independence in *P. aquilinum* and *P. vittata* to a promotion of the development velocity in *P. lonchitis*.

As far as is known to date, research on effects of brassinosteroids over germination has been done only in gymnosperm and angiosperm seeds and pollen. It has been stated that brassinosteroids are able to enhance seed germination by controlling the inhibitory effects of abscisic acid – ABA (Finkelstein et al. 2008; Zhang et al. 2009). Positive effects of brassinosteroids on seed germination have been reported previously:

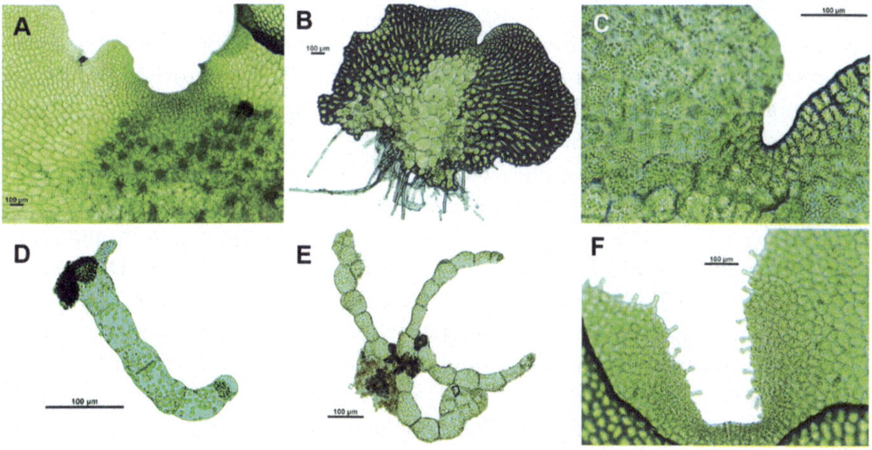

Fig. 21.6 Gametophytes of *Pteridium aquilinum* (**a–c**) and *Polystichum lonchitis* (**d–f**), 30 days after sowing. (**a**, **d**): gametophytes grown at control cultures; (**b**, **c**, **e**, **f**) gametophytes grown from spores exposed to 10^{-7} M BR

Table 21.1 Summary of the significant detected effects of brassinosteroids over critical aspects of germination and gametophyte development in three ferns from temperate latitudes. Additional information on biological and ecological aspects of the species is shown

	Bioecological information			Effects of brassinosteroids (60′ 10^{-7} M)		
	Ecological	Biological	Antheridiogen		Development	
Species	aspects	cycle	system	Germination	Velocity	Other aspects
Pteridium aquilinum	Very ample distribution, mainly open places	Sexual cycle rare, with hardly spore formation. Apogamy can be induced	Present, functional	↓*	—	High % of female gametophytes (unchecked if due to brassinosteroids)
Pteris vittata	Ample distribution, rocks in lowlands	Sexual cycle but apogamy can be induced	Present, functional	↓	—	—
Polystichum lonchitis	Comparatively limited distribution, rocks and similar in highlands	Normal sexual cycle	Present, but degenerate, non-functional	↑↑*	↑↑	Some odd branching forms at early stages

↓ inhibition, ↑ promotion, — independent, no detected effect, * statistical significance

promoted growth of embryos in *Nicotiana tabacum* L. (Leubner-Metzger 2001), accelerated germination in *Hordeum vulgare* L. (Kartal et al. 2009), stimulated germination in *Arabidopsis* Heinh. (Steber and McCourt 2001), improved germination index of aged *Oryza sativa* L. (Yamaguchi et al. 1987), increased germination rate in *Phaseolus aureus* Rox. (Srivastava et al. 2011) and a favourable effect observed in *Picea abies* (L.) H.Karst. (Kuneš et al. 2016). Nevertheless, a detrimental effect over germination has been observed in *Pinus sylvestris* L. (Kuneš et al. 2016). Brassinosteroids were also reported to enhance pollen germination in numerous species (*Prunus avium* (L.) L., Hewitt et al. 1985; *Prunus dulcis* (Mill.) D.A. Webb, Sotomayor et al. 2012; *Solanum lycopersicum* L., Singh and Shono 2005). However, in tobacco and rice, the growth-promoting effects of brassinosteroids on pollen remain contradictory. Hewitt et al. (1985) found a pronounced effect on *Nicotiana tabacum* pollen growth, but Ylstra et al. (1995) could not detect any influence. Brassinosteroid treatment greatly reduced the percentage of pollen viability in rice (Thussagunpanit et al. 2012). Our results with fern species point in the same direction, i.e. there is an effect of brassinosteroids over spore germination that is species-dependent.

Steroid hormone actions in plants reveal a regulatory network that integrates hormonal and light signalling pathways for plant growth regulation. There is an extensive crosstalk between BR and other hormonal and light-signalling pathways at multiple levels (Sun et al. 2010). Regarding the physiological connection between brassinosteroids and ABA, the endogenous hormone signalling networks mediating germination in response to the environment may have evolved independently in spores and seeds (Vesty et al. 2016). Thus, and contrary to what usually occurs in other plant groups, ABA has little effect on fern spore germination, and they are almost unaffected under ABA treatment (Suo et al. 2015), although it can affect posterior elongation of protonema (Chia and Raghavan 1982).

Steroid hormones play important roles in growth and development of various organisms. These include the sex hormones glucocorticoids and mineral corticoids in animals, the moulting hormones ecdysteroids in insects and crustaceans and, what it is important here in relation to brassinosteroids, antheridiogens in plants. Brassinosteroid (BR) and gibberellin (GA), and thus the antheridiogen, are two predominant hormones regulating germination and other developmental processes in plants. Genetic and biochemical studies of the antheridiogen response in ferns reveal many similarities and interesting differences to GA signalling and biosynthetic pathways in angiosperms.

The antheridiogen gibberellin (GA) A9 methyl ester (GA9-Me) is C3 hydroxylated and de-methylated to become an active GA (GA4). This antheridiogen is produced via the conserved gibberellin biosynthetic pathway in plants (Tanaka et al. 2014). Antheridiogens in ferns have been extensively studied (Keklowski and Lloyd 1968; Raghavan 1989; Schneller 2008). Two main effects have been detected: first, for most fern species with antheridiogen system, GA liberated from female gametophytes promotes the germination of their own non-germinated spores but also, sometimes, affects spores of other species (Prada et al. 2008); this is related also to the quick formation of antheridia at very early stages of development of the newly germinated gametophytes (Prada et al. 2008; Seral and Gabriel y Galán 2016). Second, for some species, GA liberated from light-grown gametophytes can substitute the light requirement in germination of non-germinated spores (Gemmrich 1986; Prada et al. 2008).

Interestingly, Tong et al. (2014) demonstrated in angiosperms that, when excessive active BR is applied, this hormone mostly induces GA inactivation through upregulation of the GA inactivation gene GA2ox-3 and also represses BR biosynthesis, resulting in decreased hormone levels. In this way, our present results might be explained. Both studied species which are antheridiogen sensitive (*P. aquilinum* and *P. vittata*) have shown a detrimental effect of the BR treatment over spore germination rate (Fig. 21.2). Regarding *P. vittata*, a slightly negative response to the BR treatment was found for the germination of the spores of this species, though this was not statistically supported. This may occur just due to a variation in sensitivity to BR. Indeed, spores of different species may react differently to similar BR concentrations, and, in this way, increasing BR concentrations could cause a more pronounced effect over *P. vittata* spore germination.

On the other hand, a favourable effect of the BR treatment has been observed in *P. lonchitis* spore germination (Figs. 21.3 and 21.4), a species with a non-functional, degenerate antheridiogen system (Pangua et al. 2003). Plant hormones characteristically exhibit wide activity ranges, often spanning four orders of magnitude in concentration (Trewavas 1991). In this way, we have tested three treatments: 10^{-6}, 10^{-7} and 10^{-8} M BR concentrations. A non-monotonic dose-response curve (NMDRC) has been demonstrated for brassinosteroid. At the lowest BR concentration (10^{-8} M), there were not statistically significant differences between BR treatment and control, but with increasing concentrations, more spores are capable of initiating germination (Fig. 21.3; 10^{-7} M). Nevertheless, qualitatively different effects between 10^{-7} and 10^{-6} M BR dose exposures may occur for several reasons and might explain quantitative differences in spore germination rate. A concentration of 10^{-6} M did not show any effect over *P. lonchitis* spore germination. According to the previously cited hormone-balance hypothesis between BR and GA (or antheridiogen), both plant hormones induced germination but with diverse modes of action. Whilst GA induces germination by removing the spore from its dormant state, BR increases germination by inducing growth. In fact, it is noteworthy the differences in growth observed between control and BR treatment for *P. lonchitis* (Fig. 21.2d, f). Furthermore, exposure to the same concentration (10^{-7} M BR) over 60 min significantly enhanced germination (Fig. 21.4).

In conclusion, diverse responses have been observed for the BR treatment to spore germination according to the fern species analysed. For *P. aquilinum*, which is antheridiogen sensitive and in which rhizome reproduction predominates, brassinosteroid treatment caused a negative effect over spore germination. For *P. vittata*, also antheridiogen sensitive and which forms large populations with sexual origin, BR treatment causes a slight negative effect on spore germination. And finally, for *P. lonchitis*, antheridiogen insensitive, the effect of BR treatment was positive.

P. lonchitis has been found to present a non-functional, degenerate antheridiogen system (Pangua et al. 2003), and the loss of the antheridiogen effectivity was attributed to its particular (almost) rupicolous habitat. In our opinion, this fact could not be invoked alone to explain GA loss effectiveness over spore germination, since there are other rupicolous ferns with such an antheridiogen system, for example, *P. vittata*. Bioclimatic differences due to altitude (*P. lonchitis* occurs mainly in mountains, whilst *P. vittata* does in lowlands) are perhaps better candidates to an ecological explanation of this phenomenon. Anyway, it seems that *P. lonchitis* can substitute a degenerate GA system by a more effective BR system to promote germination in order to quickly occupy favourable sites. The observed accelerated development (from filament to adult prothalli) mediated by BR would aid the species in the same sense: have sexual contacts and new sexually formed sporophytes before weather conditions are hostile to reproduction. On the other hand, the less hard environmental conditions in which *P. aquilinum* and *P. vittata* occur, along with the potent antheridiogen system exhibited by both and with the additional possibility of producing apogamous sporophytes, can explain the absence or even the inhibitory effect of BR over spore germination and gametophyte development. It seems that these species do not need BR, this seen from an ecological point of view.

Further research will be done in order to elucidate in detail the relationship between brassinosteroids, sexual organ expression and sporophyte production in the three species, considering at the time the use of antheridiogen system and the formation of apogamous plants.

Acknowledgements This study has been partially funded by Universidad Complutense Research Group Programme (Biodiversity and Taxonomy of Cryptogamic Plants, UCM 910801).

References

Banks JA (1999) Gametophyte development in ferns. Annu Rev Plant Physiol Plant Mol Biol 50:163–186

Baskin CC, Baskin JM (1998) Seeds: ecology, biogeography, and evolution of dormancy and germination. Academic Press, San Diego

Beri A, Bir SS (1993) Germination of stored spores of Pteris Vittata L. Am Fern J 83:73–78

Castroviejo S, Laínz M, López G, Montserrat P, Muñoz F, Paiva J, Villar L (eds) (1986) Flora Iberica I. CSIC, Madrid

Chia S-SE, Raghavan V (1982) Abscisic acid effects on spore germination and protonemal growth in the fern Mohria Caffrorum. New Phytol 92:31–37

Choudhary SP, Yu JQ, Yamaguchi-Shinozaki K, Shinozaki K, Tran LSP (2012) Benefits of brassinosteroid crosstalk. Trends Plant Sci 17:594–605

Clouse SD (2002) Brassinosteroid signal transduction: clarifying the pathway from ligand perception to gene expression. Mol Cell 10:973–982

Conway E (1949) The autecology of bracken Pteridium aquilinum L. Kuhn the germination of the spore, and the development of the prothallus and the young sporophyte. Proc Roy Soc Edinb 63 B:325–342

Dyer A (1979) The experimental biology of ferns. Academic Press, London

Fernández H, Bertrand AM, Sánchez-Tames R (1999) Biological and nutritional aspects involved in fern multiplication. Plant Cell Tissue Organ Cult 563:211–214

Finkelstein R, Reeves W, Ariizumi T, Steber C (2008) Molecular aspects of seed dormancy. Annu Rev Plant Biol 59:387–415

Gabriel y Galan JM, Prada C (2010) Pteridophyte spores viability. In: Fernández H, Kumar A, Revilla MA (eds) Working with ferns: issues and applications. Springer, New York, pp 193–205

Gemmrich AR (1986) Antheridiogenesis in the fern Pteris Vittata II. Hormonal control of antheridium formation. J Plant Physiol 125:157–166

Gomez-Garay A, Pintos B, Manzanera JA, Prada C, Martin L, Gabriel y Galan JM (2016) Nanoceria and bulk cerium oxide effects on the germination of asplenium adiantum-nigrum spores. Forest Systems 253:e067

Hewitt FR, Hough T, O'Neill P, Sasse JM, Williams EG, Rowan KS (1985) Effect of brassinolide and other growth regulators on the germination and growth of pollen tubes of Prunus Avium using a multiple hanging-drop assay. Funct Plant Biol 12(2):201–211

Huang Y, Hsu S, Hsieh T, Chou H, Chiou W (2011) Three Pteris species (Pteridaceae, Pteridophyta) reproduce by apogamy. Bot Stud 52:79–87

Kartal G, Temel A, Arican E, Gozukirmizi N (2009) Effects of brassinosteroids on barley root growth, antioxidant system and cell division. Plant Growth Regul 58:261–267

Klekowski EJ, Lloyd RM (1968) Reproductive biology of the Pteridophyta I. General considerations and a study of Onoclea Sensibilis L. J Linn Soc Bot 60:315–324

Korpelainen H (1997) Comparison of gametophyte growth, sex determination and reproduction in three fern species from the tropics. Nord J Bot 17:133–143

Kuneš I, Baláš M, Linda R, Gallo J, Nováková O (2016) Effects of brassinosteroid application on seed germination of Norway spruce, Scots pine, Douglas fir and English oak. iForest-Biogeoscience For 10:121

Li KR, Zhang WB, Li HK (2005) Effect of natural brassinolide on germination of Ailanthus Altissima seeds. Forestry Stud China 7:12–14

Leubner-Metzger G (2001) Brassinosteroids and gibberellins promote tobacco seed germination by distinct pathways. Planta 213:758–763

Martinez OG (2010) Gametophytes and young sporophytes of four species of the fern genus Pteris (Pteridaceae) naturalized in the American continent. Rev Biol Trop 58:89–102

Mehltreter K, Walker LR, Sharpe JM (2010) Fern ecology. University Press, Cambridge

Munroe MH, Sussex IM (1969) Gametophyte formation in bracken fern-p root cultures. Can J Bot 47:617–621

Nomura T, Bishop GJ (2006) Cytochrome P450s in plant steroid hormone synthesis and metabolism. Phytochem Rev 5:421–432

Nomura T, Kushiro T, Yokota T, Kamiya Y, Bishop GJ, Yamaguchi S (2005) The last reaction producing brassinolide is catalyzed by cytochrome P-450s, CYP85A3 in tomato and CYP85A2 in Arabidopsis. J Biol Chem 280:17873–17879

Pangua E, Quintanilla LG, Sancho A, Pajarón S (2003) A comparative study of the gametophytic generation in the Polystichum Aculeatum group Pteridophyta. Int J Plant Sci 164:295–303

Prada C, Moreno V, Gabriel y Galan JM (2008) Gametophyte development, sex expression and antheridiogen system in Pteris Incompleta Cav. Pteridaceae. Am Fern J 98:14–25

Raghavan V (1989) Developmental biology of fern gametophytes. University Press, Cambridge

Regalado L, Prada C, Gabriel y Galán JM (2010) Sexuality and apogamy in the Cuban *Asplenium auritum-monodon* Complex Aspleniaceae. Plant Syst Evol 289:137–146

Ross JJ, Reid JB (2010) Evolution of growth-promoting plant hormones. Funct Plant Biol 37:795–805

Salvo E (1990) Guía de helechos de la Península Ibérica y Baleares. Pirámide, Madrid

Schneller J (2008) Antheridiogens. In: Ranker TA, Hauffler CH (eds) Biology and evolution of ferns and lycophytes. University Press, Cambridge, pp 134–158

Seral A, Gabriel y Galán JM (2016) Gametophytic phase of Doryopteris triphylla (Pteridaceae, Polypodiopsida). Botanica Complutensis 40:63–70

Sheffield E (1996) From pteridophyte spore to sporophyte in the natural environment. In: Camus JM, Gibby M, Johns R (eds) Pteridology in perspective. Royal Botanic Gardens, Kew, pp 541–549

Shorina NI (2001) Population biology of gametophytes in homosporous polypodiophyta. Russ J Ecol 323:164–169

Singh I, Shono M (2005) Physiological and molecular effects of 24-epibrassinolide, a brassinosteroid on thermotolerance of tomato. Plant Growth Regul 47:111–119
Sotomayor C, Castro J, Velasco N, Toro R (2012) Influence of seven growth regulators on fruit set, pollen germination and pollen tube growth of almonds. J Agric Sci Technol B 2(9B):1051
Srivastava K, Raghavan N, Raghavan RP (2011) Brassinosteroids stimulate seed germination parameters and chlorophyll content in moongbean. Indian J Scientific Res 2:89
Steber CM, McCourt P (2001) A role for brassinosteroids in germination in Arabidopsis. Plant Physiol 125:763–769
Sun Y, Fan XY, Cao DM, Tang W, He K, Zhu JY et al (2010) Integration of brassinosteroid signal transduction with the transcription network for plant growth regulation in Arabidopsis. Dev Cell 19:765–777
Suo J, Zhao Q, Zhang Z, Chen S, Cao J, Liu G, Wei X, Dai S (2015) Cytological and proteomic analyses of Osmunda Cinnamomea germinating spores reveal characteristics of fern spore germination and rhizoid tip-growth. Mol Cell Proteomics 14:2510–2534. https://doi.org/10.1074/mcp.M114.047225
Takatsuto S (1994) Brassinosteroids: distribution in plants, bioassays and microanalysts by gas chromatography—mass spectrometry. J Chromatogr A 658:3–15
Tanaka J, Yano K, Aya K, Hirano K, Takehara S, Koketsu E, Ordonio RL, Park SH, Nakajima M, Ueguchi-Tanaka M, Matsuoka M (2014) Antheridiogen determines sex in ferns via a spatiotemporally split gibberellin synthesis pathway. Science 346:469–473
Thussagunpanit J, Jutamanee K, Chai-arree W, Kaveeta L (2012) Increasing photosynthetic efficiency and pollen germination with 24-Epibrassinolide in rice (Oryza Sativa L.) under heat stress. Thai J Botany 4:135–143
Tong H, Xiao Y, Liu D, Gao S, Liu L, Yin Y et al (2014) Brassinosteroid regulates cell elongation by modulating gibberellin metabolism in rice. Plant Cell 26:4376–4393
Trewavas A (1991) How do plant growth substances work? II. Plant Cell Environ 14:1–12
Vardhini BV, Rao SSR (1997) Effect of brassinosteroids on growth, metabolite content and yield of Arachis Hypogaea. Phytochemistry 48:927–930
Vesty EF, Saidi Y, Moody LA, Holloway D, Whitbread A, Needs S, Choudhary A, Bae H (2016) The decision to germinate is regulated by divergent molecular networks in spores and seeds. New Phytol 211:952–966
Whittier DP (1966) Induced apogamy in diploid gametophytes of Pteridium. Can J Bot 44:1717–1721
Yamaguchi T, Wakizuka T, Hirai K, Fujii S, Fujita A (1987) Stimulation of germination in aged rice seed by pre-treatment with brassinolide. In: Proceedings of the 14th annual plant growth regulator Society of America meeting, Hawaii, pp 26–27
Ylstra B, Touraev A, Brinkmann AO, Heberle-Bors E, Tunen AJV (1995) Steroid hormones stimulate germination and tube growth of in vitro matured tobacco pollen. Plant Physiol 107:639–643
Yokota T, Takahashi N (1986) Chemistry, physiology and agricultural application of brassinolide and related steroids. In: Plant growth substances. Springer, Berlin, pp 129–138
Yokota T, Ohnishi T, Shibata K, Asahina M, Nomura T, Fujita T, Kohchi T (2017) Occurrence of brassinosteroids in non-flowering land plants, liverwort, moss, lycophyte and fern. Phytochemistry 136:46–55
Zhang S, Cai Z, Wang X (2009) The primary signaling outputs of brassinosteroids are regulated by abscisic acid signaling. Proc Natl Acad Sci 106:4543–4548

Chapter 22
Ecomorphology of Stomata in Temperate Ferns Under Contrasting Environments

Jose Maria Gabriel y Galán, Andrea Seral, Antonio Murciano, María do Rosario Anjos, Francisco B. Cuevas-Fernández, Pablo Fernández, and Teresa Pinto

22.1 Introduction

Various traits of the vascular plants can be adduced to explain the role of these organisms in their ecosystems. One of the most interesting and fundamental are the stomata, because on them rely the ultimate control between photosynthesis and water balance (Jones 1998). For this reason, stomatal morphology responds to many environmental variables, as light, water availability and gases pressure: plants can detect these variables and send signals to the developing new leaves, which in term adjust their stomata to the detected conditions (Brownlee 2001; Hetherington and Woodward 2003). Thus, understanding morphology and physiology of stomata is necessary to know how plants fit in their ecosystems and the role of these organisms in the different terrestrial biomes (Ackerly et al. 2000; Haworth et al. 2013; McAdam and Brodribb 2012; Reich et al. 1999).

Morphological and functional ecology of stomata has been studied for a long time, essentially by observing flowering plants. Thus, a well-developed corpus of

J. M. Gabriel y Galán (✉) · A. Seral · F. B. Cuevas-Fernández · P. Fernández
Department of Plant Sciences I, Universidad Complutense, Biodiversity and Taxonomy of Cryptogamic Plants Research Group, UCM, Madrid, Spain
e-mail: jmgabrie@ucm.es

A. Murciano
Department of Applied Mathematics (Biomathematics), Faculty of Biology, Universidad Complutense, Madrid, Spain

Neural Plasticity Research Group, IdISSC" and "Neuro-computing and Neuro-robotics Research Groups, UCM, Madrid, Spain

M. R. Anjos · T. Pinto
University of Trás-os-Montes and Alto Douro, Department of Biology and Environment, Centre for the Research and Technology of Agro-Environmental and Biological Sciences (CITAB), Vila Real, Portugal

H. Fernández (ed.), *Current Advances in Fern Research*,
https://doi.org/10.1007/978-3-319-75103-0_22

knowledge is available on how angiosperm stoma is constructed and responds to the environment (Brodribb et al. 2009; Cowan 1977; Evert and Eichhorn 2006; Franks and Farquhar 2007; Hetherington and Woodward 2003; Liu et al. 2006; Maherali et al. 2002; Martinez et al. 2007; Meidner and Mansfield 1968; Öpik and Rolfe 2005; Prabhakar 2003; Quarrie and Jones 1977; Salisbury 1927; Schulze et al. 1973; Spence et al. 1986; Xu and Zhou 2008). Briefly, although different effects of abiotic factors on stomatal morphology and density may depend on plant species, the following general rules are largely accepted: the stomatal size is more genetically controlled than density, which can be dynamically modified in new leaves to adjust plants to their environment; stomatal size and density are generally correlated; and density is expected to increase with light and to decrease with water availability.

Ferns are vascular plants that often act as important biological elements in their natural ecosystems. Although their major occurrence is in tropical areas, they can be easily found in almost all terrestrial plant communities in temperate regions (Mehltreter et al. 2010). Interestingly, some evidence is appearing about certain differences in the morphology and physiology of fern stomata compared to those of angiosperms and gymnosperms, especially regarding signalling pattern (Doi et al. 2006; Franks and Farquhar 2007; Kawai et al. 2003; McAdam and Brodribb 2013). To date, however, ecological studies of fern stomatal morphology have been much less developed (Atala et al. 2012; Gabriel y Galán et al. 2011; Hunt et al. 2002; Kato and Imaichi 1992; Kessler et al. 2007; Kluge et al. 2008; Kluge and Kessler 2007; Riano and Briones 2013). In general, these few studies on the ecomorphology of fern stomata confirm the general rules pointed before. However, most of these works dealt with tropical fern species, whereas research on temperate ferns is really scarce. There is a need to deepen in the ecology of morphological traits of temperate ferns stomata before generating full models for the vascular plants in its whole.

The main aim of this work is to increase the knowledge about morphological ecology of stomata in the understudied group of temperate ferns. On one hand, we want to characterize the stomatal morphological traits in the selected species and carry on a study of biometrical correlations between these traits. On the other hand, we also want to investigate if the environmental conditions of light and water availability are influencing the size and density of stomata in these organisms, following the general rules above commented.

22.2 Materials and Methods

The following 15 species were selected: *Adiantum capillus-veneris* L. (Pteridaceae), *Asplenium marinum* L. (Aspleniaceae), *Asplenium scolopendrium* L. (Aspleniaceae), *Asplenium trichomanes* L. (Aspleniaceae), *Asplenium viride* Huds. (Aspleniaceae), *Athyrium filix-femina* (L.) Roth (Athyriaceae), *Botrychium lunaria* (L.) Sw.

(Ophioglossaceae), *Cystopteris fragilis* (L.) Berhn. (Cystopteridaceae), *Dryopteris affinis* (Lowe) Fraser-Jenk. (Dryopteridaceae), *Dryopteris filix-mas* (L.) Schott (Dryopteridaceae), *Osmunda regalis* L. (Osmundaceae), *Polystichum aculeatum* (L.) Roth ex Mert. (Dryopteridaceae), *Polystichum lonchitis* (L.) Roth (Dryopteridaceae), *Polystichum setiferum* (Forssk.) T. Moore ex Woyn. (Dryopteridaceae) and *Struthiopteris spicant* (L.) Weiss (Blechnaceae). The samples came either from extraction of herbarium material deposited in MACB (Faculty of Biology, Universidad Complutense, Madrid) or from new field collections; in these cases, voucher specimens were deposited in MACB. Detailed information on the source of samples is shown in Appendix 22.1.

Of each species, 15 to 20 different individuals were sampled. Stomata were observed in adult, undamaged medium leaf segments (pinna or pinnule) from three different fronds of each individual (one segment per frond). Dry herbarium material was treated with 6% sodium hydroxide for tissue softening followed by 50% sodium hypochlorite for tissue whitening (Rolleri et al. 2013; Ruzin 1999). Recently collected material was directly whitened without softening. Pinnae were mounted in water and observed under a compound light microscope Nikon Labophot-2 with a camera Coolpix MDC. One field from the medium area of each pinna was photographed at 20X. Each photograph was processed using the software Piximètre (Henriot and Cheype 2012), measuring the following data: length (L) of five stomata randomly selected and number of stomata present in the field. Stomata with half or more of its surface outside the observation field were excluded. Stomatal density (D) is expressed as stomata per mm^2 (st/mm^2). For each frond, we have calculated one value of potential conductance index (PCI) as follows (Holland and Richardson 2009): $PCI = L^2 \cdot D \cdot 10^{-4}$. The total number of stomata observed per species is shown in Table 22.1.

Assignment of habitat types was simplified to express essential, contrasting ecological differences (Table 22.1): regarding water availability, riverbanks (water surplus) vs forests/pastures (medium water availability) vs rocks (water deficiency), and regarding light environment, shaded forests, riverbanks and rocks (low availability) vs opened pastures and exposed rocks (high availability). These assignments come from our field observations and from previously reported information on the ecological preferences of the species (Castroviejo et al. 1986; Ferrer-Castan and Vetaas 2005; Moreno and Lobo 2008; Pichi Sermolli 1979; Salvo 1990; Salvo et al. 1984).

The normality of the data was evaluated through a Lilliefors (Kolmogorov-Smirnov) test, resulting that in all the three traits follow approximately a normal distribution. Therefore, subsequent analyses were made using parametric test: t-Student, ANOVA and Pearson correlations. Linear regression techniques were applied in order to investigate the relationships between stomatal traits and environmental conditions. All the analyses were conducted with the software SPSS (IBM).

Table 22.1 Species sampled, followed in parenthesis by the total number of individuals observed, total measurements of stomatal length (5 per frond, 3 fronds per individual) and total number of measurements of stomatal density (1 per frond, 3 fronds per individual). Habitat characterization and classification of light and water environments come from personal observations and previous research (see text)

Species (# individuals, # stomata length data, # density data)	Habitat description	Light availability	Water availability
Adiantum capillus-veneris (16, 240, 48)	Humid shadowed rocks	Low	High
Asplenium marinum (19, 285, 57)	Marine cliffs; aero-halophyte	High	Low
Asplenium scolopendrium (18, 270, 54)	Shadowed riverbanks	Low	High
Asplenium trichomanes (20, 300, 60)	Exposed rock crevices	High	Low
Asplenium viride (15, 225, 45)	Montane shadowed rock crevices	Low	Low
Athyrium filix-femina (15, 225, 45)	Shadowed riverbanks	Low	High
Botrychium lunaria (15, 225, 45)	Montane pastures, seasonally affected by snow melting	High	Medium
Cystopteris fragilis (16, 240, 48)	Shadowed rock crevices	Low	Low
Dryopteris affinis (20, 300, 60)	Humid forests	Low	Medium
Dryopteris filix-mas (15, 225, 45)	Humid forests	Low	Medium
Osmunda regalis (15, 225, 45)	Shadowed riverbanks	Low	High
Polystichum aculeatum (20, 300, 60)	Humid forests	Low	Medium
Polystichum lonchitis (20, 300, 60)	Montane exposed rocks	High	Low
Polystichum setiferum (15, 225, 45)	Humid forests	Low	Medium
Struthiopteris spicant (16, 240, 48)	Humid forests	Low	Medium

22.3 Results

22.3.1 Stomatal Traits in 15 Species of Temperate Ferns

Table 22.2 and Fig. 22.1 show the statistical values (mean value ± standard deviation) for the variables considered (stomatal length, density and PCI) for all the selected species and the statistically supported groups as resulted from the ANOVA analyses for each trait. The three species with the longest stomata are *A. scolopendrium*, *S. spicant* and *B. lunaria* (57.58, 57.47 and 57.11 μm, respectively), which form a group statistically differentiated from the rest of species. On the other hand, *A. capillus-veneris*, with a mean length of 32.91 μm, is the species with the shorter stomata and statistically isolated from the rest; it is followed by a group formed by *P. lonchitis* (38.88 μm), *A. marinum* (38.96 μm) and *A. filix-femina* (40.88 μm). Regarding stomatal density, the group formed by *A. scolopendrium*, *A. capillus-veneris* and *B. lunaria* presents the lowest values, with 18.33, 23.92 and 24.44 st/mm^2, respectively. *Asplenium marinum* (85.23 st/mm^2) and *A. viride* (79.96 st/mm^2)

Table 22.2 Biometric characterization of the sample (mean ± sd), ordered by species. The extreme values for each trait have been highlighted in bold. Species with the same lettering form a statistically supported group for each trait (Tukey test, $\alpha < 0.05$)

Species	Length(μm)	Density(st/mm²)	PCI
Adiantum capillus-veneris	**32.91 ± 2.06** a	23.92 ± 5.26 a	**2.70 ± 0.68** a
Asplenium marinum	38.96 ± 1.79 b	**85.23 ± 10.59** i	12.96 ± 1.72 h, i
Asplenium scolopendrium	**57.58 ± 3.25** i	**18.33 ± 3.29** a	6.08 ± 1.14 b
Asplenium trichomanes	46.23 ± 2.56 e, f	42.73 ± 7.33 c, d	9.18 ± 1.32 d, e
Asplenium viride	42.88 ± 1.52 c, d	79.96 ± 9.23 h	14.84 ± 2.39 i, j
Athyrium filix-femina	40.88 ± 1.82 b, c	64.55 ± 3.77 g	10.85 ± 1.02 e, f, g
Botrychium lunaria	57.11 ± 4.33 i	24.44 ± 1.83 a	8.00 ± 1.07 b, c, d
Cystopteris fragilis	45.35 ± 3.63 d, e, f	33.38 ± 4.47 b	6.99 ± 1.68 b, c
Dryopteris affinis	45.19 ± 1.05 d, e, f	52.42 ± 6.64 e, f	10.77 ± 1.24 e, f, g
Dryopteris filix-mas	50.54 ± 4.47 g, h	45.24 ± 2.54 d, e	11.75 ± 2.35 g, h
Osmunda regalis	52.41 ± 1.23 h	55.02 ± 4.47 f	**15.38 ± 1.55** j
Polystichum aculeatum	48.01 ± 2.48 f, g	37.03 ± 6.52 b, c	8.57 ± 1.56 c, d
Polystichum lonchitis	38.88 ± 3.66 b	72.77 ± 7.98 h, i	11.25 ± 2.53 f, g, h
Polystichum setiferum	44.09 ± 2.65 c, d, e	50.00 ± 4.28 d, e, f	9.76 ± 0.99 d, e, f
Struthiopteris spicant	57.47 ± 2.54 i	44.50 ± 4.50 d	14.79 ± 1.70 i, j

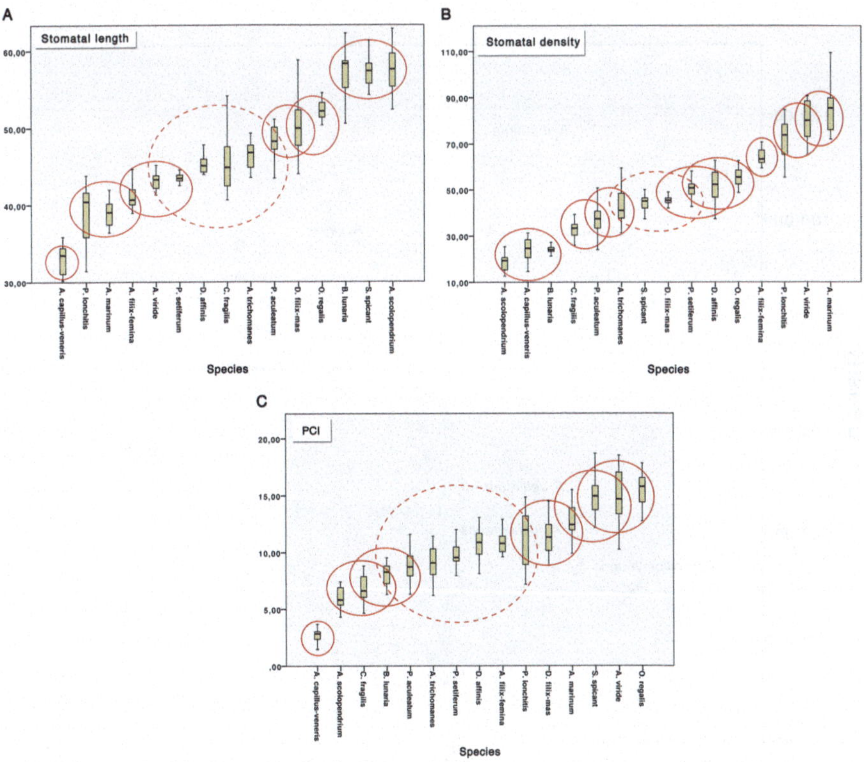

Fig. 22.1 Box and whisker plots of each stomatal trait: (**a**) stomatal length (**b**) stomatal density (**c**) potential conductance index (PCI). In each graph, species are ordered by trait value. Red solid circles indicate statistically supported groups of species, with three or less members. Dotted circles indicate groups with more than three members and not statistically supported. Groups have been established after an ANOVA and post hoc Tukey test, with p-value < 0.05

form a statistically supported group with the highest values of density. Finally, the lowest value of PCI is found in *A. capillus-veneris* (2.7), which appeared to be statistically well differentiated, followed by a group which contains the species *A. scolopendrium* (6.08), *C. fragilis* (6.99) and *B. lunaria* (8.00). The highest PCI appears in *O. regalis* (15.38), which forms a statistically supported group with *A. viride* (14.84) and *S. spicant* (14.79).

22.3.2 Biometrical Correlations Between Stomatal Traits

We have made a test to study the correlation between the stomatal length and density of stomata of the selected species (Fig. 22.2). For the total of data, the Pearson test resulted in a statistically supported inverse correlation ($\rho = -0.438$; p-value < 0.01). The same test applied over the data within each species (Table 22.3) shows that only two cases have a correlation L•D with statistical support: *Asplenium trichomanes* and *Asplenium viride*. Noteworthy, the sign of the correlation is different in both species, negative in *A. trichomanes* (longer stomata implies lower

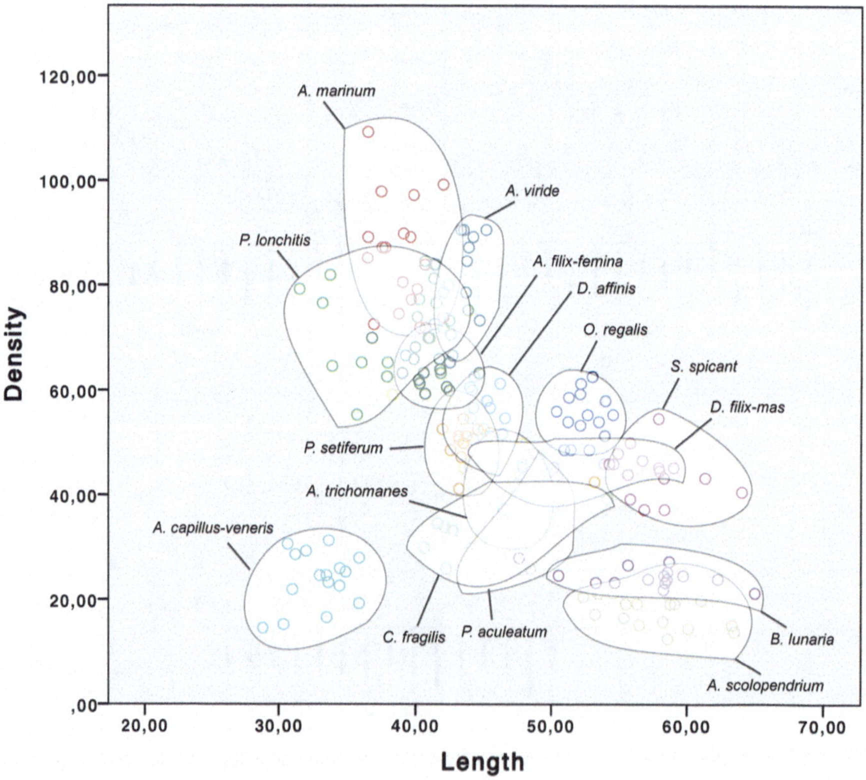

Fig. 22.2 Dispersion plot showing the correlations between length and density for the 15 selected species

Table 22.3 Results of the Pearson bivariate correlation test between pairs of stomatal traits, with the associate p-value. The cases with statistical support have been highlighted

	(L • D)		(L • PCI)		(D • PCI)	
Species	ρ	p-value	ρ	p-value	ρ	p-value
Adiantum capillus-veneris	0.239	0.373	**0.670****	0.004	**0.865****	0.000
Asplenium marinum	−0.293	0.223	0.442	0.058	**0.717****	0.001
Asplenium scolopendrium	−0.352	0.152	0.253	0.310	**0.799****	0.000
Asplenium trichomanes	**−0.553****	0.011	0.145	0.541	**0.730****	0.000
Asplenium viride	**0.557***	0.031	**0.802****	0.000	**0.940****	0.000
Athyrium filix-femina	−0.269	0.332	**0.780****	0.001	0.388	0.153
Botrychium lunaria	−0.510	0.052	**0.872****	0.000	−0.027	0.924
Cystopteris fragilis	0.230	0.392	**0.827****	0.000	**0.726****	0.001
Dryopteris affinis	−0.407	0.075	−0.046	0.849	**0.923****	0.000
Dryopteris filix-mas	0.147	0.602	**0.957****	0.000	0.420	0.119
Osmunda regalis	0.118	0.675	**0.575***	0.025	**0.878****	0.000
Polystichum aculeatum	−0.122	0.608	**0.447***	0.048	**0.825****	0.000
Polystichum lonchitis	0.147	0.535	**0.890****	0.000	**0.572****	0.008
Polystichum setiferum	−0.451	0.091	**0.713****	0.003	0.300	0.277
Struthiopteris spicant	−0.267	0.318	**0.517***	0.040	**0.684****	0.003

** and * indicate significant correlation in $\alpha = 0.01$ and $\alpha = 0.05$ levels, respectively

densities) and positive in *A. viride* (longer stomata implies higher densities). For the rest of species, both traits seem to be independent. PCI is correlated either with length or density of stomata, always in a direct relationship. In seven species, PCI is correlated with both length and density. *Asplenium viride* is the only species in which all the three traits (length, density and PCI) are statistically correlated.

22.3.3 Relationships Between Stomatal Traits and Habitat

We found significant differences between the water (ANOVA) and light (T-Student) environments of temperate fern species and all the variables studied, except for the case PCI/light, for which the test did not found statistical support (Table 22.4, Fig. 22.3). In other words, we can state that water and light environments are responsible for most of the differences measured in biometrical stomatal variables (length, density and PCI) for the temperate ferns.

The post hoc test of the ANOVA between water and length (F = 30.958; p-value < 0.01) segregated two groups: in one hand, those species that inhabit low water availability environments, which developed shorter stomata (42.345 μm), and on the other hand, species that occur in medium and high water availability, which presented longer stomata (50.096 μm and 46.286 μm, respectively). Regarding the test density*water (F = 43.600, p-value < 0.01), we discriminated again two groups: species that inhabit low water environments had higher density values (62.918 st/

Table 22.4 Mean ± standard deviation values of biometric stomatal traits of the groups established by water and light availability categories. Groups whose differences are not statistically supported are indicated with the same letter

	Water availability			Light availability	
	Low	Medium	High	Low	High
Length (μm)	42.34 ± 4.2(a)	50.09 ± 5.99(b)	46.28 ± 10.13(b)	47.15 ± 7.35(a)	44.58 ± 7.73(b)
Density (st/mm²)	62.91 ± 22.13(a)	42.53 ± 10.40(b)	39.16 ± 20.33(b)	45.15 ± 17.52(a)	58.05 ± 24.63(b)
PCI	10.99 ± 3.26(a)	10.55 ± 2.68(b)	8.53 ± 4.9(c)	10.08 ± 4.04(a)	10.47 ± 2.55(a)

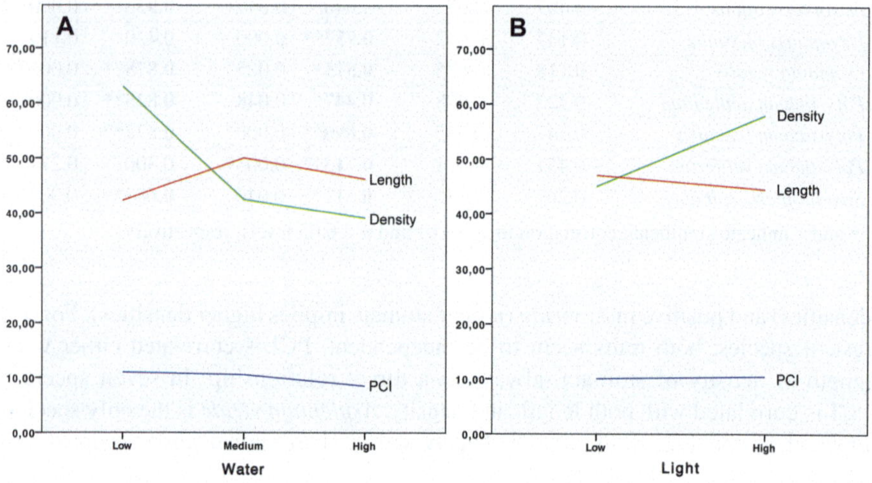

Fig. 22.3 Values of length, density and PCI for the different environments of water availability (**a**) and light availability (**b**). For water environments, stomatal length and density did not show statistical differences between medium and high water availability habitats. In the case of the light environments, differences in PCI had no statistical support

mm²) versus species of medium and high water availability places, with lower density values (42.538 st/mm² and 39.160 st/mm², respectively). Finally, about PCI*water ($F = 9.773$; p-value < 0.01), the test established also two groups: species of high water availability showed lower PCI values (8.533) against species from medium and low water availability, which rendered higher values (10.555 and 10.993, respectively).

Light is also related to biometrical stomatal traits. We encountered significant differences between light environments and stomatal length ($t = 2.44$; p-value = 0.016), with longer stomata developed by species that inhabit low light regimes (47.15 μm vs 44.58 μm of the group of high light habitats). Also, differences in light environments affected stomatal density ($t = -4.107$; p-value < 0.01), having more density those species from high light regimes (58.05 st/mm² vs 45.15 st/mm² of the group of low light habitats). Finally, as it was stated above, light environments seemed to not affect the PCI of our sample (t = −0.902; p-value = 0.368).

In summary, species that inhabit low water availability environments (exposed rocks, crevices and cliffs) developed significant shorter stomata, but at higher densities and with higher potential conductance, species from low light availability environments (shadowed rocks and forests) have longer stomata and lower density.

22.4 Discussion

22.4.1 *Variation and Correlation of Stomatal Traits*

Derived from our results, we can state that there are enough statistical differences in stomatal length to correctly assign to concrete species most of the measured stomatal variation. This phenomenon let us affirm that stomatal characters are good diagnostics of fern temperate species. The fact that the individuals could be assigned to species through these stomatal traits is related with the classic appreciation that stomatal characters (specifically those affecting its size) are genetically controlled. Some experimental data support this idea, for example: (a) the very low stomatal variability detected in some species, as *Huperzia saururus* (Lam.) Trevis. (Rolleri 1977); (b) the statement that, at least in angiosperms, size of stomata is less influenced by external factors than size of ordinary epidermal cells (Metcalfe and Chalk 1979); and (c) the known fact that the ploidy level is generally related to stomatal size and to sizes of various other structures, as spores or epidermal cells (Stebbins 1950). Two other temperate fern species studied (*Blechnum chilense* (Kaulf.) Mett. and *B. mochaenum* G. Kunkel) seemed to present its stomatal size under low external control (Atala et al. 2012). As in the temperate species here studied, stomata of tropical species previously known also show a similar behaviour (Gabriel y Galán et al. 2011).

Length and stomatal density appears to be quite independent variables for the temperate ferns studied, since they show correlation only in two species (*Asplenium trichomanes* and *A. viride*). For the angiosperms, it has been stated a strong and general relationship between density and size (Hetherington and Woodward 2003). The absence of correlation between these variables in almost all of the investigated fern species suggests that these ferns are more flexible in the adaptation of stomatal traits to the environment than other plant groups. Indeed, as length seems to be genetically controlled, one way to modify stomatal density is modifying the size of ordinary epidermal cells instead of the size of guard cells. This operative ontogenetical strategy has been suggested to be functioning also in some angiosperms (Pompelli et al. 2010; Salisbury 1927) and in some tropical ferns (Gabriel y Galán et al. 2011).

In one of the species (*A. trichomanes*), the correlation density-length is inverse, i.e. greater lengths imply lower densities. This seems to be an extended rule for the plants in general, as it is known for some tropical ferns (Gabriel y Galán et al. 2011) and angiosperms (Camargo and Marenco 2011; Hetherington and Woodward 2003). However, we have detected the remarkable exception to this rule in *A. viride*, species in which higher densities were accompanied by higher lengths. This should be interpreted in the way that density is increased by reducing the size of the ordinary epidermal cells, otherwise no increase in length could have been observed.

With respect to PCI, the ferns studied here showed a logic relation with the rest of stomatal traits, in the sense that higher densities and higher lengths promote a higher potential conductance. This has been observed in the majority of the species.

22.4.2 Ecological Significance of Stomatal Variation in Relation to Water Environment

Regarding water environment, it is generally accepted that water availability inversely correlates with stomatal density and directly with stomatal size (Xu and Zhou 2008). In agreement with these general expectations, the temperate fern species studied here present higher density and PCI and shorter stomata under water-stressed habitats (rocks) compared to habitats without water stress (humid rocks, forests, riverbanks). Strikingly, previous studies point out that tropical ferns from humid soils present high stomata numbers (Gabriel y Galán et al. 2011; Kato and Imaichi 1992). Two reasons could be adduced to explain such a contrasted difference. First, the tropical fern studied are perfect rheophytes, i.e. the rhizome is constantly under the water level in swamps and streams, whilst this is not occurring in the species here studied; second, some other uncontrolled environmental fact could be operating, as a general high radiation and a higher and constant air temperature in the tropics.

22.4.3 Ecological Significance of Stomatal Variation in Relation to Light Environment

With respect to the light environment, it has been proposed that plants that live under open sunny conditions are not light-limited so they can produce more stomata to maximize photosynthetic rates; on the other hand, shaded habitats are light-limited, and plants tend to present low stomatal densities (Gay and Hurd 1975; Givnish 1988). This has been proposed in general for deep-shade fern species, which seem to possess large stomata at low densities (Hetherington and Woodward 2003; Meidner and Mansfield 1968). Previously studied temperate fern species (*B. chilense* and *B. mochaenum*) appear to behave also in this way (Atala et al. 2012). Nevertheless, these same authors concluded that the observed differences in size and density between these two species could be attributed to phylogenetic differences better than to ecological adaptation. Our temperate fern species follow this general pattern: the light-influenced species from open places present shorter stomata but at higher densities values, whilst the light-limited species from forests and shaded crevices show longer stomata at lower densities. This could mean that the temperate fern species have adapted their stomatal apparatus to shade. Even though, since we have not detected differences in PCI, the relationship between length and density is balanced in the two light environments.

Acknowledgements This work has been partially funded by Universidad Complutense Research Groups Programme ("Biodiversity and Taxonomy of Cryptogamic Plants", UCM 910801) and by the EU-ERASMUS mobility programme. We thank the curator of MACB herbarium for her willingness to facilitate us the use of the material.

Appendix 22.1

Samples used in the study, with voucher information. All samples came from Spain except those specifically quoted. The information included is ordered by country/Spanish province, followed by location, MACB (Madrid Ciencias Biológicas) herbarium number and, in parenthesis, the individuals.

***Adiantum capillus-veneris* L.** ASTURIAS: Cangas de Onís, MACB 36194 (ind. 9–12). CASTELLÓN: Peñíscola, MACB 104250 (ind. 1–8). VALENCIA: Tuéjar, MACB 43430 (ind. 13–16). ***Asplenium marinum* L.** ASTURIAS: Luarca, MACB 36191 (ind. 8), MACB 107532 (ind. 13–16); Gozón, MACB 3646 (ind. 10, 11); Castropol, MACB 55943 (ind. 12), MACB 107533 (ind. 17–19). GUIPÚZCOA: Donostia, MACB 36078 (ind. 6). LA CORUÑA: Corme, MACB 24302 (ind. 1); Porto do Son, MACB 43332 (ind. 2); Cayón, MACB 4373 (ind. 3); Malpica, MACB 26154 (ind. 5). PONTEVEDRA: La Guardia, MACB 4773 (ind. 4); Bueu, MACB 18655 (ind. 7); Islas Cies, MACB 36119 (ind. 9). ***Asplenium scolopendrium* L.** ASTURIAS: Brieves, MACB 104248 (ind. 1–7). LEÓN: Villasecino, MACB 107538 (ind. 8–18). ***Asplenium trichomanes* L.** ASTURIAS: Luarca, MACB 107542 (ind. 17–20). BURGOS. Crespos, MACB 107543 (ind. 13–16). GUADALAJARA: río Bornova, MACB 22462 (ind. 3). LEÓN: Santalavilla, MACB 62938 (ind. 1). MADRID: Torrelodones, MACB 104525 (ind. 5–12). PALENCIA: Velilla del rio Carrión, MACB 30953 (ind. 2). SALAMANCA: Linares de Riofrío, MACB 1768 (ind. 4). ***Asplenium viride* Huds.** ASTURIAS: Teverga, MACB 107544 (ind. 9–15). GUIPÚZCOA: Aralar, MACB 43217 (ind. 5–7). HUESCA: Aínsa, MACB 36183 (ind. 4). PALENCIA: Velilla del Río Carrión, MACB 39484 (ind. 8). VIZCAYA: Zeanuri, MACB 62929 (ind. 1–3). ***Athyrium filix-femina* (L.) Roth.** ASTURIAS: Luarca, MACB 107531 (ind. 8–15). ÁVILA: Candeleda, MACB 53208 (ind. 1–2); El Barranco, MACB 57772 (ind. 5–7). CÁCERES. Viandar de la Vera, MACB 59103 (ind. 3). TERUEL: Orihuela, MACB 43319 (ind. 4). ***Botrychium lunaria* (L.) Sw.** ASTURIAS: Somiedo, MACB 104249 (ind. 1–15). ***Cystopteris fragilis* (L.) Berhn.** BURGOS: Sierra de la Demanda, MACB 56770 (ind. 1–4). LA RIOJA: Ezcaray, MACB 57500 (ind. 5); Villoslada de Cameros, MACB 107545 (ind. 9–16). LEÓN: Truchillas, MACB 44460 (ind. 6–7). TENERIFE: Tágara, MACB 36110 (ind. 8). ***Dryopteris affinis* (Lowe) Fraser-Jenk.** **ITALY**. TOSCANA: Massa, MACB 58804 (ind. 3). LUACCA: Versilia, MACB 58807 (ind. 4); Massa y Carrara, MACB 58801 (ind. 5). Massa, MACB 58802 (ind. 6). **SPAIN**. ASTURIAS:

Valgrande, MACB 31115 (ind. 11). CANTABRIA: Resconorio, MACB 31010 (ind. 12); Cosgaya, MACB 109253 (ind. 13–20); Torcollano, MACB 62921 (ind. 7). LA CORUÑA: Betanzos, MACB 92416 (ind. 1); Somozas, MACB 92391 (ind. 2); Cerdedo, MACB 43412 (ind. 9); Caveiro, MACB 77197 (ind. 8). ORENSE: Vilarino de Conso, MACB 59119 (ind. 10). ***Dryopteris filix-mas*** **(L.) Schott. ITALY**. TOSCANA: A Cansoli, MACB 58812 (ind. 4). **SPAIN**. ÁLAVA: Vitoria, MACB 59150 (ind. 2). CANTABRIA: Cosgaya, MACB 109257 (ind. 5–15). CIUDAD REAL: Puebla de Don Rodrigo, MACB 66633 (ind. 3). LÉRIDA: Vall Ferrera, MACB 62915 (ind. 1). ***Osmunda regalis*** **L.** ASTURIAS: Luarca, MACB 104246 (ind. 1–9). CANTABRIA: Liendo, MACB 107332 (ind. 12–15). CÁCERES: Villanueva de la Vera, MACB 109360 (ind. 10–11). ***Polystichum aculeatum*** **(L.) Roth ex Mert. FRANCE**. HAUTE GARONNE: Pirineos, MACB 42766 (ind. 19). **ITALY**. LUACCA: Garfagnana, MACB 58833 (ind. 18). **SPAIN**. ALAVA: Alto Iturrieta, MACB 18834 (ind. 1). BURGOS: Los Altos, MACB 56768 (ind. 5–6); Monte Cervera, MACB 59157 (ind. 12–14). GRANADA: Güejar-Sierra, MACB 62969 (ind. 2); Sierra Nevada, MACB 9546 (ind. 20). HUESCA: Panticosa, MACB 19040 (ind. 3). LEÓN: Salamón, MACB 43327 (ind. 4); Palacios del Sil, MACB 74457 (ind. 7–11); La Cueta, MACB 59213 (ind. 15). NAVARRA: Irañeta, MACB 43218 (ind. 16–17). ***Polystichum lonchitis*** **(L.) Roth.** ASTURIAS: Somiedo, MACB 104247 (ind. 1–11). CANTABRIA: Camaleño, MACB 109629 (ind. 12–20). ***Polystichum setiferum*** **(Forssk.) T. Moore ex Woyn.** LA CORUÑA: Abaña, MACB 92414 (ind. 2); Caaveiro, MACB 60505 (ind. 4). CANTABRIA: Luena, MACB 55630 (ind. 3); Cosgaya, MACB 109256 (ind. 5–15). LEÓN: Balboa, MACB 60520 (ind. 1). ***Struthiopteris spicant*** **(L.) Weiss. FRANCE**. BRETAGNE: Monfort, MACB 109254 (ind. 1–10). **SPAIN**. ÁLAVA: Ulibarri-Olleros, MACB 30899 (ind. 15–16). CIUDAD REAL: Navas de Estena, MACB 36083 (ind. 12–13). LEÓN: Burón, MACB 17528 (ind. 11). LUGO: Murás, MACB 30875 (ind. 14).

References

Ackerly DD, Dudley SA, Sultan SE, Schmitt J, Coleman JS, Linder CR, Sandquist DR, Geber MA, Evans AS, Dawson TE, Lachowicz MJ (2000) The evolution of plant ecophysiological traits: recent advances and future directions. Bioscience 50:979–995

Atala C, Saldaña A, Navarrete E (2012) Stomatal frequency and gas exchange differs in two Blechnum species (Pteridophyta, Blechnaceae) with contrasting ecological breadth. Gayana Botanica 69:161–166

Brodribb TJ, McAdam SAM, Jordan GJ, Feild TS (2009) Evolution of stomatal responsiveness to CO_2 and optimization of water-use efficiency among land plants. New Phytol 183:839–847

Brownlee C (2001) The long and the short of stomatal density signals. Trends Plant Sci 6:441–442

Camargo MAB, Marenco RA (2011) Density, size and distribution of stomata in 35 rainforest tree species in Central Amazonia. Acta Amazon 41:205–212

Castroviejo S, Laínz M, López G, Montserrat P, Muñoz F, Paiva J, Villar L (1986) Flora Iberica I. CSIC, Madrid

Cowan IR (1977) Stomatal behaviour and environment. Adv Bot Res 4:117–228

Doi M, Wada M, Shimazaki K (2006) The fern Adiantum Capillus-Veneris lacks stomatal responses to blue light. Plant Cell Physiol 47:748–755

Evert RF, Eichhorn SE (2006) Esau's plant anatomy. Wiley, New York

Ferrer-Castan D, Vetaas OR (2005) Pteridophyte richness, climate and topography in the Iberian peninsula: comparing spatial and nonspatial models of richness patterns. Glob Ecol Biogeogr 14:155–165

Franks PJ, Farquhar GD (2007) The mechanical diversity of stomata and its significance in gas-exchange control. Plant Physiol 143:78–87

Gabriel y Galán JM, Prada C, Rolleri CH, Lahoz-Beltrá R, Martínez-Calvo C (2011) Biometry of stomata in Blechnum species (Blechnaceae) with some taxonomic and ecological implications for the ferns. Revista De Biologia Tropical 59:403–415

Gay AP, Hurd RG (1975) Influence of light on stomatal density in tomato. New Phytol 75:37–46

Givnish TJ (1988) Adaptation to sun and shade - a whole-plant perspective. Aust J Plant Physiol 15:63–92

Haworth M, Elliott-Kingston C, McElwain JC (2013) Co-ordination of physiological and morphological responses of stomata to elevated [CO_2] in vascular plants. Oecologia 171:71–82

Henriot A, Cheype JL (2012) Piximètre, la measure des dimensions sur images

Hetherington AM, Woodward FI (2003) The role of stomata in sensing and driving environmental change. Nature 424:901–908

Holland N, Richardson AD (2009) Stomatal length correlates with elevation of growth in four temperate species. J Sustain For 28:63–73

Hunt MA, Davidson NJ, Unwin GL, Close DC (2002) Ecophysiology of the soft tree fern, Dicksonia Antarctica labill. Austral Ecol 27:360–368

Jones HG (1998) Stomatal control of photosynthesis and transpiration. J Exp Bot 49:387–398

Kato M, Imaichi R (1992) Leaf anatomy of tropical fern rheophytes, with its evolutionary and ecological implications. Can J Bot 70:165–174

Kawai H, Kanegae T, Christensen S, Kiyosue T, Sato Y, Imaizumi T, Kadota A, Wada M (2003) Responses of ferns to red light are mediated by an unconventional photoreceptor. Nature 421:287–290

Kessler M, Siorak Y, Wunderlich M, Wegner C (2007) Patterns of morphological leaf traits among pteridophytes along humidity and temperature gradients in the Bolivian Andes. Funct Plant Biol 34:963–971

Kluge J, Kessler M (2007) Morphological characteristics of fern assemblages along an elevational gradient, patterns and causes. Ecotropica 13:27–43

Kluge J, Bach K, essler M (2008) Elevational distribution and zonation of tropical pteridophyte assemblages in Costa Rica. BasicAppl Ecol 9:35–43

Liu S, Liu J, Cao J, Bai C, Shi R (2006) Stomatal distribution and character analysis of leaf epidermis of jujube under drought stress. J Anhui Agric Sci 34:1315–1318

Maherali H, Reid CD, Polley HW, Johnson HB, Jackson RB (2002) Stomatal acclimation over a subambient to elevated CO_2 gradient in a C3/C4 grassland. Plant Cell Environ 25:557–566

Martinez JP, Silva H, Ledent JF, Pinto M (2007) Effect of drought stress on the osmotic adjustment, cell wall elasticity and cell volume of six cultivars of common beans (Phaseolus vulgaris L.) Eur J Agron 26:30–38

McAdam SAM, Brodribb TJ (2012) Stomatal innovation and the rise of seed plants. Ecol Lett 15:1–8

McAdam SAM, Brodribb TJ (2013) Ancestral stomatal control results in a canalization of fern and lycophyte adaptation to drought. New Phytol 198:429–441

Mehltreter K, Walker LR, Sharpe JM (2010) Fern ecology. Cambridge University Press, Cambridge

Meidner H, Mansfield T (1968) Physiology of stomata. McGraw-Hill, London

Metcalfe CR, Chalk L (1979) Anatomy of the dicotyledons. Clarendon, London

Moreno JC, Lobo JM (2008) Iberian–Balearic fern regions and their explanatory variables. Plant Ecol 198:149–167

Öpik H, Rolfe SA (2005) The physiology of flowering plants. Cambridge University Press, Cambridge

Pichi Sermolli REG (1979) A survey of pteridological flora of the Mediterranean region. Webbia 34:175–242
Pompelli MF, Martins SCV, Celin EF, Ventrella MC, DaMatta FM (2010) What is the influence of ordinary epidermal cells and stomata on the leaf plasticity of coffee plants grown under full-sun and shady conditions? Braz J Biol 70:1083–1088
Prabhakar M (2003) Structure, delimitation, nomenclature and classification of stomata. Acta Bot Sin 462:242–252
Quarrie SA, Jones HG (1977) Effects of abscisic acid and water stress on development and morphology of wheat. J Exp Bot 28:192–203
Reich PB, Ellsworth DS, Walters MB, Vose JM, Gresham C, Volin JC, Bowman WD (1999) Generality of leaf trait relationships: a test across six biomes. Ecology 80:1955–1969
Riano K, Briones O (2013) Leaf physiological response to light environment of three tree fern species in a Mexican cloud forest. J Trop Ecol 29:217–228
Rolleri C (1977) Correlation of morphology and geographical distribution in lycopodium saururus. Am Fern J 67:109–120
Rolleri C, Prada C, Gabriel y Galán JM, Passarelli L (2013) Especies arborescentes del género Blechnum (Blechnaceae-Polypodiophyta). Revista De Biologia Tropical 61:377–408
Ruzin SE (1999) Plant microtechnique and microscopy. University Press, Oxford
Salisbury EJ (1927) On the causes and ecological significance of stomatal frequency, with special reference to the woodland flora. Philos Trans R Soc B 216:1–65
Salvo E (1990) Guía de helechos de la Península Ibérica y Baleares. Pirámide, Madrid
Salvo E, Cabezudo B, España L (1984) Atlas de la pteridoflora ibérica y balear. Acta Botanica Malacitana 9:105–128
Schulze ED, Lange OL, Kappen L, Buschbom U, Evenari M (1973) Stomatal responses to changes in temperature at increasing water stress. Planta 110:29–42
Spence RD, Wu H, Sharpe PJH, Clark KG (1986) Water stress effects on guard cell anatomy and the mechanical advantage of the epidermal cells. Plant Cell Environ 9:197–202
Stebbins GL (1950) Variation and evolution in plants. Columbia University Press, New York
Xu Z, Zhou G (2008) Responses of leaf stomatal density to water status and its relationship with photosynthesis in a grass. J Exp Bot 59:3317–3325

Chapter 23
Recent Advances in the Use of Mitochondrial Activity of Fern Spores for the Evaluation of Acute Toxicity

Alexis Joseph Rodríguez-Romero, Jacinto Elías Sedeño-Díaz, Eugenia López-López, Marta Esteban, Luis G. Quintanilla, and Myriam Catalá

23.1 Introduction

Ecotoxicology can be described as a vast and complex discipline which employs various approaches for investigations, including residue analyses, field surveys and toxicity testing (Wang and Freemark 1995). In order to evaluate the real effect of chemicals on the environment or the efficacy of decontamination technologies, chemical quantifications must be completed with toxicological studies on ecologically relevant organisms (Smolders et al. 2004; Dorne et al. 2007). Higher plants are an essential part of a healthy and balanced ecosystem, and new plant models are essential in the evaluation of potential impacts of chemicals on nontarget species (Gong et al. 2001). Many important environmental legislation and guidelines developed under different authorities during the past 30 years have included phytotoxicity tests as a part of chemical safety evaluation (FDA, USEPA, OECD, Environment Canada). The EU produced in 2006 an ambitious regulation involving the Registration, Evaluation, Authorization and Restriction of Chemicals (REACH).

A. J. Rodríguez-Romero · E. López-López
Laboratorio de Evaluación de la Salud de los Ecosistemas Acuáticos, Escuela Nacional de Ciencias Biológicas, Instituto Politécnico Nacional, Ciudad de México, Mexico

J. E. Sedeño-Díaz
Coordinación Politécnica para la Sustentabilidad. Instituto Politécnico Nacional. Av. Instituto Politécnico Nacional s/n, Ciudad de México, Mexico

M. Esteban
Area of Environmental Toxicity, National Center of Environmental Health, Instituto de Salud Carlos III, Majadahonda, Spain

L. G. Quintanilla · M. Catalá (✉)
Department of Biology and Geology, Physics and Inorganic Chemistry, University Rey Juan Carlos, Móstoles, Spain
e-mail: myriam.catala@urjc.es

H. Fernández (ed.), *Current Advances in Fern Research*,
https://doi.org/10.1007/978-3-319-75103-0_23

REACH requires the submission of safety data in support of the protection of human and environmental health of all chemicals produced or imported above 1 ton (around 30,000 substances), and new cost-efficient and reliable methods are needed in a scenario of budget restrictions.

Besides chemical safety evaluation, phytotoxicity testing is also warranted in the environmental assessment of complex effluents or contaminated sites. Toxic components that may be practically nontoxic to fish, crustaceans and daphnids can injure and kill aquatic or riparian vegetation endangering a whole ecosystem. Algal toxicity has often been used as a surrogate for higher plant toxicity testing. However, the sensitivity to toxicants of algae and plants is qualitatively different. For example, algae are more sensitive to metal ions but less sensitive to herbicides than vascular plants (Wang and Freemark 1995).

23.2 Characteristics of Toxicity Tests

Landis and Yu (1998) in their classic "Introduction to Environmental Toxicology" already point out that one of the most crucial aspects of a toxicity test is the suitability of the test organisms. Toxicity tests, they argue, are performed to gain an overall picture of the toxicity of a compound to a variety of species. They listed some of the criteria when choosing a species for a toxicity test:

1. The test organism should be widely available through laboratory culture, procurement from a hatchery or other culture facility or collection from the field.
2. The organism should be successfully maintained in the laboratory environment and available in sufficient quantities.
3. The genetics, genetic composition and history of the culture should be known. However, the majority of toxicity tests in environmental toxicology are conducted with organisms of unknown origin or field collection.
4. The relative sensitivities to various classes of toxicants of the test species should be known regarding the endpoints to be measured. This criterion is not often realized in environmental toxicology. The invertebrate *Daphnia magna* is one of the most commonly used organisms, yet only the results for approximately 500 compounds had been published by 1999. In contrast, by that year, already 2000 compounds had been examined using the Norway rat as the test species.
5. The sensitivity of the test species should be representative of the particular class or phylum that the species represents. Again this is an ideal criterion, not often met in the case of most test species. The limiting factor here is the lack of information on the sensitivity of poorly investigated classes or phyla.

Test protocols of most terrestrial plants are developed for crop species (cabbage, lettuce, oat, ryegrass, etc.) and evaluate seed germination, root elongation and early seedling and vegetative growth. These tests require special facilities, like greenhouses, and the measurement of the aforementioned endpoints is laborious and cannot be easily automated resulting in high costs. This is an important drawback for their extensive use in toxicology screening of new substances by the chemical industry or environmental monitoring of contaminated sites or discharges.

23.3 Current Challenges of Ecotoxicology

In the last years ecotoxicologists have expressed several concerns about standard phytotoxicity bioassays: low number of represented taxa, low ecological relevance and low sensitiveness of used endpoints. The use of an adequate range of taxa is a key point for the achievement of ecologically relevant results, and regulatory agencies are increasingly more interested in requiring nontarget phytotoxicity testing. The selection of more ecologically relevant species and robust test endpoints remains to be important challenges for ecotoxicologists. Moreover, the sensitiveness of a species to a given toxicant may vary along its life cycle. The occurrence of a toxic stress during the critical stages such as development may cause the failure of the organism to mature correctly, infertility or even subsequent death. The evaluation of lethality as unique endpoint has also aroused criticism. The use of biological markers (biomarkers), as for human health, could make evident early deleterious effects leading to death or reduced fitness. Biochemical and physiological biomarkers are especially adequate for automation, which is an important goal for massive bioassay application in toxicity screening or environmental monitoring.

Environmental monitoring and assessment, therefore, faces at present the challenge of developing new tools, more sensitive and reliable, with increased biological and ecological relevance than those currently used, able to detect, in a cost-effective way, early impacts before an irreversible disturbance of the ecosystem may occur.

The respiratory chain that takes place in mitochondria is the main energy source in eukaryotic cells providing for the necessary amounts of ATP. Alterations in this process can yield important consequences, and any toxicant affecting mitochondrial activity and cell energy budget will alter normal plant development, even jeopardizing plant survival. Measurements of reduction rates of the respiratory chain can be employed as viability assays in eukaryotic cells. The reduction of tetrazolium salts to water-insoluble coloured formazan salts by the respiratory chain has been used in the last 50 years for this aim. Several tetrazolium salts are available for this assay. The reduction of 2,3,5-triphenyltetrazolium chloride (TTC) to the corresponding triphenyl formazan (TPF) salt has successfully been used in the analysis of the viability of seeds or vegetal cells since 1951 (Smith 1951; Kalina and Palmer 1968).

23.4 Ferns and Ecotoxicology

Ferns are important components of numerous plant communities, especially temperate and tropical forests, where they play an important ecological role. Ferns are locally abundant and even dominant both in the understorey and forming a tree-fern layer (Brock et al. 2016). Forest ferns are the habitat for other species (Roberts et al. 2005), a filter that control forest regeneration (Song et al. 2012) and a food resource for insects and other animals (Sawamura et al. 2009; Arosa et al. 2010). Moreover, tree ferns are key species, as they are capable to suppress conifer and angiosperm woody species and to facilitate later stages of succession by stabilizing slopes and improving soils, thus controlling the ultimate structure of the upper canopy (Drake and Pratt 2001; Walker et al. 2010). Consequently, tree ferns are considered as ecosystem engineers (Large and Braggins 2004).

Fern spores and gametophytes have long been recognized as useful models for plant research in important areas, namely, plant development, sex determination, gamete production and fertilization, response to environmental factors and evolution of plant complex traits (Banks 1999). However, they have been neglected as subjects for ecotoxicology studies. One of the main advantages of this model is its naturally miniature size. Fern spores are single meiotic cells which range between 15 μm and 150 μm (Tryon and Lugardon 1991). After germination, spores develop into multicellular gametophytes, which are miniature gamete-producing individuals, typically a few millimetres long. In this case, the use of microtubes and microplates is imposed by the natural model. The development of rapid and reliable methods to test phytotoxicity with fern spores and gametophytes could dramatically reduce standard test costs maintaining the biological relevance of whole plant testing.

The first bioassay of acute phytotoxicity based on fern spores using mitochondrial activity as a biomarker of toxicity was only published in 2009 (Catala et al. 2009). The tetrazolium salt used was TTC. Here we will focus on some practical issues of this bioassay. This testing method based on fern spores is naturally miniaturized and combines biological and ecological relevance together with sensitivity and simplicity, thus making it a promising cost-effective tool for high-throughput toxicity screening and monitoring.

Fern spores possess physicochemical characteristics that hampered the direct use of the TTC method on them. For example, the presence of highly hydrophobic substances in the cellular wall makes the spores difficult to suspend in aqueous media. Other difficulty is that the spore wall is extremely resistant, and therefore the methods used for sample homogenization and TTC extraction must be especially vigorous. Moreover, pigments from plant tissues (i.e. chlorophyll) can interfere in the spectrophotometric measurement of TPF when it is extracted with ethanol. All these subjects must be taken into account and imply modifications of the TTC method for the study of spores.

Probably the most problematic adaptation of the method for studying fern spores is the aforesaid extraction of the formazan red. In that sense, different homogenization methods can be employed, but results in extraction ratios are different even when the order of the treatment changes. In the case of the fern *Dryopteris guanchica*, best results are obtained with a first chemical treatment with KOH/EtOH followed by thermal incubation at 65 °C and mechanical homogenization. Mechanical homogenization is a bottleneck step, since most of the methods are designed for single samples and are time consuming. We have developed an efficient method using strong agitation of the sample with a small volume of glass fragments, i.e.

smashed Pasteur pipettes. This allows the automated simultaneous homogenization of a high number of samples.

Other parameters of this bioassay must be optimized in order to achieve optimal results. Spores of different fern species have different characteristics, such as presence or absence of chlorophyll, wall structure, etc. Thus, the number of spores per sample, pH, TTC concentration, incubation time, homogenization and extraction should be tested to obtain optimal conditions of the bioassay with different species. We have successfully applied the TTC assay to the tree-fern *Cyathea costaricensis* (Cyatheaceae) and to several understorey ferns: *Dryopteris aemula*, *D. affinis*, *D. corleyi*, *D. filix-mas*, *D. oreades*, *Polystichum setiferum* (Dryopteridaceae) and *Osmunda regalis* (Osmundaceae). The most appropriate species were *D. guanchica*, *P. setiferum*, *O. regalis* and *C. costaricensis*.

23.5 Candidate Ferns for Toxicity Testing

D. guanchica is located in some Canary Islands and in northwestern Iberian Peninsula and grows in several habitats: laurel forest (Canary Islands), oak forest, alder forest and heathland (Iberian Peninsula). This species is an allotetraploid derived from *D. aemula* and *D. intermedia*. We choose *D. guanchica* as model species for the initial studies on mitochondrial activity of fern spores and gametophytes based on several advantages. First, leaves and spores are produced throughout the year, and thus mature spores can be collected in different seasons. Second, the germination rate of *D. guanchica* is faster than those of related diploids (Quintanilla and Escudero 2006), which shortens assay times. Third, spore viability can be maintained for at least 1 year with a simple technique (dry storage at 5 °C; Quintanilla et al. 2002). In addition, despite the hybrid (allopolyploid) origin of *D. guanchica*, spore abortion percentages are low and similar to those of related diploids (Quintanilla and Escudero 2006).

P. setiferum is mainly present in western and southern Europe. Its ecological amplitude is narrower than that of *D. guanchica*. *Polystichum setiferum* occurs in river-valley forest. Thus, this fern is a very good option for environmental monitoring of fluvial ecosystems. Moreover, its spore production per leaf is very high, albeit spore maturation is limited to a short period of time (early summer). As an additional advantage, the spores are smaller than those of *D. guanchica*, which is convenient for manipulation. Small spore size allows more spores per volume unit, and thus sample size can be increased without increasing assay cost. Smaller spores sink slower and give more homogenous liquid suspensions for longer times. This characteristic is important in our TTC assay, as spore suspensions are used in several steps of the protocol. *Polystichum setiferum* is commercialized as garden plant in temperate climates, providing an alternative source of spores.

O. regalis is cosmopolitan and, like *P. setiferum*, inhabits riparian forests. *O. regalis* lives in riverbanks and is sensitive to water pollution. Sporangia do not form definite sori but a massive panicle-like outgrowth to the top of the leaf (Fig. 23.1a). This is the only studied species that produces green (i.e. chlorophyll-containing)

Fig. 23.1 *Osmunda regalis*: (**a**) panicle-like cluster of sporangia at the apex of a fertile leaf, (**b**) spore with a high concentration of chloroplasts in the centre (differential interference contrast microscopy, 1000×)

spores (Fig. 23.1b). Green spores show an elevated metabolic activity from the very moment of collection, without a latency period. The main advantages we have found with these spores are easiness of suspension in aqueous solution, even without detergents, elevated mitochondrial activity levels and high sensitiveness to environmental stress. Despite the aforesaid advantages, *O. regalis* spores also show some drawbacks, since they must be either used immediately after collection or kept in an ultrafreezer (−80 °C) to maintain a high viability (Ballesteros et al. 2012).

C. costaricensis is located in Mexico, Guatemala, Honduras, El Salvador, Nicaragua, Costa Rica and Panama, mainly in humid and cloud forests, but also close to semiarid zones. The importance of *C. costaricensis* as well as other ferns in the Mexican cloud forest has been widely described (Williams-Linera et al. 2005). *Cyathea costaricensis* has stems up to 8 m in height and leaves up to 3 m in length (Fig. 23.2a), whereas its spores are exceptionally small (Fig. 23.2b; Gómez-Noguez et al. 2016). *Cyathea costaricensis* is an endangered species included in CITES Appendix II. However, it should be noted that only small leaf fragments with mature sori are necessary for toxicity tests, with no, or negligible, damage to the plant.

23.6 Toxicity Testing with Fern Spores

The fern spore bioassay for acute toxicity testing has been used to produce valuable toxicological data regarding vascular plant reproduction and development as well as for environmental monitoring and environmental technology assessment.

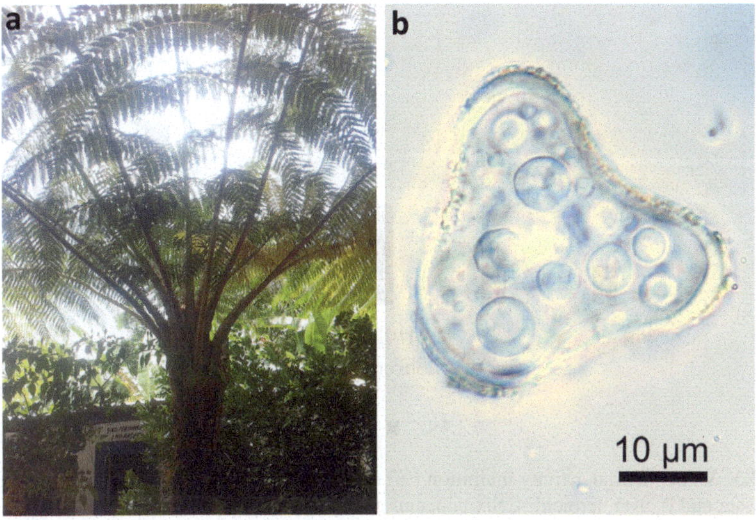

Fig. 23.2 *Cyathea costaricensis*: (**a**) trunk with leaf bases, (**b**) spore full of oil drops, which are the main energy storage (differential interference contrast microscopy, 1000×)

23.6.1 Toxicological Sensitiveness and Dose-Response Curves to Pure Substances or Mixtures

Polycyclic aromatic hydrocarbons (PAHs) constitute a family of toxic pollutants derived mainly from fuel use, transport or storage. The evaluation of the toxicity of contaminated soils or surface waters is of environmental concern. The spore mitochondrial activity bioassay was shown to be sensitive to very low concentrations of PAHs, either in known mixtures or in complex PAH-contaminated soil extracts (Catalá et al. 2009). Fern spores have also been demonstrated to be sensitive to other industrial pollutants as the comparative evaluation with the luminescent bacterium *Vibrio fischeri* (Marugán et al. 2012). In both cases, EC_{50} values differ significantly for *V. fischeri* commercial kit, *V. fischeri* lab cultures and *P. setiferum* fern spores (1.9, 16, and 101 mg cyanide L^{-1} and 27.0, 49.3, and 1440 mg phenol L^{-1}, respectively). Whereas *V. fischeri* bioassays are extremely sensitive and high dilutions must be prepared, toxicant solutions can be directly applied to spores.

The widespread use of herbicides causes major problems such as water pollution, due to their high solubility and the appearance of resistant weeds (Mariani et al. 2014). But pesticides have also been detected in the atmosphere in different areas (Reston 1999), and they can injure nearby trees and shrubs if they drift or volatilize onto leaves (Dreistadt et al. 1994). Glyphosate [N- (phosphonomethyl) glycine] is a broad-spectrum herbicide used to control a wide range of pests (weeds) and is the active ingredient of a variety of commercial herbicides. 2,4-Dichlorophenoxyacetic acid, also known as 2,4-D, is a synthetic auxin, a class

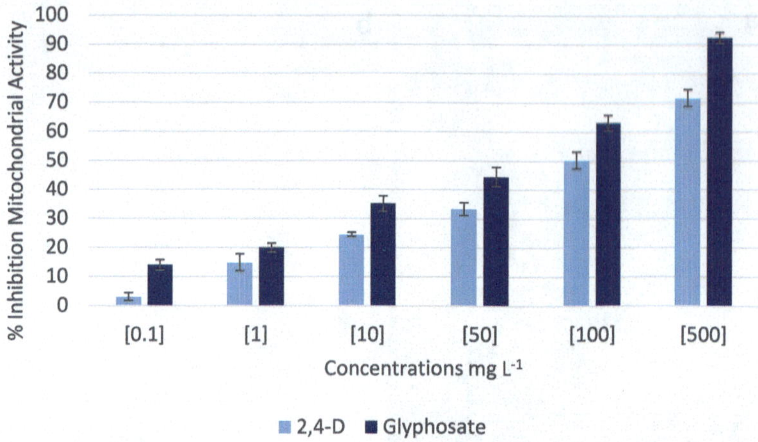

Fig. 23.3 Mitochondrial activity inhibition (%) of *C. costaricensis* at different concentrations of glyphosate and 2, 4-D herbicides. Six concentrations (0.1, 1, 10, 50, 100, and 500 mg/L) of the standard of the two herbicides tested were used (glyphosate 45,521 Sigma-Aldrich; 2, 4-D 49083 Sigma-Aldrich). The exposure period was 24 h (12 h light and 12 h darkness). Each assay was performed in triplicate using as a control group the exposure of the spores to Dyer-Tween medium, which showed the highest mitochondrial activity

of plant hormones. It is absorbed through the leaves and translocated to the meristems of the plant. Uncontrolled, unsustainable growth ensues, causing stem curl-over, leaf withering and eventual plant death (Song 2014). Both herbicides are among the world's most extensively used (Myers et al. 2016), and during application pesticide sprays can directly hit nontarget vegetation or can drift or volatilize from the treated area and contaminate air, soil and nontarget plants. Exposure to the herbicide glyphosate can severely reduce seed quality (Barnea et al. 2000) and increase the susceptibility of certain plants to disease (Vannini et al. 2015). This confers a special threat to endangered plant species and a special interest to assess the effect of the most used herbicides on nontarget native species as the case of the tree ferns.

The CE_{50} of these two herbicides was assessed on mitochondrial activity of *Cyathea costaricensis* spores. Inhibition of mitochondrial activity rated from 14.02% to 92.35% for glyphosate and from 2.97% to 71.38% for 2,4-D with the tested concentrations of the herbicides (0.1, 1, 10, 50, 100 and 500 mg L^{-1}) (Fig. 23.3, unpublished results).

Mitochondrial activity of spores of *C. costaricensis* was highly susceptible to both herbicides tested but particularly to glyphosate (Table 24.6.1.1). The 50% inhibition of mitochondrial activity was detected with 42.19 and 106.96 mg L^{-1} for glyphosate and 2,4-D, respectively (Table 23.1).

These results show that glyphosate exerts a more toxic effect for the mitochondrial activity of *C. costaricensis* than 2,4-D. Mitochondrial activity is a very sensitive biomarker to detect the negative effect of the herbicides tested on fern spores. Differences in toxicity can be related to the mechanism of action of the herbicides

Table 23.1 CE_{50} for glyphosate and 2, 4-D with confidence limits (95%)

Tested substance	CE_{50} (mg L^{-1})	Lower limit (95%)	Upper limit (95%)
Glyphosate	42.19	20.84	64.49
2,4-D	106.96	67.56	185.55

tested. In the case of glyphosate, the mode of action is inhibiting the enzyme EPSPS (5-enolpyruvylshikimate-3-phosphate), causing a reduction of the biosynthesis of amino acids (tryptophan, tyrosine and phenylalanine), which are essential for protein synthesis and the accumulation of shikimic acid and some hydroxybenzoic acids, which are processes that can trigger major toxicity effects.

Pharmaceutical products (PPs) are currently not considered in water-quality regulations; however, evidences of their potential effect on the environment have become more and more common in recent years, leading to the inclusion of these chemicals among the so-called emerging pollutants (la Farre et al. 2008). The anti-inflammatory drug diclofenac has been widely reported in the environment, but ecotoxicological data regarding ecological relevant species are still scarce, similarly to other pharmaceuticals. DNA quantification showed that diclofenac induced acute lethal phytotoxicity at 24 and 48 h (lowest observed effect concentrations, LOECs 30 and 0.3 μg L^{-1}, respectively). Hormetic effects in mitochondrial activity of spores of *P. setiferum* masked lethality, and adverse effects were only observed at 48 h (LOEC 0.3 μg L^{-1}). In subchronic exposure (1 week), LOEC for DNA was 0.03 μg L^{-1}. Mitochondrial activity showed a strong hormetic stimulation of the surviving spore population (LOEC 0.3 μg L^{-1}). Little changes were observed in chlorophyll autofluorescence (LOEC 0.3 μg L^{-1}). The results indicate that environmental concentrations of diclofenac can be deleterious for plant development (Feito et al. 2012).

Venlafaxine hydrochloride is a structurally novel antidepressant. Its occurrence in surface waters and drinking water has been repeatedly reported in the last years. Spores of *P. setiferum* were used to assess its potential acute and chronic sublethal toxicity of venlafaxine in vascular plants. The authors concluded that environmental concentrations of venlafaxine can be deleterious for the development of significant populations of sensitive individuals (Feito et al. 2013). Results showed that venlafaxine induced acute lethal toxicity at 24 and 48 h (LOECs, of 1 μg L^{-1} and 0.1 μg L^{-1}, respectively) by assessing DNA levels. At 24 h, hormetic effects in spores of *P. setiferum* mitochondrial activity masked lethality and adverse effects were observed (LOEC 1 μg L^{-1}) similarly to what was observed with diclofenac. At 48 h a reduction in the mitochondrial activity happened (LOEC 10 μg L^{-1}). In chronic exposure of 1 week, LOEC for DNA was 0.1 μg L^{-1}. Mitochondrial activity showed a strong hormetic stimulation of a surviving spore population (LOEC 10 μg L^{-1}). Changes were not observed in chlorophyll autofluorescence.

Several studies have found cocaine and its main active metabolite benzoylecgonine (BE) in the aquatic environment, derived from its consumption by humans as well as the inability of water treatment processes to eliminate it. A few studies exist on the ecotoxicology of BE to aquatic animals, but the effects of environmental anthropogenic BE on vascular plants have been neglected, and the potential risk of this substance for the riparian communities is unknown. BE induced alterations in mitochondrial activity and DNA levels of fern spores at environmental concentrations (1 ng L^{-1}), which could disrupt gametophyte germination (García-Cambero et al. 2015). Adverse effects on ferns agree with the allelopathic role described for alkaloids and their unspecific interference with plant germination. The authors conclude that the anthropogenic dispersion of alkaloid allelochemicals may pose a risk for biodiversity and irrigated food production that should be further investigated.

In recent years, there is a growing concern for endocrine-disrupting compounds (EDCs) that have the ability to alter the hormonal system. One of the substances found almost ubiquitously and in higher concentrations is the alkylphenol nonylphenol. Albeit this compound is included in priority lists as a probable risk for human health and the environment, little is known about its effects on developing plants. Esteban et al. (2016) showed that nonylphenol induced acute and subchronic toxicity at 48 h and after 1 week, regarding the effects on mitochondrial activity and DNA content, respectively. Significant effects are observed in both parameters in *P. setiferum* spores at ng L^{-1} levels, but chlorophyll autofluorescence shows little changes. Results presented in this study suggest that environmental concentrations of nonylphenol could interfere with higher plant germination development by mimicking natural allelochemicals and/or phytohormones acting as a "phytoendocrine disruptor" likely posing ecophysiological risks.

23.6.2 Environmental Technology Assessment and Development

The utility of bioassays is not limited to the determination of the toxicity of pure compounds or environmental samples, but can also be extended to the assessment of environmental technologies. The assessment of developing technologies for industrial or urban effluent decontamination is especially important. Traditionally, the efficacy of these technologies has been measured by means of chemical quantification of the elimination of specific toxicants. Nonetheless, the products resulting from a decontamination treatment may also be toxic.

Cyanide and phenol are highly toxic substances found in some industrial effluents that must be eliminated before their discharge into the environment. The analytical results of the chemical monitoring of a process of advanced oxidation for cyanide elimination reported by Marugán et al. (2012) showed that the possibility of direct application of *P. setiferum* bioassay makes it a more suitable option for the evaluation of processes for the treatment of highly contaminated water than currently used *V. fischeri* kits. Spore microbioassay was useful in the follow-up of photoxidation processes of cyanide and phenol, also reflecting the formation of intermediate degradation by-products even more toxic than the parent compound. Ferns include a series of widespread riverbank species which adds ecological relevance to the bioassay results, especially regarding the impact of discharges into surface waters. Both for cyanide and phenol, significant differences can be observed in the performance of each toxicological test. Whereas *P. setiferum* bioassay allows the direct follow-up of the process without dilution, *V. fischeri* bioassays require an important and very precise range of dilution of the water samples to achieve the sensitivity of the response to the treatment. The toxicity bioassay using fern spores has been shown to be really useful for the direct evaluation of decontamination processes such as advanced oxidation processes, in which the analysis of individual chemical species would fail in assessing the environmental risk of the effluent.

Photo-Fenton treatment is a promising advanced oxidation process (AOP) approach for the degradation of drug residues which is based on the generation of hydroxyl radicals that act as powerful oxidizing agents with a high reactivity and low selectivity to dissolved organic compounds. Fern spores were employed for the evaluation of acute and chronic toxicity based on mitochondrial activity, DNA and chlorophyll quantifications of natural river waters and photo-Fenton-treated samples. Photo-Fenton treatment provided a high degree of total organic carbon mineralization with up to 70% reduction for river water samples containing at least 56 pharmaceuticals of various chemical groups, such as anti-inflammatory, antibacterial and cardiovascular drugs. The elimination of most of the studied pharmaceutical compounds was confirmed yielding a significant decrease in toxicity, even though some residual toxicity remained after treatment. This fact, revealed by the spore bioassay, seemed to be related to the presence of toxicants in the water matrix, probably of inorganic nature, rather than the toxic effect of the studied pharmaceutical compounds, since they were effectively removed (Rodríguez-Gil et al. 2010). Similar results were obtained for drugs of abuse from six different chemical classes and their metabolites in natural fluvial waters. Photo-Fenton technology was able to significantly reduce the concentration of the drugs of abuse in several tested conditions for the adjustment of hydrogen peroxide and catalyst loadings during the process. However, toxicological analyses measured as inhibition of fern spore mitochondrial activity showed only a complete elimination of acute and chronic toxicity when a higher solid catalyst loading was used (0.6 g L^{-1}). A lower catalyst loading of 0.2 g L^{-1} was not enough for toxicity elimination. These results evidence the need for combining toxicological tests and chemical analyses in order to establish the effectiveness of the water treatment technologies (Catalá et al. 2015).

23.6.3 Environmental Monitoring of Land Use and Ecotoxicology

According to currently applied monitoring programmes, water quality and ecosystem health are to be studied by traditional water chemistry analysis and/or standard ecological assessments. Traditional chemistry gives no information on the ecological effects of the pollutants, and complete ecological assessments provide a precise diagnosis but are not suitable for routine monitoring due to the need of highly qualified human resources, time consumption and low predictive capabilities (the ecological damage may be irreversible by the time it is detected). With these limitations in mind, the development of effective environmental monitoring methodologies faces the challenge of incorporating new techniques able to merge cost-effectiveness, reliability, prognostic power and ecological relevance. A series of reliable and cost-effective microbioassay-based techniques for routine water quality monitoring including *P. setiferum* spore bioassay were developed and tested on a newly developed irrigation area in western Spain (Rodriguez-Gil et al. 2013). Mitochondrial activity, chlorophyll autofluorescence and total amount of DNA were evaluated in *P. setiferum* spores and gametophytes as markers of effects on plant development in response to fluvial and irrigation channel waters. Lipid peroxidation was assessed as a measure for acute animal toxicity in zebrafish embryos (*Danio rerio*). Pollution by pesticides (atrazine, diuron, molinate and oxadiazon) and pharmaceuticals (caffeine, cotinine, ivermectine, nicotine and paraxanthine) was detected in water courses receiving irrigation drainage. Results from these bioassays were compared to a screening-level risk characterization based on pesticide concentrations in water samples from the study area. Preliminary risk assessment described potential moderate or high risk in the lower waters of both studied rivers. The battery of bioassays was able to detect toxicity in the waters of one of the rivers as well as toxic effects on the waters from the irrigation channel. The authors conclude that such methods could both cut costs and improve the prognostic capability of current monitoring programmes.

The mitochondrial activity assay has been recently used to test the water condition of a river in a cloud forest in the intertropical area of the state of Veracruz, México (Rodríguez-Romero et al., under revision). For this study the aim was to evaluate the response of mitochondrial activity in spores of the autochthonous fern *Cyathea costaricensis* exposed to water samples of the Bobos River (Veracruz, Mexico) as a bioassessment tool concurrently with land use and physicochemical evaluation of water samples along the river. Bobos River is located in the Nautla basin, northeastern Veracruz. The upper Bobos River and its tributary, Jalacingo River, are surrounded by a cloud forest, where *C. costaricensis* is present; this portion of the river runs through a protected natural area (Filobobos River and adjacent areas), and the soil type of the basin is dominated by humic material. The land use in the basin shows remarkable differences: in the upper portion (study sites PI and HU), the basin displayed predominantly natural vegetation, while those sites located in the middle portion (MA, FI, JL and EN) are dominated by agriculture (banana, citrus and sugar cane). Finally, study sites at the lower portion (PA, RG and PL) showed a combination of three land uses, some patches of natural vegetation, agriculture and urban areas (Fig. 23.4).

The study involved three monitoring periods including the dry and cold season, February, and the rainy and warm season, June and September 2014. In each study site, the Water Quality Index (WQI) was assessed. Additionally, bioassays were carried on with spores of *C. costaricensis* that were exposed to various water samples from the Bobos River to assess mitochondrial activity and were compared to a positive control. A principal component analysis (PCA) was performed to correlate land-use attributes with environmental variables and mitochondrial activity. Three river sections were identified (Fig. 23.5):

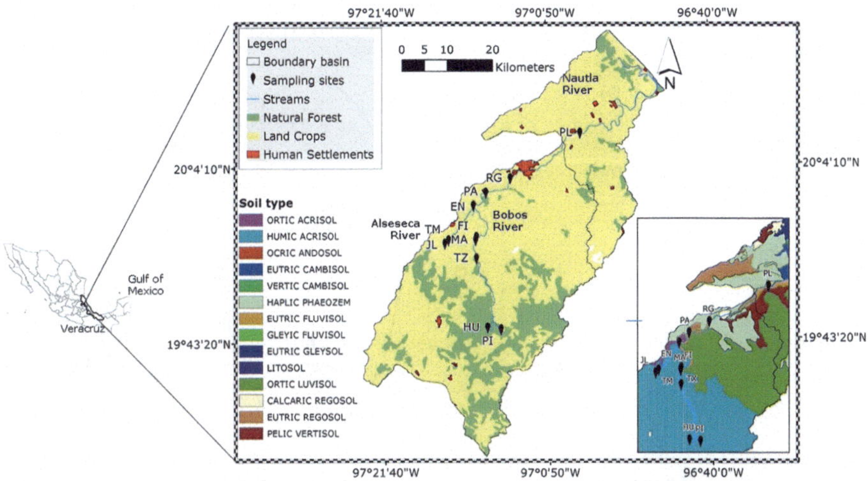

Fig. 23.4 Study area in Bobos River, main land use in the basin and soil types

1. The upper portion, characterized by the dominance of native vegetation (cloud forest), the highest scores of WQI and the lowest mitochondrial activity (63.87–77.47%), related to high hardness. The geological nature of this portion of the river is dominated by rock masses of limestone from the Early Cretaceous. These limestones are exposed in the riverbed of the upper portion, which is free from sediment due to the high slope of the river; the water of this portion of the river was rich in $CaCO_3$.
2. Study sites located at the middle portion (FI, JL and TM) were characterized by the highest levels of orthophosphate, nitrite, chloride and faecal coliforms. These sites are closely related to agricultural land use and showed the highest rates of mitochondrial activity in *C. costaricensis* spores.
3. Finally, study sites located at the lower portion of the river basin (PA, RG and PL) showed the highest water temperature, conductivity, ammonia, colour, total suspended solids, phosphorus and total nitrogen, as well as the highest concentration of total coliforms, associated with urban areas and, consequently, with lower WQI in the three periods of study. The results regarding mitochondrial activity suggest that there are stressors that inhibit the biological response in the water samples assessed from the upper portion of the river (PI, HU, TZ and MA); results oppose to those of WQI scores which in turns are in concordance with the presence of natural vegetation. In this section of the river, large $CaCO_3$–rich limestone deposits have been detected in the geological basement; during the three study seasons, $CaCO_3$ levels ranged from 62.33 to 85.49 mg L^{-1}, which correspond to moderately hard water (USGS 2016).

The dysfunction of the mitochondrial activity by the effect of calcium in excess has been described by Brooks et al. (2004), who report that excess calcium in intra- and intercellular space triggers a pathological stimulus in mitochondria that blocks the electron transport mechanism in the first complex of the inner membrane.

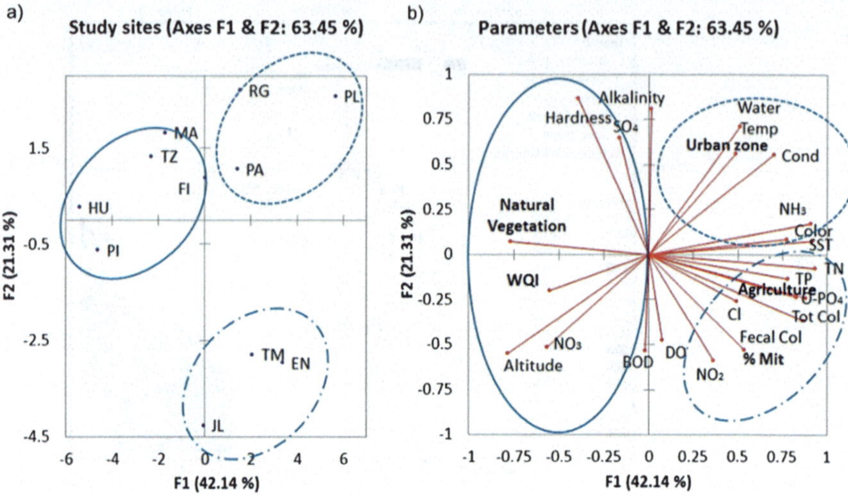

Fig. 23.5 Biplot of the principal component analysis between study sites and the physicochemical environmental variables, land use and mitochondrial activity: (**a**) study sites, (**b**) vectors of the physicochemical environmental variables

Blocking electron transport leads to the production of ROS and the inhibition of ATP production, leading to cell death. When the electron transport chain is inhibited, the reduction reaction of the tetrazolium salt to form formazan is interrupted, and this is detected by the technique. This offers the possibility of having a test method capable of detecting changes in the physicochemical composition related to the geological nature of the basin (limestone rich in $CaCO_3$), as well as to changes related to the incorporation of nutrients or contaminants derived from the activities carried out in the basin.

Mitochondrial activity also showed seasonal variations; it showed a peak in September (98.32% ± 9.01), likely resulting from nutrient enrichment in the rainy season, and was lowest in February (74.54% ± 1.60) ($p < 0.05$). Mitochondrial activity was found to be a good benchmark for the assessment of water quality, reflecting the effects of the physicochemical characteristics of the river and the effect of land use. The mitochondrial activity showed changes along the river and between seasons, associated with environmental characteristics (land use and the geological nature of the basin, as well as with those related to human impacts).

23.7 Conclusion

Our bioassay presents several practical advantages: easy collection and manipulation of test organism, low cost, only basic lab equipment needed, relatively short execution time and easiness of adaptation to high-throughput technologies (i.e.

automated analyser, microplate reader, etc.). Despite the simplicity of the method, when we compare the fern spore bioassay based on mitochondrial activity with other bioassays, we find further advantages:

1. Biological relevance: ferns are higher plants. The conclusions obtained are highly relevant for eukaryotic organisms, especially other higher plants such as crops or wild plants. Comparable bioassays based on prokaryote organisms (e.g. *Vibrio fischeri*) render results that are not easily extrapolated to eukaryotic organisms.
2. Ecological relevance: we can use spores of ferns belonging to different habitats including temperate riparian and tropical ecosystem engineer species.
3. Versatility. With the same organism and methodology, different types of bioassays are possible: acute toxicity (24–72 h) or chronic toxicity (weeks). This is also so regarding the nature of the sample to be tested: aqueous, soil or gas samples. Incubation of gametophytes can be performed in liquid samples: environmental waters, soil or gas leachates. Although still not explored, spores could also be sown on solid substrates such as soils or directly exposed to gaseous samples.
4. Fern spores can be easily transported, and some species preserved at 5 °C for months or even years and do not need continuous growth or feeding such as other organisms.
5. Very low volume of test sample needed. If the samples can be analysed in a microplate reader, less than 1 mL of test sample is needed, and automation of the procedure requires only punctual adjustments.
6. The fern spore bioassay allows the screening of a high number of samples at a minimum cost. Plant material and chemical substances are inexpensive, the latter not being especially toxic; therefore not extraordinary health security measures need to be adopted. The laboratory equipment needed is usually found in regular laboratories.
7. Spores and gametophytes of some species tolerate wide ranges of pH. This characteristic minimizes the problems related with sample pH or acidification during bioassay realization. In the case of *D. guanchica*, we have shown that mitochondrial activity is not altered in a pH range of 5–10 (Catala et al. 2009). At the same time spores may be very sensitive to water hardness as has been shown for *C. costaricensis*.

Finally, this model sums up the advantages of methods based on single cells with those of pluricellular organisms. In a short time, spores give rise to gametophytes that may be used as a pluricellular plant model with no further modifications of the method. These features make this method suitable for commercialization in the shape of a kit.

The method developed for the measurement of mitochondrial activity in fern spores allows obtaining results of cell viability, development and physiological state in short periods of time. Then, besides its use in bioassays, it can be applied to multiple fields: (a) conservation programmes, as an alternative to germination assays; (b) research in plant physiology and biochemistry, i.e. for the investigation of oxidative pathways; (c) as a tool in the study of biotic and abiotic stress; and (d) as a tool in the study of genetic and genomic factors affecting spore germination and gametophyte development.

References

Arosa ML, Ramos JA, Quintanilla LG, Brown D (2010) First report of fern (*Culcita macrocarpa*) spore consumption by a small mammal (*Apodemus sylvaticus*). Mamm Biol 75:115–121. https://doi.org/10.1016/j.mambio.2009.05.009

Ballesteros D, Estrelles E, Walters C, Ibars AM (2012) Effects of temperature and desiccation on ex situ conservation of nongreen fern spores. Am J Bot 99:721–729. https://doi.org/10.3732/ajb.1100257

Banks JA (1999) Gametophyte developtment in ferns. Annu Rev Plant Physiol Plant Mol Biol 50:163–186

Barnea N, Alon Y, Sibony M, Rubin B (2000) The effect of late season application of glyphosate on cotton plants and weeds. In: 2000 Proceedings Beltwide Cotton Conferences, San Antonio, pp 663–665

Brock JMR, Perry GLW, Lee WG, Burns BR (2016) Tree fern ecology in New Zealand: a model for southern temperate rainforests. For Ecol Manag 375:112–126

Brookes PS, Yoon YS, Robotham JL et al (2004) Calcium, ATP, and ROS: a mitochondrial love-hate triangle. Am J Physiol Cell Physiol 287:C817–C833. https://doi.org/10.1152/ajpcell.00139.2004

Catalá M, Esteban M, Rodríguez-Gil JL, Quintanilla LG (2009) Development of a naturally miniaturised testing method based on the mitochondrial activity of fern spores: a new higher plant bioassay. Chemosphere 77:983–988

Catalá M, Pagani R, Portolés M-T (2009) Regulation of hepatocyte glutathione content by hepatic sinusoidal cells activated with LPS: anatomical restrictions. Histol Histopathol 24:1541–1550

Catalá M, Domínguez-Morueco N, Migens A et al (2015) Elimination of drugs of abuse and their toxicity from natural waters by photo-Fenton treatment. Sci Total Environ 520:198–205. https://doi.org/10.1016/j.scitotenv.2015.03.042

Dorne JL, Skinner L, Frampton GK et al (2007) Human and environmental risk assessment of pharmaceuticals: differences, similarities, lessons from toxicology. Anal Bioanal Chem 387:1259–1268

Drake DR, Pratt LW (2001) Seedling mortality in Hawaiian rain forest: the role of small-scale physical disturbance. Biotropica 33:319–323. https://doi.org/10.1646/0006-3606(2001)033[0319:SMIHRF]2.0.CO;2

Dreistadt SH, Clark JK, Flint ML (1994) Pests of landscape trees and shrubs: an integrated pest management guide. University of California, Oakland

Esteban S, Llamas PM, García-Cortés H, Catalá M (2016) The endocrine disruptor nonylphenol induces sublethal toxicity in vascular plant development at environmental concentrations: a risk for riparian plants and irrigated crops? Environ Pollut 216:480–486. https://doi.org/10.1016/j.envpol.2016.05.086

Feito R, Valcárcel Y, Catalá M (2012) Biomarker assessment of toxicity with miniaturised bioassays: diclofenac as a case study. Ecotoxicology 21(1):289–296

Feito R, Valcárcel Y, Catalá M (2013) Preliminary data suggest that venlafaxine environmental concentrations could be toxic to plants. Chemosphere 90:2065–2069

García-Cambero JP, García-Cortés H, Valcárcel Y, Catalá M (2015) Environmental concentrations of the cocaine metabolite benzoylecgonine induced sublethal toxicity in the development of plants but not in a zebrafish embryo-larval model. J Hazard Mater 300:866–872. https://doi.org/10.1016/j.jhazmat.2015.08.019

Gómez-Noguez F, Pérez-García B, Mehltreter K et al (2016) Spore mass and morphometry of some fern species. Flora Morphol Distrib Funct Ecol Plants 223:99–105. https://doi.org/10.1016/j.flora.2016.05.003

Gong P, Wilke BM, Strozzi E, Fleischmann S (2001) Evaluation and refinement of a continuous seed germination and early seedling growth test for the use in the ecotoxicological assessment of soils. Chemosphere 44:491–500

Kalina M, Palmer JM (1968) Reduction of tetrazolium salts by plant mitochondria. Histochemie 14:366–695

la Farre M, Perez S, Kantiani L, Barcelo D (2008) Fate and toxicity of emerging pollutants, their metabolites and transformation products in the aquatic environment. Trac-Trends Anal Chem 27:991–1007. https://doi.org/10.1016/j.trac.2008.09.010ER

Landis WG, Yu M-H (1998) An introduction to toxicity testing. In: Introduction to environmental toxicology: impacts of chemicals upon ecological systems. CRC Press, Boca Raton, pp 21–53

Large MF, Braggins JE (2004) Tree ferns. Timber Press, Portland

Mariani M, Romero R, Cassano A, Zalazar C (2014) Degradation of a mixture of glyphosate and 2,4-D in water solution employing the UV/H2O2 process, including toxicity evaluation. In: Advanced oxidation technologies. CRC Press, Leiden, pp 99–117

Marugán J, Bru D, Pablos C, Catalá M (2012) Comparative evaluation of acute toxicity by *Vibrio fischeri* and fern spore based bioassays in the follow-up of toxic chemicals degradation by photocatalysis. J Hazard Mater 213–214:117–122

Myers JP, Antoniou MN, Blumberg B et al (2016) Concerns over use of glyphosate-based herbicides and risks associated with exposures: a consensus statement. Environ Health 15:19. https://doi.org/10.1186/s12940-016-0117-0

Quintanilla LG, Escudero A (2006) Spore fitness components do not differ between diploid and allotetraploid species of *Dryopteris* (Dryopteridaceae). Ann Bot 98:609–618. https://doi.org/10.1093/aob/mcl137

Quintanilla LG, Amigo J, Pangua E, Pajaron S (2002) Effect of storage method on spore viability in five globally threatened fern species. Ann Bot 90:461–467

Reston VA (1999) The quality of our nation's waters – nutrients and pesticides. USGS, Denver

Roberts NR, Dalton PJ, Jordan GJ (2005) Epiphytic ferns and bryophytes of Tasmanian tree-ferns: a comparison of diversity and composition between two host species. Austral Ecol 30:146–154. https://doi.org/10.1111/j.1442-9993.2005.01440.x

Rodríguez-Gil JL, Catalá M, Alonso SG et al (2010) Heterogeneous photo-Fenton treatment for the reduction of pharmaceutical contamination in Madrid rivers and ecotoxicological evaluation by a miniaturized fern spores bioassay. Chemosphere 80:381–388

Rodríguez-Gil JL, San Sebastián Sauto J, González-Alonso S et al (2013) Development of cost-effective strategies for environmental monitoring of irrigated areas in Mediterranean regions: traditional and new approaches in a changing world. Agric Ecosyst Environ 181:41–49. https://doi.org/10.1016/j.agee.2013.09.007

Sawamura M, Kawakita A, Kato M (2009) Fern – spore-feeder interaction in temperate forests in Japan: Sporing phenology and spore-feeding insect community. Am J Bot 96:594–604. https://doi.org/10.3732/ajb.0800256

Smith FE (1951) Tetrazolium Salt. Science 113:751–754

Smolders R, Bervoets L, Blust R (2004) In situ and laboratory bioassays to evaluate the impact of effluent discharges on receiving aquatic ecosystems. Environ Pollut 132:231–243

Song Y (2014) Insight into the mode of action of 2,4-dichlorophenoxyacetic acid (2,4-D) as an herbicide. J Integr Plant Biol 56:106–113. https://doi.org/10.1111/jipb.12131

Song GZM, Yates DJ, Doley D (2012) Rain forest understorey ferns facilitate tree seedling survival under animal non-trophic stress. J Veg Sci 23:847–857. https://doi.org/10.1111/j.1654-1103.2012.01398.x

Tryon AF, Lugardon B (1991) Spores of the Pteridophyta. Springer, New York

USGS (2016) Water-Quality information, water hardness and alkalinity. http://water.usgs.gov/owq/hardness-alkalinity.html

Vannini A, Guarnieri M, Bačkor M et al (2015) Uptake and toxicity of glyphosate in the lichen *Xanthoria parietina* (L.) Th. Fr. Ecotoxicol Environ Saf 122:193–197. https://doi.org/10.1016/j.ecoenv.2015.07.030

Walker LR, Landau FH, Velázquez E et al (2010) Early successional woody plants facilitate and ferns inhibit forest development on Puerto Rican landslides. J Ecol 98:625–635. https://doi.org/10.1111/j.1365-2745.2010.01641.x

Wang WC, Freemark K (1995) The use of plants for environmental monitoring and assessment. Ecotoxicol Environ Saf 30:289–301

Williams-Linera G, Palacios-Rios M, Hernández-Gómez R (2005) Fern richness, tree species surrogacy, and fragment complementarity in a Mexican tropical montane cloud forest. Biodivers Conserv 14:119–133. https://doi.org/10.1007/s10531-005-4053-5

Chapter 24
Update on the Assessment of Chronic Phytotoxicity Using Fern Spore Biomarkers

Helena García-Cortés, Myriam Catalá, and José Luis Rodríguez-Gil

24.1 Introduction

For many years, acute lethality tests have been the dominant method for both single compound and whole effluent toxicity studies due to their simplicity and economic and temporal efficiencies. The United States Environmental Protection Agency (USEPA) defines chronic toxicity as "the ability of a substance or mixture of substances to cause harmful effects over an extended period, usually upon repeated or continuous exposure sometimes lasting for the entire life of the exposed organism" (USEPA 1997). By looking at such a definition, most of us can agree that regardless of a few punctual events, organisms in nature are mainly subjected to chronic exposures. With acute exposure being a rare event, and lethality a most certainly unwanted outcome, it soon became obvious to toxicologists that new, relevant, chronic toxicity tests needed to be developed. By evaluating more subtle endpoints like growth or reproduction, and lengthening in the tests, more accurate, direct and relevant estimates of the toxicity of the studied chemical or sample could be obtained. The most conservative tests of this kind, and those that offer a better confidence on the obtained data, are full life-cycle tests. However, these are long and costly and have been limited to animal toxicology testing of compounds with a direct human health association such as drugs. More recently, subchronic studies evaluating early life stages (ELS tests) of the organisms have been implemented to provide comparable results to full life-cycle test for a fraction of the time and cost

H. García-Cortés
National Center for Environmental Health, National Health Institute Carlos III, Madrid, Spain

M. Catalá (✉)
Department of Biology and Geology, Physics and Inorganic Chemistry, ESCET, Rey Juan Carlos University, Madrid, Spain
e-mail: myriam.catala@urjc.es

J. L. Rodríguez-Gil
Department of Biology, University of Ottawa, Ottawa K1N 6N5, Ontario, Canada

H. Fernández (ed.), *Current Advances in Fern Research*,
https://doi.org/10.1007/978-3-319-75103-0_24

Table 24.1 Examples of the most commonly applied standard test for terrestrial plant toxicity testing (OECD 2006; USEPA 2012)

Issuer	Guideline num.	Guideline	Latest version
OECD	OECD 208	Terrestrial Plant Test, (seedling emergence and growth)	2006
OECD	OECD 227	Terrestrial Plant Test, (vegetative vigour)	2006
USEPA	OCSPP 850.4100	Seedling Emergence and Seedling Growth	2012
USEPA	OCSPP 850.4150	Vegetative Vigour	2012
USEPA	OCSPP 850.4230	Early Seedling Growth Toxicity Test	2012
USEPA	OCSPP 850.4300	Terrestrial Plants Field Study	2012

(Landis et al. 1993). This is, however, not such a novel approach in the field of phytotoxicity. Ever since the beginning of the development of the pesticide industry, early life stages of both crops and weeds have been tested as relatively long-lasting germination and seedling growth tests to evaluate the success of the new plant protection products.

24.2 Current State of Phytotoxicity Testing

Currently several standard guidelines are available for the evaluation of subchronic terrestrial plant toxicity. Of special relevance are those developed by the United States Environmental Protection Agency's Office of Chemical Safety and Pollution Prevention (USEPA, OCSPP) and the Organisation for Economic Co-operation and Development (OECD). A summary of the main guidelines produced by these two entities for the assessment of ecological effects on terrestrial plants is listed in Table 24.1.

Due to their primarily agronomic origin (i.e. assessment and registration of plant protection products), the list of species suggested in these protocols has traditionally been limited to a number of common crops. However, over the last few years (with great improvements since the first edition of this book was published), both organizations have increased the number of nontarget terrestrial plant (NTTPs) species suggested as possible test species. Currently the list of species suggested by the OECD consists of 32 crop species (25 dicots and 9 monocots) and 52 non-crop species. The USEPA list is somewhat shorter with only 19 crop species (15 dicots and 4 monocots); however they share the same list of non-crop species suggested by the OECD. The full list of species suggested by the two organizations is presented on Table 24.2. In addition to species-specific laboratory/greenhouse studies, the USEPA provides a guideline (OCSPP 850.4300 – Terrestrial Plants Field Study)

Table 24.2 Summary of the most common species used for terrestrial plant toxicity testing (OECD 2006; USEPA 2012)

Use	Group	Family	Species	Common name	OECD	USEPA
Crop	Dicot	Apiaceae	*Daucus carota*	Carrot	Yes	Yes
Crop	Dicot	Asteraceae	*Helianthus annuus*	Sunflower	Yes	Yes
Crop	Dicot	Asteraceae	*Lactuca sativa*	Lettuce	Yes	Yes
Crop	Dicot	Brassicaceae	*Brassica campestris var. chinensis*	Chinese Cabbage	Yes	No
Crop	Dicot	Brassicaceae	*Brassica napus*	Turnip, Rape	Yes	Yes
Crop	Dicot	Brassicaceae	*Brassica oleracea var. capitata*	Cabbage	Yes	Yes
Crop	Dicot	Brassicaceae	*Brassica rapa*	Field Mustard, Canola	Yes	Yes
Crop	Dicot	Brassicaceae	*Lepidium sativum*	Garden Cress	Yes	No
Crop	Dicot	Brassicaceae	*Raphanus sativum*	Radish	Yes	No
Crop	Dicot	Brassicaceae	*Sinapis alba*	White Mustard	Yes	No
Crop	Dicot	Chenopodiaceae	*Beta vulgaris*	Sugar Beet or Table Beet	Yes	Yes
Crop	Dicot	Cucurbitaceae	*Cucumis sativus*	Cucumber	Yes	Yes
Crop	Dicot	Fabaceae	*Glycine max*	Soybean	Yes	Yes
Crop	Dicot	Fabaceae	*Lotus corniculatus*	Birdsfoot Trefoil	Yes	No
Crop	Dicot	Fabaceae	*Phaseolus aureus*	Mung Bean	Yes	No
Crop	Dicot	Fabaceae	*Phaseolus vulgaris*	Garden Bean	Yes	Yes
Crop	Dicot	Fabaceae	*Pisum sativum*	Pea	Yes	Yes
Crop	Dicot	Fabaceae	*Trifolium pratense*	Red Clover	Yes	No
Crop	Dicot	Fabaceae	*Trigonella foenum-graecum*	Fenugreek	Yes	No
Crop	Dicot	Fabaceae	*Vicia sativa*	Vetch	Yes	No
Crop	Dicot	Linaceae	*Linum usitatissimum*	Flax	Yes	No
Crop	Dicot	Malvaceae	*Gossypium spp.*	Cotton	No	Yes
Crop	Dicot	Polygonaceae	*Fagopyrum esculentum*	Buckwheat	Yes	Yes
Crop	Dicot	Solanaceae	*Lycopersicon esculentum*	Tomato	No	Yes

(continued)

Table 24.2 (continued)

Use	Group	Family	Species	Common name	OECD	USEPA
Crop	Dicot	Solanaceae	*Solanum lycopersicon*	Tomato	Yes	Yes
Crop	Monocot	Liliaceae	*Allium cepa*	Onion	Yes	Yes
Crop	Monocot	Poaceae	*Avena sativa*	Oat	Yes	Yes
Crop	Monocot	Poaceae	*Hordeum vulgare*	Barley	Yes	No
Crop	Monocot	Poaceae	*Lolium perenne*	Perennial ryegrass	Yes	Yes
Crop	Monocot	Poaceae	*Oryza sativa*	Rice	Yes	No
Crop	Monocot	Poaceae	*Secale cereale*	Ryes	Yes	No
Crop	Monocot	Poaceae	*Sorghum bicolor*	Grain sorghum	Yes	No
Crop	Monocot	Poaceae	*Triticum aestivum*	Wheat	Yes	No
Crop	Monocot	Poaceae	*Zea mays*	Corn	Yes	Yes
Non-crop	Dicot	Apiaceae	*Torilis japonica*	Japanese Hedge-parsley	Yes	Yes
Non-crop	Dicot	Asteraceae	*Bellis perennis*	English Daisy	Yes	Yes
Non-crop	Dicot	Asteraceae	*Centaurea cyanus*	Cornflower	Yes	Yes
Non-crop	Dicot	Asteraceae	*Centaurea nigra*	Black Knapweed	Yes	Yes
Non-crop	Dicot	Asteraceae	*Inula helenium*	Elecampane	Yes	Yes
Non-crop	Dicot	Asteraceae	*Leontodon hispidus*	Big Hawkbit	Yes	Yes
Non-crop	Dicot	Asteraceae	*Rudbeckia hirta*	Black-eyed Susan	Yes	Yes
Non-crop	Dicot	Asteraceae	*Solidago canadensis*	Canada Goldenrod	Yes	Yes
Non-crop	Dicot	Asteraceae	*Xanthium pensylvanicum*	Common Cocklebur	Yes	Yes
Non-crop	Dicot	Asteraceae	*Xanthium spinosum*	Spiny Cocklebur	Yes	Yes
Non-crop	Dicot	Asteraceae	*Xanthium strumarium*	Italian Cocklebur	Yes	Yes
Non-crop	Dicot	Brasicaceae	*Cardamine pratensis*	Cuckoo Flower	Yes	Yes
Non-crop	Dicot	Caryophyllaceae	*Lychnis flos-cuculi*	Ragged Robin	Yes	Yes
Non-crop	Dicot	Chenopodiaceae	*Chenopodium album*	Lamb's Quarters	Yes	Yes
Non-crop	Dicot	Clusiaceae	*Hypericum perforatum*	Common St. John's Wort	Yes	Yes

(continued)

Table 24.2 (continued)

Use	Group	Family	Species	Common name	OECD	USEPA
Non-crop	Dicot	Convolvulaceae	*Ipomoea hederacea*	Purple Morning Glory	Yes	Yes
Non-crop	Dicot	Fabaceae	*Lotus corniculatus*	Bird's-foot Trefoil	Yes	Yes
Non-crop	Dicot	Fabaceae	*Senna obtusifolia*	Cassia, Sicklepod	Yes	Yes
Non-crop	Dicot	Fabaceae	*Sesbania exaltata*	Hemp	Yes	Yes
Non-crop	Dicot	Fabaceae	*Trifolium pratense*	Red Clover	Yes	Yes
Non-crop	Dicot	Lamiaceae	*Leonurus cardiaca*	Motherwort	Yes	Yes
Non-crop	Dicot	Lamiaceae	*Mentha spicata*	Spearmint	Yes	Yes
Non-crop	Dicot	Lamiaceae	*Nepeta cataria*	Catnip	Yes	Yes
Non-crop	Dicot	Lamiaceae	*Prunella vulgaris*	Self-heal	Yes	Yes
Non-crop	Dicot	Lamiaceae	*Stachys officinalis*	Hedge-nettle	Yes	Yes
Non-crop	Dicot	Malvaceae	*Abutilon theophrasti*	Velvetleaf	Yes	Yes
Non-crop	Dicot	Malvaceae	*Sida spinosa*	Prickly Sida	Yes	Yes
Non-crop	Dicot	Papaveraceae	*Papaver rhoeas*	Poppy	Yes	Yes
Non-crop	Dicot	Polygonaceae	*Polygonum convolvulus*	Black Bindweed	Yes	Yes
Non-crop	Dicot	Polygonaceae	*Polygonum lapathifolium*	Pale Persicaria	Yes	Yes
Non-crop	Dicot	Polygonaceae	*Polygonum pennsylvanicum*	Pennsylvania Smartweed	Yes	Yes
Non-crop	Dicot	Polygonaceae	*Polygonum persicaria*	Smartweed	Yes	Yes
Non-crop	Dicot	Polygonaceae	*Rumex crispus*	Curly Dock	Yes	Yes
Non-crop	Dicot	Primulaceae	*Anagallis arvensis*	Scarlett Pimpernel	Yes	Yes
Non-crop	Dicot	Ranunculaceae	*Ranunculus acris*	Common Buttercup	Yes	Yes
Non-crop	Dicot	Rosaceae	*Geum urbanum*	Yellow Avens	Yes	Yes
Non-crop	Dicot	Rubiaceae	*Galium aparine*	Cleavers	Yes	Yes
Non-crop	Dicot	Rubiaceae	*Galium mollugo*	Hedge Bedstraw	Yes	Yes
Non-crop	Dicot	Scrophulariaceae	*Digitalis purpurea*	Foxglove	Yes	Yes
Non-crop	Dicot	Scrophulariaceae	*Veronica persica*	Speedwell	Yes	Yes

(continued)

Table 24.2 (continued)

Use	Group	Family	Species	Common name	OECD	USEPA
Non-crop	Monocot	Cyperaceae	*Cyperus rotundus*	Purple Nutsedge	Yes	Yes
Non-crop	Monocot	Poaceae	*Agrostis tenuis*	Common Bentgrass	Yes	Yes
Non-crop	Monocot	Poaceae	*Alopecurus myosuroides*	Foxtail	Yes	Yes
Non-crop	Monocot	Poaceae	*Avena fatua*	Wild Oats	Yes	Yes
Non-crop	Monocot	Poaceae	*Bromus tectorum*	Downy Brome	Yes	Yes
Non-crop	Monocot	Poaceae	*Cynosurus cristatus*	Dog's-tail Grass	Yes	Yes
Non-crop	Monocot	Poaceae	*Digitaria sanguinalis*	Crabgrass	Yes	Yes
Non-crop	Monocot	Poaceae	*Echinochloa crusgalli*	Barnyard Grass	Yes	Yes
Non-crop	Monocot	Poaceae	*Elymus canadensis*	Canada Wild Rye	Yes	Yes
Non-crop	Monocot	Poaceae	*Festuca pratensis*	Fescue	Yes	Yes
Non-crop	Monocot	Poaceae	*Hordeum pusillum*	Little Barley	Yes	Yes
Non-crop	Monocot	Poaceae	*Phleum pratense*	Timothy	Yes	Yes

aimed at the assessment of the effects of chemicals in whole plant populations and communities. Given the site- and objective-specific nature of these tests, no detail is provided on the specific components of these plant assemblages, but importance is given to the maintenance of ecosystem services.

Some of these changes (especially those that occurred since the first edition of this book) have come to address a growing interest on the potential effects of agrochemicals in NTTPs, a field of research (together with the study of their potential effects on species at risk) that has grown tremendously over the last decade (*EFSA Journal* 2014; 12(7): 3800). At the same time, other changes come from the expansion of the applicability of terrestrial plant toxicity tests outside of their traditional agronomical niche. This was, in fact, the motivation for some of the main changes and harmonization of the USEPA guidelines, which was carried out in order to unify, and harmonize, different guidelines used under different regulations including those involved in the registration of plant protection products, as well as more general ones such as the Toxic Substances Control Act (TSCA) (15 U.S.C. 2601 et seq.) which recently went through a large reform and update process.

The situation of phytotoxicity testing in riverine and aquatic ecosystems is not better, in terms of diversity of covered organisms, than that of their terrestrial counterparts. The presence of phytotoxicity tests in aquatic toxicity guidelines is limited to a number of selected green algae species, such as *Raphidocelis subcapitata* (formerly classified as *Pseudokirchneriella subcapitata* and *Selenastrum capricorniatum), Chlorella vulgaris* or *Desmodesmus subspicatus* (formerly known as *Scenedesmus subspicatus*), the diatom *Navicula Pelliculosa* or the marine diatom *Skeletonema costatum* (Guidelines OCSPP 850.4500 and OECD 201). Cyanobacteria, such as *Anabaena flos-aquae* or *Synechococcus leopoliensis*, are often considered separately (Guidelines OCSPP 850.4550 and OECD 201). Additionally, a small number of aquatic macrophytes, namely, duckweed (*Lemna spp.*) (Guidelines OCSPP 850.4400 and OECD 221) and *Myriophyllum spicatum* (OECD 238 and OECD 239), are also considered. Wang (1991) pointed out that the archaic view of algae and plants being members of the same kingdom should not lead us to the error of accepting algae toxicity data as a surrogate of the potential effects on plants; however and as an example, only a growth inhibition test in *R. subcapitata* is required under the USEPA Whole Effluent Toxicity (WET) test guidelines (USEPA 2002). Particular attempts to apply aquatic macrophytes and aquatic environment-related species can be found in the literature (Doust et al. 1994; Ferrat et al. 2003; Lewis 1995; Mohan and Hosetti 1999), but their inclusion in regulatory documentation is far from ideal.

In both cases (terrestrial and water-related ecosystems), and even with the great expansion in the number of species considered that occurred over the last few years, ferns are still consistently ignored. This is even when ferns are the second largest group of vascular plants with more than 10,000 living species and are, in many cases, key components of numerous plant communities like tree fern tropical forests. Boutin et al. (2012) reviewed the methods employed to perform phytotoxicity tests and risk assessments and found that there are limitations related to biodiversity and the number of tested species, because most of the tests are focused on assessing the toxicity of herbicides to a narrow range of species (usually crops). They propose a different approach based on biological traits instead of on the plant phylogeny which, to date, has not been implemented. Recently, the European Food Safety Authority (EFSA), who is in charge of the registration and re-registration of plant protection products in Europe, published a scientific opinion (EFSA 2014) addressing the state of the science on risk assessment of plant protection products for nontarget terrestrial plants. In this piece, they point to incomplete species representation in currently available guidelines as one of the main issues concerning the appropriate evaluation of the risk posed by plant protection products to nontarget terrestrial plants. They specifically name cryptogams (together with mosses, liverworts, hornworts, horsetails and woody species) as one of the groups where data is most dramatically lacking, and where research should focus on the coming years.

According to EFSA (2014), fern species and other cryptogams are not used in phytotoxicity testing for regulatory purposes, although many species comprise main components of some of the moist and shaded habitats adjacent to crop fields in North America and Europe. In addition, some studies pointed out that ferns are quite sensitive to herbicides. As an example, Boutin et al. (2012) assessed the sensitivity from the two fern species, *Onoclea sensibilis* and *Dennstaedtia punctilobula*, to two herbicides and demonstrated a high sensitivity to metsulfuron methyl and to a lesser extent to glyphosate of both species when tested at the early sporophyte stage, causing a biomass reduction in both species. Newmaster and Bell (2002) carried out a study under field conditions, showing that two herbicides commonly used in forestry for conifer release, had a significant initial effect on the richness, abundance and diversity of pteridophytes, bryophytes and lichens. Rowntree and Sheffield (2005) investigated effects of the herbicide asulam on eight fern species tested at the mature sporophyte stage, assessing damage over two seasons, occurring the maximum damage one year after treatment with slight signs of recovery only observed by the second season. Four of the studied species were affected by the high and medium doses, and three of them were more sensitive than the flowering species *Rumex acetosa*. It is important to highlight that the fern species tested were exposed after their maturation, so they were probably not at their most sensitive growth stage.

24.3 Limitations of Traditional Acute Toxicity Tests

Acute toxicity refers to the effects of a single exposure to a toxicant or mixture of toxicants over a short period of time, and it is commonly presented as the lethal concentration 50 or LC_{50}, which represents the dose (or concentration) of a substance that is lethal to 50% of the tested organisms. They have been the predominant toxicity tests for many years due to their multiple advantages such as lower costs, shorter test times and easy quantification of the endpoints (usually mortality) in comparison with the commonly available chronic tests. However, they present numerous disadvantages as well. As mentioned in the introduction, most of the potentially toxic compounds that could end up in natural ecosystems will do so under chronic exposures. This is especially relevant in the case of plants whose lack of motility prevents them from reducing exposure by avoidance as animals would do. Other than in singular dramatic events, pollution usually occurs in non-lethal concentrations. Sublethal effects such as reproduction inhibition or slow growth/development are to be the most affected in this kind of scenarios, and yet acute LC_{50}'s are, in many cases, still the standard measurement in toxicology.

The lack of data about long-term effects, plant recovery or on effects on reproductive stages (Boutin et al. 2012) shows the need for chronic or subchronic bioassays that can provide relevant information, especially when the tested organisms are in a critical stage of its development.

24.4 Limitations of Traditional Terrestrial Plant Toxicity Tests

One of the main limitations of current environmental toxicology is the low number of taxa used for standard bioassays. Currently, the most commonly applied standard tests for terrestrial plant toxicity are based on crop species, and this hinders the extrapolation of the effects to wild species diversity, ecosystem services and functions (Boutin et al. 2012). Although some non-crop species are considered, ferns are not included in any list (Tables 24.1 and 24.2), despite their relevance within vascular plants.

The popularity of acute toxicity testing was not induced by the ignorance of its associated problems; however, chronic toxicity was a logistically difficult alternative to apply. Traditional subchronic phytotoxicity evaluations are based on seed germination or root elongation tests (Benenati 1990) and, as presented in the standard test guidelines, are to be performed in pots filled with soil to which the tested compound is applied (unless testing the effects of direct plant spray) (OECD 2006; USEPA 2012). These traditional tests are not only long but also require specialized facilities able to host large numbers of pots in specific environmental conditions and where the handling and disposal of large volumes of treated soil should be possible. Together with this, the large amounts of soil to be treated increase the need for large volumes of the compound or environmental sample to be tested, as well as the amount of contaminated waste to be dealt with. These characteristics not only increase the test cost dramatically but can, in many cases, make the test unfeasible. Methodologies to reduce sample volume and space requirements have been attempted, from the traditional germination tests in petri dishes to more elaborated phytotoxicity kits (Czerniawska-Kusza et al. 2006). However, all of them are still based on seed germination, and thus, space and sample volume reductions are limited by seedling size and water requirements of the tested plant. Also, the relationship between the sensitivity of the species tested in vitro with the responses of plants in the field is quite difficult to assess. The application of such an approach to non-target wild species is, therefore, complex and will involve basic plant research with the participation of experts in botany and plant ecology together with the environmental toxicologists.

24.5 Usefulness of Ferns in Phytotoxicity Testing

A series of criteria are provided for the selection of nontarget species to be tested under the OECD guidelines, in their test number 208 (Terrestrial Plants [seedling emergence and growth]) (OECD 2007). Among these:

- The species selected should be reasonably broad, e.g. considering their taxonomic diversity in the plant kingdom, their distribution, abundance, species-specific life-cycle characteristics and region of natural occurrence.
- The species have uniform seeds that are readily available from reliable standard seed source(s) and that produce consistent, reliable and even germination, as well as uniform seedling growth.
- They have been used to some extent in previous toxicity tests, and their use in, for example, herbicide bioassays, heavy metal screening, salinity or mineral stress tests or allelopathy studies indicates sensitivity to a wide variety of stressors.

- Plant is amenable to testing in the laboratory and can give reliable and reproducible results within and across testing facilities.
- The sensitivity of the species tested should be consistent with the responses of plants found in the environment exposed to the substance.

As mentioned earlier, ferns, with more than 9000 living species, are the second largest group of vascular plants and are main components of numerous plant communities. Therefore, the inclusion of this group in standard ecotoxicological tests is an important goal regarding ecological relevance. The application of model species representative of the ecosystem studied would considerably help interpret and extrapolate laboratory results (Chung et al. 2007). Fern spores are produced in large quantities from species, in many cases, such as *Polystichum setiferum*, *Osmunda regalis* or many *Dryopteris*, commonly found in plant nurseries around the world. Produced spores are uniform, with very low abortive particles which present even germination, above 90% in many cases (Quintanilla and Escudero 2006). Fern spores, and spore-developed gametophytes, have long been recognized as useful models for plant research in important areas, namely, plant development, sex determination, gamete production and fertilization, response to environmental factors and evolution of plant complex traits (Banks 1999), in part for the convenience of their small size and ease to use in aqueous suspension, which substantially simplifies laboratory procedures and decrease standard tests costs maintaining the biological relevance of whole plant testing (Catalá et al. 2009, see Chap. 23). It is important to note that due to the riparian character of some of the families/species of ferns, they can be chronically exposed to micropollutants from both terrestrial and water sources and could help evaluate water quality, elucidate the potential toxicity of priority and emerging pollutants and assess potential ecological risks.

Our team has been working on the application of a miniaturized bioassay based on the use of biomarkers of *Polystichum setiferum* fern spores that provides the use of an organism with ecological relevance, in an early phase of life cycle and the capacity to assess the impact on natural riparian ecosystem (Catalá et al. 2009). This microbioassay has been successful in the determination of the phytotoxicity and ecotoxicological risks of micropollutants such as the pharmaceutical products diclofenac and venlafaxine (Feito et al. 2012, 2013), the endocrine disruptor nonylphenol (Esteban et al. 2016), the cocaine metabolite benzoylecgonine (Garcia-Cambero et al. 2015), as well as mixtures, showing increased sensitivity compared to validated bioassays (algae growth inhibition and daphnia mobility inhibition) (Esteban et al. 2013). The "biomarker approach" of this bioassay, by assessing sensitive endpoints of phytotoxicity, helps elucidate the underlying biological mechanisms at cellular, biochemical or physiological levels, as well as infer possible effects in higher levels of complexity such as communities or ecosystems. It has also been helpful in evaluating the effectiveness of water treatment technologies based on advanced oxidation processes (photo-Fenton) for the elimination of micropollutants such as pharmaceuticals (Rodríguez-Gil et al. 2010) or drugs of abuse and their metabolites (Catalá et al. 2015). The bioassay has also constituted a useful tool for environmental monitoring of pharmaceutically active compounds of irrigated areas in Mediterranean regions, enabling a preliminary risk assessment based on the detected toxicity, decreasing costs and improving the prognostic capability of current monitoring programmes (Rodríguez-Gil et al. 2013).

24.6 DNA as a Measurement of Cellular Proliferation

The amount of DNA is a biomarker of lethality or cell proliferation during development, and it can be easily quantified by using fluorescent labelling, inserting a specific fluorophore into the DNA. Fluorescence intensity is directly proportional to fluorophore levels and DNA quantity.

The spore germination and development, as for any other pluricellular organism, is characterized by a rapid succession of cell cycles where embryonic cells duplicate their genetic material and divide giving birth to two daughter cells. Microscopic assessment is laborious and presents technical difficulties in the initial phases due to coloured cell walls. In every cell division, DNA content augments twofold; consequently, gametophyte DNA quantification can clearly indicate the stage of plant development of a certain sample of spores.

Fluorescence is the result of a process that occurs in certain molecules (generally polyaromatic hydrocarbons or heterocycles) called fluorophores. A fluorescent probe is a fluorophore designed to respond to a specific stimulus or to localize within a specific region of a biological specimen. The fact that a single fluorophore can generate many thousands of detectable photons is fundamental to the high sensitivity of fluorescence detection techniques. Fluorescence intensity is quantitatively dependent on the same parameters as absorbance—defined by the Beer-Lambert law as the product of the molar extinction coefficient, optical path length and solute concentration—as well as on the fluorescence quantum yield of the dye and the excitation source intensity and fluorescence collection efficiency of the instrument. The use of epifluorescent probes specific for nucleic acids is common on the field of cell biology and embryology as a means of an estimate of cell number.

Fluorometry allows specific and sensitive measurement of DNA concentration by use of common dyes including Hoechst dyes and PicoGreen. Unlike other nucleic acid stains (ethidium bromide, propidium iodide, etc.), the blue fluorescent Hoechst dyes are cell permeable and bind preferentially to DNA in the presence of RNA (Mocharla et al. 1987). Together with DAPI, Hoechst is included in the class of DNA minor-groove binders. The dye, weakly fluorescent itself in solution, binds to all nucleic acids, but AT-rich double-strand DNA (dsDNA) enhances fluorescence twofold greater than GC-rich strands resulting in an increase in fluorescence and a shift in the emission maximum from 500 to 460 nm (Daxhelet et al. 1989; Labarca and Paigen 1980). Recently, a similar behaviour has been fully demonstrated for the interaction of this probe either with animal or plant DNA (Maiti et al. 2009).

PicoGreen is a highly sensitive measure of double-stranded DNA (dsDNA) and is useful for very little volumes. The assay is optimized to minimize the fluorescence contributions of RNA and single-stranded DNA (ssDNA), such that dsDNA can be precisely quantified in the presence of equimolar concentrations of ssDNA and RNA with minimal influence on the quantitative results.

24.7 Chlorophyll *a* Autofluorescence as Surrogate of Plant Physiological State

Environmental pollutants can easily interact with key plant metabolic processes, leading to impairment in chlorophyll *a* (Chl *a*) production or function and thus, in the whole photosynthetic activity and in consequence, in plant fitness. In such scenario variations in Chl *a* content can easily serve as a simple and sensitive surrogate of the physiological state of the plant, in toxicity testing protocols. Precise quantification of chlorophyll a, as applied in many algal toxicity tests, is a destructive method that requires the homogenization of the sample and solvent extraction of the pigment prior to the measurement.

Chlorophyll a is itself a fluorophore and emits red fluorescence when excited at certain wavelengths. Chlorophyll *a* autofluorescence is a biomarker of plant vitality, with which the physiological and functional state of development can be assessed by estimating the total amount of chlorophyll through fluorimetric methods. In vivo fluorimetric methods allow for the measurement of potential quantum efficiency of photosynthesis, providing detailed pictures of the photosynthetic capacity of the studied plant. Studies of chlorophyll *a* (Chl *a*) fluorescence modulation in vivo have been proposed as a useful non-invasive method to evaluate the physiological state of plants (Krause and Weis 1991). However, these methods are complex and, in many cases, not practical for a toxicity bioassay approach. It has been long demonstrated that the direct measurement of the red autofluorescence from plant cells can be used to report the total amount of the cellular chlorophyll pigment with reasonable accuracy (Galbraith et al. 1988), especially when applied to treatment-comparison studies, such as those typically used in toxicology. This approach allows for rapid determinations of total chlorophyll *a* by flow cytometry or fluorometry. Optimal fluorometric parameters for Chl *a* detection are λ_{exc} = 420 nm and λ_{em} = 685 nm (Agati 1998), but thanks to the width of excitation and emission peaks, standard blue and red filters (e.g. λ_{exc} = 485–8 nm and λ_{em} = 635 nm) can be used in the fluorometric estimation of Chl *a* content in plant material with satisfactory sensibility.

The method for in vivo fluorometric quantification of Chl *a* presented in this chapter allows for a significant reduction in sampling preparation time, as spore dilutions can be measured directly on the growth medium or test solution. At the same time, the use of high-throughput techniques, such as multiwell plate fluorescence reading, allows for simple detection of relative variations in Chl *a* content between treatments as both, controls and samples, are measured virtually at the same time, and thus in the same photosynthetic conditions.

24.8 Applications of the Bioassay of Chronic Toxicity Based on Fern Spores

In recent years, a reduction in the amount of "conventional" pollutants in surface water (nitrates, heavy metals, sulphur and nitrogen oxides, etc.) has occurred worldwide, but we are witnessing the occurrence of the so-called "emerging" pollutants, which include, among others, pharmaceuticals and their metabolites (Fent et al. 2006). Very low concentrations of these substances have been detected in several countries, either in sewage treatment plants (STPs), surface waters, sea water, groundwater, sediments or drinking water. Some of the published studies state that such low concentrations of the pharmaceutical drugs found do not constitute an imminent risk for public health but also point to probable ecotoxicological chronic toxicity for aquatic organisms due to a continuous exposure to low doses (Khetan and Collins 2007). However, the information of the toxicity of these micropollutants on nontarget species at environmental concentrations is almost non-existent, in part due to low sensitivity of standard chronic tests.

The development of a bioassay to become a standard method requires the collection of a solid range of previous data on different classes of toxicants, for example, herbicide bioassays, heavy metal screening, salinity or mineral stress tests. In order to characterize the sensitivity of a chronic toxicity bioassay based on fern spores, we have performed studies on different substances and environmental samples. Our team studied the response of the spores of the riparian *Polystichum setiferum* to different classes of toxicants throughout a microbioassay based on the use of biomarkers (Catalá et al. 2009) from several approaches: environmental toxicology/ecotoxicology, assessment of environmental technology and environmental monitoring.

This assay has shown to be successful at assessing the phytotoxicity of waterborne contaminants such as pharmaceuticals (Feito et al. 2012, 2013), environmental water samples (Esteban et al. 2013), the main cocaine metabolite (benzoylecgonine, García-Cambero et al. 2015) and the endocrine disruptor nonylphenol (Esteban et al. 2016). The anti-inflammatory diclofenac showed chronic toxicity on fern spore germination (Feito et al. 2012), causing a strong decrease of DNA as well as a slight decrease in chlorophyll autofluorescence in exposed gametophytes for all concentrations studied. The antidepressant venlafaxine also exerted chronic sublethal toxicity (decrease on DNA levels) at environmental concentrations (Feito et al. 2013) and showed a biphasic response known as "hormesis" in the mitochondrial activity (see Chap. 23), a compensatory effect caused by a moderate toxic insult (Calabrese 2008). This hormetic effect has also been identified for benzoylecgonine (main cocaine metabolite) (García-Cambero et al. 2015) and the endocrine disruptor nonylphenol (Esteban et al. 2016). For these two substances, alkaloid and phenolic compounds respectively, an allelopathic role has been proposed. This behaviour, termed "allelopathy", is an ecophysiological phenomenon by which an organism produces metabolites that affect its survival or development and that of other organisms (Evenari 1949). It is important to note that hormesis has usually been described for allelochemicals, pesticides, hormones (including hormone disruptors) and pollutants in general. Also, many of the emerging pollutants are alkaloids or benzoic acid derivatives and therefore probably allelochemicals, and due to their unspecific interference with plant germination, their anthropogenic dispersion may pose a risk to biodiversity and irrigated agriculture that should be further investigated.

In order to evaluate the efficacy of water treatment technologies, we performed acute and chronic toxicity tests with *Polystichum setiferum* spores in environmental samples. Chemical analysis detected 56 pharmaceuticals in four selected river waters receiving sewage effluents in the Community of Madrid (Spain) (Rodríguez-Gil et al. 2010). Photo-Fenton oxidation provided a high degree of total organic carbon (TOC) mineralization with up to 70% reduction in total organic carbon (TOC), and a 99% elimination of the studied pharmaceuticals. The chronic bioassay with *Polystichum setiferum* demonstrated a significant reduction of the hormetic effect for DNA in all the samples and for chlorophyll in only two of the samples. Despite chemical elimination of pharmaceuticals and high reduction in TOC, a significant hormetic effect persisted when the gametophytes were exposed to supposedly decontaminated water samples. This fact seems to be related to the presence of other toxicants in the water matrix and warns against the sole application of chemical analysis for the monitoring of environmental pollution. Another study was performed in order to evaluate the effectiveness of Photo-Fenton oxidation on waters containing drugs of abuse and their metabolites (Catalá et al. 2015), finding a significant hormetic effect on mitochondrial activity (see Chap. 23).

The bioassay was also used as a technique for water quality monitoring in an irrigation area in western Spain (Rodríguez-Gil et al. 2013). The levels of 147 pesticides were measured in irrigation channels, fluvial water and selected animal and plant tissue as well as the presence of 59 human and veterinary pharmaceuticals in the watershed. Results from these assays were compared to a screening-level risk characterization based on pesticide concentrations in water samples from the study area. The battery of bioassays, in which the fern spores-based assay was included, detected toxicity in the waters as well as toxic effects on the waters from the irrigation channel. An inhibition of fern spore germination when these are exposed to irrigation channel waters polluted with pharmaceuticals and stimulants was found. Considering the complex mixture of pollutants in river waters and its possible additive/synergic effect, and the fact that this water may be used for irrigation of many crops and orchards, land productivity may be affected. These methods could both decrease costs and improve the prognostic capability of current monitoring programmes.

24.9 Conclusions

Higher plants are an essential part of a healthy and balanced ecosystem, and new plant models are essential in the evaluation of potential impacts of pollutants. In this context, ferns must be included within relevant taxa in standard bioassays. The newly developed bioassay of chronic and subchronic toxicity based on fern spores and gametophytes is a promising cost-effective tool for high-throughput toxicity screening and monitoring. The use of biomarkers in toxicity tests can provide relevant information determining the effects at cellular level as well as predicting changes in levels of more complex organization enabling corrective measures to prevent damage to the environment and human health. This higher plant testing method is naturally miniaturized and combines biological and ecological relevance together with sensitivity and simplicity. With this bioassay, we have assessed the subchronic phytotoxicity of several individual chemicals and mixtures. It has also been used to test the performance of advanced tertiary treatment technologies applied to wastewater, as well as for environmental monitoring and water quality assessment. In addition, the "biomarker approach" can lead to the discovery of interesting mechanistic responses at cellular, biochemical or physiological levels (i.e. hormesis and allelopathy, respectively).

References

15 U.S.C. §2601, et seq (1976) The toxic substances control act

Agati G (1998) Response of the in vivo chlorophyll fluorescence spectrum to environmental factors and laser excitation wavelength. Pure Appl Opt 7:797–807

Banks JA (1999) Gametophyte developement in ferns. Annu Rev Plant Phys 50:163–186

Benenati FE (1990) Keynote address: plants - Kesystone to risk assessment. In: Wang W-c, Gorsuch JW, Lower WR (eds) Plants for toxicity assessment. ASTM, Philadelphia, pp 5–14

Boutin C, Aya KL, Carpenter D, Thomas PJ, Rowland O (2012) Phytotoxicity testing for herbicide regulation: shortcomings in relation to biodiversity and ecosystem services in agrarian systems. Sci Total Environ 415:79–92

Calabrese EJ (2008) Hormesis: why it is important to toxicology and toxicologists. Environ Toxicol Chem 27:1451–1474

Catala M, Esteban M, Rodríguez-Gil JL, Quintanilla LG (2009) Development of a naturally miniaturised testing method based on the mitochondrial activity of fern spores: a new higher plant bioassay. Chemosphere 77:983–988

Catalá M, Domínguez-Morueco N, Migens A, Molina R, Martínez F, Valcárcel Y, Mastroianni N, López de Alda M, Barceló D, Segura Y (2015) Elimination of drugs of abuse and their toxicity from natural waters by photo-Fenton treatment. Sci Total Environ 520:198–205

Chung KW, Fulton MH, Scott GI (2007) Use of juvenile clam Mercenaria Mercenaria, as a sensitive indicator of aqueus and sediment toxicity. Ecotoxicol Environ Saf 67:333–340

Czerniawska-Kusza I, Ciesielczuk T, Kusza G, Cichon A (2006) Comparison of the Phytotoxkit microbiotest and chemical variables for toxicity evaluation of sediments. Environ Toxicol 21:367–372

Daxhelet GA, Coene MM, Hoet PP, Cocito CG (1989) Spectrofluorometry of dyes with Dnas of Different Base composition and conformation. Anal Biochem 179:401–403

Doust JL, Schmidt M, Doust LL (1994) Biological assessment of aquatic pollution - a review, with emphasis on plants as biomonitors. Biol Rev 69:147–186

EFSA PPR Panel (EFSA Panel on Plant Protection Products and their Residues) (2014) Scientific opinion addressing the state of the science on risk assessment of plant protection products for non-target terrestrial plants. EFSA J 12(7):3800. 163 pp. https://doi.org/10.2903/j.efsa.2014.3800

Esteban S, Fernández Rodríguez J, Díaz López G, Nuñez M, Valcárcel Y, Catalá M (2013) New microbioassays based on biomarkers are more sensitive to fluvial water micropollution than standard testing methods. Ecotoxicol Environ Saf 93:52–59
Esteban S, Llamas PM, García-Cortés H, Catalá M (2016) The endocrine disruptor nonylphenol induces sublethal toxicity in vascular plant development at environmental concentrations: a risk for riparian plants and irrigated crops? Environ Pollut 216:480–486
Evenari M (1949) Germination inhibitors. Bot Rev 15:153–194
Feito R, Valcárcel Y, Catalá M (2012) Biomarker assessment of toxicity with miniaturised bioassays: diclofenac as a case study. Ecotoxicology 21:289–296
Feito R, Valcárcel Y, Catalá M (2013) Preliminary data suggest that venlafaxine environmental concentrations could be toxic to plants. Chemosphere 90:2065–2069
Fent K, Weston AA, Caminada D (2006) Ecotoxicology of human pharmaceuticals. Aquat Toxicol 76:122–159
Ferrat L, Pergent-Martini C, Romeo M (2003) Assessment of the use of biomarkers in aquatic plants for the evaluation of environmental quality: application to seagrasses. Aquat Toxicol 65:187–204
Galbraith DW, Harkins KR, Jefferson RA (1988) Flow cytometric characterization of the chlorophyll contents and size distributions of plant protoplasts. Cytometry 9:75–83
García-Cambero JP, García-Cortés H, Valcárcel Y, Catalá M (2015) Environmental concentrations of the cocaine metabolite benzoylecgonine induced sublethal toxicity in the development of plants but not in a zebrafish embryo-larval model. J Hazard Mater 300:866–872
Khetan SK, Collins TJ (2007) Human pharmaceuticals in the aquatic environment: a challenge to green chemistry. Chem Rev 107:2319–2364
Krause GH, Weis E (1991) Chlorophyll fluorescence and photosynthesis - the basics. Annu Rev Plant Phys 42:313–349
Labarca C, Paigen K (1980) Simple, rapid, and sensitive DNA assay procedure. Anal Biochem 102:344–352
Landis WG, Hughes JS, Lewis MA (1993) Environmental toxicology and risk assessment. ASTM, Philadelphia, 431 p
Lewis MA (1995) Use of fresh-water plants for Phytotoxicity testing - a review. Environ Pollut 87:319–336
Maiti S, Maiti P, Sinha SS, Mitra RK, Pal SK (2009) Molecular recognition of plant DNA: does it differ from conventional animal DNA? Int J Biol Macromol 44:133–137
Mocharla R, Mocharla H, Hodes ME (1987) A novel, sensitive fluorometric staining technique for the detection of DNA in RNA preparations. Nucleic Acids Res 15:133–137
Mohan BS, Hosetti BB (1999) Aquatic plants for toxicity assessment. Environ Res 81:259–274
Newmaster SJ, Bell FW (2002) The effects of silvicultural disturbances on cryptogam diversity in the boreal-mixedwood forest. Can J For Res 32:38–51
OECD (2006), Test No. 208: Terrestrial Plant Test: Seedling Emergence and Seedling Growth Test, OECD Publishing, Paris
OECD (2006), Test No. 227: Terrestrial Plant Test: Vegetative Vigour Test, OECD Publishing, Paris
OECD (2007) Section 2 – effects on biotic systems (draft). In: Guidelines for the testing of chemicals. Organisation for Economic Co-Operation and Development, Paris
Quintanilla LG, Escudero A (2006) Spore fitness components do not differ between diploid and allotetraploid species of Dryopteris (Dryopteridaceae). Ann Bot 98:609–618
Rodriguez-Gil JL, Catala M, Alonso SG, Maroto RR, Valcarcel Y, Segura Y, Molina R, Melero JA, Martinez F (2010) Heterogeneous photo-Fenton treatment for the reduction of pharmaceutical contamination in Madrid rivers and ecotoxicological evaluation by a miniaturized fern spores bioassay. Chemosphere 181:41–49
Rodriguez-Gil JL, San Sebastián Sauto J, González-Alonso S, Sánchez Sánchez P, Valcarcel Y, Catalá M (2013) Development of cost-effective strategies for environmental monitoring of irrigated areas in Mediterranean regions: traditional and new approaches in a changing world. Agric Ecosyst Environ 181:41–49

Rowntree JK, Sheffield E (2005) The effects of Asulam spraying on non-target ferns. Canadian Journal of Botany, 83:1622–1629

USEPA (1997) Terms of environment: glossary, abbreviations and acronyms (EPA publication no.175-B-97-001). U.S. Environmental Protection Agency, Washington DC

USEPA (2002) Method 1003.0: Green Alga, Selenastrum capricornutum, growth test; chronic toxicity

USEPA (2012) Series 850 – ecological effects test guidelines –terrestrial plants field study. U.S. Environmental Protection Agency, Office of Chemical Safety and Pollution Prevention (OCSPP), Washington DC

Wang WC (1991) Literature-review on higher-plants for toxicity testing. Water Air Soil Poll 59:381–400

Chapter 25
Role of Ferns in Environmental Cleanup

Bhupinder Dhir

25.1 Introduction

Plants possess capacity for removing contaminants from environmental components such as soil and water. Plant species including ferns have been exploited for capacity to remove various organic and inorganic contaminants from the environment (Alexandra et al. 2014; Drăghiceanu et al. 2014). Fast growth rate, easy reproduction, nitrogen-fixing ability, easy culture, and wide occurrence establish ferns as ideal candidates to be used as remediators of environmental contamination. Both aquatic and terrestrial fern species have shown potential for removing contaminants such as heavy metals, organic compounds, and radionuclides from the environment (Olguín et al. 2002; Benaroya et al. 2004, Stepniewska et al. 2005; Sune et al. 2007). *Azolla* and *Salvinia* are the major aquatic species, and *Pteris* and *Pityrogramma* are the major terrestrial species well exploited for removing various contaminants including heavy metals from the environment (Sood et al. 2011). Fern species have been successful in treating wastewater from various sources such as industrial and municipal operations, sugar refinery, olive mill, and soil from mining sites.

Ferns possess the capacity for removing contaminants via accumulation, chelation, and detoxification mechanism. The fibrous root system supports pollutant degradation by providing favorable environmental conditions for microbial activities and pollutant metabolism (Nie et al. 2011). Plants also degrade contaminants by forming symbiotic associations with rhizospheric microbes (Glick 2010).

B. Dhir (✉)
Department of Genetics, University of Delhi South Campus, New Delhi, India

H. Fernández (ed.), *Current Advances in Fern Research*,
https://doi.org/10.1007/978-3-319-75103-0_25

25.2 Removal of Contaminants by Ferns

Fern species remove various inorganic and organic contaminants from the environment. Some of the major ones have been discussed in this section.

25.2.1 Organic Contaminants

Organic contaminants such as volatile compounds, dyes, explosives, and hydrocarbons have been removed/treated by fern species. Species such as *Azolla* show biofiltration capacity because it can alter pH and remove dissolved oxygen (DO), biological oxygen demand (BOD), chemical oxygen demand (COD), polyphenols, total organic matter from olive mill wastewater (OMWW), and sugar refinery wastewater (Ena et al. 2007).

Volatile Compounds

Fern species including *Osmunda japonica*, *Davallia mariesii*, *Polypodium formosanum*, *Polypodium dispar*, *Polypodium multifida*, *Pteris dispar*, *Pteris multifida*, and *Pelargonium* spp. have shown efficacy for removing volatile organic compounds such as formaldehyde (Forns et al. 2010) (Table 25.1). Depending upon the capacity for removal, plant species have been divided into three groups: excellent (greater than 1.2 μgm^{-3} per cm^2 of leaf area over 5 h), intermediate (1.2 or less to 0.6), and poor (less than 0.6). An average removal rate of 1.87 $\mu gm^{-3}\ cm^{-2}$ has been reported for these species. *O. japonica* was found to be the most effective species with the removal capacity of 6.64 $\mu gm^{-3}\ cm^{-2}$ over 5 h. Another species, namely, *Nephrolepis exaltata* (Boston fern), has also reported formaldehyde removal when exposed to high concentration (6200–49,600 $\mu g/m^3$) (Dela Cruz et al. 2014).

VOC removal in plants generally occurs via various pathways: (1) removal by the aboveground plant part, (2) removal by the microorganisms residing in the soil, (3) removal by the roots, and (4) removal by the growing media. Stomata and cuticle are suggested to be pathways for removal of volatile organic compounds (VOCs) by the aboveground plant parts. Diffusion of VOCs occurs through the stomata in the light and through the cuticle in the dark (Schmitz et al. 2000; Kim et al. 2008; Treesubsuntorn and Thiravetyan 2012). The studies have shown absorption of benzene, toluene, ethylbenzene, and xylene by wax layer. Uptake by plants is highly dependent on the properties of the VOCs. A hydrophilic VOC such as formaldehyde will not diffuse easily through the cuticle, whereas a lipophilic VOC such as benzene is more likely to penetrate through the cuticle. After entry these compounds are readily absorbed by the abaxial surface of leaves and processed in leaf tissues (Giese et al. 1994; Ugrekhelidze et al. 1997). The compound can undergo degradation, storage, or excretion, either at site of uptake or after translocation to other parts of the plant (Dela Cruz et al. 2014). These compounds undergo glycosylation and glutathionylation resulting in their conversion to nonvolatile compounds which play important ecological roles such as defense in plants. According to some reports,

Table 25.1 Formaldehyde removal reported in fern species after exposure duration of 5 h

Scientific name	(mg m^{-3} cm^{-2} leaf area)
Adiantum capillus-veneris	0.86
Botrychium ternatum	1.42
Coniogramme japonica	0.76
Cyrtomium caryotideum	1.09
Cyrtomium falcatum	0.67
Davallia mariesii	3.16
Dryopteris nipponensis	0.91
Microlepia strigosa	1.49
Osmunda japonica	6.64
Polypodium formosanum	3.62
Polystichum tripteron	0.92
Pteris dispar	1.95
Pteris ensiformis	1.01
Pteris multifida	1.92
Selaginella tamariscina	4.84
Thelypteris acuminate	0.51
Thelypteris decursive-pinnata	0.47
Thelypteris esquirolii	0.84
Thelypteris torresiana	0.40

VOCs enter the Calvin cycle after a two-step enzymatic oxidation to CO_2 after absorption. It is also supposed that some of them get converted to S-methylmethionine and further translocated in the phloem from where it is transported to various organs (e.g., seed, roots) (Hanson and Roje 2001). Microorganisms present in the rhizospheric zone assist in removal of volatile compounds. Plants excrete significant amount of carbon into the root zone. This stimulates the growth of microbes. Rhizospheric and phyllospheric microorganisms, as well as stomata-mediated absorption, are the major means of biofiltration of VOC (Kim et al. 2008). The reduction of VOCs (such as formaldehyde) also occurs due to microbial utilization of them as a carbon source.

Microorganisms present in the soil play an important role in removal of VOCs. When VOCs are not degraded in leaf, they are transported to roots where microbial degradation results in their breakdown and hence removal (De Kempeneer et al. 2004). Endophytic and rhizospheric bacteria present in the phyllosphere zone assist plants in removing toxic compounds from soil (McGuinness and Dowling 2009). Monooxygenases and dioxygenases are the major bacterial enzymatic systems that assist in degradation of benzene, toluene, ethylbenzene, and xylene present in the environment (Jindrova et al. 2002).

Pyrocatechol

Azolla filiculoides show high ability to remove pyrocatechol (1, 2-dihydroxybenzene, or 2-hydroxyphenol) from aqueous solutions. The removal efficiency of 60–90%

has been noted at temperature of 25–30 °C. Efficiency of removal increased with increasing amount of biomass. Removal efficiency of 85% has been noted when biomass of 1.2 g was used for removing 5ppmconcentration of pyrocatechol. Lower concentrations of the compound (less than 10 ppm) support growth of *Azolla*, while higher concentrations (50 ppm) inhibit growth. Uptake of pyrocatechol is generally followed by its degradation in plants (also known as phytotransformation). The degradation is supported by metabolic processes occurring within the plant or through the effect of enzymes produced within the plants. Microbial breakdown (rhizodegradation) of contaminant has also been reported as one of the major routes of its removal (Zazouli et al. 2013).

Hydrocarbons

Aquatic ferns such as *Azolla pinnata* possess the capacity to degrade hydrocarbon present in the growth medium (Al-Baldawi et al. 2012). The plant showed survival at low concentrations of diesel (0.5–3 mL/L) but could not tolerate high concentrations (10–50 mL/L) due to toxicity when exposed for 10 days. Biodegradation of petroleum hydrocarbons has been facilitated by bacteria present in the rhizospheric zone (Cohen et al. 2002). Bacterial growth is supported by nitrogen and phosphorous released during decomposition of dead plant biomass (Cohen et al. 2002).

Oils

Biomass of Salvinia (an aquatic fern) show greater oil sorption capacity; hence, it can be used as material for removing oil from emulsions/wastewater. The potential of *Salvinia* (approx. 90% of the oil removed by biomass) is attributed to properties such as larger surface area, hydrophobicity, and hairlike surfaces which assist in holding oil (Ribeiro et al. 2003).

Dyes

Biosorption of dyes is followed by biotransformation, and detoxification into non-toxic metabolites (decolorization) has been reported in fern species (Rizwana et al. 2014). According to literature reports, *Azolla filiculoides* exhibit efficiency to remove C.I. Acid Blue 92 dye (Vafaei et al. 2013; Khataee et al. 2013; Torbati et al. 2015). The dye is removed via absorption. Capacity of tree fern species for removal of dye Basic Red 13 via biosorption mechanism has also been reported. The sorption of the dye depends upon the particle size and temperature. The dye sorption capacity increases with the decrease in particle size of the sorbent. Maximum sorption capacity of 408 mg g^{-1} has been noted (Ho et al. 2004).

Plants break the dye molecules into smaller ones so that they can easily be translocated across the semipermeable membranes of plants (Khandare et al. 2013; Watharkar et al. 2013). Transformation and degradation of dyes are supported by intracellular enzymes such as tyrosinases, laccases, NADH-DCIP reductases, lignin peroxidases, etc. (Ghodake et al. 2009, Patil et al. 2009, Kagalkar et al. 2009, 2010). Binding of dye molecules with ions such as calcium and metals such as silicon results in dye–metal complexes which prevent harm caused due to toxic ions (avoidance mechanism). Amide and siloxane groups are found to be mainly involved in the removal of dyes (Nilratnisakorn et al. 2007, 2008).

Explosives

Salvinia species particularly *S. rotundifolia* have shown capacity of treating groundwater contaminated with explosives. The plant showed the capacity to degrade explosives such as trinitrotoluene (TNT) to generate metabolic products such as aminodinitrotoluene (ADNT) (Jacobson et al. 2003).

25.2.2 Inorganic Contaminants

Nutrients

Aquatic fern species such as *Salvinia minima* possess capacity to treat high-strength synthetic organic wastewater (HSWW). The plants show potential for removing inorganic nutrients such as ammonium and nitrate nitrogen. A significant increase in relative growth rate (RGR) and productivity (2.3-fold increase) has been reported in plants exposed to ammonium–nitrogen concentration of 70 mg L^{-1} (Olguín et al. 2007). *Azolla* is another species which acts as a biofilter and removes nutrients such as P, K, Mg, and ammonia from wastewater (Costa et al. 1999; Forni et al. 2001). The ammonia removal rate (ARR) of 6.394 h^{-1} has been reported (Carlozzi and Padovani 2016). Species such as *Azolla caroliniana* showed reduction in growth when exposed to concentration of 2 mg L^{-1} NH_4Cl. The nitrogenase activity of plants showed a significant reduction (about 33%) when exposed to 10 mg l^{-1} NH_4Cl. High concentration of nitrogenous compounds caused significant reduction in the nitrogenase activity. High nitrogen removal has been noted in *Azolla filiculoides* growing in media containing high amounts of nitrogenous compounds. Phosphorus uptake noted an increase in plants exposed to heavy metals such as Pb, Cd, and Cu. Highest values of nitrogen uptake and accumulation, viz., 9572.31, 8721.32, 7174.12, and 7088.41 mg/m^2, have been noted in *Azolla* fronds grown in the presence of 2 ppm Cu, 0.5 ppm Pb, and 0.5 ppm Cd after exposure of 30 days. *Azolla* species also exhibited potential for uptake of potassium growing in the media having presence of metals such as Pb, Cd, and Cu. Potassium uptake of 6361.60 and 6229.06 mg/m^2 has been noted in plants applied with 2 ppm Pb or Cu (Carlozzi and Padovani 2016).

Radionuclides

Azolla caroliniana have shown potential for removing radioactive ions, viz., ^{137}Cs, ^{60}Co, ^{210}Po, and ^{238}U, from radioactive wastewaters. Greater absorption of ^{210}Po followed by ^{230}Th, ^{234}U, ^{238}U, ^{232}Th, and ^{235}U has been reported. Removal of radioactive ions from dilute solutions occurs by two pathways – rapid and slow absorption. High concentration of radionuclides is removed by active process, i.e., bioaccumulation which involves living biomass. Biochemical components including polysaccharide and lipid fractions assist in binding of radionuclides (Popa et al. 2004).

Heavy Metals

Heavy metal removal and accumulation in ferns are well reported (Ho et al. 2004; Gonzaga et al. 2006; Kubicka et al. 2015; Prabhu et al. 2016) (Table 25.2).

Table 25.2 Metal accumulation reported in fern species

Metal	Rate of accumulation	Plant species
Zn	655 mg Kg^{-1}	*Adiantum aethiopicum*
As	1240 mg Kg^{-1}	*Asplenium australasicum*
As	2630 mg Kg^{-1}	*Asplenium bulbiferum*
Cu	5500 mg Kg^{-1}	*Athyrium yokoscense*
As	814 mg Kg^{-1}	*Athyrium yokoscense*
Cd	1095 mg Kg^{-1}	*Athyrium yokoscense*
Pb	2040–3464 mg Kg^{-1}	*Athyrium yokoscense*
Zn	1221–2422 mg Kg^{-1}	*Athyrium yokoscense*
As	284 mg Kg^{-1}	*Azolla caroliniana*
Cd	259 mg Kg^{-1}	*Azolla caroliniana*
Cr	963 mg Kg^{-1}	*Azolla caroliniana*
Hg	578 mg Kg^{-1}	*Azolla caroliniana*
Pb	416 mg Kg^{-1}	*Azolla caroliniana*
As	54 mg Kg^{-1}	*Azolla filiculoides*
Cd, Cr, Cu, Zn	6500–9000 mg Kg^{-1}	*Azolla filiculoides*
Pb	9300 mg Kg^{-1}	*Azolla filiculoides*
Cr	209 mg Kg^{-1}	*Azolla pinnata*
Cu	255 mg Kg^{-1}	*Azolla pinnata*
Hg, Cd	310–740 mg Kg^{-1}	*Azolla pinnata*
Pb	2701.1 mg Kg^{-1}	*Azolla pinnata*
Cd	259 mg Kg^{-1}	*Azolla* sp.
Sr	23,000 mg Kg^{-1}	*Azolla* sp.
Hg	940 mg Kg^{-1}	*Azolla* sp.
Ni	121 mg Kg^{-1}	*Blechnum cartilagineum*
Zn	202 mg Kg^{-1}	*Blechnum cartilagineum*
Zn	216 mg Kg^{-1}	*Blechnum nudum*
Cd	168–368 mg Kg^{-1}	*Dennstaedtia davallioides*
Ni	161 mg Kg^{-1}	*Doodia aspera*
Pb	469 mg Kg^{-1}	*Doodia aspera*
Cu	500 mg Kg^{-1}	*Hypolepis muelleri*
Zn	233–248 mg Kg^{-1}	*Hypolepis muelleri*
Cd	4.1 mg Kg^{-1}	*Nephrolepis cordifolia*
Ni	131 mg Kg^{-1}	*Nephrolepis cordifolia*
As	8000 mg Kg^{-1}	*Pellaea calomelanos*
Cd	87.9 mg Kg^{-1}	*Pteris falcata*
Ni	196 mg Kg^{-1}	*Pteris falcata*
Pb	62 mg Kg^{-1}	*Pteris falcata*
As	7230–14,500 mg Kg^{-1}	*Pteris vittata*
Zn	271–2000 mg Kg^{-1}	*Pteris vittata*
Cd	4134–17,170 mg Kg^{-1}	*Salvinia minima*
Pb	2000–2600 mg Kg^{-1}	*Salvinia minima*
Hg	18,575 mg Kg^{-1}	*Salvinia molesta*
Ni	18,875 mg Kg^{-1}	*Salvinia molesta*
Pb	18,275 mg Kg^{-1}	*Salvinia molesta*

Many species of ferns are known to be hyperaccumulators. These species accumulate more than 100 mg kg^{-1} Cd, 1000 mg kg^{-1} As/Co/Cu/Pb/Ni, and 10,000 mg kg^{-1} Mn/Zn in aboveground plant parts. Both living and dead biomass of *Azolla* have been exploited for removing heavy metals such as Cs, Sr, Pb, Zn, Ni, Cu, Au, Cd, and Cr from industrial effluents and sewage water (Cohen-Shoel et al. 2002; Bennicelli et al. 2004; Stepniewska et al. 2005; Rakhshaee et al. 2006; Rai 2008; Mashkani and Ghazvini 2009). *Azolla filiculoides* is also known to accumulate high concentrations of Pb, Zn, Cu, Ni, Cr, and Cd (Zhao and Duncan 1997; Asbchin et al. 2012; Sheel et al. 2015). Other species showing high accumulation of Pb, Cu, Mn, and Zn include *A. microphylla* and *A. caroliniana* (Cohen-Shoel et al. 2002; Bennicelli et al. 2004, 2005; Stepniewska et al. 2005; Jafari et al. 2010). Metal uptake has been facilitated by hydroxyl and/or carboxyl groups (TanghiGanji et al. 2005). The carboxyl groups present in the galacturonic acid, the principal constituent of pectin (an important polysaccharide constituent of cell walls), help in binding of heavy metals in *Azolla* (Cohen-Shoel et al. 2002).

Salvinia is another potential fern species with an inherent capacity to accumulate heavy metals. Accumulation of Cr, Fe, Ni, Cu, Pb, and Cd in the range of 6–9 mg g^{-1} dry weight and Co, Zn, and Mn in the range of 4 mgg^{-1} dry weight has been reported (Banerjee and Sarker 1997; Dhir 2009; Dhir et al. 2011). *Salvinia natans*, *S. molesta*, *S. herzogii*, *S. minima*, *S. auriculata*, and *S. rotundifolia* accumulate high amounts of heavy metals, hence removing them from wastewaters (Nichols et al. 2000; Hoffman et al. 2004; Olguín et al. 2002, 2005; Espinoza-quinones et al. 2005; Molisani et al. 2006; Sune et al. 2007; Sánchez-Galván et al. 2008; Xu et al. 2009; Fuentes et al. 2014; Kumari et al. 2016). Bioaccumulation of heavy metals increased with increasing concentration of heavy metals. The metal uptake occurs via fast physical processes such as adsorption, ionic exchange, and chelation, while slow biological processes such as intracellular uptake (transported through plasmalemma into cells) also aid in metal uptake. It has been postulated that uptake of heavy metals is driven by secondary transport proteins, viz., channel proteins or H^+-coupled carrier proteins. Uptake subsequently leads to translocation of metals from roots to leaves (Sune et al. 2007). Free carboxylic groups present on the cell surface provide the sites for metal binding (Olguin et al. 2005). Proteins behave as important ligand atoms and also play an important role in metal absorption. High concentration of lipids and carbohydrates present on the plant surface acts as the cationic weak exchanger groups that contribute to metal sorption by ion exchange reactions (Sánchez-Galván et al. 2008). Metal accumulation caused considerable reduction in the fresh weight and increase in dry weight.

Several species of *Pteris* including *P. vittata*, *P. cretica*, *P. longifolia*, *P. umbrosa*, *P. argyraea*, *P. quadriaurita*, *P. ryiunkensis*, and *P. biaurita* show hyperaccumulation of heavy metals such as As and Cr (Ma et al. 2001; Francesconi et al. 2002; Meharg and Hartley-Whitaker 2002; Zhao et al. 2002; Srivastava et al. 2006; Jadia and Fulekar 2008; Qing et al. 2009; Feng et al. 2010; Garcia et al. 2010). *Pteris vittata* has shown capacity to take up both inorganic and organic As species including arsenate, arsenite, and monomethylarsonic acid (MMA) (Ma et al. 2001; Kertulis et al. 2005). The plant possesses excellent abilities to take up and transport As in fronds (Zheng et al. 2010). *P. vittata* accumulate 438–755 mg kg^{-1} As when grown in an uncontaminated soil and 3525–6805 mg kg^{-1} As when grown in a contaminated soil (Ma et al. 2001). Arsenic and Cu concentration of 3510 and 7.0 mg kg^{-1} has been observed in *P. vittata* grown in soil containing 475 mg kg^{-1}As and 370 mg kg^{-1} Cu. The concentration of As ranged from 1770 to 3650 mg kg^{-1} DW in the fronds and 182 to 507 mg kg^{-1} DW in the roots of *Pteris* species *P. cretica, P. biaurita, P. quadriaurita, and P. ryukyuensis* when exposed to a concentration of 100 mg As kg^{-1} soil (Tu et al. 2002; Wang et al. 2002; Zhang et al. 2002). Structural similarity between phosphate and As(V) leads to uptake of As(V) via phosphate transport systems in the plant (Tu et al. 2004a, b). Arsenate reductase in roots aids in the reduction of As(V) (Duan et al. 2005). Majority (30–40%) of arsenate taken up by roots gets reduced to arsenite. Reduction of arsenate to arsenite inside the plant occurs in the fronds (Pickering et al. 2000; Ma et al. 2001). The activity of glutathione-dependent arsenate reductase is responsible for reduction of arsenate to arsenite. Arsenite (accounting for 93–98% of the total As) is transported via xylem sap and sequestered into vacuoles of the fronds and trichome cells (Fig. 25.1).

Pityrogramma calomelanos also show accumulation of As in fronds (~8350 μg g^{-1} dry mass) and rhizoids (lower concentration of 88–350 μg g^{-1} dry mass). Accumulation as high as 2760 to 8350 mg kg^{-1} has been noted in fronds when plants are grown in soil containing 135 to 510 mg kg^{-1} As. High content of As(III) has been reported in fronds in comparison to As(V). Heavy metal accumulation in roots is an exclusion mechanism which imparts tolerance in plants (Francesconi et al. 2002; Zhao et al. 2002; Meharg 2003).

Athyrium yokoscense accumulate metals in roots and fronds. Arsenic accumulation in the roots ranged from 506 to 2192 mg kg^{-1}, while in fronds range of 162 mg kg^{-1} to 562 mg kg^{-1} has been noted. Copper and Zn accumulation of 375 and 1165 mg kg^{-1} has been noted in roots, while 88 and 973 mg kg^{-1} has been reported in fronds. Plant showed accumulation of 242 mg kg^{-1} As and 88 mg kg^{-1} Cu when grown in soil containing 814 mg kg^{-1} As and 343 mg kg^{-1} Cu. Significantly greater heavy metal accumulation occurred in roots than fronds, suggesting a limited mobility and translocation of heavy metals once absorbed by ferns. Compartmentalization of excess Cu in the cell wall appears to contribute to Cu tolerance and accumulation. Proanthocyanidins (condensed tannins) have been proposed to have the ability to complex with heavy metals such as Pb and play an important role in Pb tolerance (Kamachi et al. 2005).

Accumulation of metals such as Cd, Cr, Cu, Ni, Pb, and Zn has also been noted in fern species including *Adiantum aethiopicum, Blechnum cartilagineum, Blechnum nudum, Calochlaena dubia, Dennstaedtia davallioides, Doodia aspera, Hypolepis muelleri, Nephrolepis cordifolia*, and *Pellaea falcata* (Francesconi et al. 2002).

Polysaccharides in the cell walls of *Lygodium japonicum* prothallium facilitate rapid accumulation of Cu. Metal accumulated into the cell wall gets tightly bound to the homogalacturonan of the cell wall pectin (Konno et al. 2005). Carboxyl, phosphate, amine, and sulfonate groups present on cell wall are responsible for

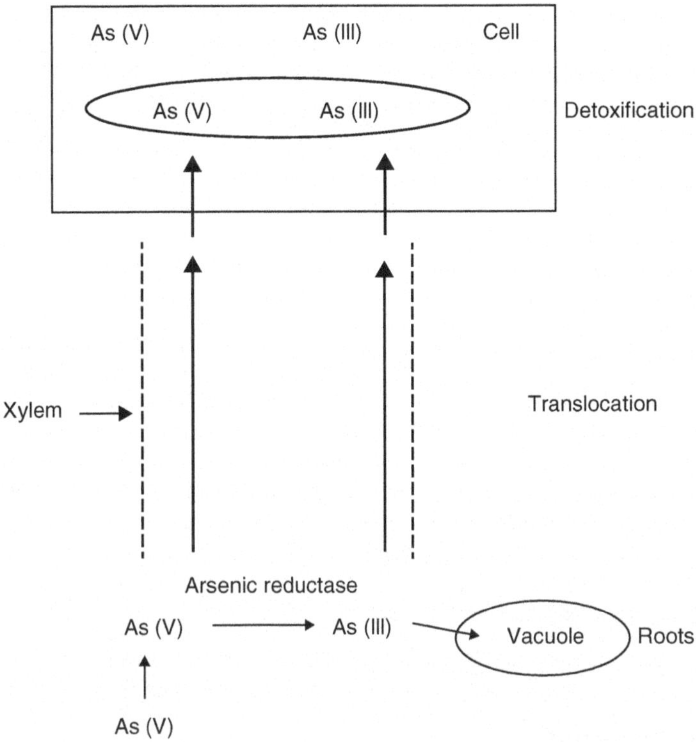

Fig. 25.1 Schematic diagram of arsenic uptake, translocation, detoxification, and sequestration in *P. vittata*

binding metal ions to the biosorbent (Tanghiganji et al. 2005). The polar functional groups of lignin-like alcohols, aldehydes, ketones, acids, phenolic hydroxides, and ethers act as the chemical binding agents (Ho 2003).

Plant tolerance capacity for heavy metals includes accumulation, compartmentalization, or exclusion. Exclusion restricts the uptake of the contaminant into biomass. Accumulation resists high concentration of the contaminants in their aerial tissues and transforms the contaminant into an inert form. Metal detoxification mechanism mainly includes chelation, compartmentalization, biotransformation, and cellular repair (Salt et al. 1998). Heavy metals are transported and deposited in a vacuole as metal chelates. The chelation is assisted by high-affinity ligands (such as oxygen-donor, sulfur-donor, or nitrogen-donor ligands) (Baker et al. 2000). In the vacuoles, sulfur-donor ligands (such as metallothioneins and phytochelatins) form highly stable complexes with heavy metals (Zhao et al. 2003; Zhang et al. 2004). Transporter proteins and intracellular high-affinity binding sites mediate the heavy metal uptake across the plasma membrane. Classes of proteins involved in heavy metal uptake include ATPases, natural resistance-associated macrophage protein (NRAMP), cation diffusion facilitator (CDF), and ZIP family. ATPases present in the plasma membrane are mainly responsible for sequestration of toxic heavy metals in vacuoles, Golgi, or endoplasmic reticulum. Uptake of As occurs via phosphate cycle and Zn via transporters, zinc-regulated transporter/iron-regulated transporter proteins (ZIP) (Muchhal et al. 1996). Proteins act as ligands for metal binding (Sharma and Sachdeva 2015). Enzymatic and nonenzymatic antioxidants play an important role in metal detoxification and recovering oxidative stress (Tiwari and Sarangi 2017). Enzymes mainly include superoxide dismutase, catalase, and ascorbate peroxidase (Kanchenko et al. 2007; Srivastava et al. 2005). Superoxide dismutase (SOD) has been known to play an important role in accumulation and detoxification of As in *Pteris vittata* and *P. multifida*.

Genes encoding for heavy metal tolerance have been identified in ferns. Gene encoding for phytochelatin synthase (PvPCS1) has been characterized. Expression of *PvPCS1* increased their Cd tolerance by mediating As chelation. Arsenate reductase gene (*PvACR2*) which plays an important role in reduction of As(V) to As(III) has been characterized. PvGrx5 which plays a role in regulating intracellular arsenite levels has been characterized. Two distinct gene families *Pht1* and *Pht2* involved in phosphate uptake in plants have been identified (Bucher et al. 2001). *Pht1* genes are root transporters, and *Pht2* genes are involved in the transport in vascular tissue of the leaf and shoot (Shin et al. 2004). Enhanced arsenate resistance has been reported in organisms expressing *PV4–8*. The role of *Pht1* and *Pht2* genes in As(V) transport has been revealed.

25.3 Conclusion

Ferns have emerged as a plant group with exorbitant potential to remove inorganic and organic contaminants from the environment. Fern species such as *Pteris*, *Azolla*, and *Salvinia* exhibit high heavy metal uptake, tolerance, and detoxification potential. Terrestrial and aquatic fern species remove/treat contaminants such as radionuclides, hydrocarbons, and volatile organic compounds; hence their role in developing phytoremediation technology for environmental cleanup needs to be emphasized.

References

Al-Baldawi IA, Sheikh Abdullah SR, Suja F, Anuar N, Idris M (2012) Preliminary test of hydrocarbon exposure on *Azolla pinnata* in phytoremediation process. In: International conference on environment, energy and biotechnology, vol 33. IACSIT Press, Singapore

Alexandra DO, Mihaela DC, Cristina SL (2014) Applications of pteridophytes in phytoremediation. Curr Trends Nat Sci 3:68–73

Asbchin SA, Omran AN, Jafari N (2012) Potential of *Azolla filiculoides* in the removal of Ni and Cu from wastewaters. Afr J Biotechnol 11(95):16158–16164

Baker AJM, McGrath SP, Reeves RD, Smith JAC (2000) Metal hyperaccumulator plants: a review of the ecology and physiology of a biochemical resource for phytoremediation of metal-

polluted soils. In: Terry N, Banuelos G (eds) Phytoremediation of contaminated soil and water. Lewis Publishers, Boca Raton, pp 85–107
Banerjee G, Sarker S (1997) The role of *Salvinia rotundifolia* in scavenging aquatic Pb (II) pollution: a case study. Bioprocess Eng 17:295–300
Benaroya RO, Tzin V, Tel-Or E, Zamski E (2004) Lead accumulation in the aquatic fern *Azolla filiculoides*. Plant Physiol Biochem 42:639–645
Bennicelli R, Stezpniewska Z, Banach A, Szajnocha K, Ostrowski J (2004) The ability of *Azolla caroliniana* to remove heavy metals (Hg(II), Cr(III), Cr(VI)) from municipal waste water. Chemosphere 55:141–146
Bennicelli RP, Balakhnina TI, Szajnocha K, Banach A, Wolinska A (2005) Potential of *Azolla caroliniana* for the removal of Pb and Cd from wastewaters. Int Agrophys 19:251–255
Bucher M, Rausch C, Daram P (2001) Molecular and biochemical mechanisms of phosphorus uptake into plants. J Plant Nutr Soil Sci 164:209–217
Carlozzi P, Padovani G (2016) These plants purify the air, removing VOC's and more via NASA study. Environ Sci Pollut Res 23:8749–8755
Cohen MF, Williams J, Yamasaki H (2002) Biodegradation of diesel fuel by an *Azolla* derived bacterial consortium. J Environ Sci Health Part A Tox Hazard Subst Environ Eng A 37(9):1593–1606
Cohen-Shoel N, Barkay Z, Ilzycer D, Gilath I, Tel-Or E (2002) Biofiltration of toxic elements by *Azolla* biomass. Water Air Soil Pollut 135:93–104
Costa ML, Santos MC, Carrapiço F (1999) Biomass characterization of *Azolla filiculoides* grown in natural ecosystems and wastewater. Hydrobiologia 415:323–327
De Kempeneer L, Sercu B, Vanbrabant W, Van Langenhove H, Verstraete W (2004) Bioaugmentation of the phyllosphere for the removal of toluene from indoor air. Appl Microbiol Biotechnol 64:284–288
Dela Cruz M, Christensen JH, Thomsen JD, Müller R (2014) Can ornamental potted plants remove volatile organic compounds from indoor air? A review. Environ Sci Pollut Res 21:13909–13928
Dhir B (2009) *Salvinia*: an aquatic fern with potential use in phytoremediation. Environ We Int J Sci Technol 4:23–27
Dhir B, Sharmila P, Saradhi PP, Sharma S, Kumar R, Mehta D (2011) Heavy metal induced physiological alterations in *Salvinia natans*. Ecotoxicol Environ Saf 74:1678–1684
Drăghiceanu OA, Bobrescu CM, Soare LC (2014) Application of pteridophytes in phytoremediation. Curr Trends Natl Sci 3(6):68–73
Duan GL, Zhu YG, Tong YP, Cai C, Kneer R (2005) Characterization of arsenate reductase in the extract of root and fronds of chinese brake fern, an arsenic hyperaccumulator. Plant Physiol 138:461–469
Ena A, Carlozzi P, Pushparaj B, Paperi R, Carnevale S, Angelo S (2007) Ability of the aquatic fern *Azolla* to remove chemical oxygen demand and polyphenols from olive mill wastewater. Grasas Aceites 58(1):34–39
Espinoza-Quinones FR, Zacarlein CE, Palacio SM, Obregon CL, Zenatti DC, Galante RM, Rossi N, Rossi FL, Pereira RA, Welter RA, Rizzulto MA (2005) Removal of heavy metals from polluted river using aquatic macrophytes *Salvinia* sp. Braz J Plant Physiol 35:744–746
Feng R, Wei C, Tu S, Tang S, Wu F (2010) Simultaneous hyperaccumulation of arsenic and antimony in Cretan brake fern: evidence of plant uptake and subcellular distributions. Microchem J 97:38–43
Fons F, Froissard D, Bessière JM, Buatois B, Rapior S (2010) Biodiversity of volatile organic compounds from five French ferns. Nat Prod Commun 5(10):1655–1658
Forni C, Nicolai MA, D'Egidio DG (2001) Potential of the small aquatic plants *Azoll*a and *Lemna* for nitrogenous compounds removal from wastewater. Trans Ecol Environ 49:315–324
Francesconi K, Visoottiviseth P, Sridockhan W, Goessler W (2002) Arsenic species in an hyperaccumulating fern, *Pityrogramma calomelanos*: a potential phytoremediator. Sci Total Environ 284:27–35

Fuentes II, Espadas-Gil F, Talavera-May C, Fuentes G, Santamaría JM (2014) Capacity of the aquatic fern (*Salvinia minima* Baker) to accumulate high concentrations of nickel in its tissues, and its effect on plant physiological processes. Aquat Toxicol 155:142–150

García ML, Lodeiro PL, Barriada JL, Herrero R, de Vicente MES (2010) Reduction of Cr(VI) levels in solution using bracken fern biomass: batch and column studies. Chem Eng J 165:517–523

Ghodake GS, Telke AA, Jadhav JP, Govindwar SP (2009) Potential of *Brassica juncea* in order to treat textile effluent contaminated sites. Int J Phytoremediation 11:297–312

Giese M, Bauer-Doranth U, Langebartels C, Sandermann H (1994) Detoxification of formaldehyde by the spider plant (*Chlorophytum comosum* L.) and by soybean (*Glycine max* L.) cell-suspension cultures. Plant Physiol 104:1301–1309

Glick BR (2010) Using soil bacteria to facilitate phytoremediation. Biotechnol Adv 28:367–374

Gonzaga MIS, Gonzaga SJA, Ma LQ (2006) Arsenic phytoextraction and hyperaccumulation by fern species. Sci Agric 63:90–101

Hanson AD, Roje S (2001) One-carbon metabolism in higher plants. Annu Rev Plant Physiol Plant Mol Biol 52:119–137

Ho YS (2003) Removal of copper ions from aqueous solution by tree fern. Water Res 37:2323–2330

Ho YS, Chiu WT, Hsu CS, Huang CT (2004) Sorption of lead ions from aqueous solution using tree fern as a sorbent. Hydrometallurgy 73:55–61

Hoffmann T, Kutter C, Santamaria JM (2004) Capacity of *Salvinia minima* Baker to tolerate and accumulate As and Pb. Eng Life Sci 4:61–65

Jacobson ME, Chiang SY, Gueriguian L, Westholm LR, Pierson J, Zhu G, Saunders FM (2003) Transformation kinetics of trinitrotoluene conversion in aquatic plants. In: McCutcheon SC, Schnoor JL (eds) Phytoremediation. Wiley, New York

Jadia CD, Fulekar MH (2008) Phytotoxicity and remediation of heavy metals by fibrous root grass (sorghum). J Appl Biosci 10:491–499

Jafari N, Senobari Z, Pathak RK (2010) Biotechnological potential of *Azolla filiculoides, Azolla microphylla* and *Azolla pinnata* for biosorption of Pb(II), Mn(II), cu (II) and Zn(II). Ecol Environ Conserv 16:443–449

Jindrova E, Chocova M, Demnerova K, Brenner V (2002) Bacterial aerobic degradation of benzene, toluene, ethylbenzene and xylene. Folia Microbiol 47:83–93

Kagalkar AN, Jagtap UB, Jadhav JP, Bapat VA, Govindwar SP (2009) Biotechnological strategies for phytoremediation of the sulphonated azo dye Direct Red 5B using *Blumea malcolmii* Hook. Bioresour Technol 100:4104–4110

Kagalkar AN, Jagtap UB, Jadhav JP, GovindwarSP BSA (2010) Studies on phytoremediation potentiality of *Typhonium flagelliforme* for the degradation of Brilliant Blue R. Planta 232(1):271–285

Kamachi H, Komori I, Tamura H, Sawa Y, Karahara I, Honma Y, Wada N, Kawabata T, Matsuda K, Ikeno S, Noguchi M, Inoue H (2005) Lead tolerance and accumulation in the gametophytes of the fern *Athyrium yokoscense*. J Plant Res 118:137–145

Kanchenko AG, Singh B, Bhatia NP (2007) Heavy metal tolerance in common fern species. Aust J Bot 55:63–73

Kertulis GM, Ma LQ, MacDonald GE, Chen R, Winefordner JD, Cai Y (2005) Arsenic speciation and transport in *Pteris vittata* L. and the effects on phosphorus in the xylem sap. Environ Exp Bot 54:239–247

Khandare RV, Kabra AN, Awate AV, Govindvar SP (2013) Synergistic degradation of diazo dye direct red 5B by *Portulaca grandiflora* and *Pseudomonas putida*. Int J Environ SciTechnol 10:1039–1050

Khataee AR, Movafeghi A, Vafaei F, Lisar SSY, Zarei M (2013) Potential of the aquatic fern *Azolla filiculoides* in biodegradation of an azo dye: modeling of experimental results by artificial neural networks. Int J Phytoremediation 15:729–742

Kim KJ, Kil MJ, Song JS, Yoo EH, Son K, Kays SJ (2008) Efficiency of volatile formaldehyde removal by indoor plants: contribution of aerial plant parts versus the root zone. J Am Soc Hortic Sci 133(4):521–526

Konno H, Nakato T, Nakashima S, Katoh K (2005) *Lygodium japonicum* fern accumulates copper in the cell wall pectin. J Exp Bot 56:1923–1931

Kubicka K, Samecka-Cymerman A, Kolon K, Kosiba P, Kempers AJ (2015) Chromium and nickel in *Pteridium aquilinum* from environments with various levels of these metals. Environ Sci Pollut Res Int 22:527–534

Kumari S, Kumar B, Sheel R (2016) Bioremediation of heavy metals by serious aquatic weed, *Salvinia*. Int J Curr Microbiol Appl Sci 5:355–368

Ma LQ, Komar KM, Tu C, Zhang WH, Cai Y, Kennelley ED (2001) A fern that hyperaccumulate arsenic: a hardy, versatile, fast-growing plant helps to remove arsenic from contaminated soils. Nature 409:579–579

Mashkani SG, Ghazvini PTM (2009) Biotechnological potential of *Azolla filiculoides* for biosorption of Cs and Sr: application of micro-PIXE for measurement of biosorption. Bioresour Technol 100:1915–1921

McGuinness M, Dowling D (2009) Plant-associated bacterial degradation of toxic organic compounds in soil. Int J Environ Res Public Health 6:2226–2247

Meharg AA (2003) Variation in arsenic accumulation hyperaccumulation in ferns and their allies. New Phytol 157:25–31

Meharg AA, Hartley-Whitaker J (2002) Arsenic uptake and metabolism in arsenic resistant and non resistant plant species. New Phytol 154:29–43

Molisani MM, Rocha R, Machado W, Barreto RC, Lacerda LD (2006) Mercury contents in aquatic macrophytes from two reservoirs in the paraiba do sul: Guandu river system, SE Brazil. Braz J Biol 66:101–107

Muchhal US, Pardo JM, Raghothama KG (1996) Phosphate transporters from the higher plant *Arabidopsis thaliana*. Proc Nat Acad Sci USA 93:10519–10523

Nichols PB, Couch JD, Al-Hamdani SH (2000) Selected physiological responses of *Salvinia minima* to different chromium concentrations. Aquat Bot 68:313–319

Nie M, Wang Y, Yu J, Xiao M, Jiang L, Yang J, Fang C, Chen J, Li B (2011) Understanding plant-microbe interactions for phytoremediation of petroleum polluted soil. PLoS One 6(3):1–8

Nilratnisakorn S, Thiravetyan P, Nakbanpote W (2007) Synthetic reactive dye wastewater treatment by narrow-leaved cattails (*Typha angustifolia* Linn.): effects of dye, salinity and metals. Sci Total Environ 384:67–76

Nilratnisakorn S, Thiravetyan P, Nakbanpote W (2008) Synthetic reactive dye wastewater treatment by narrow-leaved cattail: studied by XRD and FTIR. Asian J Energ Environ 9:231–252

Olguin J, Hernandez E, Ramos I (2002) The effect of both different light conditions and the pH value on the capacity of *Salvinia minima* BAKER for removing cadmium, lead and chromium. Acta Biotechnol 22:121–131

Olguin EJ, Sánchez-Galván G, Pérez-Pérez T, Pérez-Orozco A (2005) Surface adsorption, intracellular accumulation and compartmentalization of Pb(II) in batch-operated lagoons with *Salvinia minima* as affected by environmental conditions, EDTA and nutrients. J Ind Microbiol Biotechnol 32:577–586

Olguín EJ, Sánchez-Galván G, Pérez-Pérez P (2007) Assessment of the phytoremediation potential of *Salvinia minima* Baker compared to *Spirodela polyrrhiza* in high-strength organic wastewater. Water Air Soil Pollut 181:135–147

Patil P, Desai N, Govindwar S, Jadhav JP, Bapat V (2009) Degradation analysis of reactive red 198 by hairy roots of *Tagetes patula* L. (marigold). Planta 230(4):725–735

Pickering IJ, Prince RC, George MJ, Smith RD, George GN, Salt DE (2000) Reduction and coordination of arsenic in Indian mustard. Plant Physiol 122:1171–1177

Popa K, CecalA HD, Caraus I, Draghici CL (2004) Removal of $^{60}Co^{2+}$ and $^{137}Cs^{+}$ ions from low radioactive solutions using *Azolla caroliniana* Willd. Water fern. Central Euro J Chem 2:434–445

Prabhu SG, Srinikethan G, Hegde S (2016) Potential of pteridophytes in heavy metal phytoremediation. Int J Res Eng Technol 5:1–9

Qing E, Xiu L, Xiao Y (2009) The arsenic hyperaccumulator fern *Pteris vittata* L. Environ Sci Technol 43:8488–8495

Rai PK (2008) Phytoremediation of Hg and Cd from industrial effluent using an aquatic free floating macrophyte. *Azolla pinnata*. Int J Phytoremediation 10:430–439

Rakhshaee R, Khosravi M, TaghiGanji MT (2006) Kinetic modeling and thermodynamic study to remove Pb(II), Cd(II), Ni(II) and Zn(II) from aqueous solution using dead and living *Azolla filiculoides*. J Hazard Mater B134:120–129

Ribeiro TH, Rubio J, Smith RW (2003) A dried hydrophobic aquaphyte as an oil filter for oil/water emulsions. Spill Sci Technol Bull 8:483–489

Rizwana M, Darshan M, Nilesh D (2014) Phytoremediation of textile waste water using potential wetland plant: eco-sustainable approach. Int J Interdisciplinary Multidisciplinary Stud 1(4):130–138

Salt DE, Smith RD, Raskin I (1998) Phytoremediation. Ann Rev Plant Physiol Plant Mol Biol 49:643–668

Sánchez-Galván G, Monroy O, Gómez G, Olguín EJ (2008) Assessment of the hyperaccumulating lead capacity of *Salvinia minima* using bioadsorption and intracellular accumulation factors. Water Air Soil Pollut 194:77–90

Schmitz H, Hilgers U, Weidner M (2000) Assimilation and metabolism of formaldehyde by leaves appear unlikely to be of value for indoor air purification. New Phytol 147(2):307–315

Sharma A, Sachdeva S (2015) Cadmium toxicity and its phytoremediation a review. Int J Sci Eng Res 6(9):395–405

Sheel R, Anand M, Nisha K (2015) Phytoremediation of heavy metals (Zn and Pb) and its toxicity on *Azolla filiculoides*. Int J Sci Res 4(7):1238–1241

Shin H, Dewbre GR, Harrison MJ (2004) Phosphate transport in *Arabidopsis*: Pht1;1 and Pht1;4 play a major role in phosphate acquisition from both low- and high-phosphate environments. Plant J 39:629–642

Sood A, Uniyal PL, Prasanna R, Ahluwalia AS (2011) Phytoremediation potential of aquatic macrophyte, *Azolla*. Ambio 41:22–137

Srivastava M, Ma LQ, SinghN SS (2005) Antioxidant responses of hyper-accumulator and sensitive fern species to arsenic. J Exp Bot 56:1335–1342

Srivastava M, Ma LQ, Santos JAG (2006) Three new arsenic hyperaccumulating ferns. Sci Total Environ 364:24–31

Stepniewska Z, Bennicelli RP, Balakhnina TI, Szajnocha K, Banach A, Wolinska A (2005) Potential of *Azolla caroliniana* for the removal of Pb and Cd from wastewaters. Int Agrophys 19:251–255

Suñe N, Sánchez G, Caffaratti S, Maine MA (2007) Cadmium and chromium removal kinetics from solution by two aquatic macrophytes. Environ Pollut 145:467–473

Taghi-Ganji M, Khosravi M, Rakhshaee R (2005) Biosorption of Pb, Cd, Cu and Zn from the wastewater by treated *A. filiculoides* with $H_2O_2/MgCl_2$. Int J Environ Sci Technol 1:265–271

Tiwari S, Sarangi BK (2017) Comparative analysis of antioxidant response by *Pteris vittata* and *Vetiveria zizanioides* towards arsenic stress. Ecol Eng 100:211–218

Torbati S, Movafeghi A, Khataee AR (2015) Biodegradation of C.I. acid blue 92 by *Nasturtium officinale*: study of some physiological responses and metabolic fate of dye. Int J Phytoremediation 17:322–329

Treesubsuntorn C, Thiravetyan P (2012) Removal of benzene from indoor air by *Dracaena sanderiana*: effect of wax and stomata. Atmos Environ 57:317–321

Tu C, Ma LQ, Bondada B (2002) Arsenic accumulation in the hyperaccumulator Chinese brake and its utilization potential for phytoremediation. J Environ Qual 31:1671–1675

Tu S, Ma LQ, Luongo T (2004a) Root exudates and arsenic accumulation in arsenic hyperaccumulating *Pteris vittata* and non-hyperaccumulating *Nephrolepis exaltata*. Plant Soil 258:9–19

Tu S, Ma LQ, MacDonald GE (2004b) Arsenic absorption, speciation and thiol formation in excised parts of *Pteris vittata* in the presence of phosphorus. Environ Exp Bot 51:121–131

Ugrekhelidze D, Korte F, Kvesitadze G (1997) Uptake and transformation of benzene and toluene by plant leaves. Ecotoxicol Environ Saf 6:24–29

Vafaei F, Movafeghi A, Khataee AR, Zarei M, Lisar SSY (2013) Potential of *Hydrocotyle vulgaris* for phytoremediation of a textile dye: inducing antioxidant response in roots and leaves. Ecotoxicol Environ Saf 93:128–134

Wang J, Zhao FJ, Meharg AA, Raab A, Feldman J, McGrath SP (2002) Mechanisms of arsenic hyperaccumulation in *Pteris vittata*. Uptake kinetics, interactions with phosphate, and arsenic speciation. Plant Physiol 130:1552–1156

Wathakar AD, Khandare RV, Kamble AA, Mulla AY, Govindwar SP, Jadhav JP (2013) Phytoremediation potential of *Petunia grandiflora* Juss., an ornamental plant to degrade a disperse, disulfonate triphenylmethane textile dye Brilliant Blue G. Environ Sci Pollut Res Int 20:939–949

Xu QS, Ji WD, Yang HY, Wang HX, Xu Y, Zhao J, Shi GX (2009) Cadmium accumulation and phytotoxicity in an aquatic fern, *Salvinia natans* (Linn.) Acta Ecol Sin 29:3019–3027

Zazouli MA, Balarak D, Mahdavi Y (2013) Pyrocatechol removal from aqueous solutions by using *Azolla filiculoides*. Health Scope 2(1):25–30

Zhang W, Cai Y, Tu C, Ma LQ (2002) Arsenic speciation and distribution in an arsenic hyperaccumulating plant. Sci Total Environ 300:167–177

Zhang W, Cai Y, Downum KR, Ma LQ (2004) Thiol synthesis and arsenic hyperaccumulation in *Pteris vittata* (Chinese brake fern). Environ Pollut 131:337–345

Zhao M, Duncun JR (1997) Removal and recovery of nickel from aqueous solution and electroplating rinse effluent using *Azolla filiculoides*. Process Biochem 33(3):249–255

Zhao FJ, Dunham SJ, SP MG (2002) Arsenic hyperaccumulation by different fern species. New Phytol 156:27–31

Zhao FJ, Wang JR, Barker JHA, Schat H, Bleeker PM, McGrath SP (2003) The role of phytochelatins in arsenic tolerance in the hyperaccumulator *Pteris vittata*. New Phytol 159:403–410

Zheng J, Niu T, Wu G, Chen W (2010) One magic pteridophyte (*Pteris vittata* L.): application in remediating arsenic contaminated soils and mechanism of arsenic hyperaccumulation. Front Agric 4:293–298

Index

H. Fernández (ed.), *Current Advances in Fern Research*,
https://doi.org/10.1007/978-3-319-75103-0

B

G

H

R